Barley Science
Recent Advances
from Molecular Biology
to Agronomy of Yield and Quality

Pre-publication
REVIEWS,
COMMENTARIES,
EVALUATIONS . . .

"This book brings together internationally recognized scientists to provide broad coverage of today's knowledge on barley science. The authors comprehensively introduce genetics, molecular biology, crop physiology, and agronomy of barley crop. The main thread of this book is the linking of scientific understanding on molecular biology and crop physiology to practical breeding. By neatly synthesizing the versatile, interacting processes of yield and quality formation, this ambitious goal is skillfully achieved.

I expect that this reference book will come in handy and wear through in the hands of undergraduate and postgraduate students, cereal scientists, agronomists, and plant breeders all over the world."

Dr. Pirjo Peltonen-Sainio
Professor, MTT Agrifood Research,
Finland

"This comprehensive and wide-ranging account of barley science and the current status of research will be welcomed by all who are concerned with the crop. It will be a valuable reference source for barley scientists, for lecturers at the graduate level, and especially for those beginning research on the crop. It should have a place in all relevant department and institute libraries.

Of particular interest are the chapters on molecular mapping of the barley genome, on crop development, simulation modeling, genotype by environment interaction, and yield physiology, contributed by foremost experts in their fields. Other chapters contain critical assessments of the prospects for marker-assisted selection and physiologically based methods for augmenting yield-based selection. Quality for malting and for animal feeding is also thoroughly reviewed. Plant breeders will find much of interest in the book as it gives a foretaste of future directions for crop improvement."

R. B. Austin, DSc
Formerly Head,
Plant Physiology Department,
Plant Breeding Institute,
Cambridge, United Kingdom

More pre-publication
REVIEWS, COMMENTARIES, EVALUATIONS . . .

"This book presents the latest findings in spearheading areas in the forefront of barley research. The information presented will greatly impact the future of precompetitive and applied research and has great relevance for devising more efficient breeding strategies and production schemes. The principles of the main genomic tools are clearly outlined. At the same time, the book does not neglect more traditional, but certainly just as important, topics such as the history of the crop and its physiology from both crop physiologist and agronomist perspectives, making it particularly attractive for those mainly involved in field studies and operations. Detailed descriptions of the plant morphology and physiology coupled with the presentation of growth models allow for a clear understanding of the metabolic processes regulating the translocation and accumulation of assimilates in the kernel. Indications are provided on how to best exploit wild and unadapted germplasm, using the newer and more sophisticated molecular tools. The theory behind the analysis of `genotype x environment' interaction, a fundamental factor for yield stability and its quality, has been analyzed in great depth providing a valuable update of the latest statistical tools. Great attention has been devoted to the biochemical and physiological processes influencing the formation of reserve products in the kernel, particularly in relation to feeding quality, malting quality, and brewing."

Roberto Tuberosa, PhD
*Professor in Biotechnology
Applied to Plant Breeding,
Department of Agroenvironmental
Science and Technology,
University of Bologna, Italy*

Food Products Press®
An Imprint of The Haworth Press, Inc.
New York • London • Oxford

NOTES FOR PROFESSIONAL LIBRARIANS AND LIBRARY USERS

This is an original book title published by Food Products Press®, an imprint of The Haworth Press, Inc. Unless otherwise noted in specific chapters with attribution, materials in this book have not been previously published elsewhere in any format or language.

CONSERVATION AND PRESERVATION NOTES

All books published by The Haworth Press, Inc. and its imprints are printed on certified pH neutral, acid free book grade paper. This paper meets the minimum requirements of American National Standard for Information Sciences-Permanence of Paper for Printed Material, ANSI Z39.48-1984.

Barley Science
*Recent Advances
from Molecular Biology
to Agronomy of Yield and Quality*

FOOD PRODUCTS PRESS®
Crop Science
Amarjit S. Basra, PhD
Senior Editor

New, Recent, and Forthcoming Titles of Related Interest:

Dictionary of Plant Genetics and Molecular Biology by Gurbachan S. Miglani

Advances in Hemp Research by Paolo Ranalli

Wheat: Ecology and Physiology of Yield Determination by Emilio H. Satorre and Gustavo A. Slafer

Mineral Nutrition of Crops: Fundamental Mechanisms and Implications by Zdenko Rengel

Conservation Tillage in U.S. Agriculture: Environmental, Economic, and Policy Issues by Noel D. Uri

Cotton Fibers: Developmental Biology, Quality Improvement, and Textile Processing edited by Amarjit S. Basra

Heterosis and Hybrid Seed Production in Agronomic Crops edited by Amarjit S. Basra

Intensive Cropping: Efficient Use of Water, Nutrients, and Tillage by S. S. Prihar, P. R. Gajri, D. K. Benbi, and V. K. Arora

Physiological Bases for Maize Improvement edited by María E. Otegui and Gustavo A. Slafer

Plant Growth Regulators in Agriculture and Horticulture: Their Role and Commercial Uses edited by Amarjit S. Basra

Crop Responses and Adaptations to Temperature Stress edited by Amarjit S. Basra

Plant Viruses As Molecular Pathogens by Jawaid A. Khan and Jeanne Dijkstra

In Vitro Plant Breeding by Acram Taji, Prakash P. Kumar, and Prakash Lakshmanan

Crop Improvement: Challenges in the Twenty-First Century edited by Manjit S. Kang

Barley Science: Recent Advances from Molecular Biology to Agronomy of Yield and Quality edited by Gustavo A. Slafer, José Luis Molina-Cano, Roxana Savin, Josi Luis Araus, and Ignacio Romagosa

Tillage for Sustainable Cropping by P. R. Gajri, V. K. Arora, and S. S. Prihar

Bacterial Disease Resistance in Plants: Molecular Biology and Biotechnological Applications by P. Vidhyasekaran

Barley Science
Recent Advances from Molecular Biology to Agronomy of Yield and Quality

Gustavo A. Slafer
José Luis Molina-Cano
Roxana Savin
José Luis Araus
Ignacio Romagosa
Editors

Food Products Press®
An Imprint of The Haworth Press, Inc.
New York • London • Oxford

Published by

Food Products Press®, an imprint of The Haworth Press, Inc., 10 Alice Street, Binghamton, NY 13904-1580.

© 2002 by The Haworth Press, Inc. All rights reserved. No part of this work may be reproduced or utilized in any form or by any means, electronic or mechanical, including photocopying, microfilm, and recording, or by any information storage and retrieval system, without permission in writing from the publisher. Printed in the United States of America.

Cover design by Anastasia Litwak.

Library of Congress Cataloging-in-Publication Data

Barley science : recent advances from molecular biology to agronomy of yield and quality / Gustavo Slafer . . . [et al.] editors.
    p. cm.
    Includes bibliographical references and index.
    ISBN 1-56022-909-8 (hard : alk. paper)—ISBN 1-56022-910-1 (soft : alk. paper)
    1. Barley. I. Slafer, Gustavo A., 1960-
    [DNLM: 1. Barley—genetics. 2. Breeding—methods. 3. Genes, Plant—physiology.]
SB191.B2 B27 2001
633.1'6233—dc21

00-140222

CONTENTS

About the Editors — xiii

Contributors — xv

Preface — xix

Chapter 1. Contribution of Barley to Agriculture: A Brief Overview — 1
    Gerhard Fischbeck

History of Barley Cultivation — 1
Utilization of Barley — 5
Share of Barley in Today's Cereal Production — 6
Utilization of the Barley Crop — 10

Chapter 2. New Views on the Origin of Cultivated Barley — 15
    José Luis Molina-Cano
    Ernesto Igartua
    Ana-María Casas
    Marian Moralejo

Historical Introduction—The Quest for the Ancestors of Cultivated Barley — 15
Proposed Centers of Origin for Barley — 17
Distinctiveness of Moroccan *H. spontaneum* — 20
Distinctiveness of Western Mediterranean *H. vulgare* — 21
Similarities Between Western Mediterranean *H. spontaneum* and *H. vulgare* — 23
Phylogenetic Conclusions: A Western Mediterranean Center of Origin for Barley? — 24

Chapter 3. Molecular Mapping of the Barley Genome — 31
    Andris Kleinhofs
    Feng Han

Introduction — 31
Genetic and Physical Mapping — 32
Mapping of Quantitative Trait Loci — 35
Mapping Disease Resistance Genes — 40
Comparative Mapping and Synteny — 43
Concluding Statement — 45

Chapter 4. Wild Barley As a Source of Genes for Crop Improvement   65
Roger P. Ellis

Introduction   65
Origin and Taxonomy   67
Disease Resistance   68
Abiotic Stress Tolerance   71
Grain Quality   75
Genetic Diversity   76
Future Trends   77

Chapter 5. Genetics and Breeding of Malt Quality Attributes   85
J. Stuart Swanston
Roger P. Ellis

Introduction   85
Components of Malting Quality   87
Effects of Breeding   101
Future Directions   102

Chapter 6. Genetics and Breeding of Barley Feed Quality Attributes   115
Steven E. Ullrich

Introduction   115
Feed Quality Characteristics   116
Genetics of Feed Quality Characteristics   121
Breeding for Feed Quality   128
Projections for the Future   132

Chapter 7. Experiences with Genetic Transformation of Barley and Characteristics of Transgenic Plants   143
Henriette Horvath
Jintai Huang
Oi T. Wong
Diter von Wettstein

Introduction   143
Barley Transformation of Immature Zygotic Embryos by the Biolistic Method   144
Barley Transformation by Cocultivation of Immature Zygotic Embryos with *Agrobacterium Tumefaciens*   147

Marker-Free Transgenic Plants	150
Testing the Quality of the Target Gene, Its Product, and Tissue-Specific Expression	151
Identification Assays for Transformants and Selection of Homozygous Plants	155
Studying Organ Development with Barley Plants Transgenic for Homeotic Genes	160
The Production of Recombinant Proteins in Maturing and Germinating Barley Grains	162
Advantages of Tailoring a Heat-Tolerant (1,3-1,4)-β-Glucanase and Its Expression in Transgenic Barley	164
Field Testing	166
Micromalting of Transgenic Lines	167
Agronomic Characteristics of the Transgenic Lines	169

Chapter 8. Molecular Marker-Assisted versus Conventional Selection in Barley Breeding 177
William T. B. Thomas

Barley Breeding—Problems and Success	177
Conventional Breeding Methods	179
Marker-Assisted Selection	181

Chapter 9. Genotype by Environment Interaction and Adaptation in Barley Breeding: Basic Concepts and Methods of Analysis 205
Jordi Voltas
Fred van Eeuwijk
Ernesto Igartua
Luis F. García del Moral
José Luis Molina-Cano
Ignacio Romagosa

Introduction	205
Genotype by Environment Interaction in Barley	206
An Example Data Set: Yield of a Series of Barley Isogenic Lines in Spain	209
Breeding Implications	231
Appendix: Examples of SAS Programs for the Models Described	234

Chapter 10. Initiation and Appearance of Vegetative and Reproductive Structures Throughout Barley Development 243
Luis F. García del Moral
Daniel J. Miralles
Gustavo A. Slafer

Introduction 243
Identification of Different Stages and Developmental Phases 243
Association Between Time Course of Organogenesis and Yield Components Generation 244
Initiation and Appearance of Vegetative and Reproductive Organs 246
Developmental Responses to Environmental Factors 254
Genetic Systems Regulating Flowering Time 259

Chapter 11. Physiological Basis of the Processes Determining Barley Yield Under Potential and Stress Conditions: Current Research Trends on Carbon Assimilation 269
José Luis Araus

Introductory Remarks 269
Physiological Factors Determining Yield 270
Potential Yield versus Stress Adaptation: Breeding Implications 272
Incidence of Stresses on Yield 272
Physiological Mechanisms to Cope with Stresses 282
Final Remarks 294

Chapter 12. Genetic Bases of Barley Physiological Response to Stressful Conditions 307
Luigi Cattivelli
Paolo Baldi
Cristina Crosatti
Maria Grossi
Giampiero Valè
Antonio Michele Stanca

Adaptation to Stress Environments 307
Improving Abiotic Stress Tolerance in Barley 328
Molecular Response to Pathogens 330
Conclusions 343

Chapter 13. Physiological Changes Associated with Genetic Improvement of Grain Yield in Barley   361
    L. Gabriela Abeledo
    Daniel F. Calderini
    Gustavo A. Slafer

Introduction	361
Genetic Gains in Grain Yield	362
Physiological Changes Associated with Genetic Gains in Grain Yield	366
Comparison Between Breeding Progress Reached in Barley and Other Small-Grain Cereals	376
Conclusions	381

Chapter 14. Spectral Vegetation Indices As Nondestructive Indicators of Barley Yield in Mediterranean Rain-Fed Conditions   387
    Jordi Bort
    Jaume Casadesús
    José Luis Araus
    Stefania Grando
    Salvatore Ceccarelli

Introduction	387
Material and Methods	390
Results	393
Discussion	401
Conclusions	408

Chapter 15. Choosing Genotype, Sowing Date, and Plant Density for Malting Barley   413
    Daniel F. Calderini
    M. Fernanda Dreccer

Introduction	413
Choosing Genotype	413
Choosing Sowing Date	422
Choosing Plant Density	429
Concluding Remarks	434

Chapter 16. Computer Simulation of Barley Growth, Development, and Yield   **445**
Joe T. Ritchie
Gopalsamy Alagarswamy

Introduction	445
Phasic Development	448
Growth and Organ Development	453
Potential Model Improvement	466
Evaluation of the Model	467
Appendix: Glossary of Terms	474

Chapter 17. Bases of Preharvest Sprouting Resistance in Barley: Physiology, Molecular Biology, and Environmental Control of Dormancy in the Barley Grain   **481**
Roberto L. Benech-Arnold

Introduction	481
Physiology of Dormancy in Barley	482
Genetics and Molecular Biology of Dormancy in Barley	489
Environmental Control of Dormancy in Barley	493
Concluding Remarks	495

Chapter 18. The Proteins of the Mature Barley Grain and Their Role in Determining Malting Performance   **503**
Peter R. Shewry
Helen Darlington

Introduction	503
Barley Grain Structure in Relation to Protein Distribution	504
Barley Grain Proteins	506
Barley Grain Proteins and Malting Quality	511
Conclusions	515

Chapter 19. Changes in Malting Quality and Its Determinants in Response to Abiotic Stresses   **523**
Roxana Savin
José Luis Molina-Cano

Introduction	523
Phases of Grain Growth	524
Barley Grain Composition	530
Quality Specifications	536

Response of Some Aspects of Grain Quality to High Temperature and Drought	538
Future Research	544
Implications	544

Index   **551**

ABOUT THE EDITORS

Gustavo A. Slafer, PhD, is Associate Professor of Grain Crops Production at the Department of Plant Production and Chairperson of the Plant Production Program of the School for Graduate Studies, both at the University of Buenos Aires, Argentina. He is the author or co-author of more than 150 professional papers, abstracts, and proceedings on various aspects of crop physiology and is the editor of *Genetic Improvement of Field Crops,* and co-editor of *Wheat: Ecology and Physiology of Yield Determination* (Haworth, 1999). Dr. Slafer is an editorial board member of the *Journal of Crop Production* (Haworth) and speaks internationally on crop physiology, particularly in relation to yield generation and physiology/breeding relationships. In addition, he serves as Independent Research Scientist for the National Council for Scientific and Technical Research in Argentina at IFEVA—the Agricultural Plant Physiology and Ecology Research Institute, as well as being a member of many scientific and professional organizations.

José Luis Molina-Cano, PhD, is Scientist and Head of the Cereal Division at the Centre University of Lleida Institute of Food and Agriculture Research and Technology. He has developed 11 varieties of barley released in Spain. Since 1970 he has been directing barley seed productions and breeding programs in both academia and industry. Dr. Molina-Cano has authored more than 70 publications, including a book on barley. He is a member of the European Association for Research in Plant Breeding and a number of other scientific and professional organizations.

Roxana Savin, PhD, is Adjunct Professor of Grain Crops Production at the Department of Plant Production and a member of the staff of the School for Graduate Studies, both at the University of Buenos Aires, Argentina. She is the author or co-author of more than 45 professional papers, abstracts, and proceedings on various aspects of crop physiology. Dr. Savin is a member of the National Council for Scientific and Technical Research in Argentina, as well as other scientific and professional organizations.

José Luis Araus, PhD, is Professor of Plant Physiology at the University of Barcelona and currently Seconded National Expert at the Direction General of Research, European Community in Brussels. He is the author or co-author of more than 50 professional papers in refereed journals in addition to numerous book chapters, abstracts, and proceedings in research fields ranging from photosynthetic metabolism to paleoenvironmental reconstructions of agriculture in antiquity using ecophysiological tools. During the past decade he has served as a referee for various international journals of crop and plant physiology as well as research funding proposals. Dr. Araus is a member of many scientific and professional organizations.

Ignacio Romagosa, PhD, is Professor of Barley Genetics and Breeding at the Department of Crop Production and Forestry Science at the University of Lleida Institute of Food and Agriculture Research and Technology, Spain. He is the author or co-author of more than 50 professional papers in refereed journals in addition to five chapters in international reference books. He is also the scientific coordinator of the Mediterranean Agronomic Institute of Zaragoza of the International Center for Advanced Mediterranean Agronomic Studies. He is currently a member of the board of European Association for Research in Plant Breeding.

CONTRIBUTORS

L. Gabriela Abeledo, is Research and Teaching Agronomist, Departamento de Producción Vegetal, Facultad de Agronomía, Universidad de Buenos Aires, Buenos Aires, Argentina. E-mail: <abeledo@agro.uba.ar>.

Gopalsamy Alagarswamy, PhD, Department of Crop and Soil Sciences, Michigan State University, East Lansing, Michigan. E-mail: <alagarswamy @hotmail.com>.

Paolo Baldi, PhD, is Postdoctoral Research Associate, Experimental Institute for Cereal Research, Foggia, Italy. E-mail: <paolo.baldi@mail.ismaa.it>.

Roberto L. Benech-Arnold, PhD, is Associate Professor, Departamento de Producción Vegetal, Facultad de Agronomía, Universidad de Buenos Aires, Buenos Aires, Argentina. E-mail: <benech@agro.uba.ar>.

Jordi Bort, PhD, Unitat de Fisiologia Vegetal, Facultat de Biologia, Universitat de Barcelona, Barcelona, Spain. E-mail: <jordib@ porthos.bio.ub.es>.

Daniel F. Calderini, DrSci, is Research and Teaching Agronomist, Departamento de Producción Vegetal, Facultad de Agronomía, Universidad de Buenos Aires, Buenos Aires, Argentina. E-mail: <dfcalder@agro.uba.ar>.

Jaume Casadesús, Servei de Camps Experimentals, Universitat de Barcelona, Barcelona, Spain. E-mail: <csdsus@ porthos.bio.ub.es>.

Ana-María Casas, PhD, is Tenured Scientist, Consejo Superior de Investigaciones Científicas (CSIC). Estación Experimental de Aula Dei, Zaragoza, Spain. E-mail: <acasas@eead.csic.es>.

Luigi Cattivelli, Experimental Institute for Cereal Research, Foggia, Italy. E-mail: <l.cattivelli@iol.it>.

Salvatore Ceccarelli, PhD, International Center for Agricultural Research in the Dry Areas, Aleppo, Syria. E-mail: <s.ceccarelli@cgiar.org>.

Cristina Crosatti, PhD, is Postdoctoral Research Associate, Experimental Institute for Cereal Research, Foggia, Italy. E-mail: <c.crosatti@iol.it>.

Helen Darlington, PhD, is Postdoctoral Research Scientist, Institute of Arable Crops Research (IACR)–Long Ashton Research Station, Department of Agricultural Sciences, University of Bristol, Long Ashton, Bristol, England. E-mail: <helen.darlington@bbsrc.ac.uk>.

M. Fernanda Dreccer, PhD, is Research and Teaching Agronomist, Departamento de Producción Vegetal, Facultad de Agronomía, Universidad de Buenos Aires, Buenos Aires, Argentina. E-mail: <dreccer@agro.uba.ar>.

Fred van Eeuwijk, PhD, Department of Plant Sciences, Laboratory of Plant Breeding, Wageningen University, Wageningen, The Netherlands.

Roger P. Ellis, PhD, Scottish Crop Research Institute, Dundee, Scotland. E-mail: <rellis@scri.sari.ac.uk>.

Gerhard Fischbeck, PhD, is Professor, TU-München, Lehrstuhl für Pflanzenbau und Pflanzenzüchtung Lange Point 51, Freising-Weihenstephan, Germany. E-mail: <fischbeck@pollux.edu.agrar.tu-muenchen.de>.

Luis F. García del Moral, Dr, is Professor, Departamento de Biología Vegetal, Facultad de Ciencias, Universidad de Granada, Granada, Spain. E-mail: <lfgm@goliat.ugr.es>.

Stefania Grando, International Center for Agricultural Research in the Dry Areas, Aleppo, Syria. E-mail: <s.grando@cgiar.org>.

Maria Grossi, PhD, is Postdoctoral Research Associate, Experimental Institute for Cereal Research, Foggia, Italy. E-mail: <m.grossi@iol.it>.

Feng Han, Pioneer Hi-Bred International, Inc., Johnston, Iowa. E-mail: <hanfe@phibred.com>.

Henriette Horvath, PhD, is Postdoctoral Research Associate, Department of Crop and Soil Sciences, Washington State University, Pullman, Washington; Institute of Biochemistry and Microbiology, Technical University of Vienna, Wien, Austria. E-mail: <henny@mail.wsu.edu>.

Jintai Huang, PhD, is Postdoctoral Research Associate, Department of Crop and Soil Sciences, Washington State University, Pullman, Washington; currently at Monsanto Company, Chesterfield, Missouri. E-mail: <jintai.huang@stl.monsanto.com>.

Ernesto Igartua, PhD, is Tenured Scientist, Consejo Superior de Investigaciones Científicas (CSIC), Estación Experimental de Aula Dei, Zaragoza, Spain.

Andris Kleinhofs, PhD, is Professor, Departments of Crop and Soil Sciences and Department of Genetics and Cell Biology, Washington State University, Pullman, Washington. E-mail: <andyk@ewsu.edu>.

Daniel J. Miralles, DrSci, is Adjunct Professor, Departamento de Producción Vegetal, Facultad de Agronomía, Universidad de Buenos Aires, Buenos Aires, Argentina. E-mail: <miralles@agro.uba.ar>.

Marian Moralejo, PhD, Centro Universitat de Lleida-Institut de Recerca I Tecnologia Agroalimentaries (UdL-IRTA), Lleida, Spain. E-mail: <moralejo @pvcf.udl.es>.

Joe T. Ritchie, PhD, is Professor, Department of Crop and Soil Sciences, Michigan State University, East Lansing, Michigan. E-mail: <ritchie@pilot.msu.edu>.

Peter R. Shewry, PhD, is Professor, Institute of Arable Crops Research (IACR)-Long Ashton Research Station, Department of Agricultural Sciences, University of Bristol, Long Ashton, Bristol, England. E-mail: <Peter.Shewry@bbsrc.ac.uk>.

Antonio Michele Stanca, Experimental Institute for Cereal Research, Foggia, Italy. E-mail: <m.stanca@iol.it>.

J. Stuart Swanston, PhD, Scottish Crop Research Institute, Mylnefield, Invergowrie, Dundee, Scotland. E-mail: <jswans@scri.sari.ac.uk>.

William T. B. Thomas, PhD, Scottish Crop Research Institute, Mylnefield, Invergowrie, Dundee, Scotland. E-mail: <wthoma@scri.sari.ac.uk>.

Steven E. Ullrich, PhD, is Professor, Department of Crop and Soil Sciences, Washington State University, Pullman, Washington. E-mail: <ullrich@wsu.edu>.

Giampiero Valè, PhD, is Postdoctoral Research Associate, Experimental Institute for Cereal Research, Foggia, Italy. E-mail: <gp.vale@iol.it>

Jordi Voltas, PhD, Departmento de Producció Vegetal i Ciència Forestal, Universitat de Lleida, Llieda, Spain. E-mail: <jvoltas@pvcf.udl.es>.

Diter von Wettstein, PhD, is Professor, Department of Crop and Soil Sciences and Department of Genetics and Cell Biology, Washington State University, Pullman, Washington. E-mail: <diter@wsu.edu>.

Oi T. Wong, MS, is Technical Assistant, Departments of Crop and Soil Sciences and Department of Genetics and Cell Biology, Washington State University, Pullman, Washington.

Preface

New discoveries and novel interpretations of previous findings have produced changes in barley science at a continuously increasing pace. Barley is the world's fourth most important cereal crop and is as ancient as the origin of agriculture itself. Archaeological studies have found barley grains in preagricultural layers older than 10,000 years. Similar to other cereals, it has utility as a feed and food grain and has always been one of the preferred raw materials for malting. Both the importance and antiquity of barley as a major crop explain why it is one of the plant species that has received strongest attention by researchers in various fields of knowledge, from molecular biology to agronomy, related to its two dominant traits of interest: yield and quality.

Although books covering different aspects of barley science have been published in the past, there does not appear to be a recent volume updating and condensing the information on most fields of barley yield and quality. In the 1960s, an important book was published on malting and brewing issues: *Barley and Malt: Biology, Biochemistry, Technology,* by A. H. Cook. Almost two decades later, *Brewing Science,* a book with an even more specific scope, was published by J. R. A. Pollock (1979). Later on, two books more general in scope and closer to the scope of the present book appeared: *Barley,* edited by D. E. Briggs (1978), and *Barley,* edited by D. C. Rasmusson (1985), followed by a similar sort of book, *La cebada,* written in Spanish by J. L. Molina-Cano in 1989. Finally, within the past decade, three other internationally relevant books were published, but on far more restricted fields than those covered in the present text: *Barley: Genetics, Biochemistry, Molecular Biology and Biotechnology,* edited by P. R. Shewry (1993); *Barley: Chemistry and Technology,* edited by A. W. MacGregor and R. S. Bhatty (1993); and *Le varietá di orzo coltivate in Italia,* edited by J. Delogu, V. Terzi, L. Cattivelli, and A. M. Stanca (1993).

In this book, *Barley Science: Recent Advances from Molecular Biology to Agronomy of Yield and Quality,* we have attempted to expand and update the coverage given by the general revisions published by Briggs, Rasmusson, and Molina-Cano. In addition, Shewry, Cattivelli, and Stanca are contributors to this book, and Molina-Cano is both co-author of two chapters and co-editor of the book.

Barley Science is mainly addressed to agronomists, breeders, geneticists, and physiologists working in different fields related to barley. It is clear that the knowledge reviewed in this book has been mostly gained using barley as a crop species, but scientists interested in other crops, particularly other cereals, will also find updated information that we trust will be valuable and useful. Although the present book has not been written as a textbook, based on the prestige and well-gained worldwide recognition of most authors, we expect it to become a useful reference for advanced undergraduate and postgraduate students in courses that center their objectives in subjects covered in some of the chapters. Each author has intended his or her chapter to be a summarized description and critical update of the knowledge in each field.

The nineteen chapters contain information on all aspects of barley, from molecular biology to agronomy of yield and quality. The chapters were ordered to direct readers interested in the overall field from an introduction to discussion of end uses, passing through chapters on aspects of genetics and breeding, physiology, and agronomy. All chapters, with the exception of Chapter 14 by Bort and colleagues, consist of critical reviews of the literature for the subjects covered; Chapter 14 presents an original assessment of potential nondestructive, spectral reflectance indices to estimate yield differences among genotypes, a potentially strong technique to bridge the gap between physiology and breeding.

Overall, the views and ideas of scientists of different fields of expertise from ten different countries enriched the entire book, but as can be expected from multiauthored books, some repetition among chapters as well as occasional disagreement among authors emerge. We decided to allow this lack of homogeneity to persist as a way to highlight similarities and differences among the experts and, therefore, to emphasize the sometimes controversial nature of the knowledge generated by the progress in barley research.

We are indebted to the authors for agreeing to write the chapters and to put aside some of their duties or free time to accomplish this task. Despite the extra time this project required of us, we have truly enjoyed editing this work. We do hope the authors have enjoyed writing their chapters also, and that readers will share this feeling with us as well.

Chapter 1

Contribution of Barley to Agriculture: A Brief Overview

Gerhard Fischbeck

Cultivation of barley traces back to the earliest remains of agricultural activity in the Old World about 10,000 years ago. Only a few other species, such as Einkorn and Emmer wheat, lentils, peas and chickpeas, bitter vetches, and flax, have been found to have accompanied barley during the transitional phase from preferential reaping of wild plants to purposely performed cultivation. These plants offered suitable mutations from which the earliest forms of domesticated plant species within this region originated.

HISTORY OF BARLEY CULTIVATION

The most ancient remains uncovered from archaeologic research that document collection of food staples from the natural flora date back about 17,000 years B.C. (Ohalo II, south shore of the Sea of Galilee [Kislev et al., 1992]). They are composed of wild emmer *(Triticum dicoccoides)* and wild barley *(Hordeum spontaneum)*. Remains of wild barley appear again together with wild einkorn *(Triticum boeoticum)* in preagricultural layers retrieved from Tell Abu Hureyra, Syria (Hillmann, 1975), and date back to 8,000 to 10,000 B.C.

From about 8,000 B.C. and later, remains of nonbrittle rachis barley appear in an increasing number of excavations, in most cases together with tough rachis types of *Triticum monococcum* (einkorn) and *Triticum turgidum* ssp. *dicoccum* (emmer wheat), and characterize the Old World's contribution to the rise of agriculture. Compared to the time that has elapsed from the remains found at Ohalo II, a rapid spread of agricultural activities throughout the Fertile Crescent region is noted, as shown by Zohary and Hopf (1993). Tough rachis mutants certainly occurred also during earlier millennia, but a long time passed before they assumed their catalytic role in human-guided cereal production. Two-row ear types and hulled kernels characterize the early forms of cultivated barley. They sometimes represent larger percentages of the remains found at the earliest sites (e.g., see Tell Abu Hureyra, in Figure 3 of the book by Zohary and Hopf, 1993) than does either of the two wheat species, but especially emmer wheat, which generally increased in importance during the seventh millennium B.C.

Very interestingly, a few kernels of naked barley have been found among the remains from Tell Aswad (Van Zeist and Bakker Heeres, 1985) and again from Ali Kosh (see Figure 3 in Zohary and Hopf, 1993) dating back to the seventh millennium B.C. In addition, Ali Kosh (Helbaek, 1969) also yielded kernels from six-row ear types, indicating that the four major types of cultivated barley, namely, hulled versus naked kernels and two-row versus six-row ears, appeared already during the early phases of plant cultivation in the Old World, when it still was confined to the Fertile Crescent region (Zohary and Hopf, 1993). Based on additional mutations, spontaneous hybridization between different types of cultivated barley as well as with accompanying weedy forms of *Hordeum spontaneum* probably allowed a multitude of barley genotypes to survive within mixed stands that differed in morphological as well as in physiological traits, but which also were subjected to the forces of natural selection for adaptation to the prevailing growing conditions. Such forces gained even more in importance with the expansion of agriculture into new environments beyond the Fertile Crescent. In addition, drift effects from restriction in population size of the seed lots that were taken along as well as effects of mass selection for types that may have received preferential attention from different farming communities will have played a role.

As is documented from archaeologic research, it was already during the sixth and fifth millenniums B.C. that cultivated barley accompanied emmer and einkorn wheat, often together with weedy forms of wild barley, into the Aegean region and subsequently into all other regions of the eastern part of the Mediterranean Basin (Zohary and Hopf, 1993).

Early remains of cultivated barley from Egypt date back to the fifth millennium and to sites in the Nile Delta (Darby et al., 1977). Later phases of ancient civilization in the Nile Valley credited barley as a gift of the goddess Isis, and germinating barley kernels symbolized the resurrection of goddess Osiris. Most probably, the mixture of barley types that characterizes the ancient remains of cultivated barley from Egypt was carried from the Nile northward and, in the end, may have reached the highlands of Ethiopia. Continuous cultivation practices under a wide range of ecological conditions finally developed into the secondary center of genetic diversity, which still exists within the Ethiopian land races of barley (Lakev et al., 1997) that impressed and puzzled N. I. Vavilov and his scholars so much.

Expansion of agriculture in the eastern direction apparently took place at the same time, as is documented from early remains of the major crops. Not only the Caucasus and Transcaucasus regions were reached during the fifth millennium (Lisitsina, 1984); remains of cultivated barley also indicate that the Old World type of agriculture was practiced even in the highlands of the Indian subcontinent (Costantini, 1984).

Expansion into the western parts of the Mediterranean Basin is witnessed from remains that date back to the fourth millennium (Hopf, 1991), and there is an ongoing scientific debate as to whether Morocco represents a secondary center in which barley was domesticated independently from the Fertile Crescent sources (Molina-Cano et al., 1999; Zohary, 1999).

Another pathway of expanding agriculture originating from the Aegean region turned to the north, moving upward along the riverbeds of the Danube, throughout the Balkan region, and upward along the Dniester from Ukraine into Poland. With further expansion, cultivated barley reached Central and Northern Europe during the third millennium B.C. (Körber-Grohne, 1987).

From present-day knowlege, it appears that cultivation of barley reached China only during the second half of the second millennium B.C. (Ho, 1977). It may have been the exchange of seeds from the Old World type of agriculture with the rice-based agriculture in the Far East civilization that established it there and, further, on the Korean Peninsula and the islands of Japan.

Considering the changes in frequency of barley within the remains from ancient cereal cultures and the presence of different barley types among them (Zohary and Hopf, 1993), it appears that during the earliest phases of agricultural activities barley occurred in rather large percentages, and very often as a mixture of two-row and six-row ear types as well as hulled and naked kernels. A characteristic change during the later phases and development of agriculture throughout the Old World is associated with the preference given to free-threshing types of wheat. In the beginning this was restricted to the tetraploid progenies of emmer wheat (*T. turgidum* ssp. *durum* and others) but later on included the hexaploid common wheat *(T. aestivum)* that is supposed to have originated from spontaneous hybridizations of tetraploid cultivated wheat with *T. squarrosa,* a wild wheat-grass species that occured most probably within the Caucasus/Transcaucasus region when the tetraploid wheats entered there (Zohary and Hopf, 1993). With reference to barley, the exceptions from the general rule of increased preference for wheat deserve special interest, and it has already been observed (Hillman, 1975) that barley occupied the prime position mainly at locations characterized by less favorable growing conditions.

During the fourth millennium B.C., the Mesopotamian Basin was almost completely covered with agricultural activity, and it was during this period that six-row, hulled barley types outnumbered not only other types of barley but also cultivated wheats. It is assumed that tolerance against nutritional depletion of the cultivated soils and/or salinity problems was the reason for it (Zohary and Hopf, 1993).

Six-row ear types of barley also played a major role during later and more far-reaching expansions of agriculture, for example, into the western part of the Mediterranean Basin. Only six-row ear types of barley reached Central and Northern Europe during the third and fourth millenniums B.C., and it is also documented (Körber-Grohne, 1987) that barley represents only small percentages among the remains found from sites located on better soils, but its presence increases drastically in the remains from settlements located on very light, sandy soils, which are widely distributed across the more northern lowlands of Central Europe, together with the percentages of naked barley kernels. In fact, two-row ear types of barley remained virtually unknown to cereal cultivation practices in Central and Northern Europe throughout the first millennium A.C. Without well-documented proof, it is assumed that two-row ear types were introduced only with seeds brought along by crusaders who fought and lived in Near East countries during the twelfth and thirteenth centuries.

Much more recent in time and well documented is the spread of cultivated barley into the Americas and Oceania. Settlers following the Spanish conquerers introduced barley seeds into Mesoamerica and the southern parts of the United States (Poehlman, 1959). Mainly six-row, lax ear types originating from North Africa were found to grow well and gave rise to the ecotypes cultivated in this region and, later on, to cultivars that are still in use today. Furthermore, six-row barley types became naturalized members of cultivated crops in the Andean Highlands. Later introductions of two-row barley from Australia and Europe gained limited importance in the southern plains regions of Uruguay and Argentina.

During the eighteenth and nineteenth centuries, immigrants from many Western, Central, and Eastern European countries who settled in New England and the Midwest brought along barley seeds from land races that originally were grown in their European homesteads. Clear and sometimes surprising differences concerning the level of adaptation to the new environment were noted. These gave rise to the preferential distribution of regionally well-adapted types, such as Manchuria and Oderbrucker in the Midwest. Together with similar experiences with other immigrant-introduced crop plants, this eventually resulted in the U. S. Plant Introduction Service (Moseman and Smith, 1981), which implemented more systematic approaches after assembling crop plants from other countries and evaluating their potential for crop production in the United States.

Introduction of barley into Oceania was closely related, and in the beginning also restricted, to its ties with the British Empire. Later on, Australian scientists (Finlay and Wilkinson, 1963) used the results of comparative yield trials with a large set of barley introductions under Australian growing conditions to develop a method that can be applied to quantify differences in adaptation between the accessions tested and to evaluate their potential for crop improvement.

UTILIZATION OF BARLEY

During the rise of Old World agriculture, barley was grown together with einkorn and emmer wheat to provide food staples for human nutrition. Evidence on barley's place in the human diet at the onset of agriculture is strengthened with the early appearance of naked kernel types. An early preference of wheat over barley is correlated with the transition to baking and eating bread, based on the lack of gluten among the storage proteins of barley kernels. Although bread from barley is mentioned in the ancient literature, and at least admixtures from barley probably occured rather frequently in ancient breadmaking, food uses of barley are more related to soup and porridge dishes, which have survived together with cultivation of naked barley into present times. Most specifically, this applies to the need to expand or at least maintain production of cereals at levels required for growing population sizes. Stagnant or even declining cereal yields mean that plants need higher levels of stress resistance and shorter life cycles. It was on such a basis that agriculture introgressed into regions where barley almost regularly outyielded the more pretentious wheat species, and barley maintained or gained prime importance as a food staple in relevant regions and cultures, as is still the case in Tibet and Ethiopia, as well as in other high mountain areas where such conditions prevail.

Dough for breadmaking as well as porridge from cereal grains, if not baked or eaten in due time, easily allow naturally occuring yeasts to begin fermentation processes. With adequate care, these processes led to the production of alcoholic drinks from cereals during the early phases of ancient agriculture. It also may have been known early that sprouted kernels are better suited for such purposes. While in principal all kinds of cereal grains can be used, it again was learned through experience that filtration of the nonfermented wort extracted from malted grains can be performed most efficiently with barley due to the high extract yield obtained from well-modified starch during the malting process and the possibility of utilizing the remaining hull content as an effective wort filtration layer. It is on this basis that barley became the major source of raw material needed by the brewing guilds that formed in Europe during medieval times, from which the malting and brewing industries later developed, and well-shaped, uniform kernels from hulled, two-row ear types gained a preferential position in the production of malting barley.

Cultivation of barley for feed is a more recent development despite its being the major focus of contemporary barley production. This trend relates to the decreasing importance of barley in food production. In addition, increasing levels in the consumption of animal products together with more intensive methods for raising animals created the need for a high-energy feed. Especially for hog feeding and other lines of animal production in Europe, feed barley gained, and still occupies, a very important position.

SHARE OF BARLEY IN TODAY'S CEREAL PRODUCTION

Compared to its early companions, einkorn and emmer wheat, barley has a wider distribution and greater importance, as well as being maintained in contemporary agriculture despite a considerable reduction in recent years (see Table 1.1). The 1996/1998 average amounts to acreage of close to 65 million ha devoted to barley cultivation worldwide, occupying 9.1 percent of the world's cereal acreage (FAO, 1998). Following the much larger acreages of wheat, rice, and maize, each of which covers more than 20 percent of the world's cereal acreage, barley assumes the fourth position in total production acreage. Barley occupied this position during the early decades of the twentieth century (FAO, 1947) and maintains it in present times. A closer look at the geographical distribution of barley cultivation nevertheless reveals a very characteristic degree of unbalance (see Table 1.1).

With Europe in the leading position, Old World countries in Europe and Asia plant about 75 percent of the world's barley acreage, despite a notable tendency to decrease acreage and production. North and Central America contribute little more than 10 percent, whereas barley cultivation in the other continents stays below 10 percent of the total cereal acreage in these regions, despite substantial increases in Oceania and South America. Considering that barley production is close to 150 millon tons/year, its contribution to the contemporary world cereal basket amounts to only 7.2 percent and stays clearly below its share in the world's cereal acreage. The unbalanced geographical distribution of barley acreage is seen also in the respective shares of barley production. The contribution from barley fields in Europe amounts to nearly 60 percent of the world's barley production, while production in North America reaches about an equal share as in Asia, despite substantial differences in barley acreage.

Obviously, the broad scale survey given in Table 1.1 characterizes the different trends in current barley production but is not sufficient to describe precisely the contributions of barley to contemporary farming systems. A more regional approach is needed, as well as the comparison of data that indicate differences in the competitive position of barley against other cereal species. Therefore, the actual figures on barley acreage have been used to calculate the share barley takes from total cereal acreage in countries that have barley fields covering more than 100,000 ha (see Table 1.2). The results demonstrate large regional differentiations, especially among Old World countries.

In Syria and Iraq, barley occupies more than 40 percent of the cereal acreage. These countries include large parts of the Fertile Crescent region, where cultivation of barley originated at the rise of Old World agriculture about 10,000 years ago. The large barley acreage in this region is based on the high degree of adaptation of barley land races (Ceccarelli et al., 1995) that still dominate barley production within the area to the arid and drought-

TABLE 1.1. World Acreage and Production of Barley, 1996/1998 Average

	Acreage (million ha)	Share (%)	Production (million ton)	Share (%)	Difference[1] (%)
World	64.3		149.8		−12.2
Europe	32.7	50.8	90.3	60.3	(−10.3)[2]
Asia	15.0	23.3	23.6	15.7	(+9.7)[2]
North and Central America	7.4	11.5	22.6	15.1	+2.2
Oceania	3.3	5.1	6.9	4.6	+50.0
Africa	5.0	7.7	5.3	3.5	−12.2
South America	0.8	1.2	1.5	1.0	+36.4

Source: FAO, 1998.

[1] Increase/decrease of production compared to 1989/1991 average
[2] Without follower states of the USSR (follower states' 1996/1998 barley production 37.2 percent below USSR 1989/1991 average)

TABLE 1.2. Share of Barley in Total Cereal Acreage, 1996/1998 Average

>40%		25-40%		10-25%		<10%	
Asia							
Syria	46.3	Turkey	26.5	Kazakstan	19.5	Rep. Korea	7.2
Iraq	45.8			Iran	18.5	Afghanistan	7.0
				Azerbaijan	16.1	China	1.7
				Kyrgyzstan	15.9	India	0.8
				Pakistan	12.4		
				Uzbekistan	10.2		
Europe							
Ireland	63.3	Latvia	39.1	Ukraine	28.5	Romania	9.2
Spain	53.2	UK	37.8	Russ. Fed.	23.4	Italy	8.6
Norway	51.5	Czech Rep.	37.4	Bulgaria	14.8	Yugoslavia	6.0
Finland	50.9	Sweden	36.8	France	14.3		
Estonia	50.0	Belarus	36.1	Poland	13.2		
Denmark	44.6	Germany	32.0	Hungary	12.5		
		Austria	30.9	Moldova	12.0		
		Slovakia	28.6	Greece	11.4		
Africa							
Libya	86.4	Tunisia	29.0	Ethiopia	13.2	Rep. S. Africa	8.2
Morocco	40.9	Algeria	27.3				
North and South America and Oceania							
				Canada	23.9	United States	4.2
				Australia	20.0	Mexico	2.4
				Uruguay	19.3	Argentina	2.5
				Peru	14.9	Brazil	0.7

Source: FAO, 1998.

inflicted climate as well as to salinity-prone soil conditions. Barley also fits well into the traditional farming systems, which are based on raising large flocks of sheep and goats, that have developed and persisted over the centuries and prevail into present times. Depending on available moisture, barley stands may be used for grazing the animals or else to harvest grains, and in some cases, after grazing and sufficient rainfall, an additional grain crop can be harvested. It is easy to follow the expansion of similar farming systems in a western direction across the North African countries as well as in the eastern direction into countries where large regions show similar climatic and soil conditions, together with a common cultural background of the farming population that has settled there (see Table 1.2).

The lack of moisture over large areas of rain-fed agriculture also explains the high percentage that is devoted to barley cultivation in Spain. Low, but rather safe, hectare productivity outcompetes other cereal species with higher moisture requirements under such circumstances.

Another region in which barley occupies the prime position in cereal culture extends from Ireland across Scotland to the northern countries of Scandinavia, including Estonia. Barley production again serves mainly the farmers' needs for feeding their animal stocks. Different ecotypes of barley cultivars are sown over the region that differently combine a number of characters conferring a high degree of adaptation to the farming systems that are practiced within the subregions.

Although the mild Atlantic climate in Ireland allows for a long vegetation period together with high kernel yields from all cereal species, and with not much difference between fall- and spring-sown crops, barley is best adapted to combine harvesting and ripens early enough to perform this operation with a comparitively low weather risk. Similar conditions extend to Denmark and the northern parts of the United Kingdom, which has special needs related to its hog-based meat export industry. Within these regions, mainly two-row types of barley are grown, and these types are generally preferred by the malting industry, which also plays an important role in their economies.

The northern fringe bent for barley cultivation is placed in the Nordic countries (Norway, Finland, Estonia), where agricultural activities are restricted by a very short vegetation period, the disadvantages of which being somewhat alleviated by an excessively extended photoperiod. Except for in Estonia, cereal culture in this region is largely restricted to spring-sown crops of six-row barley. Ecotypes of land races and improved cultivars of barley have been developed that excel in earliness and high harvest index and, on this basis, outcompete all other cereal species within this ecoregion. Early maturing, six-row ear barley types generally are the borderline for cereal culture, not only for higher latitudes, but also for higher altitudes. This applies not only to Old World countries. Barley production for food, often based on six-row, free-threshing, naked kernel types, has played a major role not only in the mountain regions in Asia (Tibet), where it is still being practiced, but also in the Andean Highlands in Peru and in Ethiopia. Even if it is uncertain to what extent and in which regions the cultivation of such marginal lands will continue in the future, there are at least historical reasons to give credit to the high-ranking importance of barley in this respect.

As shown in Table 1.2, barley occupies at least medium to high levels of the cereal acreage in a large number of countries in North, Northwest, and Central Europe (Sweden, southern parts of the United Kingdom, Germany, Austria, the Czech Republic, Slovakia), even though many of these countries have decreased barley acreage to some extent in recent years while increasing wheat acreage. Similar conditions also apply to the northeastern part of France. In this region, contemporary barley production mainly goes for animal feed, but the production of malting barley maintains its traditional role in the brewing industry. For this industry, barley production is largely based on two-row spring barley cultivars, whereas the contribution of barley to the overall cereal acreage is more often based on two- and six-row winter barley cultivars. Winter barley acreage increased over several decades as a consequence of fully mechanized cereal production, which allowed early fall sowing of winter barley as a major prerequisite for realization of its high yield potential. Cultivation of winter barley in Central Europe for a long time relied on six-row, lax-ear types, but an important sector of two-row, lodging-resistant winter barley cultivars was planted for the production of not only feed grain but also malting barley.

Starting from Latvia and Belarus, medium to low cereal acreages are sown to barley into Eastern Europe across Poland and the Ukraine and further into the vast territories of the Russian Federation, including its trans-Ural provinces. Two-row spring barley cultivars prevail, except for regions surrounding the Black Sea, which have a milder winter climate. A substantial decrease in barley production in the USSR follower states has been noted before (see Table 1.1). This does not reflect a similar loss in competitive power against other cereal crops; it is explained by a general decrease in cereal production that apparently goes along with the difficulties these countries have faced during their economic transition.

A decline in the competitive power of barley cultivation in comparison to other cereal crops is noted for most countries in Southeastern Europe. Since drought reductions are less important for yield formation of cereals under rain-fed conditions than hot temperature spells during the second half of the vegetation period, maize gains greater competitive power as compared to barley and, consequently, takes over the prime position for feed grain production.

Apart from the Old World countries, in Asia, barley contributes only small percentages to the cereal acreage of this continent, since food grains from other cereal species, mainly rice and wheat, are of prime importance (namely, in China and India, respectively). Barley production, therefore, is confined to cooler climates in mountain regions or else provides the basis for special food products, but to some extent, barley for feed grain production has also gained importance in double-cropping systems.

Low percentages of barley acreage are documented for the United States, Mexico, Argentina, Brazil, and the Republic of South Africa. Barley is often planted as a cash crop in these countries, and larger percentages of the barley crop are sold to the malting industry than in other countries.

Whereas in Ethiopia as well as in Peru and Uruguay barley has occupied medium positions in cereal culture for a long period of time and for reasons that have been discussed before, a more recent development has been increased barley production, to occupy 20 percent or more of the cereal acreage, in Canada and Australia. In both countries, export of agricultural products is an important sector of the economy. Since large parts of the arable land are not suited for maize production, mainly for climatic reasons, and the world market for wheat in recent times has suffered from surplus, barley has replaced to some extent the wheat acreage in both countries, which mainly seek export markets in Asia for either feed or malting barley.

UTILIZATION OF THE BARLEY CROP

Apart from wheat and rice, the major cereal food staples in today's agriculture, barley is listed in the category of "coarse grains" together with all other cultivated cereal species. Although the share of barley out of the 1996/1998 coarse grain acreage amounts to 19.8 percent, barley contributed only 16.5 percent to the world's production of coarse grain cereals, way behind the 65 percent share of maize. Although barley acreage has decreased for several reasons by close to 14 percent since 1989/1991, a further increase in acreage of about 6 percent is reported for maize for the same time period. Even if a large part of the recent decrease in barley acreage is associated with economic disturbances within the USSR follower states, a trend that probably will not extend forever, it cannot be overlooked that maize will be a very strong competitor of barley in the future coarse grain market. Available data about world trade in barley indicate for the period 1994/1996 an export/import exchange of about 14 millon tons/year (about 9 percent of the world's barley production during this period). Close to 80 percent of the barley exports originated from the European Union (32 percent), Australia (23 percent), and Canada (22 percent).

Incorporated with the term "coarse grain" is the understanding that major parts of these commodities are used for feeding animals. Notwithstanding that considerable differences exist between countries and regions, it can be estimated that, in fact, about 85 percent of today's world barley production is used for feeding animals.

Very often barley production for feeding grains is based on the farmer's preference for a high degree of self-sufficiency in this respect, and there still are large areas where barley outcompetes other "coarse grain" species, especially maize, due to relevant limitations caused by the prevailing growing conditions (e.g., arid climate, short vegetation period), but also in areas with more and sometimes even very favorable growing conditions that do not allow maize to mature (e.g., Central and Northwest Europe). Barley will have to withstand competition even in such regions from favorable offers of feed mixtures from the feed industry that may take advantage of low-price imports of other feed grains. Another line of competition that cannot be overlooked arises from surpluses of bread wheat that, in the end, also are used for feeding animals. One of the defense lines to keep barley competitive goes with specific quality characteristics in favor of feeding barley, as has been noted for hog feeding, and may be extended if future breeding efforts succeed in improving specific quality traits (Newman and McGuire, 1985; Campbell, 1996; Ullrich, Chapter 6 in this book).

Following the leading sector of feed barley, it is the malting industry that mainly relies on the supply of suitable charges of malting barley. In this case, the term "coarse grain" certainly is not justified anymore if the highly specific quality traits are considered that need to be met for malting barley qualifications (MacGregor, 1996). Although malt from barley can be used for a number of other purposes (Briggs, 1978), it is mainly the brewing industry that consumes the barley malt production. With a world production that approaches 1.3 billion liters of beer/year it can be estimated that about 18 million tons of barley will be needed for this purpose, with some uncertainty about the amount of starch from unmalted cereal grains (maize, rice) that often is added to the mash from barley malt. Barley malt for the production of the famous Scottish whiskeys should also be mentioned; it certainly adds to the need for malting barley supplies in the United Kingdom, but this is offset by the offer from the countries with high barley production potential.

Although beer consumption appears to be decreasing in major beer-consuming countries in Europe, this is overcompensated for by rising beer consumption in other European countries and in parts of Asia and South America, so that total world beer production increases steadily. One of the consequences of such developments relates to the expansion of the world's malt market. In the mid-1990s, about 3 millon tons of malt were exported/imported, which amounts to about 17 percent of total malt production, and the European Union contributed more than 50 percent to this trade. Special interest in malting barley, therefore, will most probably increase its share in future barley production.

Except for special conditions, as has been mentioned for Tibet, Ethiopia, and Peru, barley food consumption has lost most of its earlier position in human nutrition, at least in industrialized countries. With recent efforts and progress in human nutrition research (Oakenfull, 1996), the level of nutritional quality of new or traditional foodstuffs from barley is considered as being equal to oat products. To some extent barley food consumption profits from trends to diversify mixtures in breakfast cereals and other newly developed food products. But even with further increases, barley will play only a minor role as a food crop, and this probably will not change in the foreseeable future.

Finally, possible uses for vegetative plant organs of the barley plant should also be mentioned (Briggs, 1978). In certain regions, mainly in dryland farming systems in Near East countries, barley stands provide possibilities for grazing animals in years when growth in pasture grounds remains insufficient. This may explain the preference for two-row barley types, because of their high tillering capacity. Occasionally barley stands are also used for producing hay or silage.

Very often the straw from harvested barley grains has been, and in part is still being, used for feeding ruminants during seasonal interruptions of the growth period. In many traditional farming systems that need to keep animals in stables, straw from barley and other cereal species is utilized for providing a straw shed that takes up the animal waste, providing also farmyard manure for maintaining soil fertility. With fully mechanized systems of animal production, the straw shed is often avoided, and combine harvesting at the same time cuts the straw into small pieces that are left on the ground for mulching and to restore organic matter content of the topsoil.

More recently (Munck, 1991), it has been suggested to consider whole-plant harvesting of barley and other cereals, to include all plant organs by suitable transformations, for use in feed and as food and also in producing intermediates for use in different industries. The suggestion is based on the argument that industrial whole-plant processing would not only contribute to alleviating surplus problems in cereal production but could also be important in encouraging a general shift toward the utilization of renewable resources from agricultural products.

REFERENCES

Briggs, D. E. (1978). *Barley*. London: Chapman and Hall, pp. 481-525, 560-586.

Campbell, G. L. (1996). Oat and barley as livestock feed—The future. In *V International and VII International Barley Genetic Symposium Proceedings, Invited Papers* (eds. G. Scoles and B. Rossnagel). Saskatoon: University of Saskatchewan Press, pp. 50-57.

Ceccarelli, S., S. Grando, and I. A. G. van Leur (1995). Barley landraces offer new breeding options for stress environments. *Diversity,* 11: 112-113.

Costantini, L. (1984).The beginning of agriculture in the Kachi Plain: The evidence of Mehrgarts. South Asian Archaeology 1981. *Proceedings 6th International Conference Association of South Asian Archeologists in Western Europe* (ed. B. Allchin). Cambridge: Cambridge University Press, pp. 29-33.

Darby, W. J., P. Ghaliounugi, and L. Grivetti (1977). *Food—The gift of Osiris,* Volumes 1 and 2. London: Academic Press.

FAO (1947). *Yearbook of Food and Agricultural Statistics—1947.* Washington, DC: FAO.

FAO (1998). *Production Yearbook,* Volume 52. Rome: FAO.

Finlay, K. W. and G. N. Wilkinson (1963). The analysis of adaptation in a plant breeding programme. *Australian Journal of Agricultural Research* 14: 742-754.

Helbaek, H. (1969). Plant collecting, dry farming and irrigation agriculture in prehistoric Deh Luran. In *Prehistoric and human ecology of the Deh Luran Plain* (eds. F. Hole, K. V. Flannery, and J. A. Neely). Memoirs Museum of Anthropology. No. 1. Ann Arbor: University of Michigan, p. 383.

Hillman, G. (1975). The plant remains from Tell Abu Hureyra: A preliminary report. *Proc. Prehist. Soc.* 41: 70-73.

Ho, P. T. (1977). The indigenous origins of Chinese agriculture. In *Origins of agriculture* (ed. C. A. Reed). The Hague: Mouton, pp. 413-418.

Hopf, M. (1991). South and Southwest Europe. In *Progress in Old World paleoethnobotany* (eds. W. van Zeist, K. Wasilikowa, and K. E. Behre). Rotterdam: Balkema, pp. 241-277.

Kislev, M. E., D. Nadel, and I. Carmi (1992). Grain and fruit diet 19,000 years old at Ohalo II, Sea of Galilee. *Israel. Rev. Paleobot. Palynol.* 73: 161-166.

Körber-Grohne, U. (1987). *Nutzpflanzen in Deutschland.* Stuttgart: Theiss, pp. 48-52.

Lakev B., Y. Semane, F. Alemayehu, H. Gehre, S. Grando, A. J. van Leur, and S. Ceccarelli (1997). Exploiting and diversity in barley landraces in Ethiopia. *Genetic Resources and Crop Evolution* 44: 2.

Lisitsina, G. N. (1984). The Caucasus—A centre of ancient farming in Eurasia. In *Plants and ancient man* (eds. W. van Zeist and W. A. Casparie). Rotterdam: Balkema, pp. 285-292.

MacGregor, A. W. (1996). Biochemistry of malting—The way forward. In *V International and VII International Barley Genetic Symposium Proceedings, Invited Papers* (eds. G. Scoles and B. Rossnagel). Saskatoon: University of Sasketchewan Press, pp. 1-6.

Molina-Cano, J. L., M. Moralejo, E. Igartua, and I. Romagosa (1999). Further evidence supporting Morocco as a center of origin of barley. *Theoretical and Applied Genetics* 98: 912-918.

Moseman, J. G. and D. H. Smith Jr. (1981). Purpose, development, utilization maintenance and status of the USDA barley collection. In *Barley Genetics IV* (ed. R. N. H. Whitehouse). Edinburgh: University of Edinburgh Press, pp. 67-70.

Munck, L. (1991). The contribution of barley to agriculture today and in the future. In *Barley Genetics VI,* Volume II (ed. L. Munck). Copenhagen: Munksgaard Intern Publishers, pp. 1099-1109.

Newman, C. W. and C. F. McGuire (1985). Nutritional quality of barley. In *Barley* (ed. D. C. Rasmusson). Madison, WI: ASA, CSSA Series Agronomy No. 26, pp.403-439.

Oakenfull, D. (1996). Food applications for barley. In *Barley Genetics VII* (eds. G. Scoles and B. Rossnagel). Saskatoon: University of Saskatchewan, Press. pp. 50-57.

Poehlman, J. M. (1959). *Breeding field crops*. New York: H. Holl, pp. 22-50, 150-173.

Van Zeist, W. and J. A. H. Bakker Heeres (1985). Archeological studies in the Levant 1 Neolithic sites in the Damascus Basin: Aswad, Ghoarife, Ramad. *Paleohistoria* 24: 165-256.

Zohary, D. (1999). Monophyletic vs. polyphyletic origin of the crops on which agriculture was founded in the Near East. *Genet. Res. Crop Evol.* 46: 133-142.

Zohary, D. and M. Hopf. (1993). *Domestication of plants in the Old World,* Second edition. Oxford: Oxford Science Publications, Clarendon Press, pp. 33-64.

Chapter 2

New Views on the Origin of Cultivated Barley

José Luis Molina-Cano
Ernesto Igartua
Ana-María Casas
Marian Moralejo

HISTORICAL INTRODUCTION—THE QUEST FOR THE ANCESTORS OF CULTIVATED BARLEY

For this introductory section we substantially follow the excellent historical overview presented by Russian botanist F. Kh. Bakhteyev at the First International Barley Genetics Symposium (Bakhteyev, 1964).

In investigating the origin of cultivated barley, different periods can be distinguished, during which answers to several questions were sought: Can any existing species of *Hordeum* L. be considered the ancestor of its cultivated forms, or is the initial species already extinct? What phylogenetic relationships led to the appearance of cultivated barley? What are the main routes of distribution of cultivated barley? Last but not least—Where is the center of origin of cultivated barley?

The first period started when Carl Koch published, in 1848, the first description of the wild barley *Hordeum spontaneum* C. Koch. Later, in 1855, De Candolle, in his book La Géographie Botanique Raissonnée (De Candolle, 1959), investigated the origin of cultivated plants, including barley. In a later book, *Origine des Plantes Cultivées,* published in 1882 (De Candolle, 1959), he again treated the topic of crop domestication, in depth. In addition to these workers, other scholars made substantial contributions with their investigations during this period, for example, Körnicke (1885, 1895), Rimpau (1891, 1892), Beaven (1902), Schulz (1911, 1912), and others (see the summary published by Åberg, 1940). We can summarize the two different views that arose during this period: either *H. spontaneum* was considered as the common ancestor of all cultivated forms, or, alternatively, it was the an-

---

We wish to thank Dr. J. Stuart Swanston, Scottish Crop Research Institute (Dundee, Scotland) for critically reading and improving the manuscript.

cestor of only the two-rowed types, with the origin of six-rowed barleys remaining open. The finding of an appropriate ancestor was central to the definition of the center of origin put forward by De Candolle, as that was based on the existence of an ancestral wild species growing in that territory, either *H. spontaneum* or some other, or even several connected species.

The second period started at the beginning of the twentieth century with the work of Nikolai Ivanovich Vavilov. This encompassed a very fruitful time that brought about the collection and characterization of many accessions of land races of different cultivated plants, including barley, and their wild ancestors, by Vavilov and his co-workers at the All-Union Institute V. I. Lenin Academy of Agricultural Sciences. By 1940, the world collection of barley contained almost 17,000 entries thanks mainly to this effort (see, for example, the book Five Continents, Vavilov, 1997).

The main ideas of Vavilov were published in "The Law of Homologous Series in Variation" (Vavilov, 1922), "Studies on the Origin of Cultivated Plants," "Phytogeographic Basis of Plant Breeding" (Vavilov, 1926, 1949/1950), and also Vavilov (1931, 1940) and others. There are English (Vavilov, 1992) and Spanish (Vavilov, 1951) translations of his theories. Vavilov was somewhat skeptical about considering *H. spontaneum* as the unique ancestor of cultivated barleys. Rather, he considered it as a wild type genetically close to the cultivated forms, undoubtedly having a common origin, perhaps even some common ancestors. By no means, however, did he attribute to *H. spontaneum* the role of ancestor of cultivated barley. Meanwhile, during this second period, a number of new works appeared in addition to those of Vavilov, including Arzt (1926), Becker (1927), Tschermack (1923, 1928), Larionow (1929), Schiemann (1922, 1932, 1939), Åberg (1938, 1940), and others.

We can consider the key event of the third period to be the discovery, by Åberg, of three kernels of husked barley, which turned out to be six-rowed with brittle rachis, within a mixed sample of wheat and oats collected by Smith (1947) in eastern Tibet. This husked barley was named *Hordeum agriocrithon* and considered a wild species as well as the ancestral parent of cultivated six-rowed barleys. Following that, Freisleben (1943) proposed the southeastern Himalayas as the center of origin for cultivated six-rowed barley forms.

This theory was challenged by Bakhteyev (Bakhteyev, 1947), who discovered among carbonized barley samples from archaeological excavations in different USSR republics (Armenia, Azerbaijan) grains with an elongated basal part and a pedicel resembling the neck of a bottle. This new species, first named *Hordeum lagunculiforme* (Bakhteyev, 1957), had pedicelled spikelets unlike *H. spontaneum* and *H. agriocrithon,* both of which had sessile spikelets. Bakhteyev's view, formulated after studying the crossing behavior of the aforementioned species, can be summarized as follows: *H. agriocrithon* was not a genuine wild species, but a hybrid product arising from crosses between *H. spontaneum* and cultivated six-rowed types. *Hordeum spontaneum* C. Koch emend Bacht. was defined as a new species containing both two- and six-rowed forms *(H. spontaneum* v. *lagunculiforme)*, which were, respectively, the ancestral parents of two- and six-rowed cultivated forms. He strongly recommended the barley scientists who met at Wageningen (the Netherlands) in 1964, at the First International Barley Genetics Symposium, to organize further expeditions to collect new material that could help finally to settle the origin of barley.

The theories of Bakhteyev did not meet with general acceptance, and this third period ended with the almost general acceptance of what would appear to be the current "official theory" on the origin of barley, being the one put forward by Harlan and Zohary (1966). They considered the Fertile Crescent as the unique center of origin of barley and *H. spontaneum* C. Koch as the ancestral parent of both two- and six-rowed forms. The sequence of the domestication process (see Nevo, 1992) was thus thought to be as follows: *H. spontaneum* gave rise to cultivated (nonbrittle rachis) two-rowed forms, by mutation at either the Bt or Bt_2 locus, and then the six-rowed and naked forms arose, during domestication, through mutations at the V and N loci, respectively.

The fourth period, with its general agreement, or what we would rather term *Pax Romana,* with regard to a monocentric origin of barley, has lasted up to the present. The accepted view was, however, challenged at the beginning of the 1980s, with the finding of wild barley in Tibet and Morocco, enlarging the area of distribution of the species. This provided material evidence of a multicentric origin of the crop, as we will discuss in the next section. Recently, one of the leading supporters of the Fertile Crescent monocentric theory has proposed that barley was domesticated more than once (Zohary, 1999). Surprisingly, this change of opinion is based not on our findings in Morocco nor those of the Chinese workers in Tibet, but rather on the old "eastern-western" theory about the diversity of brittle rachis genes (Takahashi, 1955).

PROPOSED CENTERS OF ORIGIN FOR BARLEY

The center of origin of a crop, as defined by De Candolle (see previous section), is a region where domestication took place, and in which the wild ancestor and the derived cultivated species coexist. Thus, the center of origin of barley should be located within the distribution range of its wild ancestor, *H. spontaneum*. At different times, several regions of the world have been proposed as centers of origin of barley: the Near East, Tibet, Ethiopia, and the Western Mediterranean region.

The Near East or, more precisely, the Fertile Crescent, is one place where domestication of barley, together with other crops and animal species, took place about 10,000 B.P. Extensive evidence from archaeology, genetics, and distribution patterns of wild and cultivated species supports this hypothesis, which has been thoroughly reviewed in the literature (e.g., Harlan, 1975, 1979; Briggs, 1978; Nevo, 1992; Smith, 1995). The authors referred to, however, also present this region as the only center of origin of barley, which is at first a questionable hypothesis.

Ethiopia was first proposed as a center of origin by Vavilov (1926) because of the vast diversity of morphological types of cultivated barley existing in the region. However, he later accepted that barley was unlikely to have been domesticated in an area in which no plausible wild ancestor is currently found (Vavilov, 1940). The presence of a high level of genetic diversity is not enough to qualify a region as a center of origin for a crop. Currently, the most accepted view considers Ethiopia as a center of diversity, though some studies still claim its status as a center of origin (Bekele, 1983). Significantly, as has been recently discovered, the USDA World Barley Collection contains entries of *H. spontaneum* from Ethiopia (H. E. Bockelman, personal communication).

The case for Tibet as a center of origin has provoked controversy for over seven decades (Vavilov, 1926; Freisleben, 1940). Though the role of *H. agriocrithon* as a possible ancestor has been ruled out, the discovery of *H. spontaneum* in the Qinghai-Xizang plateau (Xu, 1982) brought renewed interest to the debate. The Tibetan wild forms showed distinct morphological and biochemical characteristics, which set them apart from Near East forms. Takahashi (1955) had already pointed out a remarkable feature in the distribution of nonbrittle rachis genes in cultivated barley (an issue central to the domestication process). All Chinese barleys he studied carried a $Bt\ Bt\ bt_2\ bt_2$ genotype, whereas all barleys from neighboring India had the $Bt_2\ Bt_2$ or "western" genotype. It seems that selection for nonbrittle rachis during domestication acted on different loci in Oriental and Western barleys. This genetic differentiation between the two groups of cultivated barleys was also observed in a worldwide survey of barley analyzed for isozyme polymorphism (Kahler and Allard, 1981) and for ribosomal DNA polymorphism (Zhang, Saghai-Maroof, and Yang 1992), and in a recent study using random amplified polymorphic DNAs (RAPDs) (Strelchenko, Kavalyova, and Okuno, 1999). Though this domestication may have occurred rather recently (5000 B.P., according to Xu, 1982), it would have produced a genetic diversity level comparable to that seen in the Ethiopian barleys (Zhang, Dai, and Saghai-Maroof, 1992). This genetic variability, however, showed a nondifferentiated geographic pattern, unlike the Ethiopian barleys, probably due to their recent radiation from a nuclear region (Zhang et al., 1994).

The most important discovery in the Western Mediterranean region was that of the wild ancestor of barley in Morocco. *Hordeum spontaneum*

C. Koch was thought to have a geographical distribution whose western limits, according to the maps provided by Harlan and Zohary (1966) and Harlan (1968, 1979), were some scattered collection places in Cyrenaica (west of Libya). There have been few specific references to the presence of *H. spontaneum* in Northern Africa. Humphries (1980) mentioned the North of Africa without precise definition of the location, although Maire (1955) referred to Cyrenaica in his detailed flora of Northern Africa.

The occurrence of *H. spontaneum* in southern Morocco (Molina-Cano and Conde, 1980) changed opinions about the distribution range of the species, previously thought to have its western limit in Libya. The true *H. spontaneum* nature of those wild barley plants collected in Morocco was confirmed at the Vavilov Institute (F. Kh. Bakhteyev, personal communication). This initial finding led to a second trip, where more populations were found and some of them collected (Molina-Cano, Gómez-Campo, and Conde, 1982). The region explored comprised the surroundings of the Djebel Siroua mountain range. It covers a surface of about 7,000 km^2. The Djebel Siroua itself is of volcanic origin and is located between the Great Atlas and the Anti-Atlas mountains. Its highest peak has an altitude of 3,340 m. Cultivated lands around the range are sometimes irrigated near the villages, where supplies of water can be drawn from small rivers and wells. Dryland growing areas are always found on the *Artemisia* steppe, an arid environment named after its predominant species *Artemisia herba-alba*. Soils outside the *Artemisia* steppe are mainly covered by *Stipa tenacissima*. These soils are lighter and with lower water retention capacity than those of the steppe (Emberger, 1971). Crops are never cultivated on the *Stipa* lands, except in the cases where water supply from wells is guaranteed. Rainfall is very scarce in the whole region, with a yearly average of 250 mm, and it is distributed irregularly throughout winter and early spring. All 23 *H. spontaneum* populations found came from cultivated fields in the *Artemisia* steppe.

From interviews with a number of farmers from the region, we learned that wild barley was widely known there during the time of their predecessors, but nobody was able to make any comment on its origin. The weed did not worry them much because they knew that its spikes would shatter at maturity prior to ripening of common barley. This is probably the reason why the species maintains itself, as the farmers do not weed it from their barley fields.

In addition to the populations described previously, we found two more in the southern hills of the Middle Atlas, on the road from Midelt to Azrou, and a third in the very center of the Middle Atlas, in a barley field surrounded by *Cedrus atlantica* forests, near to the so-called Gounod's Cedar. This is very distant from Djebel Siroua (about 400 km north), and the habitats are completely different.

The occurrence of *H. spontaneum* in Morocco could be explained as follows: (1) Naturally occurring wild barleys would disappear at an undetermined moment after domestication had taken place. This situation might be due to the progressive encroachment of the Sahara from the third millennium B.C. onward (Harlan, 1975). Under these more arid conditions, the only suitable land for wild barley would be restricted to the more humid and favorable environment of barley fields, where soil moisture was conserved by cultivation. Away from these fields, on the very dry and hard *Artemisia* steppe, only xerophytes could survive. (2) *Hordeum spontaneum* was adapted to survive as a weed in cultivated barley fields, then introgression began, leading to its present level of similarity with its cultivated companion. A similar situation in Libya was described by Hammer, Lehmann, and Perrino (1985).

DISTINCTIVENESS OF MOROCCAN H. SPONTANEUM

How these Moroccan populations compare with the *H. spontaneum* collected in the Eastern Mediterranean region and Southwest Asia was addressed by examining a varied array of agronomic, morphological, and biochemical features of the *H. spontaneum* germplasm pool (Salcedo et al., 1984; Molina-Cano et al., 1987, 1999; Moralejo et al., 1994). Genetic variation both within and between all the observed populations of *H. spontaneum* in their natural settings was observed. The most striking observations were in the following characters: earliness, height of the plants, size of sterile and fertile florets, and color of the ear. Moroccan populations were different from those from Afghanistan, Iraq, Israel, and Libya. In particular, they had a more profuse foliar development, more erect habit at heading, significantly shorter and thicker tillers, higher kernel weight and number of kernels per spike, and higher tillering ability than populations of other origins.

Interesting results were also obtained when analyzing the CM-proteins of the grain (Salcedo et al., 1980; Salcedo, Sánchez-Monge, and Aragoncillo, 1982), which are soluble in chloroform-methanol mixtures (Salcedo, Sánchez-Monge, and Aragoncillo, 1982) and belong to the a-amylase-trypsin inhibitor family (Garcia-Olmedo et al., 1987, 1992). They are less variable both in the cultivated and in the wild species than B, C, and D hordeins (Doll and Brown 1979; Shewry et al., 1979, 1983) and are, therefore, more suitable biochemical markers for phylogenetic studies. For example, wheat CM-proteins were among those used by Johnson and co-workers in their extensive studies on the origin and evolution of the *Aegilops-Triticum* species (Johnson, 1972, 1975; Johnson and Hall, 1965).

The variants CMb-3 and CMe-2.1-2.2 (see Table 2.1) were almost exclusively found in Moroccan accessions in the *H. spontaneum* pool, with just one exception: CMe-2.1-2.2 was present in some plants of a population from Crete. The CMe-2.1-2.2 pattern, with both components inherited jointly, was prevalent among Moroccan accessions.

These results were supported by those from a wider survey of restriction fragment length polymorphism (RFLP) carried out in 35 *H. spontaneum*

TABLE 2.1. Diversity of CM-Proteins in Barley Germplasm Pools

Species	Origin	CMb			CMe		
		1	2	3	1	2.1,2.2	3
H. spontaneum	Morocco (8)	6	1	1	1	7	—
	Other origins (11)*	11	—	—	6	1	4
H. vulgare	Morocco (14)	11	—	3	2	12	—
	Spain (49)	12	37	—	29	14	3
	Other origins (63)**	59	4	—	33	18	12

Source: Adapted from Molina-Cano et al., 1987; Moralejo et al., 1994.
Note: CMb and CMe variants and frequencies present in selected groups of accessions of H. spontaneum and H. vulgare.
*Afghanistan, Crete, Greece, Iraq, Iran, Libya
**Austria, Bulgaria, Czechoslovakia, Denmark, Finland, France, Germany, Great Britain, Hungary, Ireland, Romania, Sweden, Turkey

populations from nine different geographic origins, including eight populations from Morocco (see Figure 2.1). It was concluded that diversity within the Moroccan accessions was not large, and these accessions clustered together in a compact group, clearly and significantly different from those from other origins (Molina-Cano et al., 1999). The geographic consistency of other groups was low, except for the populations from Cyprus, which formed as compact a group as their Moroccan counterparts.

Thus, the biochemical, agronomic, and morphological data seem to indicate that Moroccan weed barleys are clearly different from those found elsewhere. These results are in agreement with the findings of Giles and Lefkovitch (1984, 1985), who reported clear differences between *H. spontaneum* gene pools from Morocco and Iran, from a classification based on 46 quantitative and qualitative agronomic characters.

DISTINCTIVENESS OF WESTERN MEDITERRANEAN H. VULGARE

Several authors have pointed out the similarities between cultivated barleys growing in the Western Mediterranean region, on either side of the Straits of Gibraltar. Harlan (1957) suggested that Spanish six-rowed barleys were of North African origin, while Wiebe (1968) stated clearly that their origin was Moroccan. Peeters (1988) classified close to 5,000 barley varieties from all over the world according to several quantitative and qualitative characters. In a three-dimensional plot of relationships among barley types from different countries, he found that those from Morocco and Spain, though demonstrating a low level of diversity, clustered together and clearly diverged from those in other, especially European, countries. Further evi-

FIGURE 2.1. Dendrogram Showing the Grouping of 35 *H. spontaneum* Accessions Based on Cluster Analysis (UPGMA) of RFLP Variation Detected with 21 Probes and 3 Restriction Enzymes

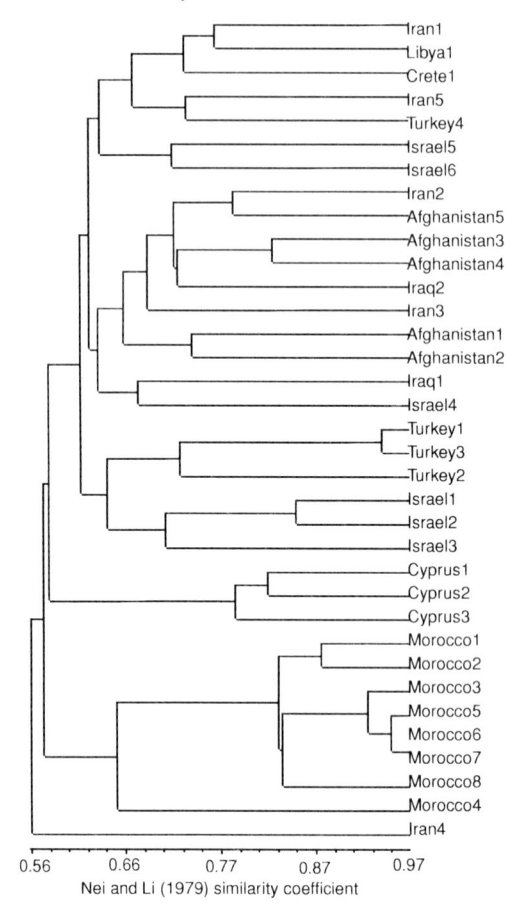

Note: Accessions are labeled according to their geographic origin and numbered sequentially within each origin.

dence of the distinctiveness of Spanish barleys was found from polymorphism at the *Est1* isozyme locus (Kahler and Allard, 1981), and from RFLP variation (Casas et al., 1998). In this last report, the two old Spanish varieties studied showed, by far, the most distinctive RFLP patterns, including several unique alleles, in comparison to 35 mostly European barley varieties of diverse pedigree.

With regard to the variation found in studies of CM-proteins or hordeins (B, C, and D) (Moralejo et al., 1994), cluster analysis led to the formation of groups composed chiefly or totally of Spanish accessions. It was concluded, consequently, that there were protein patterns typical of barleys from Spain. The CMb-2 variant (Table 2.1) was almost confined to Spanish (two-row and six-row types) and French entries (the four accessions in the "other origins" row of Table 2.1 were from France). The occurrence of CMe-2.1-2.2 was also largely restricted to the Western Mediterranean region. Five of the eight accessions from "other origins" were from France. More specifically, the combination CMb-2, CMe-1 was restricted to Spanish entries (and one entry from Turkey), whereas the combination CMb-2, CMe-2.1-2.2 was shared by Spanish, French, and Moroccan entries. These two protein combinations can be considered of Western Mediterranean origin.

The geographical significance of these groups allows us to formulate some hypotheses on the origin of the Spanish barleys studied. The near exclusivity of the Spanish gene combinations and the absence of any *H. spontaneum* with this genotype suggest an ancient in situ domestication with a subsequent disappearance of the putative wild barley ancestor. The present evidence does not contradict this hypothesis, and the ecological similarity between Greece (with actual survival of *H. spontaneum*) and Spain suggests the existence of wild barley populations in the Iberian Peninsula in Neolithic times. The disappearance of *H. spontaneum* from the Iberian Peninsula could have happened after hundreds of years of overgrazing, in a similar way as is postulated for the lowlands of Cyprus by Hadjichristodoulou (1992).

SIMILARITIES BETWEEN WESTERN MEDITERRANEAN H. SPONTANEUM AND H. VULGARE

The results in Table 2.1 show a striking genotypic similarity between *H. vulgare* and *H. spontaneum* from Morocco. The variant CMb-3 appeared only in Moroccan accessions of both species, whereas CMe-2.1-2.2 was equally prevalent in Moroccan accessions of both species. The distribution of the less frequent protein variant, CMb-3, could be explained in terms of introgression. However, the prevalence of CMe-2.1-2.2 in wild and cultivated Moroccan barleys is not easily explained by introgression. This allele probably originated in situ, suggesting that a domestication process might have occurred in Morocco, a situation similar to that postulated for the Qinghai-Xizang region (Tibet) of China (Xu, 1982).

The occurrence of the same genotype (based on CM-proteins) in Spanish and French barleys and in *H. spontaneum* from Morocco (Molina-Cano et al., 1987) suggests the introduction, to the Iberian Peninsula, of barleys that had already been domesticated in the Maghreb. This hypothesis is in agreement with historical records. Alternatively, the occurrence of wild barleys with that genotype, in the Iberian Peninsula during historical times, cannot be ruled out, so an alternative hypothesis is possible.

PHYLOGENETIC CONCLUSIONS: A WESTERN MEDITERRANEAN CENTER OF ORIGIN FOR BARLEY?

The results presented earlier and our previous work (Molina-Cano et al., 1987, 1999; Moralejo et al., 1994) have demonstrated clear genetic differences between the Moroccan populations of *H. spontaneum* and those of other origins within the distribution range of the species. These differences were consistent for all types of markers used: morphological, physiological, CM-proteins, and, currently, RFLPs.

There are some similarities between our results on Moroccan *H. spontaneum* and the situation described for Tibet. The *H. spontaneum* material from Tibet, analyzed by Ma et al. (1987), proved to be very different from that from Israel, particularly with regard to spike color, esterase isozymes, habit type, and rachilla hairs. These workers supported the hypothesis of Tibet as another center of origin of cultivated barley, distinct from the Fertile Crescent.

We could thus conceive the domestication of barley as a process carried out over a continuum of very different environments, i.e., as a noncentric crop, following Harlan's (1975) terminology. Other authors also put forth the hypothesis of a possible center of domestication of barley in Europe (Dennell, 1983; Barker, 1985)

In situ domestication of barley in the Iberian Peninsula is not inconceivable. Barley is a very ancient crop in this region, as the earliest remains of cultivated barley from the Mediterranean coast of Spain were dated with ^{14}C to 6500 and 7590 B.P. (Buxó, 1997). Although the first historical records of Phoenician and Greek visits to the Iberian Peninsula are dated at 800 B.C. (Tuñón de Lara, Tarradell, and Mangas, 1980), earlier contacts cannot be ruled out. There is good, indirect, archaeological evidence that people along the Mediterranean coast had considerable seafaring capabilities before the first crops were domesticated (Smith, 1995). Archaeological evidence clearly shows that domestication of barley (together with emmer and einkorn wheat) took place around 10,000 B.P. in the Fertile Crescent. Archaeological remains by themselves, however, cannot elucidate whether barley cultivation in other areas across the Mediterranean was due to independent domestication events or to contacts with Near East civilizations. Some of the evidence presented in this chapter suggests a possible domestication of wild barley in the Western Mediterranean region, which would thus make it a center of origin for barley, providing we make allowance for the limited evidence based on (1) the restricted number of populations analyzed and (2) the small number of markers used.

Our results, gathered over the past fifteen years (Molina-Cano and Conde, 1980; Molina-Cano, Gómez-Campo, and Conde, 1982; Molina-Cano et al., 1987, 1999; Moralejo et al., 1994), and, particularly, the ones presented here, together with those from Chinese workers (Xu, 1982; Ma et al., 1987), confirm the presence of *H. spontaneum* in regions very distant from the classically accepted distribution range of the species. The Moroccan and Tibetan populations of *H. spontaneum* have also proved to be genetically distinct from those from the Fertile Crescent. It was the presence of an immediate wild ancestor that provided one of the main reasons for proposing the Fertile Crescent as the center of origin for cultivated barley. We suggest, therefore, the hypothesis of a multicentric origin for this crop, which could have been domesticated along a broad arch, starting in Morocco and ending in Tibet.

REFERENCES

Åberg, E. (1938). *Hordeum agriocrithon* nova sp., a wild six-rowed barley. *Annals Agricultural College of Sweden* 6.

Åberg, E. (1940). The taxonomy and phylogeny of *Hordeum* L. Sect. *Cerealia* Ands. with special reference to Tibetan barleys. *Symbolae Botanicae Upsalienses* IV: 2.

Arzt, H. (1926). Serologische untersuchungen über die verwandtschaftsverhältnisse der gerste mit besonderer berücksichtigung des eiweissausgleichs innerhalb der präzipitierenden lösungen. *Botanisches Archiv* 13.

Bakhteyev, F.Kh. (1947). Is *Hordeum agriocrithon* Åberg a species? *Proceedings of the Academy of Sciences of the U.S.S.R.* 57: 195-196 (in Russian).

Bakhteyev, F.Kh. (1957). A fossil form of cultivated barley: *Hordeum lagunculiforme* Mihi. *Annals Royal Agricultural College of Sweden* 23: 309-314.

Bakhteyev, F.Kh. (1964). Origin and phylogeny of barley. *Barley Genetics* I, Wageningen: Pudoc, pp. 1-18.

Barker, G. (1985). *Prehistoric Farming in Europe*. Cambridge: Cambridge University Press.

Beaven, E.S. (1902). Varieties of barley. *J. Fed. Inst. Brew.* 8.

Becker, E.S. (1927). *Handbuch des Gesamten Pflazenbaues* I. Berlin.

Bekele, E. (1983). A differential rate of regional distribution of barley flavonoid patterns in Ethiopia, and a view on the center of origin of barley. *Hereditas* 98: 269-280.

Briggs, D.E. (1978). *Barley*. London: Chapman and Hall, 612 pp.

Buxó, R. (1997). *Arqueología de las Plantas. La explotación económica de las semillas y los frutos en el marco mediterráneo de la Península Ibérica*. Barcelona: Crítica, Grijalbo Mondadori.

Casas, A.M., E. Igartua, M.P. Vallés, and J.L. Molina-Cano (1998). Genetic diversity of barley cultivars grown in Spain, estimated by RFLP, similarity and coancestry coefficients. *Plant Breeding* 117: 429-435.

De Candolle, A.Q. (1959). *Origin of Cultivated Plants,* Second Reprint Edition (1886). New York: Noble Offset Printers, 486 pp.
Dennell, R. (1983). *European Economic Prehistory, an Economic Approach.* London: Academic Press.
Doll, H. and A.D. Brown (1979). Hordein variation of wild (*Hordeum spontaneum* C. Koch) and cultivated (*H. vulgare* L.) barley. *Canadian Journal of Genetics and Cytology* 21: 391-404.
Emberger, L. (1971). *Travaux de Botanique et Ecologie.* Paris: Masson et Cie, 520 pp.
Freisleben, R. (1940). Die Gersten der deutschen Hindukusch Expedition 1935. *Kühn Archiv* 54: 295-368.
Freisleben, R. (1943). Ein neuer fund von *Hordeum agriocrithon. Züchter* 15: 49-63.
García-Olmedo, F., G. Salcedo, R. Sánchez-Monge, L. Gómez, J. Royo, and P. Carbonero (1987). Plant proteinaceous inhibitors of proteinases and α-amylases. *Oxford Surveys of Plant Molecular and Cell Biology* 4: 275-334.
García-Olmedo, F., G. Salcedo, R. Sánchez-Monge, C. Hernández-Lucas, M.J. Carmona, J.J. López-Fando, J.A. Fernández, L. Gómez, J. Royo, F. García-Maroto, A. Castagnaro, and P. Carbonero (1992). Trypsin/α-amylase inhibitors and thionins: Possible defence proteins from barley. In *Barley: Genetics, Biochemistry, Molecular Biology and Biotechnology,* ed. P.R. Shewry. Wallingford, UK: CAB International, pp. 335-350.
Giles, B.E. and L.P. Lefkovitch (1984). Differential germination in *Hordeum spontaneum* from Iran and Morocco. *Zeitschrift für Pflanzenzüchtung* 92: 234-238.
Giles, B.E. and L.P. Lefkovitch (1985). Agronomic differences in *Hordeum spontaneum* from Iran and Morocco. *Zeitschrift für Pflanzenzüchtung* 94: 25-40.
Hadjichristodoulou, A. (1992). A new domestication of the "wild" brittle rachis gene of barley (*Hordeum vulgare* L.). *Plant Genetic Resources Newsletter* 90: 46.
Hammer, K., Chr.O. Lehmann, and P. Perrino (1985). Character variability and evolutionary trends in a barley hybrid swarm—A case study. *Biologisches Zentralblatt* 104: 511-517.
Harlan, H.V. (1957). *One Man's Life with Barley.* New York: Exposition Press, 223 pp.
Harlan, J.R. (1968). On the origin of barley. *U.S.D.A. Agriculture Handbook* 338: 9-31.
Harlan, J.R. (1975). *Crops and Man.* Madison, WI: American Society of Agronomy-Crop Science Society of America, pp. 6-9.
Harlan, J.R. (1979). Barley. In *Evolution of Crop Plants,* ed. N.W. Simmonds. London: Longman, pp. 93-98.
Harlan, J.R. and D. Zohary (1966). Distribution of wild wheats and barley. *Science* 153: 1074-1080.
Humphries, C.J. (1980). Hordeum. In *Flora Europaea,* eds. T.G. Tutin, V.H. Heywood, N.A. Burges, and D.H. Valentine. London: Cambridge University Press, Volume 5, p. 204.
Johnson, B.L. (1972). Protein electrophoretic profiles and the origin of the B genome of wheat. *Proceedings of the National Academy of Sciences, USA* 69: 1398-1402.

Johnson, B.L. (1975). Identification of the apparent B genome donor of wheat. *Canadian Journal of Genetics and Cytology* 17: 21-39.
Johnson, B.L. and O. Hall (1965). Analysis of phylogenetic affinities in the Triticinae by protein electrophoresis. *American Journal of Botany* 52: 506-513.
Kahler, A.L. and R.W. Allard (1981). Worldwide patterns of genetic variation among four esterase loci in barley (*Hordeum vulgare* L.). *Theoretical and Applied Genetics* 59: 101-111.
Koch, C. (1848). Beiträge zu einer Flora des Orientes. *Linnaea* 21.
Körnicke, F. (1885). Die arten und varietäten des getreides. *Handbuch des Getreidebaues* I, Berlin.
Körnicke, F. (1895). *Die Hauptsächlichsten Formen der Saatgerste.* Bonn.
Larionow, D. (1929). Zur frage über den phylogenetischen zusammenhang zwischen zweizeiliger und vielzeiliger gersten. *Angewandte Botanik* II.
Ma, D.Q., T.W. Xu, M.Z. Gu, S. B. Wu, and Y. C. Kang (1987). The classification and distribution of wild barley in the Tibet Autonomous Region. *Scientia Agriculturae Sinica* 20: 1-6.
Maire, R. (1955). *Flore de l'Afrique du Nord.* Paris: P. Lechevalier, Volume III, pp. 373-386.
Molina-Cano, J.L. and J. Conde (1980). *Hordeum spontaneum* C. Koch emend Bacht. collected in southern Morocco. *Barley Genetics Newsletter* 10: 44-47.
Molina-Cano, J.L., P. Fra-Mon, C. Aragoncillo, F. Roca de Togores, and F. García-Olmedo (1987). Morocco as a possible domestication center for barley: Biochemical and agromorphological evidence. *Theoretical and Applied Genetics* 73: 531-536.
Molina-Cano, J.L., C. Gómez-Campo, and J. Conde (1982). *Hordeum spontaneum* C. Koch as a weed of barley fields in Morocco. *Zeitschrift für Pflanzenzüchtung* 88: 161-167.
Molina-Cano, J.L., M. Moralejo, E. Igartua, and I. Romagosa (1999). Further evidence supporting Morocco as a centre of origin of barley. *Theoretical and Applied Genetics* 98: 913-918.
Moralejo, M., I. Romagosa, G. Salcedo, R. Sánchez-Monge, and J.L. Molina-Cano (1994). On the origin of Spanish two-rowed barleys. *Theoretical and Applied Genetics* 87: 829-836.
Nevo, E. (1992). Origin, evolution, population genetics and resources for breeding of wild barley, *Hordeum spontaneum,* in the Fertile Crescent. In *Barley: Genetics, Biochemistry, Molecular Biology and Biotechnology,* ed. P.R. Shewry. Wallingford, UK: CAB International, pp. 19-44.
Peeters, J.P. (1988). The emergence of new centres of diversity: Evidence from barley. *Theoretical and Applied Genetics* 76: 17-24.
Rimpau, W. (1891). Kreuzungsprodukte landwirtschaftlichter kulturpflanzen. *Landwirtschaftliche Jahrbuch* 20.
Rimpau, W. (1892). Die genetische entwickelung der verschiedenen formen unserer saatgerste. *Landwirtschaftliche Jahrbuch 21.*
Salcedo, G., P. Fra-Mon, J.L. Molina-Cano, C. Aragoncillo, and F. García-Olmedo (1984). Genetics of CM-proteins (A-hordeins) in barley. *Theoretical and Applied Genetics* 68: 53-59.

Salcedo, G., R. Sánchez-Monge, and C. Aragoncillo (1982). The isolation and characterization of low molecular weight hydrophobic salt soluble proteins from barley. *Journal of Experimental Botany* 33: 1325-1331.

Salcedo, G., R. Sánchez-Monge, A. Argamenteria, and C. Aragoncillo (1980). The A-hordeins as a group of salt soluble hydrophobic proteins. *Plant Science Letters* 19: 109-119.

Schiemann, E. (1922). Die phylogenie der gerste. *Zeitschrift für Naturwissenschaften* X: 6.

Schiemann, E. (1932). Entstehung der kulturpflanzen. *Handbuch für Vererbungswissenschaft*. III, Berlin.

Schiemann, E. (1939). Neue probleme der gerstenphylogenie. *Züchter* 11.

Schulz, A. (1911). Die abstammung der saatgerste *Hordeum sativum*. *Mitt. Naturf. Ges. Zu Halle*.

Schulz, A. (1912). Die geschichte der saatgerste. *Zeitschrift für Naturwissenschaften* 83.

Shewry, P.R., R.A. Finch, S. Parmar, J. Franklin, and B.J. Miflin (1983). Chromosomal location of *Hor-3*, a new locus governing storage proteins in barley. *Heredity* 50: 179-190.

Shewry, P.R., H.M. Pratt, A.J. Falks, S. Parmar, and B.J. Miflin (1979). The storage protein (hordein) polypeptide pattern of barley (*Hordeum vulgare* L.) in relation to varietal identification and disease resistance. *Journal of the National Institute of Agricultural Botany* 15: 34-50.

Smith, B.D. (1995). *The Emergence of Agriculture*. New York: Scientific American Library, 231 pp.

Smith, H. (1947). Om korn i Ost-Tibet. *Sveriges Utsädesforening Tidskrift* 57.

Strelchenko, P., O. Kavalyova, and K. Okuno (1999). Genetic differentiation and geographical distribution of barley germplasm based on RAPD markers. *Genetic Resources and Crop Evolution* 46: 193-205.

Takahashi, R. (1955). The origin of cultivated barley. *Advances in Genetics* 7: 227-266.

Tschermack, E. (1923). Zweizeilige gerste; Bastardierung. *Handbuch für Landwirtschaftliche Pflanzenzüchtung*. Berlin.

Tschermack, E. (1928). Die stammeltern unserer getreidearten. *Fortschrift für Landwirtschaft* 3.

Tuñón de Lara, M., M. Tarradell, and J. Mangas (1980). Primeras culturas e Hispania romana. In *Historia de España*, ed. M. Tuñón de Lara. Barcelona, Volume 1, pp. 65-90.

Vavilov, N.I. (1922). The law of homologous series in variation. *Journal of Genetics* 12: 47-89.

Vavilov, N.I. (1926). Studies on the origin of cultivated plants. *Bulletin of Applied Botany* 16: 1-248.

Vavilov, N.I. (1931). The Linnean species as a system. *5th International Botanical Congress*. Report of Proceedings, Cambridge, pp. 213-216.

Vavilov, N.I. (1940). The new systematics of cultivated plants. *The New Systematics*, Oxford University Press, pp. 549-566.

Vavilov, N.I. (1949/1950). Phytogeographic basis of plant breeding. *The Origin, Variation, Immunity and Breeding of Cultivated Plants. Chronica Botanica* 13: 14-44.
Vavilov, N.I. (1951). *Estudios sobre el origen de las plantas cultivadas.* Buenos Aires: Ediciones ACME Agency, 186 pp.
Vavilov, N.I. (1992). *Origin and Geography of Cultivated Plants.* Cambridge: Cambridge University Press.
Vavilov, N.I. (1997). *Five Continents.* Rome: IPGRI, 198 pp.
Wiebe, G.A. (1968). Introduction of barley into the New World. In *Barley: Origin, Botany, Culture, Winter Hardiness, Genetics, Utilization, Pests.* Washington DC: USDA Agriculture Handbook No. 338, p. 3.
Xu, T.W. (1982). Origin and evolution of cultivated barley in China. *Acta Genet. Sin.* 9: 440-446.
Zhang, Q., X. Dai, and M.A. Saghai-Maroof (1992). Comparative assessment of genetic variation at 6 isozyme loci in barley from two centers of diversity: Ethiopia and Tibet. *Chinese Journal of Genetics* 19: 119-126.
Zhang, Q., M.A. Saghai-Maroof, and P.G. Yang (1992). Ribosomal DNA polymorphisms and the Oriental-Occidental genetic differentiation in cultivated barley. *Theoretical and Applied Genetics* 84: 862-687.
Zhang, Q., G.P. Yang, X. Dai, and J.Z. Sun (1994). A comparative analysis of genetic polymorphism in wild and cultivated barley from Tibet using isozyme and ribosomal DNA markers. *Genome* 37: 631-638.
Zohary, D. (1999). Monophyletic vs. polyphyletic origin of crops on which agriculture was founded in the Near East. *Genet. Res. Crop Evol.* 46: 133-142.

Chapter 3

Molecular Mapping of the Barley Genome

Andris Kleinhofs
Feng Han

INTRODUCTION

Barley, *Hordeum vulgare,* has been a favorite genetic experimental organism since the rediscovery of Mendel's laws of heredity (Tschermak, 1901, cited from Smith, 1951). The widespread use of barley is attributable to its diploid nature ($2n = 2x = 14$), self-fertility, large chromosomes (6 to 8 µmmm), high degree of natural and easily inducible variation, ease of hybridization, wide adaptability, and relatively limited space requirements. The genetics of barley and related information have been reviewed (Smith, 1951; Nilan, 1964, 1974; Briggs, 1978; Rasmusson, 1985; Shewry, 1992). Other major sources of barley genetics information are the *Proceedings of the International Barley Genetics Symposia* (Lamberts et al., 1964; Nilan, 1971; Gaul, 1976; Asher et al., 1981; Yasuda and Konishi, 1987; Munck, 1992; Scoles and Rossnagel, 1996) and the *Barley Genetics Newsletter* (since 1971).

Barley has seven pairs of distinct chromosomes (Nilan, 1964; Ramage, 1985) containing approximately 5×10^9 bp DNA (Bennett and Smith, 1976; Arumuganathan and Earle, 1991). The seven barley chromosomes were defined based on their sizes and characteristics. Chromosomes 1 through 5 differ in their sizes measured at mitotic metaphase, with chromosome 1 being the longest and chromosome 5 being the shortest. Chromosomes 6 and 7 have satellites, with chromosome 6 having the larger satellite and chromosome 7 having the smaller satellite. The barley genome is well characterized with respect to classical genetics and cytogenetics. Over 1,000 genes and 500 translocation stocks are known (Sogaard and von Wettstein-Knowles, 1987; von Wettstein-Knowles, 1992). Conventional genetic mapping in the past 50 years has placed over 200 loci to the barley chromosomes (Franckowiak, 1997).

Research in Dr. Kleinhofs' laboratory is supported by USDA-NRI plant research project grant 9600794, the North American Barley Genome Mapping Project, Washington Barley Commission, and the American Malting Barley Association. Special thanks are due to the large number of students, postdocs, and technicians in my laboratory who have contributed to the mapping of the barley genome, particularly Dave Kudrna.

Comparative mapping has revealed that barley chromosomes 1, 2, 3, 4, 5, 6, and 7 are homeologous to wheat chromosomes 7H, 2H, 3H, 4H, 1H, 6H, and 5H, respectively. It has been recommended that barley chromosomes be designated according to their homeologous relationships with chromosomes of other Triticeae species (Linde-Laursen, 1997). In this chapter we use the barley chromosome designations with the Triticeae designations in parentheses to avoid confusion with what has been published previously. When using the Triticeae nomenclature, it is extremely important to include the genome H symbol to indicate clearly which nomenclature is being used.

Use of molecular techniques as diagnostic tools to assist the conventional breeding process demands the construction of linkage maps. Markers evenly spaced on the chromosomes are then used to scan the genome to identify associations of markers and traits. Molecular mapping of the barley genome has been facilitated by the development of molecular markers, the ability to develop doubled haploid (DH) lines, the availability of numerous mutants and cytogenetic stocks, particularly the barley-wheat addition lines, and the recent development of large insert libraries. The comprehensive molecular marker linkage maps have provided a powerful and important tool in barley for identifying quantitative trait loci (QTL) for agronomic and quality traits, determining QTL effects and action, and facilitating genetic engineering (Nilan, 1990).

GENETIC AND PHYSICAL MAPPING

Linkage Maps

Molecular mapping of the barley genome started with the first restriction fragment length polymorphism (RFLP) map for chromosome 6 published by Kleinhofs, Chao, and Sharp (1988), followed by a partial map of the whole genome by Shin et al. (1990). Since then, extensive and numerous molecular maps of the barley genome have been constructed (Graner et al., 1991; Heun et al., 1991; Kleinhofs et al., 1993; Becker et al., 1995; Kasha et al., 1995; Langridge et al., 1995; Sherman et al., 1995; Meszaros and Hayes, 1997; Waugh et al., 1997; Qi, Stam, and Lindhout, 1998). Various studies with interest in specific chromosomes, regions, genes, or traits have generated a number of additional maps (Hinze et al., 1991; Barua et al., 1993; Devos, Millan, and Gale, 1993; Giese et al., 1994; Becker and Heun, 1995; Komatsuda et al., 1995; Laurie et al., 1995; Bezant et al., 1996; Laurie et al., 1996; Schonfeld et al., 1996; Ellis et al., 1997). All these mapping efforts have resulted in locating over 2,000 different molecular markers to the barley genome. Cross mapping of markers between the Steptoe x Morex and Igri x Franka maps has generated a common platform comprising about 100 common markers (Kleinhofs and Graner, 2001). These data have been used to merge several maps and accumulate more valuable information (Langridge et al., 1995; Qi, Stam, and Lindhout, 1996). With more extensive effort to merge mapping information from different mapping populations, the North American Barley Genome Mapping Project (NABGMP) introduced the Bin Map concept (Kleinhofs and Graner, 2001). Using the Steptoe x Morex (SM) map as a base, the barley genome was divided into approximately 10 cM intervals or Bins, allowing the placement of many markers mapped on different maps in their appropriate Bins. Here we present additional data in map format (see Figures 3.1-3.7). The maps contain a single marker identifying each locus in the SM map, with cosegregating markers shown at the bottom of each figure and markers mapped in other crosses, and located to individual Bins, shown alongside the map (see appendix). The SM map contains 60, 64, 58, 41, 50, 49, and 62 unique loci on chromosomes 1through 7, respectively. An additional 235 markers cosegregate with the 384 total unique loci. A total of 952 different molecular markers or genes have been located to individual Bins on the seven barley chromosomes. Thus, the total number of different markers or genes that have been placed to the SM map chromosome Bins is 1,571. This includes most of the unique RFLP markers that have been mapped in barley, very few of the morphological/physiological genes, and practically none of the hundreds of random amplified polymorphic DNA (RAPD) or amplified fragment length polymorphism (AFLP) markers. The descriptions of individual markers and the maps are accessible at <http://barleygenomics.wsu.edu>.

High-resolution maps have been prepared for several barley genome regions. The earliest such maps involved the *lig* (previously *li*), *mlo,* and *wx* (previously *wax* or *glx*) loci (Jorgensen and Jensen, 1979; Rosichan et al., 1979; Konishi, 1981). Molecular marker-based high-resolution maps have been produced for the disease resistance loci *Rpg1, Mla,* and *mlo* (DeScenzo, Wise, and Mahadevappa, 1994; Kilian et al., 1997; Simons et al., 1997).

Updates on barley genome mapping are available at GrainGenes, <http://wheat.pw.usda.gov>; at the North American Barley Genome Mapping Project home page, <http://www.css.orst.edu/Barley/NABGMP/nabgmp.htm>; and at the Washington State University barley genomics home page, <http://barleygenomics.wsu.edu>.

Mapping Centromeres and Telomeres

Barley centromeres are readily mapped due to the availability of the ditelosomic addition lines (Islam, 1983). Markers are mapped on specific chromosome arms until the switch from one arm to the other is found (Kleinhofs et al., 1993). Physical mapping of barley chromosomes has also localized the centromeres (Kunzel and Korzun, 1996; Kunzel, Korzun, and Meister, 1999). Approximate centromere locations are indicated in the maps shown in Figures 3.1-3.7.

All short arms and 4(4H), 5(1H), and 7(5H) long arms of the 14 barley telomeres have been mapped by combined RFLP/PCR (polymerase chain reaction) methods using telomere-associated sequences (Kilian and Kleinhofs, 1992; Kilian, Kudrna, and Kleinhofs, 1999). Telomeres that have been clearly established in the SM map are shown in Figures 3.1-3.7, while telomeres mapped in other crosses are shown only in the Bins. Seven putative telomeric markers were also mapped using polymorphisms generated by pulsed field gel electrophoresis (PFGE) and hybridization with telomere repeat sequences from *Arabidopsis* and an HvRT repeat sequence from barley (Roder et al., 1993). The telomeric markers mapped by PFGE for the short arm of chromosomes 2(2H) and 5(1H) are in good agreement with the telomeric markers mapped by RFLP/PCR (Kilian, Kudrna, and Kleinhofs, 1999). Combining the results of Kilian, Kudrna, and Kleinhofs (1999) and Roder et al. (1993), only barley chromosome 2L(2HL) is without a telomeric marker, and the genetic distance of the barley genome is established at about 1,500 cM, varying slightly with different parents.

Physical Mapping

Although genetic linkage maps provide a powerful tool for narrowing the search for genes, it is not sufficient to clone a specific gene. This objective can only be achieved using a physical mapping approach. A physical map provides the real DNA distance in bp between landmarks on the chromosomes and is used as the basis for isolation and characterization of individual genes or genome regions of interest, as well as for genome sequencing. Low-resolution physical maps of the barley genome were produced by Giemsa C and N banding patterns (Linde-Laursen, 1988). Higher-resolution physical mapping has been accomplished for the *Hor1* and *Hor2* loci by PFGE (Sorensen, 1989; Siedler and Graner, 1991). Due to the large genome size of barley, high-resolution physical mapping of the whole barley genome by fingerprinting large insert clones would be very difficult and expensive with current technology. Fortunately, the barley genome, similar to other cereals, appears to be organized into gene-rich and gene-poor regions (Buschges et al., 1997; Civardi et al., 1994; Carels, Barakat, and Bernardi, 1995; DeScenzo and Wise, 1996; Gill et al., 1996). Only 12 percent of the barley genome is estimated to contain nearly all of the genes (Barakat, Carels, and Bernardi, 1997); this 12 percent is roughly equivalent to the genome size of rice and, thus, of manageable size.

Physical mapping also provides significant insights into the relationships between genetic distance and physical distance. The earliest comparisons of genetic and physical distances were based on comparing C-bands to marker genes (Linde-Laursen, 1979) and in situ hybridization of ribosomal RNA (Appels et al., 1980), B-hordein (Clark, Karp, and Archer, 1989), and ribosomal 5S RNA (Kolchinsky et al., 1990) loci to RFLP markers (Kleinhofs,

Chao, and Sharp, 1988; Kleinhofs et al., 1993). Integration of the in situ hybridization physical maps with genetic maps has been done (Fukui and Kakeda, 1990; Pedersen, Giese, and Linde-Laursen, 1995). A comprehensive comparison of the barley physical and genetic RFLP maps has been completed (Kunzel and Korzun, 1996; Kunzel, Korzun, and Meister, 1999). This work involves PCR mapping of single-copy RFLP probes on microisolated individual translocation and control chromosomes. The data are in agreement with earlier studies showing suppressed recombination in the chromosome centromeric regions, but the quantitative data show significant variation within and among individual chromosomes. The physical/genetic distances varied between 0.1 to 298 Mb/cM. The tables and map images included in this chapter are online at GrainGenes, <http://wheat.pw.usda.gov/ggpages/Barley_physical/> or <http://grain.jouy.inra.fr/ggpages/Barley_physical/>. In addition, a pictorial documentation showing the idiograms of the 120 T-lines used in the PCR mapping by Kunzel and co-workers is also online at <http://wheat.pw.usda.gov/ggpages/Barley_physical/ Idiograms/>.

Detailed comparisons of physical and genetic distances have been made at the *mlo* and *Mla* loci (DeScenzo and Wise, 1996; Simons et al., 1997). These data clearly illustrate the extreme variation in the barley physical/genetic distances even within short chromosome regions.

MAPPING OF QUANTITATIVE TRAIT LOCI

The observation that most important agronomic traits exhibit continuous phenotypic variation can be explained by the independent action and potential interaction of many discrete genes, each of which has a relatively small effect on the overall phenotype, together with the effect of environment. These "polygenes," occurring at discrete "quantitative trait loci" (QTL) (Gelderman, 1975), are expected to exhibit the fundamental Mendelian properties of segregation and recombination. Therefore, it is possible to identify linkages between genetic markers and QTL and thus determine the locations of individual QTL.

Classical quantitative genetics has provided methods for estimating the minimum number and the average gene action of QTL influencing a trait (Wright, 1968; Mather and Jinks, 1971; Falconer, 1981). However, these methods are useful in describing average properties of a group of QTL, rather than in defining the location and specific phenotypic effects of individual QTL. Linkage between a morphological marker and a QTL was first demonstrated by Sax (1923), followed by work in peas (Rasmusson, 1935), barley (Everson and Schaller, 1955), and several other species (early work reviewed by Smith, 1937). Isozyme markers were used to identify QTL in maize, soybeans [*Glycine max* (L.) Merr.], tomato (*Lycopersicon* spp.), and wild oats (*Avena fatua* L.) (reviewed by Stuber, 1992). However, the lack of suitable marker loci hampered further development.

The development of molecular markers has made systematic and accurate mapping of QTL possible by the construction of complete molecular marker linkage maps. Associations of RFLP markers with QTL have been found in several species (reviewed by Paterson, Tanksley, and Sorrells, 1991). Further, new statistical methods and software for QTL mapping have made the mapping procedure more efficient (Soller and Beckmann, 1983; Weller, 1986; Edwards, Stuber, and Wendel, 1987; Cowen, 1988; Jensen, 1989; Lander and Botstein, 1989; Luo and Kearsey, 1989; Simpson, 1989; Knapp, Bridges, and Birkes, 1990; Knapp, 1991; Jansen, 1992; Lincoln, Daly, and Lander, 1992; Moreno-Gonzalez, 1992; von Ooijen, 1992; Zeng, 1993, 1994; Zeng, 1994; Tinker and Mather, 1995).

Mapping QTL for Agronomic Traits

QTL for barley grain yield have been mapped in several populations (Hayes and Iyamabo, 1993; Hayes, Liu, et al., 1993; Backes et al., 1995; Iyamabo and Hayes, 1995; Langridge et al., 1995; Thomas et al., 1995; Tinker and Mather, 1995; Kjaer and Jensen, 1996; Romagosa et al., 1996; Tinker et al., 1996; Bezant et al., 1997; Powell et al., 1997). In the six-row SM cross, Hayes, Liu, et al. (1993) detected 14 QTL for yield with data from 16 environments. Analyzing the same data set using an additive main effect and multiplicative interaction model, Romagosa et al. (1996) found four QTL for grain yield across environments. These four QTL were verified through realized molecular marker-assisted selection responses (Romagosa, Han, Ullrich, et al., 1999). A QTL on chromosome 3(3H) was confirmed as the most important and consistent locus to determine yield across environments (Larson, Kadyrzhanova, et al., 1996; Romagosa, Han, Ullrich, et al., 1999). In a two-row Harrington/TR306 cross, Tinker et al. (1996) found five primary QTL for yield with data from 17 environments, of which only the QTL × E (environment) interaction on chromosome 1(7H) may have corresponded to QTL detected in the SM cross. Many other QTL studies for yield have detected different numbers of QTL with different populations and environments (Kjaer, Haahr, and Jensen, 1991; Backes et al., 1995; Thomas et al., 1995). Only limited coincidence, in terms of genome locations, was found among these QTL.

Hayes, Liu, et al. (1993) reported ten QTL for plant height in the six-row SM cross. However, only three primary QTL were detected in the two-row Harrington/TR306 cross (Tinker et al., 1996). QTL on chromosome 1(7H) and 7(5H) in the Harrington/TR306 cross correspond closely with QTL detected in the SM cross. Backes et al. (1995) also detected three QTL for plant height on chromosomes 1(7H), 4(4H), and 6(6H). Three QTL were also found on chromosomes 1(7H), 3(3H), and 7(5H) by Thomas et al.

(1995). Barua et al. (1993) and Thomas et al. (1995) found the *denso* dwarfing gene affecting many traits. The plant height QTL at the *denso* locus consistently showed an additive effect of 10 cm for height (Thomas et al., 1995). Comparison of QTL for plant height detected in different crosses indicated that the QTL on chromosome 7(5H) are in similar genome positions in Blenheim/E224-3, Harrington/TR305, and SM crosses.

QTL for heading date were studied (Hayes, Liu, et al., 1993; Backes et al., 1995; Kjaer, Jensen, and Giese, 1995; Laurie et al., 1995; Thomas et al., 1995; Tinker et al., 1996). None of the five major genes that affected vernalization or photoperiod response reported in Laurie et al. (1995) corresponded with positions of heading QTL in Harrington/TR306 (Tinker et al., 1996). However, three of five primary QTL for heading date on chromosomes 1(7H), 3(3H), and 4(4H) in Harrington/TR306 were detected in the vicinity of three QTL reported by Laurie et al. (1995) for earliness-per-se QTL. Thomas et al. (1995) found that three QTL for heading date, two on chromosome 3(3H) and one on chromosome 4(4H), were at similar genome locations between the Blenheim/E224-3 and SM crosses. Kjaer, Jensen, and Giese, (1995) identified two QTL for heading date on chromosome 2(2H), with one that may correspond to the QTL identified by Hayes, Liu, et al. (1993).

QTL analyses for other important agronomic traits, such as lodging, maturity, kernel weight, test weight, etc., were reported by several groups (Hayes, Liu, et al., 1993; Backes et al., 1995; Laurie et al., 1995; Bezant et al., 1996; Tinker et al., 1996). Hayes, Blake, et al. (1993) also reported barley chromosome 7 associated with components of winterhardiness. These QTL studies have accumulated valuable information in terms of number of loci involved in each trait, QTL organization in the genome, magnitude of each QTL, etc. QTL for different traits are often found in similar genome regions. This overlapping QTL phenomenon has posed interesting questions regarding whether these overlapping QTL are due to pleiotropic effects or multigene families or random linkage. Further resolution, and thus understanding of the first-order QTL analyses, is necessary.

Mapping QTL for Quality Traits

Malting quality has been a major breeding effort in barley. During malting and mashing, the endosperm components are modified and solubilized by the interaction of enzymes and their substrates (Pollock, 1962). Malting quality, attributed to these enzymes and their substrates, is composed of a number of traits, including malt extract percentage, total grain protein, soluble protein, ratio of soluble/total protein, β-glucan content, plump barley, α-amylase activity, diastatic power, and malt β-glucanase (Burger and LaBerge, 1985). Malting quality is controlled by genetic factors but also is influenced by environmental conditions such as soil, climate, and fertilizers (Hunter, 1962).

QTL analysis for malting quality in barley was first reported by Hayes, Liu, et al. (1993) using the NABGMP six-row SM cross. Some, but not all, of the QTL for malting quality were found near the amylase loci *Amy2*, *Bmy1*, and *Bmy2* (Hayes, Liu, et al., 1993). More detailed analyses with data from multiple years and locations for the same cross detected 12 loci for kernel weight, 8 for kernel plumpness, 8 for grain protein content, 8 for wort protein content, 14 for diastatic power, 9 for α-amylase, and 9 for malt extract (Han and Ullrich, 1994). QTL for barley and malt β-glucan content and for green and finished malt β-glucanase activity were mapped. Three QTL for barley β-glucan, 6 for malt β-glucan, 3 for β-glucanase in green malt, and 5 for β-glucanase in finished malt were detected (Han et al., 1995). The QTL with the largest effects on barley β-glucan, malt β-glucan, green malt β-glucanase, and finished malt β-glucanase were identified on chromosomes 2(2H), 1(7H), 4(4H), and 7(5H), respectively. These studies have identified an important chromosome 1(7H) region near the centromere containing two putative adjacent overlapping QTL for malt extract content, α-amylase activity, diastatic power, malt β-glucan content, and malt β-glucanase activity and/or dormancy (Hayes, Liu, et al., 1993; Ullrich et al., 1993; Han et al., 1995). This complex QTL region also showed the largest and the most consistent effects for these traits over multiple locations and years. QTL for malting quality in two-row barley were reported by Thomas et al. (1996) and Mather et al. (1997). Comparison of QTL between six-row barley and two-row barley and between different crosses of two-row barley revealed only a few QTL with similar genome positions. This phenomenon might be attributed to parental polymorphism, different genetic backgrounds, QTL × E interactions different methods of QTL analysis, sampling errors, etc.

QTL for malting quality involving winter barley germplasm were reported by Oziel et al. (1996). They identified a number of QTL for grain protein, soluble/total protein ratio, diastatic power, α-amylase, malt extract, fine-coarse difference, and wort β-glucan. The most important regions were chromosomes 1(7H) and 7(5H), where QTL for a number of traits overlapped. QTL for diastatic power, α-amylase, and fine-coarse difference on chromosome 1(7H) and for diastatic power and α-amylase on chromosome 7(5H) (Oziel et al., 1996) may correspond to the QTL identified in spring barley (Hayes, Lui, et al., 1993; Han et al., 1995).

Four regions of barley genome on chromosomes 1(7H), 4(4H), and 7(5H) were found to be associated with most of the differential phenotypic expression for dormancy (Ullrich et al., 1993; Oberthur et al., 1995; Larson, Bryan, et al., 1996). Major dormancy loci on chromosome 7(5H) and 1(7H) were verified via linked molecular markers (Han et al., 1996). Individual locus effects on dormancy during seed development and after-ripening were also studied (Romagosa, Han, Clancy, et al., 1999).

The concentration of the various carbohydrate components in barley grain impact both agronomic and end-use characteristics. Starch contributes positively to grain yield and other physical kernel traits, malt extract, brewer's wort, and food/feed energy content. Acid detergent fiber (ADF) contributes positively or negatively to physical grain characteristics, embryo protection, beer filtration, and nutrient utilization and intestinal health (MacGregor and Fincher, 1993). QTL mapping using the SM cross identified two QTL for starch and five QTL for ADF (Ullrich et al., 1996). The two loci for starch account for about 21 percent of the total variation. The small number of genes and low multilocus r^2 are probably due to the relatively low polymorphism between the parents. Five QTL for ADF explain nearly 65 percent of the total variation.

Fine-Structure Mapping of QTL

Current QTL analysis procedures provide only approximate localization of QTL. More accurate localization can be obtained by performing additional backcrosses to produce isogenic lines differing only in the region containing a QTL. This approach eliminates the majority of the genetic variance and makes it possible to dissect the remaining interval by examining various recombinants for flanking markers (Paterson et al., 1990). This strategy, termed substitution mapping, is based on the following principles: (1) different chromosome segments carrying molecular markers are identified, and their regions of overlap determined using all available genetic markers; (2) phenotypic effects of each chromosomal segment are determined by QTL analysis; and (3) effects shared by different segments are attributed to QTL in regions shared by those segments, while effects unique to a segment are attributed to QTL in a region unique to that segment (Paterson et al., 1990). The efficiency of substitution mapping for QTL in a specific region depends on molecular marker availability in that region.

The substitution mapping approach has been employed for fine mapping of a complex malting quality QTL region containing two putative adjacent overlapping QTL affecting malt extract, α-amylase, diastatic power, malt β-glucan, malt β-glucanase, and dormancy in the centromere region of chromosome 1 (Han et al., 1997). Isogenic lines carrying different chromosome segments only in the region containing the QTL were constructed by molecular marker-assisted backcrossing. By comparing the significant additive effects among isogenic lines, two QTL each for malt extract and for α-amylase and two to three for diastatic power were identified in certain environments and resolved into 1 to 8 cM genome fragments. Significant QTL $\times$ E interactions and epistatic interactions were observed by high resolution mapping.

Paterson et al. (1990) and Han et al. (1997) used different strategies of substitute mapping. Paterson et al. (1990) performed phenotyping and QTL analyses on BC_2F_2 plants (i.e., segregating populations) so that both additive and dominant effects could be determined. However, since they were locating several QTL at the same time, interactions among the QT might actually interfere with the detection of a specific QTL effect. Moreover, two backcrosses are not sufficient to remove all donor parent alleles to achieve a uniform genetic background. Constructing isogenic lines that differ only in the target region provides another approach to assay recombinant chromosomes. Han et al. (1997) employed this second approach for high-resolution mapping of one chromosome region at a time by (1) performing three backcrosses to obtain a more uniform genetic background and (2) selecting homozygous recombinants in the BC_3F_2. The advantages of this approach are as follows: (1) with only one chromosome region under study and a uniform genetic background, QTL effects identified from recombinants will clearly come from the target region; (2) homozygous recombinants can be tested extensively in different environments to detect small effects; and (3) further dissection of QTL fragments can be achieved by further mating to the recurrent parent and checking only the target region for recombinants. The disadvantages of the isogenic line approach are that (1) the effect of a target QTL may not be detected if there are epistatic interactions with QTL on other genome regions, and (2) some very small donor chromosome segments may persist in isogenic lines to interfere with target QTL effects, but such segments should occur at random in large segregating populations.

MAPPING DISEASE RESISTANCE GENES

Disease resistance has been an integral part of plant breeding efforts along with grain yield. Although substantial progress has been made in breeding for disease-resistant cultivars, screening for disease-resistant phenotypes is time-consuming and resource intensive in many cases. The development of molecular markers, and their subsequent use in mapping and tagging disease resistance genes, provides a more efficient alternative to select and manipulate disease resistance genes without extensive field and greenhouse tests. A wealth of studies related to major or quantitative trait gene mapping and tagging for disease resistance has been published. Graner (1996) reviewed molecular mapping of qualitative and quantitative disease resistance genes. This section will summarize additional literature, in a chronological order, and mapping and map-based cloning of several major disease resistance genes.

Thomas et al. (1995) reported QTL mapping for disease resistance to mildew, *Rhynchosporium;* brown rust; and yellow rust. For powdery mildew, only one QTL was detected in more than one environment and this QTL was probably at the *Mla* locus in the region of *Hor1* and *Hor2* on chromosome 5(1H). One QTL was detected for brown rust on the long arm of chromosome 7(5H), while three QTL for yellow rust were detected on chromosomes 1(7H), 5(1H), and 7(5H). One to two QTL were detected for *Rhynchosporium* resistance on chromosome 3(3H). Hayes et al. (1996) mapped leaf rust *(Rphx)* resistance to chromosome 1(7H), and resistance to barley yellow dwarf virus *(Ryd2)* and scald to chromosome 3(3H). Jin et al. (1993) reported that *Rph3* is located on chromosome 1(7H). Allelism tests will be required to determine the relationship between *Rphx* and *Rph3*. The *Ryd2* and scald resistance loci are 7.3 cM apart (Hayes et al., 1996). The resistance to stripe rust was mapped as QTL to chromosomes 4(4H) and 7(5H) in greenhouse and field tests (Hayes et al., 1996). An incompletely dominant gene, *RphQ,* for resistance to leaf rust caused by *Puccinia hordei* was mapped to the centromeric region of chromosome 7(5H) (Borovkova et al., 1997). Allelism tests indicated that *RhpQ* is either allelic with or closely linked to the *Rph2* locus. QTL controlling resistance to cereal aphids were mapped with the largest consistent effect locus on the short arm of chromosome 1(7H) (Moharramipour et al., 1997). Three QTL, *Rphq1*, *Rphq2,* and *Rphq3*, were effective at the seedling stage; and five QTL, *Rph2*, *Rphq3*, *Rphq4*, *Rphq5,* and/or *Rphq6,* were effective at the adult stage for partial resistance to leaf rust *(Puccinia hordei)* (Qi et al., 1998).

Stem Rust

Stem rust, caused by the fungus *Pucccinia graminis* f. sp. *tritici,* has been one of the most devastating diseases of barley and wheat in the northern Great Plains of the United States and Canada. In the past, durable resistance in barley has been achieved by the widespread use of a single resistance gene, *Rpg1*. Recently, a new virulent *P. graminis* f. sp. *tritici* pathotype, Pgt-QCC, was identified (Roelfs et al., 1991). Resistance to pathotype QCC is conferred by a single gene, *rpg4* (Jin, Steffenson, and Miller, 1994). *Rpg1* is located very close to the telomere on barley chromosome 1S(7HS) (Jin et al., 1993; Kleinhofs et al., 1993). Closely linked markers, ABG704 and ABG077, have been identified, which can be used for molecular marker-assisted selection (Kilian et al., 1994; Horvath et al., 1995). The *rpg4* has been mapped to the barley chromosome 7(5H) subtelomeric region (Borovkova et al., 1995). Two RFLP markers, MWG740 and ABG391, were found to be closely linked to *rpg4*. The *Rpg1* and *rpg4* genes have been the targets for map-based cloning (Kilian et al., 1995, 1997; Han et al., 1998). Map-based cloning of *Rpg1* is being carried out using rice as an intergenomic cloning vehicle. Rice yeast artificial chromosome (YAC), bacterial artificial chromosome (BAC), and cosmid clones were used to isolate probes mapping to the barley *Rpg1* region. The rice BAC isolated with the pM13 probe was a particularly excellent source of probes. A detailed physical map and plasmid contig was established for the rice BAC fragment containing the *Rpg1* flanking markers pM13 and B24. This fragment covers a rice DNA physical distance of ca. 90 kb and a barley genetic distance of 0.6 cM, based on 1,400 segregating gametes. Further high-resolution mapping with 3,000 gametes yielded a 30 kb rice fragment containing the *Rpg1* flanking markers B172 and B24. Complete sequencing of this 30 kb fragment resulted in the identification of an open reading frame coding a putative membrane protein, a potential candidate for *Rpg1* (Han et al., 1999). However, this membrane protein did not show homology with published plant disease resistance genes. A barley BAC clone contig covering this region is being established. Rice and barley BAC contigs were established across the *rpg4* region (Druka et al., 1999). The contig data suggest that the flanking markers are less than 100 kb apart in rice and less than 200 kb apart in barley.

Powdery Mildew

Erysiphe graminis f. sp. *hordei* is the causal agent of powdery mildew disease in barley. A series of loci conferring resistance to powdery mildew has been identified and mapped (reviewed by Graner, 1996). Among these loci, *mlo* has two distinct features: resistance alleles are recessive, and it provides resistance to all known races of the fungus. The *mlo* gene maps to the long arm of chromosome 4(4H). RFLP markers closely linked to *mlo* were identified (Hinze et al., 1991). The *mlo* gene has been isolated using a positional cloning approach (Buschges et al., 1997). The deduced 60 kDa protein is predicted to be membrane anchored by at least six transmembrane helices, with a putative nuclear localization motif and two casein kinase II motifs. The sequence feature of this gene is clearly different from that of other isolated resistance genes. Analysis of 11 mutagen-induced *mlo* alleles revealed that mutations in each case resulted in alterations of the deduced *Mlo* wild-type amino acid sequence. Susceptible intragenic recombinants, isolated from *mlo* heteroallelic crosses, could restore the *Mlo* wild-type sequence. Efforts to isolate the *Mla* gene are also in progress (DeScenzo, Wise, and Mahadevappa, 1994).

Barley Yellow Mosaic Virus

Barley mild mosaic virus (BaMMV) and barley yellow mosaic virus (BaYMV) are two major strains of the barley mosaic virus complex. Two related strains (BaYMV-1, BaYMV-2) of BaYMV have been identified that differ in their virulence pattern. At least seven loci conferring either partial resistance or immunity to the viruses have been mapped on chromosomes 3L(3HL), 4L(4HL), and 5S(1HS). Linked molecular markers have allowed effective and rapid incorporation of resistance genes without field or greenhouse tests. The single recessive gene *ym4*, chromosome 3L(3HL), which confers complete resistance to BaMMV and BaYMV-1, is a major target for positional cloning. High-resolution mapping and rice synteny around the *ym4* locus has been established (Bauer et al., 1996).

COMPARATIVE MAPPING AND SYNTENY

Comparative Maps Within Barley

It is now reasonably well established that the barley genome contains few, if any, rearrangements, at least among the parents used for extensive mapping. Mapping approximately 100 common probes on the SM and Igri x Franka maps did not find any contradictions in their locations (Kleinhofs and Graner, 2001). Probes identifying a single locus have been mapped on as many as five different maps and always appear to map in the same position on all maps, as judged by their position relative to other markers. There are a few reports of single-copy probes not mapping to the same position in different crosses. Although real rearrangements cannot be completely excluded, human error is the most likely explanation in the majority of cases. A few probes have been identified whose map location seems to have changed in relatively recent times. These probes may produce a hybridization band in one parent that is missing in another parent. In most cases, no allelic bands are found, although other bands hybridizing to the same probe may occur, representing different loci. Some examples are BCD351, ABG078, and *Caf2*. The explanation for these apparent deletion and/or insertion events is not known.

Comparative Mapping Among Grass Species

Using molecular marker systems in genome mapping has led to direct comparisons of the relative order of homologous sequences along the chromosomes of closely and distantly related species and development of the concept of genome synteny. Comparative genome mapping provides insights into genome evolution and genetic information between divergent species.

The comparisons among major grass species (barley, rice, maize, sorghum, wheat) revealed very good marker order conservation over large chromosomal regions (Devos et al., 1992; Ahn and Tanksley, 1993; Devos and Gale, 1993; Devos et al., 1993; Devos, Millan, and Gale, 1993; Kurata, Moore, et al., 1994; Moore et al., 1995). Devos and Gale (1997) provided an extensive review on comparative genetics in the grasses, including rice, foxtail millet, sugar cane, maize, the Triticeae crops, and oats. The barley genome is highly colinear, with a few minor inversions, with that of the wheat genomes A, B, and D (Devos et al., 1993; Namuth et al., 1994; Hohmann et al., 1995; Van Deynze et al., 1995; Dubcovsky et al., 1996). Thus, the comparisons of the wheat genomes with other species apply to the barley genome as well. Colinearity of the barley and rice genomes has been reported in general (Saghai Maroof et al., 1996) and in specific regions (Dunford et al., 1995; Kilian et al., 1997; Han et al., 1998). These studies indicate regions of very high gene order colinearity between barley and rice. However, there are also rearrangements and probable transpositions. Nevertheless, rice has proven to be a very good model and source of probes for saturation of specific barley chromosome regions.

Establishment of syntenic relationships among major grass species, especially at the micro level, has made possible map-based cloning of genes from a large genome (such as in barley) by utilizing the results of genetic and physical mapping of the small genome. The genomes of modern cereals differ substantially in size. For example, the barley haploid genome size is ~4900 Mb (Arumuganathan and Earle, 1991). In contrast, the rice haploid genome size is ~400 Mb (Arumuganathan and Earle, 1991). The most comprehensive rice genetic map is 1575 cM distributed over 12 linkage groups (Kurata, Nagamura, et al., 1994). By comparison, the most comprehensive barley genetic map is 1245 cM distributed over seven chromosomes (Kleinhofs, 1994). Therefore, the average DNA sizes are ~250 kb and ~4 Mb per centiMorgan in rice and barley, respectively. The rice genome also consists of less repetitive DNA than the barley genome: ~50 percent and ~80 percent in rice and barley, respectively (Bennett and Smith, 1991; Flavell, Rimpau, and Smith, 1977). The smaller size and lower repetitive sequences of the rice genome clearly indicate the advantages of using the rice genome as a tool for map-based cloning of genes from barley and other large-genome cereals (Kurata, Nagamura, et al., 1994).

Comparative Mapping of QTL

Since emphases on quantitative traits in different species could be different, such as no emphasis on malting quality traits in rice, comparative mapping of QTL among species may not be very useful and informative. However, many studies on comparative mapping of plant phenotypes across related species have been undertaken (reviewed by Paterson, 1997). The most interesting question for plant breeders will be whether QTL are conserved in the different crosses of the same species. This question is especially important in molecular marker-assisted selection. If a QTL is not in a conserved location on a genome, then tagging of the QTL becomes very difficult because QTL mapping must be done for each specific cross to identify linked molecular markers.

Thomas et al. (1996) compared heading date QTL across different barley crosses and found that QTL on chromosome 7(5H) are located at similar positions. However, no QTL for malt extract were found to be similar in the two North American barley crosses, SM and Harrington/TR306. We have compared QTL for α-amylase in three different crosses: SM, Harrington/

TR306 (H/T), and Harrington/Morex (H/M). Four, three, and three QTL for α-amylase were identified in SM, H/T, and H/M, respectively. Among these QTL, there was one common QTL (in the same or a similar marker interval) between any two of the three crosses, but no common QTL was present across all three crosses (S. E. Ullrich, F. Han, and J. A. Clancy, unpublished data). Three possible explanations for this phenomenon are as follows: (1) different genes are indeed involved in controlling quantitative traits in different germplasm or different types (six-row versus two-row); (2) no polymorphism exists for some QTL in some crosses; and (3) QTL, as with other genes or markers, are conserved in the genome within a species, but the effects and magnitude of these QTL may not be the same in different germplasm, and thus detection of QTL among crosses would be variable.

CONCLUDING STATEMENT

The advent of DNA markers has greatly enhanced our ability to solve many of the problems associated with breeding programs, such as environmental effects, quantitative inheritance, etc. The DNA marker technology is also providing, and will continue to provide, us with significant insights into genome organization and evolution. Structural genomics developed along with molecular markers has established a basis for rapid development of functional genomics research. QTL analysis provides a powerful tool to study and manipulate the genes controlling quantitative traits, but it has not solved the complexity of QTL, such as their functions, interactions, and organizations. Confirmation of identified QTL is necessary for further manipulation. High-resolution mapping is required to resolve QTL as accurately as qualitative trait loci and is essential for reducing linkage drag in marker-assisted selection, elucidating QTL function (pleiotropic and epistatic effects) and organization (gene cluster, random linkage), and cloning QTL based on their map locations. The genetic basis of QTL will ultimately be elucidated by QTL cloning. Plant breeding will enter a new era with increased knowledge of genes and QTL.

FIGURES 3.1-3.7. Barley chromosome 1-7 linkage maps based on the Steptoe x Morex (SM) 150 doubled haploid line mapping population. Markers representing each locus are shown to the right of the chromosome while the cumulative distance, starting from the short arm telomere, is shown in centiMorgans (Kosambi formula) to the left of the chromosome. The maps are subdivided in approximately 10 cM Bins and each new Bin marker is shown in bold. Markers mapped in other crosses, which could be located with reasonable accuracy to a Bin, are listed. Approximate centromere (Cen) locations are indicated.

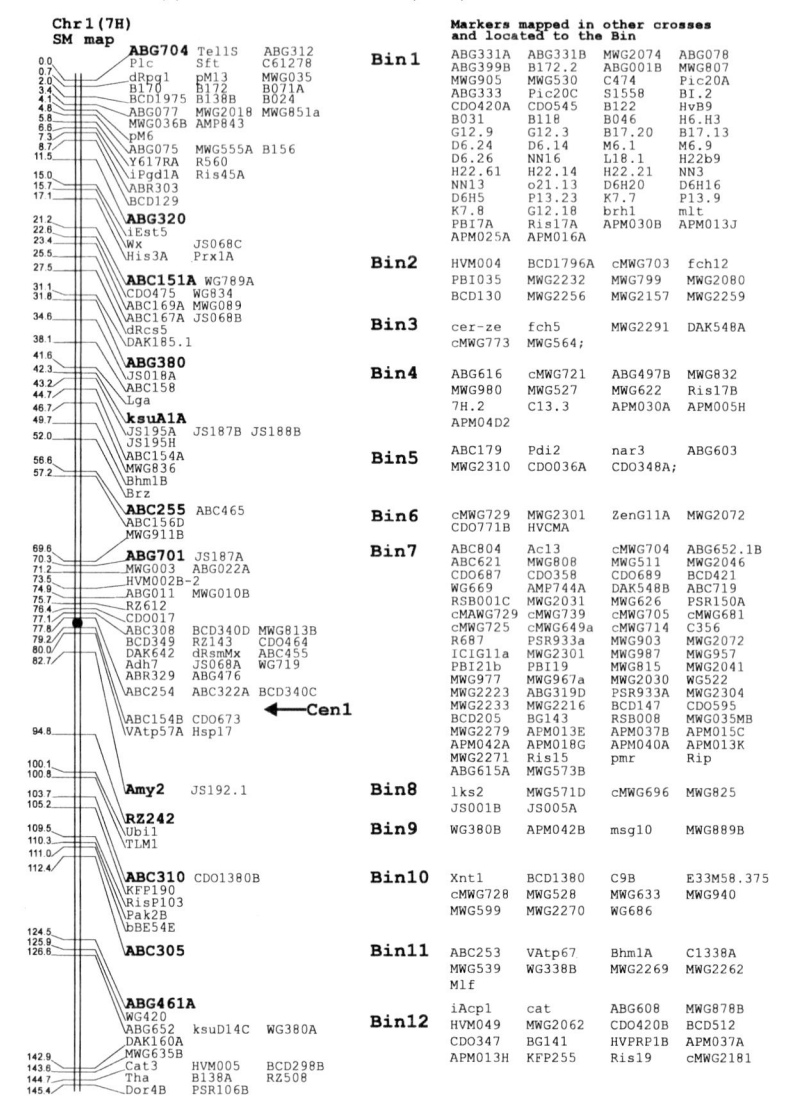

Chr1 (7H) SM map

cM	Markers on chromosome
0.0	**ABG704** Tel1S ABG312
0.7	Plc Sft C61278
3.4	dRpg1 pM13 MWG035
4.1	B170 B172 B071A
4.8	BCD1975 B138B B024
5.8	ABG077 MWG2018 MWG851a
6.8	MWG036B AMP843
7.3	pM6
8.7	ABG075 MWG555A B156
11.5	Y617RA R560
15.0	iPgd1A Ris45A
15.7	ABR303
17.1	BCD129
21.2	**ABG320**
22.8	iEst5
23.4	Wx JS068C
25.5	His3A Prx1A
27.5	
31.1	**ABC151A** WG789A
31.8	CDO475 WG834
	ABC169A MWG089
34.6	ABC167A JS068B
	dRcs5
38.1	DAK185.1
41.6	**ABG380**
42.3	JS018A
43.2	ABC158
44.7	Lga
46.7	**ksuA1A**
49.7	JS195A JS187B JS188B
52.0	JS195H
56.6	ABC154A
57.2	MWG836
	Bhm1B
	Brz
	ABC255 ABC465
	ABC156D
	MWG911B
69.8	**ABG701** JS187A
71.2	MWG003 ABG022A
73.5	HVM002B-2
74.9	ABG011 MWG010B
75.7	RZ612
76.4	CDO017
77.1	ABC308 BCD340D MWG813B
79.2	BCD349 RZ143 CDO464
80.0	DAK642 dRsmMx ABC455
82.7	Adh7 JS068A WG719
	ABR329 ABG476
	ABC254 ABC322A BCD340C
	ABC154B CDO673 ←Cen1
94.8	VAtp57A Hsp17
100.1	
100.8	
103.7	**Amy2** JS192.1
105.2	
109.5	**RZ242**
110.3	Ubi1
111.0	TLM1
112.4	
	ABC310 CDO1380B
	KFP190
	RisP103
	Pak2B
	bBE54E
124.5	
125.9	**ABC305**
126.8	
	ABG461A
	WG420
	ABG652 ksuD14C WG380A
	DAK160A
	MWG635B
142.9	Cat3 HVM005 BCD298B
143.6	Tha B138A RZ508
144.7	Dor4B PSR106B
145.4	

Markers mapped in other crosses and located to the Bin

Bin1: ABG331A ABG331B MWG2074 ABG078; ABG399B B172.2 ABG001B MWG807; MWG905 MWG530 C474 Pic20A; ABG333 Pic20C S1558 B1.2; CDO420A CDO545 B122 HvB9; B031 B118 B046 H6.H3; G12.9 G12.3 B17.20 B17.13; D6.24 D6.14 M6.1 M6.9; D6.26 NN16 L18.1 H22b9; H22.61 H22.14 H22.21 NN3; NN13 o21.13 D6H20 D6H16; D6H5 P13.23 K7.7 P13.9; K7.8 G12.18 brh1 mlt; PBI7A Ris17A APM030B APM013J; APM025A APM016A

Bin2: HVM004 BCD1796A cMWG703 fch12; PBI1035 MWG2232 MWG799 MWG2080; BCD130 MWG2256 MWG2157 MWG2259

Bin3: cer-ze fch5 MWG2291 DAK548A; cMWG773 MWG564;

Bin4: ABG616 cMWG721 ABG497B MWG832; MWG980 MWG527 MWG622 Ris17B; 7H.2 C13.3 APM030A APM005H; APM04D2

Bin5: ABC179 Pdi2 nar3 ABG603; MWG2310 CDO036A CDO348A;

Bin6: cMWG729 MWG2301 ZenG11A MWG2072; CDO771B HVCMA

Bin7: ABC804 Ac13 cMWG704 ABG652.1B; ABC621 MWG808 MWG511 MWG2046; CDO687 CDO358 CDO689 BCD421; WG669 AMP744A DAK548B ABC719; RSB001C MWG2031 MWG626 PSR150A; cMWG729 cMWG739 cMWG705 cMWG681; cMWG725 cMWG649a cMWG714 C356; R6B7 PSR93a MWG903 MWG2072; ICIG11a MWG2301 MWG987 MWG957; PBI21b PBI19 MWG815 MWG2041; Adh7 JS068A WG719; MWG977 MWG967a MWG2030 WG522; MWG2223 ABG319D PSR933A MWG2304; MWG2233 MWG2216 BCD147 CDO595; BCD205 BG143 RSB008 MWG035MB; MWG2279 APM013E APM037B APM015C; APM042A APM018G APM040A APM013K; MWG2271 Ris15 pmr Rip; ABG615A MWG573B

Bin8: lks2 MWG571D cMWG696 MWG825; JS001B JS005A

Bin9: WG380B APM042B msg10 MWG889B

Bin10: Xnt1 BCD1380 C9B E33M58.375; cMWG728 MWG528 MWG633 MWG940; MWG599 MWG2270 WG686

Bin11: ABC253 VAtp67 Bhm1A C1338A; MWG539 WG338B MWG2269 MWG2262; Mlf

Bin12: iAcp1 cat ABG608 MWG878B; HVM049 MWG2062 CDO420B BCD512; CDO347 BG141 HVPRP1B APM037A; APM013H KFP255 Ris19 cMWG2181

Chromosome2 (2H) SM Map

```
  0.0 ─ Tel2S
  1.8 ─ MWG844A
  2.6 ─ ABG313A
  4.3 ─ cMWG682
  5.1 ─ ABG058
  9.9 ─ ABG703B
        Chs1B
 18.8 ─ MWG878A  Chs1A  DAK605
 19.5 ─ WG516
 20.3 ─ ABG008
        RbcS
 23.7 ─ HVM056
 27.6 ─ DAK595A
 28.6 ─ BCD351F
 29.7 ─ ABG318
 30.5 ─ ABC156A
 31.2 ─ ABG002  MWG858
 32.7 ─ Tef4
 38.9
 39.8 ─ ABG358  R1690
 41.4 ─ CDO064
 44.2 ─ ABG459
 45.6
 47.0 ─ MWG520A
 47.7 ─ ABG005
 48.6 ─ DAK213A
 51.8
 53.2
 53.8 ─ Pox
 55.3 ─ ABC454
 56.0 ─ cMWG663  JS187C
        MWG950   JS195B   JS188A
 63.3 ─ Adh8
 66.9 ─ Bgq60
 68.3 ─ B15C   MWG567
 69.2 ─ CDO537 ABG019
 70.8
 71.5 ─ ABC468  Bmy2    ABG316C
 72.7 ─ ABC306  ABR338  P40FA           ← Cen2
 79.0
 79.8 ─ ABC162  WG996    Hvex1
 80.7 ─ ABG356  HVBKASI  HVM023
 83.3 ─ ABC167B
 84.8 ─ bBE54D
 91.2 ─ ABC451  ABG014  CDO588
 94.7 ─ RZ740
        TLM3
 98.0 ─ His3C
101.3 ─ ABC152D
        MWG865
        Rrn5S1
        ksuF15
        MWG503
        ABG072
        CDO244
121.0
122.4
123.1 ─ ksuD22  Crg3A
        ABC252  MWG520B  F3hA
138.7 ─ pKABA1
139.4 ─ Gln2   BCD135  CDO373
140.1 ─ KG004.2 KG004.1
        ABC157
145.4
        ABC165  ABC153  RZ590
150.3 ─ ABG317A
151.9 ─ ABG317B
152.6 ─ ABG316E
        Pcr1   ABG316D
156.7
        MWG844C
160.8 ─ CDO036  cMWG663A
165.0 ─ BCD410  WG645
165.5 ─ Prx2/7
166.5 ─ BG123A
168.0 ─ ABA005
169.4
170.9
172.4 ─ bBE54C
```

	Markers mapped in other crosses and located to the Bin			
Bin1	APM013A	ABG707B	cMWG684C	MWG2262B
	CDO057A			
Bin2	BCD175			
Bin3	ABC311	gsh6	cMWG655E	PSR150B
	Adh9	MWG2171	MWG2086	PBI21A
	ksuD18	BG140B	H22B9A	
Bin4	RZ069	msg2	C213	MWG578
	MWG553B	ABG328	BCD221B	WG222A
Bin5	MWG035C	cMWG763	PSR933B	MWG578
	MWG553B	MWG2067	MWG889A	CDO665A
Bin6	ABC803	ABC256	ABG461C	cMWG647
	cMWG646A	C198	MWG887A	MWG874
	MWG2133	MWG065	MWG2146	MWG2054
	CDO370	WG405A		
Bin7	Kas1.2B	yst4	ABG602	ABG716
	MWG636B	MWG2287	MWG949B	CDO474B
	WG180C	MWG2058	BCD334	CDO675A
	CDO770	BCD351B	Hot1	DAK232
Bin8	eog	Az94	CDO1328	f
	HVM026	CDO667	CDO366	BCD355
	ksuE12	WG223A	BCD111	L843.2B
	cMWG658.2	MWG2240		
Bin9	ABG619	BCD386	CDO474C	
Bin10	mVrs1	MWG876	MWG581	MWG2123
	MWG882A	MWG2212	mAnt2	MWG892
	cMWG699	MWG2211	MWG2081	MWG801
Bin11	fol-a	PSR901	AWBma21	Ha2
	MWG694	ABC620	MWG2236	PSR331
	Ole2			
Bin12	CDO665B	ABC177	ABC180	MWG882B
	fsh	Rph15		
Bin13	cMWG655B	CDO680	cnx1	lig1
	nar4	nar6	Az86	R11301
	DAK213B	HVM054	HVCSG	BCD266
	BCD453A			
Bin14	Zeo1	ABG609A	ABG613	ksuF41
	ksuG43			
Bin15	wst7	gpa	ABC171C	MWG949A
	C808	MWG636A	MWG829	MWG2068
	MWG090	MWG866	MWG989	MWG080
	MWG2076	MWG2303	MWG2200	MWG2225
	cMWH660	BCD339C	CDO678	WG338A
	ksuH16	RSB001B		

Chromosome 3 (3H) SM Map

```
0.0    Tel3S
0.8    ABG070
4.9    ABG316A  MWG571C  cMWG691
5.8    ABG655A
6.8    JS195C
7.6    JS195F

       ABC171A
19.2   ABG321
22.9   ABA303
       MWG798B
29.3   MWG584  MWG974  ABG010
31.0   ABG460  ABG057A
       ABG471  Y704L
35.1   MWG014
38.0   BCD1532  CDO1435
41.2   ABG396  ABG399A
42.0   ABG398
       KO1                        ← Cen3
       Dfr     P40FB    DAK160B
       CDO118  aCDO113  Dor4A
       ABR334  DAK202   ksuA3C
       PSR626B ksuF2B   Adh5
59.3   ABC323  ABG462   ABC156C
60.0   HVM027  HVM044
60.9   BCD828  ABC325   AtpbB
62.7
65.0   BCD1145 cMWG680
68.2   ABG703B
70.6   MWG571B
72.2   DD1.1B
       PSR156
78.0   ABC176
79.3   HVM033
81.6
86.4   ABG377  Rrn5S2  JS68D
88.6   ksuG59
       MWG555B
92.3   ABG315  HVM060
95.7   ABG453  Y704R   Amyb
96.8   MWG571A
102.3  ABR320
103.8  ABC307B
106.2
106.9  CDO345  PSR078B
110.7  Crg3B   ABG499
117.6  CDO113B
127.0  His4B
       ABG004
       WG110
141.7  mPub
146.6  ABG389
148.9  ABC161
150.9  Y617RB
153.0  ABA302
155.5  MWG902  Csu838A
156.4  ABG654
158.6  ABG655B
159.4
160.2
162.2  ABC174  BCD131
162.9  Glb4
164.3  Glb3
165.8  iBgl
169.5  ABG495B
170.2  MWG041
171.2
171.9  ABC166
174.3  Prx1B   MWG838
       TLM2
181.4  ABG319B
       ABG172
```

Markers mapped in other crosses and located to the Bin

Bin 1 BCD907 MWG848 MWG2021B MWG2158 E33M58-M002

Bin 2 BCD1532C ABC171B

Bin 3 CDO395 BCD1532B

Bin 4 alm Ugp2 R106 C131 MWG2266 MWG2234 MWG497 MWG2021B MWG848 MWG577

Bin 5 MWG022 BCD706 E33M54.128

Bin 6 msg5 R1613 C1271 E33M51.370 Rpt.a MWG2187 MWG2065 E33M59-M001 HVM009 KFP216 C122 Pt,,a RRrs1 MWG582 MWG2189 MWG595 MWG561A MWG985 MWG802 MWG852 MWG2064 MWG844B MWG952 MWG2105 MWG2138 PBI5 PBI21C BCD263 WG889B uzu CDO1174 CDO684 AWBma1 glk223 BCD127 PSR116 Yd2 Ylp E39M48.297

Bin 7 AMP751A C911 C854 C1459 BCD452 C250 MWG812 MWG904 MWG986 wst6 BCD339A BCD809 WG178 glk80 PSR543

Bin 8 MWG975 MWG2026 DAK496 WG405B WG940 ksuG59 Glb36 BCD1796B C1623A C836 MWG2167B Pic20C CDO1406 R37

Bin 9 Ugp1 MWG2013 MWG2132 E33M60-M004 MWG2281B MWG2290 CDO419B CDO718 FG71 Act8C

Bin 10 wst1 C191 C949 E33M51.316 MWG973 cMWG784 MWG961

Bin 11 als MWG973 E35M54.103

Bin 12 MWG847 ABG607 MWG961 BCD298C RZ444B

Bin 13 R1545 MWG2190 MWG803 MWG2025 MWG2273 sdw1

Bin 14 ABC178 Glb31 C742 E33M59.119 MWG932 MWG632A MWG570 E33M59.485 MWG883 L843.2C

Bin 15 ABC805 Glb32 Glb33 E35M49-M002 Glb37 ABC706B ABG709 Glb34 ABG710A HVM062 HVM070 cMWG693 R2816 R1014 Adh10 MWG549 MWG2099 MWG2040 MWG2162 MWG972 MWG546 MWG969 MWG2149 MWG2167 MWG2099A CDO105A

Bin 16 iEst2 iEst4 ABG609 iEstY ym4 MWG010A MWG085 CDO394A CDO474A BG131 Glb35 ABG609B WG222C

Molecular Mapping of the Barley Genome

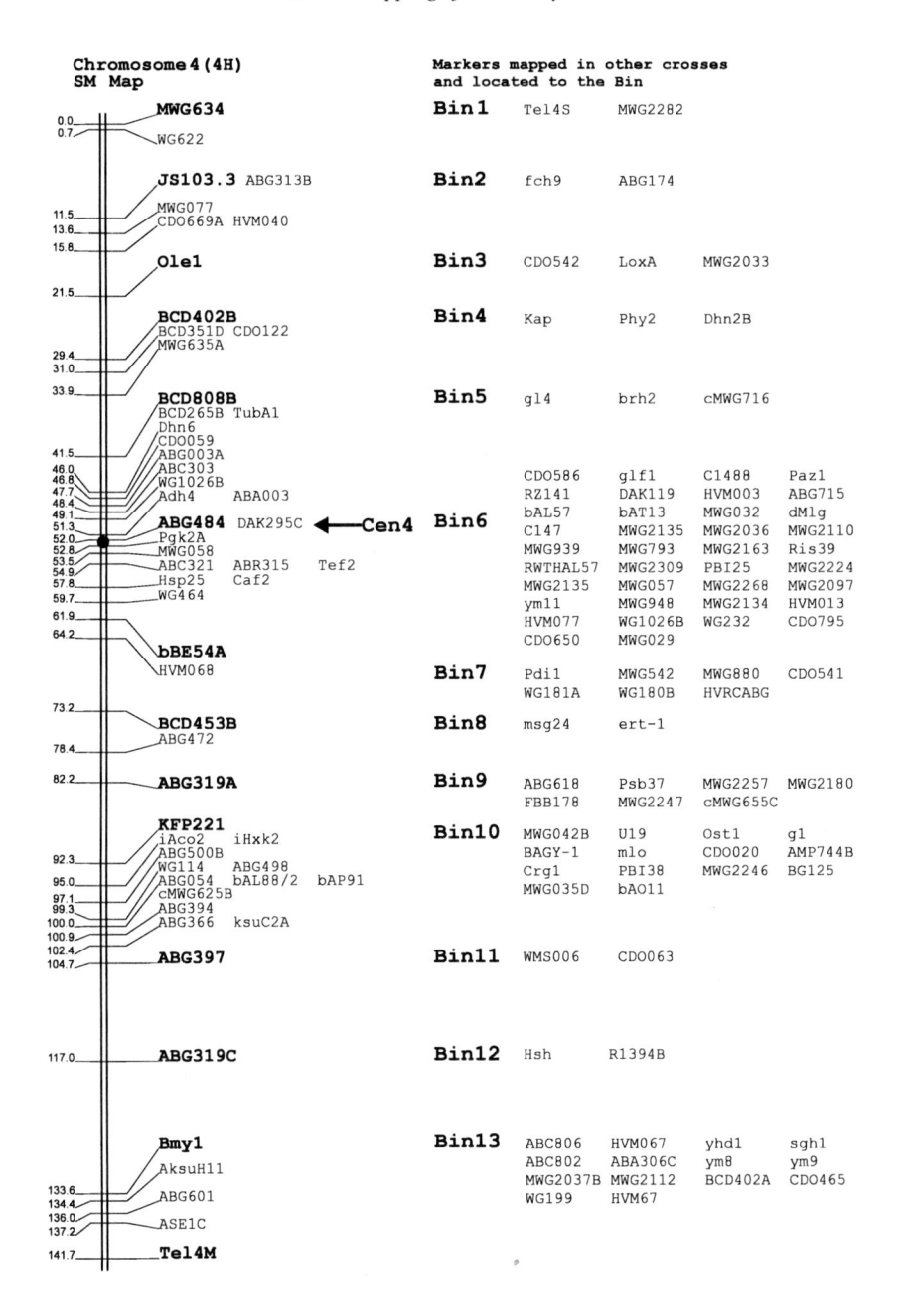

Chromosome 4 (4H) SM Map

```
 0.0 ── MWG634
 0.7 ── WG622

        JS103.3  ABG313B
11.5 ── MWG077
13.6 ── CDO669A  HVM040
15.8 ── Ole1

21.5

        BCD402B
        BCD351D  CDO122
        MWG635A
29.4
31.0
33.9 ── BCD808B
        BCD265B  TubA1
        Dhn6
        CDO059
41.5 ── ABG003A
46.0 ── ABC303
46.8 ── WG1026B
47.7 ── Adh4  ABA003
48.4
49.1 ── ABG484  DAK295C  ◄── Cen4
51.3 ── Pgk2A
52.0
52.8 ── MWG058
53.5 ── ABC321  ABR315  Tef2
54.9
57.8 ── Hsp25  Caf2
59.7 ── WG464
61.9
64.2
        bBE54A
        HVM068

73.2
        BCD453B
78.4 ── ABG472

82.2 ── ABG319A
        KFP221
        iAco2    iHxk2
        ABG500B
92.3 ── WG114    ABG498
95.0 ── ABG054   bAL88/2  bAP91
97.1 ── cMWG625B
99.3 ── ABG394
100.0
100.9── ABG366   ksuC2A
102.4
104.7── ABG397

117.0── ABG319C

        Bmy1
        AksuH11
133.6
134.4── ABG601
136.0
137.2── ASE1C
141.7── Tel4M
```

Markers mapped in other crosses and located to the Bin

Bin	Markers
Bin1	Tel4S MWG2282
Bin2	fch9 ABG174
Bin3	CDO542 LoxA MWG2033
Bin4	Kap Phy2 Dhn2B
Bin5	gl4 brh2 cMWG716
Bin6	CDO586 glf1 C1488 Paz1 RZ141 DAK119 HVM003 ABG715 bAL57 bAT13 MWG032 dM1g C147 MWG2135 MWG2036 MWG2110 MWG939 MWG793 MWG2163 Ris39 RWTHAL57 MWG2309 PBI25 MWG2224 MWG2135 MWG057 MWG2268 MWG2097 ym11 MWG948 MWG2134 HVM013 HVM077 WG1026B WG232 CDO795 CDO650 MWG029
Bin7	Pdi1 MWG542 MWG880 CDO541 WG181A WG180B HVRCABG
Bin8	msg24 ert-1
Bin9	ABG618 Psb37 MWG2257 MWG2180 FBB178 MWG2247 cMWG655C
Bin10	MWG042B U19 Ost1 gl BAGY-1 mlo CDO020 AMP744B Crg1 PBI38 MWG2246 BG125 MWG035D bAO11
Bin11	WMS006 CDO063
Bin12	Hsh R1394B
Bin13	ABC806 HVM067 yhd1 sgh1 ABC802 ABA306C ym8 ym9 MWG2037B MWG2112 BCD402A CDO465 WG199 HVM67

Chromosome 5 (1H)
SM Map

cM	Marker
0.0	Tel5P
	Aga6
7.8	Hor5
9.1	
9.8	MWG938 Act8A DAK152.1
11.3	MWG920.1B
	ABG059 Hor2
16.2	MWG835A
	MWG036A
21.0	MWG837
23.1	Hor1 Chs3
28.6	ABA004
	DD1.1A
35.5	BCD098 CDO099
	ABG053
39.7	
	Ica1
45.4	ABG074
	RZ166
51.3	ABG500A
52.3	End1
55.3	
56.9	JS074 ←—Cen5
58.3	Caf1
61.9	ABC164
64.8	ABG452 ksuF2A ABG494
65.5	BCD351C WG789B
67.2	ABC52B
67.9	Ris17C HVM043
69.2	
70.4	Pcr2 ABG022B
73.8	ABR337
	CDO105B MWG800
77.8	
79.2	Glb1 PSR626A Dor2
	BCD454
	ABC160
88.4	DAK123B
92.9	Phy1
95.3	
	His4A ABG464
101.7	PSR330 RZ444
103.4	
108.1	His3B
110.3	ABC307A
	cMWG706A
120.2	ABC257
125.3	iPgd2 BCD442
126.0	
130.2	BCD1930 cMWG733A
	AtpbA
136.4	ABG702A
138.5	ABG322B
143.6	
145.0	ABC261 ABA002
145.8	MWG635C
147.7	Cab2
151.2	Aga7
152.7	CDO400
154.0	MWG912 ABG373 CDO202
154.8	BCD340A
156.0	ABG055
159.6	ABG387A

Markers mapped in other crosses and located to Bin

Bin	Markers
Bin 1	ABC165C ABC326 OP06 R622 Rti ksuC2B ABG330
Bin 2	Sex76 BCD1434 ABC801 ksuD14A Mla14 Mla13 ciwS10 BCD135B AWBMA35 RZ783 DAK152.1 aHor2 MWG2148 MWG2021A MWG2048 MWG2197 MWG2083 MWG2245 ABG316F ABG319E Mla6 cMWG645.1 Mla12 AWBMA5
Bin 3	MWG837B KFP257A BCD249.1 BCD249.2 MWG068 BCD402C ABC156B ZENG11B RWTHAT13 Ris45B 22Epr8 BCD249
Bin 4	ksuE19 MWG896 ksuG9 ksuE18
Bin 5	CDO442 WG180A ksuD14B 102o4
Bin 6	cMWG758 MWG913 MWG2262C MWG2056 CDO473 amo1
Bin 7	msg1 HVM020 cMWG758 R210 R758 MWG2056 MWG506 Ris13 KinT8
Bin 8	C1338B WG983 MWG943
Bin 9	MWG2077
Bin 10	
Bin 11	R1394A Dor3 Kas1.2A cMWG676 cMWG649B His4C
Bin 12	TubA2 R2374B R178 MWG947 MWG954 CDO394B ksuE2B CDO989 BCD304 BCD265A
Bin 13	Blp MWG2028 cMWG701B MWG984B ABG332
Bin 14	KFP257B fch7 Aga7 drun8 ABG710B Tel5L C6 MWG632B CDO393 WG241 ksuE11

Molecular Mapping of the Barley Genome 51

Chromosome 6 (6H) SM Map

Position	Marker(s)
0.0	Tel6S
2.1	**ABG062** Lth PSR167A
6.7	MWG620
7.4	ABG466
8.1	Nar1 cMWG663B
10.9	B071B
11.7	PSR325B
13.4	
16.4	**ABG378B**
18.2	Crg2
	Apt
	Cxp3
23.5	ABC152A
24.2	His3D
24.9	
26.4	**cMWG625A**
	DD1.1C
	PSR106A
40.1	
42.2	
46.0	**ABG387B** HVM041
	Ldh1
	ABG458 PSR167B Ubi4
	JS068E ABR331
	HVM011
	ABC169B Ubi5
60.4	Rrn1 ABA006 B12DA
61.9	CDO524 DAK233 Lox
63.1	CDO534 HVM014
64.3	DAK595B
65.0	HVM022
65.7	ABG020 DAK123A ksuA3B
66.8	CDO497 HVM074 ABR335
68.1	WG223B
68.9	
72.5	**ABG474** BCD340E ←Cen6
75.6	ksuD17
77.1	G57
77.8	ABG379
78.6	ABC163 ABG388 CDO507
79.3	ABC175
80.6	RZ323
81.3	MWG820
82.0	**ABC170B**
82.7	ABG001A ksuA3D
84.2	cMWG684B
	JS195D
92.7	
	Nar7
96.9	Amy1 Nir JS195E
99.2	JS188C JS187D
	bBE54B
103.9	**MWG934**
	cMWG684C
	ABC170A PSR154
112.3	
113.1	
113.8	**Tef1**
	DAK148
	ABC807A
	ABG713
151.6	MWG798A
153.0	DAK213C

Markers mapped in other crosses and located to the Bin

Bin	Markers
Bin1	Lox1B MWG2202
Bin2	Nar8 MWG966 MWG573A MWG2218 ABG654B
Bin3	Hsp70 MWG602B Rrs13 CDO1081 ABC159A
Bin4	MWG988 MWG887B JS005B
Bin5	MWG2065 MWG916 C597
Bin6	BCD102 rob Gst1 E148A; Crg4 HVM065 ABG705B C174; cMWG679 MWG2264 MWG2231 MWG2227A; HVM34 MWG561C HVM034
Bin7	Sex1 Dip1 C601 ABC164B; MWG2029 PBI9 Ris6 WG286; MWG2313 MWG872 MWG2061 MWG2043; MWG984A
Bin8	ABC165D ABC623 ABG001C MWG2205; MWG2297 ksuF2 MWG967B MWG951; MWG2100 MWG2137 MWG2141
Bin9	BCD453A ABG711 cMWG669 MWG549A; CDO419A BG140A WG282 BCD269
Bin10	ABC807B ABC264 Dhn5 ABC302B; cMWG655D Dhn2C Dhn4 Dhn3; C74 cMWG716B BCD339B WG222B; ABC154C
Bin11	ABG461B cMWG658.1 cMWG684A DAK200B
Bin12	DAK199 BCD221A
Bin13	iAco1 UBC018 MWG897 MWG911A; MWG2196 MWG2317
Bin14	PSR156B MWG2053 MWG514 MWG2176; MWG010C

Chromosome 7(5H) SM Map

```
         Gsp
 0.0  ── DAK133   Mta9        JS018B
 1.7  ── MWG502
 2.4  ── ABC483   ABR313
 7.8
 9.9  ── MWG920.1A
10.6  ── MWG835B
         ABG316B
         ABG705A
         CDO669B
         Dor5     ABG708
         Adh6
         ABR336
36.5     ABG064   ABG497
41.9
42.8     ABG395
43.6     CDO749
44.4     FBA232*
45.4     Rrn2
46.2     Ubi2
47.5     MWG2040B ◄──── Cen7
         DAK123C
50.1     B12DB    HVM002B-1  HVM030
50.1
56.2
57.0  ── Ltp1     ksuA3A
58.4  ── WG541
59.9
64.7  ── WG530
68.0     ABA001   BG123B
68.8     ABC706A  S6F
69.6     WG889
71.4     Ale
72.6     CDO348B
74.0
75.5  ── ABC324
83.2
         ABC302
         PSR128
94.7
98.2  ── BCD926
         CDO57B
104.0    mSrh     ksuA1B    ABC168
105.6    MWG522
106.3    ABG069   BCD351E   WG364
107.1    MWG583
         ABC717
         ABG473
         FBA332*
133.0
133.9    MWG514B
138.3    CDO504
139.9    WG644
141.3    ABG712
142.8    BCD265C  iEst9
143.5    Tef3     R3226
144.3
149.2 ── WG908
152.5    FBA351*
155.5    MWG877
         ABG495A
161.8    ABG496   C606      Cab1
168.3    ABC155
173.3    ABC482
174.5    R273
175.4    FBB213A*
176.9    ABG391   DAK156
186.0    ABG390   DAK049    ABG707
188.8    ABC310A  CDO1380A
191.6    CDO484   PBI7B     R1925
193.0
193.7
194.4    ABG463
197.1    MWG813A
200.1    ABC309
201.1    ABG314B
201.8    ABG314A
202.5    MWG851C
         HVM006
         MWG851D
         MWG851B
         Tel7L
```

Markers mapped in other crosses and located to the Bin

Bin1	PSR1204	ksuD4	Tel7S	MWG618
	Act8B	Lox1A		
Bin2	ABG610	AWBMA12	CDO1335A	CDO665C
	ABG615B			
Bin3	cud1	PSR326	RphTR	BG140C
Bin4	lax-a	AWBMA1	AWBMA33	AWBMA32
	Acl1	OPR3	BCD410B	CDO460
	Ym3			
Bin5	CDO460			
Bin6	nar2	nar5	cMWG770	cMWG717
	MWG596	MWG526	MWG561B	
Bin7	ABG461D	Mlj	WG564	RSB001A
Bin8				
Bin9	var1	lj	MWG923	MWG956
	MWG2121	MWG624	MWG604	PBI39
	MWG2237	MWG2191	CDO675B	CDO771A
	ksuE2A	WG364	WG1026	
Bin10	MWG894	RZ404	Ugp3	Xa21like
	HVDHN009	MWG914	MWG549B	MWG553A
	MWG550	WG583	WG1026	MWG2230
	mRaw1			
Bin11	cMWG781	R3226	ABC159B	ABG003B
	ABG702B	MWG533	cMWG646B	cMWG716A
	MWG2092	MWG2227B	MWG2224	MWG862
	MWG933	MWG900	Sgh2	Dhn2A
Bin12	DAK111	Xyl	TubA3	MWG922
	CDO583			
Bin13	AMP744C	ARD065	Aga5	ARD578.3
	ARD188.3	CDO419C	PSR370	ABC170C
	DAK200A	cMWG740	ABC622	cMWG654
	C1257	cMWG701A	MWG2020	MWG2088
	MWG827	MWG850	MWG2037A	CDO113A
	CDO213	BCD298A	Ch1	ARD032
	E10757A	rpg4	JS005C	
Bin14	cMWG650	C9A	MWG2193	CDO484;
Bin15	MWG632C	ABC718	ABG057B	MWG891
	MWG2249	CDO506		

REFERENCES

Ahn, S. and S.D. Tanksley (1993). Comparative linkage maps of the rice and maize genomes. *Proceedings of the National Academy of Sciences, USA* 90: 7980-7984.

Appels, R., Gerlach, W.L., Dennis, E.S., Swift, H., and W.J. Peacock (1980). Molecular and chromosomal organization of DNA sequences coding for the ribosomal RNAs in cereals. *Chromosoma* 78: 293-311.

Arumuganathan, K. and E.D. Earle (1991). Nuclear DNA content of some important plant species. *Plant Molecular Biology Reporter* 9: 208-218.

Asher, M.J.C., Ellis, R.P., Hayter, A.M., and R.N.H. Whitehouse (eds.) (1981). *Barley Genetics IV*. Edinburgh: Edinburgh University Press.

Backes, G., Granner, A., Foroughi-Wehr, B., Fischbeck, G., Wenzel, G., and A. Jahoor (1995). Localization of quantitative trait loci (QTL) for agronomic important characters by the use of a RFLP map in barley (*Hordeum vulgare* L.). *Theoretical and Applied Genetics* 90: 294-302.

Barakat, A., Carels, N., and G. Bernardi (1997). The distribution of genes in the genomes of Gramineae. *Proceedings of the National Academy of Sciences, USA* 94: 6857-6861.

Barua, U.M., Chalmers, K.J., Thomas, W.T.B., Hackett, C.A., Lea, V., Jack, P., Forster, B.P., Waugh, R., and W. Powell (1993). Molecular mapping of genes determining height, time to heading, and growth habit in barley (*Hordeum vulgare*). *Genome* 36: 1080-1087.

Bauer, E., Lahaye, T., Schulze-Lefert, P., Sasaki, T., and A. Graner (1996). High resolution mapping and rice synteny around the *Ym4* virus resistance locus on chromosome 3L. In *Proceedings of the V International Oat Conference and the VII International Barley Genetics Symposium. Poster Sessions, Volume 1*, eds. A. Slinkard, G. Scoles, and B. Rossnagel. Saskatoon: University of Saskatchewan Extension Press, pp. 317-319.

Becker, J. and M. Heun (1995). Barley microsatellites: Allele variation and mapping. *Plant Molecular Biology* 27: 835-845.

Becker, J., Vos, P., Kuiper, M., Salamini, F., and M. Heun (1995). Combined mapping of AFLP and RFLP markers in barley. *Molecular and General Genetics* 249: 65-73.

Bennett, M.D. and J.B. Smith (1976). Nuclear DNA amounts in angiosperms. *Philosophical Transactions of the Royal Society of London (Biology)* 274: 227-274.

Bennett, M.D. and J.B. Smith (1991). Nuclear DNA amounts in angiosperms. *Philosophical Transactions of the Royal Society of London (Biology)* 334: 309-345.

Bezant, J., Laurie, D., Pratchett, N., and J. Chojecki (1996). Marker regression mapping of QTL controlling flowering time and plant height in a spring barley (*Hordeum vulgare* L.) cross. *Heredity* 77: 64-73.

Bezant, J., Laurie, D., Pratchett, N., Chojecki, J., and M. Kearsey (1997). Mapping QTL controlling yield and yield components in a spring barley (*Hordeum vulgare* L.) cross using marker regression. *Molecular Breeding* 3: 29-38.

Borovkova, I.G., Jin, Y., Steffenson, B.J., Kilian, A., Blake, T.K., and A. Kleinhofs (1997). Identification and mapping of a leaf rust resistance gene in barley line Q21861. *Genome* 40: 236-241.

Borovkova, I.G., Steffenson, B.J., Jin, Y., Rasmussen, J.B., Kilian, A., Kleinhofs, A., Rossnagel, B.G., and K.N. Kao (1995). Identification of molecular markers linked to the stem rust resistance gene *rpg4* in barley. *Phytopathology* 85: 181-185.

Briggs, D.E. (ed.) (1978). *Barley*. London: Chapman and Hall.

Burger, W.C. and D.E. LaBerge (1985). Malting and brewing quality. In *Barley*, ed. D.C. Rasmusson. Madison, WI: American Society of Agronomy, pp. 367-401.

Buschges, R., Hollricher, K., Panstruga, R., Simons, G., Wolter, M., Frijters, A., van Daelen, R., van der Lee, T., Diergaarde, P., Groenendijk, J., et al. (1997). The barley *Mlo* gene: A novel control element of plant pathogen resistance. *Cell* 88: 695-705.

Carels, N., Barakat, A., and G. Bernardi (1995). The gene distribution of the maize genome. *Proceedings of the National Academy of Sciences, USA* 92: 11057-11060.

Civardi, L., Xia, Y., Edwards, K.J., Schnable, P.S., and B.J. Nikolau (1994). The relationship between genetic and physical distances in the cloned *a1-sh2* interval of the *Zea mays* L. genome. *Proceedings of the National Academy of Sciences, USA* 91: 8268-8272.

Clark, M., Karp, A., and S. Archer (1989). Physical mapping of the B-hordein loci on barley chromosome 5 by in situ hybridization. *Genome* 32: 925-929.

Cowen, N.M. (1988). The use of replicated progenies in marker-based mapping of QTLs. *Theoretical and Applied Genetics* 75: 857-862.

DeScenzo, R.A. and R.P. Wise (1996). Variation in the ratio of physical to genetic distance in intervals adjacent to the *Mla* locus on barley chromosome 1H. *Molecular and General Genetics* 251: 472-482.

DeScenzo, R.A., Wise, R.P., and M. Mahadevappa (1994). High-resolution mapping of the *Hor1/Mla/Hor2* region on chromosome 5S in barley. *Molecular Plant-Microbe Interactions* 7: 657-666.

Devos, K.M., Atkinson, M.D., Chinoy, C.N., Francis, H.A., Harcouvt, R.L., Koebner, R.M.D., Liu, C.J., Masojc, P., Xie, D.X., and M.D. Gale (1993). Chromosomal rearrangements in rye genome relative to that of wheat. *Theoretical and Applied Genetics* 85: 673-680.

Devos, K.M., Atkinson, M.D., Chinoy, C.N., Liu, C.J., and M.D. Gale (1992). RFLP-based genetic map of the homoeologous group 3 chromosomes of wheat and rye. *Theoretical and Applied Genetics* 83: 931-939.

Devos, K.M. and M.D. Gale (1993). Extended genetic map of the homoeologous group 3 chromosomes of wheat, rye and barley. *Theoretical and Applied Genetics* 85: 649-652.

Devos, K.M. and M.D. Gale (1997). Comparative genetics in the grasses. *Plant Molecular Biology* 35: 3-15.

Devos, K.M., Millan, T. and M.D. Gale (1993). Comparative RFLP maps of the homoeologous group-2 chromosomes of wheat, rye and barley. *Theoretical and Applied Genetics* 85: 784-792.

Druka, A., Kudrna, D., Han, F., Kilian, A., Steffenson, B., Yo, Y., Frisch, D., Tomkins, J., Wing, R., and A. Kleinhofs (1999). Map based cloning of barley *rpg4* gene. Poster presented at Plant and Animal Genome VII, the Seventh International Conference on the Plant and Animal Genome, San Diego, January 17-21, 1998. Final Program and Abstract Guide, p. 178.

Dubcovsky, J., Luo, M.-C., Zhong, G.-Y., Kilian, A., Kleinhofs, A., and J. Dvorak (1996). Genetic map of diploid wheat, *Triticum monococcum* L., and its comparison with maps of *Hordeum vulgare* L. *Genetics* 143: 983-999.

Dunford, R.P., Kurata, N., Laurie, D.A., Money, T.A., Minobe, Y., and G. Moore (1995). Conservation of fine-scale DNA marker order in the genomes of rice and the Triticeae. *Nucleic Acids Research* 23: 2724-2728.

Edwards, M.D., Stuber, C.W., and J.F. Wendel (1987). Molecular-marker-facilitated investigations of quantitative trait loci in maize. I. Numbers, genomic distribution, and types of gene action. *Genetics* 116: 113-125.

Ellis, R.P., Forster, B.P., Waugh, R., and N. Bonar (1997). Mapping physiological traits in barley. *The New Phytologist* 137: 149.

Everson, E.H. and C.W. Schaller (1955). The genetics of yield differences associated with awn barbing in the barley hybrid (Lion x Atlas 10) x Atlas. *Agronomy Journal* 47: 276-280.

Falconer, D.S. (ed.) (1981). *Introduction to Quantitative Genetics*, Second Edition. New York: Longman.

Flavell, R.B., Rimpau, J., and D.B. Smith (1977). Repeated sequence DNA relationships in four cereal genomes. *Chromosoma* 63: 205-222.

Franckowiak, J. (1997). Revised linkage maps for morphological markers in barley, *Hordeum vulgare*. *Barley Genetics Newsletter* 26: 9-21.

Fukui, K. and K. Kakeda (1990). Quantitative karyotyping of barley chromosomes by image analysis methods. *Genome* 33: 450-458.

Gaul, H. (ed.) (1976). *Barley genetics III*. Munich: Verlag Karl Thiemig.

Gelderman, H. (1975). Investigations on inheritance of quantitative characters in animals by gene markers. I. Methods. *Theoretical and Applied Genetics* 46: 319-330.

Giese, H., Holm-Jensen, A.G., Mathiassen, H., Kjaer, B., Rasmussen, S.K., Bay, H., and J. Jensen (1994). Distribution of RAPD markers on a linkage map of barley. *Hereditas* 120: 267-273.

Gill, K.S., Gill, B.S., Endo, T.R., and E.V. Boyko (1996). Identification and high-density mapping of gene-rich regions in chromosome group 5 of wheat. *Genetics* 143: 1001-1012.

Graner, A. (1996). Molecular mapping of genes conferring disease resistance: The present state and future aspects. In *Proceedings of the V International Oat Conference and the VII International Barley Genetics Symposium*. Invited Papers, eds. G. Scoles and B. Rossnagel. Saskatoon: University of Saskatchewan Extension Press, pp. 157-166.

Graner, A., Jahoor, A., Schondelmaier, J., Siedler, H., Pillen, K., Fischbeck, G., Wenzel, G., and R.G. Herrmann (1991). Construction of an RFLP map of barley. *Theoretical and Applied Genetics* 83: 250-256.

Han, F., Kilian, A., Chen, J.P., Kudrna, D., Steffenson, B., Yamamoto, K., Matsumoto, T., Sasaki, T., and A. Kleinhofs (1999). Sequencing analysis of a rice BAC covering syntenous barley *Rpg1* region. *Genome* 42:1071-1076.

Han, F., Kleinhofs, A., Ullrich, S.E., and A. Kilian (1998). Synteny with rice: Analysis of barley malting quality QTLs and *rpg4* chromosome regions. *Genome* 41: 373-380.

Han, F. and S.E. Ullrich (1994). Mapping of quantitative trait loci for malting quality traits in barley. *Barley Genetics Newsletter* 23: 84-97.

Han, F., Ullrich, S.E., Chirat, S., Menteur, S., Jestin, L., Sarrafi, A., Hayes, P.M., Jones, B.L., Blake, T.K., Wesenberg, D.M., Kleinhofs, A., and A. Kilian (1995). Mapping of β-glucan content and β-glucanase activity loci in barley grain and malt. *Theoretical and Applied Genetics* 91: 921-927.

Han, F., Ullrich, S.E., Clancy, J.A., Jitkov, V., Kilian, A., and I. Romagosa (1996). Verification of barley seed dormancy loci via linked molecular markers. *Theoretical and Applied Genetics* 92: 87-91.

Han, F., Ullrich, S.E., Kleinhofs, A., Jones, B., Hayes, P.M., and D.M. Wesenberg (1997). Fine structure mapping of the barley chromosome 1 centromere region containing malting quality QTLs. *Theoretical and Applied Genetics* 95: 903-910.

Hayes, P.M., Blake, T., Chen, T.H.H., Tragoonrung, S., Chen, F., Pan, A., and B. Liu (1993). Quantitative trait loci on barley (*Hordeum vulgare* L.) chromosome 7 associated with components of winterhardiness. *Genome* 36: 66-71.

Hayes, P.M. and I. Iyamabo (1993). Summary of QTL effects in the Steptoe x Morex population. *Barley Genetics Newsletter* 23: 98-143.

Hayes, P.M., Liu, B.H., Knapp, S.J., Chen, F., Jones, B., Blake, T., Franckowiak, J., Rasmusson, D., Sorrells, M., Ullrich, S.E., Wesenberg, D., and A. Kleinhofs (1993). Quantitative trait locus effects and environmental interaction in a sample of North American barley germplasm. *Theoretical and Applied Genetics* 87: 392-401.

Hayes, P.M., Prehn, D., Vivar, H., Blake, T., Comeau, A., Henry, I., Johnston, M., Jones, B., Steffenson, B., Pierre, C.A. St., and F.Q. Chen (1996). Multiple disease resistance loci and their relationship to agronomic and quality loci in a spring barley population. *Journal of Quantitative Trait Loci* [online], <http://probe.nalusda.gov:8000/otherdocs/jqtl1996-02/jqtl22.html>.

Heun, M., Kennedy, A.E., Anderson, J.A., Lapitan, N.L.V., Sorrells, M.E., and S.D. Tanksley (1991). Construction of a restriction fragment length polymorphism map for barley *(Hordeum vulgare)*. *Genome* 34: 437-447.

Hinze, K., Thompson, R.D., Ritter, E., Salamini, F., and P. Schulze-Lefert (1991). Restriction fragment length polymorphism-mediated targeting of the *ml-o* resistance locus in barley *(Hordeum vulgare)*. *Proceedings of the National Academy of Sciences, USA* 88: 3691-3695.

Hohmann, U., Graner, A., Endo, T.R., Gill, B.S., and R.G. Herrmann (1995). Comparison of wheat physical maps with barley linkage maps for group 7 chromosomes. *Theoretical and Applied Genetics* 91: 618-626.

Horvath, D.P., Dahleen, L.S., Stebbing, J.A., and G. Penner (1995). A co-dominant PCR-based marker for assisted selection of durable stem rust resistance in barley. *Crop Science* 35: 1445-1450.

Hunter, H. (1962). The science of malting barley production. In *Barley and Malt: Biology, Biochemistry, Technology*, ed. A.H. Cook. New York, London: Academic Press, pp. 25-44.

Islam, A.K.M.R. (1983). Ditelosomic additions of barley chromosomes to wheat. In *Proceedings of the 6th International Wheat Genetics Symposium*, ed. S. Sakamoto. Kyoto, Japan: Kyoto University Press, pp. 233-238.

Iyamabo, I. and P.M. Hayes (1995). Effects of plot types on detection of quantitative-trait-locus effects in barley (*Hordeum vulgare* L.). *Plant Breeding* 114: 55-60.

Jansen, R.C. (1992). A general mixture model for mapping quantitative trait loci by using molecular markers. *Theoretical and Applied Genetics* 85: 252-260.

Jensen, J. (1989). Estimation of recombination parameters between a quantitative trait locus (QTL) and two marker gene loci. *Theoretical and Applied Genetics* 78: 613-618.

Jin, Y., Statler, G.D., Franckowiak, J.D., and B.J. Steffenson (1993). Linkage between leaf rust resistance genes and morphological markers in barley. *Phytopathology* 83: 220-223.

Jin, Y., Steffenson, B.J., and J.D. Miller (1994). Inheritance of resistance to pathotypes QCC and MCC of *Puccinia graminis* f. sp. *tritici* in barley line Q21861 and temperature effects on the expression of resistance. *Phytopathology* 84: 452-455.

Jorgensen, J.H. and H.P. Jensen (1979). Inter-allelic recombination in the *ml-o* locus in barley. *Barley Genetics Newsletter* 9: 37-39.

Kasha, K.J., Kleinhofs, A., Kilian, A., Saghai Maroof, M., Scoles, G.J., Hayes, P.M., Chen, F.Q., Xia, X., Li, X.-Z., Biyashev, R.M., et al. (1995). The North American Barley Genome Map on the cross HT and its comparison to the map on cross SM. In *Plant Genome and Plastome: Their Structure and Evolution*, ed. Koichiro Tsunewaki, Tokyo: Kodansha Scientific LT, Tokyo Press, pp. 73-88.

Kilian, A., Chen, J., Han, F., Steffenson, B., and A. Kleinhofs (1997). Towards map-based cloning of the barley stem rust resistance genes *Rpg1* and *rpg4* using rice as an intergenomic cloning vehicle. *Plant Molecular Biology* 35: 187-195.

Kilian, A. and A. Kleinhofs (1992). Cloning and mapping of telomere-associated sequences from *Hordeum vulgare* L. *Molecular General Genetics* 235: 153-156.

Kilian, A., Kudrna, D., and A. Kleinhofs (1999). Genetic and molecular characterization of barley chromosome telomeres. *Genome* 42: 412-419.

Kilian, A., Kudrna, D.A., Kleinhofs, A., Yano, M., Kurata, N., Steffenson, B., and T. Sasaki (1995). Rice-barley synteny and its application to saturation mapping of the barley *Rpg1* region. *Nucleic Acids Research* 23: 2729-2733.

Kilian, A., Steffenson, B.J., Maroof, M.A.S., and A. Kleinhofs (1994). RFLP markers linked to the durable stem rust resistance gene *Rpg1* in barley. *Molecular Plant-Microbe Interaction* 7: 298-301.

Kjaer, B., Haahr, V., and J. Jensen (1991). Associations between 23 quantitative traits and 10 genetic markers in a barley cross. *Plant Breeding* 106: 261-274.

Kjaer, B. and J. Jensen (1996). Quantitative trait loci for grain yield and yield components in a cross between a six-rowed and a two-rowed barley. *Euphytica* 90: 39-48.

Kjaer, B., Jensen, J., and H. Giese (1995). Quantitative trait loci for heading date and straw characters in barley. *Genome* 38: 1098-1104.

Kleinhofs, A. (1994). Barley Steptoe x Morex map. File available via Internet gopher, host: <greengenes.cit.cornell.edu>; menu: "Grainsfiles" to browse "Barley Steptoe x Morex map."

Kleinhofs, A., Chao, S., and P.J. Sharp (1988). Mapping of nitrate reductase genes in barley and wheat. In *Proceedings of the Seventh International Wheat Genetics Symposium*, Volume 1, eds. T.E. Miller and R.M.D. Koebner. Cambridge, England: Institute of Plant Science Research, pp. 541-546.

Kleinhofs, A. and A. Graner (2001). An integrated map of the barley genome. In *DNA-Based Markers in Plants*, Second Edition, eds. R. L. Phillips and I. Vasil. Dordrecht, the Netherlands: Kluwer Academic Publishers.

Kleinhofs, A., Kilian, A., Maroof, M.A.S., Biyashev, R.M., Hayes, P., Chen, F.Q., Lapitan, N., Fenwick, A., Blake, T.K., Kanazin, V., et al. (1993). A molecular, isozyme and morphological map of the barley *(Hordeum vulgare)* genome. *Theoretical and Applied Genetics* 86: 705-712.

Knapp, S.J. (1991). Using molecular markers to map multiple quantitative trait loci: Models for backcross, recombinant inbred, and doubled haploid progeny. *Theoretical and Applied Genetics* 81: 333-338.

Knapp, S.J., Bridges, W.C., and D. Birkes (1990). Mapping quantitative trait loci using molecular marker linkage maps. *Theoretical and Applied Genetics* 79: 583-592.

Kolchinsky, A., Kanazin, V., Yakovleva, E., Gazumyan, A., Kole, C., and E. Ananiev (1990). 5S-RNA genes of barley are located on the second chromosome. *Theoretical and Applied Genetics* 8: 333-336.

Komatsuda, T., Taguchi-Shiobara, F., Oka, S., Takaiwa, F., Annaka, T., and H.J. Jacobsen (1995). Transfer and mapping of the shoot-differentiation locus *Shd1* in barley chromosome 2. *Genome* 38: 1009-1014.

Konishi, T. (1981). Reverse mutation and interallelic recombination at the liguleless locus of barley. In *Barley Genetics IV*, ed. R. N. H. Whitehouse. Edinburgh, Scotland: Fourth International Barley Genetics Symposium, pp. 838-845.

Kunzel, G. and L. Korzun (1996). Physical mapping of cereal chromosomes, with special emphasis on barley. In *Proceedings of the V International Oat Conference and the VII International Barley Genetics Symposium*, Invited Papers, eds. G. Scoles and B. Rossnagel. Saskatoon: University of Saskatchewan, Extension Press, pp. 197-206.

Kunzel, G., Korzun, L., and A. Meister (1999). Cytologically integrated physical RFLP maps for the barley genome based on translocation breakpoints. *Genetics*.

Kurata, N., Moore, G., Nagamura, Y., Foote, T., Yano, M., Minobe, Y., and M. Gale (1994). Conservation of genome structure between rice and wheat. *Bio/Technology* 12: 276-278.

Kurata, N., Nagamura, Y., Yamamoto, K., Harushima, Y., Sue, N., Wu, J., Antonio, B.A., Shomura, A., Shimizu, T., Lin, S.Y., et al. (1994). A 300 kilobase interval

genetic amp of rice including 883 expressed sequences. *Nature Genetics* 8: 365-372.
Lamberts, H., Broekhuizen, S., Dantuma, G., and R.A. Martienssen (eds.) (1964). *Barley Genetics I*. Wageningen: Pudoc.
Lander, E.S. and D. Botstein (1989). Mapping Mendelian factors underlying quantitative traits using RFLP linkage maps. *Genetics* 121: 185-199.
Langridge, P., Karakousis, A., Collins, N., Kretschmer, J., and S. Manning (1995). A consensus linkage map of barley. *Molecular Breeding* 1: 389-395.
Larson, S.R., Bryan, W., Dyer, W., and T. Blake (1996). Evaluation gene effects of a major barley seed dormancy QTL in reciprocal backcross populations. *Journal of Quantitative Trait Loci* [online], <http://probe.nalusda.gov:8000/otherdocs/jqtl1996-04/larson15a.htm>.
Larson, S.R., Kadyrzhanova, D., McDonald, C., Sorrels, M., and T.K. Blake (1996). Evaluation of barley chromosome 3 yield QTL in a backcross F_2 population using STS-PCR. *Theoretical and Applied Genetics* 93: 618-625.
Laurie, D.A., Pratchett, N., Allen, R.L., and S.S. Hantke (1996). RFLP mapping of the barley homeotic mutant *lax-a*. *Theoretical and Applied Genetics* 93: 81-85.
Laurie, D.A., Pratchett, N., Bezant, J.H., and J.W. Snape (1995). RFLP mapping of five major genes and eight quantitative trait loci controlling flowering time in a winter x spring barley (*Hordeum vulgare* L.) cross. *Genome* 38: 575-585.
Lincoln, S.E., Daly, M., and E.S. Lander (1992). Mapping genes controlling quantitative traits with MAPMAKER/QTL 1.1. *Technical Report,* Second Edition. Cambridge, MA: Whitehead Institute for Biomedical Research.
Linde-Laursen, I. (1979). Giemsa C-banding of barley chromosomes. III. Segregation and linkage of C-bands on chromosomes 3, 6, and 7. *Hereditas* 91: 73-77.
Linde-Laursen, I. (1988). Giemsa C-banding of barley chromosomes. V. Localization of breakpoints in 70 reciprocal translocations. *Hereditas* 108: 65-76.
Linde-Laursen, I. (1997). Recommendations for the designation of the barley chromosomes and their arms. *Barley Genetics Newsletter* 26: 1-3.
Luo, Z.W. and M.J. Kearsey (1989). Maximum likelihood estimation of linkage between a marker gene and a quantitative trait locus. *Heredity* 63: 401-408.
MacGregor, A.W. and G.B. Fincher (1993). Carbohydrates of the barley grain. In *Barley: Chemistry and Technology*, eds. A.W. MacGregor and R.S. Bhatty. St. Paul, MN: American Association of Cereal Chemists, pp. 73-130.
Mather, D.E., Tinker, N.A., LaBerge, D.E., Edney, M., Jones, B.L., Rossnagel, B.G., Legge, W.G., Briggs, K.G., Irvine, R.B., Falk, D.E., and K.J. Kasha (1997). Regions of the genome that affect grain and malt quality in a North American two-row barley cross. *Crop Science* 37: 544-554.
Mather, K. and J.L. Jinks (eds.) (1971). *Biometrical Genetics*. Ithaca, NY: Cornell University Press.
Meszaros, K. and P.M. Hayes (1997). The Dicktoo x Morex barley mapping population. Electronic report: <http://wheat.pw.usda.gov/ggpages/DxM/>.
Moharramipour, S., Tsumuki, H., Sato, K., and H. Yoshida (1997). Mapping resistance to cereal aphids in barley. *Theoretical and Applied Genetics* 94: 592-596.
Moore, G., Devos, K.M., Wang, Z., and M.D. Gale (1995). Grasses, line up and form a circle. *Current Biology* 5: 737-739.

Moreno-Gonzalez, J. (1992). Estimates of marker-associated QTL effects in Monte Carlo backcross generations using multiple regression. *Theoretical and Applied Genetics* 85: 423-434.

Munck, L. (ed.) (1992). *Barley Genetics VI.* Copenhagen: Munksgaard International Publishers Ltd.

Namuth, D.M., Lapitan, N.L.V., Gill, K.S., and B.S. Gill (1994). Comparative RFLP mapping of *Hordeum vulgare* and *Triticum tauschii*. *Theoretical and Applied Genetics* 89: 865-872.

Nilan, R.A. (ed.) (1964). *The Cytology and Genetics of Barley, 1951-1962*. Pullman: Washington State University Press.

Nilan, R.A. (ed.) (1971). *Barley Genetics II*. Pullman: Washington State University Press.

Nilan, R.A. (1974). Barley *(Hordeum vulgare)*. In *Handbook of Genetics*, ed. R. C. King. New York: Plenum Press, pp. 93-110.

Nilan, R.A. (1990). The North American Barley Genome Mapping Project. *Barley Newsletter* 33: 112.

Oberthur, L.E., Dyer, W., Blake, T.K., and S.E. Ullrich (1995). Genetic analysis of seed dormancy in barley (*Hordeum vulgare* L.). *Journal of Quantitative Trait Loci* [online], <http://probe.nalusda.gov:8000/otherdocs/jqtl1995-05/dormancy.html>.

Oziel, A., Hayes, P.M., Chen, F.Q., and B. Jones (1996). Application of quantitative trait locus mapping to the development of winter-habit malting barley. *Plant Breeding* 115: 43-51.

Paterson, A.H. (1997). Comparative mapping of plant phenotypes. *Plant Breeding Reviews* 14: 2-37.

Paterson, A.H., DeVerna, J.W., Lanini, B., and S.D. Tanksley (1990). Fine mapping of quantitative trait loci using selected overlapping recombinant chromosome, in an interspecies cross of tomato. *Genetics* 124: 735-742.

Paterson, A.H., Tanksley, S.D., and M.E. Sorrells (1991). DNA markers in plant improvement. *Advances in Agronomy* 46: 39-90.

Pedersen, C., Giese, H., and I. Linde-Laursen (1995). Towards an integration of the physical and the genetic chromosome maps of barley by in situ hybridization. *Hereditas* 123: 77-88.

Pollock, J.R.A. (1962). The nature of the malting process. In *Barley and Malt: Biology, Biochemistry, Technology*, ed. A.H. Cook. London: Academic Press, pp. 303-399.

Powell, W., Thomas, W.T.B., Baird, E., Lawrence, P., Booth, A., Harrower, C., McNicol, J.W., and R. Waugh (1997). Analysis of quantitative traits in barley by the use of amplified fragment length polymorphisms. *Heredity* 79: 48-59.

Qi, X., Niks, R.E., Stam, P., and P. Lindhout (1998). Identification of QTLs for partial resistance to leaf rust *(Puccinia hordei)* in barley. *Theoretical and Applied Genetics* 96:1205-1215.

Qi, X., Stam, P., and P. Lindhout (1996). Comparison and integration of four barley genetic maps. *Genome* 39: 379-394.

Qi, X., Stam, P., and P. Lindhout (1998). Use of locus-specific AFLP markers to construct a high-density molecular map in barley. *Theoretical and Applied Genetics* 96: 376-384.

Ramage, R.T. (1985). Cytogenetics. In *Barley*, ed. D. C. Rasmusson. Madison, WI: American Society of Agronomy, pp. 127-154.

Rasmusson, D.C. (ed.) (1985). *Barley*. Madison, WI: American Society of Agronomy.

Rasmusson, J.M. (1935). Studies on the inheritance of quantitative characters in *Pisum. I*. Preliminary note on the genetics of flowering. *Hereditas* 20: 161-180.

Roder, M.S., Lapitan, N.L.V., Sorrells, M.E., and S.D. Tanksley (1993). Genetic and physical mapping of barley telomeres. *Molecular and General Genetics* 238: 294-303.

Roelfs, A.P., Casper, D.H., Long, D.L., and J.J. Roberts (1991). Races of *Puccinia graminis* in the United States in 1989. *Plant Disease* 75: 1127-1130.

Romagosa, I., Han, F., Clancy, J.A., and S.E. Ullrich (1999). Individual locus effects on dormancy during seed development and after ripening in barley. *Crop Science* 39: 74-79.

Romagosa, I., Han, F., Ullrich, S.E., Hayes, P.M., and D.M. Wesenberg (1999). Verification of yield QTL through realized molecular marker-assisted selection responses in a barley cross. *Molecular Breeding* 5: 143-152.

Romagosa, I., Ullrich, S.E., Han, F., and P.M. Hayes (1996). Use of the additive main effects and multiplicative model in QTL mapping for adaptation in barley. *Theoretical and Applied Genetics* 96: 30-37.

Rosichan, J., Nilan, R.A., Arenaz, P., and A. Kleinhofs (1979). Intragenic recombination at the waxy locus in *Hordeum vulgare*. *Barley Genetics Newsletter* 9: 79-85.

Saghai Maroof, M.A., Yang, G.P., Biyashev, R.M., Maughan, P.J., and Q. Zhang (1996). Analysis of the barley and rice genomes by comparative RFLP linkage mapping. *Theoretical and Applied Genetics* 92: 541-551.

Sax, K. (1923). The association of size differences with seed-coat pattern and pigmentation in *Phaseolus vulgaris*. *Genetics* 8: 552-560.

Schonfeld, M., Ragni, A., Fischbeck, G., and A. Jahoor (1996). RFLP mapping of three new loci for resistance genes to powdery mildew *(Erysiphe graminis* f. sp. *hordei)* in barley. *Theoretical and Applied Genetics* 93: 48-56.

Scoles, G. and B. Rossnagel (eds.) (1996). *Proceedings of the V International Oat Conference and the VII International Barley Genetics Symposium,* Invited Papers, Saskatoon: University of Saskatchewan Extension Press.

Sherman, J.D., Fenwick, A.L., Namuth, D.M., and N.L.V. Lapitan (1995). A barley RFLP map: Alignment of three barley maps and comparisons to Gramineae species. *Theoretical and Applied Genetics* 91: 681-690.

Shewry, P.R. (ed.) (1992). *Barley: Genetics, Biochemistry, Molecular Biology and Biotechnology*. Wallingford, UK: CAB International.

Shin, J.S., Corpuz, L., Chao, S., and T.K. Blake (1990). A partial map of the barley genome. *Genome* 33: 803-808.

Siedler, H. and A. Graner (1991). Construction of physical maps of the *Hor1* locus of two barley cultivars by pulsed field gel electrophoresis. *Molecular and General Genetics* 226: 177-181.

Simons, G., van der Lee, T., Diergaarde, P., van Daelen, R., Groenendijk, J., Frijters, A., Buschges, R., Hollricher, K., Topsch, S., Schulze-Lefert, P., et al.

(1997). AFLP-based fine mapping of the *Mlo* gene to a 30-kb DNA segment of the barley genome. *Genomics* 44: 61-70.

Simpson, S.P. (1989). Detection of linkage between quantitative trait loci and restriction fragment polymorphism using inbred lines. *Theoretical and Applied Genetics* 77: 815-819.

Smith, H.H. (1937). The relation between genes affecting size and color in certain species of *Nicotiana*. *Genetics* 22: 361.

Smith, L. (1951). Cytology and genetics of barley. *Botany Review* 17: 1-51, 133-202, 285-355.

Sogaard, B. and P. von Wettstein-Knowles (1987). Barley: Genes and chromosomes. *Carlsberg Research Communications* 52: 123-196.

Soller, M. and J. Beckmann (1983). Genetic polymorphism in varietal identification and genetic improvement. *Theoretical and Applied Genetics* 67: 25-33.

Sorensen, M.B. (1989). Mapping of the *Hor2* locus in barley by pulsed field gel electrophoresis. *Carlsberg Research Communications* 54: 109-120.

Stuber, C.W. (1992). Biochemical and molecular markers in plant breeding. In *Plant Breeding Reviews*, ed. J. Janick. New York: John Wiley and Sons, pp. 37-61.

Thomas, W.T.B., Powell, W., Waugh, R., Baird, E., Booth, A., Lawrence, P., Harrower, B., and N. Bonar (1996). Comparison of heading date QTLs on chromosome 7 across 4 barley crosses. In *Proceedings of the V International Oat Conference and the VII International Genetics Symposium,* Poster Sessions, Volume 1, eds. A. Slinkard, G. Scoles, and B. Rossnagel. Saskatoon, University of Saskatchewan Extension Press, pp. 388-390.

Thomas, W.T.B., Powell, W., Waugh, R., Chalmers, K.J., Barua, U.M., Jack, P., Lea, V., Forster, B.P., Swanston, J.S., Ellis, R.P., et al. (1995). Detection of quantitative trait loci for agronomic, yield, grain and disease characters in spring barley (*Hordeum vulgare* L.). *Theoretical and Applied Genetics* 91: 1037-1047.

Tinker, N.A. and D.E. Mather (1995). MQTL: Software for simplified composite interval mapping of QTL in multiple environments. *Journal of Quantitative Trait Loci* [online], <http://probe.nalusda.gov:8000/otherdocs/jqtl/jqtl1995-02/jqtl16r2.html>.

Tinker, N.A., Mather, D.E., Rossnagel, B.G., Kasha, K.J., Kleinhofs, A., Hayes, P.M., Falk, D.E., Ferguson, T., Shugar, L.P., Legge, W.G., et al. (1996). Regions of the genome that affect agronomic performance in two-row barley. *Crop Science* 36: 1053-1062.

Ullrich, S.E., Han, F., Froseth, J.A., Jones, B.L., Newman, C.W., and D.M. Wesenberg (1996). Mapping of loci that affect carbohydrate content in barley grain. In *Proceedings of the International Oat Conference and the VII International Barley Genetics Symposium,* Poster Sessions, Volume 1, eds. A. Slinkard, G. Scoles, and B. Rossnagel. Saskatoon: University of Saskatchewan Extension Press, pp. 141-143.

Ullrich, S.E., Hayes, P.M., Dyer, W.E., Blake, T.K., and J.A. Clancy (1993). Quantitative trait locus analysis of seed dormancy in 'Steptoe' barley. In *Pre-harvest Sprouting in Cereals 1992*, eds. M.K. Walker-Simmons and J.L. Ried. St. Paul, MN: American Association of Cereal Chemists, pp. 136-145.

Van Deynze, A.E., Nelson, J.C., Yglesias, E.S., Harrington, S.E., Braga, D.P., McCouch, S.R., and M.E. Sorrells (1995). Comparative mapping in grasses: Wheat relationships. *Molecular and General Genetics* 248: 744-754.

von Ooijen, J.W. (1992). Accuracy of mapping quantitative trait loci in autogamous species. *Theoretical and Applied Genetics* 84: 803-811.

von Wettstein-Knowles, P. (1992). Cloned and mapped genes: Current status. In *Barley: Genetics, Biochemistry, Molecular Biology and Biotechnology*, ed. P.R. Shewry. Wallingford, UK: CAB International, pp. 73-98.

Waugh, R., Bonar, N., Baird, E., Thomas, B., Graner, A., Hayes, P., and W. Powell (1997). Homology of AFLP products in three mapping populations of barley. *Molecular and General Genetics* 255: 311-321.

Weller, J.I. (1986). Maximum likelihood techniques for the mapping and analysis of quantitative trait loci with the aid of genetic markers. *Biometrics* 42: 627-640.

Wright, S. (ed.) (1968). *Evolution and the Genetics of Populations*. Chicago, IL: University of Chicago Press.

Yasuda, S. and T. Konishi (eds.) (1987). *Barley Genetics V*. Okayama, Japan: Sanyo Press.

Zeng, Z.B. (1993). Theoretical basis for separation of multiple linked gene effects in mapping quantitative trait loci. *Proceedings of the National Academy of Sciences, USA* 90: 10972-10976.

Zeng, Z.B. (1994). Precision mapping of quantitative trait loci. *Genetics* 136: 1457-1468.

Chapter 4

Wild Barley As a Source of Genes for Crop Improvement

Roger P. Ellis

INTRODUCTION

This chapter examines the most recent progress in the use of new genes from wild barley, i.e., directly from *Hordeum vulgare* L. ssp. *spontaneum* C. Koch, with less weight paid to the use of land races. This aim could not be achieved throughout as research efforts have been unevenly distributed. Loci with major effects, often interpreted as Mendelian genes, have been exploited, whereas quantitative traits have hardly been explored. The development of techniques based on the use of mapped DNA-based markers has greatly facilitated the exploration and exploitation of genetic variation in wide relatives (Tanksley and Nelson, 1996; Tanksley and McCouch, 1997).

Barley improvement programs, whether by breeding or genetic manipulation, aim to match adaptation to the local environment and to enhance quality for processing. The most significant factor determining adaptation is time of flowering. In barley, time of flowering is affected by genes controlling response to vernalization and daylength as well as earliness per se (Laurie et al., 1995; Bezant et al., 1996; Laurie, 1997). Early flowering, conditioned by daylength responses and low vernalization response, is appropriate to Mediterranean latitudes. Typically, in dry, rain-fed situations, early flowering is important to exploit winter rainfall and to avoid drought stress associated with high temperatures in midsummer (van Oosterom and Acevedo, 1992). In contrast, the highest crop yields occur in moist, cool environments in which long days allow slow plant development (Ellis and Kirby, 1980; Kirby and Ellis, 1980; Ellis and Russell, 1984; Russell and Ellis, 1988). The availability of water limits the growth and yield potential of both crops and wild populations; thus, much attention has been paid to aspects of drought tolerance (Grando, Falistocco, and Ceccarelli, 1985; Gunasekera et al., 1994; Grando and Ceccarelli, 1995; Ellis, Forster, et al., 1997; Forster et al., 1997). In dry, hot climates an additional problem is that, even when crop growth is maintained by irrigation, grain growth and maturation can be adversely affected by high air temperatures (Coles, Jamieson, and Haslemore, 1991; Ellis, Rubino, et al., 1997). In addition to daylength and rainfall, soil composition has a dramatic effect on barley growth. Although barley is more tolerant of salt than is wheat (Harlan, 1995), acidic soils cause greater stresses in barley because metal ions such as Mn^{++} and Al^{+++} are more toxic to barley than to wheat or oats (Essen and Dantuma, 1962).

In the broadest sense, processing quality in barley can be considered to relate to animal fodder (Moharramipour et al., 1999), human food (Bhatty, 1996), and malting. Much research relates to malting, brewing, and distilling (see Swanston and Ellis, Chapter 5, this book), with less attention paid to food uses of barley (see Ullrich, Chapter 6, this book). This is essentially because other cereals, such as maize, sorghum, wheat, and oats, are preferred to barley when they can be cultivated.

For a wild ancestor to be of use in cultivar improvement programs, it must offer novel genetic variability for economically important characters. Although it is obvious that new sources of disease resistance have great potential, it is not so evident that a weedy species (Tanksley and McCouch, 1997) can contribute useful genetic variation for yield and processing characters (Tanksley and Nelson, 1996; Grandillo, Zamir, and Tanksley, 1999; Zamir, Grandillo, and Tanksley, 1999). It has required wide-ranging research to describe the differences between *H. vulgare* ssp. *vulgare* and H. vulgare ssp. *spontaneum** in root characteristics (Grando and Ceccarelli, 1995), floral structure (Giles and Bengtsson, 1988), photosynthetic capacity (Burkey, 1994), salt tolerance (Nevo, Krugman, and Beiles, 1993; Pakniyat et al., 1997), grain size (Giles, 1990), milling energy (Ellis, Nevo, and Beiles, 1993), and grain protein content (Jaradat, 1991). However, this in itself is not enough because, despite current views of the taxonomic relationship between the taxa, a massive phenotypic difference exists between modern cultivars and wild barley populations. This gap can be bridged only by the application of a refined system of genetic markers to isolate characters in *H. spontaneum* and to transfer them efficiently into *H. vulgare* (Forster et al., 1997). Historically, such methods have not been available, so success in the use of *H. spontaneum* for cultivar improvement was limited by the capability of contemporary techniques. By the same token, it was not possible to carry out refined genetical analysis with the same facility as is now possible (Forster et al., 1997).

---

*Referred to henceforth as *H. vulgare* and *H. spontaneum,* respectively.

ORIGIN AND TAXONOMY

In a review published more than thirty years ago, Nilan (1964) commented, "Cultivated barley does not have a clearly traceable origin or path of descent" (p. 278).

Since then studies have not progressed to the point that a clear explanation is available for the existence of all the distinct cultivated forms of barley. Although this may not matter at the practical breeding level, there are examples of barriers to cultivar improvement, in particular the difficulty of recovering high-yielding recombinants from crosses between two-row and six-row types (Kirby and Riggs, 1978; Riggs and Kirby, 1978). *Hordeum spontaneum* has a two-row ear type, so this is not a problem for the improvement of European cultivars but could be for North African ones, and for those in other countries where six-row barley is the normal type.

Whereas the barley crop is distributed widely, its supposed progenitor is more restricted, with *H. spontaneum* occurring in the Middle East and adjacent regions of North Africa. According to Bothmer et al. (1991), *H. spontaneum* is found in the Fertile Crescent eastward to Pakistan, westward to Libya, and north to Turkey and Greece. Molina-Cano and colleagues (Molina-Cano et al., 1999; Molina-Cano et al., Chapter 2, this book), who suggest that the distribution of *H. spontaneum* extends westward to Morocco and east to Tibet, have challenged this view.

In Syria, introgression still takes place between *H. spontaneum* and cultivated barley (S. Ceccarelli, personal communication), indicating that domestication is unlikely to have been a single event. Similar observations are recorded by Molina-Cano et al. (1999) for barley fields in Morocco, where *H. spontaneum* occurs as a weed, including evidence for introgression. The extent of introgression will depend on factors such as the time of flowering, the degree of out pollination, and the fertility and vigor of the hybrids. In Scottish plots of *H. vulgare,* levels of outcrossing as high as 5 percent have been recorded (Giles, McConnell, and Fyfe, 1974), much higher than in situ estimates for *H. spontaneum* (Brown, Zohary, and Nevo, 1978).

The possibility of multiple origins and the length of the domestication process are queried by calculations of the large number of tough rachis mutants necessary to barley domestication (Ladizinsky, 1988). Surveys of chloroplast DNA may shed light on the origin of domesticated barley. The lack of diversity in the chloroplast DNA of *H. vulgare* by comparison with *H. spontaneum* (Clegg, Brown, and Whitfeld, 1984) may be evidence for a bottleneck during domestication. A similar lack of diversity has been reported for chloroplast microsatellites in *H. vulgare* (Provan et al., 1999), with the possible implication of a single domestication event. This conclusion, however, has to be tempered by consideration of the restricted sample studied. It may also be necessary to revise the general picture of the genetical changes produced by domestication (Paterson et al., 1995) to consider the particular situation of an animal-dispersed species such as barley. Seed shed from animal coats in overnight corrals could have generated grazing for nomadic herds. It is likely that the initiation of domestication was through the selective pressure of grazing by domesticated animals rather than changes in rachis fragility.

After domestication, unrecorded migration and trade would have rapidly distributed the barley crop outside the region of its origin. The result is the development of land races adapted to northern and western European environments and then North America, Australia, and Southern Africa. The recorded exchange throughout Europe of cultivars and breeding material has been extensive (Plarre and Hoffman, 1963), with widespread development of cultivars selected from land races. More recently, knowledge of plant biology and genetics has resulted in the widespread use of cultivars that are phenotypically and genetically uniform. The current situation, then, is the extreme end of a spectrum that started with *H. spontaneum* in the Middle East. This focuses on the need for in situ conservation of *H. spontaneum* and *H. vulgare* land races (Zeller, 1998).

DISEASE RESISTANCE

Hordeum spontaneum is a rich source of new disease resistances to bolster susceptible crops. Many barley diseases occur in wild populations, so it is possible to select resistance sources from observations in the field. In this chapter, a critical distinction is made between *H. spontaneum* and *H. vulgare* land races because the use of wild barley is a more formidable task for the plant breeder.

Ivandic, Walther, and Graner (1998) reported the use of molecular markers to map a new gene, *Rph16*, for resistance to leaf rust (*Puccinia hordei* Otth) on chromosome 2H. The origin of two *H. spontaneum* accessions from the gene bank of the Institut für Pflanzengenetik und Kulturpflanzenforschung (IPK), Gatersleben, Germany, used in the study was unknown, so it was not possible to trace the resistance back to "site of origin." The allele detected was dominant, and the markers employed in the mapping could equally be used for marker-assisted selection (MAS) in the development of breeding lines. The need for MAS is reinforced by comparison of the map of Ivandic, Walther, and Graner (1998) with that developed by the North American Barley Genome Mapping Project (<http://gnome.agrenv.mcgill.ca>, <http://www.css.orst.edu/research/barley/nabgmp.htm >) (see Figure 4.1), as it would appear that the *Rph16* locus is adjacent to a locus controlling attributes of malting quality.

FIGURE 4.1. Comparison of Barley Chromosome 2H Maps

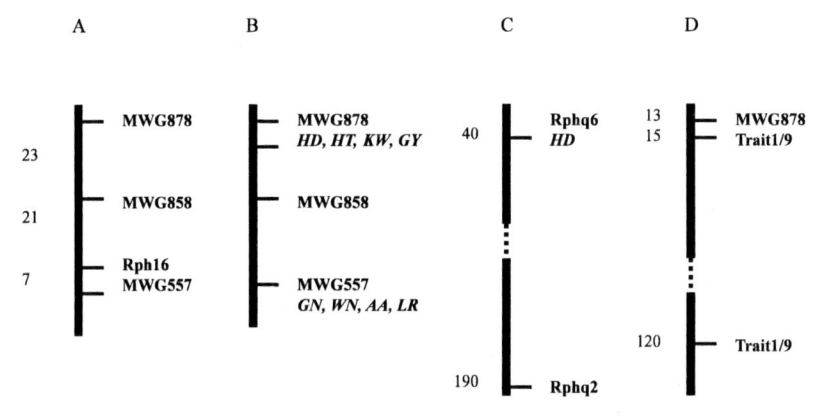

```
       A              B                    C              D

       ├ MWG878       ├ MWG878              Rphq6   13    ├ MWG878
 23    │              ├ HD, HT, KW, GY  40 ├ HD     15    ├ Trait1/9
       │              │
 21    ├ MWG858       ├ MWG858
       │              │
       ├ Rph16        │
  7    ├ MWG557       ├ MWG557
                      │ GN, WN, AA, LR
                                                    120   ├ Trait1/9
                                           190 ├ Rphq2
```

Source: (A) Ivandic, Walther, and Graner (1998); (B) Tinker and Mather (1996); (C) Qi et al. (1998); (D) Richter, Schondelmaier, and Jung (1998).

Note: Map alignments are by (A to B) positions of RFLPs drawn to the same scale as used by Ivandic, Walther, and Graner (1998) and by (B to C) positions of QTLs for heading date. Map distances in cM are given between loci (for A) or directly from the published map (for C and D). Rph and Rphq are loci for leaf rust resistance, while Trait1/9 is a locus for net blotch resistance. The QTLs are for heading date (HD), plant height (HT), kernel weight (KW), grain yield (GY), grain nitrogen (GN), wort nitrogen (WN), alpha-amylase (AA), and lodging resistance (LR).

In contrast, Qi et al. (1998) analyzed partial resistance to leaf rust found in the cultivar Vada. They argued that this form of resistance would be more durable than that based on dominant genes. If this view is substantiated empirically, then it will be necessary to use markers to select for resistances that result in this form of incomplete disease control. Qi et al. (1998) carried out a particularly sophisticated study in the form of quantitative trait loci (QTL) analysis of seedling and adult plant responses. They found six QTL for leaf rust resistance; three were effective in seedlings [located on chromosomes 2(2H), 6(6H), and 1(7H)], and five effective in adult plants [located on chromosomes 2(2H){x2}, 4(4H), 7(5H), and 6(6H)]. Two QTL—Rphq2, located distally on chromosome 2(2H), and Rphq3, located centrally on chromosome 6(6H)—conferred resistance at all growth stages.

Jana and Bailey (1995) surveyed resistance to two fungal diseases that produce similar leaf symptoms (spot blotch due to *Cochliobolus sativus* and net blotch due to *Drechlera teres* [*Pyrenospora teres*]). They tested some 290 wild barleys from 15 sites in Jordan and 5 sites in Turkey and 350 cultivated barleys (land races) from the same countries. The highest levels of resistance to the diseases were in wild barleys, but some of the land races exceeded the resistant controls. Resistance to seedling stage net blotch was resolved to 12 QTL in a cross between a susceptible cultivar and an Ethiopian land race (Hor9088) by Richter, Schondelmaier, and Jung (1998). Again, as in the case of partial resistance to leaf rust, the complexity of host-pathogen interaction was revealed only by a sophisticated analysis of disease development. This resulted in the detection of different loci for symptoms scored successively after seven and nine days of incubation. Inspection of the map location for the QTL on chromosome 2(2H) shows the proximity of the net blotch resistance Trait1/9 to the MGW878 locus, resulting in a concentration of disease resistances and important agronomic traits (see Figure 4.1) on the short arm.

In Europe the barley disease powdery mildew *(Erysiphe graminis)* has received considerable attention from breeders in attempts to alleviate yield losses that could be as high as 50 percent. Schonfeld et al. (1996) provided a genetic map that acts as a foundation for knowledge of mildew resistance manipulation. Three resistances from *H. spontaneum* were mapped in F_2 populations developed from crosses with both cv. Pallas and cv. Gitte. In addition to the mapping study, test crosses were made to examine allelism between the new resistances and lines carrying the *Mla9, Mla13, Mlp3,* and *Mlp* alleles. A recessive gene was mapped to chromosome 1S(7HS), and two semidominant genes were mapped to new resistance loci on chromosomes 7L(5HL) and 1L(7HL).

This study was a "follow up" to the exploration of mildew resistance in *H. spontaneum* initiated by authors such as Moseman (1955) and Jahoor and Fischbeck (1987a,b). It was notable that the three resistance lines were selected from an F_7 bulk derived by selfing crosses between the *H. spontaneum* accessions and cv. Elgina, cv. Oriol, cv. Aramir, and cv. Diamant and selecting for mildew resistance at each generation. In this process, many of the typical traits of wild barley, such as brittle rachis and poor adaptation, would have been modified, resulting in material closer to cultivars. The development of the F_7 lines by intercrossing with cultivars mirrors the development of land races. However, even where land races are used instead of *H. spontaneum,* the development of novel resistances can be difficult and prolonged. The *mlo* mildew resistance was discovered in the 1940s, in an Ethiopian land race, but not released in a cultivar until the 1980s. The gene was located on the long arm of chromosome 4(4H) but was associated with severe leaf flecking. This was alleviated by breeding, and eventually recombination reduced the effect of a QTL for low grain number per plant to give high-yielding cultivars (Thomas et al., 1998). The effort invested in the development of the *mlo* resistance was repaid by its durability and widespread deployment in cultivars (Thomas et al., 1998).

Jonsson and Lehmann (1999) reported an extensive program to locate and develop genes from 6,500 lines of *H. spontaneum* for mildew, leaf rust, net blotch, *Rhynchosporium,* and barley yellow dwarf virus resistance. Efforts to locate new sources of mildew resistance in *H. spontaneum* continue with, for example, the screening of the U.S. Department of Agriculture Small Grain Collection (Dreiseitl and Bockelman, <http://wheat.pw.usda.gov/ggpages/BarleyNewsletter/41/dreiseitl.html>, <http://wheat. pw.usda.gov/ggpages/BarleyNewsletter/42/dreiseitl.html>). The collection of some 1,500 accessions of *H. spontaneum* was screened in the greenhouse and field to identify over 100 accessions with good mildew resistance for further study.

Rhynchosporium secalis (Oud.) Davis causes scald in winter barley but can be particularly severe when conditions are favorable for spore spread by rain splash. In these conditions, control by fungicides is also difficult. Abbott, Lagudah, and Brown (1995) located a new gene for leaf blotch resistance, originally from *H. spontaneum* (Abbott, Brown, and Burdon, 1992), on chromosome 6H.

ABIOTIC STRESS TOLERANCE

The abiotic stress tolerances of *H. spontaneum* have been extensively explored in attempts to locate genetical variation for characters such as photoperiod response and tolerance to salt, drought, cold, and waterlogging. An essential area of research has been the description of the relationship between genetic variation and site of origin ecogeography (reviewed by Forster, 1999).

This theme was continued by the examination of genetical variation for root characters in lines from selected populations of *H. spontaneum* in hydroponic culture (Ellis et al., 2000). In this medium, root length was reduced by a "drought" treatment but increased by a "no-nitrogen" treatment. The relationship between root length and the environmental variables listed by Pakniyat et al. (1997) is summarized in Table 4.1 and Figure 4.2. Although the correlation of the root lengths with ecogeographical parameters was low, the correlation between "control" root length and mean temperature in January (Tj) is interesting. It is by no means certain that results from hydroponic culture reflect those from field conditions, but in January plants would be actively growing, while in August summer drought will have caused desiccation, reflected in a negative correlation of root length with evapotranspiration (Ev). The no-nitrogen and drought treatments show a significant relationship with latitude that reflects cooler and moister conditions in more northerly sites. In turn, this underlines the importance of plant development patterns to abiotic stress tolerance (van Oosterom and Acevedo, 1992).

TABLE 4.1. Statistically Significant Correlations Between Root Length in *H. spontaneum* Populations Given Control, Drought, and No-Nitrogen Treatments in Hydroponic Culture and Ecogeographical Parameters

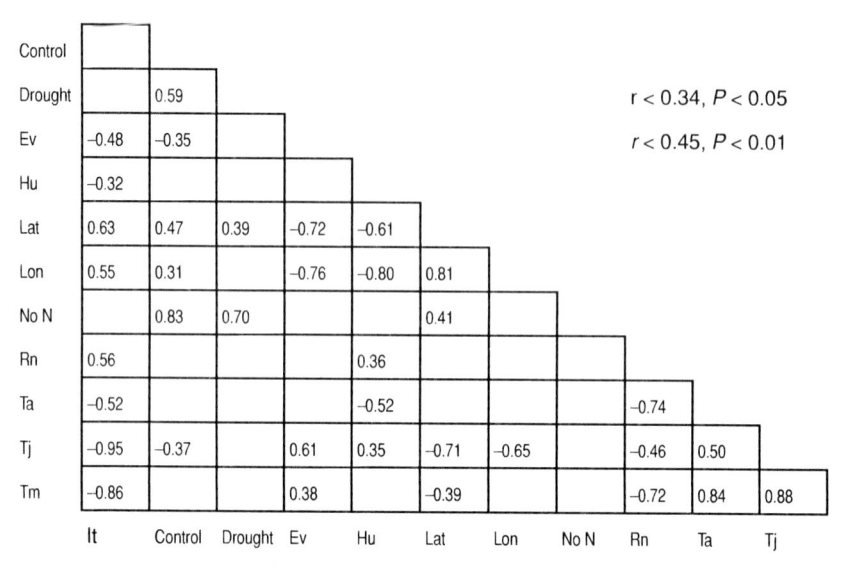

	It	Control	Drought	Ev	Hu	Lat	Lon	No N	Rn	Ta	Tj
Control											
Drought		0.59									
Ev	-0.48	-0.35									
Hu	-0.32										
Lat	0.63	0.47	0.39	-0.72	-0.61						
Lon	0.55	0.31		-0.76	-0.80	0.81					
No N		0.83	0.70			0.41					
Rn	0.56			0.36							
Ta	-0.52			-0.52					-0.74		
Tj	-0.95	-0.37		0.61	0.35	-0.71	-0.65		-0.46	0.50	
Tm	-0.86			0.38		-0.39			-0.72	0.84	0.88

$r < 0.34, P < 0.05$
$r < 0.45, P < 0.01$

Source: Ellis et al. (2000) (hydroponic culture); Pakniyat et al. (1997) (ecogeographical parameters).

Note: The ecogeographical parameters are Ev = evapotranspiration, Hu = humidity, Lat = latitude, Lon = longitude, Rn = annual rainfall, Ta = mean temperature in August, Tj = mean temperature in January, Tm = mean annual temperature.

The genetic control of drought tolerance has been explored by Teulat and colleagues (Teulat et al., 1997) in recombinant inbred lines from the cross Tadmor x Er/Amp. They reported positions of QTLs for leaf water status and related traits, namely, relative water content (RWC), the number of leaves on the main stem (NL), and total shoot fresh mass (TSFM). Drought stress reduced the RWC of Er/Amp but not Tadmor. Er/Amp had more leaves on the main stem and higher fresh weight when unstressed, but Tadmor and Er/Amp fresh weights were not different under drought stress. Only one QTL was found for fresh weight [on chromosome 2(SH), while three were reported for RWC [6(6H), 1S(7HS), and 1L(7HL)], and five for number of leaves on the main stem [5(1H), 7(5H), 6(6H), 1(7HS), and 1L(7HL)]. This study was extended to include osmotic adjustment (OA), i.e., the decrease of osmotic potential within cells due to solute accumulation during the decline of leaf water (Teulat et al., 1998). A most interesting

FIGURE 4.2. The Relationship Between Root Length in Hydroponic Culture and Ecogeographic Data from the Site of Origin

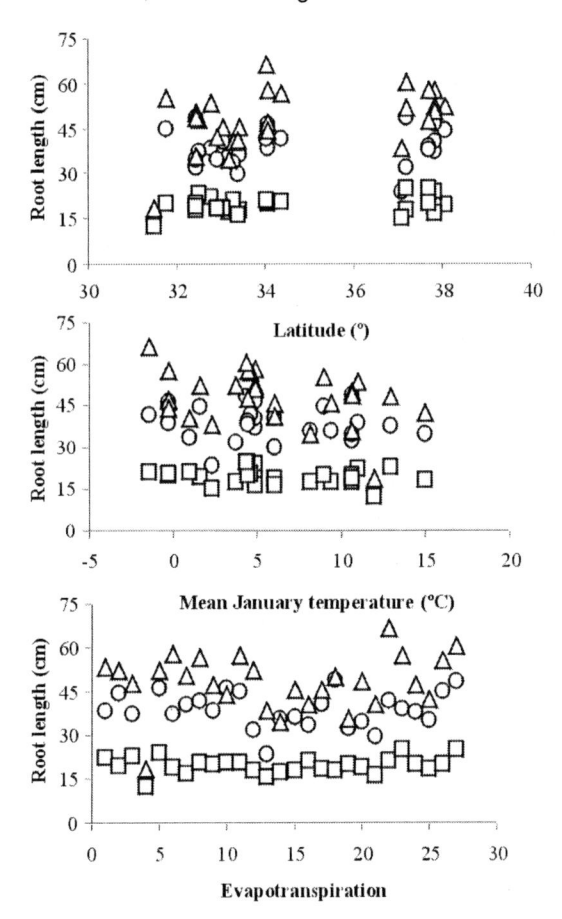

Source: Pakniyat et al. (1997).

Note: Symbols stand for control (circles), drought (squares), and no-nitrogen (triangles).

result was that a leaf osmotic potential QTL mapped to the same position as the QTL for RWC on chromosome 1(7H). In total, seven QTLs were found for leaf osmotic potential and located to chromosomes 2(2H), 7(5H), 6(6H), and 1(7H). A major problem of this study was the lack of polymorphic markers for chromosome 4(4H), so no map of this chromosome was available for QTL mapping. This restricts comparison between work in the Tadmor x Er/Amp population and Lina x HS92 (Ellis, Forster, et al., 1997).

Root length was investigated in a population of doubled haploids from the cross Clipper x Sahara 3771 (Jefferies et al., 1999) in a study of boron tolerance. Seedlings were exposed to a range of boron solutions in filter paper rolls. Root length was affected by loci on three chromosomes, namely, at the position of AWBMA125 on chromosome 3(3H), WG114 on chromosome 4(4H), and WG564 on chromosome 7(5H).

Attempts to locate genes controlling tolerance to salt were hampered by difficulties in mapping amplified fragment length polymorphisms (AFLPs) (Pakniyat et al., 1997). Despite this problem, it was possible to associate genetic diversity with ecogeographic data from the sites of origin of the original populations (Pakniyat et al., 1997.) A diverse population used for gene mapping at the Scottish Crop Research Institute was derived from the cross between the spring barley Lina and an *H. spontaneum* line collected from Canada Park, Israel. Doubled haploids from this population were tested for salt tolerance, and a number of candidate loci identified (Ellis, Forster, et al., 1997) on chromosomes 5(1H), 4(4H), 6(6H), and 1(7H). Of the characters measured, seedling dry weight and $\delta^{13}C$ are particularly interesting and were associated with six loci (see Figure 4.3). Seedling dry weight (DWC) in the

FIGURE 4.3. AFLP Marked Loci Associated with Seedling Dry Weight in Control (DWC) and Salt (DWS) Treatments and $\delta^{13}C$ Relative to Mapped Microsatellites and Amylase Loci

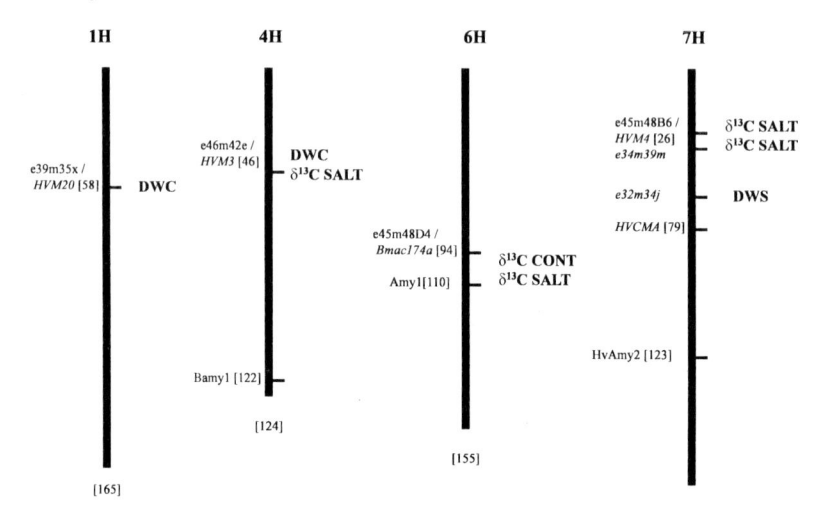

Source: Ellis, Forster, et al. (1997).

Note: Map distances in cM are given in brackets adjacent to each locus and for each chromosome.

control was associated with the AFLP e39m35x on chromosome 5(1H) and, in subsequent mapping with microsatellites (Ramsay et al., 2000), with the HVM20 microsatellite. DWC was also associated with the AFLP e46m42e, 1.5 cM distal to HVM3, on chromosome 4H some 2 cM from the locus associated with carbon isotope discrimination in salt (δ^{13}C SALT). Seedling dry weight in salt treatment (DWS) was associated with AFLP e32m36t on chromosome 1(7H) some 8 cM distal to the microsatellite HVCMA. The remaining loci on chromosome 6(6H) and 1(7H) were concerned with carbon isotope discrimination; that on 6H, located near Bmac174a, was active in both control and salt treatments, while those on 1(7H), near to the position of HVM4, were detected only in the salt treatment. These results contrast with those of Mano and Takeda (1997) because they also examined the effect of salt on germination as well as seedling growth. Three QTL contributed to the effect of salt on germination on chromosomes 4(4H), 7(5H), and 6(6H), while two on chromosomes 5(1H) and 7 (5H) affected salt tolerance in seedlings.

These studies show that there is great potential to change the tolerance of abiotic stress by the use of *H. spontaneum* genes, but there is not yet an example of successful deployment in a cultivar. Phenotypic studies indicate a wide range of variation for important characters such as time of heading (Ellis et al., 2000), drought tolerance (B. P. Forster, personal communication), salt tolerance (Nevo, Krugman, and Beiles, 1993), and tolerance to toxic minerals such as Mn^{++} and Al^{+++}.

GRAIN QUALITY

In the agriculture of the developed world, barley is used for malting and animal feed. These topics are reviewed elsewhere in this book so only a brief review is presented here. The changes that have been considered common to the domestication (Paterson et al., 1995) of many cereal species obviously understate the differences between *H. vulgare* and *H. spontaneum*. A particular area that has been neglected is that of grain size variation in relation to ear position. This has an obvious advantage in a wild species, for which flexibility in response to the environment is a prime requirement for ecological success (Giles, 1990). The pattern of grain size varies widely in *H. vulgare* and is related to the size of the rachis internodes. In a lax ear, typical of many modern cultivars, the lemma awns are parallel and grain size and time of maturity show minimal divergence from the mean time of maturity. This ear form contrasts with that typical of cultivars in the United Kingdom in the 1930s, when Spratt Archer, with a fan-shaped ear of large grain with relatively short rachis internodes, was widely grown.

At anthesis, the spikelets of an ear of a typical modern cultivar show anther dehiscence first at the middle positions, while lower and higher positions reach anthesis one to two days later. This positional difference reflects the course of preanthesis differentiation at the stem apex (Kirby and Ellis, 1980), and together these processes result in a 50 percent variation in grain size within an ear. If careful methodology is employed, it is possible to demonstrate that the rate of germination depends on grain position, with the most rapid rate shown by grains from the base of the ear rather than the tip (Ellis and Marshall, 1998). In *H. spontaneum* the developmental pattern is critically different, as the anthers are fully extruded before anthesis and grains are smaller than those of *H. vulgare*. Strikingly, in contrast to *H. vulgare*, there is a large difference in the time spikelets take to reach maturity. In both types the spikelets at the tip of the ear mature before those at the base, but this difference is more marked in *H. spontaneum*, and the rachis disarticulates before the basal spikelets mature, which has obvious adaptive consequences. Postharvest dormancy can be considerable in *H. spontaneum*, as reviewed by Gozlan and Gutterman (1999), another obvious adaptation to dry environments.

In addition to germination, a prime requirement for malting is optimal grain composition, e.g., a maximization of starch content with a reduction of complex cell wall glucans and storage proteins. Low grain milling energy is associated with the highest levels of malting quality and is an assay that can be applied to *H. spontaneum*. Ellis, Nevo, and Beiles (1993) found a wide range of variation for milling energy in *H. spontaneum*, with low levels being associated with populations from desert sites. The development of appropriate levels of enzyme during germination (Ahokas and Naskali, 1990; Ahokas and Erkkila, 1992) and the adaptation of enzymes to industrial processes are highlighted by the report that higher thermostability of β-amylase can be found in *H. spontaneum* (Eglinton, Langridge, and Evans, 1998) as well as in non-European cultivars (Kihara et al., 1998).

GENETIC DIVERSITY

Genetic diversity is necessary for any crop improvement program, and that of *H. spontaneum* has been extensively explored in the work of Professor E. Nevo and colleagues. In a series of studies, many aspects of *H. spontaneum* have been elicited, including the high level of isozyme polymorphism (Nevo, Brown, and Zohary, 1979), the rates of outcrossing (Brown, Zohary, and Nevo, 1978), plant nitrogen uptake (Corke, Nevo, and Atsmon, 1988), and a range of disease resistances, e.g., to powdery mildew (Moseman, Nevo, and El-Morshidy, 1990). Other research groups have explored variation in both *H. spontaneum* and land races, and here the efforts of Professor S. Ceccarelli's group, International Center for Agricultural Research in the Dry Areas (ICARDA), Syria, are notable, as they have made major contributions, particularly to the exploration of genetic diversity in Syria, Jordan, and Ethiopia (Leur, Ceccarelli, and Grando, 1989; Ceccarelli, Grando, and Leur, 1987; Lakew et al., 1997) and to plant breeding (van Oosterom and Ceccarelli, 1993; van Oosterom, Ceccarelli, and Peacock, 1993).

FUTURE TRENDS

Barley research offers the prospect of an exciting cascade of knowledge from the genome, through wild populations into cultivars with unique properties. To offer optimal progress, programs need to balance the use of unadapted germplasm with fundamental genetic research, phenotyping, and cultivar development. Historically this span of activities has required the involvement of publicly funded programs, and given appropriate interfaces with the plant breeding industry, this remains a certain requirement for the future health of agriculture worldwide. A publicly funded program such as at the Scottish Crop Research Institute can embrace a wide range of techniques and interact with other public and industry programs in a unique and highly productive manner.

The need exists for a more complete alignment of genetic maps constructed with RFLPs, AFLPs, and single sequence repeats (SSRs) to make a direct comparison of QTL. The rapid progress reported (Powell, Machray, and Provan, 1996) in genome mapping with SSRs will continue, and as expressed sequence tags (ESTs) are deployed, QTL will be associated with known genes. New QTL were found through phenotypic assays in barley or by synteny with other cereals, such as QTL for useful root characteristics in rice and maize (Champoux et al., 1995; Lebreton et al., 1995; Price and Tomos, 1997; Yadav et al., 1997). This knowledge, as concepts of synteny (Moore et al., 1995) are extended, will be directly applied to barley, both cultivars and wild populations, to follow up initial studies of plant genotype, such as those of Chalmers et al. (1992) and Dawson et al. (1993).

A vital step in the utilization of wild barley lines for cultivar improvement is the elucidation of the effects of domestication. Past studies have shown how simple characteristics differ between *H. vulgare* and *H. spontaneum*, so the prime requirement is to understand the effects of changes in quantitatively inherited traits. Malting quality improvement with new alleles for enzyme composition will not succeed unless marker-assisted selection allows separation of the desired character from possible adverse characters. The historical example of the failure of all attempts to develop high-lysine barley shows the complexity of the interactions involved. The speed of developments in barley genetics will ensure the successful achievement of these objectives and others that arise unexpectedly, e.g., from global climate change.

REFERENCES

Abbott, D.C., Brown, A.H.D., and Burdon, J.J. (1992). Genes for scald resistance from wild barley *(Hordeum vulgare* ssp. *spontaneum)* and their linkage to isozyme markers. *Euphytica* 61: 225-231.

Abbott, D.C., Lagudah, E.S., and Brown, A.H.D. (1995). Identification of RFLPs flanking a scald resistance gene on barley chromosome 6. *Journal of Heredity* 86: 152-154.

Ahokas, H. and Erkkila, M.J. (1992). Barley α-amylase and β–glucanase activities at germination in vulgare-type lines from backcrosses of wild, spontaneum strains with cv. Adorra. *Agricultural Science in Finland* 1: 339-350.

Ahokas, H. and Naskali, L. (1990). Geographic variation of α-amylase, β-glucanase, pullulanase and chitinase activity in germinating *Hordeum spontaneum* barley from Israel and Jordan. *Genetica* 82: 73-78.

Bezant, J., Laurie, D., Pratchett, N., Chojecki, J., and Kearsey, M. (1996). Marker regression mapping of QTL controlling flowering time and plant height in a spring barley cross *(Hordeum vulgare* L.). *Heredity* 77: 64-73.

Bhatty, R.S. (1996). Production of food malt from hull-less barley. *Cereal Chemistry* 73: 75-80.

Bothmer, von R., Jacobsen, N., Baden C., Jorgensen, R., and Linde-Laursen, I. (1991). *An ecogeographical study of the genus* Hordeum. Rome: International Board for Plant Genetic Resources, 127 pp.

Brown, A.H.D., Zohary, D., and Nevo, E. (1978). Outcrossing rates and heterozygosity in natural populations of *Hordeum spontaneum* Koch in Israel. *Heredity* 41: 49-62.

Burkey, K.O. (1994). Genetic variation of photosynthetic electron transport in barley: Identification of plastocyanin as a potential limiting factor. *Plant Science* 98: 177-187.

Ceccarelli, S., Grando, S., and Leur, J.A.G. van (1987). Genetic diversity in barley landraces from Syria and Jordan. *Euphytica* 36: 389-405.

Chalmers, K.J., Waugh, R., Watters, J., Forster, B.P., Nevo, E., Abbott, R.J., and Powell, W. (1992). Grain isozyme and ribosomal DNA variability in *Hordeum spontaneum* populations from Israel. *Theoretical and Applied Genetics* 84: 313-322.

Champoux, M.C., Wang, G., Sarkarung, S., Mackill, D.J., O'Toole, J.C., Huang, N., and McCouch, S.R. (1995). Locating genes associated with root morphology and drought avoidance in rice via linkage to molecular markers. *Theoretical and Applied Genetics* 90: 969-981.

Clegg, M.T., Brown, A.H.D, and Whitfeld, P.R. (1984). Chloroplast DNA diversity in wild and cultivated barley: Implications for genetic conservation. *Genetical Research, Cambridge* 43: 339-343.

Coles, G.D., Jamieson, P.D., and Haslemore, R.M. (1991). Effect of moisture stress on malt quality in Triumph barley. *Journal of Cereal Science* 14: 161-177.

Corke, H., Nevo, E., and Atsmon, D. (1988). Variation in vegetative parameters related to the nitrogen economy of wild barley, *Hordeum spontaneum,* in Israel. *Euphytica* 39: 227-232.

Dawson, I.K., Chalmers, K.J., Waugh, R., and Powell, W. (1993). Detection and analysis of genetic variation in *Hordeum spontaneum* populations from Israel using RAPD markers. *Molecular Ecology* 2: 151-159.

Eglinton, J.K., Langridge, P., and Evans, D.E. (1998). Thermostability variation in alleles of barley beta-amylase. *Journal of Cereal Science* 28: 301-309.

Ellis, R.P., Forster, B.P., Robinson, D., Handley, L. L., Gordon, D. C., Russell J.R., and Powell, W. (2000). Wild barley: A source of genes for crop improvement in the 21st century? *Journal of Experimental Botany* 51: 9-17.

Ellis, R.P., Forster, B.P., Waugh, R., Bonar, N., Handley, L.L., Robinson, D., Gordon, D.C., and Powell, W. (1997). Mapping physiological traits in barley. *New Phytologist* 137: 149-157.

Ellis, R.P. and Kirby, E.J.M. (1980). A comparison of spring barley grown in England and Scotland. 2. Yield and its components. *Journal of Agricultural Science, Cambridge* 95: 111-115.

Ellis, R.P. and Marshall, B. (1998). The relationship between plant development in barley and the subsequent rate of seed germination. *Journal of Experimental Botany* 49: 1021-1029.

Ellis, R.P., Nevo, E., and Beiles, A. (1993). Milling energy polymorphism in *Hordeum spontaneum* Koch in Israel and its potential utilization in breeding for malting quality. *Plant Breeding* 111: 78-81.

Ellis, R.P. and Russell, G. (1984). Plant development and grain yield in spring and winter barley. *Journal of Agricultural Science, Cambridge* 102: 85-95.

Ellis, R.P., Rubio, A., Perez-Vendrell, A.M., Romagosa, I., Swanston, J.S., and Molina-Cano, J.L. (1997). The development of β-glucanase and degradation of β-glucan in barley grown in Scotland and Spain. *Journal of Cereal Science* 26: 75-82.

Essen, van A. and Dantuma, G. (1962). Tolerance to acid soil conditions in barley. *Euphytica* 11: 282-286.

Forster, B. P. (1999). Studies on wild barley, *Hordeum spontaneum* C. Koch, at the Scottish Crop Research Institute. In *Evolutionary theory and processes: Modern perspectives, papers in honour of Eviatar Nevo,* ed. S.P. Wasser. Dordrecht: Kluwer, pp. 325-341.

Forster, B.P., Russell, J.R., Ellis, R.P., Handley, L.L., Robinson, D., Hackett, C.A., Nevo, E., Waugh, R., Gordon, D.C., Keith, R., and Powell, W. (1997). Locating genotypes and genes for abiotic stress tolerance in barley: A strategy using maps, markers and the wild species. *New Phytologist* 137: 141-147.

Giles, B.E. (1990). The effects of variation in seed size on growth and reproduction in the wild barley *Hordeum vulgare* ssp. *spontaneum*. *Heredity* 64: 239-250.

Giles, B.E. and Bengtsson, B.O. (1988). Variation in anther size in wild barley *(Hordeum vulgare* ssp. *spontaneum). Hereditas* 108: 199-205.

Giles, R.J. McConnell, G., and Fyfe, J.L. (1974). The frequency of crossing in a composite cross grown in Scotland. *Journal of Agricultural Science, Cambridge* 83: 447-450.

Gozlan, S. and Gutterman, Y. (1999). Dry storage temperatures, duration, and salt concentrations affect germination of local and edaphic ecotypes of *Hordeum*

spontaneum (Poaceae) from Israel. *Biological Journal of the Linnean Society* 67: 163-180.
Grandillo, S., Zamir, D., and Tanksley, S.D. (1999). Genetic improvement of processing tomatoes: A 20 years perspective. *Euphytica* 110: 85-97.
Grando, S. and Ceccarelli, S. (1995). Seminal root morphology and coleoptile length in wild *(Hordeum vulgare* ssp. *spontaneum)* and cultivated *(Hordeum vulgare* ssp. *vulgare)* barley. *Euphytica* 86: 73-80.
Grando, S., Falistocco, E., and Ceccarelli, S. (1985). Use of wild relatives in barley breeding. *Genetica Agaria* 3: 65-76.
Gunasekera, D., Santakumari, M., Glinka, Z., and Berkowitz, G.A. (1994). Wild and cultivated barley genotypes demonstrate varying ability to acclimate to plant water deficits. *Plant Science* 99: 125-134.
Harlan, J.R. (1995). Barley. In *Evolution of crop plants*, eds. J. Smartt and N.W. Simmonds. London: Longman, pp. 140-147.
Ivandic, V., Walther, U., and Graner, A. (1998). Molecular mapping of a new gene in wild barley conferring complete resistance to leaf rust (*Puccinia hordei* Otth). *Theoretical and Applied Genetics* 97: 1235-1239.
Jahoor, A. and Fischbeck, G. (1987a). Genetical studies of resistance of powdery mildew in barley lines derived from *Hordeum spontaneum* collected from Israel. *Plant Breeding* 99: 265-273.
Jahoor, A. and Fischbeck, G. (1987b). Sources of resistance to powdery mildew in barley lines derived from *Hordeum spontaneum* collected from Israel. *Plant Breeding* 99: 274-281.
Jana, S. and Bailey, K.L. (1995). Responses of wild and cultivated barley from West Asia to net blotch and spot blotch. *Crop Science* 35: 242-246.
Jaradat, A.A. (1991). Grain protein variability among populations of wild barley (*Hordeum spontaneum* C. Koch.) from Jordan. *Theoretical and Applied Genetics* 83: 164-168.
Jefferies, S.P., Barr, A.R., Karakousis, A., Kretschner, J.M., Manning, S., Chalmers, K.J., Nelson, J.C., Islam, A.K.M.R., and Langridge, P. (1999). Mapping of chromosome regions conferring boron toxicity tolerance in barley (*Hordeum vulgare* L.). *Theoretical and Applied Genetics* 98: 1293-1303.
Jonsson, R. and Lehmann, L. (1999). Use of new gene sources for resistance in barley breeding. *Sveriges UtsadesforHening Tidskrift* 109: 146-159.
Kihara, M., Kaneko, T., and Ito, K. (1998). Genetic variation of beta-amylase thermostability among varieties of barley, *Hordeum vulgare* L., and relation to malting quality. *Plant Breeding* 117: 425-428.
Kirby, E.J.M. and Ellis, R.P. (1980). A comparison of spring barley grown in England and Scotland. 1. Shoot apex development. *Journal of Agricultural Science, Cambridge* 95: 101-110.
Kirby, E.J.M. and Riggs, T.J. (1978). Developmental consequences of two-row and six-row ear type in spring barley. 2. Shoot apex, leaf and tiller development. *Journal of Agricultural Science, Cambridge* 91: 207-216.
Ladizinsky, G. (1998). How many tough-rachis mutants gave rise to domesticated barley? *Genetic Resources and Crop Evolution* 45: 411-414.

Lakew, B. Semeane, Y. Alemayehu, F. Gebre, H. Grando, S. Leur, J. A. G. van, and Ceccarelli, S. (1997). Exploiting the diversity of barley landraces in Ethiopia. *Genetic Resources and Crop Evolution* 44: 109-116.

Laurie, D.A. (1997). Comparative genetics of flowering time. *Plant Molecular Biology* 35: 167-177.

Laurie, D.A., Pratchett, N., Bezant, J.H., and Snape, J.W. (1995). RFLP mapping of five major genes and eight QTL controlling flowering time in a winter x spring barley (*Hordeum vulgare* L.) cross. *Genome* 38: 575-585.

Lebreton, C., Lazic-Jancic, V., Steed, A., Pekic, S., and Quarrie, S.A. (1995). Identification of QTL for drought responses in maize and their use in testing causal relationships between traits. *Journal of Experimental Botany* 46: 853-865.

Leur, J.A.G. van, Ceccarelli, S., and Grando, S. (1989). Diversity for disease resistance in barley land races from Syria and Jordan. *Plant Breeding* 103: 324-335.

Mano, Y. and Takeda, K. (1997). Mapping quantitative trait loci for salt tolerance at germination and the seedling stage in barley (*Hordeum vulgare* L.). *Euphytica* 94: 263-272.

Moharramipour, S., Takeda, K., Sato, K., Yoshida, H., and Tsumuki, H. (1999). Inheritance of gramine content in barley. *Euphytica* 106: 181-185.

Molina-Cano, J.L., Moralejo, M., Igartua, E., and Romagosa, I. (1999). Further evidence supporting Morocco as a centre of origin of barley. *Theoretical and Applied Genetics* 98: 913-918.

Moore, G., Devos, K.M., Wang, Z., and Gale, M.D. (1995). Grasses, line up and form a circle. *Current Biology* 5: 737-739.

Moseman, J.G. (1955). Sources of resistance to powdery mildew of barley. *Plant Disease Reporter* 39: 967-972.

Moseman, J.G., Nevo, E., and El-Morshidy, M.A. (1990). Reactions of *Hordeum spontaneum* to infection with two cultures of *Puccinia hordei* from Israel and United States. *Euphytica* 49: 169-175.

Nevo, E., Brown, A. H. D., and Zohary, D. (1979). Genetic diversity in the wild progenitor of barley in Israel. *Experientia* 35: 1027-1029.

Nevo, E., Krugman, T., and Beiles, A. (1993). Genetic resources for salt tolerance in the wild progenitors of wheat *(Triticum dicoccoides)* and barley *(Hordeum spontaneum)* in Israel. *Plant Breeding* 110: 338-341.

Nilan, R.A. (1964). *The cytology and genetics of barley*. Pullman: Washington State University Press, 278 pp.

Pakniyat, H., Powell, W., Baird, E., Handley, L.L., Robinson, D., Scrimgeour, C.M., Nevo, E., Hackett, C.A., Caligari, P.D.S., and Forster, B.P. (1997). AFLP variation in wild barley (*Hordeum spontaneum* C. Koch) with reference to salt tolerance and associated ecogeography. *Genome* 40: 332-341.

Paterson A.H., Lin, Y.-R., Li, Z., Schertz, K.F., Doebley, J.F., Pinson, S.R.M., Liu, S.-C., Stansel, J.W., and Irvine, J.E. (1995). Convergent evolution of cereal crops by independent mutations at corresponding genetic loci. *Science* 269: 1714-1718.

Plarre, W. and Hoffman, W. (1963). The development of barley growing and breeding in Europe. *Barley Genetics I, Proceedings of the 1st International Barley*

Genetics Symposium, eds. S. Broekhuizen, G. Dantuma, H. Lamberts, and W. Lange. Wageningen: Pudoc, pp. 7-57.

Powell, W., Machray, G., and Provan, J. (1996). Polymorphism revealed by simple sequence repeats. *Trends in Plant Science* 1: 215-222.

Price, A.H. and Tomos, A.D. (1997). Genetic dissection of root growth in rice (*Oryza sativa* L.). II. Mapping quantitative trait loci using molecular markers. *Theoretical and Applied Genetics* 95: 143-152.

Provan, J., Russell, J.R., Booth, A., and Powell, W. (1999). Polymorphic chloroplast simple sequence repeat primers for systematic and population studies in the genus *Hordeum. Molecular Ecology* 8: 505-511.

Qi, X, Niks, R.E., Stam, P., and Lindhout, P. (1998). Identification of QTLs for partial resistance to leaf rust *(Puccinia hordei)* in barley. *Theoretical and Applied Genetics* 96: 1205-1215.

Ramsay, L., Maccaulay, M., Ivanissevich, S., MacLean, K., Cardle, L., Fuller, J., Edwards, K.J., Tuvesson, S., Morgante, M., Massari, A., et al. (2000). A simple sequence repeat-based linkage map of barley. *Genetics* 156: 1997-2005.

Richter, K., Schondelmaier, L., and Jung, C. (1998). Mapping of quantitative trait loci affecting *Drechlera teres* resistance in barley with molecular markers. *Theoretical and Applied Genetics* 97: 1225-1243.

Riggs, T.J. and Kirby, E.J.M. (1978). Developmental consequences of two-row and six-row ear type in barley. 1. Genetical analysis and comparison of mature plant characters. *Journal of Agricultural Science, Cambridge* 91: 199-205.

Russell, G. and Ellis, R.P. (1988). The relationship between leaf canopy development and yield of barley. *Annals of Applied Biology* 113: 357-374.

Schonfeld, M., Ragni, A., Fischbeck, G., and Jahoor, A. (1996). RFLP mapping of three new loci for resistance genes to powdery mildew *(Erysiphe graminis* f. sp. *hordei)* in barley. *Theoretical and Applied Genetics* 93: 48-56.

Tanksley, S.D. and McCouch, S.R. (1997). Seed banks and molecular maps: Unlocking genetic potential from the wild. *Science* 277: 1063-1066.

Tanksley, S.D. and Nelson, J.C. (1996). Advanced backcross QTL analysis: A method for the simultaneous discovery and transfer of valuable QTLs from unadapted germplasm into elite breeding lines. *Theoretical and Applied Genetics* 92: 191-203.

Teulat, B., Monneveux, P., Wery, J., Borries, C., Souyris, I., Charrier, A., and This, D. (1997). Relationships between relative water content and growth parameters under water stress in barley: A QTL study. *New Phytologist* 137: 99-107.

Teulat, B., This, D., Khairallah, M., Borries, C., Ragot, C., Sourdille, P., Leroy, P., Monneveux, P., and Charrier, A. (1998). Several QTLs involved in osmotic-adjustment trait variation in barley (*Hordeum vulgare* L.). *Theoretical and Applied Genetics* 96: 688-698.

Thomas, W.T.B., Baird, E., Fuller, J.D., Lawrence, P., Young, G.R., Russell, J., Ramsay, L., Waugh, R., and Powell, W. (1998). Identification of a QTL decreasing yield in barley linked to *Mlo* powdery mildew resistance. *Molecular Breeding* 4: 381-393.

van Oosterom, E.J. and Acevedo, E. (1992). Adaptation of barley (*Hordeum vulgare* L.) to harsh Mediterranean environments. III. Plant ideotype and grain yield. *Euphytica* 62: 29-38.

van Oosterom, E.J. and Ceccarelli, S. (1993). Indirect selection for grain yield of barley in harsh Mediterranean environments. *Crop Science* 33: 1127-1131.

van Oosterom, E. J., Ceccarelli, S., and Peacock, J. M. (1993). Yield response of barley to rainfall and temperature in Mediterranean environments. *Journal of Agricultural Science* 121: 307-313.

Yadav, R., Courtois, B., Huang, N., and McLaren, G. (1997). Mapping genes controlling root morphology and root distribution on a double-haploid population of rice. *Theoretical and Applied Genetics* 94: 619-632.

Zamir, D., Grandillo, S., and Tanksley, S. D. (1999). Genes from wild species for the improvement of yield and quality of processing tomatoes. *Acta Horticulturae* 487: 285-288.

Zeller, F.J. (1998). Improving cultivated barley (*Hordeum vulgare* L.) by making use of the genetic potential of wild *Hordeum* species. *Journal of Applied Botany-Angewandte Botanik* 72: 162-167.

Chapter 5

Genetics and Breeding of Malt Quality Attributes

J. Stuart Swanston
Roger P. Ellis

INTRODUCTION

The use of barley in the production of fermented beverages probably originated 3,000 to 5,000 years B.C. (Edney, 1996), although barley had been domesticated long before that time as a food crop. Early maltsters would have followed a scientific approach, based on experimentation and observation, to develop their skills (MacLeod, 1977), and it is likely that a similar approach based on observation would have been used to select the type of grain best suited for malting. Archaeological evidence from barley grain found in Egypt and examined by scanning electron microscopy (Palmer, 1995) revealed an internal structure and a pattern of modification readily recognizable to a modern maltster.

Until the nineteenth century, it is likely that selection within locally adapted barley was the only practiced method of improvement. Pure breeding varieties of the type cultivated in the present day were unknown, and the land races farmers then grew would have been highly heterogeneous. Lein (1964) indicated that these land races were developed in isolation, throughout much of Europe, although certain geographical areas became renowned for the brewing quality of their local barley. Selection was based on physical appearance of the grain, and plump, well-filled grains with a fine husk were favored (Meredith, Anderson, and Hudson, 1962). Visual observation could be complemented by biting or chewing to determine the mealiness of the grain, an approach ultimately quantified in the development of the milling energy test (Allison, Cowe, et al., 1979).

Central European areas most favored for malting-quality barleys included Moravia, Bavaria, and Frankonia (Lein, 1964), although it was not until the 1930s that crosses were regularly made between types representative of different areas. The Hanna types from Moravia were particularly popular as parents, being also incorporated into breeding programs in other parts of

Europe, including Scandinavia, where the Hanna derivative cv. Binder was used to produce the sister lines Kenia, Opal, and Maja (Allison, 1986). Kenia was subsequently crossed with an English barley to produce the cultivar Proctor (Bell, 1951), which became the dominant English malting variety for many years and by 1960 occupied 70 percent of the English spring barley growing area (Riggs, Hanson, and Start, 1981).

Laboratory malting systems were first designed toward the end of the nineteenth century (Meredith, Anderson, and Hudson, 1962; Glennie-Holmes, 1997) to define the optimal malting conditions for a particular barley sample. As commercial barley breeding programs began to evolve, laboratory malting became a means of progeny evaluation (Whitmore and Sparrow, 1957). It was therefore possible to mimic the commercial process in the laboratory and identify the barley lines that were most suited for use, but the number of samples that could be processed and the grain quantity required made this impossible until the later stages of breeding programs (Ellis, Swanston, and Bruce, 1979). Breeders had many other objectives in addition to malting quality, e.g., yield, straw strength, and resistance to a range of diseases. Some of these characters, such as mildew resistances, were generally controlled by single dominant genes, but others, e.g., yield and partial resistance to brown rust (Jones and Clifford, 1983), were polygenic. There was little or no information about the chromosomal location of many of these genes.

Selection in early generations for characters that were easy to observe led to inadvertent selection of closely linked characteristics. The use of mildew (Swanston, 1987) and brown rust resistances from *Hordeum laevigatum* introduced a number of deleterious malting-quality properties, which persisted through several cycles of crossing, leading to the production of cultivars that were high yielding but unsuited to malting. Hot water extracts plotted against α-amylase activity for genotypes with *H. laevigatum* in their pedigrees, compared with malting cultivars grown in the same trial (see Figure 5.1), demonstrate very low levels for both characters in certain genotypes. It was clear that breeding for malting quality required different approaches, with better understanding of the underlying genetics, and testing procedures that could be introduced at earlier stages in breeding programs.

Sparrow (1970) defined malting as "the commercial exploitation of those processes that lead to germination." Problems in defining and analyzing malting quality derive, therefore, from the complex interactions of a very wide range of characters. There are, however, four major aspects of the malting process that can be considered (Sparrow, 1970), i.e., the hormone control system, the production of the enzymes, the substrate of starch, protein and cell walls within the endosperm, and the transport of soluble material to the embryo. These four components will each be considered in greater detail to determine how more recent studies have enhanced our un-

FIGURE 5.1. Hot Water Extract (L°/kg) Plotted Against α-Amylase Activity (Enzyme Units) for Eight Genotypes Derived from *H. laevigatum* (Black) and Eight Malting Cultivars (White)

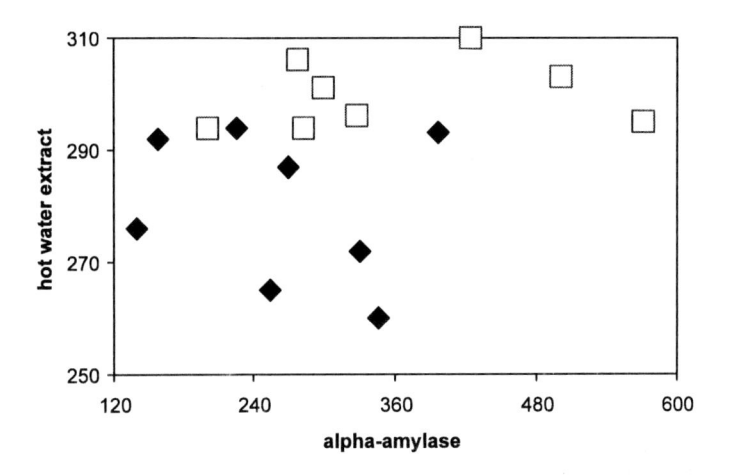

derstanding of their genetic control and how this understanding may be exploited to improve the efficiency of breeding programs.

COMPONENTS OF MALTING QUALITY

Hormone Control System

Research from the 1950s and 1960s, reviewed by MacLeod (1977), showed that addition of gibberellic acid (GA) to malting barley increased the levels of a number of hydrolytic enzymes, in particular, α-amylase and protease. Swanston (1990b) observed that addition of GA during steeping did not affect water uptake or rate of initial germination, but it appeared to increase endosperm modification and hot water extract, although some varieties were more responsive than others. This was demonstrated by measuring hot water extract after four days of germination following steeping, either with or without GA, and after five days of germination with no GA in the steep (see Figure 5.2). For five genotypes the enhancement in extract due to GA was greater than the enhancement from an additional 24 hours of germination. The cv. Tyne, however, appeared to be less responsive to added GA.

The *ari-e.GP* gene in cv. Golden Promise, which was mapped to barley chromosome 7(5H) (Thomas, Powell, and Wood, 1984), is also present in cv. Tyne (Camm et al., 1990). Franckowiak and Pecio (1991) studied the effects of GA on a large number of semidwarf genotypes in barley by deter-

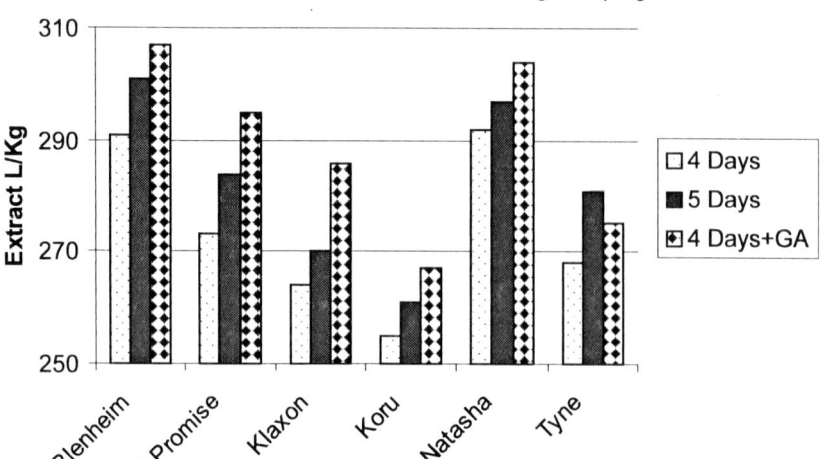

FIGURE 5.2. Hot Water Extracts of Six Cultivars After Four or Five Days of Germination and After Four Days with GA Added During Steeping

mining the elongation of coleoptile and subcrown internodes. They showed the *ari-e.GP* gene to be GA constitutive; that is, similar elongation was observed in both GA-treated and untreated seedlings.

Initially, gibberellins were regarded as malting additives (MacLeod, 1977), but it was shown that, under suitable conditions for germination, GA was synthesized in the barley embryo (Yomo and Iinuma, 1966) and moved from there to the aleurone layer (Newman and Briggs, 1976). Hayter and Allison (1976) interpreted cultivar differences in capacity to germinate, in a given concentration of abscisic acid (ABA), as indicative of variation in gibberellic acid production. They considered that it would be possible to develop lines with higher levels of starch-degrading enzymes by selecting for enhanced levels of gibberellic acid in a mutation breeding program. Allison, Ellis, et al. (1979) described five such mutants that were characterized by an increased capacity for uptake of nitrogen. Grown under levels of high fertility, these mutants demonstrated significantly higher levels of grain protein content and starch-degrading enzyme activity than the wild type. One, studied in greater detail, was shown to carry an erectoides dwarfing gene (Allison, Ellis, et al., 1979) with reduction in both neck and ear length compared to the parent variety (Ellis et al., 1986). However, there are no published data on its response to GA.

There are also differences in the ABA content of barley grains, due both to cultivar and to temperature during ripening (Goldbach and Michael, 1976). Higher levels of both endogenous ABA and ABA sensitivity are characteristic of dormant barley grains (Wang, Heimovaara-Dijkstra, and vanDuijn, 1995). Postharvest dormancy (i.e., an inability of a viable seed to germinate under conditions adequate for germination) can persist for a considerable period in certain cultivars and environments. It is a particular problem for maltsters when grain is grown under Northern European conditions, as high relative humidity, when combined with low temperatures during grain development, is readily conducive to dormancy (Corbineau and Come, 1980; Barre, 1983).

The genetics of dormancy is generally poorly understood (Molina-Cano et al., 1999), although Buraas and Skinnes (1984) suggested that dormancy may be controlled by several recessive genes, expressed at different preharvest and after-ripening stages. With recent advances in molecular mapping, it has been possible to detect four quantitative trait loci (QTL) associated with dormancy (Ullrich et al., 1993), designated SD through 4 (Han et al., 1996). Two loci on chromosome 7(5H) (one more centromeric, the other distal), one on chromosome 1(7H), and one on chromosome 4(4H) accounted for 50, 15, 5, and 5 percent, respectively, of the phenotypic variability in dormancy (Ullrich et al., 1993). Han et al. (1996) used molecular markers closely linked to these loci to select within reciprocal crosses and confirmed the large effect associated with the more centromeric locus on chromosome 7(5H). In contrast to the findings of Buraas and Skinnes (1984), Han et al. (1996) also suggested a cytoplasmic effect, although interaction with dormancy alleles in the nucleus was necessary before this was observed.

ABA acts to prevent precocious germination in a wide range of species (Koornneef, Reuling, and Karssen, 1984; Walker-Simmons, 1987; Groot and Karssen, 1992). Visser et al. (1996) indicated that the action of ABA in inhibiting germination occurs outside the embryo and reported on a barley mutant able to germinate in a concentration of ABA that was inhibitory for the wild type, which they suggested was due to an increased ability to degrade extracellular ABA. A barley mutant with a reduced basal level of ABA has also been reported (Walker-Simmons, Kurdna, and Warner, 1989), and selection of mutants capable of germination in ABA has been carried out (Allison, Ellis, et al., 1979; Molina-Cano et al., 1999). A mutant in cv. Triumph demonstrated a more rapid recovery from postharvest dormancy than its parent (Molina-Cano et al., 1999), and this was attributed to reduced ABA sensitivity, as ABA contents were very similar in the two genotypes.

Production of Enzymes

Four main types of β-glucanase detected in germinated or malted barley were reviewed by MacGregor (1990), although endo-(1,4)-β-glucanase, which was purified from germinated barley (Yin and MacGregor, 1988), was shown to be produced by fungi on the barley husk. The role of endo-(1,3)-β-glucanase during malting is unclear, as it appears to be produced in excessive amounts for the level of substrate present (Fulcher et al., 1977). Hoj et al. (1988), however, supported the suggestion that it was associated with resistance to pathogens, although they observed synthesis of the enzyme in the absence of fungal invasion.

Treatment of endosperm cell walls with a proteolytic enzyme was shown to release β-glucan (Forrest and Wainwright, 1977). Beta-glucan solubilase, an enzyme which showed carboxypeptidase activity and which increased in content during germination, was shown to release insoluble β-glucan from barley grist (Bamforth, Martin, and Wainwright, 1979). However, it was also demonstrated that mixed-linkage β-glucanases were able to hydrolyze a high proportion of β-glucan from endosperm cell walls without prior solubilization (Brunswick, Manners, and Stark, 1988).

Most interest in recent years has focused on endo-(1,3) (1,4)-β-glucanase, generally called β-glucanase, for simplicity (MacGregor, 1990). Its activity in germinating barley is controlled, in part, by genotype (Bendelow, 1976); (Smith and Gill, 1986) and also by growing conditions (Morgan, Gill, and Smith, 1983a; Stuart, Loi, and Fincher, 1988). Addition of gibberellic acid to germinating grain increases activity in the resulting malt (Brunswick, Manners, and Stark, 1987). Ellis, Swanston, et al. (1997) indicated that the rate of synthesis during malting was more rapid for barley grown under Spanish compared to Scottish growing conditions. These authors were also able to detect β-glucanase in the mature grain, which may have been synthesized during grain development, with the high preharvest temperatures in Spain causing rapid maturation and leaving a residue of enzymes (Swanston et al., 1997).

Two isoenzymes of β-glucanase have been characterized (Woodward and Fincher, 1982a,b). Isoenzyme II is controlled by a gene on barley chromosome 1(7H) (Loi, Ahluwalia, and Fincher, 1988), while the gene for isoenzyme I is on chromosome 5(1H) (Slakesi et al., 1990). More recently, Han et al. (1995) identified five QTL for β-glucanase in finished malt. Two of these were in similar areas to the locations proposed for isoenzymes I and II, and the other loci could be additional isoenzymes (Han et al., 1995). However, these authors also indicated that a major QTL for β-glucanase mapped to the same location as the major QTL for dormancy on chromosome 7(5H) (Ullrich et al., 1993). It is therefore possible that this locus is coding for content of, or response to, one of the hormones affecting enzyme synthesis.

Both β-glucanase isoenzymes are heat labile, with very little of isoenzyme I surviving kilning (Brunswick, Manners, and Stark, 1987). Residual activity, including that of isoenzyme II, is destroyed very rapidly by mashing temperatures of 60 to 65°C (Loi, Barton, and Fincher, 1987). It is therefore important that extensive β-glucan degradation occurs during malting, as the potential for further breakdown during mashing appears to be very limited (MacGregor, 1990). Recently, it has been possible to produce stable transgenic barley by incorporating a gene from a microorganism that codes for a more heat-stable β-glucanase (Jensen et al., 1998). However, although the DNA from the transgene will not be present in the final product, it remains to be seen whether beer made from genetically modified barley will be acceptable to consumers.

The most important enzymes of protein degradation are the endopeptidases and the exopeptidases (Wallace and Lance, 1988). Although better malting varieties were shown to have higher endopeptidase activity, there was only a weak correlation between endopeptidase activity and hot water extract (Morgan, Gill, and Smith, 1983b), and little other information is available on the relationship between peptidase activity and malting quality (Wallace and Lance, 1988). Smith and Simpson (1983) suggested that proteolysis was very rapid in barleys of good malting quality, while Swanston and Taylor (1988) considered that disruption of the protein matrix in a cultivar such as Triumph could be completed while cell wall modification was still continuing. Although protein modification may be a limiting factor in barleys grown under Spanish conditions, which have high grain nitrogen contents (Swanston et al., 1997), selection for peptidase activity has not been a major objective of malting quality breeding programs in northwestern Europe.

Malted barley contains three major starch-degrading enzymes, α-amylase, β-amylase, and limit dextrinase, the functions of which were reviewed by MacGregor (1990). It is known that α-amylase is the product of two gene families, with the synthesis of α-amylase 1 controlled by a gene on chromosome 1(7H) (MacGregor, 1990). The other gene family is on chromosome 6(6H) and controls the synthesis of α-amylase 2 (Brown and Jacobsen, 1982), which is responsible for about 90 percent of the total α-amylase activity (MacGregor and Balance, 1980b). In addition, Li et al. (1996) reported QTL for α-amylase activity associated with chromosomes 2(2H), 7(5H) and 5(1H). The locus on chromosome 2(2H) also affected the activity of β-amylase, limit dextrinase, and β-glucanase and thus was thought to be associated with a response to GA (Li et al., 1996).

At the temperature of starch hydrolysis in the mash, α-amylase is relatively stable (MacGregor and Balance 1980a). Although β-amylase is, by contrast, substantially heat labile, Arends et al. (1995) suggested that sufficient activity survives kilning for it to be the major contributor to the starch-degrading activity of the malt. MacGregor et al. (1999), however, indicated that worts could contain low levels of linear dextrins, suggesting that commercial malts contain insufficient β-amylase to fully hydrolyze these dextrins to maltose. Some Japanese barleys have been shown to contain β-amylase with a higher level of heat stability (Kihara, Kaneko, and Ito, 1998), and *Hordeum spontaneum,* the wild relative of cultivated barley, may also be a source of thermostable β-amylase (Eglington, Langridge, and Evans, 1998). Isoenzymes of β-amylase, identified by either isoelectric focusing (Forster et al., 1991) or electrophoresis (Allison, 1973), showed differences in thermostability, suggesting that these differences could be associated with alleles at the β-amylase locus on barley chromosome 4(4H) (Powling, Islam, and Shepherd, 1981) or at a very closely linked locus. All genotypes scored as either Sd1 (Eglington, Langridge, and Evans, 1998) or IEF group B (Kihara, Kancko, and Ito, 1998) had moderate levels of thermostability, but Sd2 types (isoelectric focusing [IEF] group A) could be divided into high, moderate, and low groups for thermostability, as shown by the genotypes in Figure 5.3. Selecting for improved heat stability of β-amylase rather than higher enzyme activity may be a more beneficial breeding strategy (MacGregor et al., 1999).

Malt contains low levels of active limit dextrinase (Lee and Pyler, 1984), but a significant additional portion can be activated by proteases or reducing agents, such as dithriothreitol or mercaptoethanol (Kristensen, Svensson, and Larsen, 1993; Sissons, 1996). MacGregor et al. (1999) indicated that 60 percent of the limit dextrinase activity could be inactivated, largely due to limit dextrinase inhibitor proteins from kilned malt that complex with the limit dextrinase enzyme in the mash (MacGregor et al., 1995). Because of the high levels of inhibitors, there may be little value in selecting for barleys that synthesize higher levels of limit dextrinase (MacGregor et al., 1999), and attempts should be made to lower inhibitor levels in the malt, either

FIGURE 5.3. Beta-Amylase Thermostability in Cultivars Classed As Either Sd2 (Black) or Sd1 (White)

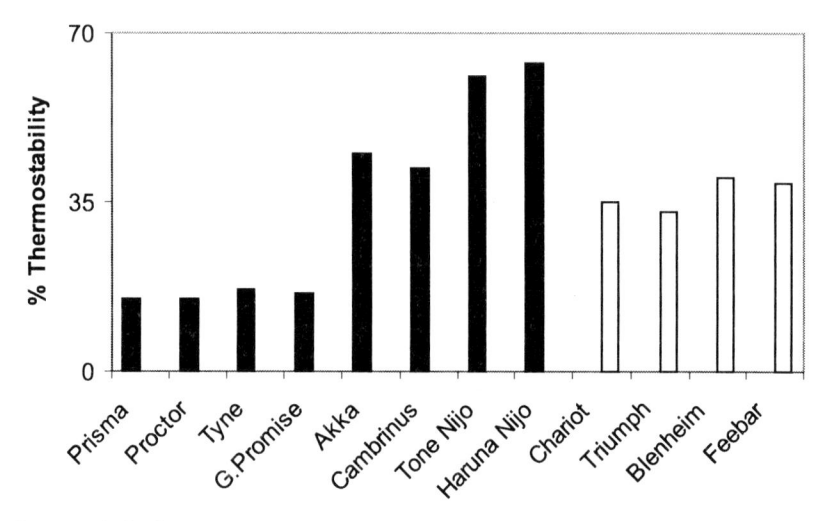

Source: J. S. Swanston, unpublished data.

through breeding or through modifying malting or mashing conditions, to improve release of the bound enzyme.

Starch, Protein, and Cell Walls

MacGregor and Fincher (1993) estimated that starch, protein, and β-glucan, the major constituent of endosperm cell walls, make up around 80 percent of the fresh weight of barley grain. Starch, which is the main seed reserve polysaccharide of green plants (Morrison and Karkalas, 1989), is by far the largest component, accounting for over 60 percent. The products of starch hydrolysis are fermented to alcohol, so any genetic factors affecting the content or ease of hydrolysis of starch would be expected to have consequences for malting quality.

In the mature barley grain, starch exists in granular form, with two distinct populations of large (A-type) and small (B-type) granules, the latter accounting for 90 percent of the total number, but only 10 percent of the total volume (Bathgate and Palmer, 1972). The starch is a combination of the two polysaccharides amylose (AM) and amylopectin (AP), with AP constituting approximately 75 percent of the starch in normal, cultivated barley. Two single gene mutations, however, alter the AM:AP ratio. The waxy mutation is located on the short arm of chromosome 1(7H) and reduces the level of, but does not eliminate, granule-bound starch synthase (Ellis et al., 1998). This may be due to partial suppression of gene expression, since the structural gene for the enzyme is preserved intact (Domon, 1996). Waxy barleys generally have amylose contents of between 2 and 10 percent, which is higher than in other waxy cereals (Tester and Morrison, 1992). The other mutation increases amylose content (Merritt, 1967), and this has been mapped to chromosome 5(1H) (Schondelmaier et al., 1992).

Although it was assumed that straight-chained AM would be more readily degraded by amylases than AP, with its highly branched molecular structure, this proved unfounded (Ellis, 1976). The high-amylose character is also associated with a reduction in the size of the A-type granules (Ellis, 1976; McDonald et al., 1991), and these granules are also more heavily embedded into the surrounding protein matrix (Swanston, 1994). This gives rise to a very compact endosperm structure that is difficult to disrupt either enzymically or mechanically (Ellis, Swanston, and Bruce, 1979; (Swanston, Ellis, and Stark, 1995). In addition, a portion of the AM forms a complex with lipid that remains stable at temperatures above 90°C (Tester and Morrison, 1993) and is not degraded during malting (Swanston, Ellis, and Stark, 1995). As there is a close relationship between the amylose and lipid contents of barley starch (Tester and Morrison, 1992), the proportion of AM inaccessible to enzymic attack is thus increased in high-amylose genotypes (Swanston, Ellis, and Stark, 1995).

Using near isogenic lines, Ullrich et al. (1986) suggested that waxy lines were of poorer malting quality than wild-type lines. This could be due to the reduced grain size and starch quantity of the waxy type (Tester and Morrison, 1993), although waxy lines also have cell walls containing higher levels of β-glucan (Ullrich et al., 1986; Que et al., 1990), which may be modified more slowly during malting (Swanston, 1995). Swanston (1996) demonstrated a range of hot water extracts in both waxy and nonwaxy inbred lines and indicated that some waxy lines with high hot water extracts could be obtained. Waxy lines were, however, affected to a greater extent by environmental conditions, with poor grain fill reducing starch content, while large grains did not modify adequately during malting (Swanston, 1996).

The waxy and high-amylose mutations are not the only sources of variation in A- and B-type granule size and number. However, those associated with mutations, e.g., the Risø mutants in cv. Bomi, tend to be characterized by smaller grain size and reduced starch content (Tester, Morrison, and Schulman, 1993). They may thus be of limited value for malting, in addition to being very low yielding (Doll, 1976), although some appear to combine high amylase activity with a more friable endosperm structure (Allison, 1978). Among wild-type barleys, genetic differences in granule traits occur, but there are also large environmental differences (Oliveira, Rasmusson, and Fulcher, 1994). Effects of this variation on malting properties are, however, unclear. Oliveira, Rasmusson, and Fulcher (1994) suggested that both A- and B-type granule volumes were higher in malting cultivars than in feed types, although overlap existed between the two populations. Borem et al. (1999) mapped QTL affecting starch granule traits in a cross between the cultivars Steptoe and Morex. They indicated that a small section of chromosome 2(2H) appeared to have a significant effect on several traits, but found no QTL in areas previously shown to affect malting-quality characters, although Steptoe had both lower granule volume and poorer malting quality than Morex (Oliveira, Rasmusson, and Fulcher, 1994). However, there appeared to be an association between starch properties and heading date (Borem et al., 1999), so differences in the length of the grain-filling period may have been a major factor in this particular cross. In addition, the malting properties considered by Borem et al. (1999) did not include fermentability, which was influenced, albeit to a limited extent, by three loci on chromosome 2(2H), as shown in Table 5.1 (Swanston, Thomas, Powell, Young, 1999). These two studies, however, utilized different germplasm and different ranges of molecular markers, so it is not easy to compare QTL locations and draw firm conclusions.

Recent research has also advanced knowledge of the inheritance of enzymes of barley starch synthesis. Kilian et al. (1994) identified five genes responsible for adenosine diphosphate (ADP)-glucose pyrophosphorylase (AGPase) in either endosperm or leaf tissue and mapped four of them. The

TABLE 5.1. QTL on Chromosome 2(2H) Influencing Fermentability in a Cross Between the Cultivar Derkado and an SCRI Breeders' Line, Showing the Effect of the 'Derkado' Allele in Increasing or Decreasing Percentage Fermentability in Trials at Dundee, Scotland, Over Two Seasons

Marker	Distance of Locus from Marker (cm)	Effect of 'Derkado' Allele	
		1995	1996
P61M51g	8	0.80	0.23
P34M34f	6	−0.87	−0.48
P61M48c	0	0.69	0.28

waxy locus in barley has also been cloned and sequenced (Rohde, Becker, and Salamini, 1988). It has not been demonstrated, however, whether the mutation conferring high amylose is in a starch branching enzyme, as it differs phenotypically from the amylose extender mutant of rice, caused by a deficiency in starch branching enzyme III (Mizuno et al., 1993), the gene for which has been mapped in both rice and maize (Harrington et al., 1997). In addition, enzymes of starch synthesis have been genetically manipulated in other crops; for example, a bacterial AGPase has been used to increase tuber starch content in transgenic potatoes (Stark et al., 1992). McElroy and Jacobsen (1995) suggested that a similar approach in barley could increase extract in addition to improving yield and grain size.

The primary storage proteins (hordeins) in the barley endosperm can be divided into three groups of subunits: sulphur-rich B hordeins, sulphur-poor C hordeins, and high molecular weight D hordeins (Shewry and Miflin, 1983). D hordeins comprise only up to 4 percent of the total hordein fraction (Miflin, Field, and Shewry, 1983). The relative proportions of the B and C hordein groups vary significantly between cultivars, although they are also influenced by environmental conditions, with increased nitrogen availability giving a higher C hordein content (Griffiths, 1987). Each of the three groups is controlled by a single gene, and all three hordein loci are on chromosome 5(1H). The B and C groups are controlled by the *Hor-2* and *Hor-1* loci, respectively, which are linked to the *Mla* mildew resistance locus (Shewry et al., 1980), and second-division segregation–polyacrylamide gel electrophoresis (SDS-PAGE) has shown the *Hor-2* locus to be highly polymorphic (Riggs et al., 1983; Shewry et al., 1979). D hordein is controlled by the *Hor-3* locus, on the long arm of chromosome 5(1H) (Miflin, Field, and Shewry, 1983).

Although it is well established that an extensive matrix of storage protein surrounding the starch granules will produce a compacted endosperm that is likely to modify poorly during malting (Palmer, 1980), the exact roles of the different hordein subunits have been subject to considerable research and discussion. Initially it was considered that malting quality was inversely related to the amount of B hordein (Baxter and Wainwright, 1979; Slack, Baxter, and Wainwright, 1979), but Benetrix, Sarrafi, and Autran (1994) suggested that this was based on the assumption that, as aggregated hordeins are associated by disulfide bonds, the B hordeins, in particular, would form a persisting martix around the starch granules. However, it was also shown that B and D hordeins could form a complex (Smith and Simpson, 1983), resulting in the formation of gel protein (Smith and Lister, 1983) that could be precipitated in greater quantity from samples of poor malting quality. In addition, Skerritt and Janes (1992) observed specific elevation of B1 and B2 subunits in poor malting samples.

By contrast, it was noted that C hordeins were more poorly modified than the other storage proteins during malting (Skerritt and Henry, 1988). Millet, Montembault, and Autran (1991) also suggested that the subaleurone layer had a large proportion of C hordein that could limit the diffusion of water during steeping. Benetrix, Sarrafi, and Autran (1994) separated protein aggregates by size exclusion high-performance liquid chromatography (HPLC) and noted changes in protein fractions due to cultivar or level of nitrogen fertilization. They indicated that the quantity of a fraction, particularly rich in C hordein, offered the best discrimination between winter feed cultivars and spring malting barleys, but their observations were based on greenhouse-grown samples, which were not subjected to malt analyses. An element of caution must therefore be employed in extrapolating these data, since varietal classification may not be the most reliable indicator of the malting potential of a particular sample.

Molina-Cano et al. (1995) compared Scottish- and Spanish-grown samples over two seasons and found that water uptake was not reduced by an increased proportion of C hordein. However, the rate of hordeins deposition during grain filling was very different between the two environments. In Scotland, the relative proportions of B and C hordeins remained fairly constant during the later stages of grain filling (Swanston et al., 1997), supporting the data of Rahman, Shewry, and Miflin (1982), who found a similar pattern from around 26 days after anthesis. In Spain, by contrast, the relative proportion of C hordein increased throughout this period and may therefore lead to differences in polypeptide composition across the endosperm, depending on the age of the cells (Swanston et al., 1997). Swanston et al. (1997) also suggested that breakdown of the protein matrix appeared to be a limiting factor in modification of the Spanish-grown malt samples, but this may reflect grain protein levels that were much higher in Spain than in Scotland (Molina-Cano et al., 1995; Swanston et al., 1997).

Chandra, Proudlove, and Baxter (1999) detected C hordein in both mealy and steely areas of the endosperm but indicated that D hordein was one of the storage proteins associated with the steely areas, and its degradation was necessary to obtain a friable malt. Discovery of a genotype lacking D hordein enabled the production of several sets of near isogenic lines, with and without D hordein, each set varying from the others in either its *Hor1* or *Hor2* alleles (Brennan et al., 1998). In contrast to previous reports (Smith and Lister, 1983; Howard et al., 1996), no relationship was found between either presence of D hordein or gel protein content and malting quality. However, grain nitrogen results reported by Brennan et al. (1998) were considerably higher than those normally acceptable for malting and extract results were all very low. In addition, wort viscosity and soluble nitrogen levels quoted suggest malting problems that may have masked any effect of the null allele at the D hordein locus.

It was suggested that the presence of a small molecular weight protein (friabilin), on the surface of the starch granules, was the basis of grain softness in wheat (Greenwell and Schofield, 1986), as it interfered with the adhesion between the starch granules and the surrounding protein matrix. The genetic factor determining the presence of friabilin appeared to be closely linked to the *Ha* gene on chromosome 7(5H) (Schofield and Greenwell, 1987), while Forster and Ellis (1990) found a major effect on milling energy in the Triticeae also to be associated with the short arm of chromosome 7(5H). Jagtap et al. (1993) detected friabilin on the surface of barley starch granules and suggested that there were differences in content between genotypes. Genes affecting endosperm texture were also located on barley chromosome 7(5H) (Jagtap et al., 1993). In addition, Brennan et al. (1992) indicated that freeze-fractured endosperms of good malting barleys showed a much cleaner separation of starch granules from surrounding protein than those of poor malting types. However Ellis et al. (1999), while confirming genotypic differences in friabilin content in barley, found no association with either grain hardness, as measured by milling energy (see Figure 5.4), or hot water extract, although it is possible that any association could be masked by allelic variation at other loci.

The major constituent of barley endosperm cell walls is the polysaccharide (1-3), (1-4) β-D-glucan (Fincher, 1975; Forrest and Wainwright, 1977). High levels of β-glucan have long been regarded as deleterious to malting quality, as they may reduce the rate of endosperm modification (Martin and Bamforth, 1980), and residual cell walls are a barrier to the enzymes responsible for protein modification and starch hydrolysis during malting and mashing. In addition, high levels of β-glucan in the wort cause increases in viscosity that may lead to filtration problems (Bamforth and Barclay, 1993). The development of enzymic (McCleary and Glennie-Holmes, 1985) and fluorimetric (Jorgensen and Aastrup, 1988) methods of determination have enabled breeders to readily select for low levels of β-glucan in barley grain. In conjunction with molecular markers, they have also permitted improved understanding of the mode of inheritance. Han

FIGURE 5.4. Milling Energy (Joules) Plotted Against Friabilin Content (Absorbance Units at 405 nm) for 20 Barley Genotypes, Grown in Two Seasons, 1997 (White) and 1998 (Black)

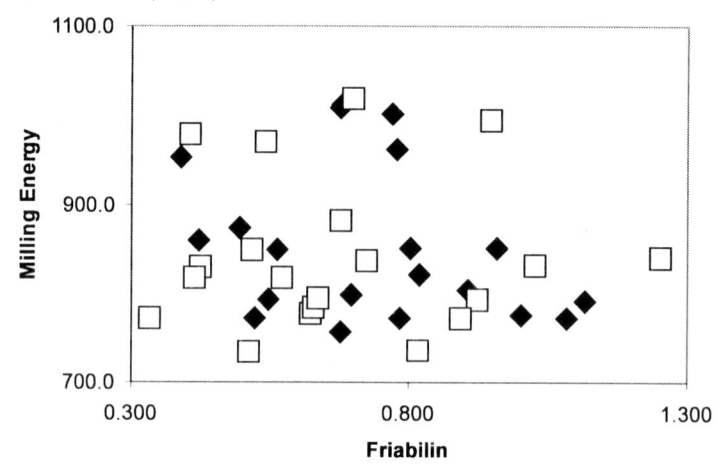

Source: Ellis et al., 1999.

et al. (1995) detected three QTL for barley β-glucan and six for β-glucan in malt, although some of the latter were coincident with QTL for malt β-glucanase activity and could be a consequence of the same genes. Powell et al. (1985) suggested that barley β-glucan was determined by the additive effects of between three and five genetic factors but did not determine their chromosomal locations.

Barley β-glucan exists in two forms, classed according to solubility or insolubility in water, which can be assessed separately by an enzymic method (Aman and Graham, 1987). Morgan and Riggs (1981) suggested that although total β-glucan content was increased when plants were grown in hotter, drier conditions, it was the soluble portion, in particular, that was raised. This was supported by Swanston et al. (1997), who found differences in the pattern of β-glucan deposition during grain filling between barleys grown in Spain and Scotland, with Spanish-grown samples having higher levels of total β-glucan, but lower levels of the insoluble portion. However, McNicol et al. (1993) suggested that the timing of the stress factor was more critical than the factor itself. They found that drought stress occurring late in the grain-filling period had no effect on β-glucan content.

Han et al. (1995) noted that the largest effect on barley β-glucan was associated with a locus on chromosome 2(2H), and under their growing conditions, use of molecular markers to select for optimum expression at this locus would be a viable breeding strategy. However, the large differences noted earlier, due to environmental conditions, could mean that the most critical loci for this character also vary between locations. Data from other environments and a better knowledge of the function of the genes identified are clearly requirements of ongoing research.

Mutants expressing reduced levels of β-glucan have been shown to have thinner endosperm cell walls (Aastrup, Erdal, and Munck, 1985), although expression of this character may be variable across the endosperm (Miller and Fulcher, 1994). Swanston (1997) demonstrated expression of a gene from one such mutant in combination with the waxy starch gene, by dividing waxy inbred lines on the basis of high or low milling energy. The waxy lines with low milling energy were shown to have similar mean levels of β-glucan and a similar level of cell wall modification, after malting, as determined by Aastrup, Gibbons, and Munck (1981), to the nonwaxy lines from the same cross (see Table 5.2). Previous research showed waxy barleys to have increased levels of β-glucan (Ullrich et al., 1986; Que et al., 1990), .while Stahl et al. (1996) suggested that the waxy character and high β-glucan levels were inherited simultaneously in a backcrossing program. This is unlikely to be an effect of tight linkage, as Han et al. (1995) did not demonstrate any effects on grain β-glucan associated with the area of barley chromosome 1 where the waxy gene is located, but the location of the gene conferring low β-glucan in waxy lines (Swanston, 1997) is not yet known.

In this section, the endosperm structure of barley has been shown to be under extremely complex biochemical and genetic control. In addition, a delicate balance exists between the various synthetic pathways, and alterations to one component can precipitate effects on many others (Swanston, 1994). Selecting for quality is, however, simplified by the fact that a desirable expression of genes coding for endosperm characteristics can lead to discernible ef-

TABLE 5.2. Mean Levels of Milling Energy, Grain β-Glucan Content, and Cell Wall Modification in High and Low Milling Energy, Waxy and Nonwaxy (Not Divided for Milling Energy) Inbred Lines from a Cross Between the Genotypes Chalky Glenn and Waxy Hector

Group	Mean Values		
	Milling Energy (Joules)	Grain β-Glucan (%)	Cell Wall Modification (%)
Waxy (Low Milling Energy)	736.6	4.19	93.6
Waxy (High Milling Energy)	857.5*	5.30*	35.3*
Nonwaxy	749.2	4.13	91.0

*Significantly different from the other two groups at the 0.1 percent level.

fects on structure and texture. The milling energy test (Allison, Cowe, et al., 1979) determines the mechanical energy required to disrupt the endosperm structure, with good malting cultivars characterized by low milling energy, and applications of the test to developing, mature, and malted grains have been reviewed (Swanston, 1990a). The degree of steeliness of the grain can also be determined by light transflectance, with steely areas randomly distributed within the endosperm and associated with higher concentrations of β-glucan and protein (Chandra, Proudlove, and Baxter, 1999). Steeliness leads to uneven distribution of water across the endosperm (Chandra, Proudlove, and Baxter, 1999), and this, in turn, affects the distribution and activity of key enzymes, leading to a lower level of modification during malting.

Transport of Soluble Material

The transport of soluble material to the embryo will promote germination and growth. However, although more rapid initial germination appears to be a feature of improved malting-quality barleys (Ulonska and Baumer, 1976; Molina-Cano et al., 1989), reducing embryo growth during malting to prevent loss of fermentable materials is a major objective of the maltster. Solving the paradox of how endosperm modification can be maximized, while loss of the products of such modification is reduced, has been predominantly a technological rather than a plant breeding challenge. By bringing the grain to a high-moisture content prior to sprouting and controlling temperature, the malting process creates a very artificial environment for germination and induces the grain to respond in a particular way. Grain germinated in petri dishes, for example, will exhibit much greater root and shoot growth than malt (Taylor and Swanston, 1987), but the pattern of endosperm modification is very different. The close correlations between grain and germinated grain milling energies suggested that substantial areas of the endosperm remained unmodified following petri dish germination, but there was extensive breakdown in areas close to the embryo (Taylor and Swanston, 1987) and loss of starch-derived material to the embryo.

Swanston and Taylor (1990) indicated that there were genotypic differences in the initial rate of germination during malting, but not all rapid-germinating cultivars had good malting quality. There was considerable genotype × environment interaction, due to factors such as duration of steep and inclusion of air rests, which had large effects on some genotypes but did not affect others. However, water uptake during steeping did not correlate significantly with the initial germination rate (Swanston and Taylor, 1990).

The length of the germination period is also critical for the transport of soluble material. Bathgate, Martinez-Frias, and Stark (1978) considered samples germinated from four to seven days and demonstrated a linear relationship between loss of fermentable material to the embryo (malting loss) and time of germination. Germination time and water uptake during steep are also both significant determinants of the measurable cyanide potential of malting barley (Cook, 1990). The cyanogenic glycoside epi-heterodendrin (EPH) (Erb et al., 1979), which is the precursor of an undesirable trace component of Scotch whiskey (Cook et al., 1990), is formed in the acrospires of malting barley. Forslund and Jonsson (1997) indicated that varying the supply of nitrogen to barley seedlings caused small, but significant differences in cyanogenic glycoside production. Differences in nitrogen supply to the acrospires of germinating grain reflect grain nitrogen content and the rate and extent of solubilization (Swanston, 1999), but, clearly, the rate of transport into the embryo is also critical.

EFFECTS OF BREEDING

Barley breeders are now able to exploit the knowledge that has been gained on factors affecting quality and to use DNA markers to select for desired traits. Although breeding and selection were carried out for many years, without such benefits, to produce cultivars that could best serve the needs of malting, brewing, and distilling companies that were introducing new technology and seeking to increase throughput, this objective can now be pursued with far greater precision. A further advantage of DNA marker technology is that it is now possible to observe, at the genotypic level, the changes that have occurred in the cultivated barley gene pool as a result of the breeding process and to compare these changes with observed phenotypic differences.

Ellis, McNicol, et al. (1997) considered changes in U.K.-recommended spring barley cultivars, since the 1950s, by using principal coordinate analysis of genetic similarities, based on genetic information determined by amplified fragment length polymorphism (AFLP) (Zabeau and Vos, 1993). Pedigree analysis cannot take account of factors such as selection, mutation, or genetic drift (Melchinger et al., 1994), unlike direct DNA assays such as AFLP (Ellis, McNicol, et al., 1997). This can lead to widely different genetic distances when genetic similarity, derived by AFLP, is compared to the pedigree kinship coefficient, as illustrated for the parental genotypes and their progeny in Table 5.3. These data clearly illustrate a closer relationship between the parental cultivars than was estimated by pedigree analysis, which may overestimate divergence, particularly when one parent has an "exotic" genotype in its own lineage (Ellis, McNicol, et al., 1997).

From a plot derived from estimates of genetic similarity, among more than 40 cultivars, Ellis, McNicol, et al. (1997) observed three broad groupings, one of which included feed cultivars with disease resistance genes pre-

TABLE 5.3. Estimates of Relatedness (%) Between Parental Cultivars Vada and Zephyr and the Cultivars Georgie and Sundance, Derived from a Cross Between Them, by Both Pedigree and AFLP Analysis

Parents	Vada	Zephy	Vada	Zephyr
Progeny:	Pedigree		AFLP Analysis	
Georgie	53	53	91	92
Sundance	53	53	89	93

Source: Ellis, Swanston, et al., 1997.

dominantly from *Hordeum laevigatum*. The other two groups were of malting cultivars, one of genotypes closely related to cv. Proctor and the other of genotypes similar to cv. Triumph. These three groups correspond to the three endosperm types classified on the basis of grain milling energy (Allison, 1986). However, Allison (1986) also pointed out that the mealy endosperm structure, with very low milling energy, found in cv. Triumph was originally present in the old European malting barley Kneifel, so it is unlikely that improvements in malting performance can be explained simply by changes in the physical structure of the endosperm. This was demonstrated by Taylor and Swanston (1987), who ranked 18 cultivars on the basis of both milling energy and hot water extract (see Figure 5.5). Although the best malting cultivars were ranked in the top third for both characters, and those ranked in the bottom third for both characters, were all feed types, many genotypes diverged considerably in their rankings for the two tests. Similarly, Ellis et al. (1999), considering over 90 genotypes, spanning many years of barley breeding, did not find a high correlation between grain milling energy and hot water extract.

Thomas et al. (1991) indicated that hot water extract and milling energy were both affected by genetic factors on the long arm of barley chromosome 4(4H), with the effect on milling energy slightly more distal. Later studies on a different population revealed a factor affecting both milling energy and hot water extract on chromosome 7(5H) (Thomas et al., 1996), but hot water extract was also influenced by factors on chromosomes 2(2H) and 3(3H). Although low milling energy appears to be a component of good malting quality, its association with extract may therefore be, at least in part, due to linkage.

FUTURE DIRECTIONS

Barley breeders have made significant changes and improvements in the crop, meaning the malting industry has access to a high-quality raw mate-

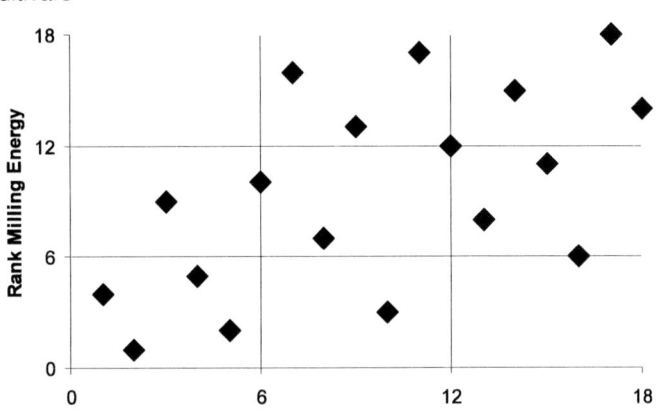

FIGURE 5.5. Comparison of Rankings for Hot Water Extract and Milling Energy of 18 Cultivars

[Scatter plot: x-axis "Rank Hot Water Extract" (0 to 18), y-axis "Rank Milling Energy" (0 to 18)]

Source: Taylor and Swanston, 1987.

rial. This has been achieved largely as a result of phenotypic selection, making use of increasingly sophisticated testing procedures. Malting, however, involves a highly complex series of interrelated biochemical pathways occurring simultaneously, so the underlying genetic control is equally complex. In recent years, major steps have been taken to locate some of the critical genes and to assess their role. This is likely to continue, making use of techniques such as expressed sequence tags (ESTs), where the messenger RNA associated with activated genes is extracted to clone sequences of complementary (c) DNA. One possible application of ESTs to malting barley has already been outlined briefly (Swanston, Thomas, Powell, Meyer, et al., 1999), in which RNA was extracted from barley malted for two days after steeping. The vast improvements in information technology of recent years also make it possible to compare cDNA sequences with those already known and held in databases, and around 70 percent of the ESTs from malted barley show homology with known sequences (Swanston, Thomas, Powell, Meyer, et al., 1999). Not all sequences that have been determined have a known function, but it was possible to detect sequences associated with both carbohydrate and amino acid metabolism. Future developments are likely to include mapping of ESTs and comparison of map locations with those of known genes or QTL.

These developments are likely to have significant effects on breeding methodology. Marker-assisted breeding has already been demonstrated in barley (Toojinda et al., 1998), and the advantages of selecting directly for the desired genotype are obvious. However, the number of traits that can be selected in this way and the cost implications of the methodology will also have to be taken into consideration. Traditional barley breeding skills have served the agricultural industry very well over many years and are likely to retain a significant role in the future, but with the speed and accuracy of selection enhanced through employment of novel technology as it becomes available.

REFERENCES

Aastrup, S., Erdal, K., and L. Munck (1985). Low β-glucan barley mutants and their malting behaviour. In *Proceedings of the 20th Congress of the European Brewery Convention*. Oxford: Oxford University Press, pp. 387-393.

Aastrup, S., Gibbons, G.C., and L. Munck (1981). A rapid method for estimating the degree of modification in barley by measurement of cell wall breakdown. *Carlsberg Research Communications* 46: 77-86.

Allison, M.J. (1973). Genetic studies on the β-amylase isozymes of barley malt. *Genetica* 44: 1-15.

Allison, M.J. (1978). Amylase activity and endosperm hardness of high lysine barleys. *Journal of the Institute of Brewing* 84: 231-232.

Allison, M.J. (1986). Relationships between milling energy and hot water extract values of malts from some modern barleys and their parental cultivars. *Journal of the Institute of Brewing* 92: 604-607.

Allison, M.J., Cowe, I.A., Borzucki, R., Bruce, F.M., and R. McHale (1979). Milling energy of barley. *Journal of the Institute of Brewing* 85: 262-265.

Allison, M.J., Ellis, R.P., Hayter, A.M., and J.S. Swanston (1979). Breeding for malting quality at the Scottish Plant Breeding Station. *Scottish Plant Breeding Station Annual Report* 58: 92-139.

Aman, P. and H. Graham (1987). Analysis of total and insoluble mixed-linked (1-3), (1-4)β-D-glucan in barley and oats. *Journal of Agricultural and Food Chemistry* 35: 704-709.

Arends, A.M., Fox, G.P., Henry, R.J., Marschke, R.J., and M.H. Symons (1995). Genetic and environmental variation in the diastatic power of Australian barley. *Journal of Cereal Science* 21: 63-70.

Bamforth, C.W. and A.H.P. Barclay (1993). Malting technology and the uses of malt. In *Barley: Chemistry and Technology*, eds. A.W. MacGregor and R.S. Bhatty. St. Paul, MN: American Association of Cereal Chemists, pp. 297-354.

Bamforth, C.W., Martin, H.L., and T. Wainwright (1979). A role for carboxypeptidase in the solubilization of barley β-glucan. *Journal of the Institute of Brewing* 85: 334-338.

Barre B. (1983). La dormance de l'orge, 2e partie: Causes des variations de l'aptitude a germer des grains d'orge au moment de leur recolte. *Bios* 14: 65-80.

Bathgate, G.N., Martinez-Frias, J., and J.R. Stark (1978). Factors controlling the fermentable extract in distillers malt. *Journal of the Institute of Brewing* 84: 22-29.

Bathgate, G.N. and G.H. Palmer (1972). A reassessment of the chemical structure of barley and wheat starch granules. *Die Starke* 24: 336-341.

Baxter, E.D. and T. Wainwright (1979). Hordein and malting quality. *Journal of the American Society of Brewing Chemists* 37: 8-12.

Bell, G.D.H. (1951). Barley breeding and related researches. *Journal of the Institute of Brewing* 57: 247-260.

Bendelow, V.M. (1976). Selection for malt β-glucanase activity in barley breeding programmes. *Canadian Journal of Plant Science* 56: 225-229.

Benetrix, F., Sarrafi, A., and J.-C. Autran (1994). Effects of genotype and nitrogen nutrition on protein aggregates in barley. *Cereal Chemistry* 71: 75-82.

Borem, A., Mather, D.E., Rasmussen, D.C., Fulcher, R.G., and P.M. Hayes (1999). Mapping quantitative trait loci for starch granule traits in barley. *Journal of Cereal Science* 29: 153-160.

Brennan, C.S., Harris, N., Smith, D.B., and P.R. Shewry (1992). Structural differences in the mature endosperms of good and poor malting barley cultivars. *Journal of Cereal Science* 24: 171-178.

Brennan, C.S., Smith D.B., Harris, N., and P.R. Shewry (1998). The production and characterisation of *Hor3* null lines of barley provides new information on the relationship of D hordein to malting performance. *Journal of Cereal Science* 28: 291-299.

Brown, A.H.D. and J.V. Jacobsen (1982). Genetic basis and natural variation of α-amylase isozymes in barley. *Genetic Research, Cambridge* 40: 315-324.

Brunswick, P., Manners, D.J., and J.R. Stark (1987). The development of β-D-glucanases during germination of barley and the effect of kilning on individual isoenzymes. *Journal of the Institute of Brewing* 93: 181-186.

Brunswick, P., Manners, D.J., and J.R. Stark (1988). Degradation of isolated barley endosperm cell walls by purified endo (1-3) (1-4)-β-D-glucanases and malt extracts. *Journal of Cereal Science* 7: 153-168.

Buraas T. and H. Skinnes (1984). Genetic investigations on seed dormancy in barley. *Hereditas* 101: 235-244.

Camm, J.-P., Ellis, R.P., Swanston, J.S., and W.T.B. Thomas (1990). Malting quality of barley. In *Scottish Crop Research Institute Annual Report*, eds. D.A. Perry and T.D. Heilbronn. Dundee, Scotland: Burns and Harris, pp. 16-20.

Chandra, G.S., Proudlove, M.O., and E.D. Baxter (1999). The structure of barley endosperm—An important determinant of malt modification. *Journal of the Science of Food and Agriculture* 79: 37-46.

Cook, R. (1990). The formation of ethyl carbamate in Scotch whisky. In *Proceedings of the Third Aviemore Conference on Malting, Brewing and Distilling*, ed. I. Campbell. London: Institute of Brewing, pp. 237-243.

Cook, R.H., McCaig, N., McMillan, J.M.B., and W.B. Lumsden (1990). Ethyl carbamate formation in grain-based spirits. Part III. The primary source. *Journal of the Institute of Brewing* 96: 233-244.

Corbineau, F. and D. Come (1980). Quelques caracteristiques de la dormance du caryopse d'orge (*Hordeum vulgare* L. variete Sonja). *Compes Rendus de l'Academie des Sciences de Paris* 290: 547-550.

Doll, H. (1976). Genetic studies of high lysine barley mutants. In *Proceedings of the Third International Barley Genetics Symposium*, ed. H. Gaul. Munich: Verlag Karl Thiemig, pp. 542-546.

Domon, E. (1996). Polymorphisms within waxy gene in indigenous barley cultivars revealed by the polymerase chain reaction. In *Proceedings of the Fifth International Oat Conference and the Seventh International Barley Genetics Symposium, Poster Session,* Volume 1, eds. A. Slinkard, G. Scoles, and B. Rossnagel. Saskatoon: University of Saskatchewan Extension Press, pp. 60-61.

Edney, M.J. (1996). Barley. In *Cereal Grain Quality*, eds. R.J. Henry and P.S. Kettlewell. London: Chapman and Hall, pp. 113-151.

Eglington, J.K., Langridge, P. and D.E. Evans (1998). Thermostability variation in alleles of barley beta-amylase. *Journal of Cereal Science* 28: 301-309.

Ellis, R.P. (1976). The use of high amylose barley for the production of whisky worts. *Journal of the Institute of Brewing* 82: 280-281.

Ellis, R.P., Cochrane, M.P., Dale, M.F.B., Duffus, C.M., Lynn, A., Morrison, I.M., Prentice, R.D.M., Swanston, J.S., and S.A. Tiller (1998). Starch production and industrial use. *Journal of the Science of Food and Agriculture* 77: 289-311.

Ellis, R.P., Ferguson, E., Swanston, J.S., Forrest, J.M.S., Fuller, J.D., Lawrence, P.E., Powell, W., Russell, J., Tester, R.F., Thomas, W.T.B., and G.R. Young (1999). Use of DNA marker-based assays to define and select malting characteristics in barley. In *HGCA Research Report 183*. London: Home-Grown Cereals Authority, 33 pp.

Ellis, R.P., McNicol, J.W., Baird, E., Booth, A., Lawrence, P., Thomas, B., and W. Powell (1997). The use of AFLPs to examine genetic relatedness in barley. *Molecular Breeding* 3: 359-369.

Ellis, R.P., Powell, W., Swanston, J.S., and C.E. Thomas (1986). Genetical analysis of a barley mutant with reduced height and increased diastatic power. *Journal of Agricultural Science, Cambridge* 106: 619-621.

Ellis, R.P., Swanston, J.S., and F.M. Bruce (1979). A comparison of some rapid screening tests for malting quality. *Journal of the Institute of Brewing* 85: 282-285.

Ellis, R.P., Swanston, J.S., Rubio, A., Perez-Vendrell, A., Romagosa, I., and J.-L. Molina-Cano (1997). The development of β-glucanase and degradation of β-glucan in barley grown in Scotland and Spain. *Journal of Cereal Science* 26: 75-82.

Erb, N., Zinsmeister, H.D., Lehmann, G., and A. Nahrstedt (1979). A new cyanogenic glycoside from *Hordeum vulgare*. *Phytochemistry* 18: 1515-1517.

Fincher, G.B. (1975). Morphology and chemical composition of barley endosperm cell walls. *Journal of the Institute of Brewing* 81: 116-122.

Forrest, I.S. and T. Wainwright (1977). The mode of binding of β-glucans and pentosans in barley endosperm cell walls. *Journal of the Institute of Brewing* 83: 279-286.

Forslund, K. and L. Jonsson (1997). Cyanogenic glycosides and their metabolic enzymes in barley, in relation to nitrogen levels. *Physiologia Plantarum* 101: 367-372.

Forster, B.P. and R.P. Ellis (1990). Milling energy requirement of the aneuploid stocks of common wheat, including alien addition lines. *Theoretical and Applied Genetics* 80: 806-809.

Forster, B.P., Thompson, D.M., Watters, J., and W. Powell (1991). Water-soluble proteins of mature barley endosperm: Genetic control, polymorphism and linkage with β-amylase and spring/winter habit. *Theoretical and Applied Genetics* 81: 787-792.

Franckowiak, J.D. and A. Pecio (1991). Coordinator's report: Semidwarf genes: A listing of genetic stocks. *Barley Genetics Newsletter* 21: 116-127.

Fulcher, R.G., Setterfield, G., McCully M.E., and P.J. Wood (1977). Observations on the aleurone layer. II. Fluorescence microscopy of the aleurone-sub-aleurone junction with emphasis on possible β-1,3-glucan deposits in barley. *Australian Journal of Plant Physiology* 4: 917-928.

Glennie-Holmes, M. (1997). The misdevelopment of the modern micromalter. *Ferment* 10: 51-55.

Goldbach, H. and G. Michael (1976). Abscisic acid content of barley grains during ripening as affected by temperature and variety. *Crop Science* 16: 797-799.

Greenwell, P. and J.D. Schofield (1986). A starch granule protein associated with endosperm softness in wheat. *Chemistry* 63: 379-380.

Griffiths, D.W. (1987). The ratio of B to C hordein as estimated by high performance liquid chromatography. *Journal of the Science of Food and Agriculture* 38: 229-235.

Groot, S.P.C. and C.M. Karssen (1992). Dormancy and germination of abscisic acid-deficient tomato seeds. *Plant Physiology* 99: 952-958.

Han, F., Ullrich, S.E., Chirat, S., Menteur, S., Jestin, L., Sarrafi, A., Hayes, P.M., Jones, B.L., Blake, T.K., Wesenberg, D.M., et al. (1995). Mapping of β-glucan content and β-glucanase activity loci in barley grain and malt. *Theoretical and Applied Genetics* 91: 921-927.

Han, F., Ullrich, S.E., Clancy J.A., Jitkov, V., Kilian, A., and I. Romagosa (1996). Verification of barley seed dormancy loci via linked molecular markers. *Theoretical and Applied Genetics* 92: 87-91.

Harrington, S.E., Bligh, H.F.J., Park, W.D., Jones, C.A., and S.R. McCouch (1997). Linkage mapping of starch branching enzyme III in rice (*Oryza sativa* L.) and prediction of location of orthologous genes in other grasses. *Theoretical and Applied Genetics* 94: 564-568.

Hayter, A.M. and M.J. Allison (1976). Breeding for high diastatic power. In *Proceedings of the Third International Barley Genetics Symposium,* ed. H. Gaul. Munich: Verlag Karl Thiemig, pp. 612-619.

Hoj, P.B., Slade, A.M., Wettenhall, R.E.H., and G.B. Fincher (1988). Isolation and characterization of a (1-3)-β-glucan hydrolase from germinating barley *(Hordeum vulgare)* amino acid sequence similiarity with barley (1-3) (1-4) –β-glucanases. *FEBS Letters* 230: 67-71.

Howard, K.A., Gayler, K.R., Eagles, H., and G.M. Halloran (1996). The relationship between D hordein and malting quality in barley. *Journal of Cereal Science* 24: 47-53.

Jagtap, S.S., Beardsley, A., Forrest, J.M.S., and R.P. Ellis (1993). Protein composition and grain quality in barley. In *Aspects of Applied Biology 36 (Cereal Quality III)*. Wellesbourne, Warwick, UK: Association of Applied Biologists, pp. 51-60.

Jensen, L.G., Politz, O., Olsen, O., Thomsen, K.K., and D. von Wettstein (1998). Inheritance of a codon optimized transgene expressing heat stable (1,3-1,4)-β-glucanase in scutellum and aleurone of germinating barley. *Hereditas* 129: 215-225.

Jones, D.G. and B.C. Clifford (1983). *Cereal Diseases, Their Pathology and Control* (Second Edition). Chichester: John Wiley and Sons.

Jorgensen, K.G. and S. Aastrup (1988). Determination of β-glucan in barley, malt, wort and beer. In *Modern Methods of Plant Analysis, New Series: Beer Analysis*, Volume 12, eds. H.F. Linslens and J.F. Jackson. Berlin, Heidelberg, New York: Springer, pp. 88-108.

Kihara M., Kaneko, T., and K. Ito (1998). Genetic variation of β-amylase thermostability among varieties of barley, *Hordeum vulgare* L., and relation to malting quality. *Plant Breeding* 117: 425-428.

Kilian, A., Kleinhofs, A., Villand, P., Thorbjornsen, T., Olsen, O.-A., and L.A. Kleczkowski (1994). Mapping of the ADP-glucose pyrophosphorylase genes in barley. *Theoretical and Applied Genetics* 87: 869-871.

Koornneef M., Reuling, G., and C.M. Karssen (1984). The isolation and characterization of abscisic acid insensitive mutants of *Arabidopsis thaliana*. *Physiologia Plantarum* 61: 377-383.

Kristensen, M., Svensson, B., and J. Larsen (1993). Purification and characterization of barley limit dextrinase during malting. In *Proceedings of the 24th Congress of the European Brewery Convention, Oslo*. Oxford: Oxford University Press, pp. 37-43.

Lee, W.J. and R.E. Pyler (1984). Barley malt limit dextrinase: Varietal, environmental and malting effects. *Journal of the American Society of Brewing Chemists* 42: 11-17.

Lein, A. (1964). Breeding for malting quality. In *Barley Genetics I, 1st Proceedings of the First International Barley Genetics Symposium*, eds. S. Broekhuizen, G. Dantuma, H. Lamberts, and W. Lange. Pudoc, Wageningen: Centre for Agricultural Publications and Documentation, pp. 310-324.

Li, C.D., Lance, R.C.M., Fincher, G.B., and P. Langridge (1996). Molecular mapping of loci controlling hydrolytic enzyme activity. In *Proceedings of the Fifth International Oat Conference and the Seventh International Barley Genetics Symposium, Poster Papers* Volume 1, eds. A. Slinkard, G. Scoles, and B. Rossnagel, Saskatoon: University of Saskatchewan Extension Press, pp. 344-346.

Loi, L., Ahluwalia, B., and G.B. Fincher (1988). Chromosomal location of genes encoding barley (1-3, 1-4)-β-glucan 4-glucanohydrolases. *Plant Physiology* 87: 300-302.

Loi, L., Barton, P.A., and G.B. Fincher (1987). Survival of barley (1-3) (1-4)-β-glucanase isoenzymes during kilning and mashing. *Journal of Cereal Science* 5: 45-50.

MacGregor, A.W. (1990). Current state of research into barley carbohydrates and enzymes. In *Proceedings of the Third Aviemore Conference on Malting, Brewing and Distilling*, ed. I. Campbell. London: Institute of Brewing, pp. 10-33.

MacGregor, A.W. and D.L. Balance (1980a). Hydrolysis of large and small starch granules from normal and waxy barley cultivars by alpha-amylases from barley malt. *Cereal Chemistry* 57: 397-402.

MacGregor, A.W. and D. L. Balance (1980b). Quantitative determination of α-amylase enzymes in germinated barley after separation by isoelectricfocusing. *Journal of the Institute of Brewing* 86: 131-133.

MacGregor, A.W., Bazin, S.L., Macri, L.J., and J.C. Babb (1999). Modelling the contribution of alpha-amylase, beta-amylase and limit dextrinase to starch degradation during mashing. *Journal of Cereal Science* 29: 161-169.

MacGregor, A.W. and G.B. Fincher (1993). Carbohydrates of the barley grain. In *Barley: Chemistry and Technology*, eds. A.W. MacGregor and R.S. Bhatty. St. Paul, MN: American Association of Cereal Chemists, pp. 73-130.

MacGregor, A.W., Macri, L.J., Bazin, S.L., and G.W. Sadler (1995). Limit dextrinase inhibitor in barley and malt and its possible role in malting and brewing. In *Proceedings of the 25th Congress of the European Brewery Convention, Brussels*. Oxford: Oxford University Press, pp. 185-192.

MacLeod, A.M. (1977). The impact of science on malting technology. In *Proceedings of the 16th Congress of the European Brewery Convention, Amsterdam*. Oxford: Oxford University Press, pp. 63-75.

Martin, H.L. and C.W. Bamforth (1980). The relationship between β-glucan solubilase, barley autolysis and malting potential. *Journal of the Institute of Brewing* 86: 216-221.

McCleary, B.V. and M. Glennie-Holmes (1985). Enzymic quantification of (1-3), (1-4)-β-D-glucan in barley and malt. *Journal of the Institute of Brewing* 91: 285-295.

McDonald, A.M.L., Stark, J.R., Morrison, W.R., and R.P. Ellis (1991). The composition of starch granules from developing barley genotypes. *Journal of Cereal Science* 13: 93-112.

McElroy, D. and J. Jacobsen (1995). What's brewing in barley biotechnology? *Bio/Technology* 13: 245-249.

McNicol, P.K., Jacobsen, J.V., Keys, M.M., and I.M. Stuart (1993). Effects of heat and water stress on malting quality and grain parameters of Schooner barley grown in cabinets. *Journal of Cereal Science* 18: 61-68.

Melchinger, A.E., Graner, A., Singh, M., and M.M. Messmer (1994). Relationships between European barley germplasm. 1. Genetic diversity among winter and spring cultivars revealed by RFLPs. *Crop Science* 34: 1191-1199.

Meredith, W.O.S., Anderson, J.A., and L.E. Hudson (1962). Evaluation of malting barley. In *Barley and Malt: Biology, Biochemistry, Technology*, ed. A.H. Cook. New York and London: Academic Press, pp. 207-270.

Merritt, N.R. (1967). A new strain of barley with starch of high amylose content. *Journal of the Institute of Brewing* 73: 583-585.

Miflin, B.J., Field, J.M., and P.R. Shewry (1983). Cereal storage proteins and their effects on technological properties. In *Seed Proteins,* eds. J. Daussant, J. Mosse, and J. Vaughan. London: Academic Press, pp. 253-319.

Miller, S.S. and R.G. Fulcher (1994). Distribution of (1-3),(1-4)-β-D-glucan in kernels of oats and barley using microspectrofluorometry. *Cereal Chemistry* 71: 64-68.

Millet, M.O., Montembault, A., and J.-C. Autran (1991). Hordein composition differences in various anatomical regions of the kernel between two different barley types. *Sciences des Alimentations* 11: 155-161.

Mizuno, K., Kawasaki, T., Shimada, H., Satoh, H., Kobayashi, E., Okumura S., Arai, Y., and T. Baba (1993). Alteration of the structural properties of starch components by the lack of an isoform of starch branching enzyme in rice seeds. *Journal of Biological Chemistry* 268: 19084-19091.

Molina-Cano, J.-L., Ramo, T., Ellis, R.P., Swanston, J.S., Bain, H., Uribe-Escheverria, T., and A.M. Perez Vendrell (1995). Effect of grain composition on water uptake by malting barley: A genetic and environmental study. *Journal of the Institute of Brewing* 101: 79-83.

Molina-Cano, J.-L., Roca de Togores, F., Royo, C., and A. Perez (1989). Fast germinating low β-glucan mutants induced in barley with improved malting quality and yield. *Theoretical and Applied Genetics* 78: 748-754.

Molina-Cano, J.-L., Sopena, A., Swanston, J.S., Casas, A.M., Moralejo, M.A., Ubieto, A., Lara, I., Perez Vendrell, A.M., and I. Romagosa (1999). A mutant induced in the malting barley cv. Triumph with reduced dormancy and ABA response. *Theoretical and Applied Genetics* 98: 347-355.

Morgan, A.G., Gill, A.A., and D.B. Smith (1983a). Some barley grain and green malt properties and their influence on malt hot-water extract. I. β-glucan, β-glucan solubilase and endo-β-glucanase. *Journal of the Institute of Brewing* 89: 283-291.

Morgan, A.G., Gill, A.A., and D.B. Smith (1983b). Some barley grain and green malt properties and their influence on malt hot-water extract. II. Protein, proteinase and moisture. *Journal of the Institute of Brewing* 89: 292-298.

Morgan, A.G. and T.J. Riggs (1981). Effects of drought on yield and on grain and malt characters in spring barley. *Journal of the Science of Food and Agriculture* 32: 339-346.

Morrison, W.R. and J. Karkalas (1989). Starch. In *Methods in Plant Biochemistry,* Volume 1: *Carbohydrates,* ed. P.M. Dey. London: Academic Press.

Newman, J.C. and D.E. Briggs (1976). Glyceride metabolism and gluconeogenesis in barley endosperm. *Phytochemistry* 15: 1453-1458.

Oliveira, A.B., Rasmusson, D.C., and R.G. Fulcher (1994). Genetic aspects of starch granule traits in barley. *Crop Science* 34: 1176-1180.

Palmer, G.H. (1980). The morphology and physiology of malting barleys. In *Cereals for Food and Beverages,* eds. G. Inglett and L. Munck. New York and London: Academic Press, pp. 301-338.

Palmer, G.H. (1995). Structure of ancient cereal grains. *Journal of the Institute of Brewing* 101: 103-112.

Powell, W., Caligari, P.D.S., Swanston, J.S., and J.L. Jinks (1985). Genetical investigations into beta-glucan content in barley. *Theoretical and Applied Genetics* 71: 461-466.

Powling, A., Islam, A.K.M.R., and K.W. Shepherd (1981). Isozymes in wheat-barley hybrid derivative lines. *Biochemical Genetics* 19: 237-254.

Que, X., Newman R.K., Newman, C.W., and D.F. McGuire (1990). Waxy gene effects on dietary fiber, beta glucan content and viscosity of barleys. *Barley Newsletter* 34: 89-92.

Rahman, S., Shewry, P.R., and B.J. Miflin (1982). Differential protein accumulation during barley grain development. *Journal of Experimental Botany* 33: 717-728.

Riggs, T.J., Hanson, P.R., and N.D. Start (1981). Genetic improvements in yield of spring barley and associated changes in plant phenotype. In *Proceedings of the Fourth International Barley Genetics Symposium*, eds. M.J.C. Asher, R.P. Ellis, A.M. Hayter, and R.N.H. Whitehouse. Edinburgh, Scotland: Edinburgh University Press, pp. 97-103.

Riggs, T.J., Sanada, M., Morgan, A.G., and D.B. Smith (1983). Use of acid gel electrophoresis in the characterisation of 'B' hordein in relation to malting quality and mildew resistance in barley. *Journal of the Science of Food and Agriculture* 34: 576-586.

Rohde, W., Becker, D., and F. Salamini (1988). Structural analysis of the *waxy* locus from *Hordeum vulgare*. *Nucleic Acids Research* 16: 7185-7186.

Schofield, J.D. and P. Greenwell (1987). Wheat starch granule proteins and their technological significance. In *Cereals in a European Context*, ed. I.D. Morton, Chichester, England: Ellis Horwood, pp. 407-420.

Schondelmaier, J., Jacobi, A., Fischbeck, G., and A. Jahoor (1992). Genetical studies on the mode of inheritance and localization of the *amo1* (high amylose) gene in barley. *Plant Breeding* 109: 274-280.

Shewry, P.R., Field, J.M., Kirkman, M.A., Faulks, A.J., and B.J. Miflin (1980). The extraction, solubility and characterisation of two groups of barley storage proteins. *Journal of Experimental Botany* 121: 393-407.

Shewry, P.R. and B.J. Miflin (1983). Characterization and synthesis of barley seed proteins. In *Seed Proteins, Biochemistry, Genetics, Nutritive Value*, eds. W. Gottschalk and H.P. Muller. The Hague: Martinus Nijhoff, pp. 143-205.

Shewry, P.R., Pratt, H.M., Faulks, A.J., Parmar, S., and B.J. Miflin (1979). The storage protein (hordein) polypeptide pattern of barley (*Hordeum vulgare* L.) in relation to varietal identification and disease resistance. *Journal of the National Institute of Agricultural Botany* 15: 34-50.

Sissons, M.J. (1996). Studies on the activation and release of bound limit dextrinase in malted barley. *Journal of the American Society of Brewing Chemists* 54: 19-25.

Skerritt, J.H. and R.J. Henry (1988). Hydrolysis of barley endosperm storage proteins during malting. II. Quantification by enzyme- and radio-immunoassay. *Journal of Cereal Science* 7: 265-281.

Skerritt, J.H. and P.W. Janes (1992). Disulphide-bonded "gel protein" aggregates in barley: Quality-related differences in composition and reductive dissociation. *Journal of Cereal Science* 16: 219-235.

Slack, P.T., Baxter, E.D., and T. Wainwright (1979). Inhibition by hordein of starch degradation. *Journal of the Institute of Brewing* 85: 112-114.

Slakesi, N., Baulcombe, D.C., Devos, K.M., Ahluwalia, B., Doan, D.N.P., and G.B. Fincher (1990). Structure and tissue-specific regulation of genes encoding barley-(1-3, 1-4)-β-glucan endohydrolases. *Molecular and General Genetics* 224: 437-439.

Smith, D.B. and A.A. Gill (1986). Quantitative changes in the concentrations of the major components of four genotypes of barley grains during malting and mashing. *Journal of the Institute of Brewing* 92: 360-366.

Smith, D.B. and P.R. Lister (1983). Gel forming proteins in barley grain and their relationship with malting quality. *Journal of Cereal Science* 1: 229-239.

Smith, D.B. and P.A. Simpson (1983). Relationships of barley proteins soluble in sodium dodecyl sulphate to malting quality and varietal identification. *Journal of Cereal Science* 1: 185-197.

Sparrow, D.H.B. (1970). Some genetical aspects of malting quality. In *Proceedings of the Second International Barley Genetics Symposium*, ed. R.A. Nilan. Pullman: Washington State University Press, pp. 559-574.

Stahl, A., Persson, G., Johansson, L.-A., and H. Johansson (1996). Breeding barley for functional food starch. In *Proceedings of the Fifth International Oat Conference and the Seventh International Barley Genetics Symposium, Poster Papers*, Volume 1, eds. A. Slinkard, G. Scoles, and B. Rossnagel. Saskatoon: University of Saskatchewan Extenshion Press, pp. 95-97.

Stark, D.M., Timmerman, K.P., Barry, G.F., Preiss, J., and G. Kishore (1992). Regulation of the amount of starch in plant tissues by ADP glucose pyrophosphorylase. *Science* 258: 287-292.

Stuart, I.M., Loi, L., and G.B. Fincher (1988). Varietal and environmental variations in (1-3, 1-4)-β-glucan levels and (1-3, 1-4)-β-glucanase potential in barley: Relationships to malting quality. *Journal of Cereal Science* 7: 61-71.

Swanston, J.S. (1987). The consequences, for malting quality, of *Hordeum laevigatum* as a source of mildew resistance in barley breeding. *Annals of Applied Biology* 110: 351-355.

Swanston, J.S. (1990a). Applications of milling energy to barley and malt assessment. In *Advances in Malting Barley 1990*, ed. J.S. Pierce. Exeter: European Brewery Convention, pp. 31-33.

Swanston, J.S. (1990b). The use of milling energy to predict increases in hot water extract in response to the addition of gibberellic acid during steeping. *Journal of the Institute of Brewing* 96: 209-212.

Swanston, J.S. (1994). Malting performance of barleys with altered starch composition. PhD Thesis. Edinburgh, Scotland: Heriot Watt University.

Swanston, J.S. (1995). Effects on barley grain size, texture and modification during malting associated with three genes on chromosome 1. *Journal of Cereal Science* 22: 157-161.

Swanston, J.S. (1996). Associations of the waxy *(wx)* gene with malting quality characters in random inbred lines of barley. *Journal of the Institute of Brewing* 102: 355-358.

Swanston, J.S. (1997). Waxy starch barley genotypes with reduced β-glucan contents. *Cereal Chemistry* 74: 452-455.
Swanston, J.S. (1999). Quantifying cyanogenic glycoside production in the acrospires of germinating barley grains. *Journal of the Science of Food and Agriculture* 79: 745-749.
Swanston, J.S., Ellis, R.P., Perez-Vendrell, A., Voltas, J., and J.-L. Molina-Cano (1997). Patterns of barley grain development in Spain and Scotland and their implications for malting quality. *Cereal Chemistry* 74: 456-461.
Swanston, J.S., Ellis, R.P., and J.R. Stark (1995). Effects on grain and malting quality of genes altering barley starch composition. *Journal of Cereal Science* 22: 265-273.
Swanston, J.S. and K. Taylor (1988). A comparison of some rapid tests to assess extent of modification during a laboratory scale malting procedure. *Journal of the Institute of Brewing* 94: 311-314.
Swanston, J.S. and K. Taylor (1990). The effects of different steeping regimes on water uptake, germination rate, milling energy and hot water extract. *Journal of the Institute of Brewing* 96: 3-6.
Swanston, J.S., Thomas, W.T.B., Powell, W., Meyer, R., Machray, G.C., and P.E. Hedley (1999). Breeding barley for malt whisky distilling—A genome based approach. In *Proceedings of the Fifth Aviemore Conference on Malting, Brewing and Distilling*, ed. I. Campbell. London: Institute of Brewing.
Swanston, J.S., Thomas, W.T.B., Powell, W., Young, G.R., Lawrence, P.E., Ramsay, L., and R. Waugh (1999). Using molecular markers to determine barleys most suitable for malt whisky distilling. *Molecular Breeding* 5: 103-109.
Taylor, K. and J.S. Swanston (1987). Malting quality assessment in a petri-dish. In *Aspects of Applied Biology 15 (Cereal Quality)*. Wellesbourne, Warwick, UK: Association of Applied Biologists, pp. 523-528.
Tester, R.F. and W.R. Morrison (1992). Swelling and gelatinization of cereal starches. III. Some properties of waxy and normal nonwaxy barley starches. *Cereal Chemistry* 69: 654-657.
Tester, R.F. and W.R. Morrison (1993). Swelling and gelatinization of cereal starches. VI. Starches from waxy Hector and Hector barleys at four stages of grain development. *Journal of Cereal Science* 17: 11-18.
Tester, R.F., Morrison, I.M., and A.H. Schulman (1993). Swelling and gelatinization of cereal starches. V. Risø mutants of Bomi and Carlsberg II barley cultivars. *Journal of Cereal Science* 17: 1-9.
Thomas, W.T.B., Powell, W., Swanston, J.S., Ellis, R.P., Chalmers, K.J., Barua, U.M., Jack, P., Lea, V., Forster, B.P., Waugh, R., and D.B. Smith (1996). Quantitative trait loci for germination and malting quality characters in a spring barley cross. *Crop Science* 36: 265-273.
Thomas, W.T.B., Powell, W., Swanston, J.S., and B.P. Forster (1991). The associations between the linked loci mlo, β-Amy-1 and WSP-3 and quantitative characters in barley. In *Proceedings of the EUCARPIA (Cereal Section) Meeting, Schwerin*. Wageningen: EUCARPIA, pp. 255-260.
Thomas, W.T.B., Powell, W., and W. Wood (1984). The chromosomal location of the dwarfing gene present in the spring barley variety Golden Promise. *Heredity* 53: 177-183.

Toojinda, T., Baird, E., Booth, A., Broers, L., Hayes, P., Powell, W., Thomas, W., Vivar, H., and G. Young (1998). Introgression of quantitative trait loci (QTLs) determining stripe rust resistance in barley: An example of marker-assisted line development. *Theoretical and Applied Genetics* 96: 123-131.

Ullrich, S.E., Clancy, J.A., Eslick, R.F., and R.C.M. Lance (1986). β-glucan content and viscosity of extracts from waxy barley. *Journal of Cereal Science* 4: 279-285.

Ullrich, S.E., Hayes, P.M., Dyer, W.E., Blake, T.K., and J.A. Clancy (1993). Quantitative trait locus analysis of seed dormancy in 'Steptoe' barley. In *Pre-Harvest Sprouting in Cereals, 1992*, eds. W.K. Walker-Simmons and J.L. Rieds. St. Paul, MN: American Association of Cereal Chemists.

Ulonska, E. and M. Baumer (1976). Investigation of the value of water uptake and germination for the estimation of malting quality in barley. In *Proceedings of the Third International Barley Genetics Symposium*, ed. H. Gaul, Munich: Verlag Karl Thiemig, pp. 579-593.

Visser, K., Vissers, A.P.A., Cagirgan, M.I., Kijne, J.W., and M. Wang (1996). Rapid germination of a barley mutant is correlated with a rapid turnover of abscisic acid outside the embryo. *Plant Physiology* 111: 1127-1133.

Walker-Simmons, M. (1987). ABA levels and sensitivity in developing wheat embryos of sprouting resistant and susceptible cultivars. *Plant Physiology* 84: 61-66.

Walker-Simmons, M., Kurdna, D.A., and R.L. Warner (1989). Reduced accumulation of ABA during water stress in a molybdenum cofactor mutant of barley. *Plant Physiology* 90: 728-733.

Wallace, W. and R.C.M. Lance (1988). The protein reserves of the barley grain and their degradation during malting and brewing. *Journal of the Institute of Brewing* 96: 379-386.

Wang, M., Heimovaara-Dijkstra, S., and B. Van Duijn (1995). Modulation of germination of embryos isolated from dormant and nondormant grains by manipulation of endogenous abscisic acid. *Planta* 195: 586-592.

Whitmore, E.T. and D.H.B. Sparrow (1957). Laboratory micro-malting techniques. *Journal of the Institute of Brewing* 63: 397-398.

Woodward, J.R. and G.B. Fincher (1982a). Purification and chemical properties of two (1-3,1-4)-β-D-glucan endohydrolases from germinating barley. *European Journal of Biochemistry* 121: 663-669.

Woodward, J.R. and G.B. Fincher (1982b). Substrate specificities and kinetic properties of two (1-3, 1-4)-β-D-glucan endohydrolases from germinating barley *(Hordeum vulgare)*. *Carbohydrate Research* 106: 111-122.

Yin, X.S. and A.W. MacGregor (1988). An approach to the identification of a beta-glucan solubilase from barley. *Journal of the Institute of Brewing* 94: 327-330.

Yomo, H. and H. Iinuma (1966). Production of a gibberellin-like substance in the embryo of barley during germination. *Planta* 71: 113-118.

Zabeau, M. and P. Vos (1993). Selective restriction fragment amplification: a general method for DNA fingerprinting. European Patent Application Number 92402629.7/Publication Number 0 534858A1.

Chapter 6

Genetics and Breeding of Barley Feed Quality Attributes

Steven E. Ullrich

INTRODUCTION

Barley (*Hordeum vulgare* L.), as an ancient crop, has undergone considerable genetic change through the domestication process, and subsequently as barley adaptation and use has expanded, and more recently as formal breeding has occurred. Over the thousands of years of its existence as an agricultural crop, barley use has shifted from a major food grain to a major feed grain. Along the way, the use of barley for malting, brewing, and distilling has been a major factor in its evolution and development, as well. Today barley is a very widely adapted and widely used cereal grain crop with very wide genetic diversity. Feed is the primary use of barley grain around the world. On a global basis, approximately 67 percent of the crop is used for feed, 28 percent for malt and directly for food (Bamforth and Barclay, 1993), and 5 percent for seed. Barley cultivars are often categorized as to feed or malting type and sometimes food type.

Most barley quality improvement research and breeding activity has been directed toward malting barley in spite of its minority use. Malting barley generally commands a premium price over feed barley, and the malting, brewing, and distilling industries have been particularly supportive of research and breeding efforts. Sometimes national or regional agricultural policies favor the production of malting barley over feed barley. By comparison, relatively little effort has gone into feed or food improvement, and little support has come from those industries. In fact, there has been a relative lack of interest in barley nutritional improvement by the feed and livestock and poultry industries. Often it is just physical characteristics, such as high test or volume weight or plump uniform kernels, that influence the feed barley price. Frequently in barley breeding programs, yield and adaptation, with or without malting quality, are emphasized. Most often when a new barley cultivar is released, it is designated a malting type or a feed type. It is designated a malting type if it has passed rigorous industry tests and is accepted by the industry. On the other hand, feed types are so designated because they do not possess acceptable malting-quality characteristics, not because they possess particular feed or nutritional characteristics. The associated industries do not oversee the development or designation of feed barley cultivars. There are several notable exceptions to this situation, which will be explained in the next section. Not until the 1960s was there much recognition that "barley is not just barley" as far as feed value is concerned. Whereas there has been relatively little breeding progress toward nutritional quality improvement per se, significant research effort has focused on nutritional quality and feed and food value of barley, and significant variation in nutritional quality traits has been identified within the past 30 to 40 years. Recent reviews on these subjects include those by Bhatty (1986, 1993, 1996), Munck (1992), Newman and Newman (1992a,b), and Oakenfull (1996).

The issue of malting versus feed quality in barley has been researched and debated over the years. It is frequently inferred that the best feed barleys are malting barleys. It is argued that feeding yeast in the brewing process and feeding animals, especially monogastrics, may not be that different. Correlation studies involving both malting and feed varieties and malting and feed quality parameters have generally resulted in the establishment of positive associations between the two types of quality parameters (Hockett and White, 1981; Molina-Cano et al., 1997) or no association (Ullrich, Coon, and Sever, 1981). The general conclusion then is that breeding for one quality type or end use of barley is compatible with breeding for the other. Perhaps these comparisons are somewhat flawed in that, although high-quality malting types are typically used in the research, high-quality feed types are not necessarily used for the reasons described earlier. Furthermore, since high-quality feed types are not typically developed by design, the high end for feed barley likely has not been established. There seems to be great potential for the development of truly high-quality feed barley, and this will be pursued in the next section.

The primary objectives of this chapter are to (1) review barley feed quality characteristics, (2) consider the genetics of the various feed quality traits, and (3) discuss the realized and potential selection and breeding successes for improved feed quality. Although barley is utilized for forage through grazing, hay, straw, and silage, this chapter will concentrate on grain attributes and use.

FEED QUALITY CHARACTERISTICS

Considerable research effort has gone into analyzing barley, describing its grain and grain components, and measuring the performance of animals fed barley-based diets to determine the important attributes of barley that make it a good or poor feed. The identification of important feed quality traits is paramount to improving barley's feed or nutritional value through breeding. The categories of grain traits include physical or morphological characteristics, chemical composition (quality and antiquality factors), and animal performance parameters.

Physical Characteristics

Physical or morphological characteristics influence or are indicative of the nutritional quality of barley. A common measurement taken on grain for marketing purposes is test or volume weight, which may be expressed in kilograms per hectoliter ($kg \cdot hl^{-1}$); it is a measure of density. The standard test weight of barley by which grain quality is compared is approximately 62 $kg \cdot hl^{-1}$. Test weight may range from about 52 to 72 $kg \cdot hl^{-1}$ (up to 80 $kg \cdot hl^{-1}$ for hull-less types), depending upon growing conditions and genotype. Generally, the denser the kernel, the higher the nutrient content (e.g., starch and protein versus air spaces and fiber). The measurement and interpretation of test weight values can be difficult, however. Factors such as the presence of awns, chaff, broken kernels, and foreign materials; kernel size and shape; and the presence of hulls or partial hulls and hull thickness can influence the measure of test weight, as well as internal kernel anatomy and composition. Kernel plumpness and uniformity are also characteristics that may be important. Plumpness is an indication of endosperm size and starch content. The higher the plump kernel percentage, the higher the starch content, generally. Kernel uniformity is important for feed processing such as dry and steam rolling. A major simply inherited trait is hulled versus hull-less kernels. The hull makes up 10 to 15 percent of the dry weight of the barley kernel and is composed primarily of cellulose, hemicellulose, pectins, lignin, and a small amount of protein. The hull is the major contributor of crude or acid detergent fiber in the barley kernel, which is largely nutritionally unavailable to nonruminant livestock and poultry. Absence or removal of the hull reduces the fiber content of barley to that of maize and wheat (Bhatty, 1993). Grain harvested from two-row and six-row spikes will often differ in terms of physical traits, specifically test weight, kernel weight, and kernel plumpness, all of which tend to be higher in grain from two-row spikes. Most of the physical traits are interrelated, as may be inferred from the earlier descriptions. Another physical aspect is particle size of cracked barley (after dry rolling or coarse grinding in feed preparation), which has recently received attention as a feed quality factor for ruminants (Bowman and Blake, 1996). Grain digestibility by ruminants is inversely related to particle size, and feed intake by steers is increased with larger particle size (Bowman and Blake, 1996). Therefore, a compromise in particle size is usually necessary to balance feed intake and digestibility.

Chemical Composition

Chemical composition is critical in terms of nutrient content. The barley kernel, as is the case with cereals in general, is a complex integration of carbohydrates, proteins, lipids, minerals, and other compounds. Barley grain chemical components and their nutritional implications were thoroughly reviewed recently by Newman and Newman (1992b) and Bhatty (1993). Compositional data (whole kernel dry weight basis) listed in the following come from these two references unless otherwise noted. Although giving percent composition for the various kernel components is valuable, it must be remembered that considerable variation exists due to effects of genetics and production environment. Means given for a species can be particularly misleading.

Carbohydrates are the major components (~80 percent) of barley grain, with starch as the principal compound (~40 to 65 percent; mean ~55 percent). Starch occurs primarily in the endosperm in two forms: amylopectin (~75 percent), a branched chain molecule, and amylose (~25 percent), an unbranched straight-chain molecule. Genetic variants create a range from very low to no amylose (waxy types) to amylose levels up to about 45 percent (high-amylose types). Given the relatively high level of starch in the kernel, barley is considered primarily as an energy source in livestock and poultry diets. However, nonstarch polysaccharides and lignin (~10 to 20 percent), sometimes designated total dietary fiber, and primarily the structural components of cell walls, negatively modify the energy value of barley grain (reduce digestible energy), primarily in monogastric animals. Acid detergent fiber (ADF) (~2 to 12 percent, with hull-less types on the low end) is one commonly used measure of insoluble fiber and includes primarily cellulose and lignin. Other fiber components that are somewhat soluble are mixed-linked ($1{\rightarrow}3$, $1{\rightarrow}4$) β-glucans (1.4 to 11.5 percent; mean 4.8 percent based on analyses of 10,978 accessions from the USDA national barley germplasm collection; D. M. Peterson and D. M. Wesenberg, personal communication, 1999) and arabinoxylans (~4 to 8 percent), which have a tendency to increase the viscosity of solutions and colloids. These compounds are generally considered to lower digestibility, especially in poultry. Beta-glucans, which are the primary component in the starchy endosperm cell walls, also cause modification, filtration, and haze problems in malting and brewing as well (Bamforth, 1982). On the other hand, β-glucans have been positively implicated in hypocholesterolemic effects in humans, other mammals, and poultry (Newman, Newman, and Graham, 1989). Arabinoxylans, which are also known as pentosans and found primarily in the cell walls of the hull and aleurone (Han and Froseth, 1992), similar to β-glucans, increase the viscosity in the gastrointestinal tract, which decreases digestibilty. However, recent research suggests that arabinoxylans may be the greater antinutritional factor of the two, especially in swine (Han, 1994).

Protein (6.9 to 25.0 percent; mean 14.8 percent based on analyses of 10,978 accessions from the USDA national barley germplasm collection; D. M. Peterson and D. M. Wesenberg, personal communication, 1999) and amino acid levels are significant in barley and impact feed value and diet formulation. Protein and amino acid levels are influenced by both genetic and environmental factors. Protein content particularly is influenced strongly by the production environment. As an illustration, Torp, Doll, and Haahr (1981) found that protein content in barley grain ranged from 8.1 to 14.7 percent for the same genotype grown at different locations, even at similar nitrogen fertilizer levels. Lysine and threonine are the first and second limiting amino acids, respectively, in barley grain for swine (Bhatty, 1993). Much emphasis in barley research has been on grain lysine content, which averages about 0.45 percent or less of the whole kernel in normal barley compared to nutritional requirements of 0.60 to 1.0 percent in growing pigs of various ages. Considerable effort in the study and improvement of lysine in barley occurred in the 1970s and 1980s after the discovery of the first high-lysine mutant in barley, 'Hiproly' (Munck et al., 1970). Newman and McGuire (1985) cited numerous studies that clearly indicated the nutritional advantages of high-lysine grain over normal-lysine barley in nonruminant diets. High-lysine barley will be discussed further in the genetics and breeding sections.

Lipid levels are relatively low, with a narrow range (0.9 to 3.2 percent; mean 1.6 percent based on analyses of 3,275 accessions from the USDA national barley germplasm collection; D. M. Peterson and D. M. Wesenberg, personal communication, 1999) in barley grain, and lipid content distribution is three-quarters in the endosperm and one-quarter in the embryo (Price and Parsons, 1979). Several mutants of barley have up to 7 percent lipid in the grain (Newman and Newman, 1992b). However, in general, there has been relatively little success in finding high-fat barley genotypes, whether it be through screening germplasm collections or induced mutagenesis. All in all, there has been relatively little work on barley lipids and apparently no inheritance studies. Barley grain does contain some important lipid components, such as antioxidant tocotrienols, which have been shown to inhibit cholesterol synthesis. This is more of a human nutrition than an animal nutrition factor.

Polyphenolic or flavonoid compounds such as anthocyanin are found in numerous plant tissues. Only barley and sorghum contain proanthocyanidins, which occur in the testa of barley. Proanthocyanidins have various effects on plant pest resistance and feed, malting, and brewing quality. Proanthocyanidins are recognized to cause haze problems in beer due to the formation of insoluble phenol-protein complexes (Wettstein et al., 1985). Likewise, some documentation indicates that similar complexes form in the gut of animals, limiting protein digestibility (Newman et al., 1984; Øverland, 1994). However, other research shows that the effects on digestion are rather minor due to the relatively low concentration of these phenolics (Newman et al., 1987; Newman and Newman, 1992b). Over 600 proanthocyanidin-deficient mutants have been induced (Jende-Strid, 1995), with the majority being anthocyanin free, as well. Thirty distinct *Ant* loci that affect flavonoid synthesis have been identified (Jende-Strid, 1995, 1998), some of which have been used to study the genetic control of phenolic pathways in barley (Jende-Strid, 1993). The most useful proanthocyanidin-free mutants for barley breeding purposes are proving to be those which synthesize anthocyanin. These so-called "pant" (pigmented ant) mutations involve the last steps in the flavonoid pathway and have less drastic effect on vigor and yield than the original "ant" types that are both proanthocyanidin and anthocyanin free (Jende-Strid, 1991).

Elemental minerals occur in barley grain in relatively small concentrations (~2 to 4 percent) but do impact animal nutrition. The major components are phosphorus, otassium, calcium with lesser amounts of chlorine, magnesium, sulphur, and sodium (Owen et al., 1977). The hull contains the highest proportion of minerals, followed by the embryo, then the endosperm (Newman and Newman, 1992b). Phosphorus (P) is one of the most important mineral nutrients for animals, but up to 80 percent of cereal kernel phosphorus is unavailable because it is bound to phytic acid (Newman and Newman, 1992b). In addition, phytic acid binds to several other important cations, such as calcium, iron, magnesium, and zinc. Because nonruminants cannot effectively metabolize phytate, relatively high concentrations of phosphorus end up in the feces of these animals when fed high-grain diets, which can cause waste disposal, and thereby environmental, problems. Recently, several low-phytic-acid mutants with normal P levels (~4 percent) were chemically induced in maize *(Zea mays)* (Raboy and Gerbasi, 1996) and barley (Larson et al., 1998; Rasmussen and Hatzack, 1998), providing new genetic variation for this trait and new opportunities to alleviate nutritional and environmental problems associated with grain phytic acid (Ertl, Young, and Raboy, 1998). Limited feeding trials have thus far been conducted with trout and pigs. Animals consuming barley grain of low-phytic-acid mutants excreted 50 to 55 percent less P than those consuming wild-type barley, while maintaining equal performance and bone strength (Raboy and Cook, 1999).

Animal Performance Parameters

Many attempts have been made to predict feed value by measuring grain or feed nutrient or antinutrient components, but the baseline for quality is animal performance. Nutrient composition multiple regression prediction equations, for example, can be indicative of feed value, but they are in the end tied to animal performance measurements (e.g., Campbell, Salmon, and Classen, 1986). Feeding or metabolism trials (in vivo) to measure overall digestibility or digestibility and metabolism of specific components, such as starch, protein, energy, or antiquality factors (e.g., arabinoxylans, β-glucans), ultimately are required to measure quality and to distinguish between high- and low-feed-value genotypes of given plant species such as barley. Depending upon the target animal species or age group, relatively large amounts of feed components such as grain may be required to measure quality through animal performance trials. Large animals such as cattle also require a lot of feeding trial facility space. Time required to run feeding trials can also be a factor. All of these factors can add up to a relatively high cost of conducting feed evaluation trials and can be a problem in plant breeding situations, where numbers of genotypes to be evaluated can be high. Alternatives have been developed that require greatly reduced amounts of feed materials. These include, for example, the mobile nylon bag (in situ) (Sauer, Jørgenson, and Berzins, 1983; Nocek, 1988; Honeyfield, Froseth, and Ullrich, 1990; Han, Froseth, and Ullrich, 1993; Ovenell-Roy et al., 1998) and in vitro (Tilley and Terry, 1963; Duncan et al., 1991; Boisen and Eggum, 1991) techniques in ruminants and nonruminants. In the mobile nylon bag technique, small amounts of a target feed component, such as grain or grain fraction, are placed in a small nylon mesh bag that is introduced into a test animal's digestive tract (e.g., rumen, small or large intestine) through a cannula and recovered from the rumen or the feces. The amount of the target material that disappears represents the digestible portion of the material. In vitro techniques involve incubation of target feed components with digestive enzymes obtained from an animal species of interest in the laboratory to measure disappearance or digestibility. These techniques require verification from full-blown animal digestibility trials. The use of rats, chicks, or weanling pigs for digestion or metabolism trials also requires relatively small amounts of a target feed component, but results from these trials can be highly indicative of feed value (Fadel et al., 1987; Newman and Newman, 1990). Just as grain compositional characteristics are heritable in barley, so too, apparently, are animal performance characteristics, as will be illustrated in the next section.

GENETICS OF FEED QUALITY CHARACTERISTICS

As described earlier, barley contains a host of feed quality determinants, some of which have been characterized genetically and many of which have not. Some of the barley traits are quite unique, while most are commonly shared with other cereal grain species. Some of the traits (e.g., anatomical and simple chemical composition traits) are simply inherited, controlled by one or two genes, while most (e.g., traits that involve complex physiological processes) are quantitatively inherited, controlled by several to many genes with varying effects. The expression of quantitatively inherited traits is commonly affected by the environment and by environment × genotype interaction. This means that for many of the simply inherited traits, genetic knowledge typically dates back many years (Nilan, 1964; Søgaard and Wettstein-Knowles, 1987). On the other hand, little is known about the inheritance of the quantitatively inherited traits, except that it is complex. It has only been since the availability of comprehensive molecular marker-based genetic maps and the advent of quantitative trait locus (QTL) analysis technology that significant progress in genetic information about quantitative traits has been possible. For barley, that means beginning in the late 1980s and early 1990s. QTL analysis is useful for genetically dissecting complex traits and for identifying favorable alleles in diverse germplasm for breeding purposes. The first comprehensive barley molecular map (Kleinhofs et al., 1993) and QTL analysis of agronomic and quality traits (Hayes et al., 1993) were based on the six-row cross between 'Morex' (malting type) and 'Steptoe' (feed type) through the coordinated efforts of the multinational, multiinstitutional, and multidisciplinary North American Barley Genome Mapping Project (NABGMP). Subsequently, a number of maps from other crosses have been constructed and a number of QTL analyses have been run for quality traits. Besides the Steptoe/Morex map, NABGMP core maps were made from the two-row cross between 'Harrington' (malting type) and TR306 (feed type) (Kasha et al., 1995) and the two-row/six-row cross Harrington/Morex (Hayes et al., 1997). Comprehensive barley molecular mapping projects have been developed in Europe, Australia, and Japan as well (Ullrich, 1997a). The NABGMP core and many other mapping populations consist of completely homozygous doubled haploid lines (DHLs), which make their utility essentially unlimited for genetic analysis.

Physical Characteristics

Among the physical characteristics described earlier, only the hulled/hull-less and two-/six-row traits are simply inherited. The presence or absence of the hull is controlled by the *Nud* (hulled)/*nud* (hull-less) gene located on chromosome 1(7H) (Nilan, 1964). The row number trait is controlled by two genes: the *Vrs1 vrs1* and the *Int-c int-c* genes located on chromosomes 2(2H) and 4(4H), respectively (Nilan, 1964). Commercial two-rows have the genotype *Vrs1Vrs1int-cint-c*, and commercial six-rows, *vrs1vrs1Int-cInt-c*. Both the hull and the row type traits are simply inherited, but there may be considerable pleiotropy associated with the genes involved (Powell, Ellis, and Thomas, 1990), and generally distinct germplasm subgroups have evolved for each type (Bell and Lupton, 1962)

A summary of feed quality QTL detected in eight crosses from North American, European, and Australian mapping projects is presented in Table 6.1. The Harrington/TR306 mapping population has been utilized in a number of QTL analysis projects, including comprehensive studies on agro-

TABLE 6.1. Chromosome Location of QTL for Various Barley Feed Quality Traits Based on Analysis of Eight Crosses

Cross	Chromosome Number						
	1(7H)	2(2H)	3(3H)	4(4H)	5(1H)	6(6H)	7(5H)
Steptoe/ Morex		KW, KP, GS, BG, ADF, GP	KW, KP, PS, GP, DMDC, SDC	KW, KP, ADF, GP, DMDG, SDG	KW, BG, ADF, DMDG, SDG	KW, GS, GP	KW, KP, GP
Harrington/ TR306	TW, KW, GP			TW, KW, KP, GP	GP	TW, KW	TW, KW, KP, GP
Harrington/ Morex	KP, GP	KP, GP		GP	KP		KP, GP
Blenheim/ E224-3	KW	TW, GP	TW, KW, KP, GP		TW, KP, GP	TW, KW, KP	TW
Chebec/ Harrington		KW, KP, GP		KW	KW		KP
Galleon/ Haruno N.		GP			KP		
Calicuch/ Bowman		GP					
Dicktoo/ Morex	GP			GP	GP	GP	GP

Note: In some cases there may be more than one QTL for a given trait on a given chromosome from a given cross. TW = test weight, KW = kernel weight, KP = kernel plumpness, PS = particle size of cracked grain, GS = grain starch content, BG = grain β-glucan content, ADF = grain acid detergent fiber content, GP = grain protein content, DMDC = dry matter digestibility of cracked grain, SDC = starch digestibility of cracked grain, DMDG = DMD of ground grain, SDG = SD of ground grain.

nomic (Tinker et al., 1996) and malting-quality (Mather et al., 1997) traits. Five test weight QTL were mapped to chromosomes 1(7H), 4(4H), 6(6H), and 7(5H) (Tinker et al., 1996). Thomas et al. (1995) also detected test weight QTL on chromosomes 6(6H) and 7(5H) as well as on chromosomes 2(2H), 3(3H), and 5(1H) in the mapping population from the European two-row cross 'Blenheim'/E224-3. Four kernel weight QTL were also detected in the Harrington/TR306 study on the same chromosomes that correspond to four of the Harrington/TR306 test weight QTL. Kernel weight QTL were mapped to similar regions of chromosomes 1(7H), 2(2H), 4(4H), 5(1H), 6(6H), and/or 7(5H) in several other studies as well (Jensen, 1989; Powell et al., 1990; Thomas, Powell, and Swanston, 1991; Han and Ullrich, 1994; Thomas et al., 1995; Backes et al., 1995; GrainGenes database online: <http://greengenes.cit.cornell.edu/waiteqtl/>). Kernel weight QTL were placed on the Steptoe/Morex map, two on chromosome 2(2H) and one each on chromosomes 3(3H), 4(4H), 5(1H), 6(6H), and 7(5H) (Han and Ullrich, 1994). Thomas et al. (1995) detected kernel weight QTL on chromosome 3(3H) as well. Kernel plumpness percentage QTL have been mapped on all three NABGMP core maps: six QTL from Steptoe/Morex (Han and Ullrich, 1994), three from Harrington/TR306 (Mather et al., 1997), and five from Harrington/Morex (Marquez-Cedillo et al., 1999). Four of the kernel plumpness QTL located on chromosomes 2(2H), 4(4H), and 7(5H) were in common in two of the three NABGMP crosses. Thomas et al. (1995) also found a kernel plumpness QTL in one of the similar chromosome 5 regions as well as unique QTL on chromosomes 3(3H) and 6 (6H). Kernel plumpness QTL were also detected in the Australian crosses from 'Chebec'/Harrington [chromosomes 2(2H) and 7(5H)] and 'Galleon'/'Haruna Nijo' [chromosome 5(1H)] (http://greengenes.cit.cornell.edu/waiteqtl/). A number of the QTL for the test weight, kernel weight, and kernel plumpness overlapped within and among the various crosses cited here, which is not surprising because these three traits are interrelated. Therefore, the 55 QTL independently determined for the three traits cited here most likely really represent far fewer actual barley genes. However, the exact number cannot be determined at this time, given the rather gross nature of QTL analysis and the difficulty in comparing maps from different crosses.

Particle size of cracked barley was evaluated in the Steptoe/Morex mapping population. Two apparent QTL were located on chromosome 3(3H) and were coincident with QTL for percent dry matter digestibility and percent starch digestibility of cracked barley in cattle (Bowman and Blake, 1996). This seems to indicate the importance of cracked barley particle size to digestibility in cattle.

Chemical Composition

The inheritance of the concentration of most carbohydrates important to barley feed quality tends to be complex. Most of the important carbohydrates described earlier are complex compounds or mixtures of compounds. Therefore, most genetic information about the accumulation of carbohydrates in grain is based on QTL analysis. There have been two recent genetic analyses of complex carbohydrate content. Starch content QTL were located on the Steptoe/Morex map on the long arm of chromosome 2(2H) (Ullrich et al., 1996) and near the telomere of the long arm of chromosome 6(6H) (Bowman and Blake, 1996; Ullrich et al., 1996). Genes affecting simply inherited amylose:amylopectin ratio traits in starch have been known for a long time. The waxy gene *(wax)* located on chromosome 1 conditions a high amylopectin level approaching 100 percent (Nilan, 1964). The waxy condition *(waxwax)* is recessive to the wild-type condition *(WaxWax)* of approximately 75 percent amylopectin to 25 percent amylose. The high-amylose gene *(amo1)* located on chromosome 3(3H) in the homozygous recessive state brings the amylose level to about 45 percent (Merritt, 1967; Søgaard and Wettstein-Knowles, 1987). Both the waxy and high-amylose genes may have increased β-glucan and decreased starch levels associated with them (Ullrich et al., 1986; Que et al., 1997; Swanston, Ellis, and Tiller, 1997). The high-amylose gene has elevated lysine levels and lower kernel weight associated with it, as well (Ullrich and Eslick, 1978c). Furthermore, the two genes *(wax, amo1)* interact with each other. When the two traits are together in the same genotype, the β-glucan and starch levels are increased and decreased, respectively, exceeding simple additive effects (Swanston, Ellis, and Tiller, 1997). On the other hand, waxy genotypes with normal levels of β-glucans have been identified (Swanston, 1997). The nutritional significance of both of these traits is not clear. Advantages and disadvantages have been demonstrated in feeding trials with various animal species comparing waxy and high-amylose barleys with normal wild-type barley (Newman and Newman, 1992b).

Although simply inherited, low β-glucan induced mutants have been identified (Aastrup, 1983; Molina-Cano et al., 1989), the genetics of kernel β-glucan content has proven to be rather complex. Powell et al. (1989) reported that barley β-glucan content is controlled by an additive genetic system of three to five "effective factors," but their chromosome locations could not be determined. QTL analysis of the Steptoe/Morex mapping population revealed three QTL: two located on chromosome 2(2H) and one located on chromosome 5(1H) (Han et al., 1995). However, the three QTL together accounted for only 34 percent of the total β-glucan percentage. Acid detergent fiber content inheritance has also been studied in the Steptoe/Morex DHL population. Ullrich et al. (1996) found a cluster of three QTL on chromosome 2(2H) and single QTL near the telomeres on chromosomes 4S(4HS) and 5(1HS). Bowman and Blake (1996), who evaluated the Steptoe/Morex DHs, as well, reported an apparently different QTL on chromosome 4 near the centromere.

Grain protein content inheritance in barley is rather complex but may be viewed from different perspectives. The content of some specific proteins and protein groups has monofactorial inheritance, and their genes have been mapped. For example, two isozymes of the enzyme α-amylase *(amy1* and *2)* have been mapped to chromosomes 1(7H) and 6(6H) and may influence α-amylase level, and the B, C, and D hordein gene families *(Hor 1, 2,* and *3)* that act as single loci have been mapped with close linkage to chromosome 5(1HS) (Søgaard and Wettstein-Knowles, 1987). Total grain protein content, which is often a measure of feed quality, is quantitatively inherited. Little useful genetic information about grain protein content had been determined before the advent of the molecular genetics application tool QTL analysis. Several of the QTL analysis studies on carbohydrates described above included protein content as well. Regions of all seven chromosomes have been associated with total grain nitrogen or protein content. One apparent QTL has been located near the centromere on chromosome 1(7H) from the Harrington/TR306 (Mather et al., 1997) and Harrington/Morex (Marquez-Cedillo et al., 1999) mapping populations, and another QTL on the long arm from the 'Dicktoo'/Morex cross (Oziel et al., 1996). Conservatively, four QTL have been mapped on chromosome 2(2H): one near the telomere of 2S(2HS) from the Galleon/Haruna Nijo cross (<http://greengenes.cit.cornell.edu/waiteqtl/>); another near the middle of 2S(2HS) detected in four crosses (Blenheim/E224-3 [Thomas et al., 1996]; Chebec/Harrington, <http://greengenes.cit.cornell.edu/ waiteqtl/>; Steptoe/Morex [Han and Ullrich, 1994; Marquez-Cedillo et al., 1999]; Harrington/Morex [Marquez-Cedillo et al., 1999]); a third QTL near the centromere from the Harrington/Morex (Marquez-Cedillo et al., 1999) and 'Calicuchima'-sib/'Bowman' (Hayes et al., 1997) crosses; and the fourth QTL near the end of 2L(2HL) from Steptoe/Morex (Han and Ullrich, 1994). Two grain protein QTL detected in some environments from the Steptoe/Morex population were mapped on chromosomes 3S(3HS) and 3L(3HL) (Hayes et al., 1993; Han and Ullrich, 1994; Larson et al., 1997). The chromosome 3L(3HL) QTL was mapped in the same region as one from the Blenheim/E224-3 cross (Thomas et al., 1996). It appears that at least three QTL are located on chromosome 4(4H): one QTL on 4S(4HS) was detected in the Dicktoo/Morex (Pan et al., 1994), Steptoe/Morex (Han and Ullrich, 1994), and Harrington/Morex (Marquez-Cedillo et al., 1999) crosses, and two QTL were found on 4L(4HL), with one detected from Harrington/TR306 (Mather et al., 1997) and the other from Steptoe/Morex (Hayes et al., 1993; Marquez-Cedillo et al., 1999). One apparent protein content QTL occurs on chromosome 5S(1HS), which was mapped in the same region as the hordein *(Hor)* loci from the Blenheim/E224-3 (Thomas et al., 1996), Dicktoo/Morex (Pan et al., 1994; Oziel et al., 1996), and Harrington/TR306 (Marquez-Cedillo et al., 1999) crosses. Two QTL were mapped on chromosome 6(6H): one on 6S(6HS) from Dicktoo/Morex (Pan et al., 1994; Oziel et al., 1996) and the other on 6L(6HL) from Steptoe/Morex (Han and Ullrich, 1994). At least three QTL regions were mapped on chromosome 7(5H). One QTL was detected in the 7S(5HS) telomere region from the Harrington/TR306 cross (Mather et al., 1997). Another QTL was mapped on 7S(5HS) near the centromere from the Dicktoo/Morex (Pan et al., 1994; Oziel et al., 1996), Steptoe/Morex (Hayes et al., 1993; Han and Ullrich, 1994; Marquez-Cedillo et al., 1999), and Harrington/Morex (Marquez-Cedillo et al., 1999) crosses. A third QTL was detected on 7L(5HL) from the Dicktoo/Morex (Pan et al., 1994) and Steptoe/Morex (Han and Ullrich, 1994) crosses. Considerable mapping activity has revealed concentrations of QTL on chromosomes 2(2H), 4(4H), and 7(5H) for protein content, but QTL were detected on all seven chromosomes, which is not surprising, since there are many different proteins and protein classes in barley grain. Many of the QTL described herein are visualized in a mapping review article in *Barley Genetics Newsletter* (Zale et al., 2000).

Grain lysine content in barley appears to be simply inherited, at least based on the considerable study of high-lysine mutants. After the discovery of the naturally occurring high-lysine and high-protein Ethiopian barley Hiproly (Munck et al., 1970), a plethora of high-lysine mutants were induced from chemical and radiation mutagens (Bansal, 1970; Ingversen, Køie, and Doll, 1973; Doll, Køie, and Eggum, 1974). Many of the mutants have been characterized biochemically and genetically (Doll, Køie, and Eggum, 1974; Brandt, 1976; Tallberg, 1982). The most-studied mutants are Hiproly [*lys* on chromosome 7(5H)] (Karlsson, 1976) and the 'Bomi' Risø mutant 1508 [*lys3* also on chromosome 7(5H)] (Ullrich and Eslick, 1978a). Although most of the high-lysine genes segregate as monofactorial recessives (*lys* genes) (Doll, 1976; Ullrich and Eslick, 1978a,b,c), pleiotropic effects are common, including altered endosperm protein components, reduced starch, and elevated free amino acids and lipid contents (Doll and Køie, 1978; Køie and Doll, 1979; Tallberg, 1982) compared to their progenitor varieties. The various biochemical alterations are manifested in morphological alterations as well, including increased embryo:endosperm ratio and shrunken endosperm (Tallberg, 1977; Doll and Køie, 1978), which further translate into reduced kernel weight and yield compared to their progenitors (Doll and Køie, 1978). The morphological effects of the high-lysine genes prompted investigation of the designated spontaneous shrunken endosperm mutants of barley (*seg* and *sex* genes), and it turns out that these mutants are high in lysine as well (Ullrich and Eslick, 1978b,c). Several of the designated shrunken endosperm mutants (especially *sex* types) are allelic to the designated high-lysine mutants *(lys)*, e.g., *sex3* = *lys3*, *sex5* = *lys4*, and *sex1* = *lys5* (Søgaard and Wettstein-Knowles, 1987). Despite the availability of various high-lysine gene sources, the associated shrunken endosperm and reduced yield problems have been difficult to overcome in breeding.

The proanthocyanidin-free trait in barley grain is inherited as a monofactorial recessive in all the various mutants studied genetically thus far (Jende-Strid, 1995). Only a few of the proanthocyanidin-free genes have been located to chromosomes, including two each on chromosomes 1(7H), 3(3H), and 6(6H) (Jende-Strid, 1995). The mutants most used for breeding primarily malting types have been ant 13 [chromosome 6(6H)], ant 17 [chromosome 3(3H)], and several of the pigmented mutants, including *ant* 26, 27, and 28.

Very little research has been conducted on the inheritance of barley grain mineral content with the exception of phosphorus. The discovery of low-phytic-acid mutants in barley has stimulated considerable study of their inheritance. More than 20 low-phytic-acid (lpa) mutants have been induced in barley (Raboy and Cook, 1999). These mutants appear to be one of two types. Homozygous lpa-1 forms have various levels of phytic acid P reduction, with a concomitant increase in inorganic P and no obvious change in total kernel P. In homozygous lpa-2 types, kernel phytic acid P is reduced, which is accompanied by increases in both inorganic P and lower inositol forms (organic) of P. Barley lpa-1 and lpa-2 were mapped to chromosomes 2 and 7, respectively (Larson et al., 1998).

Animal Performance Parameters

The relationships between animal performance characteristics and barley genes have received very little research attention. As indicated earlier, Bowman and Blake (1996), working with the Steptoe/Morex mapping population, determined that the same region of barley chromosome 3(3H) was associated with percent dry matter digestibility and percent starch digestibility of cracked barley grain in cattle, as was particle size of cracked or dry-rolled barley grain. In the same study, coincident QTL for percent dry matter digestibility and percent starch digestibility of ground barley grain were found on chromosomes 4(4H) and 5(1H). The chromosome 4(4H) QTL region is also coincident with a QTL for grain ADF content.

BREEDING FOR FEED QUALITY

As noted in the introduction, relatively little effort has focused on breeding barley for improved feed or nutritional quality per se compared to breeding for adaptation and malting quality. In this section, past and present feed quality breeding efforts will be reviewed and projections for the future of feed quality improvement will be made. Barriers to feed quality improvement have included a lack of consensus about high-priority traits to improve, the absence of mass screening techniques critical in evaluating large numbers of progeny, a dearth of genetic information about quality traits, and a lack of industry interest and support. Progress in overcoming at least some of these barriers is occurring, and the recent emphasis on creating value-added agricultural commodities is affecting barley feed quality improvement to some extent.

Considering the information presented here, numerous studies have established the existence of broad genetic diversity for most important feed quality characteristics, with the notable exception of grain lipid content. The surging biotechnology revolution has provided the materials and tools to begin to understand the genetics of complexly inherited traits (including those for quality), thereby allowing breeding to begin to proceed in a more efficient manner than in the past. Improvements in technology in general have allowed the development and improvement of some screening methods. Understanding trait diversity and unraveling the mystery of quality trait genetics have allowed for some prioritization of quality traits for breeders.

Screening and Selection Methods

Selection of simply measured traits can be done directly, such as with hull-less versus hulled (visual observation) and waxy versus normal nonwaxy (IKI solution). As measurements become more difficult, time-consuming, and expensive, alternative, usually indirect methods are often sought or employed. For example, starch, β-glucan, acid detergent fiber, and proanthocyanidins are quite difficult and expensive to measure. Near infrared reflectance (NIR) and near infrared transmission (NIT) have been used to measure indirectly characteristics such as starch, lipid, protein, β-glucan, and fiber content as well as some animal performance traits, such as digestibility (Helm, Sauer, and Helberg, 1996; Givens, Beaver, and Deaville, 1997; Oatway and Helm, 1999). A relatively recent breakthrough with this technology is the ability to use whole-grain samples (faster, nondestructive) versus ground samples (more time-consuming, destructive). A number of in situ and in vitro tests, including those described earlier, have been developed to circumvent full-scale feeding trials. With the development of molecular maps and the mapping of genes or chromosome regions associated with traits (QTL analysis), the identification of closely linked molecular markers allows for the development of molecular marker-assisted selection (MMAS). This perhaps is the ultimate indirect measure of a trait because actual genes are selected (genotypic selection), as opposed to conventional phenotypic selection (Paterson, 1998). Of course, as with any indirect method, thorough verification must be done in the development of the MMAS strategy. There are many types of molecular markers and most linkage maps utilize RFLPs (restriction fragment length polymorphisms). However, PCR (polymerase chain reaction)-based markers are more breeder friendly. PCR-based marker types include RAPDs (random amplified polymorphic DNAs), microsatellites or SSRs (simple sequence repeats), AFLPs (amplified fragment length polymorphisms), STSs (sequence tagged sites), ESTs (expressed sequence tags), etc. (Burow and Blake, 1998). Whereas DNA is required for MMAS, it need not necessarily be extracted and purified. PCR techniques in which plant tissue can be used directly are available (Clancy et al., 1996).

Simply Inherited Traits

Working with simply inherited traits generally means breeding and selection are simplified compared to working with complexly inherited traits. Selection for a single gene, with little or no environmental influence, generally makes life easy for the breeder, assuming that the trait is easily measured. A few examples will illustrate the point. Although the hull-less trait in barley controlled by the *nud* gene has been known since ancient times in the Eastern Hemisphere, it has been deployed relatively recently in North America, where interest and activity in hull-less barley breeding have increased in the past 25 to 30 years. During this time, a whole new industry has been developed in western Canada based on the release of hull-less barley, primarily for the swine industry (Bhatty, 1986, 1993, 1996). Although hull-less barley is established as a legitimate crop in western Canada, it is not in the western United States. One impetus for developing hull-less feed barley in western North America is a relative lack of maize production. Nevertheless, most barley breeding programs throughout North America have a hull-less barley component to their programs to exploit its potential in at least niche markets (Liu et al., 1996; Ullrich, 1997b; Paris et al., 1999). Often the hull-less trait is combined with the waxy trait, conditioned by the *wax* gene, and to some extent with the high-amylose trait, conditioned by *amo1*, giving combinations that have potential both in feed and food markets. An extensive and impressive group of "specialty starch" hull-less barley types have been developed at the Crop Development Centre of the University of Saskatchewan (B. G. Rossnagel, personal communication, 1998). The hull-less trait has also been used extensively in national and ICARDA (International Center for Agricultural Research in Dry Areas) breeding programs in and for developing nations in Latin America, Africa, and western and southern Asia and in East Asia for human food purposes (Ullrich, 1997a).

Low-phytic-acid barley breeding is already underway in the western United States and Canada. The lpa-1 type mutants are primarily used for crossing into elite hulled and hull-less germplasm (Raboy and Cook, 1999). These mutants reduce phytic acid P by 50 to 75 percent compared to the wild type, with little apparent effect on plant function. Tight linkage to two molecular markers on chromosome 2 allows a breeder to use MMAS effectively (Larson et al., 1998).

One of the greatest efforts, if not the most significant one, to improve the nutritional quality of barley through breeding has been the drive to develop high-lysine and, to a lesser extent, high-protein barley cultivars. There is plenty of justification for doing such, given the protein crisis in human nutrition, first realized in the 1960s, and the relatively high cost of protein supplements required in cereal-based livestock and poultry diets. However, after great investment of time and money in the 1970s and 1980s in a number of countries around the world, no high-lysine barley cultivar was ever released to farmers. The obstacles described in the previous genetics section could not be completely overcome. Still, considerable progress was made in breeding improved-yield, high-lysine barley (Persson and Karlsson, 1977; Seko and Kato, 1981; Ullrich et al., 1984; Bang-Olsen, Stilling, and Munck, 1987). In spite of at least ten high-lysine mutant genes available (Hiproly and Mutant 1508 were most used) and crossing to elite germplasm, reduced kernel weight and concomitant reduced grain yield could not be returned to acceptable commercial levels. Nutritional or feed quality was not the problem; it was agronomic performance. An extensive historical review of the high-lysine breeding saga was written by Munck (1992).

As can be seen in this and earlier sections, a number of important simply inherited traits have been developed via induced mutagenesis. Mutation breeding has been, and still is, a viable option for feed quality improvement and crop improvement in general. Diploid barley is particularly suited for induced mutagenesis. Konzak, Kleinhofs, and Ullrich (1984) and Ullrich (1997a) presented reviews on this subject.

Complexly Inherited Traits

The genetics of quantitative traits was largely a mystery until the advent of the application of molecular biology technology. Breeding directly for the improvement of quantitative feed quality attributes of barley, such as animal performance traits, has been largely nonexistent. There has been some limited evaluation of breeding lines by screening methods such as the mobile nylon bag technique in mid- to late generations before cultivar release (Honeyfield, Froseth, and Ullrich, 1990). The dual-purpose (malting and feed) cultivar Crest, released by Washington State University, was partially selected on the basis of mobile nylon bag digestibility data (Muir et al., 1992). However, typically, the majority of breeding programs, at most, have evaluated feed quality of advanced breeding lines just prior to cultivar release or even after release. The identification of feed quality genes, including QTL, through molecular means allows the deployment of genes heretofore unknown for the improvement of barley feed value via MMAS. Although QTL analysis data are beginning to accumulate for feed quality traits in barley, little application to breeding has occurred to date. An exception is recent molecular genetics and breeding work at Montana State University. Through the efforts of QTL mapping of barley grain yield, digestibility, and processing characteristics for cattle (Bowman and Blake, 1996), a molecular marker-assisted recombinant inbred line population was developed from a 'Lewis'/'Baronesse' cross. Both of these cultivars are well-adapted "feed" barleys in the northern Great Plains and Pacific Northwest. A range of feed quality and agronomic trait levels was expressed in the population, and from the population, a line was identified with good agronomic performance and a 10 percent increase in average daily weight gain in steers (versus Baronesse). This line was released in 1999 as the cultivar Valier, and it represents a cultivar developed using MMAS for both agronomic and feed quality traits (Blake and Bowman, 1999). Molecular marker-assisted selection may be the most feasible way of directed breeding for such bottom-line feed value traits as the animal performance traits of digestibility, rate of weight gain, and feed:gain ratio. The use of MMAS in cultivar development is beginning to pick up momentum in barley breeding and has been applied to other traits as well. The barley stripe rust (incited by *Puccinia striiformis* f. sp. *hordei*)-resistant cultivar Orca was developed by MMAS and released in 1998 (Hayes et al., 2000), but most cultivar development work is still in progress.

PROJECTIONS FOR THE FUTURE

It is always difficult and risky to look into and predict the future, but on the other hand, it is useful to make conjectures based on current trends in a work such as this. Obviously a genetics and breeding revolution is occurring, with tremendous advancements in knowledge and technology brought about in biology, chemistry, and computer sciences. Although a balance will likely be struck between the conventional and molecular approaches, molecular approaches should continue to dominate the changes and the way plant breeders work to develop new cultivars. In crop plant species in general, most economically important traits, such as adaptation, agronomic characteristics, and quality traits, are complex and quantitatively inherited. This means that plant breeders traditionally have not had very useful information about the genes that control these traits. Until the advent of molecular linkage maps and QTL analysis, breeding for quantitative traits could not be very direct, and this is certainly the case with barley feed quality traits. Now, as pointed out earlier, we are poised to make greater strides in quantitative trait crop improvement through the deployment of the genes that actually control traits of interest through MMAS. This has begun to happen in the improvement of feed quality in barley, and hopefully the trend will continue and accelerate. However, molecular genetics has not given plant breeders a "magic bullet." There is still much to learn about the location of genes, their interactions with one another and the environment, and how to combine them in the best-designed cultivars. Even though barley is a diploid, and thereby relatively easy to work with genetically, it does have a very large genome that consists of approximately 5 billion DNA base pairs and probably about 25,000 genes, but the 25,000 genes make up less than 5 percent of the barley genome. Thus far, less than 10 percent of the total number of genes, occupying perhaps 0.5 percent of the genome, have been identified (Kleinhofs, 1999).

Advancements in molecular marker technology have been moving rapidly in terms of marker type and ease of marker use, and by combining DNA and computer chip technology. Methods for analyzing hundreds or even thousands of genes at one time have recently been developed. DNA chip technology utilizes microscopic arrays (microarrays) of molecules immobilized on solid surfaces for biochemical analysis. Advanced arraying technologies such as photolithography, microspotting, and ink jetting, coupled with sophisticated fluorescence detection systems and bioinformatics, permit molecular data gathering at an unprecedented rate (Lemieux, Aharoni, and Schena, 1998). Microarrays, for example, can be constructed by arraying PCR-amplified DNA samples on a glass slide or synthesizing nucleotide oligomers of known sequences directly on a microchip. The DNA could then be hybridized by probes to identify a large number of genes or completely genotype individuals. As an illustration, all of the 6,000+ yeast genes have been placed on a single glass slide. This type of microarray could be hybridized with fluorescent-labeled messenter (m) RNA probes to determine which genes are expressed at a specific time and under specific environmental conditions in the life cycle of the organism (DeRisi, Iyer, and Brown, 1997; Lashkari, DeRisi, and McCusker, 1997; Eisen et al., 1998). Microarrays can be used for gene expression analysis, polymorphism detection, DNA resequencing, and genotyping on a genomic scale (Lemieux, Aharoni, and Schena, 1998). Microarray-based characterization of plant genomes has the potential to revolutionize plant breeding and agricultural biotechnology. This technology requires knowledge about the genes in a genome and their sequences. This knowledge can be gained one gene at a time, as is currently being done in barley, or, alternatively, by sequencing all of the barley genes in one project by physically mapping all of the barley gene-rich chromosome regions (Künzel and Korzun, 1996; Barakat, Corels, and Bernardi, 1997). This could be done by utilizing the barley bacterial artificial chromosome (BAC) library. This technology is currently expensive and requires considerably more knowledge than is presently available for barley, but already modest attempts are being made at using the library, and it will be viable for barley improvement in the near future.

Another apsect not previously considered is the role of genetic transformation. Certainly transformation technology in barley, as in the other cereals, lags behind dicot crop species, but that is changing rapidly. Great progress has been made in the years since the first routinely successful barley transformation protocol was published (Wan and Lamaux, 1994). The topic of barley transformation is competently reviewed by Horvath et al. in this book (see Chapter 7). In their chapter, the authors describe the transformation of barley with an engineered bacterial heat-stable β-glucanase gene, which has potential brewing and feed quality benefits. There is a great potential for feed quality improvement via transformation, given an essentially unlimited sounce of genes for the purpose. Of course, most of those genes have yet to be discovered, isolated, and/or cloned, but these things will happen in the future. However, at this writing, in spite of the advancements, the presence of transgenic crops and crop products in the marketplace is being hotly contested and debated. This brings sociological, cultural, and political aspects into the picture, removes somewhat the biological science involved, and does cloud the future. It is likely that this cloud will dissipate with time and the sure acquisition of new knowledge.

Newman and Newman (1992a) attempted to describe the ideal feed barley but concluded that this cannot be done, considering differences in livestock and poultry species, age groups, and production goals or requirements. However, since barley grain is usually considered an energy source, a common feed goal would be to increase available energy. The Newmans' formula included high levels of starch and free sugars (>76 percent), high fat content (4 to 5 percent), and low nonstarch polysaccharide (NSP) (<12 percent) levels. The NSPs or fiber include primarily cellulose and lignin (ADF), β-glucans, and arabinoxylans in barley grain. This formula is compatible with malting and brewing quality as well. However, there is also the matter of starch, sugar, lipid, and fiber types and their extent and rates of digestibility. The Newmans also considered protein quality in their formula, but that was when there was still some optimism about breeding commercial high-lysine barley. They did not think the impact of phenolics or phytate was critical to the ideal barley formula. This could change, especially with the induction of low-phytate mutants in barley. There is no simple answer or approach to devising or developing an ideal feed barley cultivar, but hopefully this chapter has provided some insight into the situation of breeding barley that is at least improved in feed quality. Furthermore, hopefully, the feed and animal industries will come to appreciate, encourage, and support improved feed barleys.

REFERENCES

Aastrup, S. (1983). Selection and characterization of low β-glucan mutants from barley. *Carlsberg Research Communications* 48:307-316.

Backes, G., A. Graner, B. Foroughi-Wehr, G. Fischbeck, G. Wenzel, and A. Jahoor (1995). Localization of quantitative trait loci (QTL) for agronomic important characters by use of a RFLP map in barley (*Hordeum vulgare* L.). *Theoretical and Applied Genetics* 90:294-302.

Bamforth, C.W. (1982). Barley β-glucans: Their role in malting and brewing. *Brewers Digest* 57:22-27.

Bamforth, C.W. and A.H.P. Barclay (1993). Malting technology and uses of malt. In *Barley: Chemistry and Technology,* eds. A.W. MacGregor and R.S. Bhatty. St. Paul, MN: American Association of Cereal Chemists, pp. 297-354.

Bang-Olsen, K., B. Stilling, and L. Munck (1987). Breeding for yield in high-lysine barley. In *Barley Genetics V. Proceedings of the Fifth International Barley Genetics Symposium,* eds. S. Yasuda and T. Konishi. Okayama, Japan: Sanyo Press, pp. 865-870.

Bansal, H.C. (1970). A new mutant induced in barley. *Current Science* 39:494.

Barakat, A., N. Corels, and G. Bernardi (1997). The distribution of genes in the genomes of Gramineae. *Proceedings of the National Academy of Sciences, USA* 94:6857-6861.

Bell, G.D.H. and F.G.H. Lupton (1962). The breeding of barley varieties. In *Barley and Malt,* ed. A.H. Cook. New York: Academic Press, pp. 45-96.

Bhatty, R.S. (1986). The potential of hull-less barley: A review. *Cereal Chemistry* 63:97-103.

Bhatty, R.S. (1993). Nonmalting uses of barley. In *Barley: Chemistry and Technology,* eds. A.W. MacGregor and R.S. Bhatty. St. Paul, MN: American Association of Cereal Chemists, pp. 355-417.

Bhatty, R.S. (1996). Hullless barley: Development and utilization. In *Proceedings V International Oat Conference and VII International Barley Genetics Symposium, Invited Papers,* eds. G. Scoles and B. Rossnagel. Saskatoon: University of Saskatchewan Extension Press, pp. 106-112.

Blake, T.K. and J.G.P. Bowman (1999). Building a better feed barley for beef cattle. In *Proceedings of the 32nd Barley Improvement Conference.* Milwaukee, WI: American Malting Barley Association, pp. 11-18.

Boisen, S. and B.O. Eggum (1991). Critical evaluation of in vitro methods for estimating digestibility in simple stomach animals. *Nutrition Research Review* 4:141-162.

Bowman, J. and T.K. Blake (1996). Barley feed quality for beef cattle. In *Proceedings of the V International Oat Conference and VII International Barley Genetics Symposium, Invited Papers,* eds. G. Scoles and B. Rossnagel. Saskatoon: University of Saskatchewan Extension Press, pp. 82-90.

Brandt, A. (1976). Endosperm protein formation during kernel development of wild type and a high-lysine barley mutant. *Cereal Chemistry* 53:890-901.

Burow, M.D. and T.K. Blake (1998). Molecular tools for the study of complex traits. In *Molecular Dissection of Complex Traits,* ed. A.H. Paterson. New York: CRC Press, pp. 13-30.

Campbell, Z.G.L., R.E Salmon, and H.L. Classen (1986). Prediction of metabolizable energy of broiler diets from chemical analysis. *Poultry Science* 65:2126-2134.

Clancy, J.A., V.A. Jitkov, F. Han, and S.E. Ullrich (1996). Barley tissue as direct template for PCR: A practical breeding tool. *Molecular Breeding* 2:181-183.

DeRisi, J.L., V.R. Iyer, and P.O. Brown (1997). Exploring the metabolic and genetic control of gene expression on a genomic scale. *Science* 278:680-686.

Doll, H. (1976). Genetic studies of high-lysine barley mutants. In *Barley Genetics III. Proceedings of the Third International Barley Genetics Symposium,* ed. H. Gaul. Munich: Verlag Karl Thiemig, pp. 542-546.

Doll, H. and B. Køie (1978). Influence of the high-lysine gene from barley mutant 1508 on grain carbohydrate and protein yield. In *Seed Protein Improvement by Nuclear Techniques.* Vienna: IAEA, pp. 107-114.

Doll, H., B. Køie, and B.O. Eggum (1974). Induced high-lysine mutants of barley. *Radiation Botany* 14:73-80.

Duncan, R.W., J.R. Males, M.L. Nelson, and E.L. Martin (1991). Corn and barley mixtures in finishing steer diets containing potatoe processing residue. *Journal of Production Agriculture* 4:426-432.

Eisen, M.B., P.T. Spellman, P.O. Brown, and D. Botstein (1998). Cluster analysis and display of genome-wide expresion patterns. *Proceedings of the National Academy of Sciences, USA* 95:14863-14868.

Ertl, D., K.A. Young, and V. Raboy (1998). Plant genetic approaches to phosphorus management in agricultural production. *Journal of Environmental Quality* 27:299-304.

Fadel, J.G., R.K. Newman, C.W. Newman, and A.E. Barnes (1987). Hypocholesterolemic effects of beta-glucans in different barley diets fed to broiler chicks. *Nutrition Reports International* 35:1049-1058.

Givens, D.I., J.L. Beaver, and E.R. Deaville (1997). The principles, practices, and some future applications of near infrared spectroscopy for predicting the nutritive value of foods for animals and humans. *Nutrition Research Review* 10:83-114.

Han, M.S. (1994). Functional and nutritional significance of dietary fiber in barley for pigs and the effect of enzyme supplementation. PhD Dissertation. Washington State University, Pullman, WA.

Han, M.S. and J.A. Froseth (1992). Composition of pearling fractions of barleys with normal and waxy starch. *Proceedings Western Section American Society of Animal Science* 43:155-158.

Han, M. S., J. A. Froseth, and S. E. Ullrich (1993). Simplification of a method to predict the feeding value of barley for swine. *Barley Newsletter* 36:149-152.

Han, F. and S.E. Ullrich (1994). Mapping of quantitative trait loci associated with malting quality in barley. *Barley Genetics Newsletter* 23:84-97.

Han, F., S.E. Ullrich, S. Chirat, S. Menteur, L. Jestin, A. Sarrafi, P.M. Hayes, B.L. Jones, T.K. Blake, D.M. Wesenberg, A. Kleinhofs, and A. Kilian (1995). Mapping of β-glucan content and β-glucanase activity loci in barley grain and malt. *Theoretical and Applied Genetics* 91:921-927.

Hayes, P.M., J. Cerono, H. Witsenboer, M. Kuiper, M. Zabeau, K. Sato, A. Kleinhofs, D. Kudrna, A. Kilian, M. Saghai-Maroof, et al. (1997). Characterizing and exploiting genetic diversity and quantitative traits in barley *(Hordeum vulgare)* using AFLP markers. *Journal of Quantitative Trait Loci,* online: <http://probe.nalusda.gov:8000/otherdocs/jqtl/jqtl1997-02/>.

Hayes, P.M., A.E. Corey, R. Dovel, R. Karow, C. Mundt, K. Rhinart, and H. Vivar (2000). Registration of 'Orca' barley. *Crop Science* 40:849.

Hayes, P.M., B.H. Liu, S.J. Knapp, F. Chen, B. Jones, T. Blake, J. Franckowiak, D. Rasmusson, M. Sorrells, S.E. Ullrich, et al. (1993). Quantitative trait locus effects and environmental interaction in a sample of North American barley germplasm. *Theoretical and Applied Genetics* 87:392-401.

Helm, J.H., W.C. Sauer, and L.A. Helberg (1996). The use of the mobile nylon bag technique and near infrared reflectance to determine digestible energy and protein content of hulless barley for pigs. In *Proceedings V International Oat Conference and VII International Barley Genetics Symposium, Poster Sessions,*

Volume 1, eds. A. Slinkard, G. Scoles, and B. Rossnagel. Saskatoon: University of Saskatchewan Extension Press, pp. 123-125.

Hockett, E.A. and L.M. White (1981). Simultaneous breeding for feed and malting quality. In *Barley Genetics IV. Proceeding of the Fourth International Barley Genetics Symposium,* ed. M. Asher. Edinburgh, UK: Edinburgh University Press, pp. 234-241.

Honeyfield, D. C., J. A. Froseth, and S. E. Ullrich (1990). Use of the mobile nylon bag technique to determine digestible energy of new barley breeding lines early in the WSU cultivar development program. *Barley Newsletter* 33:201-206.

Ingversen, J., B. Køie, and H. Doll (1973). Induced seed protein mutant of barley. *Experientia* 29:1151-1152.

Jende-Strid, B. (1991). A new type of proanthocyanidin-free barley. In *Barley Genetics VI. Proceedings of the Sixth International Barley Genetics Symposium,* Volume I: *Short Papers,* ed. L. Munck. Copenhagen: Muncksgaard International Publishing, pp. 504-506.

Jende-Strid, B. (1993). Genetic control of flavonoid biosynthesis in barley. *Hereditas* 119:187-204.

Jende-Strid, B. (1995). Coordinators report: Anthocyanin genes. *Barley Genetics Newsletter* 24:162-165.

Jende-Strid, B. (1998). Coordinators report: Anthocyanin genes. *Barley Genetics Newsletter* 28:103.

Jensen, J. (1989). Estimation of recombination parameters between a quantitative trait locus (QTL) and two marker loci. *Theoretical and Applied Genetics* 78:613-618.

Karlsson, K.E. (1976). Linkage studies on the *lys*-gene in relation to some marker genes and translocations. In *Barley Genetics III. Proceedings of the Third International Barley Genetics Symposium,* ed. H. Gaul. Munich: Verlag Karl Thiemig, pp. 536-541.

Kasha, K.J., A. Kleinhofs, A. Kilian, M. Saghai-Maroof, G.J. Scoles, P.M. Hayes, F.Q. Chen, X. Xia, X.-Z. Li, R.M. Biyashev, et al. (1995). The North American Barley Genome Map on the cross HT and its comparison to the map on cross SM. In *Plant Genome and Plastome: Their Structure and Evolution,* ed. K. Tsunewaki. Tokyo: Kodansha Scientific, pp. 73-88.

Kleinhofs, A. (1999). Barley genome mapping: Where are we and where are we going. In *Proceedings of the 32nd Barley Improvement Conference.* Milwaukee, WI: American Malting Barley Association, pp. 1-4.

Kleinhofs, A., A. Kilian, M.A. Saghai-Maroof, R.M. Biyashev, P. Hayes, F.Q. Chen, N. Lapitan, A. Fenwick, T.K. Blake, V. Kanazin, et al. (1993). A molecular, isozyme and morphological map of the barley *(Hordeum vulgare)* genome. *Theoretical and Applied Genetics* 86:705-712.

Køie, B. and H. Doll (1979). Protein and carbohydrate components in the Risø high-lysine barley mutants. In *Seed Protein Improvements in Cereals and Grain Legumes.* Vienna: IAEA, pp. 205-215.

Konzak, C.F., A. Kleinhofs, and S.E. Ullrich (1984). Induced mutations in seed-propagated crops. In *Plant Breeding Reviews,* Volume 2, ed. J. Janick. Westport, CT: AVI Publishing, pp. 13-72.

Künzel, G. and L. Korzun (1996). Physical mapping of cereal chromosomes, with special emphasis on barley. In *Proceedings V International Oat Conference and VII International Barley Genetics Symposium, Invited Papers,* eds. G. Scoles and B. Rossnagel. Saskatoon: University of Saskatchewan Extension Press, pp. 197-206.

Larson, S.R., D.K. Habernicht, T.K. Blake, and M. Adamson (1997). Backcross gains for six-rowed grain and malt qualities with introgression with a feed barley yield QTL. *Journal of the American Society of Brewing Chemists* 55:52-57.

Larson, S.R., K.A. Young, A. Cook, T.K. Blake, and V. Raboy (1998). Linkage mapping of two mutations that reduce phytic acid content of barley grain. *Theoretical and Applied Genetics* 97:141-146.

Lashkari, D.A., J.L. DeRisi, and J.H. McCusker (1997). Yeast microarrays for genome wide parallel genetic and gene expression analysis. *Proceedings of the Naitonal Academy of Sciences, USA* 94:13057-13062.

Lemieux, B., A. Aharoni, and M. Schena (1998). Overview of DNA chip technology. *Molecular Breeding* 4:277-289.

Liu, C.T., D.M. Wesenberg, C.W. Hunt, A.L. Branen, L.D. Robertson, D.E. Burrup, K.L. Dempster, and R.J. Haggerty (1996). *Hulless Barley: A New Look for Barley in Idaho.* Moscow: University of Idaho Cooperative Extension Current Information Series 1050.

Marquez-Cedillo, L.A., P.M. Hayes, B.L. Jones, A. Kleinhofs, W.G. Legge, B.G. Rossnagel, K. Sato, S.E. Ullrich, D.M. Wesenberg, and the NABGMP (1999). QTL analysis of malting quality in barley based on the doubled haploid progeny of two elite North American varieties representing different germplasm groups. *Theoretical and Applied Genetics* 101:173-184.

Mather, D.I., N.A. Tinker, D.E. LaBerge, M. Edney, B.L. Jones, B.G. Rossnagel, W.G. Legge, K.G. Briggs, R.B. Irvine, D.E. Falk, and K.J. Kasha (1997). Regions of the genome that affect grain and malt quality in a North American two-row barley cross. *Crop Science* 37:544-554.

Merritt, N.R. (1967). A new strain of barley with starch of high amylose content. *Journal of the Institute of Brewing* 73:583-585.

Molina-Cano, J.L., M. Francesch, A.M. Perez-Vendrell, T. Ramo, J. Voltas, and J. Brufau (1997). Genetic and environmental variation in malting and feed quality of barley. *Journal of Cereal Science* 25:37-47.

Molina-Cano, J.L., F. Roca de Togores, C. Royo, and A. Pérez (1989). Fast germinating low β-glucan mutants induced in barley with improved malting quality and yield. *Theoretical and Applied Genetics* 78:748-754.

Muir, C.E., R.A. Nilan, S.E. Ullrich, J.A. Forseth, and B.C. Miller (1992). Registration of 'Crest' barley. *Crop Science* 32:1506-1507.

Munck, L. (1992). The case of high-lysine barley breeding. In *Barley: Genetics, Biochemistry, Molecular Biology and Biotechnology,* ed. P.R. Shewry. Wallingford, UK: CAB International, pp. 573-601.

Munck, L., K.E. Karlsson, A. Hagberg, and B.O. Eggum (1970). Gene for improved nutritional value in barley seed protein. *Science* 168:985-987.

Newman, C.W. and C.F. McGuire (1985). Nutritional quality of barley. In *Barley*, ed. D.C. Rasmusson. Madison, WI: American Society of Agronomy, pp. 403-456.

Newman, C.W. and R.K. Newman (1990). Improved performance of weanling pigs fed a low β-glucan barley. *Proceedings Western Section American Society of Animal Science* 41:223-226.

Newman, C.W. and R.K. Newman (1992a). Characteristics of the ideal barley for feed. In *Barley Genetics VI. Proceedings of the Sixth International Barley Genetics Symposium*, Volume II, ed. L. Munck. Copenhagen: Muncksgaard International Publishing, pp. 925-939.

Newman, C.W. and R.K. Newman (1992b). Nutritional aspects of barley seed structure and composition. In *Barley: Genetics, Biochemistry, Molecular Biology and Biotechnology*, ed. P.R. Shewry. Wallingford, UK: CAB International, pp. 351-368.

Newman, C.W., R.K. Newman, K. Bolin-Heintzman, N.J. Roth, and E.A. Hockett (1987). Factors affecting protein utilization in proanthocyanidin-free barley. In *Barley Genetics V. Proceedings of the Fifth International Barley Genetics Symposium*, eds. S. Yasuda and T Konishi. Okayama, Japan: Sanyo Press, pp. 833-840.

Newman, R.K., C.W. Newman, A.M. El-Negoumy, and S. Aastrup (1984). Nutritional quality of proanthocyanidin-free barley. *Nutrition Reports International* 30:809-816.

Newman, R.K., C.W. Newman, and H. Graham (1989). The hypocholesterolemic function of barley β-glucans. *Cereal Foods World* 34:883-886.

Nilan, R.A. (1964). *The Cytology and Genetics of Barley 1951-1962*. Pullman: Washington State University Press.

Nocek, J.E. (1988). In situ and other methods to estimate ruminal protein and energy digestibility: A review. *Journal of Dairy Science* 71:2051-2069.

Oakenfull, D. (1996). Food applications for barley. In *Proceedings V International Oat Conference and VII International Barley Genetics Symposium, Invited Papers*, eds. G. Scoles and B. Rossnagel. Saskatoon: University of Saskatchewan Extension Press, pp. 50-57.

Oatway, L.A. and J.H. Helm (1999). The use of near infrared reflectance spectroscopy to determine quality characteristics in whole grain barley. *Barley Newsletter* 42, online: <http://wheat.pw.usda.gov/ggpages/barleynewsletter/>.

Ovenell-Roy, K.H., M.L. Nelson, J.A. Froseth, S.M. Parish, and E.L. Martin (1998). Variation in chemical composition and nutritional quality among barley cultivars for ruminants. II. Digestion, ruminal characteristics and in situ disappearance kinetics. *Canadian Journal of Animal Science* 78:377-388.

Øverland, M., K.B. Heintzman, C.W. Newman, R.K. Newman, and S.E. Ullrich (1994). Chemical composition and physical characteristics of proanthocyanidin-free and normal barley isotypes. *Journal of Cereal Science* 20:85-91.

Owen, B.D., F. Sosulski, K.K. Wu, and J.J. Farmer (1977). Variation in mineral content in Saskatchewan feed grains. *Canadian Journal of Animal Science* 57:679-687.

Oziel, A., P.M. Hayes, F.Q. Chen, and B.L. Jones (1996). Application of quantitative trait locus mapping to the development of winter-habit malting barley. *Plant Breeding* 115:43-51.

Pan, A., P.M. Hayes, F. Chen, H.H. Chen, T. Blake, S. Wright, I. Karsai, and Z. Bedo (1994). Genetic analysis of the components of winterhardiness in barley. *Theoretical and Applied Genetics* 89:900-910.

Paris, R.L., M.E. Vaughn, C.A. Griffey, and J.M. Harter-Dennis (1999). Development of hullless barley varieties as an improved feed crop. *Barley Newsletter* 42, online: <http://wheat.pw.usda.gov/ggpages/barleynewsletter/>.

Paterson, A.H. (1998). QTL mapping in DNA marker assisted plant and animal improvement. In *Molecular Dissection of Complex Traits,* ed. A.H. Paterson. New York: CRC Press, pp. 131-144.

Persson, G. and K.E. Karlsson (1977). Progress in breeding for improved nutritive value in barley. *Cereal Research Communications* 5:169-178.

Powell, W., P.D.S. Caligari, J.S. Swanston, and J.L. Jinks (1989). Genetic investigations into beta-glucan content in barley. *Theoretical and Applied Genetics* 71:461-466.

Powell, W., R.P. Ellis, M. McCaulay, J. McNichol, and R.P. Forster (1990). The effect of selection for protein and isozyme loci on quantitative traits in a doubled haploid population of barley. *Heredity* 65:115-122.

Powell, W., R.P. Ellis, and W.T.B. Thomas (1990). The effects of major genes on quantitatively varying characters in barley. III. The 2 row and 6 row locus (V-v). *Heredity* 65:259-264.

Price, P.B. and J. Parsons (1979). Distribution of lipids in embryonic axis, bran-endosperm, and full fractions of hulless barley and hullless oat grain. *Agriculture and Food Chemistry* 27:813-815.

Que, Q., L. Wang, R.K. Newman, C.W. Newman, and H. Graham (1997). Influence of the hullless, waxy starch and short-awn genes on the composition of barleys. *Journal of Cereal Science* 26:251-257.

Raboy, V. and A. Cook (1999). An update on ARS barley low phytic acid research. *Barley Genetics Newsletter* 29, online: <http://wheat.pw.usda.gov/ggpages/bgn/>.

Raboy, V. and P. Gerbasi (1996). Genetics of myo-inositol phosphate synthesis and accumulation. In *Myoinositol Phosphates, Phosphoinositides, and Signal Transduction,* ed. B.B. Biswas. New York: Plenum, pp. 257-285.

Rasmussen, S.K. and F. Hatzack (1998). Identification of two low-phytate barley (*Hordeum vulgare* L.) grain mutants by TLC and genetic analysis. *Hereditas* 129:107-112.

Sauer, W.C., H. Jørgenson, and R. Berzins (1983). A modified nylon bag technique for determining apparent digestibilities of protein in feedstuffs for pigs. *Canadian Journal of Animal Science* 63:233-237.

Seko, H. and I. Kato (1981). Breeding for high-lysine hull-less barley. In *Barley Genetics IV. Proceedings of the Fourth International Barley Genetics Symposium,* ed. M. Asher. Edinburgh, UK: Edinburgh University Press, pp. 336-340.

Søgaard, B. and P. von Wettstein-Knowles (1987). Barley: Genes and chromosomes. *Carlsberg Research Communications* 52:123-196.

Swanston, J.S. (1997). Waxy starch barley genotypes with reduced β-glucan contents. *Cereal Chemistry* 74:442-445.

Swanston, J.S., R.P. Ellis, and S.A. Tiller (1997). Effects of the waxy and high amylose genes on the total beta-glucan and extractable starch. *Barley Genetics Newsletter* 27:72-74.

Tallberg, A. (1977). The amino acid composition in endosperm and embryo of a barley variety and its high-lysine mutant. *Hereditas* 87:43-46.

Tallberg, A. (1982). Characterization of high-lysine barley genotypes. *Hereditas* 96:229-245.

Thomas, W.T.B., W. Powell, and J.S. Swanston (1991). The effects of major genes on quantitatively varying characters in barley. 4. The GP ert and denso loci and quality characters. *Heredity* 66:381-389.

Thomas, W.T.B., W. Powell, J.S. Swanston, R.P. Ellis, K.J. Chalmers, U.M. Barua, P. Jack, V. Lea, B.P. Forster, R. Waugh, and D.B. Smith (1996). Quantitative trait loci for germination and malting quality characters in a spring barley cross. *Crop Science* 36:265-273.

Thomas, W.T.B., W. Powell, R. Waugh, K.J. Chalmers, U.M. Barua, P. Jack, V. Lea, B.P. Forster, J.S. Swanston, R.P. Ellis, et al. (1995). Detection of quantitative trait loci for agronomic, yield, grain and disease characters in spring barley (*Hordeum vulgare* L.) *Theoretical and Applied Genetics* 91:1037-1947.

Tilley, J.M.A. and R.A. Terry (1963). A two-stage technique for measuring the in vitro digestibility of forage crops. *Journal of the British Grassland Society* 18:104-111.

Tinker, N.A., D.E. Mather, B.G. Rossnagel, K.J. Kasha, A. Kleinhofs, P.M. Hayes, D.E. Falk, L.P. Ferguson, L.P. Shugar, W.G. Legg, et al. (1996). Regions of the genome that affect agronomic performance in two-row barley. *Crop Science* 36:1053-1062.

Torp, J., H. Doll, and V. Haahr (1981). Genotypic and environmental influence upon the nutritional composition of barley grain. *Euphytica* 30:719-728.

Ullrich, S.E. (1997a). Barley improvement: An evolutionary perspective. In *Crop Improvement for the 21st Century,* ed. M.S. Kang. Trivandrum, India: Research Signpost, pp. 165-192.

Ullrich, S.E. (1997b). Two new niche market barley cultivars released. *Wheat Life* 40(5):45-46.

Ullrich, S.E., J.A. Clancy, R.F. Eslick, and R.C.M. Lance (1986). Beta-glucan content and viscosity of waxy barley. *Journal of Cereal Science* 4:279-285.

Ullrich, S.E., C.N. Coon, and J.M. Sever (1981). Relationships of nutritional and malting quality traits of barley. In *Barley Genetics IV. Proceedings of the Fourth International Barley Genetics Symposium,* ed. M. Asher. Edinburgh, U.K: Edinburgh University Press, pp. 225-233.

Ullrich, S. E. and R. F. Eslick (1978a). Inheritance of the associated kernel characters, high-lysine and shrunken endosperm of the barley mutant Bomi, Risø 1508. *Crop Science* 18:828-831.

Ullrich, S. E. and R. F. Eslick (1978b). Lysine and protein characterization of induced shrunken endosperm mutants of barley. *Crop Science* 18:963-966.

Ullrich, S.E. and R.F. Eslick (1978c). Lysine and protein characterization of spontaneous shrunken endosperm mutants of barley. *Crop Science* 18:809-812.

Ullrich, S.E., F. Han, J.A. Froseth, B.L. Jones, C.W. Newman, D.M. Wesenberg, and the NABGMP (1996). Mapping of loci that affect carbohydrate content in barley grain. In *Proceedings V International Oat Conference and VII International Barley Genetics Symposium, Poster Sessions,* Volume I. eds. A. Slinkard, G. Scoles, B. Rossnagel. Saskatoon: University of Saskatchewan Extension Press, pp. 141-143.

Ullrich, S.E., P.M. Hayes, W.E. Dyer, T.K. Blake, and J.A. Clancy (1993). Quantitative trait locus analysis of seed dormancy in "Steptoe" barley. eds. M.K. Walker-Simmons and J.L. Reid. In *Preharvest sprouting in cereals 1992.* St. Paul, MN:American Association of Cereal Chemists, pp. 136-145.

Ullrich, S.E., A. Kleinhofs, C.N. Coon, and R.A. Nilan (1984). Breeding for improved protein in barley. In *Cereal Grain Protein Improvement,* ST1/PUB/664. Vienna: IAEA, pp. 93-104.

Wan, Y. and P.G. Lamaux (1994). Generation of large numbers of independently transformed fertile barley plants. *Plant Physiology* 104:37-48.

Wettstein, D. von, R.A. Nilan, B. Ahrenst-Larsen, K. Erdal, J. Ingversen, B. Jende-Stride, K. Nyegaard Kristiansen, J. Larsen, H. Outtrup, and S.E. Ullrich (1985). Proanthocyanidin-free barley for brewing: Progress in breeding for high yield and research tool in polyphenol chemistry. *Technical Quarterly of the Master Brewers Association of the Americas* 22: 41-52.

Zale, J.M., J.A. Clancy, B.L. Jones, P.M. Hayes, and S.E. Ullrich (2000). Summary of barley malting quality QTLs mapped in various populations. *Barley Genetics Newsletter* 30, online: <http://wheat.pw.usda.gov/ggpages/bgn/>.

Chapter 7

Experiences with Genetic Transformation of Barley and Characteristics of Transgenic Plants

Henriette Horvath
Jintai Huang
Oi T. Wong
Diter von Wettstein

INTRODUCTION

A comprehensive review on genetic transformation of barley (*Hordeum vulgare* L.) has recently been published by Peggy G. Lemaux and colleagues in *Molecular Improvement of Cereals* (1999). It traces the history of experiments to transform barley, summarizes all successful reports of stable and transient transformations, and emphasizes the importance of achieving efficient regeneration of transformed callus tissue into somatic embryos and these into mature plants. It focuses on efforts to achieve transformation of cultivars other than 'Golden Promise,' which at present is the host genotype, giving frequencies of transformed plants (4 to 6 percent) without the requirement of testing a very large number of bombarded callus cultures. At a low frequency (less than 1 percent of bombarded scutella of immature zygotic embryos), transformants were obtained from the cultivars Haruna Nijo and Dissa (Hagio et al., 1995) or by careful adjustment of bombardment conditions from cultivars Dera, Corniche, Salome, and Femina (Koprek et al., 1996). Successful regeneration of transformants was further reported for the winter barley cultivar Igri (Lührs and Lörz, 1987), for four 'Galena' lines transformed with the genes encoding β-glucuronidase *(uidA)*

Financial support for the original research reported in this chapter by CSREES/NRICPG Grant 9502390 and by Applied Phytologics, Inc., Sacramento, California, is gratefully acknowledged. We would like to thank Dr. Norbert Wolf for supplying plasmid pAM470 (Figure 7.5). This is scientific paper 0800-07 from the College of Agriculture and Home Economics Research Center, Washington State University, Pullman, Washington.

and phosphinotricin acetyl transferase *(bar)*, as well as for two 'Harrington' lines expressing the *uidA* gene and the gene encoding hygromycin phosphotransferase *(hpt)* (Cho, Jiang, and Lemaux, 1998). However, transformation efficiencies were lower than 1 percent.

A chimaeric transgenic barley plant originated from particle bombardment of the embryonic axis of an immature embryo of 'Kymppi,' an elite barley cultivar in Finland. The plasmids used contain the gene encoding neomycin phosphotransferase *(nptI)* and the *uidA* gene (Ritala et al., 1994). One out of 277 plantlets proved to be NPTII-positive, was grown to maturity, and segregated in the T_2 progeny in a Mendelian ratio. In a different approach, protoplasts of the barley cultivar Igri were derived from embryogenic cells grown in suspension culture and transformed by bombardment. Regenerated transgenic plants carried the neomycin phosphotransferase II (NPTII) gene, which conferred resistance to kanamycin (Funatsuki et al., 1995). Protoplasts from primary calli of barley were transformed with the neomycin (G418) resistance gene using a polyethylene glycol (PEG) DNA uptake method (Kihara, Saeki, and Ito, 1998). Two plants were regenerated from these calli and NPTII expression shown. These results illustrate that protoplasts can be used as a starting material for the generation of transgenic barley.

Sonja Tingay developed a highly efficient and reliable method for the production of stable barley transformants by cocultivation of immature barley embryos with disarmed *Agrobacterium tumefaciens* (Tingay et al., 1997) (cf. Photo 7.1 and Figure 7.2).

It can thus be stated that several transformation methods that are routinely applied to dicotyledonous plants can be used with barley. One can expect improvements in the efficiency of transformation as well as a wider range of good host varieties to emerge from further experimentation.

In this chapter we present initially two protocols that are used for barley transformation—the biolistic method and the *Agrobacterium*-mediated procedure—and review techniques that are useful in analyzing transgenic barley plants at the molecular level in breeding programs as well as for basic research. In another part of our presentation, we describe what has been learned from studying the inheritance of the transgenes and their behavior in field experiments.

BARLEY TRANSFORMATION OF IMMATURE ZYGOTIC EMBRYOS BY THE BIOLISTIC METHOD

Barley and other small grain cereals were first transformed with the particle gun, as illustrated by the successful transformation of rice (Li et al., 1993), wheat (Vasil et al., 1992), oats (Somers et al., 1992; Pawlowski and Somers, 1998), rye (Castillo, Vasil, and Vasil, 1994), and barley (Wan and Lemaux, 1994). Yuechun Wan and Peggy Lemaux (1994) obtained self-fertile transgenic barley plants by microprojectile bombardment of immature zygotic embryos and of calli developed from these. Also, microspore-derived embryos were transformed. The malting barley cultivar Golden Promise turned out to be the least recalcitrant genotype for the procedure. Gold particles were coated with a plasmid containing the selectable marker gene *bar,* encoding phosphinotricin acetyl transferase from *Streptomyces hygroscopicus* (Murakami et al., 1986), and the screenable marker gene *uidA,* encoding β-glucuronidase, either alone or in combination with another plasmid containing the gene for the barley yellow dwarf virus coat protein (Wan and Lemaux, 1994). Estimated copy number for the *bar* gene integrated into the plant genome ranged between 1 and 20, revealing an 87 percent cotransformation frequency for *bar* and *uidA.* Significant progress has been made in improving the regeneration frequency of transformed barley tissue by inclusion of 6-benzyl-aminopurine (6-BAP) and the micronutrient copper sulfate in the callus induction medium as well as in the regeneration medium (Cho, Jiang, and Lemaux, 1998). It was also found that exposure to dim light during callus growth promoted greening of cells, and their differentiation into leaf and root meristems.

A successful protocol for obtaining transgenic plants with the biolistic method is presented in Figure 7.1. With the described protocol (Figure 7.1), transgenic barley plants expressing a protein-engineered heat-stable (1,3-1,4)-β-glucanase during germination were produced (Jensen et al., 1996). Fourteen fertile transgenic barley plants were obtained by bombardment of immature zygotic embryos of the cultivar Golden Promise with 1 μm gold particles that were coated with three linearized plasmids. One plasmid carried the synthetic codon-optimized gene for the heat-stable β-glucanase under the control of the barley high-pI α-amylase promoter and the code for the signal peptide necessary for the export of the enzyme from the aleurone cells to the endosperm. The second plasmid carried the reporter gene *uidA* under the control of the constitutive synthetic Emu promoter, and the third plasmid contained the *bar* (bialaphos resistance) gene driven by the maize ubiquitin promoter and the gene's first intron for selection of transformed plants (Jensen et al., 1996). This gene is commonly used to screen for transformants of both di- and monocotyledonous species (De Block et al., 1987; De Block, De Sonville, and Debrouwer, 1995; White, Chang, and Bibb, 1990). Phosphinotricin acetyl transferase inactivates the natural herbicide bialaphos produced by *Streptomyces* as well as the chemically synthesized phosphinotricin (PPT). Bialaphos was therefore supplied in the media and taken up by the regenerating embryos. Only those that inactivated the herbicide survived and developed into transgenic plants. The transgenic embryos that developed shoots and roots were transferred to soil and grown to maturity in the greenhouse. Spikes were fully fertile with

FIGURE 7.1. Timeline to Obtain Transgenic Barley Plants with the Biolistic Method

Day 1	Excise immature zygotic embryos (1.5 to 2.5 mm) from 'Golden Promise' and bisect longitudinally. Place cut embryos scutellum side down onto callus induction medium (CIM) without bialaphos and incubate at 24°C in the dark for 12 to 24 hours.
Day 2	Transfer immature embryos to CIM without bialaphos, but with 0.4 M mannitol for 4 to 6 hours, then bombard with gold particles carrying linearized plasmid DNA.
Day 3	One day after bombardment, transfer embryos to CIM containing 5 mg·L^{-1} bialaphos.
Week 1	First round of selection: Calli are kept on bialaphos containing CIM at 24°C in the dark.
Week 3	Second round of selection: Transfer calli to fresh bialaphos containing CIM and incubate at 24°C in the dark.
Week 5	Third round of selection: Transfer calli to fresh bialaphos containing CIM and incubate at 24°C in the dark.
Week 7	Fourth round of selection: Transfer calli to fresh bialaphos containing CIM and incubate at 24°C in the dark.
Week 9	Transfer calli to shoot generation media (SGM) containing 1 mg·L^{-1} bialaphos; incubate at 24°C (16 h light/ 8 h dark).
Week 13	Transfer plantlets to root generation media (RGM) containing 1 mg·L^{-1} bialaphos; incubate at 24°C (16 h light/8 h dark).
Week 15	Transfer plants to soil and grow to maturity (16 h light/16°C; 8 h dark/ 12°C).
Month 7.5	Harvest mature seeds.

Note: CIM (pH 5.8) contains Murashige and Skoog medium (Murashige and Skoog, 1962) supplemented with 30 g·L^{-1} maltose, 1.0 mg·L^{-1} thiamine-HCl, 0.25 g·L^{-1} *myo*-inositol, 1.0 g·L^{-1} casein hydrolysate, 0.69 g·L^{-1} L-proline and 2.5 mg·L^{-1} dicamba, solidified by 3.5 g·L^{-1} phytagel. SGM medium (pH 5.6) consists of Murashige and Skoog medium with the ammonium nitrate concentration changed to 165 mg·L^{-1} supplemented with 62 g·L^{-1} maltose, 0.4 mg·L^{-1} thiamine-HCl, 0.1 g·L^{-1} *myo*-inositol, 1 g·L^{-1} casein hydrolysate, 0.75 g·L^{-1} glutamine, and 1 mg·L^{-1} 6-benzyl-aminopurine, solidified with 3.5 g·L^{-1} phytagel. RGM: CIM without any dicamba added.

readily germinating grains. The construction of the plasmids is described in Jensen et al. (1998). The three transgenes showed linked Mendelian inheritance over three generations, and the heat-stable β-glucanase gene was expressed into a functional enzyme during germination (Jensen et al., 1998). Competitive polymerase chain reaction revealed that one of the primary heterozygous transformants carried six copies of the transgene, while its homozygous progeny plants contained 12 copies in the genome (Jensen et al., 1998). Secreted hybrid (1,3-1,4)-β-glucanase amounted to 20 mg·kg^{-1} of transgenic grain, corresponding to 0.3 to 0.5 µg/grain.

BARLEY TRANSFORMATION BY COCULTIVATION OF IMMATURE ZYGOTIC EMBRYOS WITH AGROBACTERIUM TUMEFACIENS

During the past few years *Agrobacterium*-mediated transformation has proven to be an efficient method for obtaining transgenic plants of rice (Hiei et al., 1994), maize (Ishida et al., 1996), barley (Tingay et al., 1997), and wheat (Cheng et al., 1997). Sonja Tingay and co-workers obtained routinely stable barley transformants by cocultivation of immature embryos with disarmed *Agrobacterium tumefaciens*. In their experiments, the explants were injured by removing the embryonic axis and bombarding the scutellum with gold particles prior to inoculation with *Agrobacterium,* thereby increasing the number of transformation events. A transformation efficiency of 4.2 percent was achieved with the binary vector containing *uidA* as the reporter gene and *bar* as the selectable marker gene. After cocultivation the explants were washed and placed on selection media containing bialaphos and timentin to inhibit *Agrobacterium* growth, allowing formation of callus containing the transgenes. Timentin is a mixture of the penicillin derivative ticarcillin and clavulanic acid, which is a competitive inhibitor of β-lactamase, an enzyme that can prevent the effect of the antibiotic. In tobacco, timentin has been shown to be a more favorable bactericide than cefotaxime, as it promotes shoot formation (Nauerby, Billing, and Wyndaele, 1997) and is more cost-effective and stable than vancomycin. Independently transformed lines had incorporated from one to ten copies of the two transferred bacterial genes. The transgenes are tightly linked and the insertion was stably inherited. Hybridization experiments revealed that the majority of independent transformants had integrated complete copies of the T-DNA.

A successful protocol for obtaining transgenic plants by cocultivation of immature embryos with *Agrobacterium* is presented in Figure 7.2. With the described protocol, transgenic barley plants were produced expressing a protein-engineered heat-stable (1,3-1,4)-β-glucanase in the endosperm of maturing grains. For transformation, a binary vector was used, consisting of the disarmed Ti plasmid with the virulence genes for mobilization of the

FIGURE 7.2. Timeline to Obtain Transgenic Barley Plants by *Agrobacterium*-Mediated Transformation

Day 1	Excise immature zygotic embryos (1.5 to 2.5 mm) from 'Golden Promise' and bisect longitudinally. Place cut embryos on callus induction medium (CIM) without bialaphos and incubate at 24°C in the dark for two days.
Day 2	Start *Agrobacterium* culture and grow at 28°C.
Day 3	Add overnight *Agrobacterium* culture dropwise to embryos and co-cultivate at 24°C in the dark for 48 hours. *Agrobacterium* as well as callus growth is visible (Figure 7.1A).
Day 5	Wash off bacterial cells with LB medium until no more bacteria are visible, then wash one more time with LB containing 200 mg·L^{-1} timentin and let excess liquid drain off on sterile filter paper. Transfer individual embryos to CIM containing 4 mg·L^{-1} bialaphos and 200 mg·L^{-1} timentin.
Week 1	First round of selection: Calli are kept on bialaphos and timentin containing CIM at 24°C in the dark.
Week 3	Second round of selection: Transfer calli to fresh bialaphos and timentin containing CIM and incubate at 24°C in the dark (Figure 7.1B).
Week 5	Third round of selection: Transfer calli to fresh bialaphos and timentin containing CIM and incubate at 24°C in the dark.
Week 7	Transfer calli to shoot generation media (SGM) containing timentin and 3 mg·L^{-1} bialaphos; incubate at 24°C (16 h light/8 h dark) (Figure 7.1C).
Week 11	Transfer plantlets to root generation media (RGM) containing timentin and 3 mg·L^{-1} bialaphos; incubate at 24°C (16 h light/8 h dark) (Figure 7.1D).
Week 15	Transfer plants to soil and grow to maturity (16 h light/16°C; 8 h dark/12°C).
Month 7	Harvest mature seeds.

Note: All media contain 5 μm copper sulfate. CIM (pH 5.8) contains Murashige and Skoog medium (Murashige and Skoog, 1962) supplemented with 30 g·L^{-1} maltose, 1.0 mg·L^{-1} thiamine-HCl, 0.25 g·L^{-1} *myo*-inositol, 1.0 g·L^{-1} casein hydrolysate, 0.69 g·L^{-1} L-proline and 2.5 mg·L^{-1} dicamba, solidified by 3.5 g·L^{-1} phytagel. SGM medium (pH 5.6) consists of Murashige and Skoog medium with the ammonium nitrate concentration changed to 165 mg·L^{-1} supplemented with 62 g·L^{-1} maltose, 0.4 mg·L^{-1} thiamine-HCl, 0.1 g·L^{-1} *myo*-inositol, 1 g·L^{-1} casein hydrolysate, 0.75 g·L^{-1} glutamine, and 1 mg·L^{-1} 6-benzyl-aminopurine, solidified with 3.5 g·L^{-1} phytagel. RGM: CIM without any dicamba added.

T-DNA and a plasmid carrying the target genes between the left and the right T-DNA borders (see Figure 7.3). *Agrobacterium tumefaciens* strain AGL-1 (Lazo, Stein, and Ludwig, 1991) contains the disarmed Ti plasmid, which is derived from the hypervirulent, attenuated, tumor-inducing plasmid pTiBo542 by precise excision of the T-region. It also has an insertion mutation in its *recA* gene that stabilizes the recombinant plasmid and renders the strain resistant to carbenicillin. AGL-1 was transformed with the vector pJH271 by electroporation based on a published protocol (Mersereau, Pazour, and Das, 1990). Vector pJH271 is derived from pBIN19 and contains the gene encoding the codon-optimized H(A12-M)ΔY13 thermotolerant (1,3-1,4)-β-glucanase (Jensen et al., 1996, 1998). The improved thermostability of the enzyme has been obtained by intragenic recombination in vitro between the genes from *Bacillus amy-*

FIGURE 7.3. *Agrobacterium* Vector pJH271 Constructed to Express the Heat-Stable

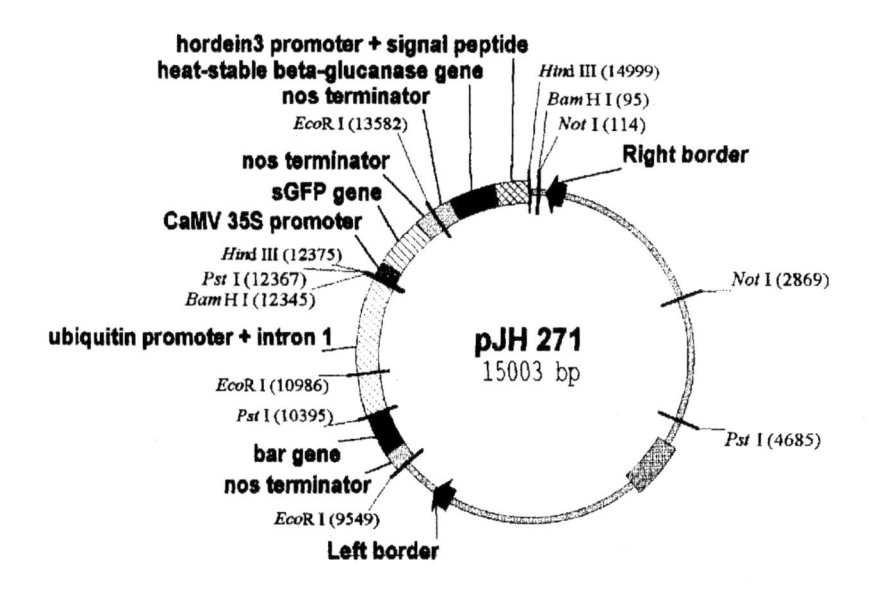

Note: The vector carries three different genes in a cassette between the left and right border sequences of the T-DNA. The hybrid β-glucanase gene is placed under the control of the D-hordein gene promoter and is provided with the code for the signal peptide, to target the enzyme into the protein bodies of the developing endosperm. The gene for the green fluorescent protein *(SGFP)* is under the control of the constitutive 35S promoter and the *bar* gene is driven by the maize ubiquitin *(Ubi-1)* promoter and the gene's first intron. All three genes integrate as one cassette into the barley genome.

loliquefaciens and *B. macerans,* specifying β-glucanases with the same substrate specificity as the barley enzyme (Olsen et al., 1991). The hybrid protein H(A12-M)ΔY13 has a polypeptide chain combining the 12 N-terminal amino acids from the *B. amyloliquefaciens* enzyme with the 202 C-terminal amino acids of the *B. macerans* enzyme, while a tyrosin in position 13 has been deleted. This enzyme exhibits a half-life of > 4 h at 70°C (pH 5.0) (Politz et al., 1993). The hybrid gene was provided with the promoter from the barley *Hor3* gene encoding the D-hordein storage prolamin (Sørensen et al., 1996) with the code for the D-hordein signal peptide, which targets the hordein into storage protein bodies deposited in the vacuoles of the developing endosperm. Additionally, the *bar* and *sGFP* genes were located within the two T-DNA borders as selectable markers. The open reading frame of the *bar* gene (White, Chang, and Bibb, 1990) was inserted between the maize ubiquitin *Ubi-1* promoter plus its first intron (Christensen, Sharrock, and Quail, 1992) and the nopaline synthetase *(nos)* terminator from *Agrobacterium tumefaciens* (Bevan, Barnes, and Chilton, 1983). The synthetic green fluorescent protein gene sequence *(sGFP)* (Chiu et al., 1996) is under control of the constitutive cauliflower mosaic virus (CaMV) 35S promoter and the *nos* terminator. Ten independent transformants were selected on bialaphos medium with this plasmid. The T_1 grains produced up to 40 μmg of the recombinant enzyme per milligram of soluble protein, corresponding to 800 mg·kg^{-1} of transgenic grain. The features of these transformants are discussed in the following section.

MARKER-FREE TRANSGENIC PLANTS

In the plasmid pJH271, the target gene and the selectable marker genes are contained in a single cassette between the left and right border sequences of the T-DNA in the binary vector. The selectable marker genes are required for the selective regeneration of transgenic plants but are unwanted because they make the derived cultivars herbicide resistant or resistant to therapeutic antibiotics used in mammals. In the latter case, large-scale growth of such plants may foster bacteria resistance to the antibiotic and render the antibiotic ineffective as a therapeuticum for humans and animals. The selectable markers in the transgenes also prevent retransformation with additional genes with the same selection procedure. It is thus desirable to create marker-free transgenic plants. This was successfully achieved in tobacco and rice with the binary "double cassette" vectors, which carry on the same plasmid two separate cassettes, each bracketed by a left and a right T-DNA border sequence (Komari et al., 1996). One cassette contained the gene encoding the β-glucuronidase *(uidA)* and the other a hygromycin or kanamycin resistance gene. Both T-DNA segments were cotransferred with a frequency of about 50 percent to rice or tobacco. As the two cassettes were frequently inserted into different chromosomes or chromosome arms, the resistance gene segregated independently from the marker gene in the T_2 generation. A "double cassette" vector pAM470 for barley transformation was constructed (see Figure 7.4) by Norbert Wolf (unpublished). It contains the structural gene for the heat-stable β-glucanase and the *bar* gene in different cassettes and is presently used to generate transgenic plants to study the co-transformation frequency and segregation pattern in barley.

TESTING THE QUALITY OF THE TARGET GENE, ITS PRODUCT, AND TISSUE-SPECIFIC EXPRESSION

It is desirable to test the possibility of expression of a gene to be used in transformation and the quality of its protein product prior to embarking on the long road of making transgenic plants. Thus, in model experiments of

FIGURE 7.4. Double-Cassette Vector pAM470 Expected to Give Rise to Barley Transformants

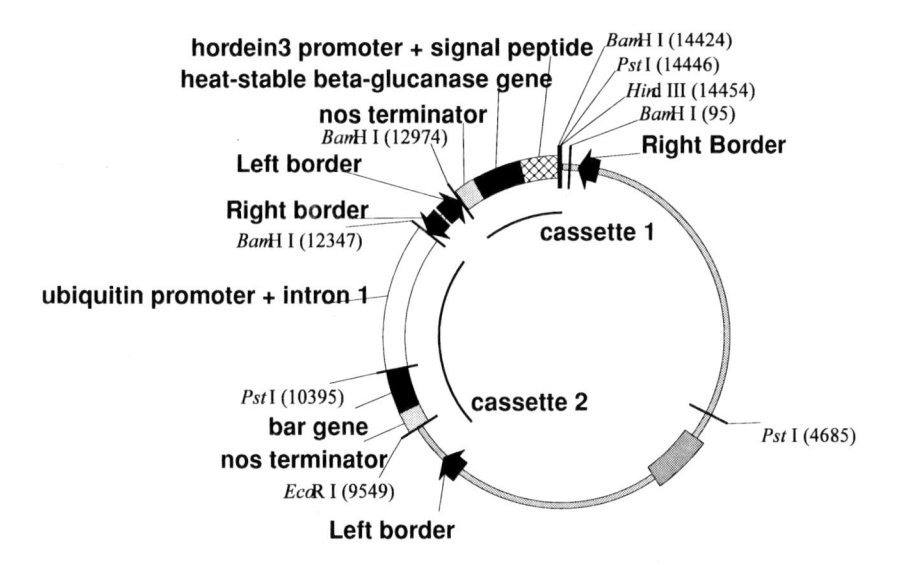

Note: The double-cassette vector pAM470 contains the *bar* gene under the control of the maize ubiquitin *(Ubi-1)* promoter and its first intron in one border set and carries the hybrid β-glucanase gene driven by the D-hordein gene promoter and its signal peptide code in the second pair of borders. This vector has the β-glucanase gene and the marker gene integrated independently.

tailoring a gene encoding a heat-stable (1,3-1,4)-β-glucanase, the gene was first expressed in *Escherichia coli,* where sufficient amounts of enzyme were produced for testing its survival at the high temperatures (70°C) used in mashing of wort (Olsen et al., 1991) and in pressing and pasteurizing feed pellets. Recombinant enzyme was subsequently produced in yeast, and its N-glycosylation pattern determined (Meldgaard and Svendsen, 1994; Meldgaard et al., 1995).

Protoplasts of the target tissue are an excellent tool for transient expression studies of foreign genes (Siemens and Schieder, 1996). Regarding barley, techniques for efficient isolation and transfection of protoplasts have been established for developing endosperm (Diaz and Carbonero, 1992) and aleurone tissue of germinating grain (Skriver et al., 1991; Wolf, 1992; Phillipson, 1993; Skjødt, Phillipson, and Simpson, 1994). They were helpful in characterizing the promoters of structural genes for α-amylase and (1,3-1,4)-β-glucanase with reporter proteins and identifying within the promoters the gibberellin- and abscisin-responsive DNA sequences interacting with transacting transcription factors.

When aleurone protoplasts were transfected with a plasmid containing the structural gene for the protein-engineered bacterial (1,3-1,4)-β-glucanase provided with a barley high-pI α-amylase gene promoter and the code for the signal peptide required for secretion of the enzyme, no recombinant β-glucanase was synthesized (see Figure 7.5). A comparable plasmid with a version of the protein reading frame that had been codon-optimized to a guanine plus cytosine (G+C) content of 63 percent gave an average production of 40 ng secreted (1,3-1,4)-β-glucanase per 2×10^5 protoplasts (Jensen et al., 1996). The aleurone protoplasts secreted two glycosylated forms of the enzyme, in addition to the unglycosylated molecules (see Figure 7.5). It was thought that the codon optimization for expression in aleurone cells was necessary because the endogenous genes for expression of α-amylases and mixed-linked β-glucanases in these cells had a high G+C content. It turns out that expression of the bacterial hybrid (1,3-1,4)-β-glucanase in endosperm cells with the D-hordein promoter likewise requires codon optimization, even though the genes for storage proteins have a high adenine plus thymine (A+T) content.

An alternative procedure for testing expression of the target gene and suitable tissue-specific promoters employs the microprojectile-mediated DNA delivery into intact cells of cell cultures or plant parts (Klein et al., 1987). Stable transformation of suspension culture cells of the barley cultivar Pokko was obtained with two separate plasmids carrying genes encoding and expressing neomycin phosphotransferase II (NPTII) and β-glucuronidase (GUS), respectively (Ritala et al., 1993). Bombardment of Himalaya barley half grains or aleurone layers is a standard procedure for the functional analysis of the high-pI and low-pI α-amylase promoters

FIGURE 7.5. Test of Transgene Expression of Heat-Stable (1,3-1,4)-β-Glucanase in Aleurone Protoplasts

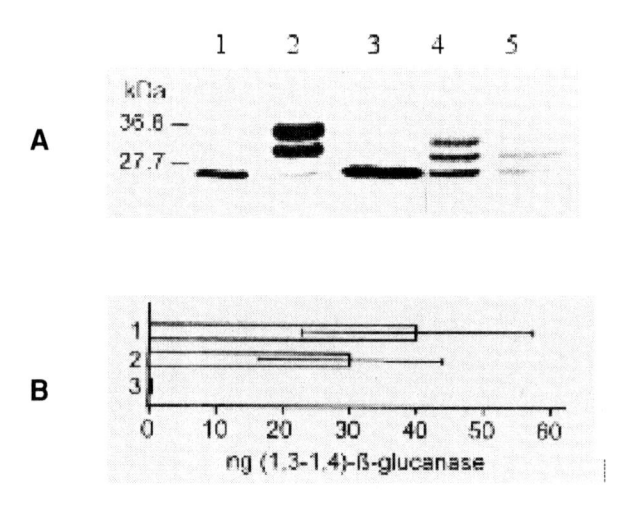

A: Purified unglycosylated enzyme from *E. coli* (lane 1), β-glucanase expressed by yeast in two N-glycosylated forms (lane 2) and after deglycosylation (lane 3). The enzyme produced and secreted by transfected aleurone protoplasts from two independent experiments (lanes 4 and 5) gives three bands, two of which are glycosylated forms. B: The amount of heat-stable (1,3-1,4)-β-glucanase produced by transfected aleurone protoplasts. Protoplasts transfected with the codon-optimized version of the gene show high expression of the enzyme after 110 hours (lane 1) and after 65 hours (lane 2) of incubation, whereas no enzyme production could be detected by protoplasts transfected with the nonoptimized version of the gene (lane 3).

(Lanahan et al., 1992; Gubler and Jacobsen, 1992; Rogers and Rogers, 1992; Rogers, Lanahan, and Rogers, 1994). With this test system, Frank Gubler and colleagues (1995) identified a Myb transcription factor that binds to the TAACAAA element in the promoters of the high-pI α-amylases. Cobombardment of aleurone layers with a construct containing the Myb transcription factor complementary (c) DNA driven by the actin 1 promoter, and a plasmid encoding the *uidA* reporter gene under the control of the amylase promoter, led to activation of transcription of the reporter gene in the absence of gibberellin. This transcription factor is therefore considered to be a component of the signal transduction chain from gibberellin reception to gene activation in aleurone cells.

Bombardment of intact cells of the developing barley endosperm revealed that the B1-hordein promoter exclusively drives expression of the *uidA* gene in starchy endosperm and subaleurone cells (Knudsen and Müller, 1991). This system is also useful to characterize the nitrogen response elements in the C-hordein promoter (Müller and Knudsen, 1993). In experiments using anthocyanin as a reporter, the barley structural gene for dihydroflavanol reductase *(Ant 18)* was delivered by microprojectile bombardment into leaf sheath tissue of the mutant *ant 18-162,* which lacks the enzyme and the capacity to synthesize anthocyanins as the result of a missense mutation. The wild-type gene complemented the mutation, and individual cells of the leaf sheath tissue formed red anthocyanin pigment (Wang, Olsen, and Knudsen, 1993). Using the same procedure, the two GT to AC base transitions at the 5' splice site of intron 3 in the mutant gene *ant 18-161* were identified among several base changes as the cause for the absence of detectable mature messenger (m) RNA in the mutant (Olsen, Wang, and von Wettstein, 1993). Transient expression of the the B-, C-, and D-hordein promoters coupled to the *uidA* reporter gene in bombarded developing endosperm revealed the D-hordein promoter to be three- to fivefold more active than B- or C-hordein promoters (Sørensen et al., 1996).

The B- and C-hordein promoters are methylated in the somatic tissues of the barley plant but demethylated during endosperm development (Sørensen, 1992). In the high-lysine mutant *lys3a* of the cultivar Bomi, transcription of the genes encoding B- and C-hordeins in the endosperm is severely reduced due to persistent hypermethylation of the promoters, while the transcription of the D-hordein gene is unaffected. A CpG island in its promoter prevents methylation of this promoter in endosperm as well as leaf tissue. Comparison of transient expression in 'Bomi' and *lys3a* endosperm demonstrated that the activities of unmethylated D-hordein and *Hor1-14* C-hordein promoters were equivalent, while in the mutant the activities of the *Hor1-17* C-hordein and *Hor2-4* B-hordein promoters were reduced two- and tenfold, respectively. Methylation of the plasmids in vitro prior to expression severely inhibited B- and D-hordein promoter activities. Using these two categories of promoters for expression of recombinant proteins in the developing barley endosperm, DNA methylation and chromatin organization are considered likely factors that cause variation of expression in different transgenic individuals.

The transient assay and stable transformation of callus was also used to demonstrate activity of the maize *Ac* transposon in barley cells (McElroy et al., 1997). Scutellar tissue of immature embryos was cobombarded with a plasmid containing the *Ac* transposase gene provided with the *Ubi1* or 35S promoter, and a plasmid with a *uidA* reporter gene that had been inactivated by insertion of a Ds element. Excision of the *Ds* element was observed by histochemical-detected GUS activity. Sequence analysis of the excision products showed the typical nucleotide deletions or mutations characteristic for transposition events in maize. Additionally, a callus line stably transformed with the *Ac* gene was bombarded with the *Ds*, containing *uidA* reporter gene, and excision of the *Ds* element was readily observed. It is concluded that barley plants transgenic for an *Ac/Ds* transposon system will be suitable for gene tagging.

IDENTIFICATION ASSAYS FOR TRANSFORMANTS AND SELECTION OF HOMOZYGOUS PLANTS

Polymerase Chain Reaction (PCR)

Polymerase chain reaction is widely used for initial screening of transgenic plants, and for following transgene segregation in subsequent generations. Traditionally, this is achieved by PCR reactions giving rise to a single band corresponding to a specific primer set (see Figure 7.6A). When screening for multiple genes in a single transformant, PCR amplification of one gene at a time is time-consuming, especially when analyzing large numbers of samples. A multiplex PCR protocol was optimized to facilitate screening for transgenic plants. In the example provided in Figure 7.6B, four pairs of primers were used in a single PCR reaction for simultaneous amplification of four different gene fragments. Genomic DNA of putative transgenic plants was used as a template, and PCR amplifications were made with specific primers for the three transgenes and with one set of primers for an endogenous gene as a quality control of the DNA preparation. Among all conditions tested, optimal primer concentration was most critical for a successful multiplex PCR.

Phosphinotricin Acetyl Transferase (PAT) Activity Assay

Single plants were tested for homozygosity using an assay developed for the presence of phosphinotricin acetyl transferase, the product of the *bar* gene (Kramer et al., 1993). Phosphinotricin (PPT) inhibits glutamine synthetase in plant cells, resulting in an accumulation and secretion of ammonia that is toxic to plant development (Murakami et al., 1986). Phosphinotricin acetyl transferase detoxifies phosphinotricin by acetylation (De Block et al., 1987). Grains are germinated on MS (Murashige-Skoog) media containing the pH indicator chlorophenol red (0.05 g·L^{-1}) and 3 mg·L^{-1} bialaphos, which is a tripeptide composed of two L-alanine residues and PPT, an L-glutamic acid analogue. Seedlings expressing the bialaphos resistance gene can be identified by color change of the medium from red to yellow as the roots secrete H$^+$ and the pH decreases from 6.0 to 4.5. Nontransformed cells secrete ammonia, making the medium alkaline and retaining its strong red color. The change in pH is evaluated after five to seven days of germination (see Photo 7.1).

Bialaphos resistance has also been tested by spraying plants with 0.5 percent Basta, containing phosphinotricin as its active compound, or by painting leaves with a bialaphos-containing solution (Wan and Lemaux, 1994).

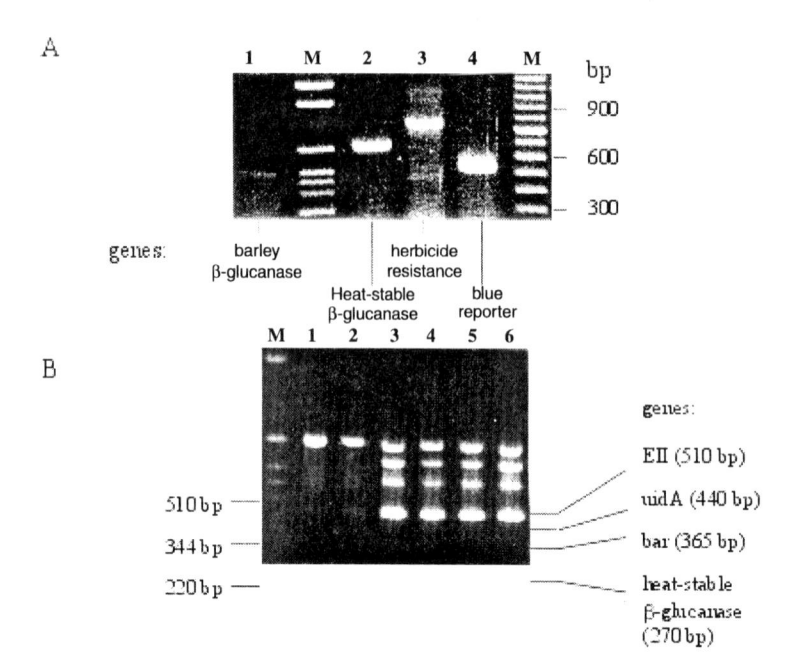

FIGURE 7.6. PCR Analysis of Transgenic Barley Plants

Note: Genomic DNA was used as a template for amplification of the three transgenes and the endogenous β-glucanase isoenzyme II *(EII)* gene to test the DNA quality.

A: The primers used amplified gene fragments of 509 bp (*EII* gene, lane 1), of the complete 705 bp sequence for the heat-stable β-glucanase, including its signal peptide (lane 2), and of the 845 bp spanning the *bar* gene (lane 3) as well as of the 539 bp *uidA* gene (lane 4). B: 1.2 percent agarose gel electrophoresis of multiplex PCR products amplified from genomic DNAs of four T_2 generation transgenic plants containing the genes for the endogenous barley (1,3-1,4)-β-glucanase, the β-glucuronidase, the phosphinotricin acetyl transferase, and the heat-stable β-glucanase (lanes 3-6). Lanes 1 and 2 are amplifications of null segregants.

b-Glucuronidase (GUS) Assay

The GUS assay tests for the presence of the selectable marker gene *uidA*, which is under the control of a constitutive promoter. GUS can be measured by fluorescence, spectrophotometry, or histochemical staining (Jefferson, Kavanagh, and Bevan, 1987). This assay is used most frequently for identification of transgenic plants. It uses X-gluc (5-bromo-4-chloro-3-indolyl β-D-glucuronide) or 4-Mug (4-methylumbelliferyl β-D-glucuronide) as a substrate. GUS cleaves the β-linkage leading to indol and the fluorochrome methylumbelliferone, respectively. The oxidative dimerization of indol is responsible for the blue insoluble indigo. Pollen and roots of transgenic germinated grains are screened individually for GUS activity by histochemical staining, thereby determining the homozygous lines.

Heat-Stable (1,3-1,4)-β-Glucanase Assay

Using lichenan, the β-glucanase assay can be a quick method to screen for transgenic plants producing this enzyme. Germinated grains are heat treated to inactivate the endogenous barley enzyme, cut in half, and placed endosperm-side down on MS plates containing 2 g·L^{-1} lichenan. Beta-glucanase diffuses out of the endosperm into the medium, thereby degrading lichenan, a (1,3-1,4)-β-glucan. Congo Red can stain the undegraded substrate, while degradation of the polysaccharide by active (1,3-1,4)-β-glucanase is visible as clear zones surrounding the grain (see Photo 7.3). The amount of enzyme produced in transgenic plants is determined with the method of McCleary (1988). Protein extracts from mature or germinated grains are incubated with the azo-barley glucan substrate at 65°C. The blue substrate is depolymerized by the heat-stable (1,3-1,4)-β-glucanase, but not by the endogenous enzyme, which is inactivated at this temperature. The degraded substrate fragments remain soluble in the presence of precipitant solution, while the undegraded long-chain molecules precipitate. Upon centrifugation, the absorbance of the supernatant is measured at 590 nm. This value is proportional to the level of (1,3-1,4)-β-glucanase activity in the sample.

Western Blot Analysis and Zymogram

Protein extracts from mature/germinated grains were separated by second-division segregation–polyacrylamide gel electrophoresis (SDS-PAGE), transferred to nitrocellulose membrane, and probed with antibodies raised against the recombinant β-glucanase made by *E. coli* (see Figure 7.7A). Alternatively, the extracts can be analyzed by a zymogram (see Figure 7.7B).

Southern Blot Analysis

Southern blot analysis of restricted DNA from transgenic plants produced by the biolistic method shows multiple bands, one of which usually can be identified as the intact transgene. Both larger and smaller restriction fragments are encountered among the multiple bands, in spite of the fact that in a majority of cases the transgenes are inherited as a single genetic locus. Wojciech Pawlowski and David Somers (1998) have analyzed the

FIGURE 7.7. Product Analysis

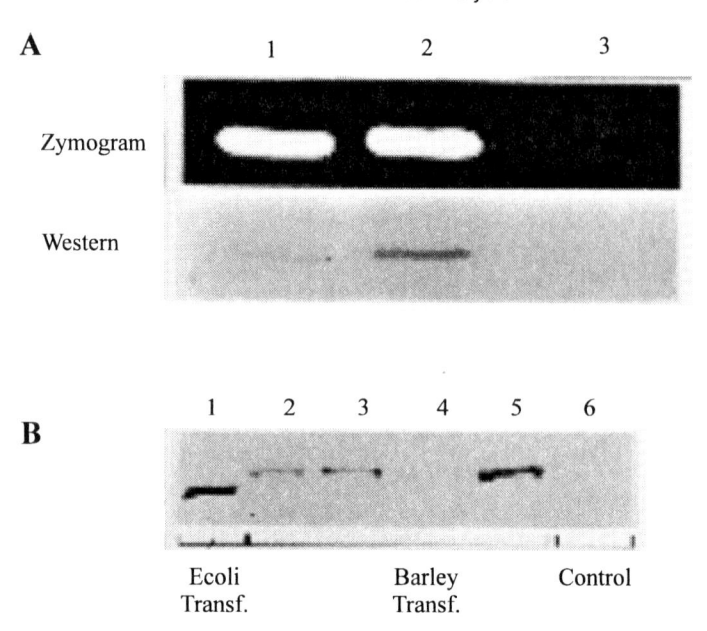

A: Shows the protein analysis of heat-stable β-glucanase expressed by *E. coli* (lane 1) and in extracts of germinated grains of a transformed barley line (lane 2) and its wild-type control (lane 3). The zymogram was obtained by isoelectric focusing in agarose. The gel was then overlaid with a lichenan-containing layer and subsequently stained with Congo Red, which reveals the active enzyme as a clear zone. The same samples were blotted onto nitrocellulose and tested with antibodies against the recombinant β-glucanase. A band is detected in the Western blot from *E. coli* and the transformed grain, but not from the untransformed grain. B: Western blot analysis of extracts containing recombinant enzyme separated by SDS-PAGE. Lane 1, *E. coli*; lanes 2-5, barley transformants expressing different amounts of heat-stable (1,3-1,4)-β-glucanase; lane 6, untransformed control. The β-glucanase produced by barley grains migrates slightly slower than the enzyme produced in *E. coli*. This is due to glycosylation of the recombinant protein.

transgene integration patterns in transgenic oat lines and demonstrated that multiple, rearranged and/or truncated transgene fragments were integrated into the genome in a single locus. Pulse field gel electrophoresis of fragments produced with a rare cutting enzyme indicates such a locus to span 36 to 280 kb. The complete or partial transgenes are interspersed with the chromatin of the host chromosome.

FIGURE 7.8. Southern Blot Analysis of Seven Homozygous Lines Transgenic for the Heat-Stable (1,3-1,4)-β-Glucanase Gene Obtained by the Ballistic Method (A) and of Ten Lines Obtained with the *Agrobacterium*-Mediated Method (B)

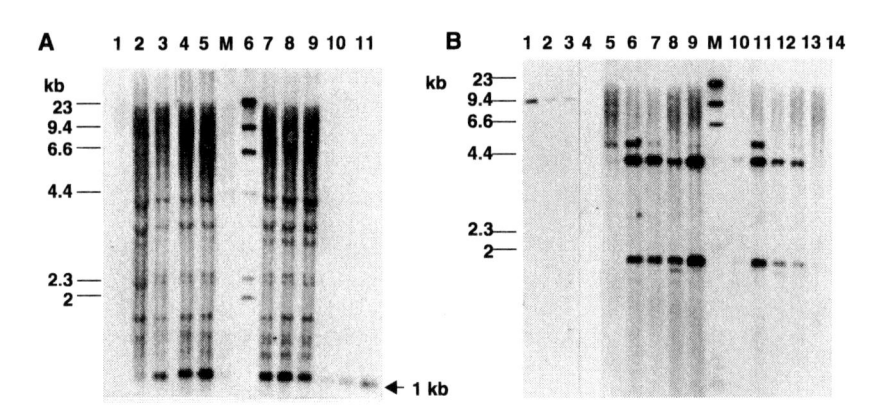

A: The genomic DNA was isolated and purified with the DNeasy Plant Mini Kit (QIAGEN), 10 μg digested with *EcoRI* and *PstI*, separated on a 0.8 percent agarose gel, blotted onto a positively charged nylon membrane and probed with the digoxygenin (DIG)-labeled coding region of the heat-stable (1,3-1,4)-β-glucanase gene (620 bp). Lanes: 1 = 'Golden Promise'; 2 = line originating from the T_0 transformant 6.2.1 with low enzyme expression; 3 = line originating from T_0 transformant 6.2.2 with low enzyme expression; 4-5 = lines from primary transformant 6.2.2 with intermediate enzyme expression; 6-8 = lines from T_0 transformant 8.2.1 with high enzyme expression; 9-11 = 1, 3, and 6 copies of the plasmid cut with *EcoRI* and *PstI* hybridized with the same probe. B: 10 μg genomic DNA digested with *EcoRI* and probed with the DIG-labeled coding region of the heat-stable (1,3-1,4)-β-glucanase gene (620 bp). Lanes: 1-3 = 20, 4, and 2 copies of the plasmid cut with *EcoRI* (10.2 kb); 4 = 'Golden Promise'; 5-14 = T_1 seedlings transgenic for the heat-stable (1,3-1,4)-β-glucanase gene.

In the Southerns prepared from the transgenic lines expressing the heat-stable (1,3-1,4)-β-glucanase (see Figure 7.8A), a complex pattern is visible. The smallest 1 kb band has the expected size of the gene with the signal peptide code, the terminator, and 95 bases of the promoter. Unexpected is the identical pattern of the larger fragments, even though lines 5607, 5609, and 5610 have arisen from a different embryo bombardment than lines 5612, 5613, 5603, 5604, and 5606. The latter lines express quite different enzyme amounts. An estimate of the copy number (cf. Figure 7.8A) would indicate that the high-expressing lines (5607-5610) contain 15 to 20 copies of the intact transgene, and a specific 1.2 kb band. The intermediate-expressing lines originating from the 6.2.2 T_0 plant had a similar copy number that is in agreement with the 12 copies determined by competitive PCR for the homozygous T_3 plant 6.2.2.4.8.21. A considerably lower copy number (~3) is apparent for the low-enzyme-expressing line 5605, originating from the 6.2.1 T_0 plant.

In the Southerns prepared from the *Agrobacterium*-mediated transformants, simpler fragment patterns are observed. Southern hybridizations of the DNA from transgenic plants obtained with the *Agrobacterium* vector pJH271 (see Figure 7.3) are presented in Figure 7.8B. The genomic plant DNA was cut with the *EcoRI* restriction enzyme and probed with the coding region of the heat-stable (1,3-1,4)-β-glucanase (620 bp), lacking an *EcoRI* restriction site. The vector plasmid used (see Figure 7.3) contains an *EcoRI* restriction site in the *nos* terminator of the transgene. Therefore, the fragments are bordered at one end by this *EcoRI* site, and at the other end by one in the chromosome near the insertion site. The different size bands thus indicate the numbers of insertions, ranging from one to four copies of the transgene.

STUDYING ORGAN DEVELOPMENT WITH BARLEY PLANTS TRANSGENIC FOR HOMEOTIC GENES

Among the homeotic mutants in barley, the dominant *Hooded* mutation (cf. Figure 7.9D and H-J) has become the focus of genetic and developmental studies because of its drastic effect on spike morphology: the florets appear to carry hoods. The *Hooded* mutation causes the development of ectopic meristems on the upper part of the lemma and awn (Stebbins and Yagil, 1966), which can lead to complete ectopic florets with palea, two lodiculi, three anthers, and an ovary with two stigmas (see Figure 7.9H). The additional ectopic florets or incomplete florets along the lemma and awn consistently show an inverse polarity to the preceding one (see Figure 7.9H-J). The *Hooded* gene has been cloned and turned out to be a homeobox gene coding an amino acid sequence with 90 percent identity to the maize *knotted 1* homeobox gene (Müller et al., 1995). In the mutant investigated, the hooded phenotype was associated with duplication in an intron of the gene. The *Ds* transposon-tagged *knotted* 1 gene of maize has been cloned by Sarah Hake, Erich Vollbrecht, and Michael Freeling (1989). The knotted leaf phenotype of mutations in this maize gene is caused by an ectopic expression in the shoot apical meristems forming leaves, inflorescences, and florets (Smith et al., 1992; Jackson, Veit, and Hake, 1994). This leads to alterations in cell differentiation around the lateral veins of the leaf ("vein clearing"), and to "knots" caused by extra cell divisions along the lateral vein, producing finger-like protrusions. Recessive mutations in the gene

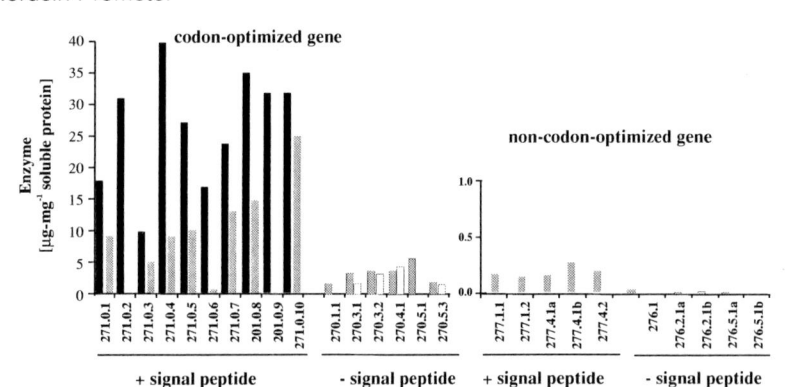

FIGURE 7.9. Expression of Heat-Stable β-Glucanase Under Control of the D-Hordein Promoter

Note: The expression of recombinant heat-stable β-glucanase in individual mature grains is shown in different transgenic barley lines obtained by *Agrobacterium*-mediated transformation. The plants, which contain the recombinant β-glucanase gene under control of the D-hordein gene promoter *(Hor3)* and its signal peptide code, produce higher levels of the enzyme than lines not carrying the signal peptide. The overall expression of the hybrid β-glucanase is 10 to 40 times higher in plants transformed with the codon-optimized gene, showing the importance of adapting the codon usage for expressing recombinant genes in barley. No heat-stable enzyme was detected in transgenic lines carrying the non-codon-optimized version of the gene without the signal peptide code.

leading to reduced KN1 mRNA and protein result in disturbances of tassel development and the formation of exra carpels and ovules (Kerstetter et al., 1997).

Rosalind Williams-Carrier and colleagues (1997) have obtained a barley transformant containing the maize *knotted 1* gene driven by the ubiquitin gene promoter. The transgenic plant developed the typical traits of the barley *Hooded* mutation, as seen in the scanning electron micrographs of the transgenic plant in Figure 7.9B-G. The phenotypes are compared to those of three different *Hooded* lines in Figure 7.9H-J.

Using affinity-purified specific antibodies against KN1 protein (Lucas et al., 1995) on paraffin sections through developing wild-type and transgenic florets showed that the protein in the wild type was restricted to the stem portions of the rachis and rachillae but absent from the differentiated organs, i.e., lemma, awn, anthers, ovaries, and ovules (see Photo 7.4A). In the transgenic plant, the protein was also present in the epidermal and subepidermal layers of the differentiating and differentiated lemma as well as in the developing filaments of the ectopic anther (see Photo 7.4C-F). The same pattern of KN1 protein was observed in the *Hooded* mutant. In situ hybridization with a digoxygenin-labeled *kn*1 cDNA probe corroborated the presence of *kn*1 mRNA in the cells and tissues containing the knotted protein (see Photo 7.4G-H).

This study illustrates that genetic transformation of barley using sense or antisense constructs of the homeotic genes of maize, *Arabidopsis,* and *Antirrhinum* can be usefully applied to identify genes determining the development of the organs in the barley floret.

THE PRODUCTION OF RECOMBINANT PROTEINS IN MATURING AND GERMINATING BARLEY GRAINS

There are two modes of producing recombinant enzymes or other proteins in the barley grain. One exploits the promoters active in the developing endosperm, such as those of the genes for the hordein storage proteins. The other uses the promoters of the genes encoding the hydrolytic enzymes that have been explored with the heat-stable (1,3-1,4)-β-glucanase. In the developing endosperm, the targeted proteins can be deposited into the storage protein bodies with the signal peptides of the hordein polypeptides. Localization of a recombinant protein in the cytosol is achieved by using a promoter for an enzyme synthesized on cytosolic ribosomes.

To investigate the importance of codon usage and the role of a signal peptide for enzyme localization and preservation in maturing grains, four different binary vectors were designed by modifying the heat-stable β-glucanase expression vector, pJH271 that uses the D-hordein promoter (cf. Figure 7.3). In the first vector, the open reading frame of the mature protein was codon-optimized to a G+C content of 63 percent, and the code for the signal peptide was included. In the second plasmid, the signal peptide code was left out. The third and fourth vector carried the non-codon-optimized version of the open reading frame for the mature protein with or without the signal peptide code. Barley plants transformed with these four constructs were obtained by *Agrobacterium*-mediated transformation, and mature T1 grains were germinated and analyzed for activity of heat-stable (1,3-1,4)-β-glucanase in the endosperm (see Figure 7.9). Up to 40 µg recombinant enzyme per milligram of soluble protein was obtained in individual kernels, when the codon-optimized open reading frame and the signal peptide code were present. This is 20 to 40 times more than the recombinant enzyme produced with the α-amylase promoter in the aleurone layer of transgenic grains during germination. The heat-stable (1,3-1,4)-β-glucanase can be readily seen not only in Western blots but also by Coomassie Blue staining (see Figure 7.10). In the absence of the signal peptide, localization of the recombinant enzyme is expected in the cytosol instead of the protein bodies.

FIGURE 7.10. Identification of Heat-Stable (1,3-1,4)-β-Glucanase in Mature Transgenic Grains (Lanes 4-6) and in Germinating Transgenic Grains (Lane 2)

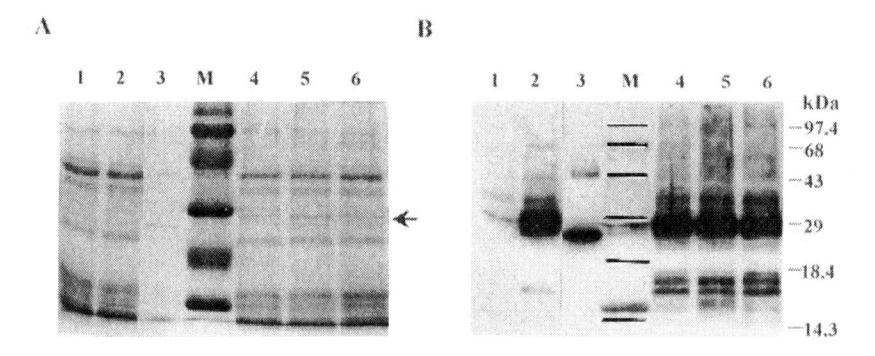

A: SDS gel. The protein is visible by Coomassie Blue staining only in lanes 4-6 after expression with the D-hordein promoter (arrow), but not after expression with the α-amylase promoter during grain germination. B: Western blot. The specific antibody recognizes the recombinant enzyme both in the mature (lanes 4-6) and, in smaller amounts, in the germinating grains. Lane 1 = extract of untransformed 'Golden Promise.' Lane 3 = purified unglycosylated recombinant enzyme synthesized in *E. coli*. The enzyme produced in barley is glycosylated and therefore has a larger apparent molecular weight.

Much lower amounts of recombinant enzyme were present in the mature grains in the absence of the signal peptide. It remains to be determined if this is due to decreased synthesis or to proteolytic degradation during programmed cell death in the last stages of grain maturation. In transgenic grains expressing the *uidA* gene with the D- or B_1-hordein gene promoter without a signal peptide code, the GUS activity per milligram of protein declined during grain filling (Cho et al. 1999). Ten to fourteen days after fertilization the B_1-hordein promoter sustained a GUS activity of 3,500 pmol·min^{-1}·mg^{-1} protein, while in mature grains an activity of only 100 pmol·min^{-1}·mg^{-1} protein was present. Similarily, the values obtained with the D-hordein promoter declined from 3,800 at 20 days after fertilization to 280 at maturity, which may indicate proteolysis of the recombinant enzyme in the cytosol during grain maturation.

In plants containing the unmodified β-glucanase gene, the enzyme could not be detected unless it was synthesized as a precursor with the N-terminal signal peptide (see Figure 7.9). Small amounts (less than 0.5 μmg·mg^{-1} soluble protein) were synthesized and present in the mature grain by the noncodon-optimized gene, when the code for the signal peptide was included. The difference is possibly due to degradation of the enzyme in the cytosol, while it remains active inside the protein bodies. It is concluded that targeting a protein into the storage bodies with the D-hordein promoter and its signal peptide code is a better way to obtain high yields of recombinant protein than is expression with the α-amylase promoter and its signal peptide code in the aleurone.

It should be mentioned that β-amylase is synthesized only during grain filling in the cytosol of the endosperm and has to be active in starch degradation during germination. This implies that certain proteins in the endosperm cytosol can withstand proteolysis during the last stages of grain maturation. In the course of producing barley lines transgenic for the green fluorescence protein (GFP) gene driven by the barley β-amylase promoter, GFP was synthesized specifically in the endosperm during grain filling.

ADVANTAGES OF TAILORING A HEAT-TOLERANT (1,3-1,4)-β-GLUCANASE AND ITS EXPRESSION IN TRANSGENIC BARLEY

Malt production and its use for food, feed, and beverages are based on two unique features of the germinating barley grain: one is the prominent heat stability of the α-amylases that survive the last step in the malt manufacturing process, i.e., the kiln drying at temperatures of up to 80°C; the second is the unique secretory system from the aleurone and scutellum tissues to the endosperm storage tissue. During germination the aleurone synthesizes the enzymes required to depolymerize the stored β-glucans, starch, and proteins and secretes these into the endosperm (Fincher, 1989). Transcription of the genes encoding the β-glucanases, the α-amylases, and the endo- and carboxypeptidases is induced by the hormone gibberellic acid, and the transcripts are translated by the newly formed endoplasmic reticulum, transferred into its lumen, moved through the Golgi apparatus, and secreted through the scutellum and aleurone cell walls into the endosperm. The barley (1,3-1,4)-β-glucanase isoenzyme II is responsible for the depolymerization of the (1,3-1,4)-β-glucans, the major component (75 percent) of the endosperm cell walls (Fincher, 1975). This has to take place before α-amylases and proteases can reach the stored starch and proteins (Gibbons, 1980). The successively produced sugars and amino acids are taken up by the scutellar epithelium and transferred through tissue layers of this organ into the growing embryo.

In contrast to the αa-amylases, the endogenous barley (1,3-1,4)-β-glucanase does not survive temperatures above 50 to 60°C and therefore is not significantly active in any process after kiln drying. This limits the use of barley varieties in which depolymerization of the (1,3-1,4)-β-D-glucan walls during malting is incomplete. Thermo-inactivation may result in an unmanageably high viscosity of the mash and reduced yield of extract (Bamforth, 1982). Survival of the enzyme in the kilning process would extend the use of malt, where the activity of the enzyme is needed or advantageous after the kiln-drying process.

Barley is an unacceptable component in chicken feed, as it is considered a grain providing low metabolic energy. The low nutritional value of barley as feed for poultry is due to the fact that birds, in their gastrointestinal tract, do not produce (1,3-1,4)-β-glucanase, the enzyme required to depolymerize the (1,3-1,4)-β-D-glucan present in the barley grain. This leads to high viscosity of the feed in the intestine, limited uptake of nutrients, and sticky droppings adhering to the chickens and the floors of the production cages (White et al., 1981, 1983). The latter complicates hygienic handling of broiler chickens. Inclusion of fungal enzyme preparations or purified (1,3-1,4)-β-glucanase from *Bacillus amyloliquefaciens* in barley diets can overcome the nutritional deficiencies of barley grain and prevent the occurrence of sticky droppings (Jensen et al., 1957; Willingham, Jensen, and McGinnis, 1959; Moscatelli, Ham, and Rickes, 1961; Rickes et al., 1962). Transgenic barley or malt-containing active-recombinant (1,3-1,4)-β-glucanase is expected to constitute a grain providing high metabolic energy in chicken diets. To be useful, the (1,3-1,4)-β-glucanase enzyme expressed in barley has to survive the heat generated when barley is pressed into feed pellets. It also has to survive pasteurization, which prevents infection of the chickens with *Salmonella typhimurium,* which causes gastroenteritis in consumers through the release of lipopolysaccharides that attack the mucous membranes of the intestine. The endogenous barley enzyme synthesized during malting for the depolymerization of the β-glucans does not survive pasteurization.

Bacillus species synthesize (1,3-1,4) β-glucanases with the same substrate specificity as the barley enzyme made in the aleurone cells during germination and secreted into the endosperm. Hybrid genes encoding the enzyme from *B. amyloliquefaciens* and *B. macerans* were constructed and expressed in *E. coli*. It was discovered that hybrid enzymes combining portions of the polypeptide from one species with portions of the polypeptide from the other species were more heat stable than both parental enzymes (Borriss et al., 1989; Olsen et al., 1991). Compared to the parent enzymes, the best hybrid enzyme increased the half-life at 70°C, pH 5.0, from six minutes to more than four hours (Politz et al., 1993). This hybrid contains 12 N-terminal amino acids from the *B. amyloliquefaciens* parent and 198 residues of the chain from the *B. macerans* parent. Pilot mashing with hybrid enzyme demonstrated the usefulness of the enzyme in degrading large amounts of β-glucans at high temperature. The hybrid enzyme will therefore survive the heat generated during pressing of feed pellets and their pasteurization.

FIELD TESTING

Transgenic barley plants obtained by the biolistic method were propagated in field trials in 1996, 1997, and 1998. Expression levels of the recombinant β-glucanase were measured by activity determination in extracts of germinated grains according to McCleary (1988), but after heating for 30 min to 65°C, a treatment eliminating activity of the endogenous barley β-glucanase. In most lines there is good correlation between the amounts of enzyme generated during germination in the three years (see Figure 7.11A). Different levels of enzyme could be observed among two sets of transgenic lines. Lines 5607 to 5610 derive from the original T_0 transformant 8.2.1 and have the same high level of enzyme expression, reaching 0.72 to 0.93 µmg·mg^{-1} protein in 1997 and 0.78 to 1.21 µmg·mg^{-1} protein in 1998. Considerable variation is observed among the lines derived from the T_0 transgenic plants 6.2.1 (5605) and 6.2.2 (5601-5604, 5606, 5611-5620). Some of them reach expressions of 0.30 to 0.40 µmg·mg^{-1} protein, while others show low production. A measurement of the endogenous barley (1,3-1,4)-β-glucanase yielded about 10 µmg/grain. Thus, there is room for improvement.

Western Blot analysis of the heat-stable β-glucanase protein of individual lines correlated with the amount of enzyme activity. The timing of synthesis and localization of the transgene product were investigated by immunofluorescence microscopy of sections through germinating grains probed with monospecific antibodies, recognizing the heat-stable (1,3-1,4)-β-glucanase, and compared to the formation of α-amylase isoenzymes (Jensen et al., 1998). Very similar patterns of synthesis of the two types of enzymes in the epidermal layer of the scutellum and in the aleurone tissue are observed, as well as a comparable time course for their secretion into the endosperm. The transgenic grains contained heat-stable β-glucanase in the coleoptile and leaf primordium. This is not expected for a recombinant enzyme driven by the barley α-amylase gene promoter.

The specific activity and V_{max} of H(A12-M)ΔY13 are 200-fold higher than those of barley (1,3-1,4)-β-glucanase, when measured under optimal conditions. In processing of barley grain material, it was found that addition of 20 µmg H(A12-M)ΔY13 is sufficient for complete depolymerization of 5 to 10 g soluble barley (1,3-1,4)-β-glucan from 1 kg of malt containing high levels of the polymer. Thus, enzyme in ~1 g (40 grains) of transgenic malt is expected to depolymerize the soluble β-glucan of 1 kg (~25,000 grains) of normal barley, while 200 µmg enzyme or 10 g of transgenic malt will depolymerize the 50 g of water-soluble and -insoluble (1,3-1,4)-β-glucan of 1 kg of normal barley cultivars.

FIGURE 7.11. The Consistent Expression of Heat-Stable β-Glucanase in High- and Low-Expressing Lines

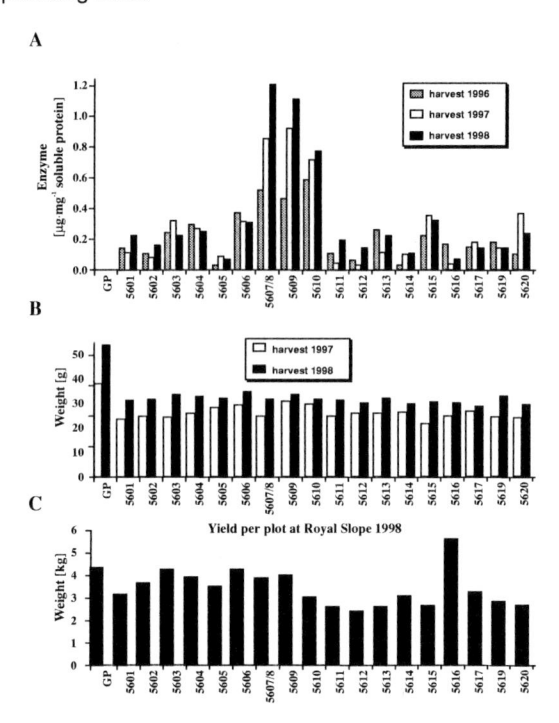

A: The expression of heat-stable β-glucanase was determined in germinated seeds from homozygous transgenic lines harvested from field trials in 1996, 1997, and 1998. The transformants carry the hybrid β-glucanase gene under the control of the high-pI α-amylase promoter and its signal peptide code for expression of the enzyme during germination/malting. The amount of enzyme produced remained similar in successive years. The increase in enzyme production seen in 1997 and 1998 compared to 1996 in the lines 5607/5608, 5609, and 5610 is probably due to better growth conditions in the field, as judged by a generally increased 1,000 grain weight (B). C: Yield per 90 ft² plots of lines transgenic for a protein-engineered, heat-stable (1,3-1,4)-β-glucanase. GP = 'Golden Promise.'

MICROMALTING OF TRANSGENIC LINES

Micromalting experiments have been carried out to investigate the synthesis of the engineered β-glucanase and its stability to high temperatures during the kilning procedure. In Figure 7.12A, results of micromalting ex-

FIGURE 7.12. Micromalting Experiment with 37 g Lots of the Barley Cultivars Golden Promise (GP), Harrington (Harr), and Four Representative Transgenic Lines Expressing Different Amounts of Heat-Stable Hybrid (1,3-1,4)-β-Glucanase

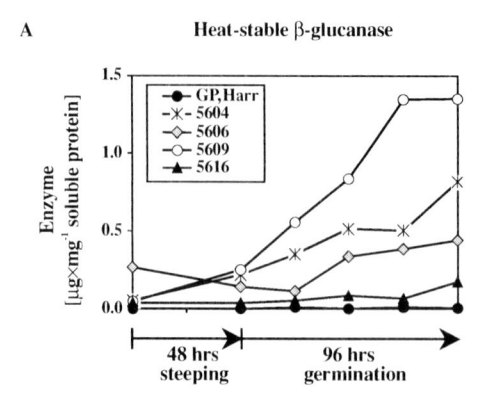

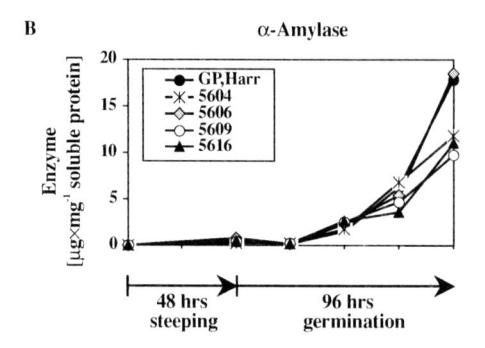

Note: Malting schedule: 8 hours steeping in aerated water at 14°C; 16 hours resting in humid air and 24 hours steeping, after which time the grains had reached a water content of 40 percent; 96 hours germination.

A: Measurements of heat-stable β-glucanase. B: Measurements of endogenous barley α-amylase.

periments for 'Golden Promise,' 'Harrington,' and four transgenic lines are presented. A time course of a strong increase in heat-stable β-glucanase during germination is apparent for the high-enzyme-expressing line 5609, as well as a substantial increase in lines 5604 and 5606, which synthesize intermediate levels of the recombinant enzyme. In agreement with the immunohistochemical analysis, the recombinant β-glucanase could consistently be detected in low amounts in the dissected embryo, but not in the endosperm of the mature grain, indicating that the α-amylase promoter is leaky in the transgenes, probably because of their chromatin location or high copy number. In the low-recombinant-enzyme expressing lines (e.g., 5616), no enzyme induction is observed during germination. The endogenous barley (1,3-1,4)-β-glucanase isoenzyme II followed the same time course of induction and synthesis in all investigated transgenic lines and untransformed controls. Induction of α-amylases (see Figure 7.12B) is frequently reduced in the transgenic grains relative to the untransformed control. After kiln drying of the green malt (see Table 7.1), a survival of 40 to 80 percent of the α-amylase activity is observed for wild types and transgenic lines. Recombinant (1,3-1,4)-β-glucanase activity displayed a survival comparable to that of α-amylase, i.e., values of 70 to 100 percent. The endogenous barley β-glucanase in the kilned malt had a remaining activity of 10 to 40 percent of that in the green malt. The barley enzyme was determined according to McCleary (1988) under conditions (pH 4.5, 30°C) in which the heat-stable enzyme has negligible activity.

AGRONOMIC CHARACTERISTICS OF THE TRANSGENIC LINES

The agronomic performance of the transgenic lines relative to that of 'Golden Promise' can be surmised from the data of Figure 7.11B and C. The 1,000 grain weight of the transgenic lines—irrespective of transgene expression levels—is reduced by 10 to 20 percent, while yield levels can reach, especially under conditions of irrigation, similar values to those of the host

TABLE 7.1. Percentage Activity of Recombinant (1,3-1,4)-β-Glucanase, Endogenous Barley (1,3-1,4)-β-Glucanase, and Barley α-Amylase Remaining After Kiln Drying of Green Malt

Lines		Survival of enzymes after 6 h at 45°C + 4 h at 80°C kilning		
		Recombinant β-glucanase (%)	Endogenous β-glucanase (%)	α-Amylase (%)
Golden Promise		—	16	80
Harrington		—	40	69
Transgene	5604	90	11	41
	5606	110	22	41
	5609	75	29	47
	5616	69	20	67

cultivar Golden Promise. Although the data are preliminary and restricted to a few transgenic barley lines, it is obvious that individual transgenic lines display similar yield or quality variations, commonly encountered in mutation breeding programs. Valuable mutation are utilized in cultivar development by fitting them into more suitable genotypes through recombination breeding. In crosses with modern barley varieties, we have obtained homozygous transgenic plants with the desired height and spike development (see Photo 7.1F).

REFERENCES

Bamforth, C.W. (1982). Barley β-glucans: Their role in malting and brewing. *Brewers Digest* 57: 22-35.

Bevan, M., Barnes, W.M., and Chilton, M.D. (1983). Structure and transcription of the nopaline synthase gene region of T-DNA. *Nucleic Acids Research* 11: 369-385.

Borriss, R., Olsen, O., Thomsen, K.K., and von Wettstein, D. (1989). Hybrid bacillus endo-(1,3-1,4)-β-glucanases: Construction of recombinant genes and molecular properties of the gene products. *Carlsberg Research Communications* 54: 41-54.

Castillo, A.M., Vasil, V., and Vasil, I.K. (1994). Rapid production of fertile transgenic plants of rye (*Secale cereale* L.). *Bio/Technology* 12: 1366-1371.

Cheng, M., Fry, J.E., Pang, S., Zhou, H., Hironaka, C.M., Duncan, D.R., Conner, T.W., and Wan, Y. (1997). Genetic transformation of wheat mediated by *Agrobacterium tumefaciens*. *Plant Physiology* 115: 971-980.

Chiu, W.L., Niwa, Y., Zeng, W., Hirano, T., Kobayashi, H., and Sheen, J. (1996). Engineered GFP as a vital reporter in plants. *Current Biology* 6: 325-330.

Cho, M.J., Choi, H.W., Buchanan, B.B., and Lemaux, P.G. (1999). Inheritance of tissue-specific expression of barley hordein promoter-*uidA* fusions in transgenic barley plants. *Theoretical and Applied Genetics* 98: 1253-1263.

Cho, M.J., Jiang, W., and Lemaux, P.G. (1998) Transformation of recalcitrant barley cultivars through improvement of regenerability and decreased albinism. *Plant Science* 138: 229-244.

Christensen, A.H., Sharrock, R.A., and Quail, P.H. (1992). Maize polyubiquitin genes: Structure, thermal perturbation of expression and transcript splicing, and promoter activity following transfer to protoplast by electroporation. *Plant Molecular Biology* 18: 675-689.

De Block, M., Bottermann, J., Vandewiele, M., Dockx, J., Thoen, C., Gosselé, V., Movva, N.R., Thompson, C., Van Montagu, M.V., and Leemans, J. (1987). Engineering herbicide resistance in plants by expression of a detoxifying enzyme. *EMBO Journal* 69: 2513-2518.

De Block, M., De Sonville, A., and Debrouwer, D. (1995). The selection mechanism of phosphinotricin is influenced by the metabolic status of the tissue. *Planta* 197: 619-626.

Diaz, I. and Carbonero, P. (1992) Isolation of protoplasts from developing barley endosperm: A tool for transient expression studies. *Plant Cell Reproduction* 10: 595-598.

Fincher, G.B. (1975). Morphology and chemical composition of barley endosperm cell walls. *Journal of the Institute of Brewing* 81: 116-122.

Fincher, G.B. (1989). Molecular and cellular biology associated with endosperm mobilization in germinating cereal grains. *Annual Review of Plant Physiology and Molecular Biology* 40: 305-346.

Funatsuki, H., Kuroda, H., Kihara, M., Lazzeri, P.A., Müller, E., Lörz, H., and Kishinami, I. (1995). Fertile transgenic barley generated by direct DNA transfer to protoplasts. *Theoretical and Applied Genetics* 91: 701-712.

Gibbons, G.C. (1980). On the sequential determination of α-amylase transport and cell wall breakdown in germinating seeds of *Hordeum vulgare*. *Carlsberg Research Communications* 45: 177-184.

Gubler, F. and Jacobsen, J.V. (1992). Gibberellin-responsive elements in the promoter of a barley high-pI α-amylase gene. *Plant Cell* 4: 1435-1441.

Gubler, F., Kalla, R., Roberts, J.K., and Jacobsen, J.V. (1995). Gibberellin-regulated expression of a *myb* gene in barley aleurone cells: Evidence for Myb transactivation of a high-pI α-amylase gene promoter. *Plant Cell* 7: 1879-1891.

Hagio, T., Hirabayashi, T., Machii, H., and Tomotsune, H. (1995). Production of fertile transgenic barley (*Hordeum vulgare* L.) plant using the hygromycin-resistance marker. *Plant Cell Reproduction* 14: 329-334.

Hake, S., Vollbrecht, E., and Freeling, M. (1989). Cloning *knotted*, the dominant morphological mutant in maize, using the *Ds2* transposon tag. *EMBO Journal* 8: 15-22.

Hiei, Y., Ohta, S., Komari, T., and Kumashiro, T. (1994). Efficient transformation of rice (*Oryza sativa* L.) mediated by *Agrobacterium* and sequence analysis of the boundaries of the T-DNA. *Plant Journal* 6: 271-282.

Ishida Y., Ohta, S., Komari., T., and Kumashiro T. (1996). High efficiency transformation of maize (*Zea mays* L.) mediated by *Agrobacterium tumefaciens*. *Nature Biotechnology* 14: 745-750.

Jackson, D., Veit, B., and Hake, S. (1994). Expression of maize *KNOTTED1* related homeobox genes in shoot apical meristem predicts patterns of morphogenesis in the vegetative shoot. *Development* 120: 405-413.

Jefferson, R.A., Kavanagh, T.A., and Bevan, M.W. (1987). Gus fusions: β–Glucuronidase as a sensitive and versatile gene fusion marker in higher plants. *EMBO Journal* 6: 3901-3907.

Jensen, L., Fry, R.E., Allred, J.B., and McGinnis (1957). Improvement in the nutritional value of barley for chicks by enzyme supplementation. *Poultry Science* 36: 919-921.

Jensen, L.G., Olsen, O., Kops, O., Wolf, N., Thomsen, K.K., and von Wettstein, D. (1996). Transgenic barley expressing a protein-engineered, thermostable (1,3-1,4)-β-glucanase during germination. *Proceedings of the National Academy of Sciences, USA* 93: 3487-3491.

Jensen, L.G., Politz, O., Olsen, O., Thomsen, K.K., and von Wettstein, D. (1998). Inheritance of a codon-optimized transgene expressing heat stable (1,3-1,4)-β-glucanase in scutellum and aleurone of germinating barley. *Hereditas* 129: 215-225.

Kerstetter, R.K., Laudencia-Chingcuanco, D., Smith, L.G., and Hake, S. (1997). Loss-of-function mutations in the maize homeobox gene, *knotted1*, are defective in shoot meristem maintenance. *Development* 124: 3045-3054.

Kihara, M., Saeki, K., and Ito, K. (1998). Rapid production of fertile transgenic barley (*Hordeum vulgare* L.) by direct gene transfer to primary callus-derived protoplasts. *Plant Cell Reproduction* 17: 937-940.

Klein, T.M., Kornstein, L., Sanford, J.C., and Fromm, M.E. (1987). Genetic transformation of maize cells by particle bombardment. *Plant Physiology* 91: 440-444.

Knudsen, S. and Müller, M. (1991). Transformation of the developing barley endosperm by particle bombardment. *Planta* 185: 330-336.

Komari, T., Hiei, Y., Saito, Y., Murai, N., and Kumashiro, T. (1996). Vectors carrying two separate T-DNAs for co-transformation of higher plants mediated by *Agrobacterium tumefaciens* and segregation of transformants free from selection markers. *Plant Journal* 10: 165-174.

Koprek, T., Hänsch, R., Nerlich, A., Mendel, R.R., and Schulze, J. (1996). Fertile transgenic barley of different cultivars obtained by adjustment of bombardment conditions to tissue response. *Plant Science* 119: 79-91.

Kramer, C., DiMaio, J., Carswell, G.K., and Shillito, R.D. (1993). Selection of transformed protoplast-derived *Zea mays* colonies with phosphinotricin and a novel assay using the pH indicator chlorophenol red. *Planta* 190: 454-458.

Lanahan, M.B., Ho, T.H., Rogers, S.W., and Rogers, J.C. (1992). A gibberellin response complex in cereal α-amylase gene promoters. *Plant Cell* 4: 203-211.

Lazo, G.R., Stein, P.A., and Ludwig, R.A. (1991). A DNA transformation-competent *Arabidopsis* genomic library in *Agrobacterium. Bio/Technology* 9: 963-967.

Lemaux, P.G., Cho, M.J., Zhang, S., and Bregitzer, P. (1999). Transgenic cereals: *Hordeum vulgare* L. (barley). In *Molecular Improvement of Cereal Crops*, ed. I.K. Vasil. Great Britain: Kluwer Academic Publishers, pp. 255-316.

Li, L., Qu, R., Kochko, A., Fauquet, C., and Beachy, R.N. (1993). An improved rice transformation system using the biolistic method. *Plant Cell Reproduction* 12: 250-255.

Lucas, W.J., Bouché-Pillon, S., Jackson, D.P., Nguyen, L., Baker, L., Ding, B., and Hake, S. (1995). Selective trafficking of *KNOTTED1* homeodomain protein and its mRNA through plasmodesmata. *Science* 270: 1980-1983.

Lührs, R. and Lörz, H. (1987). Plant regeneration in vitro from embryogenic cultures of spring and winter-type barley (*Hordeum vulgare* L.) varieties. *Theoretical and Applied Genetics* 75: 16-25.

McCleary, B.V. (1988). Soluble, dye-labeled polysaccharides for the assay of endohydrolases. *Methods in Enzymology* 160: 74-86.

McElroy, D., Louwerse J.D., McElroy, S., and Lemaux, P.G. (1997). Development of a simple transient assay for *Ac/Ds* activity in cells of intact barley tissue. *Plant Journal* 11: 157-165.

Meldgaard, M., Harthill, J., Petersen, B., and Olsen, O. (1995). Glycan modification of a thermostable recombinant (1-3,1-4)-β-glucanase secreted from *Saccharomyces cerevisiae* is determined by strain and culture conditions. *Glycoconjugate Journal* 12: 380-390.

Meldgaard, M. and Svendsen, I. (1994). Different effects of N-glycosylation on the thermostability of highly homologous bacterial (1,3-1,4)-β-glucanases secreted from yeast. *Microbiology* 140: 159-166.

Mersereau, M., Pazour, G.J., and Das, A. (1990). Efficient transformation of *Agrobacterium tumefaciens* by electroporation. *Gene* 90: 149-151.

Moscatelli, E.A., Ham, E.A., and Rickes, E.L. (1961). Enzymatic properties of a β-glucanase from *Bacillus subtilis*. *Journal of Biology and Chemistry* 236: 2858-2862.

Müller, K.J., Romano, N., Gerstner, O., Garcia-Maroto, F., Pozzi, C., Salamini, F., and Rohde, W. (1995). The barley *Hooded* mutation caused by a duplication in a homeobox gene intron. *Nature* 374: 727-730.

Müller, M. and Knudsen, S. (1993). The nitrogen response of a barley C-hordein promoter is controlled by positive and negative regulation of GCN4 and endosperm box. *Plant Journal* 4: 343-345.

Murakami, T., Anzai, H., Imai, S., Satoh, A., Nagaoka, K., and Thompson, C.J. (1986). The bialaphos biosynthetic genes of *Streptomyces hygroscopicus:* Molecular cloning and characterization of the gene cluster. *Molecular and General Genetics* 205: 42-50.

Murashige, T. and Skoog, F. (1962). A revised medium for rapid growth and bioassays with tobacco tissue cultures. *Physiology of Plants* 15: 473-497.

Nauerby, B., Billing, K., and Wyndaele, R. (1997). Influence of the antibiotic timentin on plant regeneration compared to carbenicillin and cefotaxime in concentrations suitable for elimination of *Agrobacterium tumefaciens*. *Plant Science* 123: 169-177.

Olsen, O., Borriss, R., Simon, O., and Thomsen, K.K. (1991). Hybrid Bacillus (1-3,1-4)-β-glucanases: Engineering thermostable enzymes by construction of hybrid genes. *Molecular and General Genetics* 225: 177-185.

Olsen, O., Wang, X., and von Wettstein, D. (1993). Sodium azide mutagenesis: Preferential generation of A·T → G·C transitions in the barley *Ant18* gene. *Proceedings of the National Academy of Sciences, USA* 90: 8043-8047.

Pawlowski, W.P. and Somers, D.A. (1998). Transgenic DNA integrated into the oat genome is frequently interspersed by host DNA. *Proceedings of the National Academy of Sciences, USA* 95: 12106-12110.

Phillipson, B.A. (1993). Expression of hybrid (1-3,1-4)-β-glucanase in barley protoplasts. *Plant Science* 91: 195-206.

Politz, O., Simon, O., Olsen, O., and Borriss, R. (1993). Determinants for enhanced thermostability of hybrid (1-3,1-4)-β-glucanases. *European Journal of Biochemistry* 216: 829-834.

Rickes, E.L., Ham, E.A., Moscatelli, E.A., and Ott, W.H. (1962). The isolation and biological properties of a β-glucanase from *B. subtilis*. *Archives of Biochemistry and Biophysiology* 96: 371-375.

Ritala, A., Aspegren, K., Kurtén, U., Salmenkallio-Marttila, M., Mannonen, L., Hannus, R., Kauppinen, V., Teeri, T.H., and Enari, T.M. (1994). Fertile transgenic barley by particle bombardment of immature embryos. *Plant Molecular Biology* 24: 317-325.

Ritala, A., Mannonen, L., Aspegren, K., Salmenkallio-Marttila, M., Kurtén, U., Hannus, R., Mendez Lozano, J., Teeri, T.H., and Kauppinen, V. (1993). Stable transformation of barley tissue culture by particle bombardment. *Plant Cell Reproduction* 12: 435-440.

Rogers, J.C., Lanahan, M.B., and Rogers, S.W. (1994). The *cis*-acting gibberellin response complex in high-pI α-amylase gene promoters. *Plant Physiology* 105: 151-158.

Rogers, S.W. and Rogers, J.C. (1992). The importance of DNA methylation for stability of foreign DNA in barley. *Plant Molecular Biology* 18: 945-961.

Siemens, J. and Schieder, O. (1996) Transgenic plants: Genetic transformation—Recent developments and the state of the art. *Plant Tissue Culture and Biotechnology* 2: 66-75.

Skjødt, L.T., Phillipson, B.A., and Simpson, D.J. (1994). Isolation and characterization of aleurone protoplasts from a malting variety of barley (*Hordeum vulgare* L.). *Protoplasma* 177: 132-143.

Skriver, K., Olsen, F.L., Rogers, J.C., and Mundy, J. (1991). *Cis*-acting DNA elements responsive to gibberelin and its antagonist absidic acid. *Proceedings of the National Academy of Sciences, USA* 88: 7266-7270.

Smith, L.G., Greene, B., Veit, B., and Hake, S. (1992). A dominant mutation in the maize homeobox gene, *Knotted-1*, causes a switch from determinate to indeterminate cell fates. *Genes Development* 7: 787-795.

Somers, D.A., Rines, H.W., Gu, W., Kaeppler, H.F., and Bushnell, W.R. (1992). Fertile, transgenic oat plants. *Bio/Technology* 10: 1589-1594.

Sørensen, M.B. (1992). Methylation of B-hordein genes in barley endosperm is inversely correlated with gene activity and affected by the regulatory gene *Lys3*. *Proceedings of the National Academy of Sciences, USA* 89: 4119-4123.

Sørensen, M.B., Müller, M., Skerrit, J., and Simpson, D. (1996). Hordein promoter methylation and transcriptional activity in wild-type and mutand barley endosperm. *Molecular and General Genetics* 250: 750-760.

Stebbins, G.L. and Yagil, E. (1966). The morphogenetic effects of the hooded gene in barley. I. The course of development in hooded and awned genotypes. *Genetics* 54: 727-741.

Tingay, S., McElroy, D., Kalla, R., Fieg, S., Wang, M., Thornton, S., and Bretell, R. (1997). *Agrobacterium tumefaciens*-mediated barley transformation. *Plant Journal* 11: 1369-1376.

Vasil, V., Castillo, A.M., Fromm, M.E., and Vasil, I.K. (1992). Herbicide resistant fertile transgenic wheat plants obtained by microprojectile bombardment of regenerable embryogenic callus. *Bio/Technology* 10: 662-674.

Wan, Y. and Lemaux, P.G. (1994). Generation of large numbers of independently transformed fertile barley plants. *Plant Physiology* 104: 37-48.

Wang, X., Olsen, O., and Knudsen, S. (1993). Expression of the dihydroflavanol reductase gene in an anthocyanin-free barley mutant. *Hereditas* 119: 67-75.

White, J., Chang, S.-Y.P., and Bibb, M.J. (1990). A cassette containing the *bar* gene of *Streptomyces hygroscopicus:* A selectable marker for plant transformation. *Nucleic Acids Research* 18: 1062.

White, W.B., Bird, H.R., Sunde, M.L., Marlett, J.A., Prentice, N.A., and Burger, W.C. (1983). Viscosity of β-D-glucan as a factor in the enzymatic improvement of barley for chicks. *Poultry Science* 62: 853-862.

White, W.B., Bird, H.R., Sunde, M.L., Prentice, N., Burger, W.C., and Marlett, J.A. (1981). The viscosity interaction of barley beta-glucan with *Trichoderma viride* cellulase in the chick intestine. *Poultry Science* 60: 1043-1048.

Williams-Carrier, R.E., Lie, Y.S., Hake, S., and Lemaux, P.G. (1997). Ectopic expression of the maize *kn1* gene phenocopies the *Hooded* mutant of barley. *Development* 124: 3737-3745.

Willingham, H.E., Jensen, L.S., and McGinnis, J. (1959). Studies on the role of enzyme supplements and water treatment for improving the nutritional value of barley. *Poultry Science* 38: 539-544.

Wolf, N. (1992). Structure of the genes encoding *Hordeum vulgare* (1→3,1→4)-β-glucanase isoenzymes I and II and functional analysis of their promoters in barley aleurone protoplasts. *Molecular and General Genetics* 234: 33-42.

PHOTO 7.1. Transformation of Barley. A: Bisected immature embryos during cocultivation with *Agrobacterium tumefaciens* for one day. B: Three-week-old calli in second round of selection on bialaphos-containing media. Only the ivory-colored parts of the calli have potential for plant regeneration. C: Shoot regeneration in callus on medium containing bialaphos. D: Transformants carrying the *bar* gene develop vigorous roots. E: Propagation of T_5 generation of transgenic GP lines in a field trial, 1999. F: F_3 generation from cross of lines transgenic for the heat-stable β-glucanase gene with the proanthocyanidin-free cultivar Caminant. Short recombinant plants of erect 'Golden Promise' phenotype are at left and of Caminant phenotype with drooping ears, at right.

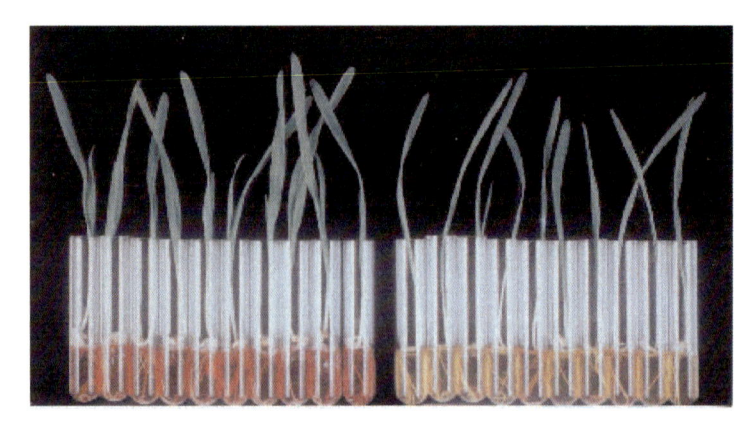

PHOTO 7.2. PAT Assay for Screening Barley Grains Containing the *bar* Gene. *Note:* Seeds are germinated on media containing bialaphos and chlorophenol red for five days. The yellow color indicates positive transformants (right), whereas media with nonexpressing seedlings remain red (left).

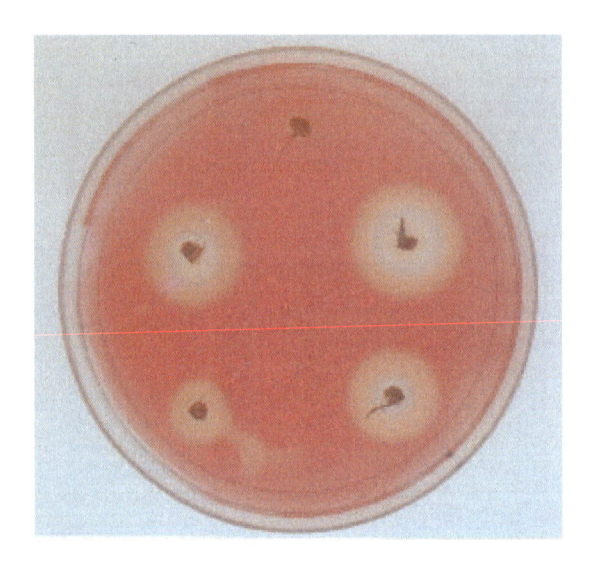

PHOTO 7.3. Germinated and Heated Half Seeds Screened on an Agar Plate Containing Lichenan for the Presence of Heat-Stable β-Glucanase, Which Degrades the Substrate. *Note:* Lichenan in the medium is stained with Congo Red, whereas the red dye does not bind to the degraded product. Four transformants and one untransformed germling have been analyzed.

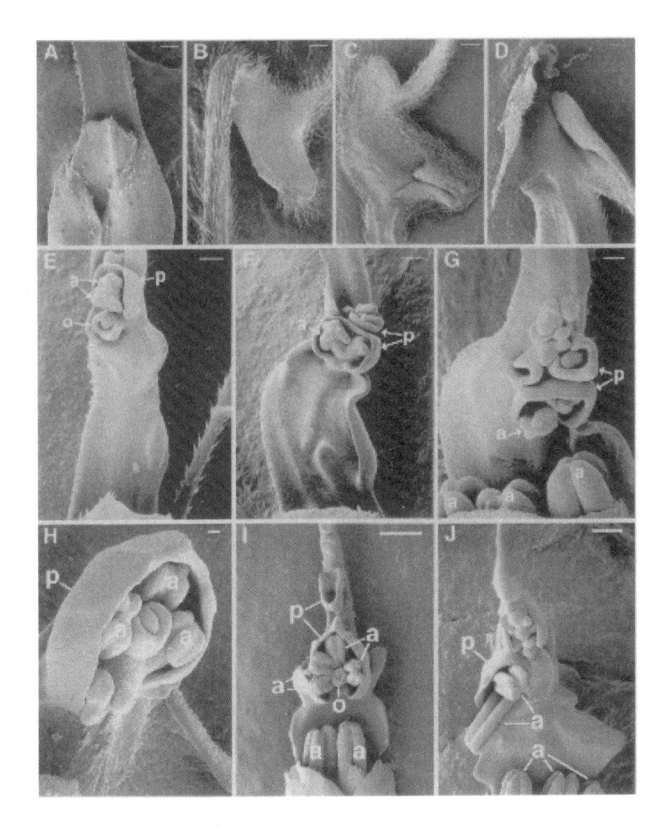

PHOTO 7.4. Morphology of Barley Transformant Expressing the Maize *knotted* 1 Homeobox Gene, Ectopic Expression of Which Results in a Phenotype Resembling the *Hooded* Mutation. *Source:* Williams-Carrier et al., 1997. Reprinted with permission of the Company of Biologists, Ltd. A: Scanning electron micrograph of the transition from the lemma (l) to the awn (aw) in a wild-type floret. B,C: Ectopic outgrowths of tissue on the awn of transformant containing the *knotted* gene driven by the ubiquitin gene promoter. D: Hood formed by an inverted lemma on the awn of the primary floret, as seen by the inverted direction of the epidermal hairs. E: An ectopic floret with palea (p), anther (a), and ovary (o) has formed on the awn in opposite orientation to the palea tip of the primary floret visible at the bottom. F: Two ectopic florets in opposite orientation have formed on the lemma of the primary floret. G: Several ectopic meristems are visible at the lemma-awn transition zone. The anthers of the primary floret are seen at the bottom. H-J: Barley *Hooded* mutant. H: Single ectopic floret. I: The palea of the primary floret was removed to visualize the first ectopic floret containing a palea, four anthers, and an ovary. A second ectopic palea in opposite orientation to the first one is seen in opposite orientation distally on the awn. J: An incomplete ectopic floret and several additional floral primordia present distally on the awn. (Scale bar: A, F = 220 µm; B = 360 µm; C = 270 µm; D = 250 µm; E = 150 µm.)

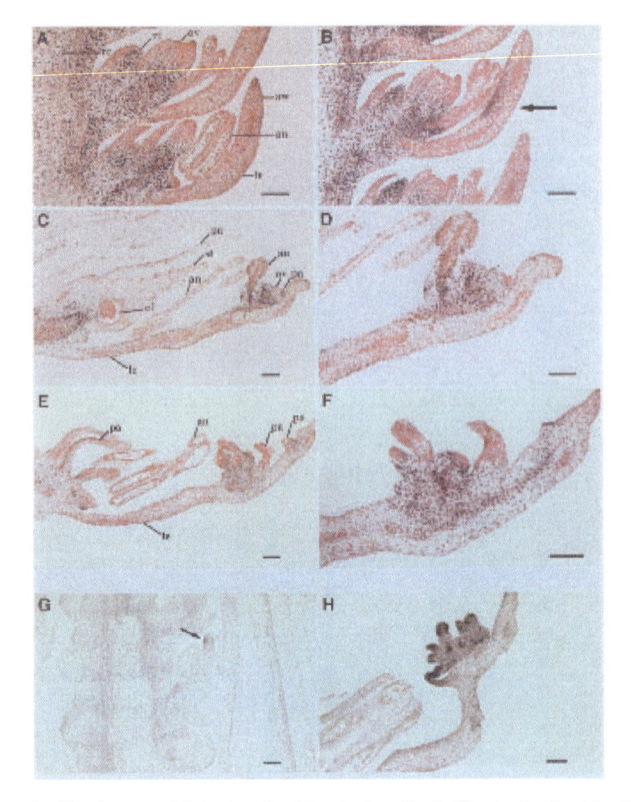

PHOTO 7.5. In Situ Immunohistochemical Analysis with Antibody Against the KNOTTED 1 Protein and Detection of mRNA with the *knotted* 1 cDNA Probe. *Source*: Williams-Carrier et al., 1997. Reprinted with permission of the Company of Biologists, Ltd. A: In an immature spike of wild-type barley the KNOTTED protein is detected only in the stem portions of the rachis (rc) of immature florets and in the rachillae (rl). B: In the immature florets of the transgenic plant expressing the *knotted* gene, the protein is also seen in the lemma-awn (le-aw) transition zone at a location that is presumed to develop into an ectopic floret (arrow). C: An individual immature floret of the transgenic barley is immunochemically stained with the KNOTTED antibody in a normal fashion at the base of the ovule (ol) of the primary floret (left). Ectopic staining is seen in the epidermal and subepidermal layers of the additional floret as well as in the lemma at the position of the developing floret. D: Higher magnification of C. E: An individual floret of the *Hooded* mutant. Typical zones of expression are seen at the base of the normal floret at the left. Expression of the homeobox protein analogous to that of the barley plant transgenic for the *knotted* gene is recorded in the ectopic flower on the lemma at right. F: A closer view of the ectopic floret (an = anther; aw = awn; le = lemma; ol = ovule; ov = ovary; pa = palea; rc = rachis; rl = rachilla; st = stigma). G: Dioxygenin labeling in a section of immature spike traces *knotted* 1 mRNA to the lemma-awn transition zone of a primary floret (arrow). This is a presumptive position for an ectopic floret. H: An individual lemma from a floret of the transgenic plant. One anther from the primary floret is at the left; *knotted* mRNA is at the base of the ectopic floret and extends up into the developing floral organs (scale bars/50 μm).

Chapter 8

Molecular Marker-Assisted versus Conventional Selection in Barley Breeding

William T. B. Thomas

BARLEY BREEDING—PROBLEMS AND SUCCESS

Conventional barley breeding involves hybridizing parents that complement each other for desirable characteristics and selecting superior recombinants from among the progeny. This relies upon the presence of genetic variation for the characters and a suitable selection screen to identify desirable lines. The effectiveness of selection then depends upon the heritability of the character being measured. Characters with high heritabilities have a greater degree of genetic control, and single selection screens are very effective in such cases. This applies particularly to characters controlled by major genes, such as the barley dwarfing genes *sdw1* and *ari-e.GP,* those conferring disease resistance, and vernalization genes. Many important characters, such as yield and malting quality, are, however, controlled by a number of genes, each of small effect. Such characters are generally more subject to modification by the environment and its interaction with genotype and are therefore of relatively low heritability. Selection in a single environment for such characters is unreliable, particularly in the early generations of a breeding program, and plant breeders generally assess such characters in multisite trials over one or more seasons. The heritabilities of established characters such as yield and malting quality have been widely studied, and breeders either choose their selection schemes in light of this knowledge or through empirical means. The genetic control of new characters will generally be unknown but can be readily established. Unbiased estimates of genetic parameters are best established through growing a population of random inbred lines from crosses that segregate for a new character. Doubled haploid, single-seed descent, or pedigree inbreeding can all be used to establish random inbred populations, with doubled haploidy being the most efficient method, provided one has access to an efficient protocol, as dominance effects are absent.

I thank SERAD for their financial support and Drs. Ellis, Forster, and Swanston for allowing me to quote results from unpublished research and for their helpful comments on this manuscript.

Barley breeding has been very successful with an estimated yield advance due to the introduction of new cultivars of 1.2 percent per annum for all barley in the United Kingdom over the period 1947 to 1983 (Silvey, 1986). Martiniello et al. (1987) observed a similar increase of 1 percent per annum due to variety for two-row winter barley in Italy since 1960 (see also Abeledo, Calderini, and Slafer, Chapter 13 in this book). The introduction of Triumph in the United Kingdom in 1980 set new standards for the combination of yield and malting quality, but breeding progress for increased yield has still been maintained in the intervening period. Data gathered from yield trials at the Scottish Crop Research Institue (SCRI) of cultivars recommended for growing in the United Kingdom from 1980 to 1997 show that new varieties have continued to contribute a significant increase of 1 percent per annum (Ellis and Thomas, unpublished data). Malting quality, as measured by hot water extract, also increased over the same period but was more gradual, so the rate of annual increase was not significant (Ellis, Thomas, and Swanston, unpublished data). Evidence to substantiate improvement in quality was, however, apparent from the gradual downgrading of Triumph's malting grade over the period that it was on the United Kingdom recommended list.

The starting point of any breeding program is choosing parental combinations. Barley breeders generally choose parents that complement each other for the overall objectives of the program. Breeders prefer to use pair crosses, if possible, but will use three-way or more complex crosses should they be required to meet the objectives. There is no obvious connection between the number of parents used to make a cross and the commercial success of the lines that are produced from it. Chariot and Optic are currently the most popular spring barley cultivars in the United Kingdom, being the results of pair and three-way crosses, respectively. In contrast Camargue, which was a popular spring barley cultivar in Scotland, was the result of a complex cross.

Most commercial barley breeding programs in the United Kingdom assess each year several hundred crosses that are made by hybridizing superior parents emerging from their own and competing programs. Often the merits of potential parents are judged on the basis of limited information, although breeders may have an idea of the disease resistance genes carried by individual lines. Breeders, thus, have little idea of the number of gene differences between the parents of their crosses, let alone the linkage relationships that they are trying to manipulate. Just one difference on each chromosome between parents in a pair cross means that there are 2^7 or 128, different homozygous combinations. To have a 75 percent chance of identifying just one of these requires a population of over 176 individuals. With a large number of crosses, this soon becomes unmanageable, and breeders compromise by selecting among progeny from crosses between similar parents, which maximizes their chances of identifying superior recombinant inbreds.

Many attempts have been made to find methods that successfully identify the best crosses in a breeding program. Diallel analysis was used to examine agronomic and yield characters (e.g., Riggs and Hayter, 1975), and Whitehouse (1969) attempted to identify good crosses for malting quality by canonical analysis. More recently, cross prediction methods have been tried. The probability of producing superior inbred lines from a cross for a character can be estimated as the normal probability integral corresponding to

$$(s-m)/\sqrt{(V_A)}$$

where s is the standard one is trying to exceed, and m and V_A are the mean and additive genetic variation, respectively, of the cross for that character (Jinks and Pooni, 1976). The only extra parameters that are required to extend univariate to multivariate predictions are the genetic correlations between the characters (Pooni and Jinks, 1978). These parameters can be estimated from random samples of doubled haploid lines derived from the F_1 of a cross or from F_3 families. Predictions based on parameters estimated from doubled haploids (Powell et al., 1985) and F_3 families (Thomas, 1987) have been used to give an accurate ranking of cross potential, although the predicted number of superior recombinants did not agree very accurately with the observed in either case. In practice, breeders are not extracting the extremes of the distribution for many of the characters that they are trying to recombine in a successful new cultivar. This means that the estimate of V_A becomes less important in determining the overall potential of a cross. From a range of predictions of cross potential gathered at SCRI over a number of seasons, it was found that the estimates of V_A for each character, from a multiple regression analysis of the predictions with the parameters as explanatory variables, were not significant, whereas the estimates of m and s were highly significant (Thomas, unpublished data). If one is therefore relying on the mean for the prediction, one might as well estimate it from the midparental value of the parents; i.e., cross the best with the best. Whitehouse (1969) reported little deviation of cross means from midparental values from canonical analysis of parents and F_3 families. The problem is successfully identifying the best parents at an early enough stage in their development. Improvement in parental selection prior to crossing would be of the greatest benefit to plant breeders.

CONVENTIONAL BREEDING METHODS

Selection of superior recombinants is normally done through pedigree inbreeding, with populations of several thousand from each cross being raised for the F_2 generation. Selection of F_2's is carried out for characters with high heritabilities, such as disease resistance, plant height, and maturity. A number of variations are possible for the succeeding generations, ranging from the F_2 bulk to full pedigree systems (Lupton and Whitehouse, 1957). The advantage of the former is that seed amounts can be bulked up quickly, enabling plot yield trials from the F_4 generation onward. The disadvantage of the F_2 bulk scheme is that by the F_6 generation, one will have assembled a reasonable amount of data, but each selection is likely to be a heterogeneous mixture of a number of homozygous components, which will need to be isolated, bulked up, and retested. The advantage of the full pedigree system is that homozygous lines are identified and can then be bulked up directly for trials and multiplication. The disadvantage is that a large number of lines have to be maintained in a program before any information on yield is obtained, which is one of the key characters in a successful cultivar. Given that one presumes selection has been successful for the highly heritable characters in the early stages of a breeding program, it does not make much sense to delay selection for the key characters such as yield and quality for too long, so breeders generally use variations of the pedigree line trial scheme (Lupton and Whitehouse, 1957), which essentially is a hybrid between the F_2 bulk and full pedigree. A subsample of the bulk is taken at around the F_4 or F_5 stage and is used to derive a homozygous multiplication stock that is gradually substituted for the bulk in the trialing scheme.

The application of the derivatives of the pedigree scheme to barley breeding has largely been derived empirically, although many studies have been carried out to examine the genetic architecture of characters under selection (e.g., Riggs and Hayter, 1975; Tapsell and Thomas, 1983; Thomas and Tapsell, 1983). The objective of the pedigree system is not only to select the best genotypes but also to produce inbred lines for marketing. As selection is more meaningfully carried out on homozygous lines (Snape, 1982), doubled haploids have been deployed in barley breeding as a way to speed up the breeding cycle. The "instant inbreds" produced by the doubled haploid process mean that all the selection stages can be carried out on homozygous material, thus improving the accuracy of the estimation of phenotype. The stage at which doubled haploidy is most usefully deployed in a breeding scheme depends to some extent upon the efficiency of the production system. Barley doubled haploids were produced initially via the *Hordeum bulbosum* method (Kasha and Kao, 1970). This was labor intensive, and the numbers that could be produced annually were in the order of a few thousand, so it could not be used as a direct substitute for the pedigree system, in which breeders assess several hundred thousand individuals at the F_2 stage. It could be applied to elite crosses, but this only gives an advantage when selecting parents from the early stages of previous cycles of the program whose phenotypic characterization is limited. Some preliminary selection at the F_2 and even the F_3 stage to identify promising selections from which doubled haploids could be produced is an alternative way of deploying doubled haploidy in plant breeding but has the disadvantage of reducing the time advantage of the method. This is especially so for spring barley, for which shuttle breeding is carried out by many breeders to rapidly develop inbred lines, and the *Hordeum bulbosum* technique has not been widely used in barley breeding, although it has led to some successful cultivars (Pickering and Devaux, 1991).

Another and microspore culture are alternative methods of producing doubled haploids, the latter being particularly attractive, as it lends itself to batch procedures. Clapham (1973) first reported successful anther culture of barley, but progress was limited until maltose was substituted for sucrose in the culture medium (Hunter, 1988; Finnie, Powell, and Dyer, 1989). The technique was initially highly genotype dependent, but gradual improvements have been made so that an average of over 25 fertile green plants per 100 anthers cultured has been achieved over a range of barley crosses (Cistue, Ramos, and Castillo, 1999). This success rate is sufficiently high to enable doubled haploid production on a large enough scale to base whole breeding programs on F_1-derived doubled haploids (Kasha, 1996). However, although doubled haploid lines are regularly entered into official trials, many breeders persist with conventional schemes. This emphasizes the fact that breeders have to be convinced that there will be a demonstrable improvement in the efficiency of their program before they are prepared to invest in alternative technologies, whether doubled haploids or marker-assisted selection (Ceccarelli and Grando, 1993).

MARKER-ASSISTED SELECTION

Major Genes

Breeders have used markers wherever possible to improve the efficiency of their breeding programs. For instance, the *Hs* gene controlling hairy leaf sheath in barley is sufficiently close to the vernalization gene *Sh1* (Sogaard and von Wettstein Knowles, 1987) for it to be used as a diagnostic marker in winter x spring barley crosses. Such examples are exceedingly rare for morphological and isozyme markers in barley, and it is only with the advent of molecular markers that sufficient polymorphism could be detected to identify closely linked markers that could be used in marker-assisted selection for plant breeding. The efficiency of molecular markers in selection for major genes has been conclusively demonstrated, particularly when introducing genes from "exotic" germplasm in a backcrossing program (e.g., Melchinger, 1990). The main benefit that the use of molecular marker-assisted selection can bring in such a scheme is that phenotypic selection is no longer necessary, so that both dominant and recessive genes can be identified in the F_1's from a backcrossing series. When marker-assisted selection for the target character is coupled with selection against donor in the remainder of the genome, then donor genome elimination is considerably accelerated (Tanksley et al., 1989). For instance, Toojinda et al. (1998) demonstrated that lines approximately equivalent to BC_3 could be obtained through a combination of marker-assisted selection in the BC_1 generation followed by doubled haploid production. The BC_3 equivalent inbred lines can be produced within a two-year time scale for both winter and spring barley, which, if one allows a further three years for multiplication and trialing means that lines can be entered into official trials five years after the initial cross. This compares very favorably with about seven years for a dominant gene and 12 for a recessive gene in a conventional backcrossing program. In a practical breeding situation, one might wish to substitute agronomically improved genotypes for the initial recurrent parent and, providing all the parental lines had been adequately genotyped, marker-assisted selection for both the target and background would greatly accelerate the deployment of new major genes, resulting in improved cultivars.

There are, however, few published examples of marker-assisted selection in barley breeding, and commercial breeders have largely persisted with their tried and trusted conventional selection programs. Table 8.1 summarizes the published examples of major genes that might be of relevance to commercial barley cultivars for which molecular markers could be deployed in their selection. It can be seen that most are disease resistance genes for which selection is inexpensive by conventional means and is also effective. The exceptions are the genes involved in malting quality traits, such as beta-glucanases and beta-amylases, and virus resistance, for which direct selection is less easy. Marker-assisted selection is of value in these cases, and it is noteworthy that a variety of marker systems have been developed to aid selection for resistance to the barley yellow mosaic virus (BaYMV) complex (Table 8.1). In some parts of the world, the BaYMV complex is a major problem, and with the difficulties in establishing an effective phenotypic screen, marker-assisted selection for resistance to it has been proposed (Tuvesson et al., 1998). This one disease has received a lot of attention with restriction fragment length polymorphism (RFLP) markers being developed for the *rym4* gene initially, followed by random amplified polymorphic DNA (RAPD), sequence-tagged site (STS), and simple sequence repeat (SSR) markers. The Bmac29 SSR is particularly interesting in that it is closely linked to the resistance locus on chromosome 3(3H) (Graner et al., 1999) and can distinguish between the *rym4* and *rym5* resistance alleles as well as the susceptible allele. The size difference between the resistant and susceptible Bmac29 alleles is sufficiently large for them to be separated on agarose gels (R. Meyer, personal communication), so the SSR marker appears to be especially suitable for marker-assisted selection. In other regions of the world, or when resistance to the BaYMV complex is not the major objective, breeders have tended to use the available molecular markers more to choose parents or between otherwise similar lines; i.e., small-scale screening has been carried out.

TABLE 8.1. List of Major Genes That Could Potentially Be Utilized in Barley Breeding, Their Homeologous Chromosomal Locations (Chrom.) and the Distance (cM), Where Known, to the Nearest Molecular Marker

Gene	Function	Marker Type	Chrom.	Distance	Reference
??	eta-xylanase	RFLP	5(1H)	0.7	Banik et al., 1997
??	limit dextrinase	SSR	7(5H)	0	Li et al., 1999
Amo1	high-amylose starch	RFLP	1(7H)	2	Schondelmaier et al., 1992
ant28-484	proanthocyanidin production	RAPD	3(3H)	0	Garvin et al., 1998
ari-e.GP	dwarfing gene	SSR	5(1H)	5	Thomas et al., 1998
Bmy1	beta-amylase activity	STS	4(4H)	0	Erkkila, 1999
Dwf2	dwarfing gene	SSR	4(4H)	5.7	Ivandic et al., 1999
ea7	photoperiod response	RFLP	6(6H)	6.7	Stracke and Borner, 1998
eph	epiheterodendrin production	SSR	1(7H)	7	Swanston et al., 1999
gai	dwarfing gene	RFLP	2(2H)	0	Borner et al., 1999
gal	dwarfing gene	RFLP	2(2H)	3.6	Borner et al., 1999
Glb (x7)	(1-3)-beta-glucanase	RFLP, PCR	3(3H)		Li et al., 1996
grm	gramine synthesis	RFLP	1(7H)	1.4	Yoshida et al., 1997
Ha2	resistance to CCN	RFLP	2(2H)	~4	Kretschmer et al., 1997
Ha4	resistance to CCN	RFLP	5(1H)	6.2	Barr et al., 1998
Ipa1	phytic acid content	STS	2(2H)	0	Larson et al., 1998
Mla	resistance to powdery mildew	RFLP	1(7H)	<1	Schuller et al., 1992
Mla6	resistance to powdery mildew	RFLP	1(7H)	<5	Graner et al., 1991
Mla25-28	resistance to powdery mildew	RFLP	1(7H)	<1	Jahoor and Fischbeck, 1993
Mla29	resistance to powdery mildew	RFLP	1(7H)	<1	Kintzios et al., 1995
Mla32	resistance to powdery mildew	RFLP	1(7H)	<1	Kintzios et al., 1995
Mlf	resistance to powdery mildew	RFLP	7(5H)	5.3	Schonfeld et al., 1996
Mlg	resistance to powdery mildew	RFLP	4(4H)	~1	Gorg et al., 1993
Mlj	resistance to powdery mildew	RFLP	5(1H)	3.5	Schonfeld et al., 1996
MlLa	resistance to powdery mildew	RFLP	2(2H)	1	Giese et al., 1993
MlLa	resistance to powdery mildew	RFLP	2(2H)	3	Hilbers et al., 1992
MlLa	resistance to powdery mildew	ASA	2(2H)	<1	Mohler and Jahoor, 1996

TABLE 8.1 *(continued)*

Gene	Trait	Marker	Chr	cM	Reference
mlo	resistance to powdery mildew	RFLP	4(4H)	~1	Hinze et al., 1991
mlo	resistance to powdery mildew	RAPD	4(4H)	1.6	Manninen et al., 1997
mlt	resistance to powdery mildew	RFLP	7(5H)	3.2	Schonfeld et al., 1996
MlTR	resistance to powdery mildew	RFLP	5(1H)	5	Falak et al., 1999
n	hull-less grain	RFLP	7(5H)	0	Heun et al., 1991
paZ	protein Z formation	RFLP	4(4H)	0	Kaneko et al., 1999
Ppd-H1	photoperiod response	RFLP	2(2H)	1.1	Laurie et al., 1995
Ppd-H2	photoperiod response	RFLP	2(2H)	3	Laurie et al., 1995
Pt,,a	resistance to net blotch (net)	STS	3(3H)	0.8	Graner et al., 1996
Rdg1a	resistance to leaf stripe	RFLP	2(2H)	0.2	Thomsen et al., 1997
Rh	resistance to scald	RAPD	3(3H)	6.8	Barua, Chalmers, Hackett, et al., 1993
Rh	resistance to scald	RFLP	3(3H)	0	Graner and Tekauz, 1996
Rh	resistance to scald	ASA	3(3H)		Penner et al., 1996
Rh2	resistance to scald	RFLP	7(5H)	0	Schweizer et al., 1995
Rpg1	resistance to stem rust	RFLP	7(5H)	0.3	Kilian et al., 1994
Rpg1	resistance to stem rust	ASA	7(5H)	0	Penner et al., 1995
rpg4	resistance to stem rust	RAPD	5(1H)	0.8	Borovkova et al., 1995
Rph9	resistance to leaf rust	SSR	5(1H)	10.2	Borovkova et al., 1998
Rph12	resistance to leaf rust	RAPD	5(1H)	17.8	Borovkova et al., 1998
Rph16	resistance to leaf rust	STS	2(2H)	0	Ivandic et al., 1998
RphQ	resistance to leaf rust	RAPD	5(1H)	12	Poulsen et al., 1995
RphQ	resistance to leaf rust	STS	5(1H)	1.6	Borovkova et al., 1997
Rpt4	resistance to net blotch (spot)	RFLP	7(5H)	6.9	Williams et al., 1999
Rrs13	resistance to scald	RFLP	6(6H)	7.3	Abbot et al., 1995
Rsm	resistance to BSMV	RFLP	7(5H)	0	Edwards and Steffenson, 1996
rym3	BaYMV complex resistance	RFLP	5(1H)	7.2	Saeki et al., 1999
rym4	BaYMV complex resistance	RFLP	3(3H)	1.2	Graner and Bauer, 1993
rym4	BaYMV complex resistance	STS	3(3H)	1.2	Bauer and Graner, 1995
rym4	BaYMV complex resistance	RAPD	3(3H)	3.2	Weyen et al., 1996
rym5	BaYMV complex resistance	SSR	3(3H)	1.3	Graner, Streng, Kellermann, Schiemann, et al., 1999

rmm7	BaMMV partial resistance	RFLP	1(7H)		Graner, Streng, Kellermann, Proseler, et al., 1999
rym9	BaYMV complex resistance	RFLP, RAPD	4(4H)	0	Bauer et al., 1997
rym11	BaYMV complex resistance	RFLP, RAPD	4(4H)	0	Bauer et al., 1997
sdw1	dwarfing gene	RFLP	3(3H)	12.8	Barua, Chalmers, Thomas, et al., 1993
sdw1	dwarfing gene	RFLP	3(3H)	7.8	Laurie et al., 1993
Sh	vernalization response	RFLP	4(4H)	1	Laurie et al., 1995
Sh2	vernalization response	RFLP	5(1H)	~5	Laurie et al., 1995
Yd2	resistance to BYDV	RFLP	3(3H)	0	Collins et al., 1996
Yd2	resistance to BYDV	STS	3(3H)	0.7	Paltridge et al., 1998

When efficient phenotypic screens are available, it is understandable that breeders are reluctant to divert resources into marker-assisted selection schemes for the same targets. This strategy is probably satisfactory when one is dealing with a gene with no deleterious associations with any other traits, which is not always the case with the introductions of new resistance genes from unadapted germplasm. The *Hordeum laevigatum* mildew and leaf rust resistance was first introduced in the spring barley cultivars Vada and Minerva in the 1960s, but due to an adverse linkage (Swanston, 1987), it was not until the release of the cultivar Doublet in the early 1980s that the resistances were coupled with good malting-quality characteristics. In such cases, it may be more efficient to use marker-assisted selection after the initial round of crossing to generate a large pool of lines carrying the target character and then rely on phenotypic selection for other characters. This would maximize the chances of finding recombinants in which the deleterious linkages of the target character had been broken, whereas relying on conventional selection for the other characters and then using markers to identify lines carrying the target character is more likely to result in their preservation. Such a strategy might not lead directly to cultivars in the first round of crossing but should produce improved lines that, following recrossing and selection, lead to introgression of the target character with minimal deleterious associations. A suggested advantage of deploying marker-assisted selection for major disease resistance genes is that a number of genes for resistance to a particular disease can easily be accumulated in one cultivar. Such resistance gene pyramids arise purely by chance through phenotypic selection but have been advocated as a means of providing more durable disease resistance than that provided by deployment of single major genes in turn. Wolfe (1984) argues that there are three reasons why this strategy might not succeed in breeding for resistance to barley powdery mildew: One is that many of the resistance gene loci are closely linked and opportunities to recombine different sources are limited. Another reason is that single gene resistances are likely to be deployed before being integrated into a pyramid, and with matching virulences for such resistances, it is likely that pathogen races virulent on the complex will soon predominate. The final reason is that possession of a resistance gene complex is no guarantee of commercial success. While marker-assisted selection offers the opportunity to break close linkages and new resistance loci have been found (Schonfeld et al., 1996), the assembly of previously defeated single major gene resistances into a complex does not appear to confer durable powdery mildew resistance (Brown, Simpson, and Wolfe, 1993).

Marker-Assisted Selection for QTL: Detection

Marker-assisted selection would be of greatest benefit in selection of yield and quality, which are the key characters necessary for a successful cultivar but are the hardest to select for efficiently. Many studies have highlighted the fact that marker-assisted selection has the greatest advantage over phenotypic selection when applied to characters of low heritability (Lande and Thompson, 1990), which would suggest that they are ideally suited to selection for yield and quality. Since the development of molecular marker maps for a range of barley crosses, some studies have revealed QTL for both these characters (see Table 8.2). Attempts to verify some of these QTL have also been made, mainly in the Steptoe x Morex population. Larson et al. (1996) attempted to transfer positive alleles from Steptoe into a Morex background but, while this approach was successful for one, it failed for another QTL. Romagosa et al. (1999), in an experiment examining four different yield QTL from Steptoe x Morex detected by Hayes et al. (1993), found that two of the four were detected consistently across a range of environments, but the other two were much more environmentally labile. In a similar experiment examining malting quality, Han et al. (1997) found that, of two QTL that accounted for over 25 percent of the phenotypic variance in hot water extract in the original Steptoe x Morex population, one was nonsignificant and the other accounted for just 6 percent of the phenotypic variance in a further sample of lines from the population. Tinker et al. (1996) detected a QTL region on the short arm of chromosome 5H in Harrington x TR306 that affected four characters. In a verification study carried out on another sample of lines from the cross, three of the four effects in the region were again detected, but the magnitude of the effects was different from that of the original study (Spaner et al., 1999).

QTL x environment interaction has also been reported for yield and quality across a wide range of environments with relatively diverse germplasm (Hayes et al., 1993; Tinker et al., 1996; Mather et al., 1997) and a relatively narrow range of environments and germplasm (Thomas et al., 1995, 1996). The existence of such interactions is not surprising since genotype x envir-

TABLE 8.2. List of Crosses, Population Type (Pop.), and Groupings of Characters for Which Genome-Wide QTL Studies Have Been Published

Cross	Pop.	Agronomic	Yield	Quality	Disease	Other
Arda x Opale	DH					1
Arena x Hor9088	F2				2	
Blenheim x Kym	DH	3	4	5		
Blenheim x E224/3	DH	6, 7	6, 7	7, 8	6, 9	10
Calchicuma sib x Bowman	DH	11	11	11	11, 12	
Chebec x Harrington	DH	13		13	13	
Chevron x M69	RIL				14	
Clipper x Sahara	DH			13		15
Derkado x B83-12/21/5	DH		16	17	9, 18	
Dicktoo x Morex	DH	19		20		19, 21, 22, 23
Dicktoo x Plaisant	DH	24				24
Galleon x Haruna Nijo	DH			13		
Harrington x TR306	DH	25	25	26, 27	28, 29	30, 31, 32
Igri x Danilo	DH	33	33		33, 34	
Igri x Triumph	DH	35				
Lina x *H. spontaneum*	DH					36
Prisma x Apex	RIL		37			37, 38
Proctor x Nudinka	DH		39		39, 40, 41	
Steptoe x Morex	DH	21	21	21, 27, 42, 43	28, 44, 45, 46	31, 47, 48, 49
Tadmor x ER/APM	RIL	50		50		50, 51
Tysofte Prentice x Vogelsanger Gold	DH	52	53			54
Vada x L94	RIL		55		56, 57	

1. Tuberosa et al., 1997; abiotic stress
2. Richter et al., 1998
3. Bezant et al., 1996
4. Bezant et al., 1997a
5. Bezant et al., 1997b
6. Thomas et al., 1995
7. Powell et al., 1997
8. Thomas et al., 1996
9. Thomas et al., 1997
10. Thomas and Young, 1996; tissue culture
11. Hayes et al., 1996
12. Chen et al., 1994
13. Langridge et al., 1996
14. de la Pena et al., 1999
15. Jeffries et al., 1999; abiotic stress
16. Thomas et al., 1998
17. Swanston et al., 1999
18. Thomas, 1999
19. Pan et al., 1994
20. Oziel et al., 1996
21. Hayes et al., 1993; abiotic stress
22. Sanguineti et al., 1994; abiotic stress

TABLE 8.2 *(continued)*

23. Hayes et al., 1997; abiotic stress
24. Karsai et al., 1997; abiotic stress
25. Tinker et al., 1996
26. Mather et al., 1997
27. Iwasa et al., 1999
28. Sato et al., 1996
29. Spaner et al., 1998
30. Kantani et al., 1998; spoilage
31. Mano and Takeda, 1997; abiotic stress
32. Moharriamipour et al., 1997; aphid resistance
33. Backes et al., 1995
34. Backes et al., 1996
35. Laurie et al., 1995
36. Ellis et al., 1997; abiotic stress
37. Yin, Stam, et al., 1999
38. Yin, Kropff, and Stam, 1999; physiological traits
39. Heun, 1992
40. Geise, 1996
41. Pecchioni et al., 1999
42. Ullrich et al., 1993
43. Han et al., 1995
44. Steffenson et al., 1996
45. Iwasa et al., 1997
46. El-Attari et al., 1998
47. Mano et al., 1996; tissue culture
48. Taketa et al., 1998; tissue culture
49. Borem et al., 1999; starch granules
50. Baum et al., 1996; physiological characters
51. Teulat et al., 1998; abiotic stress
52. Kjaer et al., 1995
53. Kjaer and Jensen, 1996
54. Kjaer and Jensen, 1995; elements
55. Stam et al., 1997
56. Qi et al., 1998
57. Qi et al., 1999

onment interactions are known to be commonplace for yield and quality, but it does pose problems in QTL detection and deployment. Software such as MQTL (Tinker and Mather, 1995) and PLABQTL (Utz and Melchinger, 1996) is available to detect and estimate the magnitude of QTL × environment interactions. The approach seems to be robust, as Romagosa et al. (1996) compared the results obtained from an MQTL analysis over the whole Steptoe x Morex data set, with mapping of PCA scores obtained from additive main effect and multiplicative interaction model (AMMI) analysis, and detected coincident peaks. There was evidence of crossover interactions, from the Steptoe x Morex data set so, in a marker-assisted selection program, a breeder would have to be aware of the suitability of selected QTL alleles for the target environment. Most QTL × environment interactions however, have been found to be magnitude effects, which emphasizes the need for testing over several environments for successful QTL detection.

While QTL of quite large effect have been found in most of the studies reported in Table 8.2, multilocus models of the multiple QTL detected for some of the characters have never been found to account for all the phenotypic variation for a character. Furthermore, the population sizes used in the QTL studies reported in Table 8.2 are not big enough to reliably detect QTL of small effect, and it is unlikely to be an efficient deployment of resources to use a mapping population to seek to identify all the QTL that may control a character. This means that marker-assisted selection will not replace phenotypic selection for quantitative traits but be an adjunct that can be deployed to increase the efficiency of selection. The confidence interval surrounding a QTL is also large, often more than 30 cM (Kearsey and Farquhar, 1998), which can cause problems in choosing flanking markers, as the interval may also encompass a region containing an undesirable allele for another important character. The development of recombinant chromosome substitution lines (Paterson et al., 1990) has led not only to refining the location of QTL but also to the discovery of new loci, some in linkage repulsion, in *Brassica oleracea* (Rae, Howell, and Kearsey, 1999). Such a resource would be of great benefit in refining QTL locations in barley. An alternative approach is the simultaneous mapping and development of improved lines through advanced backcross QTL analysis (Tanksley and Nelson, 1996). This approach has been demonstrated to have led to the introgression of useful variation from wild relatives into tomato (Tanksley et al., 1996) and rice (Xiao et al., 1998), and it therefore appears to have a lot of potential when working with wide crosses. Routine barley breeding is still making considerable advances working within the adapted gene pool (Rasmusson and Phillips, 1997), and the successful deployment of marker-assisted selection needs to be demonstrated within a relatively narrow genetic base for it to be accepted by commercial breeders.

Marker-Assisted Selection for QTL: Deployment—Problems and Opportunities

The efficiency of a number of different selection schemes in a sample of lines from the Steptoe x Morex cross has been compared for yield (Romagosa et al., 1999) and quality (Han et al., 1997). Both studies concluded that marker-assisted selection, when combined with phenotypic selection, was at least as effective as phenotypic selection alone. The real attraction of the combined approach is that one can use marker-assisted selection as a negative selection tool for the early elimination of all the lines with the "wrong" QTL alleles and then concentrate resources on phenotypic selection among an elite germplasm pool. Accumulation of favorable QTL alleles for yield has, to some extent, been tested through crossing two selections from the Steptoe x Morex mapping population that would segregate for four of the seven detected QTL, the other three being fixed for the favorable (Steptoe) allele (Zhu et al., 1999). The results were somewhat disappointing in that only 2 of the 24 lines selected by genotyping 100 F_1-derived doubled haploids from the cross between the two selections exceeded the mean yield of Steptoe across the five test environments. One of these lines was a recombination of the Steptoe alleles, and the other had just one Morex allele at the yield QTL loci. Zhu et al. (1999) concluded that phenotypic selection would have been as effective as the marker-assisted selection and that marker-assisted selection would be better targeted toward allelic combinations rather than individual loci.

There are a number of reasons why there are few examples of marker-assisted selection being used in plant breeding. Young (1999) highlights the time and effort required to develop a suitable marker screen for use in selecting for the *rhg1* major gene for resistance to soybean cyst nematode, a similar situation to the resources devoted to identifying molecular markers for use in breeding for resistance to the BaYMV complex noted earlier. The situation is worse for complex traits such as yield and quality, where accurate estimation of QTL positions and effects is highly unreliable. Beavis (1997) demonstrates this problem in a simulation study for maize, in which he tabulates the efficiency (or power) of various population sizes for detecting various numbers of simulated QTL with varying degrees of heritability. Although barley has a different breeding system, it is worth noting that, for a character with ten QTL, a heritability of 30 percent and a population size of 100, the efficiency of detection was just 9 percent and the effects were greatly over-estimated. Most of the studies reported in Table 8.2 have been carried out on populations of between 100 and 200 individuals, and yield and quality are of relatively low heritability, so it is perhaps not surprising that the QTL validation experiments noted earlier failed to redetect all the QTL. The QTL that are detected, however, are highly likely to be "real" QTL, rather than "false positives" (Beavis, 1997).

There are other reasons why marker-assisted selection has yet to be taken up on a large scale. The first is that a considerable investment in resources has to be made in order to carry out marker-assisted breeding using current technology. Using manual DNA extraction protocols, one could not expect to process more than 100 samples per person per day, i.e., a maximum output of less than 25,000 per annum. This is an order of magnitude less than each year's initial population in a commercial breeding program. When working with a single marker, there are opportunities for streamlining the process so that 5,000 samples could be handled in one day (Tuvesson et al., 1998). Even so, this would mean devoting between a three- to six-month person-year to DNA preparation for a commercial program. Increased use of robotics and/or direct PCR amplification from tissue would alleviate the bottleneck of DNA extraction but requires capital investment. Capital investment will not be made, however, until there is evidence of a demonstrable benefit in terms of breeding efficiency. This means that marker-assisted selection programs need to be demonstrably more effective than phenotypic selection on progeny from crosses between elite parents. The cost of consumable resources for marker-assisted selection has to decrease by a considerable amount as well. The entire budget for a breeding and multiplication program would be overspent just through deploying marker-assisted selection on an F_1-derived doubled haploid program based upon 100,000 individuals, let alone a conventional F_2 program, given current costs. Marker-assisted selection does, however, offer the possibility of decreasing the population size required to reach a selection target (Knapp, 1998). The development of DNA chip technology may prove a viable high-throughput method for genotyping (Hoheisel, 1997) and could provide a suitable medium upon which to carry out large-scale screening for biallelic markers, such as single nucleotide polymorphisms, that are diagnostic for target characters. Early generation screens for most major gene characters in barley are currently orders of magnitude more cost-effective than marker-assisted selection, so it would be a waste of resources to deploy marker-assisted selection at this stage of a breeding program. Phenotypic selection for yield and quality is a large component of the annual breeding budget, and it is therefore logical to carry out early generation phenotypic screening, to ensure that the agronomic phenotype largely meets requirements, followed by marker-assisted selection for yield and quality. This reduces the amount of material being entered into the yield and quality testing, which currently forms a large part of the annual breeding budget. Ideally, such a marker-assisted selection scheme should be based upon leaf material sampled from selections in the field and be rapid enough to enable genotypic screening in time for harvest.

The markers that flank the QTL reported in the studies listed in Table 8.2 are largely RFLPs and AFLPs (A = amplified), systems that are relatively monomorphic in barley. Although evidence suggests that RFLP marker-trait associations are maintained for major gene characters such as photoperiod and vernalization response (Igartua et al., 1999), no comparable study has been published for QTL. Given the problems of deploying RFLPs in mapping crosses between elite barley parents, it is unlikely that marker-QTL associations will be maintained across parental genotypes. The development of SSR markers may alleviate the problem of monomorphic marker systems, as they have been found to be very polymorphic, which, together with their multiallelic nature, makes them highly informative in discriminating between elite lines. With just four SSR markers it is possible to distinguish between all the cultivars on the U.K. recommended lists of winter and spring barley (Russell et al., 1997). SCRI has developed a library of over 500 SSRs, many of which have been placed on a reference map (Ramsay et al., 2000). These markers are available to the barley research community, and as they become incorporated onto more and more genetic maps, they can be expected to provide more discriminatory information for the tagging of QTL. Given a reasonably dense map of SSR loci, it will be possible to combine closely linked loci to define a "haplotype" that should be diagnostic for a QTL, should it not be possible to identify a single SSR for the purpose. It would then be possible for breeders to genotype their parents, identify the QTL alleles that they carry, and choose appropriate crossing and selection strategies based upon that knowledge. At a wider level, molecular markers are ideal to screen barley collections, particularly land races and *Hordeum spontaneum,* for novel alleles at loci tagging QTL. The marker alleles could then be used to introgress rapidly the donor segment into an adapted background through a backcrossing scheme, and the resultant lines used to establish the potential worth of the novel alleles in developing new cultivars.

Studies of the variation that exists within the cultivated gene pool at SCRI have shown that 17 "foundation genotypes" carry over 70 percent of the total SSR allelic variation found in a sample of spring barley cultivars and key progenitors taken over the whole of the twentieth century (Russell et al., 2000). There are 2×10^{19} different allelic combinations that can be assembled from the foundation genotypes. Even the cultivars released since 1985, which contain just 35 percent of the total allelic variation, provide enough combinations for a plant breeding program with an initial population of 500,000 to run for over 8.5 million years. SSRs are largely anonymous markers, and although they are extremely useful in genotyping, assignment of functionality is a different matter, and many of these different SSR allelic combinations may not alter phenotype. A large number of expressed sequence tags (ESTs) are being developed around the world with the International Triticeae EST Cooperative, established to act as a publicly available source of information (Langridge et al., 1999). The information that resides in such databases can be expected to be transferred to genetic maps, either through the development of single nucleotide polymorphisms (SNPs) or other mapping strategies. This will lead to the identification of candidate genes for QTLs, which will make genetic maps much more functional and enable truly targeted marker-assisted selection. Functional maps would also have the additional benefit of enabling the variation that exists in unadapted germplasm to be screened and introgressed efficiently.

REFERENCES

Abbott, D.C., E.S. Lagudah, and A.H.D. Brown (1995). Identification of RFLPs flanking a scald resistance gene on barley chromosome 6. *Journal of Heredity* 86: 152-154.

Backes, G.A., A. Graner, B. Foroughi-Wehr, G. Fischbeck, G. Wenzel, and A. Jahoor (1995). Localization of quantitative trait loci (QTL) for agronomic important characters by the use of a RFLP map in barley (*Hordeum vulgare* L.). *Theoretical and Applied Genetics* 90: 294-302.

Backes, G., G. Schwarz, G. Wenzel, and A. Jahoor (1996). Comparison between QTL analysis of powdery mildew resistance in barley based on detached primary leaves and on field data. *Plant Breeding* 115: 419-421.

Banik, M., C.D. Li, P. Langridge, and G.B. Fincher (1997). Structure, hormonal regulation, and chromosomal location of genes encoding barley $(1 \rightarrow 4)$-β-xylan endohydrolases. *Molecular and General Genetics* 253(5): 599-608.

Barr, A.R., K.J. Chalmers, A. Karakousis, J.M. Kretschmer, S. Manning, R.C.M. Lance, J. Lewis, S.P. Jeffries, and P. Langridge (1998). RFLP mapping of a new cereal cyst nematode resistance locus in barley. *Plant Breeding* 117: 185-187.

Barua, U.M., K.J. Chalmers, C.A. Hackett, W.T.B. Thomas, W. Powell, and R. Waugh (1993). Identification of RAPD markers linked to a *Rhynchosporium secalis* locus in barley using near-isogenic lines and bulked segregant analysis. *Heredity* 71: 177-184.

Barua, U.M., K.J. Chalmers, W.T.B. Thomas, C.A. Hackett, V. Lea, P. Jack, B.P. Forster, R. Waugh, and W. Powell (1993). Molecular mapping of genes determining height, time to heading, and growth habit in barley *(Hordeum vulgare)*. *Genome* 36: 1080-1087.

Bauer, E. and A. Graner (1995). Basic and applied aspects of the genetic analysis of the *ym4* virus resistance locus in barley. *Agronomie* 15: 469-473.

Bauer, E., J. Weyen, A. Schiemann, A. Graner, and F. Ordon (1997). Molecular mapping of novel resistance genes against barley mild mosaic virus (BaMMV). *Theoretical and Applied Genetics* 95(8): 1263-1269.

Baum, M., H. Sayed, J.L. Araus, S. Grando, S. Ceccarrelli, G. Backes, V. Mohler, A. Jahoor, and G. Fischbeck (1996). QTL analysis of agronomic important characters for dryland conditions in barley. In *V International Oat Conference and VII International Barley Genetics Symposium Proceedings–Poster Sessions* (1), eds. A. Slinkard, G. Scoles, B. Rossnagel. Saskatoon: University of Saskatchewan Extension Press, pp. 241-243.

Beavis, W.D. (1997). QTL analyses: Power, precision, and accuracy. In *Molecular Dissection of Complex Traits*, ed. A.H. Paterson. Boca Raton, FL: CRC Press, pp. 145-162.

Bezant, J., D. Laurie, N. Pratchett, J. Chojecki, and M. Kearsey (1996). Marker regression mapping of QTL controlling flowering time and plant height in a spring barley (*Hordeum vulgare* L.) cross. *Heredity* 77: 64-73.

Bezant, J., D. Laurie, N. Pratchett, J. Chojecki, and M. Kearsey (1997a). Mapping QTL controlling yield and yield components in a spring barley (*Hordeum vulgare* L.) cross using marker regression. *Molecular Breeding* 3: 29-38.

Bezant, J.H., D.A. Laurie, N. Pratchett, J. Chojecki, and M.J. Kearsey (1997b). Mapping of QTL controlling NIR predicted hot water extract and grain nitrogen content in a spring barley cross using marker-regression. *Plant Breeding* 116: 141-145.

Borem, A., D.E. Mather, D.C. Rasmusson, R.G. Fulcher, and P.M. Hayes (1999). Mapping quantitative trait loci for starch granule traits in barley. *Journal of Cereal Science* 29: 153-160.

Borner, A., V. Korzun, S. Malyshev, and V. Ivandic (1999). Molecular mapping of two dwarfing genes differing in their GA response on chromosome 2H of barley. *Theoretical and Applied Genetics* 99: 670-675.

Borovkova, I.G., Y. Jin, and B.J. Steffenson (1998). Chromosomal location and genetic relationship of leaf rust resistance genes *Rph9* and *Rph12* in barley. *Phytopathology* 88(1): 76-80.

Borovkova, I.G., Y. Jin, B.J. Steffenson, A. Kilian, T.K. Blake, and A. Kleinhofs (1997). Identification and mapping of a leaf rust resistance gene in barley line Q21861. *Genome* 40(2): 236-241.

Borovkova, I.G., B.J. Steffenson, Y. Jin, J.B. Rasmussen, A. Kilian, A. Kleinhofs, B.G. Rossnagel, and K.N. Kao (1995). Identification of molecular markers

linked to the stem rust resistance gene *rpg4* in barley. *Phytopathology* 85(2): 181-185.
Brown, J.K.M., C.G. Simpson, and M.S. Wolfe (1993). Adaptation of barley powdery mildew populations in England to varieties with two resistance genes. *Plant Pathology* 42: 108-115.
Ceccarelli, S. and S. Grando (1993). From conventional plant breeding to molecular biology. In *International Crop Science*, I. International Crop Science Congress, Ames, Iowa, USA, July 14-22 1992, eds. D.R. Buxton, R.A. Forsberg, B.L. Blad, H. Asay, G.M. Paulsen, and R.F. Wilson. Madison, WI: Crop Science Society of America, pp. 533-537.
Chen, F.Q., D. Prehn, P.M. Hayes, D. Mulrooney, A. Corey, and H. Vivar (1994). Mapping genes for resistance to barley stripe rust *(Puccinia striiformis* f sp *hordei)*. *Theoretical and Applied Genetics* 88: 215-219.
Cistue, L., A. Ramos, and A.M. Castillo (1999). Influence of anther pretreatment and culture medium composition on the production of barley doubled haploids from model and low responding cultivars. *Plant Cell Tissue and Organ Culture* 55: 159-166.
Clapham, D. (1973). Haploid *Hordeum* plants from anthers in vitro. *Zeitschrift fur Pflanzenzuchtung* 69: 142-155.
Collins, N.C., N.G. Paltridge, C.M. Ford, and R.H. Symons (1996). The *Yd2* gene for barley yellow dwarf virus resistance maps close to the centromere on the long arm of barley chromosome 3. *Theoretical and Applied Genetics* 92(7): 858-864.
Edwards, M.C. and B.J. Steffenson (1996). Genetics and mapping of barley stripe mosaic virus resistance in barley. *Phytopathology* 86(2): 184-187.
El-Attari, H., A. Rebai, P.M. Hayes, G. Barrault, G. Dechamp-Guillaume, and A. Sarrafi (1998). Potential of doubled-haploid lines and localization of quantitative trait loci (QTL) for partial resistance to bacterial leaf streak *(Xanthomonas campestris* pv. *hordei)* in barley. *Theoretical and Applied Genetics* 96: 95-100.
Ellis, R.P., B.P. Forster, R. Waugh, N. Bonar, L.L. Handley, D. Robinson, D.C. Gordon, and W. Powell (1997). Mapping physiological traits in barley. *New Phytologist* 137: 149-157.
Erkkila, M.J. (1999). Intron III-specific markers for screening of beta-amylase alleles in barley cultivars. *Plant Molecular Biology Reporter* 17: 139-147.
Falak, I., D.E. Falk, N.A. Tinker, and D.E. Mather (1999). Resistance to powdery mildew in a doubled haploid barley population and its association with marker loci. *Euphytica* 107(3): 185-192.
Finnie, S.J., W. Powell, and A.F. Dyer (1989). The effect of carbohydrate composition and concentration on anther culture response in barley *(Hordeum vulgare* L.). *Plant Breeding* 103: 110-118.
Garvin, D.F., J.E. Miller-Garvin, E.A. Viccars, J.V. Jacobsen, and A.H.D. Brown (1998). Identification of molecular markers linked to *ant28-484,* a mutation that eliminates proanthocyanidin production in barley seeds. *Crop Science* 38(5): 1250-1255.
Giese, H. (1996). Quantitative resistance to barley leaf stripe *(Pyrenophora graminea)* is dominated by one major locus. *Theoretical and Applied Genetics* 93: 97-101.

Giese, H., A.G. Holm-Jensen, H.P. Jensen, and J. Jensen (1993). Localization of the *laevigatum* powdery mildew resistance gene to barley chromosome 2 by the use of RFLP markers. *Theoretical and Applied Genetics* 85(6/7): 897-900.

Gorg, R., K. Hollricher, and P. Schulze-Lefert (1993). Functional analysis and RFLP-mediated mapping of the *Mlg* resistance locus in barley. *Plant Journal* 3: 857-866.

Graner, A. and E. Bauer (1993). RFLP mapping of the *ym4* virus resistance gene in barley. *Theoretical and Applied Genetics* 86: 689-693.

Graner, A., B. Foroughi-Wehr, and A. Tekauz (1996). RFLP mapping of a gene in barley conferring resistance to net blotch *(Pyrenophora teres)*. *Euphytica* 91(2): 229-234.

Graner A., A. Jahoor, J. Schondelmaier, H. Siedler, K. Pillen, G. Fischbeck, G. Wenzel, and R.G. Herrmann (1991). Construction of an RFLP map of barley. *Theoretical and Applied Genetics* 83: 250-256.

Graner, A., S. Streng, A. Kellermann, G. Proseler, A. Schiemann, H. Peterka, and F. Ordon (1999). Molecular mapping of genes conferring resistance to soil-borne viruses in barley—An approach to promote understanding of host pathogen interactions. *Journal of Plant Diseases and Protection* 106: 405-410.

Graner, A., S. Streng, A. Kellermann, A. Schiemann, E. Bauer, R. Waugh, B. Pellio, and F. Ordon (1999). Molecular mapping and genetic fine-structure of the *rym5* locus encoding resistance to different strains of the barley yellow mosaic virus complex. *Theoretical and Applied Genetics* 98(2): 285-290.

Graner, A. and A. Tekauz (1996). RFLP mapping in barley of a dominant gene conferring resistance to scald *(Rhynchosporium secalis)*. *Theoretical and Applied Genetics* 93: 421-425.

Han, F., I. Romagosa, S.E. Ullrich, B.L. Jones, P.M. Hayes, and D.M. Wesenberg (1997). Molecular marker-assisted selection for malting quality traits in barley. *Molecular Breeding* 3: 427-437.

Han, F., S.E. Ullrich, S. Chirat, S. Menteur, L. Jestin, A. Sarrafi, P.M. Hayes, B.L. Jones, T.K. Blake, D.M. Wesenberg, et al. (1995). Mapping of beta-glucan content and beta-glucanase activity loci in barley grain and malt. *Theoretical and Applied Genetics* 91: 921-927.

Hayes, P., D. Prehn, H. Vivar, T. Blake, A. Comeau, I. Henry, M. Johnston, B. Jones, B. Steffenson, C.A. St. Pierre, and F. Chen (1996). Multiple disease resistance loci and their relationship to agronomic and quality loci in a spring barley population. *Journal of Agricultural Genomics* 2: <http://www.ncgr.org/ag/jag/ papers96/paper296/indexp296.html>.

Hayes, P.M., T.H.H. Chen, W. Powell, W. Thomas, Z. Bedo, I. Kasari, and K. Meszaros (1997). Dissecting the components of winter hardiness in barley. *Acta Agronomica Hungarica* 45: 241-246.

Hayes, P. M., B.H. Liu, S.J. Knapp, F. Chen, B. Jones, T. Blake, J. Franckowiak, D. Rasmusson, M. Sorrells, S.E. Ullrich, et al. (1993). Quantitative trait locus effects and environmental interaction in a sample of North American barley germ plasm. *Theoretical and Applied Genetics* 87: 392-401.

Heun, M. (1992). Mapping quantitative powdery mildew resistance of barley using a restriction fragment length polymorphism map. *Genome* 35: 1019-1025.

Heun, M., A.E. Kennedy, J.A. Anderson, N.L.V. Lapitan, M.E. Sorrells, and S.D. Tanksley (1991). Construction of a restriction fragment length polymorphism map for barley *(Hordeum vulgare)*. *Genome* 34: 437-447.

Hilbers, S., G. Fischbeck, and A. Jahoor (1992). Localization of the *laevigatum* resistance gene *MlLa* against powdery mildew in the barley genome by the use of RFLP markers. *Plant Breeding* 109: 335-338.

Hinze, K., R.D. Thompson, E. Ritter, F. Salamini, and P. Schulze-Lefert (1991). Restriction fragment length polymorphism-mediated targeting of the ml-o resistance locus in barley *(Hordeum vulgare)*. *Proceedings of the National Academy of Sciences of the United States of America* 88(9): 3691-3695.

Hoheisel, J.D. (1997). Oligomer-chip technology. *Trends in Biotechnology* 15: 465-469.

Hunter, C.P. (1988). Plant regeneration from microspores of barley *Hordeum vulgare* L. PhD thesis. Ashford: University of London, 235 pp.

Igartua, E., A.M. Casas, F. Ciudad, J.L. Montoya, and I. Romagosa (1999). RFLP markers associated with major genes controlling heading date evaluated in a barley germ plasm pool. *Heredity* 83: 551-559.

Ivandic, V., S. Malyshev, V. Korzun, A. Graner, and A. Borner (1999). Comparative mapping of a gibberellic acid-insensitive dwarfing gene *(Dwf2)* on chromosome 4HS in barley. *Theoretical and Applied Genetics* 98(5): 728-731.

Ivandic, V., U. Walther, and A. Graner (1998). Molecular mapping of a new gene in wild barley conferring complete resistance to leaf rust *(Puccinia hordei* Otth). *Theoretical and Applied Genetics* 97(8): 1235-1239.

Iwasa, T., H. Heta, and K. Takeda (1997). QTL mapping for powdery mildew *(Erysiphe graminis* DC. f. sp. *hordei* EM Marchal) resistance in barley *(Hordeum vulgare* L.). *Bulletin of the Research Institute for Bioresources Okayama University* 5: 69-78.

Iwasa, T., H. Takahashi, and K. Takeda (1999). QTL mapping for water sensitivity in barley seeds. *Bulletin of the Research Institute for Bioresources Okayama University* 6: 21-28.

Jahoor, A. and G. Fischbeck (1993). Identification of new genes for mildew resistance of barley at the *Mla* locus in lines derived from *Hordeum spontaneum*. *Plant Breeding* 110: 116-122.

Jefferies, S.P., A.R. Barr, A. Karakousis, J.M. Kretschmer, S. Manning, K.J. Chalmers, J.C. Nelson, A.K.M.R. Islam, and P. Langridge (1999). Mapping of chromosome regions conferring boron toxicity tolerance in barley *(Hordeum vulgare* L.). *Theoretical and Applied Genetics* 98(8): 1293-1303.

Jinks, J.L. and H.S. Pooni (1976). Predicting the properties of recombinant inbred lines derived by single seed descent. *Heredity* 36: 253-266.

Kaneko, T., N. Hirota, S. Yokoi, R. Kanatani, and K. Ito (1999). Molecular marker for protein Z content in barley *(Hordeum vulgare* L.). *Breeding Science* 49(2): 69-74.

Kantani, R., H. Takahashi, and K. Takeda (1998). QTL analysis for expressivity of hull-cracked grain in two-rowed spring barley. *Bulletin of the Research Institute for Bioresources Okayama University* 5: 183-191.

Karsai, I., K. Meszaros, Z. Bedo, P.M. Hayes, A. Pan, and F. Chen (1997). Genetic analysis of the components of winterhardiness in barley (*Hordeum vulgare* L.). *Acta Biologica Hungarica* 48: 67-76.

Karsai, I., K. Meszaros, P.M. Hayes, and Z. Bedo (1997). QTL analysis of winter-hardiness-related traits in a doubled haploid population of barley developed from a cross between Dicktoo x Plaisant. In *Proceedings of the International Symposium on Cereal Adaptation to Low Temperature Stress in Controlled Environments*, eds. Z. Bedo, J. Sutka, T. Tischner, O., Veisz, and B. Koszegi. Martonvásár: Agricultural Research Institute of the Hungarian Academy of Sciences, pp. 92-96.

Kasha, K.J. (1996). Biotechnology and cereal improvement. In *V International Oat Conference and VII International Barley Genetics Symposium Proceedings–Invited Papers*, eds. G. Scoles and B. Rossnage., Saskatoon: University of Saskatchewan Extension Press, pp. 133-140.

Kasha K.J. and K.N. Kao (1970). High frequency haploid production in barley (*Hordeum vulgare* L.). *Nature* 225: 874-876.

Kearsey, M.J. and A.G.L. Farquhar (1998). QTL analysis in plants: Where are we now? *Heredity* 80: 137-142.

Kilian, A., B.J. Steffenson, M.A.S. Maroof, and A. Kleinhofs (1994). RFLP markers linked to the durable stem rust resistance gene *Rpg1* in barley. *Molecular Plant-Microbe Interactions* 7: 298-301.

Kintzios, S., A. Jahoor, and G. Fischbeck (1995). Powdery mildew resistance genes *Mla29* and *Mla32* in *H. spontaneum* derived winter barley lines. *Plant Breeding* 114: 265-266.

Kjaer, B. and J. Jensen (1995). The inheritance of nitrogen and phosphorus content in barley analysed by genetic markers. *Hereditas* 123: 109-119.

Kjaer, B. and J. Jensen (1996). Quantitative trait loci for grain yield and yield components in a cross between a six-rowed and a two-rowed barley. *Euphytica* 90: 39-48.

Kjaer, B., J. Jensen, and H. Giese (1995). Quantitative trait loci for heading date and straw characters in barley. *Genome* 38: 1098-1104.

Knapp, S.J. (1998). Marker-assisted selection as a strategy for increasing the probability of selecting superior genotypes. *Crop Science* 38: 1164-1174.

Kretschmer, J.M., K.J. Chalmers, S. Manning, A. Karakousis, A.R. Barr, A.K.M.R. Islam, S.J. Logue, Y.W. Choe, S.J. Barker, R.C.M. Lance, and P. Langridge (1997). RFLP mapping of the *Ha2* cereal cyst nematode gene in barley. *Theoretical and Applied Genetics* 94: 1060-1064.

Lande, R. and R. Thompson (1990). Efficiency of marker-assisted selection in the improvement of quantitative traits. *Genetics* 124: 743-756.

Langridge P., O. Anderson, M. Gale, P. Gustafson, P. McGuire, and C. Qualset (1999). International Triticeae EST Cooperative (ITEC): <http://wheat.pw.usda.gov/genome/>.

Langridge, P., A. Karakousis, J. Kretschmer, S. Manning, K. Chalmers, R. Boyd, C. dao Li, R. Islam, S. Logue, R. Lance, and SARDI (1996). RFLP and QTL analysis of barley mapping populations: <http://greengenes.cit.cornell.edu/WaiteQTL/>.

Larson, S.R., D. Kadyrzhanova, C. McDonald, M. Sorrells, and T.K. Blake (1996). Evaluation of barley chromosome-3 yield QTLs in a backcross F_2 population using STS-PCR. *Theoretical and Applied Genetics* 93: 618-625.

Larson, S.R., K.A. Young, A. Cook, T.K. Blake, and V. Raboy (1998). Linkage mapping of two mutations that reduce phytic acid content of barley grain. *Theoretical and Applied Genetics* 97: 141-146.

Laurie, D.A., N. Pratchett, J.H. Bezant, and J.W. Snape (1995). RFLP mapping of five major genes and eight quantitative trait loci controlling flowering time in a winter x spring barley (*Hordeum vulgare* L.) cross. *Genome* 38: 575-585.

Laurie, D.A., N. Pratchett, C. Romero, E. Simpson, and J.W. Snape (1993). Assignment of the *denso* dwarfing gene to the long arm of chromosome 3(3H) of barley by use of RFLP markers. *Plant Breeding* 111: 198-203.

Li, C.D., P. Langridge, R.C.M. Lance, P. Xu, and G.B. Fincher (1996). Seven members of the $(1 \rightarrow 3)$-β-glucanase gene family in barley *(Hordeum vulgare)* are clustered on the long arm of chromosome 3 (3HL). *Theoretical and Applied Genetics* 92(7): 791-796.

Li, C.D, X.Q. Zhang, P. Eckstein, B.G. Rossnagel, and G.J. Scoles (1999). A polymorphic microsatellite in the limit dextrinase gene of barley (*Hordeum vulgare* L.). *Molecular Breeding* 5: 569-577.

Lupton, F.G.H. and R.N.H. Whitehouse (1957). Studies on the breeding of self-pollinating cereals. I. Selection methods in breeding for yield. *Euphytica* 6: 169-184.

Manninen, O.M., T. Turpeinen, and E. Nissila (1997). Identification of RAPD markers closely linked to the mlo-locus in barley. *Plant Breeding* 116(5): 461-464.

Mano, Y., H. Takahashi, K. Sato, and K. Takeda (1996). Mapping genes for callus growth and shoot regeneration in barley (*Hordeum vulgare* L.). *Breeding Science* 46: 137-142.

Mano, Y. and K. Takeda (1997). Mapping quantitative trait loci for salt tolerance at germination and the seedling stage in barley (*Hordeum vulgare* L.). *Euphytica* 94: 263-272.

Martiniello, P., G. Delogu, G. Odoardi, and M. Stanca (1987). Breeding progress in grain yield and selected agronomic characters of winter barley (*Hordeum vulgare* L.) over the last quarter of a century. *Plant Breeding* 99: 289-294.

Mather, D.E., N.A. Tinker, D.E. LaBerge, M. Edney, B.L. Jones, B.G. Rossnagel, W.G. Legge, K.G. Briggs, R.B. Irvine, D.E. Falk, and K.J. Kasha (1997). Regions of the genome that affect grain and malt quality in a North American two-row barley cross. *Crop Science* 37: 544-554.

Melchinger, A.E. (1990). Use of molecular markers in breeding for oligogenic disease resistance. *Plant Breeding* 104: 1-19.

Moharramipour, S., H. Tsumuki, K. Sato, and H. Yoshida (1997). Mapping resistance to cereal aphids in barley. *Theoretical and Applied Genetics* 94: 592-596.

Mohler, V. and A. Jahoor (1996). Allele-specific amplification of polymorphic sites for the detection of powdery mildew resistance loci in cereals. *Theoretical and Applied Genetics* 93(7): 1078-1082.

Oziel, A., P.M. Hayes, F.Q. Chen, and B. Jones (1996). Application of quantitative trait locus mapping to the development of winter-habit malting barley. *Plant Breeding* 115: 43-51.

Paltridge, N.G., N.C. Collins, A. Bendahmane, and R.H. Symons (1998). Development of YLM, a codominant PCR marker closely linked to the *Yd2* gene for resistance to barley yellow dwarf disease. *Theoretical and Applied Genetics* 96(8): 1170-1177.

Pan, A., P.M. Hayes, F. Chen, T.H.H. Chen, T. Blake, S. Wright, I. Karsai, and Z. Bedo (1994). Genetic analysis of the components of winterhardiness in barley (*Hordeum vulgare* L.). *Theoretical and Applied Genetics* 89: 900-910.

Paterson, A.H., J.W. Deverna, B. Lanini, and S.D. Tanksley (1990). Fine mapping of quantitative trait loci using selected overlapping recombinant chromosomes in an interspecies cross of tomato. *Genetics* 124: 735-742.

Pecchioni, N., G. Vale, H. Toubia-Rahme, P. Faccioli, V. Terzi, and G. Delogu (1999). Barley-*Pyrenophora graminea* interaction: QTL analysis and gene mapping. *Plant Breeding* 118: 29-35.

de la Pena, R.C., K.P. Smith, F. Capettini, G.J. Muehlbauer, M. Gallo-Meagher, R. Dill-Macky, D.A. Somers, and D.C. Rasmusson (1999). Quantitative trait loci associated with resistance to *Fusarium* head blight and kernel discoloration in barley. *Theoretical and Applied Genetics* 99: 561-569.

Penner, G.A., J.A. Stebbing, and B. Legge (1995). Conversion of an RFLP marker for the barley stem rust resistance gene *Rpg1* to a specific PCR-amplifiable polymorphism. *Molecular Breeding* 1(4): 349-354.

Penner G.A., A. Tekauz, E. Reimer, G.J. Scoles, B.G. Rossnagel, P.E. Eckstein, W.G. Legge, P.A. Burnett, T. Ferguson, and J.F. Helm (1996). The genetic basis of scald resistance in western Canadian barley cultivars. *Euphytica* 92: 367-374.

Pickering, R.A. and P. Devaux (1991). Haploid production: Approaches and uses in plant breeding. In *Barley: Genetics, Biochemistry, Molecular Biology and Biotechnology*, ed. P.R. Shewry. Oxford: CAB International, pp. 519-547.

Pooni, H.S. and J.L. Jinks (1978). Predicting the properties of recombinant inbred lines derived by single seed descent for two or more characters simultaneously. *Heredity* 40: 349-361.

Poulsen, D.M.E., R.J. Henry, R.P. Johnston, J.A.G. Irwin, and R.G. Rees (1995). The use of bulk segregant analysis to identify a RAPD marker linked to leaf rust resistance in barley. *Theoretical and Applied Genetics* 91(2): 270-273.

Powell, W., P.D.S. Caligari, J.W. McNicol, and J.L. Jinks (1985). The use of doubled haploids in barley breeding. 3. An assessment of multivariate cross prediction methods. *Heredity* 55: 249-254.

Powell, W., W.T.B. Thomas, E. Baird, P. Lawrence, A. Booth, B. Harrower, J.W. McNicol, and R. Waugh (1997). Analysis of quantitative traits in barley by the use of amplified fragment length polymorphisms. *Heredity* 79: 48-59.

Qi, X., G. Jiang, W. Chen, R.E. Niks, P. Stam, and P. Lindhout (1999). Isolate-specific QTLs for partial resistance to *Puccinia hordei* in barley. *Theoretical and Applied Genetics* 99: 877-884.

Qi, X., R.E. Niks, P. Stam, and P. Lindhout (1998). Identification of QTLs for partial resistance to leaf rust *(Puccinia hordei)* in barley. *Theoretical and Applied Genetics* 96: 1205-1215.

Rae, A.M., E.C. Howell, and M.J. Kearsey (1999). More QTL for flowering time revealed by substitution lines in *Brassica oleracea. Heredity* 83: 586-596.

Ramsay, L., M. Macaulay, S. degli Ivanissivich, K. MacLean, L. Cardle, J. Fuller, K. Edwards, S. Tuvesson, M. Morgante, A. Massari (2000). A simple sequence repeat-based linkage map of barley. *Genetics* 156:1997-205.

Rasmusson, D.C. and R.L. Phillips (1997). Plant breeding progress and genetic diversity from de novo variation and elevated epistasis. *Crop Science* 37: 303-310.

Richter, K., J. Schondelmaier, and C. Jung (1998). Mapping of quantitative trait loci affecting *Drechslera teres* resistance in barley with molecular markers. *Theoretical and Applied Genetics* 97: 1225-1234.

Riggs, T.J. and A.M. Hayter (1975). A study of the inheritance and inter-relationships of some agronomically important characters in spring barley. *Theoretical and Applied Genetics* 46: 257-264.

Romagosa, I., F. Han, S.E. Ullrich, P.M. Hayes, and D.M. Wesenberg (1999). Verification of yield QTL through realized molecular marker-assisted selection responses in a barley cross. *Molecular Breeding* 5: 143-152.

Romagosa, I., S.E. Ullrich, F. Han, and P.M. Hayes (1996). Use of the additive main effects and multiplicative interaction model in QTL mapping for adaptation in barley. *Theoretical and Applied Genetics* 93: 30-37.

Russell, J.R., R.P. Ellis, W.T.B. Thomas, R. Waugh, J. Provan, A. Booth, J. Fuller, P. Lawrence, G. Young, and W. Powell (2000). A retrospective analysis of spring barley germplasm development from "foundation genotypes" to currently successful cultivars. *Molecular Breeding* 6: 553-568.

Russell, J., J. Fuller, G. Young, G. Taramino, W. Thomas, M. Macaulay, R. Waugh, and W. Powell (1997). Discriminating between barley genotypes using microsatellite markers. *Genome* 40: 442-450.

Saeki, K., C. Miyazaki, N. Hirota, A. Saito, K. Ito, and T. Konishi (1999). RFLP mapping of BaYMV resistance gene *rym3* in barley *(Hordeum vulgare). Theoretical and Applied Genetics* 99: 727-732.

Sanguineti, M.C., R. Tuberosa, S. Stefanelli, E. Noli, T.K. Blake, and P.M. Hayes (1994). Utilization of a recombinant inbred population to localize QTLs for abscisic acid content in leaves of drought-stressed barley (*Hordeum vulgare* L.). *Russian Journal of Plant Physiology* 41: 572-576.

Sato, K., K. Takeda, and P.M. Hayes (1996). QTL analysis for net blotch resistance in barley. In *V International Oat Conference and VII International Barley Genetics Symposium Proceedings–Poster Sessions* (1), eds. A. Slinkard, G. Scoles, and B. Rossnagel. Saskatoon: University of Saskatchewan Extension Press, pp. 298-299.

Schondelmaier, J., A. Jacobi, G. Fischbeck, and A. Jahoor (1992). Genetic studies on the mode of inheritance and localization of the *amo1* (high amylose) gene in barley. *Plant Breeding* 109: 274-280.

Schonfeld, M., A. Ragni, G. Fischbeck, and A. Jahoor (1996). RFLP mapping of three new loci for resistance genes to powdery mildew *(Erysiphe graminis* f. *hordei)* in barley. *Theoretical and Applied Genetics* 93: 48-56.

Schuller, C., G. Backes, G. Fischbeck, and A. Jahoor (1992). RFLP markers to identify the alleles at the Mla locus conferring powdery mildew resistance in barley. *Theoretical and Applied Genetics* 84: 330-338.

Schweizer, G.F., M. Baumer, G. Daniel, H. Rugel, and M.S. Roder (1995). RFLP markers linked to scald *(Rhynchosporium secalis)* resistance gene *Rh2* in barley. *Theoretical and Applied Genetics* 90(7/8): 920-924.

Silvey, V. (1986). The contribution of new varieties to cereal yield in England and Wales between 1947 and 1983. *Journal of the National Institute of Agricultural Botany* 17: 155-168.

Snape, J.W. (1982). The use of doubled haploids in plant breeding. In *Induced Variability in Plant Breeding,* ed. C. Broertjes. Wageningen, the Netherlands: Centre for Agriculture Publishing and Documentation, pp. 52-58.

Sogaard, B. and P. von Wettstein Knowles (1987). Barley: Genes and chromosomes. *Carlsberg Research Communications* 52: 123-196.

Spaner, D., B.G. Rossnagel, W.G. Legge, G.J. Scoles, P.E. Eckstein, G.A. Penner, N.A. Tinker, K.G. Briggs, D.E. Falk, J.C. Afele, et al. (1999). Verification of a quantitative trait locus affecting agronomic traits in two-row barley. *Crop Science* 39: 248-252.

Spaner, D., L.P. Shugar, T.M. Choo, I. Falak, K.G. Briggs, W.G. Legge, D.E. Falk, S.E. Ullrich, N.A. Tinker, B.J. Steffenson, and D.E. Mather (1998). Mapping of disease resistance loci in barley on the basis of visual assessment of naturally occurring symptoms. *Crop Science* 38: 843-850.

Stam, P., I. Bos, X. Qi, and R. Niks (1997). QTL mapping of yield and yield-related traits in a wide cross of spring barley *(Hordeum vulgare* L). In *Advances in Biometrical Genetics—Proceedings of the Tenth Meeting of the EUCARPIA Section Biometrics in Plant Breeding*. Poznan, Poland: Institute of Plant Genetics, pp. 277-280.

Steffenson, B. J., P.M. Hayes, and A. Kleinhofs (1996). Genetics of seedling and adult plant resistance to net blotch *(Pyrenophora teres* f. *teres)* and spot blotch *(Cochliobolus sativus)* in barley. *Theoretical and Applied Genetics* 92: 552-558.

Stracke, S. and A. Borner (1998). Molecular mapping of the photoperiod response gene *ea7* in barley. *Theoretical and Applied Genetics* 97(5/6): 797-800.

Swanston, J.S. (1987). The consequences for malting quality of *Hordeum laevigatum* as a source of mildew resistance in barley breeding. *Annals of Applied Biology* 110: 351-356.

Swanston. J.S., W.T.B. Thomas, W. Powell, G.R. Young, P.E. Lawrence, L. Ramsay, and R. Waugh (1999). Using molecular markers to determine barleys most suitable for malt whisky distilling. *Molecular Breeding* 5: 103-109.

Taketa, S., H. Takahashi, and K. Takeda (1998). Genetic variation in barley of crossability with wheat and quantitative trait loci analysis. *Euphytica* 103: 187-193.

Tanksley, S.D., S. Grandillo, T.M. Fulton, D. Zamir, Y. Eshed, V. Petiard, J. Lopez, and T. Beck-Bunn (1996). Advanced backcross QTL analysis in a cross between an elite processing line of tomato and its wild relative *L. pimpinellifolium*. *Theoretical and Applied Genetics* 92: 213-224.

Tanksley, S.D. and J.C. Nelson (1996). Advanced backcross QTL analysis: A method for the simultaneous discovery and transfer of valuable QTLs from unadapted germplasm into elite breeding lines. *Theoretical and Applied Genetics* 92: 191-203.

Tanksley, S.D., N.D. Young, A.H. Paterson, and M.W. Bonierbale (1989). RFLP mapping in plant breeding: New tools for an old science. *Bio/Technology* 7: 257-264.

Tapsell, C.R. and W.T.B. Thomas (1983). Cross prediction studies in spring barley. 2. Estimation of genetical and environmental control of yield and its component characters. *Theoretical and Applied Genetics* 64: 353-358.

Teulat, B., D. This, M. Khairallah, C. Borries, C. Ragot, P. Sourdille, P. Leroy, P. Monneveux, and A. Charrier (1998). Several QTLs involved in osmotic-adjustment trait variation in barley (*Hordeum vulgare* L.). *Theoretical and Applied Genetics* 96: 688-698.

Thomas, W.T.B. (1987). The use of random F3 families for cross prediction in spring barley. *Journal of Agricultural Science, Cambridge* 108: 431-436.

Thomas, W.T.B. (1997). Mapping disease resistance in spring barley. In *Approaches to Improving Disease Resistance to Meet Future Needs: Airborne Pathogens of Wheat and Barley*. COST Action 817 Prague, November 1997, pp. 48-51.

Thomas, W.T.B. (1999). QTL mapping of net blotch resistance in a spring barley cross. In *Disease Resistance and Cereal Leaf Pathogens Beyond the Year 2000*. COST Action 817 Martina Franca, November 1999.

Thomas, W.T.B., E. Baird, J.D. Fuller, P. Lawrence, G.R. Young, J. Russell, L. Ramsay, R. Waugh, and W. Powell (1998). Identification of a QTL decreasing yield in barley linked to Mlo powdery mildew resistance. *Molecular Breeding* 4: 381-393.

Thomas, W.T.B., W. Powell, J.S. Swanston, R.P. Ellis, K.J. Chalmers, U.M. Barua, P. Jack, V. Lea, B.P. Forster, R. Waugh, and D.B. Smith (1996). Quantitative trait loci for germination and malting quality characters in a spring barley cross. *Crop Science* 36: 265-273.

Thomas, W.T.B., W. Powell, R. Waugh, K.J. Chalmers, U.M. Barua, P. Jack, V. Lea, B.P. Forster, J.S. Swanston, R.P. Ellis, et al. (1995). Detection of quantitative trait loci for agronomic, yield, grain and disease characters in spring barley (*Hordeum vulgare* L). *Theoretical and Applied Genetics* 91: 1037-1047.

Thomas, W.T.B. and C.R. Tapsell (1983). Cross prediction studies in spring barley. 1. Estimation of genetical and environmental control of morphological and maturity characters. *Theoretical and Applied Genetics* 64: 345-352.

Thomas, W.T.B. and G.R. Young (1996). Quantitative trait loci for components of anther culture response in barley. In *Gametic Embryogenesis in Monocots*. COST Action 824 Vienna, December 1996.

Thomsen, S.B., H.P. Jensen, J. Jensen, J.P. Skou, and J.H. Jorgensen (1997). Localization of a resistance gene and identification of sources of resistance to barley leaf stripe. *Plant Breeding* 116: 455-459.

Tinker, N.A. and D.E. Mather (1995). MQTL: Software for simplified composite interval mapping of QTL in multiple environments. *Journal of Agricultural Genomics* 1: <http://www.ncgr.org/ag/jag/papers95/paper295/indexp295.html>.

Tinker, N.A., D.E. Mather, B.G. Rossnagel, K.J. Kasha, A. Kleinhofs, P.M. Hayes, D.E. Falk, T. Ferguson, L.P. Shugar, W.G. Legge, et al. (1996). Regions of the genome that affect agronomic performance in two-row barley. *Crop Science* 36: 1053-1062.

Toojinda, T., E. Baird, A. Booth, L. Broers, P. Hayes, W. Powell, W. Thomas, H. Vivar, and G. Young (1998). Introgression of quantitative trait loci (QTLs) determining stripe rust resistance in barley: An example of marker-assisted line development. *Theoretical and Applied Genetics* 96: 123-131.

Tuberosa, R., G. Galiba, M.C. Sanguineti, E. Noli, and J. Sukta (1997). Identification of QTL influencing freezing tolerance in barley. *Acta Agronomica Hungarica* 45: 413-417.

Tuvesson, S., L. von Post, R. Oumlauthlund, P. Hagberg, A. Graner, S. Svitashev, M. Schehr, and R. Elovsson (1998). Molecular breeding for the BaMMV/BaYMV resistance gene *ym4* in winter barley. *Plant Breeding* 117(1): 19-22.

Ullrich, S.E., P.M. Hayes, W.E. Dyer, T.K. Blake, and J.A. Clancy (1993). Quantitative trait locus analysis of seed dormancy in 'Steptoe' barley. In *Pre-Harvest Sprouting in Cereals 1992*, eds. M.K. Walker-Simmons and J.L. Ried. St. Paul, MN: American Association of Cereal Chemists, pp. 136-145.

Utz, H.F. and A.E. Melchinger (1996). PLABQTL: A program for composite interval mapping of QTL. *Journal of Agricultural Genomics* 2: <http://www.ncgr.org/ag/jag/papers96/paper196/indexp196.html>.

Weyen, J., E. Bauer, A. Graner, W. Friedt, and F. Ordon (1996). RAPD-mapping of the distal portion of chromosome 3 of barley, including the BaMMV/BaYMV resistance gene *ym4*. *Plant Breeding* 115(4): 285-287.

Whitehouse, R.N.H. (1969). Canonical analysis as an aid to plant beeding. In *Barley Genetics II, Proceedings of the Second International Barley Genetics Symposium*, Washington 1969, ed. R.A. Nilan. Pullman: Washington State University Press, pp. 269-282.

Williams, K.J., A. Lichon, P. Gianquitto, J.M. Kretschmer, A. Karakousis, S. Manning, P. Langridge, and H. Wallwork (1999). Identification and mapping of a gene conferring resistance to the spot form of net blotch *(Pyrenophora teres* f *maculata)* in barley. *Theoretical and Applied Genetics* 99: 323-327.

Wolfe, M.S. (1984). Trying to understand and control powdery mildew. *Plant Pathology* 33: 451-466.

Xiao, J., J. Li, S. Grandillo, S.N. Ahn, L. Yuan, S.D. Tanksley, and S.R. McCouch (1998). Identification of trait-improving quantitative trait loci alleles from a wild rice relative, *Oryza rufipogon*. *Genetics* 150: 899-909.

Yin X.Y., M.J. Kropff, and P. Stam (1999). The role of ecophysiological models in QTL analysis: The example of specific leaf area in barley. *Heredity* 82: 415-421.

Yin, X., P. Stam, C.J. Dourleijn, and M.J. Kropff (1999). AFLP mapping of quantitative trait loci for yield-determining physiological characters in spring barley. *Theoretical and Applied Genetics* 99: 244-253.

Yoshida, H., T. Iida, K. Sato, S. Moharramipour, and H. Tsumiki (1997). Mapping a gene for gramine synthesis in barley. *Barley Genetics Newsletter* 27: 22-23.

Young, N.D. (1999). A cautiously optimistic vision for marker-assisted breeding. *Molecular Breeding* 5: 505-510.

Zhu, H., G. Briceno, R. Dovel, P.M. Hayes, B.H. Liu, C.T. Liu, and S.E. Ullrich (1999). Molecular breeding for grain yield in barley: An evaluation of QTL effects in a spring barley cross. *Theoretical and Applied Genetics* 98: 772-779.

Chapter 9

Genotype by Environment Interaction and Adaptation in Barley Breeding: Basic Concepts and Methods of Analysis

Jordi Voltas
Fred van Eeuwijk
Ernesto Igartua
Luis F. García del Moral
José Luis Molina-Cano
Ignacio Romagosa

INTRODUCTION

Barley breeding is largely empirical. In the first segregating generations, breeders focus on highly heritable traits, such as height, spike morphology, and phenology, and concentrate later on grain yield and end-use quality. Extensive multilocation trials carried out during a series of years are used in the final selection cycles to identify superior genotypes. This task is not generally easy due to the frequent presence of genotype by environment interaction (GE). GE is differential genotypic expression across environments. It attenuates association between phenotype and genotype, reducing genetic progress in breeding programs.

Means across environments are adequate indicators of genotypic performance only in the absence of GE. If GE is present, use of means across environments ignores that genotypes differ in relative performance over environments. In plant breeding, the most important type of GE is crossover or qualitative, which implies changes in the rankings of genotypes across environments (Baker, 1988). With noncrossover interactions, genotypes with superior means can be recommended for all environments. Crossover GE complicates breeding, testing, and selection of superior genotypes. Identification of specifically adapted genotypes then becomes a way for increasing genetic gains.

For trials in which locations and genotypes are repeated across years, the sum of squares of GE can be partitioned into genotype by location (GL), genotype by year (GY), and genotype by location by year (GLY) interactions. This partitioning must be based on a series of trials carried out over sets of sufficient locations and years that are representative of the target area for the program. If so, it allows for an assessment of the spatial and temporal components of genotypic adaptation. If GL is the dominant portion of GE, then specific adaptation is exploitable by identifying homogeneous regions that minimize GE within regions and form uniform domains for release and recommendation of genotypes. Where GY and GLY terms dominate, as too often happens, no simplification involving spatial subdivision of regions is possible.

The statistical analysis of GE and its breeding implications are among the most recurrently addressed topics in breeding literature and have been reviewed throughout the years (most recently by van Eeuwijk [1996)] Kang and Gauch [1996], Cooper and Hammer [1996], and Kempton and Fox [1997]). Most studies are purely empirical, describing postdictively fixed genotypic performances across a fixed sample of environments. A preferred analytical approach should characterize both genotypes and environments in terms of a number of biotic and abiotic variables that directly affect performance. The objective of this work is not to present a historical review of the subject, nor to discuss in a statistical sense a subset of methods from the array of available techniques. Our approach involves the analysis of a specific barley data set to describe and illustrate current statistical models focusing on interpretable GE variation. The methodology proposed is based on the selection of independent genetic, physiological, and biophysical variables and aims at being a predictive breeding strategy.

The structure of this chapter is as follows. First, a review of the latest barley studies on GE is provided, briefly describing both the biotic and abiotic causes underlying GE. Next, a worked example of a barley trial is given. After description of the data, a series of alternative multiplicative models are developed. Finally, the implications of GE in barley breeding are discussed. As an appendix, the SAS programs (SAS Institute 1989a, 1989b, 1997) are given for the models described.

GENOTYPE BY ENVIRONMENT INTERACTION IN BARLEY

GE for grain yield in barley has been reported in numerous studies under several designations (e.g., different response patterns, adaptation, or stability of genotypes). Although it is always difficult to identify a causal relationship between genetic or environmental factors and the phenotypic expression of GE, knowledge of crop physiology and agronomy together with rational use of statistical techniques is likely to shed light on the probable causes of this phenomenon. Most often, however, no clear cause of GE has been reported in the literature due to lack of information about the environments (weather, soil) or about the genotypes themselves. In a recent paper about GE and its implications for wheat breeding in Australia, Basford and Cooper (1998), following Baker (1990), recognized two major categories of studies on GE depending upon the level of understanding of the genotypes and/or the environments. The first category listed defined causes, such as the presence of diseases, soilborne constraints, drought, etc. The second category was reserved for those studies which showed GE without an apparent biotic and/or abiotic explanation. The approach proposed by Basford and Cooper (1998) has also been followed in this study. A total of 215 hits were returned from the Commonwealth Agricultural Bureaux (Farnham Royal, U.K.) databases for the period 1973-1999 (September) when searching for "barley" and "genotype-by-environment." Out of them, 77 specifically dealt with grain yield (sometimes among other traits). The majority (52) followed in the second category, whereas 25 attempted to give some insight on the causes of GE.

Most studies in which the magnitude of GE for grain yield was measured detected a statistically significant interaction. Generally, only studies with limited genotypic and/or environmental diversity presented negligible interaction. Some of the most interesting studies in this area have been carried out by Ceccarelli and co-workers under Mediterranean conditions (Ceccarelli and Grando, 1991; Ceccarelli, 1994; 1996; Ceccarelli, Grando, and Impiglia, 1998; among others). This series of experiments consistently reported a crossover-type GE; that is, the best-performing varieties in low- and high-yielding environments differed. Other studies, however, identified a few barley varieties presenting good yield and stability across large target areas even though GE was present, as Berbigier and Denis (1981) for Europe, or Atlin, McRae, and Lu (2000) and Kong et al. (1994) for Canada. This apparent discrepancy might be explained by differences in the productivity ranges among studies, and in the nature of germplasm evaluated.

The climatic variables that most affect barley yields are temperature and rainfall. These two factors also play a primary role in the occurrence of GE. One of the two main distinguishing characteristics in the barley germplasm pool is the suitability of cultivars for autumn or spring sowing (the other would be spike row number). This suitability is governed by overall response to temperature (earliness per se), ability to hasten growth after exposure to cold temperatures (vernalization), cold tolerance, and photoperiod response (Borthwick, Parker, and Heinze, 1941; Jones and Allen, 1986; Ellis et al., 1989). All but the last of these phenomena are driven by temperature. Overall, they govern the development of the plant by triggering its growth phases and modulating their duration. The most noticeable outcome of this environmental control is the variation in heading date of the genotypes when exposed to contrasting environments. Many reports show GE for heading date and also for cold tolerance in barley collections, but few of them establish the association between GE for these traits and yield response patterns. Talamucci (1975) and Bouzerzour and Refoufi (1992) found a strong sowing date × variety interaction for yield when comparing autumn and spring sowings, being most likely caused by differential genotypic sensitivities to temperature or photoperiod.

Drought is one of the most limiting factors encountered by barley. It is a function of rainfall, temperature, and soil water-holding capacity. Differential responses of genotypes in low- and high-yielding environments often reflect the consequences of differences in rainfall regimes (e.g., Soliman and Allard, 1991; van Oosterom et al., 1993; Voltas, Romagosa, et al., 1999). In Mediterranean climates, terminal water stress is a common event, as grain filling usually develops under an increasingly higher evapotranspirative demand. In this situation, variation in heading date is usually an important cause of GE for yield: earlier cultivars generally perform better than later ones in low-yielding environments because of higher water availability at the end of the crop season (van Oosterom et al., 1993; Jackson et al., 1994; Abay and Cahalan, 1995). This advantage disappears under high-yielding, rain-fed or irrigated conditions (Jackson et al., 1994). In a few instances, true drought tolerance mechanisms have been related to yield GE in barley. For instance, Muñoz et al. (1998) found that GE between old and modern cultivars was correlated with differences in transpiration efficiency indirectly estimated by ^{13}C discrimination.

Soil characteristics have been seldom identified as responsible factors for GE in barley (Tewari, 1975; Talbot, 1984). It has to be acknowledged, however, that this is an area often overlooked by breeders, as soil data are sometimes difficult to obtain. Of the many studies focusing on the optimization of the levels of fertilizer, only a few found differences between cultivars. Not surprisingly, one of the few GE events reported for nitrogen fertilization (Lekes and Zinisceva, 1990) found that modern spring barley varieties had a better response to increased levels of nitrogen (up to 80 kg·ha^{-1}) as compared to old spring ones. Soil acidity (Barszczak and Barszczak, 1980) and salinity (Isla, Royo, and Aragüés, 1997) have also been reported as factors responsible for producing crossover-type GE for yield of barley varieties.

Studies addressing GE for disease reactions rarely extend to the effects on yield. Abo-Elenin, Morsi, and Gomaa (1977) and Soliman and Allard (1991) reported that GE for yield could be partially explained by contrasting disease tolerance to net blotch and leaf rust, and net blotch and scald, respectively, in the varieties studied. Wright and Gaunt (1992) also reported that several diseases influenced yield differentially in two barley cultivars.

Other genotypic features that are major causes of GE for yield have been revealed by the allocation of cultivars to different germplasm groups. For example, some studies reported contrasting yield responses across environments between two-row and six-row types (Talamucci, 1975; Nurminiemi,

Bjornstad, and Rognli, 1996), North American arctic versus temperate varieties (Dofing et al., 1992), local land races versus improved material (Ceccarelli and Grando, 1991), or old versus modern cultivars (Muñoz et al., 1998). In addition, the genetic structure of the varieties may also be important, as pointed out by Soliman and Allard (1991): composite crosses (a population constituted by intercrossing several cultivars) proved to be less interactive than inbred cultivars.

There have been attempts to break down GE for yield in terms of yield components in the belief that yield components may be less interactive and easier to deal with in breeding programs. Some authors have reported that high-yielding ability under favorable conditions is mostly associated with kernel number per unit area (Dofing et al., 1992; Jackson et al., 1994). Conversely, more components appear to be related to higher yields in harsher environments: kernel number, kernel weight, and also dry-matter accumulation prior to or after anthesis (Jackson et al., 1994) or spikes per unit area, kernels per spike, and kernel weight (Dofing et al., 1992).

A helpful tool to describe GE for grain yield has been the use of barley mutants (Arain, 1975; Gustafsson, Ekman, and Dormling, 1977; Molina-Cano et al., 1990; Romagosa et al., 1993). Usually, the mutant strains and their wild parents diverge for just a few traits (such as height, spike density, heading date, erectoides trait). The comparison of mutant strains to their parents provides a simple test for association of various agronomic traits and yield. This approach will be developed extensively in the following section.

In the past decade, efforts to elucidate the genetic factors causing GE have veered toward the use of molecular markers. Quantitative trait loci (QTL) responsible for adaptation have been reported in several populations: Steptoe/Morex (Hayes et al., 1993; Romagosa et al., 1996; Zhu et al., 1999), Harrington/TR306 (Tinker et al., 1996), and Blenheim/Kym (Bezant et al., 1997). The magnitude of individual QTL effects (measured as the amount of GE variance explained by a particular QTL) varied among populations, being much larger in Steptoe/Morex than in the other two populations. Some QTL for GE were coincident with those for the genotype main effect within a given population, but the agreement for loci among populations was low. Its implementation in breeding programs remains a challenge.

AN EXAMPLE DATA SET: YIELD OF A SERIES OF BARLEY ISOGENIC LINES IN SPAIN

We want to use real barley data to illustrate how available empirical (genotypic yields), genetic, and ecophysiological information can be integrated into a single model for adaptation. To that end, we propose the use of true near-isogenic lines, which share a common genetic background, as a framework to aid in understanding the magnitude and nature of GE. The available plant material consists of three mutants (M01, M02, and M03) induced in the two-rowed variety Beka and their three binary recombinants (M12, M13, and M23). These lines were grown in a multienvironment trial conducted in some of the most important Spanish barley-producing regions from 1985 to 1989: Granada (G), Sevilla (S), and the Central Plateau of Spain (C) (the latter includes the provinces of Badajoz, Palencia, Soria, and Toledo) (see Table 9.1). The experimental layout at each environment was a randomized complete block design with four replicates. Climatic records included the variables average mean temperature (T), rainfall (R), and ratio of rainfall to total evapotranspirative demand (R/ETP) for three consecutive growth periods: tillering, jointing, and grain filling. The lines were characterized by means of nine morphophysiological measurements taken for two years in two contrasting locations (see Table 9.2): number of leaves per plant and leaf area at both anthesis and physiological maturity, leaf angle, length of the main shoot, days from emergence to anthesis, days from anthesis to maturity, and spike density. Although the genetic basis of the mutations is not exactly known, most of the phenotypic variation recorded, including grain yield, is probably the effect of only three recessive Mendelian genes segregating 3:1 in the F_2 generation (Molina-Cano, 1982). Accordingly, the terms *gene* and *mutation* will be used interchangeably throughout the text. A detailed description of the seven near-isogenic barley lines, including the original variety Beka, their underlying genetic constitution, and the overall experimental conditions can be found in Molina-Cano et al. (1990) and Romagosa et al. (1993).

A previous GE analysis of a related data set (Romagosa et al., 1993) showed that the genotypic interactive pattern over environments could be explained by the morphophysiological constitution of the mutants. However, the statistical analysis in that study did not integrate in a single model external genetic and physiological data of the different lines and physical data of the environments used. The example developed hereafter aims at re-examining the usefulness of integrating ecophysiological and statistical tools in the interpretation of GE based on the joint application of two multiplicative models for interaction: the additive main effects and multiplicative interaction (AMMI) model (Gauch, 1988) and the factorial regression model (Denis, 1988). Both provide information and insight beyond the classical analysis of variance of two-way genotype by environment tables. AMMI represents an empirical approach (based on yield itself) to analyze GE. Factorial regression attempts to describe interaction by including external genetic, phenotypic, and environmental information (e.g., morphophysiological traits, climatic data, etc.) on the levels of the genotypic and environmental factors. It implies a more analytical approach to the study of GE.

TABLE 9.1. Description of the 13 Trials Used

Location					Environment code	Year	Mean grain yield (t·ha⁻¹)	Climatic characterization								
								Tillering (t)			Jointing (j)			Grain filling (gf)		
Province	City		Coordinates	Altitude				T_{mean}(°C)	R (mm)	R/ETP	T_{mean}(°C)	R (mm)	R/ETP	T_{mean}(°C)	R (mm)	R/ETP
BADAJOZ	Mérida		38°54'N 6°20'W	270	BA9	1989	3.96	8.9	28	0.30	13.6	39	0.41	12.8	47	0.49
GRANADA	Colomera		37°23'N 3°42'W	706	G18	1988	4.11	7.2	123	1.73	10.2	34	0.38	13.9	81	0.69
	Cotillar		37°30'N 3°30'W	912	G27	1987	4.44	10.5	391	2.85	16.1	48	0.35	21.4	1	0.01
					G28	1988	7.27	8.2	157	1.28	12.3	96	0.91	15.5	65	0.47
					G29	1989	3.41	8.0	162	0.98	8.5	83	0.89	15.9	20	0.12
PALENCIA	Villarramiel		42°02'N 4°50'W	900	PA8	1988	3.14	8.9	121	0.75	13.7	60	0.49	16.5	103	0.67
SEVILLA	Alcalá del Rio		37°32'N 5°58'W	10	S15	1985	5.89	8.7	135	3.12	13.6	75	1.34	13.0	5	0.05
					S16	1986	6.04	9.9	162	1.98	10.8	60	1.14	13.4	57	0.60
	Brenes		37°29'N 5°49'W	20	S28	1988	5.74	11.2	126	1.42	15.0	4	0.03	16.7	20	0.15
					S29	1989	5.51	9.9	41	0.89	12.3	63	1.03	15.3	14	0.12
SORIA	Ecija		37°32'N 5°05'W	112	S39	1989	5.15	9.5	22	0.47	11.0	44	0.74	14.8	49	0.45
	Almazán		41°29'N 2°32'W	938	SO9	1989	3.17	6.7	98	0.53	12.1	63	0.45	14.9	39	0.25
TOLEDO	Tembleque		39°41'N 3°30'W	725	TO9	1989	1.56	5.0	18	0.15	10.9	14	0.16	9.4	38	0.42

TABLE 9.2. Morphophysiological Characterization of the Genotypes Studied for Two Years in Two Contrasting Locations

Line	Code	Morphophysiological traits*								
		LPLA	LPLM	LAA	LAM	LANGLE	SHOOT	DEMAN	DANM	SPIKE
Beka	BEKA	15.2	5.8	9.7	4.9	70.4	76.9	131	31	2.8
Mutant-1	M01	13.4	4.6	6.2	2.7	71.3	60.4	124	31	2.9
Mutant-2	M02	13.7	6.2	8.7	5.5	68.5	63.7	134	30	4.4
Mutant-3	M03	14.4	5.6	10.3	5.4	70.3	70.9	135	29	3.6
Mutant-12	M12	12.5	6.0	6.2	3.2	72.1	53.1	125	31	4.4
Mutant-13	M13	13.5	4.6	6.5	3.7	72.3	59.7	128	29	3.5
Mutant-23	M23	13.7	6.3	10.4	5.3	68.2	63.1	136	29	4.3

*List of characters: LPLA = no. of leaves/plant at anthesis; LPLM = no. of leaves/plant at maturity; LAA = leaf area at anthesis (cm^2); LAM = leaf area at maturity (cm^2); LANGLE = leaf angle (degrees); SHOOT = length of the main shoot (cm); DEMAN = days emergence-anthesis; DANM = days anthesis-maturity; SPIKE = spike density (no. kernels/cm of rachis).

The following strategy of analysis will be followed in this example: AMMI will be used to get first insight into the data. External information will be superimposed on the AMMI description of the data with the aid of a graphical representation of the AMMI-fit to the GE: the biplot. Subsequently, various factorial regression models will be fitted, based upon the results of the AMMI analysis and the subsequent biplot interpretation.

Analysis of Variance of Multienvironment Trials

The classical fixed two-way analysis of variance (ANOVA) model with interaction for multienvironment trials includes additive terms for the main effects of genotypes and environments together with an extra additive term to account for the interaction effect. In this situation, the expectation of genotype i in environment j is

$$E(Y_{ij}) = \mu + g_i + e_j + ge_{ij} \qquad (1)$$

where μ is the grand mean, g_i and e_j are genotypic and environmental main effects, and ge_{ij} accounts for the specific response of genotype i in environment j.

This model was applied to our barley data (see Table 9.3). The magnitude of the sum of squares (SS) for GE was large, i.e., about three times the genotype main effect SS, and highly significant when tested against the pooled intrablock error. The information provided by this ANOVA model is restricted; we merely know that some form of GE must be present. Further ap-

TABLE 9.3. Analysis of Variance for Grain Yield ($\cdot ha^{-1}$) of the Seven Barley Genotypes Grown in 13 Environments

Source of variation	df	Sum of squares	Mean square	Variance ratio[a]	R^2 (%)[b]
Total	363	920.7			
Environment (E)	12	806.1	67.18	131.73**	87.6
Reps (E)	39	19.9	0.51		2.2
Genotype (G)	6	15.8	2.63	21.64**	1.7
G × E	72	50.3	0.70	5.74**	5.5
Error	234	28.5	0.12		3.0

[a] Environmental main effect tested against Reps (E)
[b] Fraction of sum of squares associated with each term or interaction
** Significant at 1 percent level

proaches to examine GE should dissect the information included in the term ge_{ij}. The exploration of GE starts by fitting a multiplicative model for interaction, the AMMI model.

AMMI Model

In AMMI, the interaction ge_{ij} is partitioned into successive multiplicative terms or products of the form $c_i d_j$, where c_i can be interpreted as the genotypic sensitivity of genotype i to a hypothetical environmental variable d, which has value d_j in environment j. Alternatively, d_j can be interpreted as the environmental potentiality of environment j to a hypothetical genotypic variable c, which takes value c_i for genotype i. AMMI generates a family of models as follows:

$$E(Y_{ij}) = \mu + g_i + e_j + \sum_{k-1}^{K} c_{ik} d_{jk} \qquad (2)$$

where K is the number of multiplicative terms needed to provide an adequate description of the ANOVA interaction.

The K hypothetical environmental/genotypic variables have the property of discriminating maximally between genotypes/environments and can be obtained by applying either a singular value decomposition or a principal components analysis to the interaction. The AMMI model is a very powerful tool to get insight into GE (Gauch, 1988, 1992; van Eeuwijk, 1995; van Eeuwijk and van Tienderen, 2002). It is recognized to provide parsimony; that is, it captures real structure or pattern with fewer degrees of freedom (*df*) than the standard ANOVA interaction term.

The application of AMMI to the barley data set is shown in Table 9.4. The first four terms of AMMI were significant using an approximate F-statistic (Gollob, 1968). It is important to keep in mind that Gollob's test most often retains axes of little practical relevance (with a low proportion of explained GE variation). In fact, most of the interaction (77.8 percent) was accounted for by the first and second multiplicative terms. Such a result may suggest that AMMI4 is actually overfitting the data, which means that a portion of the total noise included in the interaction term is also retained by the AMMI model. An easy inspection of this phenomenon can be done by estimating the amount of noise present in the interaction from the pooled intrablock error. This value is then compared with the SS retained in consecutive AMMI*n* models (Gauch, 1992). Such a procedure may help to fix an adequate number of multiplicative terms containing real structure. For our data set, the intrablock error was 0.12 and the interaction had 72 *df*; accordingly, the interaction contained about 0.12 × 72 = 8.65 noise SS (17.2 percent), and 50.3–8.65 = 41.65 pattern SS (82.8 percent). This last percentage,

TABLE 9.4. Summary of the AMMI Analysis for Grain Yield

Source of variation	df	Sum of squares	Mean square	Variance ratio[a]	R^2 (%)[b]
G × E	72	50.3	0.70	5.74**	
AMMI1	17	26.4	1.55	12.91**	52.5
AMMI2	15	12.7	0.85	7.08**	25.3
AMMI3	13	5.6	0.43	3.58**	11.2
AMMI4	11	4.0	0.36	3.03**	7.9
Residual	16	1.6	0.10	0.83 ns	3.1
Error	234	28.5	0.12		

[a]Environmental main effect tested against Reps (E)
[b]Fraction of sum of squares associated with each interaction term
**Significant at 1 percent level
nsNot significant

which is slightly larger than that retained by the first two multiplicative terms, shows that AMMI2 is more parsimonious and effective than the original ANOVA model; that is, it requires fewer *df* for an adequate description of the interaction and excludes most of its noise. We would tend to exclude the third and fourth (and later) axes from the AMMI model. The ANOVA interaction is then replaced by the first two multiplicative terms, which can easily be visualized with the aid of a biplot (see Figure 9.1).

Biplot Representation

In Figure 9.1 genotypes and environments are depicted as points on a plane. The position of the point for genotype i is given by the estimates for the genotypic scores (c_{i1}, c_{i2}); similarly, the point coordinates for environment j originate from the estimates for the environmental scores (d_{i1}, d_{i2}). Distances from the origin (0,0) are indicative of the amount of interaction that was exhibited by either genotypes over environments or environments over genotypes. For example, the genotype M01 showed a highly interactive behavior, whereas the environment PA8 exhibited low interaction. The nearly additive behavior of PA8 indicates that genotypic yields in that environment were highly correlated with the overall genotypic means across environments. In a vector representation, the genotype and environment points determine lines starting at the origin (0,0). The interaction effect of genotype i in environment j is approximated by projecting the genotype point (c_{i1}, c_{i2}) onto the line determined by the environmental vector, which has a slope d_{i2}/d_{i1}, where distance from the origin provides information about the magnitude of the interaction. The angle between the vectors of ge-

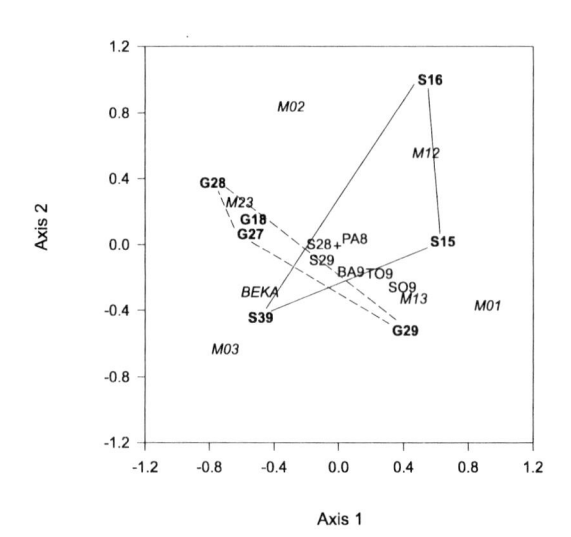

FIGURE 9.1. Biplot of the AMMI Analysis for Grain Yield

Note: Genotypes are represented in italic letters, environments in capitals. The lines connect contrasting environments (in bold) from the Sevilla region (solid lines) and the Granada region (dashed lines).

notype i and environment j tells us something about its nature: the interaction is positive for acute angles, negligible for right angles, and negative for obtuse angles.

Let us now examine the distribution of genotypes in Figure 9.1. The original variety, Beka, lay on the lower left-hand side, not very far from the origin when compared with the three original mutants (M01, M02, and M03) that exhibited larger interactions. These mutants were farther away from one another, drawing a nearly perfect equilateral triangle in the biplot. Their position relative to their binary recombinants informs us about the nonallelic combinations and/or interactions that are controlling adaptation to specific environments. For instance, M12 was intermediate in position to M01 and M02, from which an additive behavior of the nonallelic genes 1 and 2 can be inferred with respect to adaptation. An identical situation arises for M23, which can be found approximately in between M02 and M03. M13 was considerably closer to M01, suggesting that both genotypes exhibited similar adaptive patterns. In a genetic context, this means that dominance is somehow exerted by gene 1 on gene 3; that is, gene 1 is epistatic to gene 3. The genetic information included in the biplot is not exhausted yet. A further examination reveals that the first multiplicative term separates those genotypes containing gene 1, at the right-hand side, from the rest, at the left-hand side. Similarly, the second term distinguishes between genotypes incorporating gene 2, at the upper side of the biplot, from the rest, at its lower side.

A better understanding of the interaction patterns described so far is possible if we succeed (1) in identifying the physiological systems affected by the mutations playing a role in the phenotypic response to the environment and (2) the environmental factors interacting with the genetic makeup underlying the adaptive physiology.

Assessment of the Genetic and Physiological Basis of Adaptation Based on the AMMI Analysis

This step may be addressed by relating the genotypic scores of AMMI2 to additional information, such as the genetic constitution of the mutants and/or their phenotypic characterization. Directions of greatest change for genotypic variables can be included in the biplot. These directions can be obtained from a regression of the external variables on the genotypic scores for axes 1 and 2. The regression coefficients in relation to the origin (0,0) define a line that, after projection of the genotypes on it, gives an ordering of these genotypes with respect to that variable. The half line from the origin through the point defined by the regression coefficients represents the above-average value for that genotypic variable; the opposite half line represents the below-average value for the same variable.

The technique of including external genotypic information is applicable to both quantitative and qualitative variables. For example, three new variables can be defined for each of the three mutant genes characterizing the genetic constitution of Beka, its mutants, and their recombinants (see Table 9.5). These variables were included in the biplot according to the methodology described earlier (see Figure 9.2). The enriched biplot corroborates that gene 1 resembles the first multiplicative term and gene 2 the second one. Indeed, correlations between the genotypic scores of axis 1/axis 2 and gene 1/gene 2 were 0.96 and 0.93, respectively. Both mutations arise as the main genetic factors underlying GE and adaptation in this example. Gene 3 was poorly related to either axis 1 or axis 2, being better reflected in the third multiplicative term of AMMI ($r = 0.69$). In a hypothetical 3-D representation of the interaction, each of the three mutant genes would nearly represent a particular dimension in the AMMI3-derived figure.

The morphophysiological attributes of the mutants could also be incorporated in the biplot as true quantitative variables, but only those variables which explained enough variation by the regression on the genotypic scores merited imposition. For this example, the arbitrary threshold used to prescreen the external data was fixed to about 60 percent variation (Figure 9.2). The joint inclusion of genetic and phenotypic information makes the biplot extremely informative; for example, the morphophysiological traits responsible for adaptation

TABLE 9.5. Coefficients for the Identification of Single Mutant Genes

Genotype	Genetic codification		
	Gene 1	Gene 2	Gene 3
BEKA	0	0	0
M01	1	0	0
M02	0	1	0
M03	0	0	1
M12	1	1	0
M13	1	0	1
M23	0	1	1

Note: The presence/absence of a gene is indicated by a value of 1/0.

FIGURE 9.2. Biplot of the AMMI Analysis Including Genetic and Morphophysiological Information

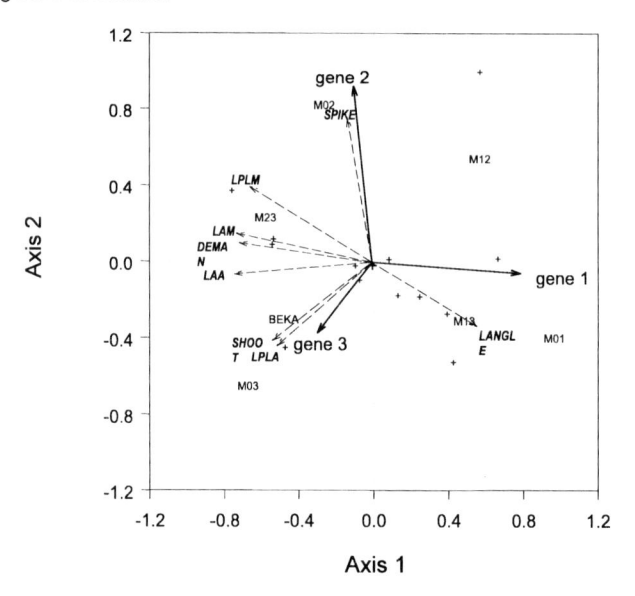

Note: The arrows indicate directions of greatest change for genetic (solid lines) and morphophysiological (dashed lines) variables. Environments are represented by crosses.

can be related rather straightforwardly with their genetic rationale. Thus, genotypes carrying gene 1 had smaller leaf area at anthesis and maturity and were earlier in heading. This information is provided in the biplot by the opposite direction of gene 1 in comparison with that for the phenotypic variables LAA, LAM, and DEMAN. Such phenotypic traits proved to be highly associated: the very early genotypes are likely to possess limited photosynthetic machinery brought about by their inferior vegetative period. Genotypes carrying gene 2 were characterized by denser spikes (note the coincident directions of both vectors), and those carrying gene 3 by shorter grain-filling periods (this last information could be inferred by the inclusion of a third multiplicative term). Additionally, both gene 1 and gene 2 seemed to influence the expression of phenotypic features such as the length of the main shoot, which was reduced in those genotypes bearing one or both genes. In summary, all the aforementioned morphophysiological traits are adaptively sound and have an explicit genetic background that could be identified according to the information revealed by the biplot.

Incorporation of Genetic and Ecophysiological Information into Factorial Regression Models

A second type of multiplicative models is the so-called factorial regression model. The main difference with AMMI is that now the ANOVA interaction, ge_{ij}, is partitioned in one or more multiplicative terms composed of explicit (rather than hypothetical) variables containing genetic, phenotypic, or environmental information. For example, suppose that we want to assess the extent to which one or more genetic (specific genes, QTL, etc.) or phenotypic (morphophysiological) features are advantageous or detrimental for grain yield (or any other response variable) in each of j environments. The factorial regression model then takes the general formulation:

$$E(Y_{ij}) = \mu + g_i + e_j + \sum_{k=1}^{K} x_{ik}\tau_{jk}$$

where x_{ik} refers to the value of any type of genetic/phenotypic variable k, either qualitative or quantitative, for genotype i; and τ_{jk} represents an environmental potentiality of environment j to the explicit genetic/phenotypic variable k.

The heterogeneity in the τ_j's for successive $x_1 \ldots x_K$ variables accounts for the interaction, while the sum of multiplicative terms

$$\sum_{k=1}^{K} x_{ik}\tau_{jk}$$

approximates the ANOVA term, ge_{ij}. To facilitate interpretation, the external variables can be centered to mean zero and standardized to unit standard deviation. The parameters τ_{jk} can easily be estimated by standard least-squares techniques, as equation 3 is a linear model.

Alternatively, we may be interested in monitoring the response of several genotypes to a set of climatic and/or edaphic factors. The multiplicative portion of equation 3 is then replaced by the summation

$$\sum_{h=1}^{H} \beta_{ih} Z_{jh}, \text{ where } z_{jh}$$

indicates the value of environmental variable h in environment j, and β_{ih} is the genotypic sensitivity of genotype i to the explicit environmental variable h. For further details on theory and application of factorial regression models, see Denis (1988); van Eeuwijk (1995); van Eeuwijk, Keizer, and Bakker (1995); and van Eeuwijk and van Tienderen (2002).

Genetic information of different types can be incorporated in the factorial regression model (equation 3). This methodology may aim at identifying the molecular genetic variation, in terms of either major genes or polygenes through QTL mapping, involved in different adaptive responses. It can readily be applied to our barley data using the genetic characterization of the mutants. The fact that most of the phenotypic variation recorded could be ascribed to the effect of only three apparently recessive mutant genes (Molina-Cano, 1982) greatly simplifies the development of a model formulation for interaction. Most often, however, the number of genes/QTL to test for a possible adaptive role may be large (Hayes et al., 1993; Tinker and Mather, 1995). Since the number of candidate models increases together with the parallel increase in external factors to be tested, a selection strategy should ordinarily be employed. Such a strategy may include the use of principal components analysis and AMMI to screen these factors for possible redundancies (Voltas, van Eeuwijk, and Sombrero, et al., 1999). For this particular example, the order of gene inclusion was based upon the explanation of GE by each individual gene, starting with the most important one, then the second one, and so forth. A factorial regression model proposal including genetic information is presented in Table 9.6. Note that the model was extended to incorporate regression of the genotypic main effect as well. In addition, significant cross product terms of two genes were included. This was done to account for possible epistatic (or interactive) effects between nonallelic genes.

Partitioning the genotype main effect (see Table 9.6) aimed at identifying mutant genes that conferred an overall productive advantage or disadvantage to the individual. The methodology is similar to that employed for fitting factorial regression models for the interaction term. In practice, the regression coefficients can easily be obtained using, for instance, the ESTIMATE option of Procedure GLM in SAS (SAS Institute, 1989b).

TABLE 9.6. Partition of Genotypic Variation and GE According to the Effect of Individual Mutant Genes and Their Epistatic Interactions (If Significant) on Grain Yield

Source of variation	df	Sum of squares	Mean square	Variance ratio[a]	R^2 (%)[b]
Total	363	920.7			
Environment (E)	12	806.1	67.18	131.73**	87.6
Reps (E)	39	19.9	0.51		2.2
Genotype (G)	6	15.8	2.63	21.64**	1.7
Gene 1	1	9.7	9.74	80.03**	61.4
Gene 2	1	0.0	0.01	0.04[ns]	0.1
Gene 3	1	2.9	2.90	23.89**	18.4
Gene 1 × Gene 3	1	3.0	2.96	23.76**	19.0
Deviations	2	0.2	0.10	0.83[ns]	4.3
G × E	72	50.3	0.70	5.74**	5.5
Gene 1 × E	12	24.6	2.05	16.82**	48.9
Gene 2 × E	12	11.5	0.96	7.88**	22.8
Gene 3 × E	12	5.9	0.49		11.7
Gene 1 × Gene 3 × E	12	4.7	0.39	3.2**	9.3
Deviations	24	3.7	0.15	1.26[ns]	7.3
Error	234	28.5	0.12		3.0

[a]Environmental main effect tested against Reps (E)
[b]Fraction of sum of squares associated with each term or interaction
**Significant at 1 percent level
[ns]Not significant

Their interpretation is rather straightforward. For example, for gene 3, a positive regression coefficient was obtained with value 0.22 ± 0.04 t·ha^{-1} (mean ± standard error [SE]). This amount quantifies the overall beneficial effect of this gene on yield. Conversely, a detrimental effect of -0.33 ± 0.04 t·ha^{-1} on yield was found for gene 1, whereas gene 2 exhibited, on average, an irrelevant effect on yield.

The genetic description of the genotype main effect is of little interest in the presence of substantial interaction. A biological interpretation should concentrate on the examination of the GE term. The model shown in Table 9.6 suggests that the most important gene accounting for specific adaptation was gene 1, followed by genes 2 and 3, while an epistatic effect of gene 1 to gene 3 seemed also present. The inclusion of all three mutant genes in the regression model could be anticipated from the AMMI analysis. Together with the epistatic effect of gene 1 to gene 3, the gene × environment terms explained about 93 percent of the GE SS with 67 percent of its degrees of freedom. The amount of interaction retained by just gene 1 and gene 2 in the factorial regression model (71.7 percent) was comparable to that of AMMI2. This observation corroborates that the two multiplicative terms of AMMI2 were closely linked to those mutant genes. It was already postulated that AMMI2 seemed to retain mostly pattern, but it was unclear whether further terms could account for something more than interaction noise. The genetic analysis suggests that AMMI2 was probably leaving out some structure hidden in the GE matrix. In fact, the third multiplicative AMMI term might deserve further attention given its relation to gene 3. The factorial regression model presented so far reveals that gene 1, gene 3, and the epistatic effect of gene 1 to gene 3 seemed simultaneously responsible for the genotype main effect and the specific adaptation of genotypes to the environments. Gene 2 had importance only for the explanation of the interaction effect.

The physiological mechanisms of adaptation underlying the three recessive mutant genes were roughly sketched after examination of the AMMI2 biplot (see Figure 9.2). A more detailed interpretation could be obtained by correlating the indicator variables for the three genes (as described in Table 9.5) with the measured morphophysiological traits (see Table 9.7). Hence, the ecophysiological basis for adaptation/misadaptation brought about by gene 1 can be summarized in extreme earliness, low leaf area index, and, to a lesser extent, erect leaves and short stems. Gene 2's influence on adaptive traits resumes in dense spikes and, to a much lesser extent, in few leaves per plant, short stems, and relatively horizontal leaves. Finally, a short grain-filling duration coupled, to some extent, with a long vegetative period is the only adaptive effect that can readily be ascribed to gene 3.

Incorporation of Environmental Information into Factorial Regression Models

Once the underlying physiology and genetics of adaptation have been established, the example may proceed focusing on the determination of those environmental factors to which the genotypes are responding differentially. To this end, a good starting point is the biplot of Figure 9.1. An initial inspection may consist in checking whether the distribution of environments throughout the biplot followed any apparent pattern. For example, is it feasible to group trials belonging to identical agroecological regions? In our data set, three different geographical regions are distinguished: Granada (G), Sevilla (S), and the Central Plateau of Spain (C). Environments of either Granada or Sevilla could not easily be clustered in the biplot. In contrast, trials of the Central Plateau grouped together around the origin, showing little or no interaction. It could therefore be expected that GE concentrated on particular environments from Sevilla and Granada. A further examination of the biplot reveals that two interactive patterns seemed mostly responsible for the interaction. These are in-

TABLE 9.7. Correlation Matrix Between Morphophysiological Traits and Gene Descriptors

				Morphophysiological traits					
Genes	LPLA	LPLM	LAA	LAM	LANGLE	SHOOT	DEMAN	DANM	SPIKE
Gene 1	−0.707	−0.927**	−0.959**	−0.955**	0.841*	−0.750*	−0.917**	0.312	−0.136
Gene 2	−0.522	0.510	0.079	0.226	−0.487	−0.480	0.238	0.000	0.906**
Gene 3	0.106	0.075	0.371	0.334	−0.101	−0.071	0.495	−0.935**	0.136

*Significant at 5 percent level
**Significant at 1 percent level

cluded in Figure 9.1. The first one was determined by a diagonal line that connected trials G28, G18, and G27 on the upper left hand side and trial G29 oppositely on the lower right-hand side. Indeed, interaction was certainly present within the Granada region. The second one came from a triangle, with trials S16 and S15 connected on its shortest side and trial S39 being the opposite extreme. These trials belong to the Sevilla region.

The interactive patterns for Granada and Sevilla defined two rather orthogonal contrasts in the biplot. This information stems from the approximate right angle formed by the lines describing both patterns (see Figure 9.2) and suggests a partition of the interaction term, ge_{ij}, based on regionalization to account for heterogeneity of environments within regions (r):

$$E(Y_{ij}) = \mu + g_i + e_j + gr_{ir} + ge(r)_{ij(r)} \qquad (4)$$

where gr_{ir} accounts for the fraction of interaction due to differences between regions, and $ge(r)_{ij(r)}$ represents that part of the interaction attributable to differences between environments within regions.

This last term can be further partitioned into three extra terms (one for each of Granada, Sevilla, and Spanish Central Plateau regions) to account for differences between environments that belong to a particular region. The same approach can be used for the environment main effect (see Table 9.8). The full model reveals that differences between regions were responsible for the most important fraction (58 percent) of the environmental variation. This result suggests that such differences were rather consistent, i.e., with little environmental variation within regions. The highest yields (mean ± standard deviation [SD]) were achieved in Sevilla (5.67 ± 0.35 t·ha^{-1}), followed by Granada (4.81 ± 1.70 t·ha^{-1}), and the Spanish Central Plateau (2.96 ± 1.01 t·ha^{-1}). On the other hand, the between-regions term of GE was of less importance than the within-region variation for Granada or Sevilla, which together captured most of the interaction (75 percent). Thus, it seemed reasonable to focus on the examination of the intraregional GE variation in order to model genotypic adaptive responses to environmental factors. To that purpose, AMMI and factorial regression were reapplied to our data set.

AMMI methodology was used for the preselection of climatic variables meriting further testing in factorial regression models. The procedure is analogous to that employed for the genetic and morphophysiological data. First, the environmental factors are regressed on the environmental scores of AMMI axes 1 and 2. The resulting regression coefficients thus determine the coordinates for the environmental variables in the biplot, provided that enough variation is explained by the regression on the scores. The application of this methodology to our barley data proved unfruitful, since none of the available climatic variables was related highly enough to the first two multiplicative terms of AMMI. A possible explanation of this apparent in-

TABLE 9.8. Partition of Environmental Variation and GE Based on Regional Division

Source of variation	df	Sum of squares	Mean square	Variance ratio[a]	R^2 (%)[b]
Total	363	920.7			
Environment (E)	12	806.1	67.18	131.73**	87.6
Between Regions	2	465.2	232.58	456.04**	57.7
Within Granada (Gr)	3	242.3	80.78	158.4**	30.1
Within Sevilla (S)	4	13.8	3.45	6.76**	1.7
Within Central Plateau (C)	3	84.8	28.28	55.45**	10.5
Reps (E)	39	19.9	0.51		2.2
Genotype (G)	6	15.8	2.63	21.64**	1.7
G × E	72	50.3	0.70	5.74**	5.5
G × Region	12	10.9	0.91	7.49**	21.7
G × Gr	18	15.5	0.86	7.07**	30.8
G × S	24	22.0	0.92	7.53**	43.7
G × C	18	1.9	0.11	0.87[ns]	3.8
Error	234	28.5	0.12		3.0

[a]Partition of the environmental main effect tested against Reps (E)
[b]Fraction of sum of squares associated with each term or interaction
**Significant at 1 percent level
[ns]Not significant

adequacy arises from the fact that, as suggested in equation 4, the adaptive variation should also be searched for within particular regions. This is ignored when AMMI is fitted to the complete GE table. The climatic variables must then be examined within each particular region responsible for interaction, such as Sevilla or Granada. This was achieved by fitting separate AMMI models for Sevilla and Granada, in order to relate later the environmental information to the interaction. Two multiplicative terms provided a good description of interaction, capturing 89.5 percent and 92.4 percent of GE for Sevilla and Granada, respectively.

Three climatic variables were closely associated to the environmental scores of the AMMI2 model for Sevilla: the ratio of rainfall to total evapotranspirative demand during tillering (R/ETP$_t$, R^2 = 86 percent) and during grain filling (R/ETP$_{gf}$, R^2 = 58 percent), and the mean temperature during grain filling (T$_{gf}$, R^2 = 61 percent). For Granada, the climatic variables that most precisely resembled the environmental scores were the mean temperature during jointing (T$_j$, R^2 = 85 percent), and the ratio of rainfall to total evapotranspirative demand during grain filling (R/ETP$_{gf}$, R^2 = 95 percent). One must be cautious when establishing an arbitrary threshold for the selection of external variables. The decision should be based upon the number of environments to which to relate this additional information. For example, the threshold must be high if relatively few environments are involved in the analysis, as is the case for Sevilla and Granada. Nonetheless, with such a limited number of environments, a tacit uncertainty exists for any relationship to be casual rather than causal. If more environments had been included in the analysis, then the threshold could have been relaxed, and the danger of erroneously selecting external variables would have decreased.

The performance of this greatly reduced set of variables in a factorial regression model was tested. As grain yield is a complex process taking place over the entire crop life cycle, it is logical that the order of inclusion of variables followed a temporal scale, beginning with variables observed during tillering (t), continuing with those for jointing (j) to assess whether additional information was added after correction for the t variables, and finally including those variables related to the grain filling period (gf). This scheme was applied separately for Sevilla and Granada according to the GE partition of equation 4 and Table 9.8. The resulting factorial regression proposal is shown in Table 9.9. The climatic variables already preselected for Granada (T_j and R/ETP_{gf}) were incorporated in the model. Together they retained about 88 percent of the within-region GE with 67 percent of its degrees of freedom. For Sevilla, two external variables were included in the model (R/ETP_t and R/ETP_{gf}), which captured about 71 percent of the within-region GE with 50 percent of its degrees of freedom. The feasibility of including all four external variables was supported by the improved parsimony of the resulting model, as it captured a high amount of GE variation with fewer df than the initial within-region terms for Granada or Sevilla. An additional check consisted in the computation of the F-ratio Ms_{Gxcov}/Ms_{resid} for each external variable, which can be calculated from Table 9.9. A significant F-value would definitely indicate an appropriate GE partitioning. For example, the F-ratio for T_j in Granada was 3.73 ($p = 0.067$); for R/ETP_{gf} in Granada, 3.86 ($p = 0.062$); for R/ETP_t in Sevilla, 3.26 ($p = 0.039$); and for R/ETP_{gf} in Sevilla, 1.64 ($p = 0.219$). The latter was the only environmental descriptor far from significance, but we decided to keep it in the model since (1) its F-ratio was still much larger than unity and (2) R/ETP_{gf}, as indicator of terminal drought, plays an important adaptive role in Mediterranean climates. We will come back later to this specific point.

Genotypic Sensitivities to Environmental Changes

The genotypic sensitivities, β_{ih}, of the factorial regression partition of GE for each region,

$$\sum_{h=1}^{H} \beta_{ih} Z_{jh},$$

are shown in Table 9.10. These sensitivities are indicative of the nature of the interaction exhibited by each mutant. For example, M01 was highly sensitive to environmental changes in Granada, being negatively affected, in

TABLE 9.9. Factorial Regression Model for the Partitioning of GE for Grain Yield

Source of variation	df	Sum of squares	Mean square	Variance ratio[a]	R^2 (%)[b]
G × E	72	50.3	0.70	5.74**	
G × Region	12	10.9	0.91	7.49**	21.7
G × Granada	18	15.5	0.86	7.07**	30.8
G × T	6	6.7	1.12	9.33**	43.2
DEMAN × T_j	1	2.8	2.84	23.67**	41.7
Deviations	5	3.9	0.77	6.42**	58.3
G × R/ETP$_{gf}$	6	7.0	1.16	9.67**	45.2
DEMAN × R/ETP$_{gf}$	1	4.8	4.75	39.58**	68.6
Deviations	5	2.2	0.44	3.67**	31.4
Deviations	6	1.8	0.30	2.50*	11.6
G × Sevilla	24	22.0	0.92	7.53**	43.7
G × R/ETP$_t$	6	10.4	1.73	14.41**	47.3
SHOOT × R/ETP$_t$	1	5.2	5.21	43.42**	50.0
Deviations	5	5.2	1.04	8.67**	50.0
G × R/ETP$_{gf}$	6	5.2	0.87	7.25**	23.6
SHOOT × R/ETP$_{gf}$	1	1.8	1.82	15.17**	34.6
Deviations	5	3.5	0.69	5.75**	65.4
Deviations	12	6.4	0.53	4.42**	29.1
G × Central Plateau	18	1.9	0.11	0.87[ns]	3.8
Error	234	28.5	0.12		

[a] Partition of the environmental main effect tested against Reps (E)
[b] Fraction of sum of squares associated to each term or interaction
*Significant at 5 percent level
**Significant at 1 percent level
[ns] Not significant

relative terms, by high temperatures during jointing and good water availability during grain filling. The performance of M13 was similar to that of M01, but, in contrast, M02 behaved in a completely different manner. Remarkably, sensitivities for recombinants 12 and 23 were intermediate to those of their parental mutants, suggesting again an additive behavior of genes 1/2 and 2/3. M03 did not show any particular response to the climatic variables of Granada, but it exhibited the largest sensitivity to the climatic variables of Sevilla, being favored by a low water availability during tillering and grain filling. Recombinant 12 displayed an opposite behavior to M03 in Sevilla, though similar to that of its parental lines M01 and M02.

Genotypes showing large sensitivities to environmental variables are also expected to exhibit a high overall interaction. The amount of interaction displayed by each genotype (ecovalence) (Wricke, 1962) is given in Table

TABLE 9.10. Genotype Mean Grain Yield, Ecovalence (W_i) Values, Contribution (%) of Genotypes to GE Sum of Squares (SS), and Estimates of the Regression Coefficients of Relevant Environmental Variables for Each Genotype According to the Factorial Regression Model

Genotype	Mean yield (t·ha^{-1})	W_i	% SS (GE)	Regression coefficients				
				Granada			Sevilla	
				T_j	R/ETP_{gf}		R/ETP_t	R/ETP_{gf}
BEKA	4.64	4.24	8.4	0.34**	0.11		−0.20*	−0.20*
M01	4.16	12.37	24.6	−0.51**	−0.63**		0.39**	0.13
M02	4.73	7.14	14.2	0.59**	0.37**		0.20*	0.28**
M03	4.75	7.91	15.7	0.11	0.12		−0.62**	−0.30**
M12	4.36	6.48	12.9	−0.17	−0.02		0.41**	0.23**
M13	4.62	6.69	13.3	−0.50**	−0.23*		0.06	−0.25**
M23	4.73	5.51	10.9	0.13	0.28**		−0.24**	0.11

*/** Regression coefficients with confidence intervals that do not include zero at the 1 percent/5 percent level, respectively

9.10. The most interactive genotype was M01, carrying gene 1, the most sensitive mutant gene to environmental changes (cf. Table 9.6). This behavior can be inferred either from the biplot (see Figure 9.2), or by the fact that, as indicated earlier, M01 was largely sensitive to most environmental variables (see Table 9.10). M03 and M02 then followed M01. M13 was the most interactive recombinant, probably because the epistatic effect of gene 1 to gene 3 approximated M13 to the performance of M01. In contrast, the additive behavior of recombinants 12 and 23 turned into a higher stability. Finally, the mother variety Beka was the least interactive genotype over environments.

The genotypic sensitivities to selected environmental variables are possibly related to one or more of the morphophysiological traits characterizing the genotypes. An easy way to verify such relationships is to calculate correlations between regression coefficients and morphophysiological traits. Alternatively, a factorial regression model can be fitted in which regression coefficients β_i to an environmental variable z_j are replaced by a constant, c, times a morphophysiological variable, x_i ($\beta_i = c \times x_i$), where the constant c is estimated from the data. A residual term, $\beta*_i = z$, can be added to check whether the morphophysiological trait leaves a significant residual sensitivity. The aim is to relate the environmental factors to which the individual is reacting with the physiological systems involved in the phenotypical response. The formulation of an extended model with one environmental variable z_j and one genotypic variable x_i is

$$E(Y_{ij}) = \mu + g_i + e_j + cx_i z_i + \beta^*_i z_i \tag{5}$$

The terms for the fully expanded factorial regression proposal for Granada and Sevilla according to equation 5 are shown in Table 9.9.

Morphophysiological traits that were best related to the environmental factors included in the model were days from emergence to anthesis (DEMAN) for Granada, and length of the main shoot (SHOOT) for Sevilla. Other traits could have been incorporated in the model instead of DEMAN and/or SHOOT with a similar overall fitting. This result is not contradictory since, as already indicated, a number of different traits share the same genetic background. Once the relevant phenotypic traits were determined, the genotypic sensitivities could be associated with the underlying physiology responsible for adaptation. For Granada, genotypes such as M02, which showed an extended vegetative period and a higher leaf area (see Table 9.2), benefited from good water availability and high temperatures in the later part of the growing cycle. These genotypes could therefore express their higher potentiality for yield under favorable conditions. In contrast, those lines characterized by a more limited preanthesis period (e.g., M01) performed relatively better under low temperatures during jointing (Tj). This may suggest that, when temperatures are suboptimal for growth, such genotypes are not penalized in excess in spite of their restricted photosynthetic machinery. Besides, the favorable response of early genotypes to low R/ETP ratios during grain filling suggests the existence of an escape strategy to the abrupt termination of the growing season caused by drought (Loss and Siddique, 1994). For Sevilla, length of the main shoot (SHOOT) was likely to be associated to differences in lodging susceptibility and/or harvest index (HI) among cultivars. Lodging susceptibility is typical of tall cultivars and causes poor grain set and deficient grain filling. In addition, a high HI, which is inversely related to shoot length, is important to maximize yield potential in barley (Riggs et al., 1981). In this example, lines with reduced shoots (i.e., M12, Table 9.2) took advantage of good water status during vegetative growth and grain filling, whereas taller cultivars performed relatively better when growing conditions were harsher. The extent to which such a behavior is dependent upon genotypic differences in HI or lodging susceptibility remains unclear due to the lack of precise field records.

An interesting point of discussion that arises from the examination of the proposed model is the rationale behind the distinctive incorporation of morphophysiological traits for the Granada and Sevilla regions. Different adaptation patterns at each region were already unmasked using an empirical model such as AMMI. It is therefore reasonable to expect different physiological systems controlling genotypic adaptive responses. However, which agroecological features distinguishing Granada from Sevilla may aid to explain such a distinct control? Essentially, the Sevilla region is characterized by relatively high rainfall during the growing cycle, mild temperatures, and deep, fertile soils. These properties led to a higher average yield in Sevilla (5.7 t·ha^{-1}) compared to Granada (4.8 t·ha^{-1}), the latter being characterized by lower temperatures, and shallower soil. Yield potential is therefore superior in the Sevilla region. In this area, larger genotypic dependence for higher yields is to be expected on traits such as HI that confer a productive advantage under near-optimal conditions. On the other hand, Granada often suffers from drought, and under these circumstances, an escape strategy is recognized to be a beneficial genotypic trait for maximizing yield.

It is always crucial to raise a word of caution on the use and misuse of analytical models for the assessment of adaptation. Factorial regression models are little more than just plain regressions, and, accordingly, it is important to be cautious about the inferences to be made from a regression analysis. Any strong association detected between a particular genetic, morphophysiological, or environmental variable and grain yield by no means implies that such an external variable is unequivocally a cause of yield variation. Previous knowledge on the subject is basic to choosing an adequate array of factors for incorporation into factorial regression models. Even so, different combinations of variables may yield similar results in terms of model fitting (Voltas, van Eeuwijk, Araus, and Romagosa, 1999) because many variables are usually highly correlated, and many sets of variables will perform equally well in different models. Therefore, the choice of proper models should be based not only on statistical considerations but also on a proper ecophysiological understanding of the phenomenon under study.

BREEDING IMPLICATIONS

The analytical models presented so far are prone to produce the erroneous impression that a single morphophysiological trait may systematically provide an adequate description of plant adaptive responses to environmental factors. Most often, yield increases associated with a particular trait have been found to be small (Loss and Siddique, 1994), and breeders are reluctant to base a selection strategy on the incorporation of a single, specific trait (Ceccarelli and Grando, 1991). The use of true near-isogenic lines in adaptation studies aims at avoiding the misleading results derived from the effect of different genetic backgrounds on yield (Rasmusson and Gengenbach, 1983; Molina-Cano et al., 1990). In our example, only three apparently recessive Mendelian genes were responsible for a broad range of associated physiological processes (Molina-Cano, 1982). The monitored responses to environmental factors are the effect of these genes and their nonallelic combinations, which demonstrates the role of major genes in adaptation (Orr and Coyne, 1992). However, what we usually observe in any segregation following a cross between two contrasting lines is the effect of many quantitative genes plus their interactions (Pérez de la Vega, 1997). The framework provided by the use of near-isogenic lines may often be of limited practical importance to breeders whose objective may be just to incorporate a target trait into genotypes of wide genetic origin (Acevedo and Ceccarelli, 1989). The value of this approach for understanding the mechanisms underlying GE and adaptation and their implications in plant breeding must also be recognized, as it has been illustrated previously.

Genotype by environment interaction has important implications in breeding programs (Fox, Crossa, and Romagosa, 1997), including (1) the goal of wide or specific adaptation and choice of locations for selection; (2) resource allocation in advanced line testing across sites and/or years; and (3) trade-off between empirical multienvironment testing of large numbers of genotypes and evaluation of a limited set of lines in detailed physiological studies.

Related to wide/specific adaptation is the question of breeding location(s): Can selection under optimum high-input environments identify genotypes adapted to more stressed environments? A significant body of literature on the issue of wide versus specific adaptation has been contributed by Salvatore Ceccarelli and co-workers of the ICARDA (International Center for Agricultural Research in the Dry Areas) barley breeding program (Ceccarelli, 1989, 1994; Ceccarelli and Grando, 1996). They have strongly advocated the exploitation of specific adaptation for optimum use of resources, particularly in marginal environments, arguing that selection for high yield potential has not increased yield under low-input conditions. Ceccarelli (1994) favored and demonstrated the benefits of selection under conditions similar to the target environment, concluding that barley genotypes bred for poor rain-fed areas should be selected under these unfavorable conditions. On the contrary, CIMMYT (International Maize and Wheat Improvement Center) wheat genotypes bred under high-input environments have shown to be superior in yield and have demonstrated better adaptation across large ecological areas than genotypes developed locally, perhaps with the exception of very marginal sites (Braun, Rajaram, and van Ginkel, 1997). Poor adaptation of CIMMYT genotypes to specific environments often reflected susceptibility to specific plant diseases.

The success of CIMMYT in releasing wheat genotypes that combine high yield potential and wide adaptation involves a completely different approach from that followed at ICARDA. Continuous selection cycles, referred to as "shuttle breeding," are carried out in alternative high-yield potential environments differing in altitude, latitude, photoperiod, temperature, rainfall, soil type, and disease spectrum. Experiments for which selection has taken place in alternate locations are scarce in barley. For example, Patel et al. (1987) found that when barley populations were alternated between diverse sites, representing different zones of adaptation, natural selection improved yields, but less than when selection was carried out in single locations. This dispute about breeding philosophies will continue until molecular studies shed light on the nature of genetic changes imposed by selection. So far, just a few studies of this kind have been published for barley. Allard (1999) found that natural selection in a composite cross II of barley (CCII) favored different allelic combinations of four esterase genes for dry and wet conditions in Davis, California. When natural selection in CCII was continued for another 20 generations in Bozeman, Montana (with a harsh continental climate), the two winning allelic combinations in California rapidly declined, and other combinations became dominant. Thus, different alleles were favored in areas with rather different climates. Pomortsev et al. (1996) also studied the fate of a barley population (cross of two cultivars) in the very distinct areas of Moscow and the Pamirs, by analyzing several morphological and protein markers. Natural selection acted against alleles at different loci in both locations, but they identified an association of hordein A and B alleles that provided adaptation for both conditions. This evidence is anecdotal, but in the years to come we will undoubtedly witness an enormous expansion of our knowledge in this area.

An appropriate allocation of available resources is inherent to the success of any breeding program. Breeders aim to cover a representative sample of spatial and temporal environmental variation. Limited resources and increasing pressure to develop new cultivars can reduce the number of years of testing below the minimum number required for representativeness. Significant GY suggests testing for many crop cycles. However, breeders often substitute year with location variation, assuming that GL interaction is of the same nature as GY (GLY). Resource allocation for multilocation-multiyear trials is discussed in extent by Talbot (1997). Decisions are based on the relative magnitude of the GL, GY, GLY, and error variance components. In this context, the issue of repeatability of GE is also important (Basford and Cooper, 1998, p. 163): "the lack of repeatability may be an inherent feature of the complex systems with which we deal or a result of the way in which we sample the system." However, little information is available about repeatability of GE. There is a need to assess the proportion of GE associated to predictable and, thus, repeatable pattern and to noise. In the absence of a sizable proportion of repeatability, exploitation of GE is not possible. A series of papers (Fox and Rosielle, 1982; Romagosa and Fox, 1993; Cooper and Fox, 1996; Fox, Crossa, and Romagosa, 1997) has suggested the use of reference and probe genotypes for the characterization of environmental variation and to assess repeatability. A breeder could define a long-term target environment using relative ranks from a common reference set of genotypes grown over locations and years. Results from each location in a given year should be considered in accordance with its across-year representativeness. Probe genotypes with differential response to known biotic and abiotic conditions could also be used to characterize environments. However, the practical use of these two concepts is still limited.

In most breeding programs based on extensive multilocation trials, empirical selection of segregating populations still enables genetic gains for unrecognized stresses. The alternative approach for the empirical testing of a large number of advanced lines is exposure to a few "key locations" with defined stresses. Once major stresses are identified, manipulation of the selection environment and selection of specific parental crosses should result in an increase in the heritability of those traits involved. Balance between both approaches should depend upon how well major stresses are defined.

Few barley breeders routinely assess genotypic adaptability and stability. Nonetheless, most of them develop a deep knowledge of their environments and of the adaptation of their barley genetic materials. This appreciation may arise from detailed field observations, often by visual comparison of newer lines to check varieties in many locations, in what frequently appears to be an intuitive process. However, we believe that statistical assessment of genotypic adaptability (and stability) is needed, aiming not to replace breeders' impressions, but to complement them (Fox, Crossa, and Romagosa, 1997). We want to emphasize that it is essential that statistical assessment runs in parallel with agroecological understanding of the genotypes and the environments. GE techniques will gain acceptance among breeders through better-documented, more user-friendly software that handles large unbalanced data, missing values, and, for certain species, multiple traits. Production of such mature software that produces real-time analyses is still a bottleneck (Fox, Crossa, and Romagosa, 1997; Basford and Cooper, 1998).

APPENDIX: EXAMPLES OF SAS PROGRAMS FOR THE MODELS DESCRIBED

All models presented in this chapter may be fitted using SAS software. The most relevant statements are presented in the following, with classification variables for genotypes, environments, and blocks labeled as gen, loc, and block, respectively. In most cases, other alternatives to the proposed SAS codes yield identical results.

AMMI Model

AMMI analysis can easily be performed using SAS/IML (Interactive Matrix Language) (SAS Institute, 1989a) software. Procedure IML requests the two-way residuals from additivity of n genotypes in m environments previously included in the SAS data set residual. These interaction effects are then transferred to a column vector x that must be reshaped to originate a new matrix (resid) with dimensions $m \times n$, prior to the singular value decomposition done by the SVD subroutine. The program output gives the squares of the singular values (the sum of squares corresponding to each axis) and the environmental and genotypic scores after being multiplied (scaled) by the square root of their corresponding singular values. This program must be adjusted for the specific number of environments and genotypes, as shown in the text.

```
PROC IML;
    USE residual;
    READ ALL VAR{env gen r_yield};
    READ ALL VAR _NUM_ INTO x;
    r_yield= x[,4];                      /* column vector x */
    m= 13;                               /* m is the number of environments */
    n= 7;                                /* n is the number of genotypes */
    resid=shape(r_yield,m,n);            /* m,n matrix containing residuals from additivity */

    CALL SVD(u,q,v,resid);               /* singular value decomposition of the 'resid' matrix */
    axes= 'Axis-1':'Axis-7';             /*7 is the lowest value of m and n */
    b=q#q;                               /* eigen values, or squares of singular values q */
    SUMb=SUM(b);
    e=(b/SUMb)*100;                      /* percentage of SS corresponding to the axes */
    sqq=SQRT(q);                         /* square root of singular values for scaling of the scores */
    d= DIAG(sqq);                        /* diagonal matrix containing the square root of singular values */
    uq=u*d;                              /* scaling of the genotypic scores */
```

```
vq=v*d;                           /* scaling of the environmental scores */
score='SCORE1':'SCORE7';  /*7 is the lowest value of n and m*/

PRINT 'Eigen Values', b[ROWNAME=axes COLNAME='SS' FORMAT=12.4];
PRINT '%SS Explained by Each Axis', e[ROWNAME=axes COLNAME='%SS' FORMAT=12.2];
PRINT 'Genotypic Scores', vq[ROWNAME=gen COLNAME=score FORMAT=12.4];
PRINT 'Environmental Scores', uq[ROWNAME=env COLNAME=score FORMAT=12.4];
```

QUIT;

Analysis of Variance Based on Regional Division of Environments

The additive model (equation 4) can be fitted using procedure GLM of SAS with fixed genotypes and environments and random blocks. The environmental main effect and its partitioning are then tested over the term region*env*block (blocks within environments within regions). The following code first creates three different variables within an SAS data set in which to allocate environments according to the previous assortment in three regions: Granada, Sevilla, and Spanish Central Plateau. Environments not belonging to a particular region are given a homogeneous value (e.g., zero). By using this methodology, the *within-regions* term of equation 4 $ge(r)_{ij(r)}$, can immediately be split into three terms accounting for the *between-environments* variation at each of Granada, Sevilla, and Central Plateau regions. This is true after correction for the *between-regions* term of the equation, gr_{ir}. Accordingly, type I sum of squares should be used. The same procedure applies to the partition of the environmental main effect.

```
DATA . . . .;
    IF region='granada' THEN env_gr=env;
    ELSE env_gr=0;
    IF region='sevilla' THEN env_se=env;
    ELSE env_se=0;
    IF region='central' THEN env_ce=env;
    ELSE env_ce=0;
RUN;

PROC GLM;
    CLASS region env env_gr env_se env_ce gen block;
    MODEL yield = region env_gr env_se env_ce
                  region*env*block
                  gen
                  gen*region gen*env_gr gen*env_se gen*env_ce/SS1 E1;
    RANDOM region*env*block/TEST;
RUN;
```

Factorial Regression Including Phenotypic and Environmental Information

As already mentioned, equation 5 needs to be developed independently for regions Granada and Sevilla. This can readily be achieved by fitting two separate factorial regression models, one for each region. Alternatively, the construction of a unique model including factorial regression per region may be attempted using procedure GLM. The following SAS code aims at the latter purpose. Similarly to the preceding ANOVA model, four different variables must be defined beforehand in which to allocate values per region of relevant environmental information (T_j, R/ETP_t, and R/ETP_{gf}). Environmental information not belonging to the implicated region is given an equal value of zero, by which to eliminate its influence on the corresponding fitting of any factorial regression submodel within the most general model. Introduction of environmental variables on the levels of the environmental effect must be carried out after inclusion of the *between-regions* (gen*region) interaction term in the model. Cross products relating specific phenotypic features (DEMAN, SHOOT) to environmental information should be included prior to the linear regression of genotypes on each environmental variable.

```
DATA . . . . ;
   IF region='granada' THEN cov_env1=Tj;
   ELSE cov_env1=0;
   IF region='granada' THEN cov_env2=RETPgf;
   ELSE cov_env2=0;
   IF region='sevilla' THEN cov_env3= RETPt;
   ELSE cov_env3=0;
   IF region='sevilla' THEN cov_env4= RETPgf;
   ELSE cov_env4=0;
RUN;

PROC GLM;
   CLASS gen env region block;
   MODEL yield =    region env*region
                    region*env*block
                    gen
                    gen*region
                    deman*cov_env1 gen*cov_env1 deman*cov_env2 gen*cov_env2
                    shoot*cov_env3 gen*cov_env3 shoot*cov_env4 gen*cov_env4 gen*env/SS1 E1;
   RANDOM region*env*block/TEST;
RUN;
```

Estimates of Genotypic Sensitivities

The estimation of genotypic sensitivities for any trait (e.g., cov_env1) can be carried out with SAS in different ways. The option SOLUTION in the MODEL statement of procedure GLM or MIXED produces estimates that are given as differences with the last genotype (fixed to a value of zero). It seems more convenient to obtain sensitivities centered around zero, for which the ESTIMATE statement must be used. One must be aware that the default F-tests provided by PROC GLM are those of the main effects and interaction against the deviations from the model. For the estimation of genotypic sensitivities, such an "error" term should preferably be replaced by the available intrablock error of the standard ANOVA. The MIXED procedure allows the user to specify an intrablock error together with its own degrees of freedom. To that end, the DDF *(denominator degrees of freedom)* option of the MODEL statement is set equal to the degrees of freedom for the intrablock error (234) for all terms of the interaction, while the intrablock error itself is defined by the statement PARMS (magnitude of the intrablock error, 0.12) /NOITER EQCONS=1.

```
PROC MIXED;
   CLASS gen env region block;
   MODEL yield =      region env*region
                      gen
                      gen*region
                      genot*cov_env1 genot*cov_env2
                      genot*cov_env3 genot*cov_env4/HTYPE=1DDF=., ., 234,234,234,234,234,234
   PARMS 0.5, 0.12/NOITER EQCONS=1 ;
   ESTIMATE      'BEKA*Tj'    GENOT*cov_env1  6 -1 -1 -1 -1 -1 -1/ DIVISOR=7;
   ESTIMATE      'M01*Tj'     GENOT*cov_env1 -1  6 -1 -1 -1 -1 -1/ DIVISOR=7;
   ....
   ESTIMATE      'M23*Tj'     GENOT*cov_env1 -1 -1 -1 -1 -1 -1  6/ DIVISOR=7;
RUN;
```

REFERENCES

Abay, F. and C. Cahalan (1995). Evaluation of response of some barley landraces in drought prone sites of Tigray (Northern Ethiopia). *Crop Improvement* 22: 125-132.

Abo-Elenin, R.A., Morsi, L.R., and A.A. Gomaa (1977). Yield stability parameters for barley cultivars in Egypt. *Agricultural Research Review* 55: 189-206.

Acevedo, E. and S. Ceccarelli (1989). Role of the physiologist breeder in a breeding program for drought resistance conditions. In *Drought Resistance in Cereals*, ed. F.W.G. Baker. Wallinford, UK: CAB International, pp. 117-139.

Allard, R.W. (1999). *Principles of Plant Breeding*. New York: J. Wiley.

Arain, A.G. (1975). Adaptation studies of radiation-induced barley mutants. *Experientia* 31: 526-528.

Atlin, G.N., McRae, K.B., and X. Lu (2000). Genotype x region interaction for two-row barley yield in Canada. *Crop Science* 40: 1-6.

Baker, R.J. (1988). Tests for crossover genotype-environmental interactions. *Canadian Journal of Plant Sciences* 68: 405-410.

Baker, R.J. (1990). Crossover genotype-environmental interaction in spring wheat. In *Genotype-by-Environment Interaction and Plant Breeding*, ed. M.S. Kang. Baton Rouge: Lousiana State University, pp. 42-51.

Barszczak, Z. and T. Barszczak (1980). Tolerance of wheat and barley varieties to soil acidity. *Wissenschaftliche Beitrage der Martin-Luther-Universitat Halle-Wittenberg* 20: 678-690.

Basford, K.E. and M. Cooper (1998). Genotype x environment interactions and some considerations of their implications for wheat breeding in Australia. *Australian Journal of Agricultural Research* 49: 153-174.

Berbigier, A. and J.B. Denis (1981). Variety x locality interaction. Analysis of spring barley yields in 1978. Comparison of 1977 with 1978. *Agronomie* 1: 641-650.

Bezant, J., Laurie, D., Pratchett, N., Chojecki, J., and M. Kearsey (1997). Mapping QTL controlling yield and yield components in a spring barley (*Hordeum vulgare* L.) cross using marker regression. *Molecular Breeding* 3: 29-38.

Borthwick, H.A., Parker, M.W., and P.H. Heinze (1941). Effect of photoperiod and temperature on development of barley. *Botanical Gazette* 103: 326-341.

Bouzerzour, H. and B. Refoufi (1992). Effect of sowing date and rate, and site environment on the performance of barley cultivars grown in the Algerian high plateaux. *Rachis* 11: 19-24.

Braun, H.J., Rajaram, S., and M. van Ginkel (1997). CIMMYT's approach to breeding for wide adaptation. In *Adaptation in Plant Breeding*, ed. P.M.A. Tigerstedt. Dordrecht, the Netherlands: Kluwer Academic Publishers, pp. 197-205.

Ceccarelli, S. (1989). Wide adaptation: How wide? *Euphytica* 40: 197-205.

Ceccarelli, S. (1994). Specific adaptation and breeding for marginal conditions. *Euphytica* 77: 205-219.

Ceccarelli, S. (1996). Positive interpretation of genotype by environment interactions in relation to sustainability and biodiversity. In *Plant Adaptation and Crop Improvement*, eds. M. Cooper and G.L. Hammer. Wallingford, UK: CAB International, pp. 467-486.

Ceccarelli, S. and S. Grando (1991). Environment of selection and type of germplasm in barley breeding for low-yielding conditions. *Euphytica* 57: 207-219.

Ceccarelli, S. and S. Grando (1996). Drought as a challenge for the plant breeder. *Plant Growth Regulation* 20: 149-155.

Ceccarelli, S., Grando, S., and A. Impiglia (1998). Choice of selection strategy in breeding barley for stress environments. *Euphytica* 103: 397-318.

Cooper, M. and P.N. Fox (1996). Environmental characterization based on probe and reference genotypes. In *Plant Adaptation and Crop Improvement*, eds. M. Cooper and G.L. Hammer. Wallingford, UK: CAB International, pp. 529-547.

Cooper, M. and G.L. Hammer (1996). *Plant Adaptation and Crop Improvement*. Wallingford, UK: CAB International.

Denis, J.B. (1988). Two-way analysis using covariates. *Statistics* 19: 123-132.

Dofing, S., Berge, T.G., Baenziger, P.S., and C.W. Knight (1992). Yield and yield component response of barley in subarctic and temperate environments. *Canadian Journal of Plant Science* 72: 663-669.

Ellis, R.H., Summerfield, R.J., Roberts, E.H., and J.P. Cooper (1989). Environmental control of flowering in barley *(Hordeum vulgare)*. III. Analysis of potential vernalization responses, and methods of screening germplasm for sensitivity to photoperiod and temperature. *Annals of Botany* 63: 687-704.

Fox, P.N., Crossa, J., and I. Romagosa (1997). Multi-environment testing and genotype × environment interaction. In Statistical Methods for Plant Variety Evaluation, eds. R.A. Kempton and P.N. Fox. London: Chapman and Hall, pp. 117-137.

Fox, P.N. and A.A. Rosielle (1982). Reducing the influence of environmental main effects in pattern analysis of plant breeding environments. *Euphytica* 31: 645-656.

Gauch, H.G. (1988). Model selection and validation for yield trials with interaction. *Biometrics* 44: 705-715.

Gauch, H.G. (1992). *Statistical Analysis of Regional Yield Trials*. Amsterdam: Elsevier.

Gollob, H.F. (1968). A statistical model that combines features of factor analysis and analysis of variance techniques. *Psycrometrika* 33: 73-115.

Gustafsson, A., Ekman, G., and I. Dormling (1977). Effects of the Pallas gene in barley: Phene analysis, overdominance, variability. *Hereditas* 86: 251-266.
Hayes, P.M., Liu, B.H., Knapp, S.J., Chen, F., Jones, B., Blake, T., Franckowiak, J., Rasmusson, D., Sorrells, M., Ullrich, S.E., et al. (1993). Quantitative trait locus effects and environmental interaction in a sample of North American barley germplasm. *Theoretical and Applied Genetics* 87: 392-401.
Isla, R., Royo, A., and R. Aragüés (1997). Field screening of barley cultivars to soil salinity using a sprinkler and drip irrigation system. *Plant and Soil* 197: 105-117.
Jackson, P.A., Byth, D.E., Fischer, K.S., and R.P. Johnston (1994). Genotype x environment interactions in progeny from a barley cross. II. Variation in grain yield, yield components and dry matter production among lines with similar times to anthesis. *Field Crops Research* 37: 11-23.
Jones, J.L. and E.J. Allen (1986). Development in barley *(Hordeum sativum)*. *Journal of Agricultural Science, Cambridge* 107: 187-213.
Kang, M.S. and H.G. Gauch (1996). *Genotype by Environment Interaction: New Perspectives.* Boca Raton, FL: CRC Press.
Kempton, R.A. and P.N. Fox (1997). *Statistical Methods for Plant Variety Evaluation.* London: Chapman and Hall.
Kong, D., Choo, T., Jui, P., Ferguson, T., Therrien, M.C., Ho, K.M., May, K.W., and P. Narasimhalu (1994). Genetic variation and adaptation of 76 Canadian barley cultivars. *Canadian Journal of Plant Science* 74: 737-744.
Lekes, J. and L. Zinisceva (1990). Selection in spring barley with a view of more efficient nitrogen utilization. Communication to the *Vereinigung osterreichischer Pflanzenzuchter,* Gumpenstein, Osterreich, 20-22 November 1990. Austria: Bundesanstalt fur Alpenlandische Landwirtschaft, Gumpenstein, pp. 107-114.
Loss, S.P. and K.H.M. Siddique (1994). Morphological and physiological traits associated with wheat yield increases in Mediterranean environments. *Advances in Agronomy* 52: 229-275.
Molina-Cano, J.L. (1982). Genetic, agronomic, and malting characteristics of a new erectoides mutant induced in barley. *Zeitshrift fur Pflanzenzüchtung* 88: 34-42.
Molina-Cano, J.L., García del Moral, L.F., Ramos, J.M., García del Moral, M.B., Jiménez-Tejada, P., Romagosa, I., and F. Roca de Togores (1990). Quantitative phenotypical expression of three mutant genes in barley and the basis for defining an ideotype for Mediterranean environments. *Theoretical and Applied Genetics* 80: 762-768.
Muñoz, P., Voltas, J., Araus, J.L., Igartua, E., and I. Romagosa (1998). Changes over time in the adaptation of barley releases in north-eastern Spain. *Plant Breeding* 117: 531-535.
Nurminiemi, M., Bjornstad, A., and O.A. Rognli (1996). Yield stability and adaptation of nordic barleys. *Euphytica* 92: 191-202.
Orr, H.A. and J.A. Coyne (1992). The genetics of adaptation: A reassessment. *American Naturalist* 140: 725-742.
Patel, J.D., Reinbergs, E., Mather, D.E., Choo, T.M., and J.D.E. Sterling (1987). Natural selection in a doubled-haploid mixture and a composite cross of barley. *Crop Science* 27: 474-479

Pérez de la Vega, M. (1997). Plant genetic adaptedness to climatic and edaphic environment. In *Adaptation in Plant Breeding,* ed. P.M.A. Tigerstedt. Dordrecht, the Netherlands: Kluwer Academic Publishers, pp. 27-38.

Pomortsev, A.A., Kalabushkin, B.A., Blank, M.L., and A. Bakhronov (1996). Investigation of natural selection in artificial hybrid populations of spring barley. *Genetika (Moskva)* 32: 1536-1544.

Rasmusson, D.C. and B.G. Gengenbach (1983). Breeding for physiological traits. In *Crop Breeding,* ed. D.R. Wood. Madison, WI: American Society of Agronomy, pp. 231-254.

Riggs, T.J., Hanson, P.R., Start, N.D., Miles, D.M., Morgan, C.L., and M.A. Ford (1981). Comparison of spring barley varieties grown in England and Wales between 1880 and 1980. *Journal of Agricultural Science, Cambridge* 97: 599-610.

Romagosa, I. and P.N. Fox (1993). Genotype-environment interaction and adaptation. In *Plant Breeding, Principles and Prospects,* eds. M.D. Hayward, N.O. Bosemark, and I. Romagosa. London: Chapman and Hall, pp. 373-390.

Romagosa, I., Fox, P.N., García del Moral, L.F., Ramos, J.M., García del Moral, M.B., Roca de Togores, F., and J.L. Molina-Cano (1993). Integration of statistical and physiological analysis of adaptation of near-isogenic barley lines. *Theoretical and Applied Genetics* 86: 822-826.

Romagosa, I., Ullrich, S.E., Han, F., and P.M. Hayes (1996). Use of the additive main effects and multiplicative interaction model in QTL mapping for adaptation in barley. *Theoretical and Applied Genetics* 93: 30-37.

SAS Institute, Inc. (1989a). *SAS/IML Software: Usage and Reference,* Version 6, First Edition. Cary, NC: SAS Institute, Inc.

SAS Institute, Inc. (1989b). *SAS/STAT User's Guide,* Version 6, First Edition. Cary, NC: SAS Institute, Inc.

SAS Institute, Inc. (1997). *SAS/STAT Software: Changes and Enhancements Through Release 6.12.* Cary, NC: SAS Institute, Inc.

Soliman, K.M. and R.W. Allard (1991). Grain yield of composite cross populations of barley: Effects of natural selection. *Crop Science* 31: 705-708.

Talamucci, P. (1975). Comportamento di 16 cultivar di orzo in semina autunnale e primaverale nella montagna pistoiese. *Sementi Elette* 21: 33-42.

Talbot, M. (1984). The identification of specific variety x environment interactions in spring barley. In *Cereal Production. Proceedings of the Second International Summer School in Agriculture,* Royal Dublin Society and WK Kellog Foundation, ed. E.J. Gallagher. UK: Butterworth and Co., Ltd., p. 110.

Talbot, M. (1997). Resource allocation for selection systems. In *Statistical Methods for Plant Variety Evaluation,* eds. R.A. Kempton and P.N. Fox. London: Chapman and Hall, pp. 162-174.

Tewari, S.N. (1975). Path coefficient analysis for grain yield and its components in a collection of barley *(Hordeum vulgare)* germplasm. In *Third International Barley Genetics Symposium,* Authors T-Z, ed. H. Gaul. München, Germany: Verlag Karl Thiemig, p. 118.

Tinker, N.A. and D.E. Mather (1995). Methods for QTL analysis with progeny replicated in multiple environments. *Journal of Quantitative Trait Loci* (online): <http://probe.nalusda.gov:8000/otherdocs/jqtl/jqtl1995-01/jqtl15.html>.

Tinker, N.A., Mather, D.E., Rossnagel, B.G., Kasha, K.J., Kleinhofs, A., Hayes, P.M., Falk, D.E., Ferguson, T., Shugar, L.P., Legge, W.G., et al. (1996). Regions of the genome that affect agronomic performance in two-row barley. *Crop Science* 36: 1053-1062.

van Eeuwijk, F.A. (1995). Linear and bilinear models for the analysis of multi-environment trials. I. An inventory of models. *Euphytica* 84: 1-7.

van Eeuwijk, F.A. (1996). Between and beyond additivity and non-additivity: The statistical modelling of genotype by environment interaction in plant breeding. PhD thesis. Wageningen, the Netherlands.

van Eeuwijk, F.A., Keizer, L.C.P., and J.J. Bakker (1995). Linear and bilinear models for the analysis of multi-environment trials. II. An application to data from the Dutch Maize Variety Trials. *Euphytica* 84: 9-22.

van Eeuwijk, F.A and P.H. van Tienderen (2002). Analysis of genotype by environment interaction: ANOVA and beyond. In *The Evolution of Phenotypic Plasticity,* eds. P.H. van Tienderen and P.M. Brakefield. Oxford: Oxford University Press (in press).

van Oosterom, E.J., Kleijn, D., Ceccarelli, S., and M.M. Nachit (1993). Genotype-by-environment interactions of barley in the Mediterrranean region. *Crop Science* 33: 669-674.

Voltas, J., Romagosa, I., Lafarga, A., Armesto, A.P., Sombrero, A., and J.L. Araus (1999). Genotype by environment interaction for grain yield and carbon isotope discrimination of barley in Mediterranean Spain. *Australian Journal of Agricultural Research* 50: 1263-1271.

Voltas, J., van Eeuwijk, F.A., Araus, J.L., and I. Romagosa (1999). Integrating statistical and ecophysiological analyses of genotype by environment interaction for grain filling of barley. II. Grain growth. *Field Crops Research* 62: 75-84.

Voltas, J., van Eeuwijk, F.A., Sombrero, A., Lafarga, A., Igartua, E., and I. Romagosa (1999). Integrating statistical and ecophysiological analyses of genotype by environment interaction for grain filling of barley. I. Individual grain weight. *Field Crops Research* 62: 63-74.

Wricke, G. (1962). Über eine Methode zur Erfassung der ökologischen Streubreite in Feldversuchen. *Zeitschrift fur Pflanzenzüchtung* 47: 92-93.

Wright, A.C. and R.E. Gaunt (1992). Disease-yield relationship in barley. I. Yield, dry matter accumulation and yield-loss models. *Plant Pathology* 41: 676-687.

Zhu, H., Briceño, G., Dovel, R., Hayes, P.M., Liu, B.H., Liu, C.T., and S.E. Ullrich (1999). Molecular breeding for grain yield in barley: An evaluation of QTL effects in a spring barley cross. *Theoretical and Applied Genetics* 98: 772-779.

Chapter 10

Initiation and Appearance of Vegetative and Reproductive Structures Throughout Barley Development

Luis F. García del Moral
Daniel J. Miralles
Gustavo A. Slafer

INTRODUCTION

Development in grain crops not only is critical in determining the adaptation of the crop to particular environments, but it also plays a central role in the dynamic generation of yield itself (Miralles and Slafer, 1999; Slafer and Whitechurch, 2001). Since certain developmental phases are more critical than others in terms of yield generation (Kirby and Appleyard, 1984; Landes and Porter, 1989), the identification of the factors that affect the length of those phases is of paramount importance to evaluate the effects on yield components when these environmental factors act. Development and growth are distinct but related processes. Growth can be defined several ways, but the most acceptable definition is an irreversible increase in physical dimensions or dry weight with time. Development can be defined as a sequence of phenological events controlled by genetic and environmental factors, determining changes in the morphology and/or function of some organs (Landsberg, 1977).

The general aim of this chapter is to provide a description of the dynamics of initiation and appearance of different structures of the plant throughout crop development. The genetic and environmental variables affecting the time course of different stages are highlighted.

IDENTIFICATION OF DIFFERENT STAGES AND DEVELOPMENTAL PHASES

Several scales have been proposed to describe and identify the major developmental stages (see a review in Landes and Porter, 1989). Some of them (Large, 1954; Haun, 1973; Zadoks, Chang, and Konzak, 1974) characterize plant growth stages through external plant appearance, without dissection of the shoot apex, while others (Andersen, 1952; Waddington, Cartwright, and Wall, 1983; Kirby and Appleyard, 1986) describe the morphological changes that occur in the apical meristem. Scales based on external features give a nondestructive identification of growth stages but provide no information about the sequence and timing of changes in the shoot apex, where development is actually occurring. The relationship between growth and developmental stages is not constant, due to the different responses of the rate of development of the shoot apex and the rate of emergence of leaves and tillers to major environmental factors. For this reason, only direct observation of the shoot apex (or using models based on the knowledge gained from dynamics of responses of morphological-developmental attributes to combinations of environmental factors [Aitken, 1971; Sadras and Villalobos, 1993]) could provide accurate information about barley development.

The development of barley can be partitioned into three major phases, i.e., vegetative, reproductive, and grain-filling phases. Figure 10.1 represents a schematic diagram of developmental progress on an arbitrary time scale during which phases are delimited by internal and external developmental indicators. The vegetative phase from sowing (actually the imbibition of the seeds) to floral initiation is characterized by leaf and tiller initiation. The reproductive phase starts with the onset of spikelet differentiation in the apex and finishes with the pollination of the ovary within the spikelets bearing fertile florets, normally coinciding with heading. During the early part of the reproductive phase, spikelets are initiated until a maximum number of spikelet primordia is reached. Later on, during the last part of this phase, the spikes and stems grow rapidly and some tillers and spikelets die, with the number of fertile florets (and virtually the number of grains) being concomitantly determined. The grain-filling phase is the last developmental phase, subsequent to pollination, during which the grain endosperm develops a variable number of cells and then dry matter is accumulated to determine the final grain weight, in the time that the embryo develops the initial vegetative primordia (roots and leaves) to be able to produce a successful seedling in the next generation. The duration of the different developmental phases varies widely, depending on the geographic area (e.g., different latitudes), time of sowing, and cultivar, in response to the interplay of genetic and environmental factors controlling developmental rates.

ASSOCIATION BETWEEN TIME COURSE OF ORGANOGENESIS AND YIELD COMPONENTS GENERATION

As illustrated in Figure 10.1, the timing of particular developmental phases is associated with times when particular yield components are being

FIGURE 10.1. Schematic Diagram of Barley Development Showing the Different Stages Throughout the Crop

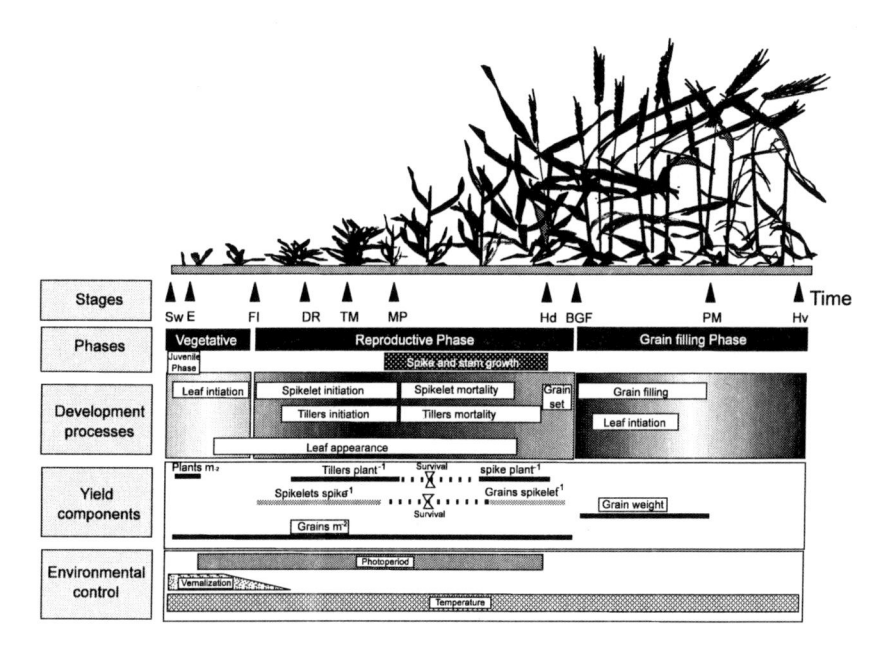

Source: Adapted from Slafer and Rawson, 1994b.
Note: Sowing (Sw), seedling emergence (E), floral initiation or "collar" stage (Fi), double ridge (DR), triple mound (TM), maximum number of total primordia initiated in the apex (MP), heading (Hd), beginning of grain-filling period (BGF), physiological maturity (PM) and harvest (Hv). Boxes indicate different phases, developmental processes, and yield components formation. Environmental factors that control the length of different vegetative and reproductive phases are also indicated.

produced, so that there are yield components being formed constantly during the life cycle of the crop. However, some developmental phases are more important in determining yield potential than others. Grain yield in cereals appears to be better associated with the number of grains per unit land area (Savin and Slafer, 1991; Magrin et al., 1993; Slafer, Calderini, and Miralles, 1996; Frederick and Bauer, 1999) than with final grain weight. The reproductive phase containing the period between spikelet initiation and heading is critical in determining yield potential (see discussion in Slafer and Rawson, 1994b). This is because, as discussed later in this chapter, final grain number seems to depend more strongly on the survival, than on the generation, of structures that may potentially produce a grain. The mortality of some of the potential structures takes place during the late reproductive phase, a few weeks immediately preceding heading, at least accepting extrapolations from wheat studies (e.g., Fischer, 1985; Thorne and Wood, 1987; Savin and Slafer, 1991; Slafer et al., 1994).

During the late part of the reproductive phase, stems and spikes grow rapidly while competing for assimilates, spikelet abortion being the result of this intense competition for metabolites, including carbohydrates and nitrogen compounds (Gallagher, Biscoe, and Scott, 1976; Kirby, 1977; Ellis and Kirby, 1980; Cottrell et al., 1985). Therefore, all crop attributes that affect the partitioning between vegetative and reproductive organs during the reproductive phase are extremely important for yield potential. As main yield components are determined during the reproductive phase, the identification of those environmental factors which modify the duration of this phase is important to understand how yield could be modified when the length of vegetative and reproductive phases changes.

The spike in barley consists of single-flowered spikelets, and the final number of grains per spike is achieved in a sequence of generation and degeneration (or lack of further developmental progress) of floral structures. At floral initiation the apical meristem starts the differentiation of spikelet structures until the maximum number of spikelet primordia is reached. From then on, some of the spikelets initiated do not progress sufficiently to produce a fertile spikelet, a process frequently known as spikelet abortion. Of both processes (spikelet initiation and abortion), it seems that the latter is far more relevant in determining final number of grains per spike than the former.

INITIATION AND APPEARANCE OF VEGETATIVE AND REPRODUCTIVE ORGANS

Leaf and Spikelet Initiation

The mature barley embryo contains the primordia of the first three to four leaves of the future main shoot (Kirby and Appleyard, 1986). The schematic transformation of the apex in the main shoot once germination was triggered and across different stages of plant development is shown in Figure 10.2. From germination to seedling emergence, the apex initiates about two (one to three) new leaf primordia; the exact number of primordia initiated is largely dependent upon sowing depth. Therefore, at seedling emergence, the shoot apex has five to seven leaf primordia. The initiation of leaf primordia in the apex, which continues until the onset of floral initiation, appears to take place at a single rate for a particular genotype under constant

FIGURE 10.2. Schematic Diagram of Morphological Changes Occurring in the Main Stem Barley Apex Throughout Different Developmental Stages

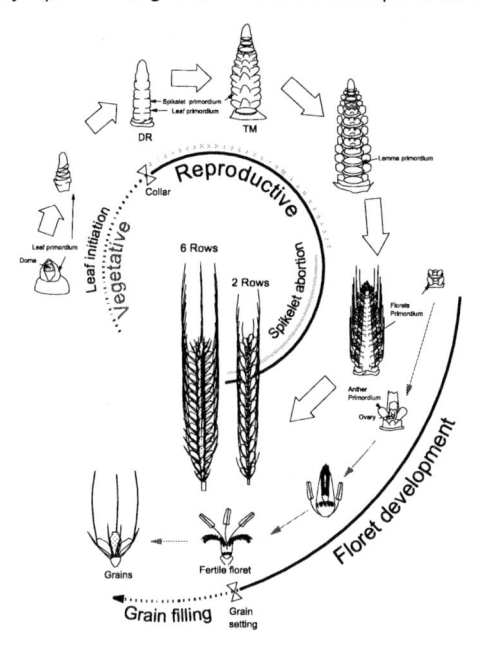

Note: Sowing (Sw), seedling emergence (E), floral initiation or "collar" stage (Fi), double ridge (DR), triple mound (TM), maximum number of total primordia initiated in the apex (MP), heading (Hd), beginning of grain-filling period (BGF), physiological maturity (PM), and harvest (Hv).

environmental conditions; that is, the interval between the initiation of two consecutive leaf primordia (viz. plastochron) is a common value regardless of the number of leaf primordia already initiated or to be initiated before the apex becomes reproductive. However, genetic variation for plastochron may be large, as has been shown for wheat (Evans and Blundell, 1994), and higher temperatures (within a wide range) increase the rate of leaf initiation (concomitantly reducing the plastochron). As the rate of leaf initiation is virtually insensitive to photoperiod and vernalization, the plastochron assumes a constant value for any environmental condition when measured on a thermal time basis (as described for wheat by Wilhelm and McMaster, 1995; Miralles and Slafer, 1999). Bearing in mind that genetic variation may be expected for phyllochron, a common value in the literature might be about 40 to 50°C day, above a base temperature of 0°C (Kirby et al., 1987; Delécolle et al., 1989).

During the vegetative phase, the apical meristem produces single ridges that will later grow up into leaves, having initially a conical shape (dome) of about 0.2 mm in length (see Figure 10.2) that later elongates. The maximum number of leaves in the main shoot is determined at the time of cessation of leaf initiation, when the apex changes from vegetative to reproductive stage, starting the initiation of spikelet primordia. This switch occurs frequently during the period when the shoot apex has an elongated shape and is initiating single ridges, as the first spikelet primordia are normally initiated as single ridges with no morphological differences from leaf primordia. Usually, the first visual evidence of the change from the vegetative to the reproductive stage in the apex is the appearance of a double ridge (see Figure 10.2). The lower ridge corresponds, to a leaf primordium that does not develop further, and the upper ridge to a spikelet primordium (Bonnett, 1966; Kirby and Appleyard, 1986). The apex at this moment is about 0.5 mm long, and this stage is usually referred to as the beginning of floral initiation, although about half of the total number of spikelet primordia have already been formed by this time under many agronomic conditions (Kirby and Appleyard, 1986). The true change from vegetative to reproductive stages is named "collar" stage, which occurs well before the appearance of the first double ridge under most agronomic conditions (see Figure 10.2).

Most of the spikelet primordia are initiated faster than the leaf primordia, and the change in the slope of the relationship between the number of primordia and thermal time generally coincides with floral initiation (see Figure 10.3). Floral initiation or collar stage occurrence can be determined only by subtracting final leaf number from the accumulated number of primordia, as there is no unequivocal morphological change in the apex to identify this stage.

The following important apical stage is known as the triple mound stage because the spikelet primordium has differentiated three protuberances that constitute the central and lateral spikelets. At this stage, spikes of two-rowed and six-rowed varieties appears similar, but subsequently the development of the lateral spikelets slows down in two-rowed varieties (Bonnet, 1966). The subsequent stages are characterized by the appearance in sequence of the glume, lemma, and stamen on the median spikelets. The number of spikelet primordia initiated in barley varies between 10 and 45, depending on genotype and environment (Kirby, 1977; Kitchen and Rasmusson, 1983; García del Moral, 1992). This spikelet initiation phase ceases when awn primordia are evident on the most advanced spikelets (see Figure 10.3). After the maximum number of spikelets is reached, a number of later-initiated primordia at the tip of the shoot apex do not progress to produce a mature spikelet. About 30 to 40 percent of the maximum number of spikelet primordia initiated abort before ear emergence (Kirby and Faris, 1972; Kernich, Halloran, and Flood, 1996). Some evidence (Appleyard, Kirby, and Fellowes, 1982;

FIGURE 10.3. Schematic Representation of the Relationship Between Cumulative Primordia in the Apex (Leaves and Spikelets) and Time After Emergence

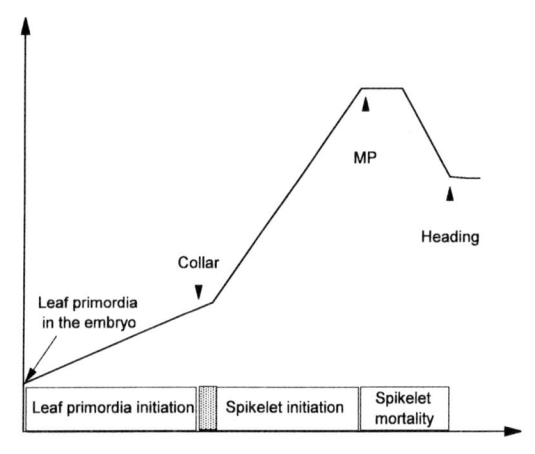

Note: Shadow zone within spikelet initiation phase indicates that some spikelets could be initiated at similar rates as leaves.

Miralles, Richards, and Slafer, 2000) reported that although the number of fertile spikelets is positively related to the maximum number of spikelets initiated, the slope of the relationship is significantly smaller than one. Thus, the larger the maximum number of spikelet primordia, the more important the proportion of spikelet abortion for grain number determination.

Coincidentally with the cessation of spikelet initiation and achievement of the maximum number of spikelet primordia, at the base of the spike, the young collar encircles the developing rachis and represents the first node (Bonnet, 1966). At this stage, which coincides with the beginning of stem elongation, the young spike is about 3 mm long and still at or below ground level (see Figure 10.1). During the spikelet initiation period, the more distal primordia exhibit the highest growth rates, but after the awn-primordium initiation stage, these differences disappear, and, therefore, the relative sizes of spikelets established during the period of primordia initiation are maintained throughout the time of spike differentiation (Kirby, 1977; Scott et al., 1983).

As described earlier, the final number of grains per spike depends upon both the maximum number of spikelets initiated in the spike (from floral initiation to awn initiation) and the proportion of those spikelets which survive (from the initiation of the awn primordium to anthesis; Ellis and Kirby, 1980; Russel et al., 1982; Ellis and Russell, 1984; Jones and Allen, 1986;

García del Moral, Jiminez-Tejada, et al., 1991). García del Moral, Jiminez-Tejada, et al. (1991) found similarities in spikelet abortion between various genotypes, suggesting that abortion may be mostly under environmental control. If this is so, even though the death of some spikelets would be inevitable, increasing the number of grains per spike by reducing spikelet abortion could be feasible through management improvement in barley. However, Kernich, Halloran, and Flood (1997) observed in a different, larger set of cultivars only little variation in the number of spikelet primordia formed by awn initiation, but a considerable variation in the number of aborted spikelets. These results suggest that to increase genetically the yield of the spike in barley, there seems to be more potential in reducing the level of spikelet abortion during the stem/spike growth phase taking place between awn initiation and heading than in increasing the maximum number of spikelet primordia per spike. This is in line with the ideas for further increasing yield potential in wheat through genetic manipulations of developmental phases (Slafer, Calderini, and Miralles, 1996; Slafer, Araus, and Richards, 1999)

Leaf Emergence

After the coleoptile emerges through the soil surface, this organ stops its growth and the tip of the first initiated leaf appears. At this time, about five to seven leaves have been already initiated. Under most conditions, further leaves are initiated before floral initiation takes place, the precise number depending on the genotype by environment (photoperiod and vernalization) interaction. The final leaf number is a critical determinant of the duration of the whole cycle to heading: all the initiated leaf primordia must appear before the last internode elongates and the spike emerges from the sheath of the last-appeared leaf (flag leaf). Therefore, time to heading strongly depends on the number of leaves initiated in the main shoot and the rate at which these leaves emerge. The phyllochron (period between the appearance of two successive leaves) is estimated as the reciprocal of the rate of leaf appearance, and, therefore, the duration from seedling emergence to heading is given by the product of the number of leaves initiated and the phyllochron (plus the time from the appearance of the flag leaf to heading, frequently assessed to last about two phyllochrons).

In fact, the length of the phyllochron not only is important for determining the total duration to heading but also, integrated with the duration of the plastochron, determines the length of the reproductive phase. Although its length is affected by genetic and environmental factors, the phyllochron is always longer than the plastochron (due to the fact that the rate of leaf initiation is faster than that of leaf emergence). Between five to seven leaf primordia are initiated before seedling emergence, which warrants a minimum duration of reproductive development, given the number of primordia that have to appear from floral initiation until flag leaf emergence. This number of leaves to appear after leaf initiation is larger when the timing of floral initiation is delayed (and the final leaf number is increased), due to the fact that the phyllochron-to-plastochron ratio, though variable (see Evans and Blundell, 1994), is always higher than one. Consequently, the longer the period from seedling emergence to floral initiation due to lack of satisfaction of photoperiod or vernalization requirements, the higher the number of leaf primordia that have to appear after floral initiation (Hay and Kirby, 1991), and therefore the longer the reproductive phase from floral initiation to anthesis.

Genetic and environmental factors affect the rate of leaf appearance. Some evidence has shown, in relation to genetic variability in phyllochron within the same experimental conditions, that the phyllochron ranged from 50 to 97° C day (Frank and Bauer, 1995; Kernich, Halloran, and Flood, 1995; Kernich, Slafer, and Halloran, 1995). The effect of environment on the rate of leaf emergence is evident when the same cultivar is sown at different sowing dates or locations (Kirby, Appleyard, and Fellowes, 1985; Kirby and Perry, 1987; Masle, Doussinault, and Sun, 1989). The phyllochron of cereals strongly depends on temperature (Klepper, Rickman, and Peterson, 1982; Cao and Moss, 1989; Slafer and Rawson, 1994a; Frank and Bauer, 1995). Due to this strong dependence, phyllochron is frequently estimated in degree-days, and after correction for temperature differences by using thermal time, the rate of leaf emergence is more or less constant from seedling emergence to flag leaf emergence (Baker, Gallagher, and Monteith, 1980; Kirby, Appleyard, and Fellowes, 1985; Cao and Moss, 1989; Slafer et al., 1994). Other factors, however, such as daylength or its rate of change at seedling emergence, carbohydrate reserves, and moderate water and nutrient stresses, have been shown to have little effect on the phyllochron of wheat and barley (Cutforth, Jame, and Jefferson, 1992; Frank and Bauer, 1995). Although phyllochron appeared to be insensitive to the rate of change of photoperiod when this factor was artificially manipulated under field conditions (Kirby, Appleyard, and Fellowes, 1982; Ritchie, 1991; Slafer et al., 1994; Kernich, Slafer, and Halloran, 1995), the absolute photoperiod appears to have a general impact on phyllochron (Slafer and Rawson, 1997). This effect tends to be rather small in most cultivars growing in the field, but marked responses can be seen when very sensitive cultivars are grown under photoperiods sufficiently short (Slafer and Rawson, 1997). In fact, in some studies, it has been reported that phyllochron may change with crop ontogeny, with a general trend for early leaves appearing faster than later leaves (Stapper and Fischer, 1990; Jamieson et al., 1995; Hotsonyame and Hunt, 1997; Slafer and Rawson, 1997). This generally occurs when the final number of leaves is increased in response to short photoperiods or no vernalization (Slafer and Rawson, 1997; Miralles and Richards, 2000).

Thus, a single genotype could show a single or two phyllochron values depending upon the final number of leaves initiated.

Dynamics of Tillering

Tillering is one of the most important developmental features for barley, since it has a decisive influence on both establishing the yield potential of the crop and compensating for wide variation in plant density (Kirby and Faris, 1972; García del Moral, Ramos, and Recalde, 1984, García del Moral, Ramos, et al., 1991; Dofing and Knight, 1992; García del Moral and García del Moral 1995). Under field conditions, a population of barley plants produces tillers in a characteristic pattern. That is, tiller numbers increase rapidly during the first few weeks after seedling emergence, reach a maximum shortly after floral initiation, diminish rapidly before spike emergence, and finally stabilize until harvest (García del Moral, Ramos, and Recalde, 1984). Furthermore, the growth of many developing tillers may be suppressed after tiller emergence, and the loss of these potentially grain-bearing tillers may also limit sink capacity later on, thus reducing grain yield. Tiller mortality often begins after floral initiation in the main shoot, as developing tillers compete with limited success for available assimilates against developing spikelets and florets on the main stem (Lauer and Simmons, 1988; García del Moral and García del Moral, 1995).

The proportion of tillers that senesce without contributing to grain yield varies with the cultivar and environment (Simmons, Rasmusson, and Wiersma, 1982; García del Moral and García del Moral, 1995). Although tillering pattern seems to be similar for diverse cultivars, barley shows pronounced differences between cultivars, both for maximum tiller production (Kirby, 1967; Cannell, 1969a; García del Moral and García del Moral, 1995) and for tiller mortality (Lauer and Simmons, 1988; García del Moral and García del Moral, 1995). In general, two-rowed barleys show higher tillering than six-rowed cultivars, and the winter forms generally produce more tillers than the spring ones (Kirby and Riggs, 1978; García del Moral and García del Moral, 1995). Long daylengths, high temperatures, and higher plant densities tend to diminish tillering (Cannell, 1969b; Kirby and Appleyard, 1980; Simmons, Rasmusson, and Wiersma, 1982; García del Moral and García del Moral, 1995), while light intensity and water and nitrogen availability promote formation and growth of secondary tillers (Cannell, 1969b; McDonald, 1990; Ramos, De la Morena, and García del Moral, 1995) or at least reduce their mortality. Competition among the shoots for nutrients, light, and water seems to be one of the principal causes for tiller mortality in barley. Thus, tillers that have at least three fully emerged leaves (Kirby and Jones, 1977) or are over one-third the height of the main stem at jointing (García del Moral, Ramos, and Recalde, 1984) are those which most likely survive. In fact, under field conditions, tiller mortality appears inversely related to maximum tiller production (García del Moral and García del Moral, 1995). According to Casal, Sanchez, and Deregibus (1987) and many other studies cited therein, tiller mortality also could be influenced by a photomorphogenesis-modulated process, depending on the changes in light quality within the crop community, tiller death being promoted especially by a lower red:far-red ratio.

Grain Filling

In two-rowed barley, with only the central spikelets being fertile, the grains are uniformly symmetrical. The lateral florets are reduced and are often raised on a short pedicel. In six-rowed varieties, all three spikelets at each node of the rachis are fertile. The median grains are symmetrical, but the lateral grains are unsymmetrical to a greater or lesser extent, each with a right-handed or left-handed bias (Briggs, 1978).

Rate of dry-matter accumulation in the barley grain is initially slow, increasing to a nearly constant rate when grain weight reaches between 4 and 6 mg (Gallagher, Biscoe, and Scott, 1976; see also Savin and Molina-Cano, Chapter 19 in this book). The length of the grain ceases to increase about seven days after pollination, but width increase continues longer, and the increase in dry weight remains for longer still. Usually, grain growth rate ranges from 0.9 to 2.2 mg/day in function of the supply of assimilates and position of the grain within the spike, with grains in the middle of the spike having the highest growth rates (Gallagher, Biscoe, and Scott, 1976; Scott et al., 1983). Final grain size is determined by both rate of growth and duration of grain filling period. The first variable is, in turn, a function of the availability of assimilates, from either photosynthesis during grain growth or contributions from dry matter accumulated in preanthesis, and the intrinsic growth capacity of the grains, whereas the second is mainly fixed by environmental conditions. Grain weight is a relatively stable feature in barley (Gallagher, Biscoe, and Scott, 1975) because mobilization of reserves from the stem and sheaths compensates for a shortage of current photoassimilates during grain filling in most cases. The contribution of preanthesis assimilates to final grain yield depends on environmental conditions during grain-filling. Estimates ranged from 2 to 74 percent (Gallagher, Biscoe, and Scott, 1975), 12 to 17 percent (Bidinger, Musgrave, and Fischer, 1977), 11 to 44 percent (Austin et al., 1980), or 60 to 80 percent (Daniels, Alcock, and Scarisbrick, 1982). This contribution is particularly high under stress conditions, when current photosynthesis is limited and grain growth depends increasingly on mobilized assimilates from the stem and other vegetative organs, including the flag leaf sheath and spike peduncle (Austin et al., 1980; Daniels, Allcock, and Scarisbrick, 1982). In fact, under Mediterranean conditions, the grain yield of barley appears strongly dependent on the crop dry weight attained at anthesis (Ramos, García del Moral, and Recalde, 1985), suggesting a higher contribution of vegetative reserves during grain filling.

Awns are apical extensions of the lemmas that carry three vascular bundles, of which the largest is centrally placed. They are roughly triangular in cross section and contain two tracts of parenchymatous photosynthetic tissue, connected with the atmosphere through two rows of stomata (Briggs, 1978). Awns senesce and cease photosynthesizing after the leaves, this attribute being particularly important under semiarid conditions (Biscoe, Littleton, and Scott, 1973). In addition to their photosynthetic role, awns may alter the sink capacity of subtended grains because they intensify the accumulation of cytokinins during grain formation, thus increasing the number of cells in the endosperm and the sink capacity of the grain (Michael and Seiler-Kelbitsch, 1972). Thus, larger grains are usually subtended by larger awns (Briggs, 1978).

The causes that determine the cessation of growth in cereal grains are not sufficiently clear, although they apparently are more related to a diminished ability for starch synthesis caused by enzyme dehydration than to a lack of carbohydrates (Biscoe et al., 1975). The rapid loss of water in the barley grain during ripening seems to be associated with an increase in the levels of abscisic acid in the endosperm, which would increase pericarp permeability, thus causing grain dehydration (King, 1976; Mounla, 1979). This effect is probably enhanced by lipid deposition in the vascular tissue at the base of the grain toward the end of the ripening period, thus disturbing the water supply to the endosperm. Renwick and Duffus (1987), however, have suggested that water content in the barley grain could be under genetic control, being dependent on the rate of carbohydrate supply during grain filling.

DEVELOPMENTAL RESPONSES TO ENVIRONMENTAL FACTORS

The major components of the environment affecting development are temperature, both temperature per se and low temperature associated with the vernalization requirements, and photoperiod (Rasmusson, 1987; Ellis et al., 1988; Slafer and Rawson, 1994b). Whereas photoperiod and vernalizing temperatures affect the rate of development of only some particular phases (i.e., even in sensitive cultivars, some of the phases are virtually insensitive to these factors), temperature per se affects all developmental phases, and there are no cultivars whose developmental rates are insensitive to this factor. Other factors related to level of nutrients in the soil, water availability, plant density, and radiation have sometimes, but not always, been reported to alter time to heading. However, the effects of these "minor" factors on development, when occurring, are quite small compared with those of temperature and photoperiod (Miralles and Slafer, 1999) and will not be discussed herein.

Temperature Responses

As mentioned previously, temperature has a universal effect on barley development: it affects all genotypes and every developmental phase, starting when the seed is imbibed and finishing at maturity. This effect for a wide range of temperatures is positive, that is, the higher the temperature, the faster the rate of development (determining a shorter period). In most cases, a linear regression between the rate of development (calculated as the reciprocal of the duration of the considered phase) and temperature describes adequately this relationship (Slafer and Rawson, 1994b). This model provides a couple of parameters of enormous empirical value in simulation models and agronomy. The intercept on the abscissa estimates a base temperature (below which the progress in development would be negligible) and a single slope (reflecting the sensitivity to temperature) for the range of temperatures in which the relationship is linear. If the range of temperatures is wide enough, the linearity is frequently lost at higher temperatures, and an optimum temperature may also be calculated as that at which the rate of development shows the maximum value. The mechanistic basis for the widely used concept of thermal time to predict the duration of different developmental phases is the linearity of this relationship. The reciprocal of the slope of the linear relationship between rate of development and temperature is the thermal time measured in degree-days (°C day) that the phase lasts at any temperature within the range between the base and optimum temperatures. Thus, the thermal time for a particular phase and genotype is invariable across a wide range of temperatures, in a determined photoperiod and under vernalization conditions. Although the effect of temperature is universal in that no cultivars are insensitive, there are genotypic differences in the sensitivity. These differences may be seen in any of the parameters of the relationship: cultivars may differ in their base and optimum temperatures and in the slope of their relationship (in thermal time). In fact, this differing sensitivity to temperature is widely mentioned in the literature as leading to differences in intrinsic earliness or earliness per se.

Some controversy exists about where temperature should be measured to calculate thermal time. Usually thermal time is calculated as the summation of the daily average of maximum and minimum air temperatures. However, as the apex during the first developmental stages remains below the soil surface, it would be more accurate to use the soil temperature for calculating thermal time during the vegetative and early reproductive phases preceding stem elongation (e.g., Jamieson et al., 1995).

Temperature affects not only the rate of development, consequently changing the duration of different developmental phases, but also the rate of leaf and spikelet initiation and that of leaf emergence (Klepper, Rickman, and Peterson, 1982; Kirby, Appleyard, and Fellowes, 1982, 1985; Jones and

Allen, 1986; Cao and Moss, 1989; Kirby, 1995; Frank and Bauer, 1995; Hay and Ellis, 1998). The rates of leaf initiation and leaf appearance are also frequently linearly related to temperature (e.g., Slafer and Rawson, 1994a), and the reciprocals of those slopes are the plastochron and phyllochron measured in degree-days with unique values for any temperature between the base and optimum thresholds. As temperature affects in a similar way the rates of vegetative development and leaf initiation, final leaf number is largely unaffected by mean temperatures (Rawson and Zajac, 1993; Slafer and Rawson, 1994b). Although similar reasoning may apply to the rates of development and primordia initiation during the spikelet initiation period, the effect of temperature on the final number of fertile spikelets may be important, as this trait depends also upon the dry-matter economy of the crop during stem elongation. For example, low temperatures diminish the rate of primordia production and lengthen the duration of developmental phases, thus diminishing competition for resources and promoting the formation of a higher number of spikelets (Gallagher, Biscoe, and Scott, 1976).

Higher temperatures reduce the final number of tillers, presumably due to an increasing demand for carbohydrates by the main stem due to a faster shift to the stem elongation phase. On the contrary, low temperatures delay the shift to the stem elongation phase, during which intraplant carbohydrate competition allows for more tillers to be initiated and to survive later on (Cannell, 1969a; Kirby and Riggs, 1978; García del Moral and García del Moral, 1995).

Grain-filling phase also is affected by temperature. Besides the well-known effect of a heat shock during this period (Savin and Nicolas, 1996), increases in air temperature during grain filling diminish individual grain weight due to a reduction in the length of the effective grain-filling period, which cannot be compensated for by increases in the rate of grain filling (Sofield et al., 1977). Higher temperatures during this period also stimulate leaf senescence (although the reduction in final grain weight due to higher temperatures may be independent of their effects on leaf senescence; Slafer and Miralles, 1992) and increase the carbohydrate losses by grain respiration (Thorne, 1974; Wiegand and Cuellar, 1981).

Vernalization Responses

Most winter barley apices (and even some spring cultivars) must be vernalized by a period of low temperature, typically in the range from 3 to 12°C (Trione and Metzger, 1970). A general model of the response to vernalization is schematically shown in Figure 10.4. Vernalizing temperatures, contrary to photoperiods, are experienced directly by the apex meristem and the stimulus perceived by the embryo of the imbibed seed. Vernalization mainly

FIGURE 10.4. Diagram of the Changes in Duration of Development to Heading for Different Photoperiod and Vernalization Treatments

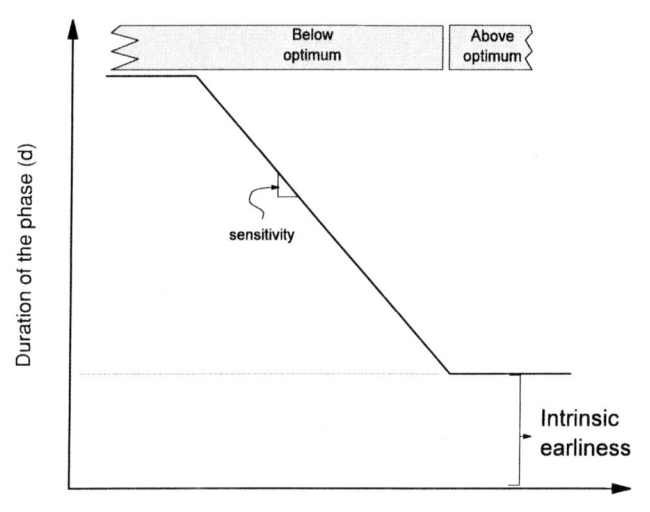

[Figure: Graph with y-axis "Duration of the phase (d)" and x-axis "Duration of photoperiod (h) and vernalization (d) treatments". Bars at top indicate "Below optimum" and "Above optimum" zones. A descending line labeled "sensitivity" drops from the below-optimum plateau to a lower plateau labeled "Intrinsic earliness".]

Note: Bars show the inductive and noninductive zones. The slope of the relationship between duration and photoperiod or vernalization requirements in the inductive zone represents the photoperiod and vernalization sensitivity.

affects the length of the vegetative phase, altering its duration. Opposite to what has been described for temperature per se, vernalization does not affect the rate of leaf initiation (or does so very slightly); thus, the effects of this factor on the length of the leaf initiation phase bring about concomitant changes in the final number of leaves. In fact, this variable is generally used as an indicator of vernalization sensitivity once the other requirements are satisfied (Kirby, Appleyard, and Fellowes, 1985; Rawson and Zajac, 1993). Although widely recognized that vernalization acts during the vegetative phase, some evidence in wheat suggests vernalization effects during the spikelet initiation phase (Rahman, 1980).

The effect of vernalization on the leaf emergence rate of barley has been little investigated. In three winter barleys, Chun (1993) found that exposition to vernalizing temperatures accelerated the rate of leaf emergence, suggesting an interaction between vernalization and thermal time in determining leaf appearance.

Vernalization has been one of the most poorly characterized aspects of cereal development, probably due to the difficulties arising out of isolating the progress toward satisfying the vernalization requirement from other concurrent processes, such as the acquisition of frost hardiness (Hughes and Dunn, 1996; Hay and Ellis, 1998).

Photoperiod Responses

Barley is a quantitative long-day species; that is, the length of the sensitive phase responds to increasing photoperiods (below those of the optimum threshold), accelerating the rate of development (consequently reducing the length of the phase), with a shape similar to that described schematically in Figure 10.4. Daylength sensitivity is, notwithstanding, very variable between cultivars, with some of them being virtually insensitive. Regarding photoperiod-sensitive cultivars, this sensitivity is not uniform throughout development.

The vegetative phase from sowing to floral initiation is sensitive when considered as a whole (the shape of the relationship described in Figure 10.4). However, as the stimulus is perceived by the leaves and the signal is transmitted to the apex (Evans, 1987), barley plants cannot respond to photoperiod until the first leaf emerges, the length of the initial part of the vegetative phase from sowing to seedling emergence being independent of the daylength. Immediately after seedling emergence, the stimulus may be perceived, provided there is no juvenile phase. Barley, as do many other crops (soybean—Collison et al., 1993; maize—Kiniry et al., 1983; sunflower—Villalobos et al., 1996), may exhibit a juvenile phase of insensitivity to photoperiod. This is a phase during which sensitive cultivars do not respond to otherwise inductive photoperiods (Takahashi and Yasuda, 1970; Yasuda, 1982; Roberts et al., 1988). The length of the juvenile phase, which together with the length of sowing-seedling emergence, determines a minimum duration of the vegetative phase and the minimum final leaf number. This preinductive, juvenile phase varies quite markedly (Boyd, 1996), and some cultivars that grow under long photoperiods exhibit only about six leaves and do not seem to experience a significant juvenile phase (Hay and Ellis, 1998).

The number of leaves formed on the main shoot is reduced by long days: although short photoperiods lengthen the period of leaf initiation, they do not seem to affect the rate of leaf initiation (or at least not in a similar proportion), thus allowing the initiation of more leaves (Kirby and Appleyard, 1980). Although some positive effects of photoperiod on the rate of spikelet initiation have been reported, they are considerably smaller than those on the rate of phasic development during spikelet initiation. Consequently, these opposing effects do not fully compensate, resulting in fewer spikelets being initiated under long days (Kirby and Appleyard, 1980).

Although it has been consistently demonstrated in the literature that photoperiod affects the rate of development of the leaf and spikelet initiation phases (from seedling emergence to awn initiation), other studies revealed that photoperiod may also affect the duration of the late reproductive phase of stem growth, during which a variable number of spikelets previously initiated die. Experiments based in reciprocal photoperiod transfers (Kernich, Halloran, and Flood, 1996; Miralles and Richards, 2000) at the time of stem growth initiation demonstrated that the duration of this late reproductive phase was largely determined by the photoperiod to which they were exposed at that time. This direct photoperiod sensitivity during the stem growth phase, in which the stem and spike grow at the same time while strongly competing for assimilates, is crucial for the determination of yield potential (Slafer, Calderini, and Miralles, 1996). It is during this phase that the fate of the previously initiated spikelets is determined. Therefore, photoperiod sensitivity could be used to modify the length of this period for improving the ability to allocate assimilates in the spikes, reducing spikelet abortion (Kernich, Halloran, and Flood, 1996). If genetic differences in photoperiod sensitivity during this late reproductive phase can be identified, this trait could be used in breeding for increased yield potential (Slafer, Calderini, and Miralles, 1996).

GENETIC SYSTEMS REGULATING FLOWERING TIME

As in many other crops, in barley, genetic control of flowering time is a complex process involving different genes that regulate response to vernalization, photoperiod, and "earliness per se" or "intrinsic earliness." Major genes responsible for the sensitivity to vernalization, named *Sh, Sh2,* and *Sh3,* are located on chromosomes 4H, 5H, and 7H, respectively (Snape, 1996). Although genotypic variability exists within winter and spring barleys, most genotypes with a winter habit, associated with the action of *Sh* alleles, need a period of growth at low temperatures before floral development can proceed at the fastest possible rate, if the response is quantitative in nature. Conversely, spring barley genotypes carrying the dominant alleles, *Sh,* have virtually no vernalization requirements. Major genes controlling responses to photoperiod, named *Ppd-H1* and *Ppd-H2,* are located on the short and long arms of chromosomes 2H and 5H, respectively (Laurie et al., 1994). These genes determine the acceleration or delay of flowering time after vernalization requirements have been satisfied. Genes determining the "earliness per se" of the genotype seem to be distributed throughout the genome, as most chromosomes appear to carry these genes, which are generally located more frequently as QTL (quantitative trait loci) than as major genes (Laurie et al., 1995; Franckowiak, 1997; Snape 1996). Genetic analyses have demonstrated that all chromosomes carry some "earliness per se" genes, named *eam, eps,* or *eak* genes (see <http://wheat.pw.usda.gov/ggpages/bgn/>). For example, while Jensen and Jorgensen (1975) found the *eak* gene in chromosome 5H, Franckowiak (1997) constructed maps showing the marker loci for the position of *eam* genes in four different chromosomes: *eam8* (1H), *Eam1* (2H), *eam10* (3H), *eam9* (4H), and *eam7* (6H). Snape (1996) described the location of genes controlling flowering time, placing the *eps* genes in all chromosomes but 1H (see Table 3 in Snape, 1996).

Barley has been used for many genetic analyses, mainly due to the facility for creating and studying mutants (Nilan, 1991). Thus, many mutants have been isolated and mapped using morphological, isozyme, and physiological markers (Franckowiak, 1995), including developmental mutants with changes in tillering pattern (*cul2;* Kirby, 1973), time of heading (*eam8;* Yasuda, 1977), vernalization requirement (*sgh, Sgh2, Sgh3;* Takashahi and Yasuda, 1970), photoperiod response (Laurie et al., 1994), and stem morphology (*ert* series; Persson, 1969). The study of barley mutants has helped to establish the basic processes that occur at the shoot apex, although fewer than half of the mutants have known loci (Hay and Ellis, 1998). This indicates the necessity for linking between classical and mutation-based studies through mapping with DNA-based markers.

Although barley is an autogamous species, there is sufficient DNA-level diversity for linkage map construction in populations derived from crosses between related genotypes. Within the past five years, as DNA marker technologies have been improved, progressively denser molecular marker linkage maps have been developed in a range of germplasm (Graner et al., 1991; Kleinhofs et al., 1993; Becker et al., 1995). Over 1,000 loci have been placed on these maps, providing a comprehensive catalog of markers and complete genome coverage (Landgridge et al., 1995; Qi, Stam, and Lindhout, 1996; Hayes et al., 1997), including several loci controlling heading time (Hayes et al., 1997). Little is known, however, about the role of genes determining the fate of cells at the apex and of those, for instance, that determine daylength or vernalization response, which control developmental rates and events (Hay and Ellis, 1998). Breeding objectives for these characters must be established in view of relationships with heading time and maturity. Otherwise, selection for agronomic performance and/or quality could have negative correlated responses in terms of maturity, which often defines the suitability of a genotype for a specific production environment (Hayes et al., 1997).

REFERENCES

Aitken, Y. (1971). Non-destructive methods for estimation of flower initiation in subterranean clover and cereals. *Journal of the Australian Institute of Agricultural Science* 37: 57-60.

Andersen, S. (1952). Methods for determining stages of development in barley and oats. *Physiologia Plantarum* 5: 199-210.

Appleyard, M., E.J.M. Kirby, and G. Fellowes (1982). Relationship between the duration of phases in the pre-anthesis life cycle of spring barley. *Australian Journal of Agricultural Research* 33: 917-925.

Austin, R.B., C.L. Morgan, M.A. Ford, and R.D. Blackwell (1980). Contributions to grain yield from pre-anthesis assimilation in tall and dwarf barley phenotypes in two contrasting seasons. *Annals of Botany* 45: 309-319.

Baker, C.K., J.N. Gallagher, and J.L. Monteith (1980). Daylength change and leaf appearance in winter wheat. *Plant Cell and Environment* 3: 285-287.

Becker, J., P. Vos, M. Kuiper, F. Salamini, and M. Heun (1995). Combined mapping of AFLP and RFLP markers in barley. *Molecular Genome Genetics* 249: 65-73.

Bidinger, F.R., R.B. Musgrave, and R.A. Fischer (1977). Contribution of stored pre-anthesis assimilate to grain yield in wheat and barley. *Nature* 270: 431-433.

Biscoe, P.V., J.N. Gallagher, E.J. Littleton, J.L. Montheith, and R.K. Scott (1975). Barley and its environment. IV. Sources of assimilate for the grain. *Journal of Applied Ecology* 12: 295-318.

Biscoe, P.V., E.J. Littleton, and R.K. Scott (1973). Stomatal control of gas exchange in barley awns. *Annals of Applied Biology* 75: 285-297.

Bonnett O.I. (1966). Inflorescences of maize, wheat, rye, barley and oats: Their initiation and development. *University of Illinois College Agriculture, Agriculture Experimental Station Bulletin* 721: 105 pp.

Boyd, W.J.R. (1996). Developmental variation, adaptation and yield determination in spring barley. *Proceedings of the VII International Barley Genetics Symposium*, eds. G. Scoles and B. Rossnagel. Saskatoon: University of Saskatchewan Extension Press, pp. 276-283.

Briggs, D.E. (1978). *Barley*. London: Chapman and Hall, 612 pp.

Cannell, R.Q. (1969a). The tillering pattern in barley varieties. I. Production, survival and contribution to yield by component tillers. *Journal of Agricultural Science, Cambridge* 72: 405-422.

Cannell, R.Q. (1969b). The tillering pattern in barley varieties. II. The effect of temperature, light intensity and daylength on the frequence of occurrence of the coleoptile node and second tillers in barley. *Journal of Agricultural Science, Cambridge* 72: 423-435.

Cao, W. and D.N. Moss (1989). Daylength effect on leaf emergence and phyllochron in wheat and barley. *Crop Science* 29: 1021-1025.

Casal, J.J., R.A. Sanchez, and V.A. Deregibus (1987). The effect of light quality on shoot extension growth in three species of grasses. *Annals of Botany* 59: 1-7.

Chun, J.U. (1993). Variation in rate of leaf emergence, initiation of primordia, stem elongation and heading time as affected by vernalization duration of barley with differing growth habits. *Field Crops Research* 32: 159-172.

Collison S.T., R.J. Summerfield, R.H. Ellis, and E.H. Roberts (1993). Durations of the photoperiod-sensitive and photoperiod-insensitive phases of development to flowering in four cultivars of soybean [*Glycine max* (L.) Merrill]. *Annals of Botany* 71: 389-394.

Cottrell, J.E., R.H. Easton, J.E. Dale, A.C. Wadsworth, J.S. Adams, R.D. Child, and G.V. Hoad (1985). A comparison of spike and spikelet survival in main stem and tillers of barley. *Annals of Applied Biology* 106: 365-377.

Cutforth, H.W., Y.W. Jame, and P.G. Jefferson (1992). Effect of temperature, vernalization and water stress on phyllochron and final main-stem leaf number of HY320 and Neepawa spring wheat. *Canadian Journal of Plant Science* 72: 1141-1151.

Daniels, R.W., M.B. Alcock, and D.H. Scarisbrick (1982). A reappraisal of stem reserve contributions to grain yield in spring barley (*Hordeum vulgare* L.). *Journal of Agricultural Science, Cambridge* 98: 347-355.

Delécolle, R., R.K.M. Hay, M. Guerif, P. Pluchard, and C. Varlet-Grancher (1989). A method of describing the progress of apical development in wheat based on the time course of organogenesis. *Field Crops Research* 21: 147-160.

Dofing, S.M. and C.W. Knight (1992). Alternative model for path-analysis of small-grain yield. *Crop Science* 32: 487-489.

Ellis, R.H. and E.J.M Kirby (1980). A comparison of spring barley grown in England and in Scotland. 2. Yield and its components. *Journal of Agricultural Science, Cambridge* 95: 111-115.

Ellis, R.H., E.H. Roberts, R.J. Summerfield, and J.P. Cooper (1988). Environmental control of flowering in barley (*Hordeum vulgare* L.). II. Rate of development as a function of temperature and photoperiod and its modification by low-temperature vernalization. *Annals of Botany* 62: 145-158.

Ellis, R.H. and G. Russell (1984). Plant development and grain yield in spring and winter barley. *Journal of Agricultural Science, Cambridge* 102: 85-95.

Evans, L.T. (1987). Short day induction of inflorescence initiation in some winter wheat varieties. *Australian Journal of Plant Physiology* 14: 277-286.

Evans, L.T. and C. Blundell (1994). Some aspects of photoperiodism in wheat and its wild relatives. *Australian Journal of Plant Physiology* 21: 551-562.

Fischer, R.A. (1985). Number of kernels in wheat crops and the influence of solar radiation and temperature. *Journal of Agricultural Science* 100: 447-461.

Franckowiack, J.D. (1995). Coordinator's report: Chromosome 2. *Barley Genetic Newsletter* 24: 132-138.

Franckowiack, J.D. (1997). Revised linkage maps for morphological markers in barley *(Hordeum vulgare)*. *Barley Genetic Newsletter* 26: 9-21.

Frank, A.B. and A. Bauer (1995). Phyllochron differences in wheat, barley, and forage grasses. *Crop Science* 35: 19-23.

Frederick, J.R. and P.J. Bauer (1999). Physiological and numerical components of wheat yield. In *Wheat: Ecology and Physiology of Yield Determination,* eds. E.H. Satorre and G.A. Slafer. Binghamton, NY: Food Products Press, pp. 45-65.

Gallagher, J.N., P.V. Biscoe, and R.K. Scott (1975). Barley and its environment. V. Stability of grain weight. *Journal of Applied Ecology* 12: 319-336.

Gallagher, J.N., P.V. Biscoe, and R.K. Scott (1976). Barley and its environment. VI. Growth and development in relation to yield. *Journal of Applied Ecology* 13: 567-583.

García del Moral, L.F., J.M. Ramos, M.B. García del Moral, and P. Jimenez-Tejada (1991). Ontogenetic approach to grain production in spring barley based on path-coefficient analysis. *Crop Science* 31: 1179-1185.

García del Moral, L.F., J.M. Ramos, and L. Recalde (1984). Tillering dynamics of winter barley as influenced by cultivar and nitrogen fertilizer: A field study. *Crop Science* 24: 179-181.

García del Moral, M.B. (1992). Fisiología del crecimiento, desarrollo y productividad de cebadas de invierno y de primavera en ambiente mediterráneo. Tesis Doctoral, Facultad de Ciencias, Universidad de Granada.

García del Moral, M.B. and L.F. García del Moral (1995). Tiller production and survival in relation to grain yield in winter and spring barley. *Field Crop Research* 44: 85-93.

García del Moral, M.B., M.P. Jimenez-Tejada, L.F. García del Moral, J.M. Ramos, F. Roca de Togores, and J.L. Molina-Cano (1991). Apex and ear development in relation to the number of grains on the main-stem ears in spring barley *(Hordeum distichon)*. *Journal of Agricultural Science, Cambridge* 117: 39-45.

Graner, A., A. Jahoor, J. Schondelmaier, H. Siedler, K. Pillen, G. Fischbeck, G. Wensel, and R. G. Herrmann (1991). Construction of an RFLP map of barley. *Theoretical and Applied Genetics* 83: 250-256.

Haun, J.R. (1973). Visual qualification of wheat development. *Agronomy Journal* 65: 116-119.

Hay, R.K.M. and R.P. Ellis (1998). The control of flowering in wheat and barley: What recent advances in molecular genetics can reveal. *Annals of Botany* 82: 541-554.

Hay, R.K.M. and E.J.M. Kirby (1991). Convergence and synchrony—A review of the coordination of development in wheat. *Australian Journal of Agricultural Research* 42: 661-700.

Hayes, P.M., J. Cerono, H. Witsenhoer, M. Kuiper, M. Zabeau, K. Sato, A. Kleinhofs, D. Kudrna, A. Kilian, M. Saghai-Maroof, et al. (1997). Characterizing and exploiting genetic diversity and quantitative traits in barley *(Hordeum vulgare)* using AFLP markers. *Journal of Quantitative Trait Loci* 3, Article 2: <http: //probe. nalusda.gov:8000/otherdocs/jqtl>.

Hotsonyame, G.K. and L.A. Hunt (1997). Sowing date and photoperiod effects on leaf appearance in field-grown wheat. *Canadian Journal of Plant Science* 77: 23-31.

Hughes, M.A. and M.A. Dunn (1996). The molecular biology of plant acclimation to low temperature. *Journal of Experimental Botany* 47: 291-305.

Jamieson, P.D., I.R. Brooking, J.R. Porter, and D.R. Wilson (1995). Prediction of leaf appearance in wheat: A question of temperature. *Field Crops Research* 41: 35-44.

Jensen, J. and J.H. Jorgensen (1975). The barley chromosome 5 linkage map. I. Literature survey and map estimation procedure. *Hereditas* 80: 5-16.

Jones, J.L. and E.J. Allen (1986). Development in barley *(Hordeum sativum)*. *Journal of Agricultural Science, Cambridge* 107: 187-213.

Kernich, G.C., G.M. Halloran, and R.G. Flood (1995).Variation in developmental patterns of wild *(Hordeum spontaneum* L.) and cultivated barley *(H. vulgare* L.). *Euphytica* 82: 105-115.

Kernich, G.C., G.M. Halloran, and R.G. Flood (1996). Constant and interchanged photoperiod effects on the rate of development in barley *(Hordeum vulgare)*. *Australian Journal of Plant Physiology* 23: 489-496.

Kernich, G.C., G.M. Halloran, and R.G. Flood (1997). Variation in duration of preanthesis phases of development in barley *(Hordeum vulgare)*. *Australian Journal of Agricultural Research* 48: 59-66.

Kernich, G.C., G.A. Slafer, and G.M. Halloran (1995). Barley development as affected by rate of change of photoperiod. *Journal of Agricultural Science, Cambridge* 124: 379-388.

King, R.W. (1976). Abscisic acid in developing grains and its relation to grain growth and maturation. *Planta* 123: 43-51.

Kiniry J.R., J.T. Ritchie, R.L. Musser, E.P. Flint, and W.C. Iwig (1983). The photoperiod sensitive interval in maize. *Agronomy Journal* 75: 687-690.

Kirby, E.J.M. (1967). The effect of plant density upon the growth and yield of barley. *Journal of Agricultural Science* 68: 317-324.

Kirby, E.J.M. (1973). Abnormalities induced in barley ears by gibberellic acid. *Journal of Experimental Botany* 24: 317-324.

Kirby, E.J.M. (1977). The growth of the shoot apex and apical dome of barley during ear initiation. *Annals of Botany* 41: 1297-1308.

Kirby, E.J.M. (1995). Factors affecting rate of leaf emergence in barley and wheat. *Crop Science* 35: 11-19.

Kirby, E.J.M. and M. Appleyard (1980). Effects of photoperiod on the relation between development and yield per plant of a range of spring barley varieties. *Zeitschrift für Pflanzenzüchtung* 85: 226-239.

Kirby, E.J.M. and M. Appleyard (1984). Cereal plant development and its relation to crop management. In *Cereal Production,* ed. E.J. Gallagher. London: Butterworth, pp. 161-173.

Kirby, E.J.M. and M. Appleyard (1986). *Cereal Development Guide.* Stoneleigh, UK: NAC Cereal Unit, 95 pp.

Kirby, E.J.M., M. Appleyard, and G. Fellowes (1982). Effect of sowing date on the temperature response of leaf emergence and leaf size in barley. *Plant Cell and Environment* 5: 477-484.

Kirby, E.J.M., M. Appleyard, and G. Fellowes (1985). Effect of sowing date and variety on main shoot leaf emergence and number of leaves of barley and wheat. *Agronomie* 5: 117-126.

Kirby, E.J.M. and D.G. Faris (1972). The effect of plant density on tiller growth and morphology in barley. *Journal of Agricultural Science* 78: 281-288.

Kirby, E.J.M. and H.G. Jones (1977). The relations between the main shoot and tillers in barley plants. *Journal of Agricultural Science, Cambridge* 88: 381-389.

Kirby, E.J.M. and M.W. Perry (1987). Leaf emergence rates of wheat in a Mediterranean environment. *Australian Journal of Agricultural Research* 38: 455-464.

Kirby, E.J.M., J.R. Porter, W. Day, J.S. Adam, M. Appleyard, S. Ayling, C.K. Baker, R.K. Belford, P.V. Biscoe, A. Chapman, et al. (1987). An analysis of primordium initiation in Avalon winter wheat crops with different sowing dates and at nine sites in England and Scotland. *Journal of Agricultural Science* 109: 107-121.

Kirby, E.J.M. and T.J. Riggs (1978). Developmental consequences of two-row and six-row ear type in spring barley. 2. Shoot apex, leaf and tiller development. *Journal of Agricultural Science* 91: 207-216.

Kitchen, B.M. and D.C. Rasmusson (1983). Duration and inheritance of leaf initiation, spike initiation, and spike growth in barley. *Crop Science* 23: 939-943.

Kleinhofs, A., A. Kilian, M. Saghai-Maroof, R. Biyashev, P. Hayes, F. Chen, N. Lapitan, A. Fenwick, T. Blake, V. Kanazin, et al. (1993). A molecular, isozyme, and morphological map of the barley *(Hordeum vulgare)* genome. *Theoretical and Applied Genetics* 86: 705-712.

Klepper, B., R.W. Rickman, and C.M. Peterson (1982). Quantitative characterization of vegetative development in small cereal grains. *Agronomy Journal* 74: 789-792.

Landes, A. and J.R. Porter (1989). Comparison of scales used for categorizing the development of wheat, barley, rye and oats. *Annals of Applied Biology* 115: 343-360.

Landgridge, P., A. Karakoussis, N. Collins, J. Kretschmer, and S. Manning (1995). A consensus linkage map of barley. *Molecular Breeding* 1: 389-395.

Landsberg, J.J. (1977). Effects of weather on plant development. In *Environmental Effects on Crop Physiology,* eds. J.J. Lansdberg and C.V. Cutting. London: Academic Press, pp. 289-307.

Large, E.C. (1954). Growth stages of cereals. Illustration of the Feekes scale. *Plant Pathology* 3: 128-129.

Lauer, J.G. and S.R. Simmons (1988). Photoassimilate partitioning by tillers and individual tiller leaves in field-grown spring barley. *Crop Science* 28: 279-282.

Laurie D.A., N. Pratchett, J.H. Bezant, and J.W. Snape (1994). Genetic analysis of a photoperiod response gene on the short arm of chromosome 2(H) of *Hordeum vulgare* (barley). *Heredity* 72: 619-627.

Laurie D.A., N. Pratchett, J.H. Bezant, and J.W. Snape (1995). RFLP mapping of five major genes and eight quantitative trait loci controlling flowering time in a winter x spring barley cross. *Genome* 38: 575-585.

Magrin, G.O., A.J. Hall, C. Baldy, and M.O. Grondona (1993). Spatial and interannual variations in the photothermal quotient: Implications for the potential kernel number of wheat crops in Argentina. *Agricultural and Forest Meterology* 67: 29-41.

Masle J., G. Doussinault, and B. Sun (1989). Response of wheat genotypes to temperature and photoperiod in natural conditions. *Crop Science* 29: 712-721.

McDonald, G.K. (1990). A differential response of a two-row barley and a six-row barley to detillering under a restricted watering regime. *Australian Journal of Agricultural Research* 41: 1065-1970.

Michael, G. and H. Seiler-Kelbitsch (1972). Cytokinin content and kernel size of barley grain as affected by environmental and genetic factors. *Crop Science* 12: 162-165.

Miralles, D.J. and R.A. Richards (2000). Responses of leaf and tiller emergence and primordium initiation in wheat and barley to interchanged photoperiod. *Annals of Botany* 85: 655-663.

Miralles, D.J., R.A. Richards, and G.A. Slafer (2000). Duration of stem elongation period influences the number of fertile florets in wheat and barley. *Australian Journal of Plant Physiology* 27: 931-940.

Miralles, D.J. and G.A. Slafer (1999). Wheat development. In *Wheat: Ecology and Physiology of Yield Determination,* eds. E.H. Satorre and G.A. Slafer. Binghamton, NY: Food Products Press, pp. 13-43.

Mounla, M.A.Kh. (1979). Phytohormones and grain growth in cereals. In *Crop Physiology and Cereal Breeding,* eds. J.H.J. Spiertz and T. Kramer. Wageningen, the Netherlands: Pudoc, pp. 20-28.

Nilan, R.A. (1991). Recent advances in barley mutagenesis. *Barley Genetics VI, Proceedings of the Sixth International Barley Genetics Symposium,* ed. L. Munck. Helsingborg, Sweden: The Organizing Committee of the Nordic Countries, pp. 823-831.

Persson, G. (1969). An attempt to find suitable genetic markers for dense ear loci in barley II. *Hereditas* 63: 1-28.

Qi, X., P. Stam, and P. Lindhout (1996). Comparison and integration of four barley genetic maps. *Genome* 39: 379-394.

Rahman, M.S. (1980). Effect of photoperiod and vernalization on the rate of development and spikelet number per spike in 30 varieties of wheat. *Journal of the Australian Institute of Agricultural Science* 46: 68-70.

Ramos, J.M., I. De la Morena, and L.F. García del Mora (1995). Barley response to nitrogen rate and timing in a Mediterranean environment. *Journal of the Agricultural Science* 125: 175-182.

Ramos, J.M., L.F. García del Moral, and L. Recalde (1985). Vegetative growth of winter barley in relation to environmental conditions and grain yield. *Journal of Agricultural Science, Cambridge* 104: 413-419.

Rasmusson, D.C. (1987). An evaluation of ideotype breeding. *Crop Science* 27: 1140-1146.

Rawson, H. M. and M. Zajac (1993). Effect of higher temperatures, photoperiod and seed vernalization on development in two spring wheats. *Australian Journal of Plant Physiology* 20: 211-222.

Renwick, F. and C.M. Duffus (1987). Factors affecting dry weight accumulation in developing barley endosperm. *Physiologia Plantarum* 69: 141-146.

Ritchie, J.T. (1991). Wheat phasic development. In *Modeling Plant and Soil Systems,* eds. J. Hanks and J.T. Richie. Madison, WI: American Society of Agronomy, pp. 31-54.

Roberts, E.H., J.P. Summerfield, J.P. Cooper, and R.H. Ellis (1988). Environmental control of flowering in barley (*Hordeum vulgare* L.). I. Photoperiod limits to long-day responses, photoperiod-insensitive phases and effects of low-temperature and short-day vernalization. *Annals of Botany* 62: 127-144.

Russell, G., R.P. Ellis, J. Brown, G.M. Milbourn, and A.M. Hayter (1982). The development and yield of autumn- and spring-sown barley in southeast Scotland. *Annals of Applied Biology* 100: 167-178.

Sadras, V.O and F.J. Villalobos (1993). Floral initiation, leaf initiation and leaf appearance in sunflower. *Field Crops Research* 33: 449-457.

Savin, R. and M.E. Nicolas (1996). Effects of short periods of drought and high temperature on grain growth and starch accumulation of two malting barley cultivars. *Australian Journal of Plant Physiology* 23: 201-210.

Savin, R. and G.A. Slafer (1991). Shading effects on the yield of an Argentinian wheat cultivar. *Journal of Agricultural Science* 116: 1-7.

Scott, W.R., M. Appleyard, G. Fellowes, and E.J.M. Kirby (1983). Effect of genotype and position in the ear on carpel and grain growth and mature grain weight of spring barley. *Journal of Agricultural Science, Cambridge* 100: 382-391.

Simmons, S.R., D.C. Rasmusson, and J.V. Wiersma (1982). Tillering in barley: Genotype, row spacing and seeding rate effects. *Crop Science* 22: 801-805.

Slafer, G.A., J.L. Araus, and R.A. Richards (1999). Physiological traits to increase the yield potential of wheat. In *Wheat: Ecology and Physiology of Yield Determination,* eds. E.H. Satorre and G.A. Slafer. Binghamton, NY: Food Products Press, pp. 379-415.

Slafer, G.A., D.F. Calderini, and D.J. Miralles (1996). Yield components and compensation in wheat: Opportunities for further increasing yield potential. In *Increasing Yield Potential in Wheat: Breaking the Barriers,* eds. M.P. Reynolds, S. Rajaram, and A. McNab. Mexico DF: CIMMYT, pp. 101-133.

Slafer, G.A., D.F. Calderini, D.J. Miralles, and M.F. Dreccer (1994). Preanthesis shading effects on the number of grains of three bread wheat cultivars of different potential number of grains. *Field Crops Research* 36: 31-39.

Slafer, G.A. and D.J. Miralles (1992). Green area duration during the grain filling period of wheat as affected by sowing date, temperature and sink strength. *Journal of Agronomy and Crop Science* 168: 191-200.

Slafer, G.A. and H.M. Rawson (1994a). Does temperature affect final numbers of primordia in wheat? *Field Crops Research* 39: 111-117.

Slafer, G.A. and H.M. Rawson (1994b). Sensitivity of wheat phasic development to major environmental factors: A re-examination of some assumptions made by physiologists and modellers. *Australian Journal of Plant Physiology* 21: 393-426.

Slafer, G.A. and H.M. Rawson (1997). Phyllochron in wheat as affected by photoperiod under two temperature regimes. *Australian Journal of Plant Physiology* 24: 151-158.

Slafer, G.A. and E.M. Whitechurch (2001). Manipulating wheat development to improve adaptation and to search for alternative opportunities to increase yield potential. In *Application of Physiology in Wheat Breeding,* eds. M.P. Reynolds, I. Ortiz-Monasterio, and A. McNab. Mexico, DF: CIMMYT, in press.

Snape, J.W. (1996). The contribution of new biotechnologies to wheat breeding. In *Increasing Yield Potential in Wheat: Breaking the Barriers,* eds. M.P. Reynolds, S. Rajaram, and A. McNab. Mexico DF: CIMMYT, pp. 167-181.

Sofield, I., L.T. Evans, M.G. Cook, and I.F. Wardlaw (1977). Factors influencing the rate and duration of grain filling in wheat. *Australian Journal of Plant Physiology* 4: 785-797.

Stapper, M. and R.A. Fischer (1990). Genotype, sowing date and planting spacing influence on high-yielding irrigated wheat in southern New South Wales. I. Phasic development, canopy growth and spike production. *Australian Journal of Agricultural Research* 41: 997-1019.

Takahashi, R. and S. Yasuda (1970). Genetic of earliness and growth habitat in barley. *Barley Genetics II. Proceedings of the Second International Barley Genetics Symposium,* ed. R.A. Nilan. Pullman: Washington State University Press, pp. 388-408.

Thorne, G.N. (1974). Physiology of grain yield of wheat and barley. *Rothamsted Experimental Station Report for 1973,* Part 2: 5-25.

Thorne, G.N. and D.W. Wood (1987). Effects of radiation and temperature on tiller survival, grain number and grain yield in winter wheat. *Annals of Botany* 59: 413-426.

Trione, E.J. and R.J. Metzger (1970). Wheat and barley vernalization in a precise range of temperatures. *Crop Science* 10: 390-392.

Villalobos, F.J., A.J. Hall, J.T. Ritchie, and F. Orgaz (1996). Oilcrop-Sun: A development, growth, and yield model of the sunflower crop. *Agronomy Journal* 88: 403-415.

Waddington, S.R., P.M. Cartwright, and P.C. Wall (1983). A quantitative scale of spike initial and pistil development in barley and wheat. *Annals of Botany* 51: 119-130.

Wiegand, C.L. and J.A. Cuellar (1981). Duration of grain filling and weight of wheat as affected by temperature. *Crop Science* 21: 95-101.

Wilhelm, W.W. and G.S. McMaster (1995). Importance of the phyllochron in studying development and growth in grasses. *Crop Science* 35: 1-3.

Yasuda, S. (1977). Linkage of the barley earliness gene *eak* and its pleiotropic effects under different growth conditions. *Bericht des Ohara Institut für Landwirtschaftliche Biologie, Okayama* 17: 15-28.

Yasuda, S. (1982). The physiology of earliness in barley. *Barley Genetics IV. Proceedings of the Fourth International Barley Genetics Symposium,* ed. M. Asher. Edinburgh, UK: Edinburgh University Press, pp. 507-517.

Zadoks, J.C., T.T. Chang, and C.F. Konzak (1974). A decimal code for the growth stage of cereals. *Weed Research* 14: 415-421.

Chapter 11

Physiological Basis of the Processes Determining Barley Yield Under Potential and Stress Conditions: Current Research Trends on Carbon Assimilation

José Luis Araus

INTRODUCTORY REMARKS

Yield potential is defined as the "yield of a cultivar grown in environments to which it is adapted, when nutrients and water are non-limiting, and when pests, diseases, weeds, lodging and other stresses are effectively controlled" (Evans, 1989). Several aspects of yield potential merit attention. For example, How is it determined? What morphological and physiological characters are involved and what combination of these would maximize its expression? What is the genetic variability in yield potential and traits related to this? How can high-yield potential genotypes be selected as early as possible in the improvement process and in an efficient and cost-effective way?

However, average crop yields fall well below their potential, due to the incidence of stresses. Crop management and breeding can reduce the incidence of stress on yield. The effect of these practices is sustained by an understanding of their physiology. The purpose of this chapter is to examine the physiological basis of the processes determining barley yields.

For crop plants, it is important to consider how the combined operation of all functions leads to the production of the harvestable component, which in cereal crops is generally grain. This chapter focuses basically on the physiological aspects of photosynthetic (carbon) metabolism that determine grain yield as well as the incidence of stresses on these processes. The physiological basis of some agronomic practices and potential selection criteria for crop breeding are also discussed.

Basic aspects related to how yield is attained, such as crop development and determination of yield components, and those related to grain quality

---

This study was supported by the research CICYT projects PB97-0865 and AGF 99-0611-C03-03 (Spain).

and mineral nutrition are developed in detail in other chapters of this book as well as in another recent publication (Wallace and Yan, 1998). The genetic bases of physiological responses to a wide array of stresses, as well as those questions concerning the genetics of yield components and selecting traits and the implementation of molecular markers for quantitative trait loci (QTL), are also discussed elsewhere in this book (see Cattivelli et al., Chapter 12) in this book. The requirements needed whereby a physiological criterion can be incorporated in a conventional breeding program have also been discussed elsewhere (Jackson et al., 1996; Araus, 1996; Araus et al., 1998, 2001). In addition, a large number of reviews on physiological approaches to increase yield potential and stress adaptation in small-grain cereals have appeared in recent years (some of which are included in this chapter's references). Therefore, the present chapter does not provide an exhaustive review of this subject. This study is limited to describing the advances made in the physiology of barley yields in the past five years.

PHYSIOLOGICAL FACTORS DETERMINING YIELD

Photosynthetic transformation of solar energy into harvestable plant parts can be clearly divided into three major processes: (1) interception of incident (solar) radiation by the canopy, (2) conversion of the intercepted (radiant) energy to chemical energy (accumulated in the plant dry matter), and (3) partitioning of the dry matter produced between the harvested parts and the rest of the plant. A wheat crop with high yield potential needs to carry out these three processes efficiently and in a balanced way. The yield or harvestable part (grain yield for barley, GY) of a crop over a given period of time can, therefore, be expressed as follows (see also Hay, 1999):

$$GY = RAD \times \%RI \times RUE \times HI \tag{1}$$

where RAD is the total quantity of incident solar radiation received by the crop over the growing period; %RI is the fraction of RAD intercepted by the canopy over the crop cycle; RUE is the radiation-use efficiency, that is, the overall photosynthetic efficiency of the crop: the total dry matter produced per unit of intercepted photosynthetic active radiation (PAR); and HI is the harvest index or fraction of the total dry matter harvested as yield: this is normally expressed in terms of aboveground production, excluding the root system.

Total biomass, the result of RAD × %RI × RUE, can be understood in a physiological sense as the result of crop photosynthesis. Thus, it is well accepted that total canopy photosynthesis during growth is closely related to yield, as reported in several species (Zelitch, 1982; Ashley and Boerma, 1989). If RAD is expressed as mean incident radiation received per day and not as total amount, we must include another independent component: green area duration, or GAD.

Although a function of crop managing practices, such as planting date, or crop breeding strategies to modify plant duration, RAD is generally outside the control of the agronomist and/or breeder (Hay, 1999). Consequently, genetic and environmental factors normally affect yield through the two "shoot" parameters: %RI and RUE. One possibility, for example, is to increase %RI by the canopy throughout the growing season by a faster approach to full cover. Another way is to increase RUE, say by improving the distribution of PAR among the various leaves by modifying the canopy structure, or by changing the assimilatory and deassimilatory characteristics of leaves. A great effort has been invested these past few years to identify the basis of canopy development, and a great deal is now known in this regard (see other chapters in this book).

Water stress is the main environmental factor limiting yield. Thus, Passioura (1996) proposed a parallel way of considering grain yield in a water-limited situation:

$$GY = W \times WUE \times HI \qquad (2)$$

where W is the water transpired by the crop, and WUE, water-use efficiency, is the biomass per unit of water evapotranspired.

The increase in atmospheric CO_2 concentrations from ca. 270 ppm before the beginning of the industrial revolution 250 years ago to the current level of ca. 370 ppm seems to have affected long-term net assimilation and water-use efficiency (Araus and Buxó, 1993), thus explaining increases in yield of nearly 50 percent under present-day conditions (Mayeux et al., 1997; Araus, Slafer, and Romagosa, 1999). Future situations are expected to involve a further steady increase in atmospheric CO_2 concentrations, which is likely to have a positive effect on yields. In a study of well-managed crops in Germany, it was concluded that barley yield would increase by 0.35 percent per ppm increase in CO_2, whereas increases in wheat would be about 25 percent lower (Manderscheid and Weigel, 1995). Considering the increase in CO_2 of about 30 ppm from 1970 to 1990, the contribution of CO_2 to the increase in yield of spring cereals during this period is estimated to be between 25 and 50 percent. Of course, in harsh environments or where management practices (particularly nitrogen fertilization) are less than optimal, the effect of increasing CO_2 on yield can be much lower or nil. Under the cool, maritime climate in Norway, total grain yield was not affected by increasing CO_2, even when biomass was increased (Saebo and Mortensen, 1996). On the other hand, the protein content of the grain decreased.

POTENTIAL YIELD VERSUS STRESS ADAPTATION: BREEDING IMPLICATIONS

Yields are often far below potential values. This is due to either episodic or long-duration stress situations. Plants can cope with stresses by means of evasion and resistance strategies, the last including both avoidance and tolerance responses (Larcher, 1995). Whatever the strategy involved, the response (either adaptative or acclimative) of plants to stress limits their yield potential.

The previous statement has obvious implications when defining breeding strategies to improve crop yield and stability in stressed regions. For barley, selection in a high-yielding location (or under high-yielding conditions) does not identify genotypes suitable for low-yielding environments, which are more representative of the production conditions of low-input agriculture. Selection in low-yielding environments appears more efficient (Ceccarelli, 1994; Ceccarelli and Grando, 1993, 1996; Bouzerzour and Dekhili, 1995). Ceccarelli and Grando (1993, 1996) pointed out that crop yield potential and stress tolerance (defined as yield under stress) were, by and large, mutually exclusive. Therefore, selection in high-yielding environments is expected to produce negative or no yield response for low-yielding environments. In the same way, Ceccarelli et al. (1998) concluded that the highest-yielding lines under stress not only were selected under stress but also were land races collected in very dry areas (< 250 mm total annual rainfall). This confirms earlier findings and supports the idea that the most effective way to improve productivity of crops grown in less-favored areas is to use adapted germplasm and to select in the target environments.

INCIDENCE OF STRESSES ON YIELD

Following equation 1 on physiological processes determining yield, the effect of stresses can be placed at two levels: one is the total amount of radiation absorbed by the crop during its cycle, and the second is the photosynthetic efficiency of radiation conversion. These two components determine the total biomass attained by the crop. Harvest index (the third component of equation 1) will reflect the amount of available assimilates (determined by the first two components of the equation) to build up the (potential size of) sink and further to fill it. Nevertheless, harvest index is comparatively less affected by stresses than the total amount of biomass produced by the crop. The component a priori more sensitive to most stresses, and that which limits more the biomass yielded, is the first, the amount of radiation absorbed. This can be diminished either by lower leaf growth and/or leaf area duration.

Photosynthetic Area

The amount of total radiation intercepted by the canopy depends on the photosynthetic area available during the crop cycle. This in turn is the result of the amount of photosynthetic area formed and its duration.

Leaf expansion depends on cell turgor, which is affected by nutrients and water status. Fricke et al., 1997, and Fricke and Flowers, 1998, conclude that leaf cell expansion in barley relies on high rates of water and solute supply, which may not be sustainable during periods of sufficient N supply (limitation by water supply: water potential gradients) or limiting N supply (limitation by solute provision: reduced osmotic pressure generation rates). To minimize the possibility of growth limitation by water and osmolyte provision, longitudinal and lateral cell expansion peak at different locations along the growth zone (Fricke and Flowers, 1998).

However, Lu and Neumann (1998) conclude that reductions in barley leaf growth due to water stress are not caused by reductions in the osmotic potential gradients between the expanding leaf cells and their external water source in the apoplast. Nevertheless, growth inhibition seems to be accompanied by significant reductions in leaf and cell wall extensibility. Moreover, there is a strong positive relationship between the reductions in leaf growth induced by water stress and the reduction in leaf extensibility (Lu and Neumann, 1998). Decrease in elongation rates of barley leaves as the soil dries is paralleled with an increase in the pH of xylem sap. Bacon et al. (1998) suggest that xylem sap pH would act as a drought signal to reduce leaf elongation rate via an abscisic acid (ABA)-dependent mechanism.

Development of photosynthetic area is related to the synthesis of chlorophyll. Chlorophyll synthesis in barley is controlled by two light-dependent NADPH-protochlorophyllide oxidoreductase enzymes. One of these is active only transiently in etiolated seedlings at the beginning of illumination, and the other also operates in green plants (Holtorf et al., 1995). Stresses can shorten the duration of photosynthetic area. Thus, photooxidative damage can induce premature senescence and thus shorter leaf duration and lower integrated radiation (Krasnovsky, 1993). Such damage by excess of radiation is usually associated with the effect of many stresses (see Lambers et al., 1998).

Energy utilization efficiency depends on the row distances at which genotypes are planted. Among row spacings, 20 cm showed the most efficient canopy architecture, with significantly higher energy utilization and yields than crops planted with smaller or larger row spaces (Gontia et al., 1994).

The most obvious way to optimize yield using the conceptual framework of equations 1 and 2 is through phenological adjustment. Thus, simultaneous selection for early maturity and a relatively long time to heading is recommended for the development of early maturing, high-yielding cultivars adapted to northern conditions (Dofing, 1995). In contrast, for Mediterranean environments, characterized by drought stress during grain filling, earlier heading and maturity are recommended and indeed have been, up until now, the most successful indirect (i.e., apart from yield itself) traits when breeding for improving yield under these conditions (Cattivelli et al., 1994; Loss and Siddique, 1994; Slafer et al., 1994; Slafer and Araus, 1998; Araus et al., 2001). However, long-duration cultivars, when sown earlier in the season and no frost damage occurs, have no yield disadvantage and can even be potentially more productive, in comparison with short-duration cultivars (Mitchell et al., 1996).

Radiation-Use Efficiency

Reductions in radiation-use efficiency are in general less important than those in total radiation absorbed as a factor limiting yield under stress conditions, but in the case of barley and other monocarpic seed crops, its contribution to the harvestable yield can be major, and not only under Mediterranean conditions (Jamieson, Martin, Francis, and Wilson, 1995). Whether, for example, drought affects light interception or light-use efficiency depends on the timing of water stress in relation to canopy development. Thus, for barley under Mediterranean conditions, development of water stress is usually slow during early stages of growth, whereas severe stress develops after maximum light interception is achieved and no more photosynthetic area is developed. In these cases, water stress has a small effect on light interception but a large effect on radiation-use efficiency (Fukai, 1995a). Nevertheless, a decrease in the duration of green area during grain filling can also limit yield.

Jamieson, Martin, Francis, and Wilson (1995) reported for barley in New Zealand that when drought was imposed from emergence, decreased growth rates were caused primarily by reductions in radiation-use efficiency. In these treatments, radiation-use efficiency was depressed even after drought was relieved, which at first would not agree with the widely accepted strong sensitivity of leaf growth to water status. In contrast, in treatments in which drought was imposed around anthesis, the primary cause of reduced biomass production was a decrease in the amount of radiation intercepted, mostly associated with more rapid leaf senescence.

Any increase in yield through increasing RUE should be achieved by increasing simultaneously the capacity for both photoassimilation and sink strength (Richards, 1996a). The way to obtain this may be to focus on improving assimilate supply during spike development, thereby increasing sink capacity, which itself would lead to higher assimilation rates during grain filling (Richards, 1996b). Indeed the higher carbon isotope discrimination of modern barley varieties compared with land races and old varieties (Voltas et al., 1999) would agree with a higher RUE of the former due to an increase in stomatal conductance.

Desassimilatory processes (dark respiration plus photorespiration in C_3 plants) decrease the RUE. Experimental evidence suggests that vegetative growth and yield of C_3 crops were enhanced by spraying methanol to leaves, and that overall crop water use was reduced by methanol sprays. Methanol might act as a carbon source for the plant and a photorespiration inhibitor. However, Albrecht et al. (1995) concluded for barley that methanol applications do not increase crop growth or yield.

Harvest Index

The HI value of modern varieties of barley falls within the range of 0.4 to 0.6. Breeding progress is illustrated by trends in the HI of old, outclassed, and recent varieties of temperate and Mediterranean barley (compared under uniform conditions). HI showed a progressive increase throughout the twentieth century, although the improvement was much slower in Australia or Canada or Spain (where, for example, drought stress is present) than in the United Kingdom. As in wheat, most frequently the improvement in barley HI has been a consequence of increased grain population density coupled with stable individual grain weight (Hay, 1995). The high heritability of HI is based on its rather weak response to variation in environmental factors (such as fertilization, population density, application of growth regulators) in the absence of severe stress. Even though the principal cereal crops appear to be approaching the upper limit of HI, and future yield gains will have to be sought in increased biomass production, there will still be a need for the concept of HI as a factor to consider when interpreting the crop response to different stress factors and climatic change.

Naylor et al. (1998) have reported that, for U.K. conditions, cultivars and management systems that result in high biomass of winter barley will also produce high grain yield. They concluded that whereas HI cannot be easily targeted for manipulation by growers, crop management systems to increase biomass are easier to handle. In addition, higher shoot uniformity generally leads to increased biomass and grain yield per shoot (Naylor et al., 1998).

Source versus Sink Limitation

Beyond mechanical limitations to support the spike by the stem, sink limitations seem to be the basic factor limiting further increases in HI under favorable conditions. Under nonstress conditions, available information strongly supports the fact that sink size limits yield (Hay, 1995). However, this is not always the case under Mediterranean conditions, for which a colimitation between sink and source seems to exist (Voltas, Romagosa, and Araus, 1998).

In six-rowed barley types, central spikelets commonly bear heavier grains than lateral spikelets. This particularity of central grains is mostly caused by a higher grain-filling rate, with these grains having higher intrinsic potential yield (Voltas, Romagosa, and Araus, 1998). On the other hand, while grain yield under Mediterranean conditions is source limited in all environments, lateral grains showed a greater degree of source limitation than central grains. In addition, Voltas, Romagosa, and Araus (1998) point out that sink reduction increased grain weight of both central and lateral grains by stimulating grain-filling rate, and also by lengthening grain-filling duration in poor rain-fed environments. Ellis and Marshall (1998) reported that although there was significant variation in nitrogen concentration by grain position, these effects were smaller than the variations in grain weight within the ear and were not related to germination rate.

Grain Number versus Grain Mass

It has been extensively reported in barley that grain yield is more strongly related to grain number than grain size, with individual grain weight being the most stable component of yield (Gallagher et al., 1975; see also Smith et al., 1999). However, this is not always the case, for example, under Mediterranean conditions, where drought stress develops during the last part of the crop cycle, coinciding with grain filling. This is due to the fact that the supply of assimilates to the developing inflorescence plays a critical role in establishing final grain number and grain size (Hendrix et al., 1986). Thus, for example, barley isolines having awnless spikes showed a higher number of grains per spike than those genotypes with normal (i.e., awned) spikes (Bort et al., 1994). On the other hand, dry mass of individual grains was higher in the awned than in the awnless isoline, this being associated with greater photosynthesis and fewer number of grains in the awned ears (Bort et al., 1994).

Individual grain mass can be negatively affected by drought, high temperature, and any other factor or treatment that may reduce grain growth duration or the growth rate of grains (Mogensen, 1992; Bulman and Smith, 1993; Savin and Nicolas, 1996; Voltas, Romagosa, and Araus, 1998). Regarding salinity, it has been concluded from an extensive long-term experiment that spike length, number of flowers, number of grains per spike, and grain specific weight are the most tolerant characters, whereas grain size, grain yield, and grain weight per spike are the most sensitive characters (Royo and Aragüés, 1995).

Cultivars developed in temperate regions tended to have slower grain-filling rates than those developed in subarctic regions. Rapid grain-filling rate in cultivars grown in subarctic regions was associated with high kernel weight, but not with a long grain-filling duration or high grain yield. Northern-adapted cultivars should have preheading periods lasting as long as possible, followed by short grain-filling periods (Dofing, 1995).

Implications for Grain Quality: Dry Mass and Nitrogen Content of Grains

It is well established that either the total grain yield or the mass of individual grains is strongly negatively related with protein content of grains (Simmonds, 1996). This evidence has obvious implications for grain quality. Individual grain dry mass can be negatively affected by drought, high temperature, and other abiotic and biotic stresses that may reduce either growth duration or growth rate of grains (Mogensen, 1992; Savin and Nicolas, 1996; Voltas, Romagosa, and Araus, 1998). For example, high temperature reduces individual grain mass, due to a reduction in number rather than size of starch granules, whereas nitrogen content increases (Savin et al., 1996). Nevertheless, mean kernel mass can be reduced as drought stress increases, this fact being evident not only under Mediterranean conditions (Jamieson, Martin, and Francis, 1995). Even when water stress and nitrogen treatment affected grain protein concentration, Bertholdsson (1998) reported that cultivars with stable grain protein concentration had longer seminal roots and higher root:shoot length ratios.

The grain quality trait most affected by nitrogen fertilization is grain nitrogen (Grundy et al., 1996). Nitrogen content in the plant at anthesis was a good predictor of grain protein content, this characteristic in turn being positively correlated with embryo size and grain volume, and negatively correlated with nonstructural carbohydrate content in the plant at anthesis (García del Moral et al., 1998). Nevertheless, the contribution of postanthesis nitrogen uptake to total grain nitrogen seems to be negatively related to grain nitrogen concentration (De Ruiter and Brooking, 1996). This indicates that late nitrogen uptake may not always be detrimental to quality, as the processes of carbon and nitrogen accumulation are closely linked during grain filling (De Ruiter and Brooking, 1996).

Manipulation of Phenology

The rapid spike growth phase, which lasts only a few weeks, is critical in determining yield potential. This is when final grain number is set, not only determining the partitioning of photoassimilates to yield, but also influencing photosynthetic assimilation rate during seed filling. Recently, the duration of the rapid spike growth phase has been manipulated using short-day photoperiod treatment, showing a highly significant relationship between phase duration and the number of fertile florets per spike (Miralles and Richards, 2000). Indeed, increasing the duration of the rapid spike growth phase may improve grain weight potential in addition to grain number (Reynolds and Pfeiffer, 2000).

It could be expected that aboveground biomass is substantially higher in early sown and/or longer duration crops. However, this is not necessarily translated into higher grain yield. Indeed, greater competition for assimilates between the growing spikes and the elongating stem in the earlier sowings could be the determining factor (Gómez-Macpherson and Richards, 1995). Kernich et al. (1995) reported that even when the number of spikelets was positively correlated with the number of leaves, the ratio of the number of spikelets to the number of leaves declined with increasing time to anthesis, indicating that each successive leaf was associated with a diminishing increase in the number of spikelets. One way of overcoming this competition is to shorten the stems genetically (Gómez-Macpherson and Richards, 1995). This should capitalize on the considerable advantages in terms of water-use efficiency that early sowing offers.

Present Trends and Retrospective Studies

In general terms, yields of newer cultivars of barley (and other small grain cereals) grown under modern management are higher and more stable, or at least they are no less stable than those of crops grown early in the twentieth century (Slafer and Kernich, 1996). Nevertheless, under Mediterranean conditions, it has been reported that the most variable yield component across genotypes and years was the number of ears per square meter, which determined final grain yield (Cantero-Martínez et al., 1995a). Thus, modern cultivars have greater plasticity and growth rates during preanthesis when the number of grains per ear is determined. In addition, they avoid terminal water stress, and their grain-filling period is less adversely affected than that of late-flowering, older cultivars. Finally, modern cultivars produce less leaf area but have a greater extinction coefficient and intercept more radiation per unit leaf area.

In a study focused on determining the changes in morphological and agronomical characters of spring barley in Germany (i.e., under absence of drought stress) over a period of 80 years, the superior grain yield of the new germplasm was attributable not only to the higher number of ears per plant but also to the larger kernel weight. In addition, for further gain in grain yield, improvement of the straw yield seems more promising than increasing HI (Schittenhelm et al., 1996).

Water Use and Its Efficiency

Water-use efficiency (WUE) may be defined as biomass produced per unit evapotranspiration and thus depends on transpiration efficiency and soil evaporation (Smith et al., 1999). Whereas transpiration efficiency is related in some way to radiation-use efficiency, soil evaporation is related to canopy development. Thus, the slower the canopy development, the greater the evaporation from soil, and so the lower the WUE of the canopy (Fukai, 1995a). Additionally, under Mediterranean conditions, WUE is usually higher in early maturing cultivars (Fukai, 1995b; see also Bort et al., 1998). This is mostly due to the differences in transpiration efficiency (TE, biomass produced per unit transpiration), the transpiration depending on the vapor pressure deficit (VPD) of the air. On the other hand, evaporation accounted for about 10 to 30 percent of total water use for barley (Fukai, 1995b).

Transpiration efficiency can be assessed in an integrated but simple way through the analysis of the carbon isotope composition of the plant material and its further transformation to carbon isotope discrimination values. The physiological basis of the relationship between TE and carbon isotope discrimination in barley was well established over a decade ago (Hubick and Farquhar, 1989). Thus, carbon isotope discrimination (Δ) is negatively correlated with TE. Since then, the usefulness of Δ as a breeding tool in barley (either under potential or stressful conditions) has been extensively studied (Craufurd et al., 1991; Romagosa and Araus, 1991; Acevedo, 1993). Genotypes showing lower Δ would be those more productive under water-limited conditions. In fact, this has been proved usually to be the case when plants are grown in pots with a given amount of water. However, under field conditions, and except for very poor environments (Acevedo, 1993; Voltas et al., 1999), this is not frequently the case, and positive relationships between Δ and yield are reported (Romagosa and Araus, 1991; Craufurd et al., 1991; see also Richards, 1996b). The basis of this positive relationship may be sustained by the fact that Δ is also strongly positively related to the total amount of water available for the crop (Araus, Febrero, et al., 1997). This could help to explain the fact that, under field conditions, Δ frequently correlates positively with yield, whereas from Passioura's equation a negative relationship might be expected. Other factors (such as phenology, etc.) can also contribute to this positive relationship; these are very pedagogically discussed in Richards (1996b).

From the carbon isotope records on fossil barley grains recovered from archaeological sites, it can be hypothesized (considering a steady VPD) that from the beginning of cultivation, coinciding with the Neolithic revolution during the first half of the Holocene, WUE has increased slowly until the past two centuries, when the increase in atmospheric CO_2 led to a high increase in the estimated WUE (Araus and Buxó, 1993). Nevertheless, from retrospective studies, it is evident that no genetic improvement in WUE during the last century has been achieved. Indeed, the available information suggests that modern varieties show consistently higher Δ (and, therefore, for a given VPD, lower TE) values than the old varieties and land races (Muñoz et al., 1998). Higher stomatal conductance in new genotypes might be responsible (see Richards, 1996a,b). Nevertheless, the earlier flowering and maturation patterns of these cultivars (Richards, 1996a) would counteract such a tendency through a lower VPD.

Early Vigor and Fast Development

In severe rain-fed environments, both aboveground biomass and grain yield are greater in barley than, for example, bread and durum wheats, triticale, and oats. Barley also has a faster leaf area development and biomass accumulation than the other species (López-Castañeda and Richards, 1994). López-Castañeda and Richards, 1994, and López-Castañeda et al., 1996, concluded that the higher yield achieved by barley in harsh environments was due to faster leaf area growth and earlier flowering. These factors ensured that water loss from the soil surface was minimized and that growth was completed before the rapid rise in temperature and VPD (López-Castañeda and Richards, 1994).

Regarding early vigor, López-Castañeda et al. (1995) reported for barley approximately 40 percent more aboveground biomass and two times greater leaf area than for wheat by the two-leaf stage. In addition, whereas the average net assimilation rate (NAR) was 25 percent greater in wheat than in barley, the average leaf ratio (LAR) was 23 percent higher in barley. Variation in specific leaf area (SLA) was largely responsible for the variation in NAR and LAR. Factors between germination and the appearance of the second main stem leaf must be responsible for the greater early vigor in barley compared with that of wheat.

The size of the embryo was the single most important factor to account for the higher vigor of barley compared with that of wheat, whereas earlier emergence in barley accounted for the remaining variation in vigor (López-Castañeda et al., 1996). On the other hand, the SLA (leaf area per unit dry weight) and its inverse, the specific leaf weight (SLW), of the first leaves of the seedling have been found to be positively and negatively correlated, respectively, with early vigor (Richards, 1996a; Bort et al., 1998). In addition, López-Castañeda et al. (1996) suggest that the breadth of the first seedling leaf is an indirect estimate of early vigor that integrates embryo size and SLA and could be used in a breeding program to increase the vigor of cereals. Febrero et al. (1992) reported a positive relationship between carbon discrimination of seedlings and early vigor for hydroponic barley seedlings cultured in a growth chamber. However, this relationship is lacking under field conditions, except when only the genotypes most resistant to low temperatures are considered (Bort et al., 1998).

Of importance, the relationship between early vigor and grain yield is not a general fact, particularly when comparisons are made only across genotypes of one species and evaluations are performed under field conditions. Thus, Bort et al. (1998) found no relationship between early vigor and grain yield in winter barley. Different aspects can be involved in the lack of effect of higher early vigor on final yield. At the early stages of crop development, a different sensitivity to low temperatures can be involved (Bort et al., 1998), with greater susceptibility being shown by the more developed genotypes. Alternatively interplant competition after the seedling stage can also be involved in this lack of relationship. Additionally, in those typically Mediterranean environments where the faster development of leaf area may result in the premature exhaustion of soil water, a greater early vigor, by itself, can be a negative attribute (Richards, 1996b). Indeed, vigorous early growth may be a good trait when combined with early flowering, which may improve WUE over the crop cycle.

Morphological and Physiological Adaptations

Among the morphological and structural adaptations that are widely accepted to confer better performance under drought stress conditions, it is worth mentioning the presence of awns and glaucousness (or wax bloom). Presence of awns in barley is reported to increase TE of spikes (Bort et al., 1994). On the other hand, the roles of glaucousness in yield and TE (evaluated indirectly using Δ) were studied using near-isogenic lines of two-rowed barley (Febrero et al., 1998). Thus, grain yield, total straw biomass, and Δ were higher in glaucous than nonglaucous lines under both irrigated and rain-fed conditions. The higher Δ of the glaucous line suggests that these plants had lower transpiration efficiency than the nonglaucous line in the long term. In contrast, no significant effect of either glaucousness or water regime was observed in cuticular conductance or in the amount of cuticular waxes (Febrero et al., 1998).

During the past decade, there has been controversy over whether there is a degree of C_4 photosynthetic metabolism in ears of C_3 cereals such as barley. This would lead to higher TE and better performance in these organs at high temperatures. In this context, CO_2 exchange and the initial products of photosynthesis were examined in flag leaf blades and various ear parts of barley. The CO_2 compensation concentration at 210 mmol·mol^{-1} CO_2 in ear parts was similar to or greater than that in flag leaves. In addition, all tissues assimilated CO_2 mainly by the Calvin (C_3) cycle, with little fixation of $^{14}CO_2$ into the C_4 acids malate and aspartate. Therefore, these collective data support the conclusion that in the ear parts of barley, C_4 photosynthetic metabolism is nil (Bort et al., 1995). However, there is still the possibility that the spikes show some degree of crassulacean acid metabolism (CAM). Indeed, the grains show photosynthetic activity, refixing the CO_2 released during respiration with much higher phosphoenolpyruvate carboxylase (PEPcase) activities than in the leaves. However PEPcase activity may be addressed to anaplerotic metabolism instead of CO_2 fixation. Thus, Macnicol and Raymon (1998) concluded that in the aleurone from maturing, rapidly acidifying grains, the flux of glycolytic carbon through the PEP branch relative to that through pyruvate kinase is three to five times greater than in young aleurone. This increase in flux can be accounted for by severalfold increases in the amount and activity of PEPcase. Irrespective of particularities in the photosynthetic C3 metabolism of the barley spikes, what seems well demonstrated is that during photosynthesis, spikes refix CO_2 released from respiratory processes, probably from filling of grains (Bort et al., 1996). This can have an obvious role in increasing the TE of these organs.

Management practices can also modify the amount of water used and its efficiency. Thus, in the semiarid areas where barley is widely grown, dryland crop production is limited by low and variable precipitation. Conservation tillage may improve soil and water conservation in these areas. However, López and Arrue (1997) reported that crop response was similar between the conventional and reduced tillage treatments at all locations, whereas poor performance with no tillage was observed at the most arid sites. Paradoxically, this unfavorable crop response to no tillage was due to lower crop water use and a larger proportion of total water lost as evaporation. Therefore, only reduced tillage would provide an alternative to conventional tillage in the dryland growing areas. However, Sharratt (1998) reported that in dry years with high evaporative demand, no tillage or removal of stubble and loose straw from the soil surface would enhance grain production and WUE of barley in the subartic.

PHYSIOLOGICAL MECHANISMS TO COPE WITH STRESSES

Plant adaptation and acclimation to the large array of potential stress factors involves many levels of response, from molecular to the whole community or canopy through the physiological level. These responses are of two kinds: (1) a common set of mechanisms, irrespective of the stress involved, and (2) particular mechanisms to cope with a given stress. As an example of the second kind of response in barley, frost hardiness was strongly correlated with messenger (m) RNA levels of different cold-inducible genes in plants grown in different temperature environments. However, this correlation did not extend to plants exposed to drought or nutrient stresses (Pearce et al., 1996).

Common Mechanisms

Protection Against Excess Radiation and Photooxidative Damage

Stress can impair the utilization of incoming light for photosynthesis. Under these photoinhibitory circumstances, the excess of resulting radiation can trigger the development of photooxidative processes. Therefore, photooxidative damage is a common response in the face of many different stresses. Photogeneration of singlet oxygen molecules ($^1O_2$) is the most likely reason for chloroplast photodamage (Krasnovsky, 1993) and premature senescence of photosynthetic organs. One of the major sources of singlet oxygen is likely to be the triplet states of newly formed pigments (mainly chlorophylls) that are not quenched by carotenoids. In addition, it has been shown for barley that an increase in the amount of active oxygen photogenerated by O_2 reduction at photosystem I (PSI) triggers ribulose-1,5-bisphosphate carboxylase/oxygenase (rubisco) degradation, with the large subunit of rubisco being the target (Desimone et al., 1996). Further results have shown that active oxygen species modify rubisco to become a substrate for stromal protease(s) (Desimone et al., 1998).

Also widely reported is the fact that photosystem II (PSII) is the most sensitive to the photoinhibitory impairment associated with numerous stresses. In addition, for barley, it has been reported that PSI can experience photodamage during low-light illumination at chilling temperatures (Tjus et al., 1998). This would arise because the defense against active oxygen species by active oxygen-scavenging enzymes is insufficient under these specific conditions. These data (Tjus et al., 1998) suggest that photoinhibition of PSI at chilling temperatures is an important phenomenon in a cold-tolerant plant species. Nevertheless, Stefanov et al. (1996) reported for barley that, compared with PSII, the electron transport of PSI was almost resistant to the effect of high temperature. By contrast, data in barley support the evidence that heat stress induces a loss in the activity of PSII, whereas the decrease in PSI activity is largely due to inactivation (or loss) of a component between the two photosystems (Takeuchi and Thornber, 1994). Indeed, the pigments that are lost from thylakoids during heat stress are mainly removed from the PSII pigment proteins. In this context, Takeuchi and Thornber (1994) reported that heat changes the quaternary structure of PSII, resulting in removal of the oxygen-evolving enhancer proteins from the thylakoid, but did essentially no damage to the PSI complex.

Perception of light stress is under the control of cellular development and differentiation. In addition, photorespiration in barley is regulated by high light also at the level of transcription and accumulation of mRNA for light-stress proteins (Poetter et al., 1996).

Photooxidative damage can accelerate and even trigger senescence. It has been proposed that oxidative damage during senescence could be favored by the inability of senescing leaves to modulate the steady-state level of superoxide dismutase mRNA, and probably those of other antioxidant enzymes, concomitant with the rate of oxyradical formation (Casano et al., 1994).

The dual role of carotenoids, either as photosynthetic or photoprotective pigment, is well established. Recent data in barley suggest that the efficiency of energy transfer from carotenoids to chlorophylls varies with the light environments both in the short and the long term, with excess light energy noticeably inhibiting the photosynthetic light-harvesting function of carotenoids. This carotenoid decoupling from the chlorophyll antennae may have photoprotective implications (Havaux et al., 1998). On the other hand, Krol et al. (1995) also suggest that, in barley, under conditions of light stress, there is induction of light-induced proteins that bind photoconvertible xanthophylls and replace "normal" light-harvesting Chl a/b-binding (LHC) proteins.

Results in barley highlight the importance of the xanthophyll cycle. It can work as an "emergency mechanism" that rapidly provides thylakoid membrane lipids with rigidifying carotenoid molecules upon sudden increase in light intensity (Tardy and Havaux, 1997). Recent results in barley indicate that the reduction state of photosynthetic electron transport chain components could be involved in light sensing for the regulation of nuclear-encoded chloroplast gene expression. Thus, the xanthophyll cycle pigments, as well as the early light-inducible stress proteins and their mRNA, accumulated with increasing PSII excitation pressure, irrespective of the way high excitation pressure was obtained (Montane et al., 1998). The previous evidence can have implications for the function and adaptation of photosynthetic membranes to stressful light and temperature.

Regarding changes associated with climatic change, an increase in CO_2 seems to decrease oxidative stress in barley, whereas soluble catalase and superoxide dismutase activities increase rapidly after plants are transferred to current CO_2 levels (Azevedo et al., 1998). On the other hand, the available data regarding the possible increase in UV-B (280-320 nm λ) associated with the depletion in stratospheric ozone do not suggest stressful effects on barley. Thus, in an experiment simulating up to 25 percent ozone depletion on the summer solstice at 40°N latitude, it was observed that the increase in UV-B irradiation did not significantly affect leaf area, fresh and dry weights, total chlorophylls and carotenoids, or photosynthetic quantum efficiency (Liu et al., 1995). Net assimilation was decreased by UV-B only at internal CO_2 levels above $250\,\mu l \cdot l^{-1}$. As an adaptive response, leaf specific fresh weight (leaf mass per leaf area) as well as the content of flavonoids (saponarin and lutonarin) increase in UV-B-irradiated plants (Liu et al., 1995; Reuber et al., 1996). Increase in flavonoids is assumed to be an efficient protective response, since no changes in variable chlorophyll fluorescence were apparent in response to increasing UV-B radiation (Reuber et al., 1996).

The Role of ABA and Related Growth Regulators

ABA is involved in the response to numerous stresses apart from drought and other widespread stresses. For example, there is evidence of the positive role of ABA as a root-to-shoot signal that assists in maintaining leaf growth in plants experiencing compaction stress (Mulholland, Taylor, et al., 1996). ABA would exert a positive role in maintaining leaf expansion under these conditions (Mulholland, Black, et al., 1996). Interaction between osmotic stress and ABA may be an important mechanism in explaining leaf growth inhibition of droughted barley plants (Dodd and Davies, 1996).

At a different level, stomatal conductance in barley has been related to the concentration of ABA in the xylem sap measured at the end of the night, although apparent stomatal sensitivity to ABA in the sap was lower in earliest anthesis lines (Borel et al., 1997). Nevertheless, stomatal control appears to have low genetic variability. In addition, stomatal control depends on evaporative demand and soil water status. In this regard, barley shows anisohydric behavior, with daytime leaf water potential markedly decreasing with evaporative demand during the day and being lower in droughted than in watered plants (Tardieu and Simonneau, 1998). Anisohydric behavior seems to be linked to an absence of interaction between hydraulic (e.g., conductivity) and chemical (ABA through the xylem) mechanisms.

Whereas a drop in activity of most photosynthetic enzymes is the common response to the application of ABA, it has been suggested that increased carbonic anhydrase activity is a photosynthetic response in barley to elevated ABA concentration (Popova et al., 1996). This might have implications in the photosynthetic metabolism during stomatal closure, favoring, for example, a higher fixation through the PEPcase. In the same way, Tsonev et al. (1998) reported a strong increase in the activities of PEPcase and carbonic anhydrase as a response to both jasmonic acid and salinity stress. Moreover, pretreatments with jasmonic acid before salinization diminished the inhibitory effect of high salt concentration on the growth and photosynthesis of barley seedlings (Tsonev et al., 1998). However, Kramell et al. (1995) reported that salt stress did not stimulate the levels of endogenous jasmonates in barley leaf segments. By contrast, treatments inducing water stress increased both jasmonic acid and its conjugates (Kramell et al., 1995). It has been suggested that jasmonic acid may affect photosynthesis indirectly, either as a stress-modulating substance or through alterations in gene expression (Metodiev et al., 1996). Leopold et al. (1996) hypothesized the existence of a stress-induced lipid-based signaling pathway in which an endogenous increase in jasmonates initiates gene expression of newly formed proteins.

Under conditions of slightly alkaline apoplastic pH, which occurs in stressed leaves, the epidermis (compared with the mesophyll cells) may serve as the main source for fast stress-dependent redistribution into the guard cell apoplast of ABA that arrives via the transpiration stream (Daeter and Hartung, 1995). Apart from roots, ABA also may be released from photosynthetic organs. Thus, structurally intact and functionally active chloroplasts would be required for drought stress to elicit an increase in ABA (Popova, 1998).

Osmotic stress or treatment with ABA led to the synthesis of novel proteins that were identical to jasmonate-induced proteins. Available results support the hypothesis that the accumulation of jasmonates, probably by de novo synthesis, is an intermediate and essential step in the signaling pathway between (osmotic) stress and activation of genes coding for polypeptides of high abundance (Lehmann et al., 1995).

Stress Proteins, Osmoprotectants, and Osmoregulators

A number of proteins have been identified that typically accumulate in plants in response to any environmental stimulus that has a dehydrative component or is temporally associated with dehydration. Among the induced proteins, dehydrins (late embryogenesis abundant [LEA] D-11 family) have been the most commonly observed. They are produced in a wide variety of plant species in response to environmental stimuli with a dehydrative component, including drought, low temperature, and salinity, in addition to developmental stages such as seed and pollen maturation. Therefore, current evidence tentatively points to dehydrins as key components of dehydration tolerance in barley (Close, 1996).

Dehydrins may be structure stabilizers with detergent and chaperone-like properties and an array of nuclear and cytoplasmic targets (Close, 1996). Nevertheless, and despite their widespread occurrence and abundance in cells under dehydrative conditions, the biochemical role of dehydrins remains elusive. The subcellular location of dehydrins would be consistent with a biochemical role as an intracellular stabilizer, possibly with surfactant characteristics (Campbell and Close, 1997).

Barley responds with an increased accumulation of osmoprotectant glycinebetaine (betaine) when subjected to different stresses inducing cell dehydration. Thus, barley plants are able to accumulate betaine at high levels in their leaves in response not only to water, salt stress, or high temperatures but also to cold stress. Accumulation of betaine during acclimation to cold is associated to some extent with freezing tolerance in leaves (Kishitami et al., 1994). In addition, the levels of betaine accumulated during cold acclimation might be also associated with the earliness or lateness of the maturity of cultivars, namely, that late cultivars accumulate more betaine than early cultivars. Nevertheless, the higher levels of betaine in the late cultivars might have resulted from coselection for lateness of maturity and freezing tolerance (Nomura et al., 1995).

The last step of betaine synthesis is catalyzed by the enzyme betaine dehydrogenase (BADH). BADH transcripts are reported to accumulate in response to drought or salt stress, indicating a common response of the plant to osmotic changes that affects its water status (Ishitani et al., 1995). The addition of ABA to plants during growth also increased the levels of BADH transcripts dramatically (Ishitani et al., 1995). Nakamura et al. (1996) reported betaine levels several times higher in young than in old leaf blades, the former being more tolerant to salinity than the latter.

Results in barley suggest that proline has a bifunctional role in the acclimation to high salt stress, an osmoregulant role in the light, and a substrate for dark respiration, to supply energy to compartmentation of ions into vacuoles in the dark (Sanada et al., 1995).

Specific Mechanisms: The Case of Salt Stress

Barley is more salt tolerant than other cereals. Nevertheless, evidence suggests that not only barley but also other species of the tribe Triticeae share some common genetic mechanisms of tolerance of sudden salt stress (Zhong and Dvorak, 1995). However, it has been reported for barley that biomass and grain yields were reduced by 2.4 to 3.1 percent for each 1 dS·m^{-1} increase in the saturated soil extract (Ece), whereas the Ece's producing 50 percent yield reduction of biomass or grain yield ranged from 17 to 21 dS·m^{-1} (Khan and Glenn, 1996).

Experimental data support the hypothesis that in barley the growth response to salinity has two phases. The first phase would show a large decrease in growth rate caused by the salt outside the roots, i.e., an "osmotic" response. The second phase would exhibit an additional decline in growth caused by salt having built up to toxic levels within plants, i.e., a "salt-specific" response (Munns et al., 1995). In addition, the salt within the plant reduces growth by causing premature senescence of old leaves, and hence a reduced supply of assimilates to the growing regions. Indeed, for barley plants exposed to salt stress, sodium and chloride ions accumulated preferentially in old rather than in young blades (Nakamura et al., 1996).

It has been reported that salt stress induces a decrease in the antioxidation activity of roots, resulting in an uncontrolled acceleration of free-radical processes (Kasumov et al., 1998). At the molecular level, adaptation to salt stress seems to involve an increase in the vacuolar hydrogen ion pumps in barley roots in order to maintain stability of tonoplast under salt stress (Zhang et al., 1998).

Implications of Stress Physiology in Breeding and Crop Management

In regions regularly exposed to abiotic stresses, plant breeding programs are being conducted to select genotypes having higher and more stable yields. Screening techniques that are able to identify desirable genotypes based on the evaluation of physiological traits related to stress evasion and stress resistance (either avoidance or tolerance) may be useful, particularly if they are rapid, simple, and inexpensive. The usefulness of physiological screening techniques can be extended in principle to the evaluation of crop management techniques.

For some of the most important abiotic stresses, such as drought and salinity, which develop progressively during crop cultivation, it is worth selecting for integrative (either in time or in level of plant organization) traits (Araus et al., 1998, 2001; Slafer and Araus, 1998; Slafer et al., 1999). Among these integrative traits, mention should be made of Δ (described earlier in this study). The utilization of remote sensing techniques at the canopy levels, such is infrared thermometry and indices based on the spectral signature of visible and near-infrared radiation reflected by the canopy (see Araus et al., 2001, and also Bort et al., Chapter 14 in this book), is also worth considering.

However, for other stresses, such as frost, cold, and high temperature, which develop more or less suddenly and usually represent short-duration episodes, the scenery can be quite different, and more instantaneous or low-level (e.g., biochemical) traits can be evaluated. These provide an indication of plant tolerance more than resistance to stresses. These traits are related to the levels of osmoprotective substances or to the photosynthetic status of the leaves. Betaine is the most known osmoprotective substance in barley. Photosynthetic status can be rapidly assessed through chlorophyll fluorescence techniques.

Chlorophyll fluorescence measurements on intact leaves are adequate techniques to assess the incidence of punctual stresses such as low or high temperature under light conditions. In this context, it is widely reported that heat stress induces an increase in the minimum (i.e., initial) chlorophyll fluorescence (F_o), whereas the maximum fluorescence (F_m) gradually decreases (Briantais et al., 1996; Pospisil and Naus, 1998; Pospisil et al., 1998). In the same way, F_o also increased in response to freezing (Pospisil et al., 1998). The rise of F_o is thought to be connected with the blocking of electron transport in the reaction center of PSII (Pospisil and Naus, 1998).

Light Stress

Barley breeders at the International Center for Agricultural Research in Dry Areas (ICARDA) have observed that genotypes adapted to dry regions have leaves that are lighter in color than those of unadapted ones, this being more evident after heading. Light leaf color was basically associated with a higher chlorophyll a:b ratio, which indicates a reduction in the amount of antenna chlorophyll compared to that in the core complex of PSII. In addition, genotypes with light green leaves had consistently less chlorophyll per unit leaf area (Watanabe et al., 1995; Tardy et al., 1998). It has been postulated that light green leaves may confer the ability to adapt to high levels of irradiance under drought conditions, reducing the risk of damage from the overexcitation of the PSII reaction center. Recently, Tardy et al. (1998) concluded that the pale green Syrian land race Tadmor, which is well suited to harsh Mediterranean environments, is equipped to survive excessive irradiance through a passive reduction of the light absorbance of its leaves (which mitigates the heating effect of strong light), and through the active protection of its photochemical apparatus by rapid xantophyll cycling. The very low chlorophyll content of Syrian land races is related to their drought adaptation. A very low chlorophyll content decreases leaf absorbance, which, in turn, reduces the potentially damaging heating effect of high solar radiation in droughted plants whose stomata are closed (Havaux and Tardy, 1999).

Genotype differences in xantophyll cycling can be evaluated in a fast and nondestructive manner through the measurement of spectroradiometrical indices such is the photochemical reflectance index (Filella et al., 1996; Araus et al., 2001; Tambussi et al., 2000). Spectroradiometrical techniques also can be useful for routine determinations of leaf color. Thus, Watanabe (1998), working with genotypes of barley taken from the breeding program at ICARDA, showed that the genotypes of light green color reflected more light that those of dark green color, especially in the green region (520 to 600 nm) of the spectrum. Thus, the light green genotypes had a lower color temperature than the dark green genotypes.

Until now, chlorophyll fluorescence has been the most widely used technique for the physiological assessment of light stress. However, alternative methodologies for chlorophyll fluorescence assessments are also already available. In barley leaves, the kinetics and magnitudes of chlorophyll fluorescence are linearly correlated with O_2 evolution and thermal dissipation using photoacoustic spectroscopy. These correlations are sustained, at least in part, by the common dependence of the previous parameters on the redox state of the acceptor side of PSII (Bukhov et al., 1997).

Other methodological approaches have been specifically developed for the evaluation of the degree of photooxidative stresses in barley. Thus, the measurement of pigment phosphorescence at 77 K would provide an index of the degree of plant photooxidative stress (Krasnovsky, 1993). On the other hand, Vallelian et al. (1998) suggested that the extracellular, acid-soluble proteins, which belong to the group of germin-like proteins, may serve as markers for oxidative stress in cereals.

Portable chlorophyll meters may also be useful for determining genotypical differences in plant phenology as well as in leaf tolerance to photooxidative damage (Bort et al., 1998), and even as a surrogate of Δ (Araus, Bort, et al., 1997). Indeed, for barley, grain yield has been positively correlated with chlorophyll content on leaf area, measured with a portable device (Bort et al., 1998). In addition, determination of leaf chlorophyll content with a chlorophyll meter accurately indicates plant nitrogen status, allowing nitrogen fertilizer requirements to be accurately determined and resulting in increased nitrogen uptake efficiency (Peltoten et al., 1995).

Osmoregulation and Osmoprotection

The protective role of betaine in barley is supported by the work of Harinasut et al. (1996). Thus, whereas the quantum yield of PSII evaluated by chlorophyll fluorescence was decreased under salt stress, such decrease was largely prevented by betaine application (Harinasut et al., 1996). In the framework of the previous evidence, it has been proposed that crop losses caused by environmental stresses might be reduced by applying osmoprotectants to the crop. However, Makela et al. (1996) did not find any significant advantage of betaine treatments in Finland, although they suggested the necessity of performing further studies in stress-prone environments.

Proline also accumulates in barley in response to different stresses, including drought and salinity. However, proline concentration, in addition to ionic (N, P, K) concentration and relative water content, in expanded leaves did not perform well for screening drought resistance in barley (Yasseen and Al Maamari, 1995).

Nevertheless, Teulat, Rekika, et al. (1997) concluded that the lowest osmotic adjustment capacities were noted in drought-susceptible varieties, while a high capacity was found in genotypes exhibiting high yield stability across contrasting environments. Relative water content, leaf osmotic potential, and accumulation of soluble sugars were found to be highly related to osmotic adjustment; they could be used as criteria for rapid evaluation of osmotic adjustment in segregating populations. Arnau et al. (1997) reported variability in osmotic adjustment capacity between genotypes: several drought-tolerant genotypes showed higher osmotic adjustment than did drought-susceptible ones. However, drought-tolerant land races with a very extensive root development did not show osmotic adjustment Therefore, barley exhibits different strategies for coping with drought stress (Arnau et al., 1997).

ABA and Other Growth Regulators

From a breeding perspective, the selection of genotypes with high endogenous ABA levels induced by stress has been proposed as desirable. To date, however, no positive results in (wheat) yield under water stress have been attained (Read et al., 1991).

Regarding management practices, ethephon, a precursor of ethylene, reduces lodging in barley by shortening straw. In the absence of lodging, however, and under conditions of environmental stress, ethephon may reduce yields, particularly of two-row barley, for which spike emergence may be severely reduced (Bridger et al., 1995).

Drought Stress

Drought, considered as the combination of water stress plus high irradiance and temperature stresses, is the main abiotic factor limiting yield. In

Mediterranean environments, drought can affect the last part of the crop cycle, coinciding with grain filling. Under these circumstances, the capacity to sustain translocation of assimilates accumulated in the stem and leaves prior to anthesis to the filling grains can be a very positive trait. In such a context, translocation capacity after chemical desiccation of the photosynthetic apparatus has been proposed as a screening approach. However, its performance in barley showing genetic differences for grain yield, kernel weight, and harvest index has not been demostrated (Gavuzzi et al., 1997). Moreover, from the published reports in wheat (Loss and Siddique, 1994) and personal experience in barley (Romagosa, Voltas, and Araus, unpublished results), this method is best conducted under irrigated conditions or in wetter conditions.

Nogués et al. (1994) proposed the determination of modulated chlorophyll fluorescence coupled to measurements of infrared thermometry and gas exchange as a screening method to score barley for drought resistance. Drought-resistant genotypes show smaller variations in photochemical quenching, net photosynthesis, transpiration efficiency, and leaf temperature than do drought-sensitive genotypes. Nevertheless, such a combined approach is time-consuming and equipment intensive and would therefore be suited only to the selection of parental lines.

Under drought conditions, seminal roots may be more important than nodal roots, and plants often reach maturity with growth of their seminal roots only. In this sense, land races do not differ from modern cultivars for number of seminal root axes, but they have the longest seminal root systems. Moreover, land races have a more vigorous seminal root system than modern cultivars (Grando and Ceccarelli, 1995). Kafawin (1998) has suggested using root length and root dry weight as selection criteria for developing drought-tolerant barley cultivars. However, Gorny (1996) pointed out the absence in barley of close genetic correlations between the seminal and adventitious root systems. Soil water resulting from episodic growing season rainfall evaporates rapidly in semiarid regions. Plants may not benefit from such water addition if near-surface roots are unable to resume water uptake rapidly following periods of soil water deficit. In this sense, Wraith et al. (1995) developed a procedure of evaluating genotypic differences in root uptake responses in the upper soil layer following rewetting after soil water deficit.

Arnau and Monneveux (1995) reported that the relative water content of the penultimate leaf as well as plant height and last and penultimate internode lengths were shown to be closely related to water stress susceptibility of grain filling and to have medium to high (broad-sense) heritability values. By contrast, even when there was genotypic variability for the photosynthesis response of the spike to drought, these photosynthetic traits did not correlate with yield.

The importance of Δ when selecting for water-limited environments has been emphasized before. Indeed, this is one of the most promising selecting traits, even when the price of analysis is one of the main factors limiting its widespread use. Total mineral accumulation (evaluated as ash content) in vegetative photosynthetic parts (mainly upper leaves) and in mature kernels has been proposed as a indirect selection criterion, either in substitution for or in addition to Δ of mature grains, to assess grain yield of barley crops in Mediterranean areas (Febrero et al., 1994; Voltas, Romagosa, Muñoz, and Araus, 1998). Thus, ash concentration in kernels is particularly adequate as a complementary criterion to Δ in poor rain-fed environments. By contrast, ash content in vegetative plant parts sampled at anthesis was not consistently associated with grain yield (Voltas, Romagosa, Muñoz, and Araus, 1998).

Populations of wild barley (*Hordeum spontaneum* Koch) can become an important source of genetic variability when breeding for improving yield under stress. In this regard, peripheral populations seem to be phenotypically more variable and more resistant to water stress than core populations. Indeed, the association of water stress resistance with high phenotypic variability gives support to the hypothesis that populations which are genetically more variable are better sources of germplasm (Volis et al., 1998).

Crop Management

Regarding crop management, water stress is reported to reduce the effectiveness of mineral fertilization. However, at least for nitrogen, this is not necessarily true in all circumstances. When plants were subjected to water stress, nitrogen content per plant decreased, with wheat appearing more severely affected than barley. However, nitrogen concentration was slightly higher for the stressed plants as compared with the controls (Pessarakli and Fardad, 1995). In the semiarid conditions (such as those of northeast Spain) where barley is cultivated in the Mediterranean region, there is no benefit for yield in high rates of nitrogen fertilization, and even during a very dry season, this can depress yield and harvest index (Cantero-Martínez et al., 1995b). These same authors concluded by indicating that with a lower amount of nitrogen fertilizer than used at present (100 kg·ha^{-1}, the normal dressing in these areas), it is possible to improve the sustainability and stability of the cropping system without reducing grain production (Cantero-Martínez et al., 1995b). It has also been reported that a larger fallowing increases the water content of the upper soil layers, thus raising average yields (Austin, Cantero-Martinez, et al., 1998). Based on barley yield/rainfall regressions for data from a dry area in the Ebro valley (Spain), Austin, Playan, and Gimeno (1998) estimated that the annual yields from a crop-fallow system would be 15 percent greater than those from annual cropping. However, for fallowing to be economic, yields per crop would need to be about twice those obtained with annual cropping. Unless there were yield benefits from fallowing apart from those resulting from extra water storage in the soil, fallowing appears to be an uneconomical alternative in most Mediterranean areas.

Fay et al. (1996) reported that mycorrhizal colonization enhanced the maximum photosynthetic rate at a low phosphorus level. However, this was associated with a higher stomatal conductance and was not related to increased leaf phosphorus or to changes in the photon yield or the ratio of variable to maximum chlorophyll fluorescence (F_v/F_m). Iron deficiency caused significant decrease in water-use efficiency of barley plants (Sharma and Tripathi, 1994). Iron-deficient plants showed a marked decrease in net photosynthesis and concentration of sugars and starch in the leaves. The decrease in photosynthesis had a nonstomatal basis; it was primarily attributable to a decrease in leaf chlorophyll concentration.

Hydrophilic superabsorbent polymers, which retain large amounts of plant-available moisture, have been promoted for use as soil amendments in drought-prone regions. However, the high amounts of polymer required to elicit a crop yield response under relatively mild water-deficit conditions limit the value of these polymers for agricultural field use in barley (Volkmar and Chang, 1995). On the other hand, application of nitrogenous substances on the plant can alleviate the effect of stresses. By application with natural amino alcohols, such as 2-aminoethanol and choline chloride, the effect of drought stress on barley plants was diminished (Bergmann et al., 1994). Treated plants showed an increase in grain yield and a decrease in the stress metabolites betaine, proline, and trigonelline, as well as in arginine (precursor of the stress metabolite putrescine).

Salinity

There is wide genotypic variability for salt stress in barley. From an extensive evaluation of genotypes, Mano et al. (1996) concluded that many of the highly salt-tolerant varieties were Chinese six-rowed ones, while most of the highly sensitive varieties were from West Asia with two-rowed, black lemma. Some of the most tolerant varieties could germinate even in sea water. Nevertheless, salt tolerance at the germination stage was independent of the salt tolerance of seedlings (Mano et al., 1996).

However, the establishment of proper agronomical practices and plant breeding programs for saline environments is limited by the lack of adequate field screening methods (Isla et al., 1997, 1998). In such a context, leaf ion concentrations should not be used as screening criteria in breeding programs for increasing salinity tolerance in barley. The highest-yielding cultivars under nonsaline conditions were also most productive under moderately saline conditions, though not under highly saline conditions (Isla et al., 1997). The same team (Isla et al., 1998) further assessed the relationship between grain yield, carbon isotope discrimination, canopy temperature, stomatal conductance, and grain ash content in a set of barley cultivars grown in a soil salinity gradient imposed by a triple-line-source sprinkler system. A linear increase in soil salinity produced significant and linear decreases in grain yield and Δ, and increases in midday differential canopy temperature. Increasing soil salinity also decreased stomatal conductance and increased grain ash content. However, Isla et al. (1998) concluded that none of the studied characters was useful in screening for high yield under saline environments. Therefore, green biomass and futher grain yield under salt stress remained the only reliable means of identifying higher salt tolerance in barley, and the same conclusion can be extended to severe drought stress conditions (Ceccarelli and Grando, 1996). In such a context, remote sensing evaluation of green biomass using, for example, spectroradiometrical indices based on the visible and near-infrared radiation reflected by the canopy may become a proper methodology for breeding purposes (Peñuelas et al., 1997). In the case of Δ, it was positively and significantly correlated with grain yield under nonsaline conditions, suggesting that Δ is a useful indicator of yield potential in barley. However, under highly saline conditions, although the relationship between Δ and grain yield was still positive, it did not attain statistical significance.

Regarding crop management practices, adding additional water during irrigation as a leaching fraction increased the amount of evaporation rather than reducing saturated soil extract (Khan and Glenn, 1996). However, other management practices can alleviate the effect of salt stress. Thus, added silicon enhances the uptake of potassium and inhibits the uptake of sodium by salt-stressed barley, thus mitigating the toxicity of salt to barley, increasing salt tolerance of the plant and therefore enhancing dry-matter yield (Liang et al., 1996). Supplemental calcium ameliorated the deleterious effects of salinity on barley growth, thereby improving its salt tolerance (Huang and Redmann, 1995, 1996). Increasing plant phosphorus supply can increase salt tolerance in barley by enhancing accumulation of mineral ions for osmotic adjustment and restricting sodium accumulation in shoots (Al-Karaki, 1997).

For an exhaustive review on the effect of salt stress in barley and other crop plants, it is worth consulting the first edition of a book on this topic, published under the somewhat misleading (at least for the first edition) title *Handbook on Plant and Crop Stress* (Pessarakli, 1994, 1999).

FINAL REMARKS

Future avenues for locating genotypes and genes for abiotic stress tolerance in barley will require a multidisciplinary approach, involving a strategy using genetic maps, molecular markers, and wild relatives (Forster et al., 1997; see also Cattivelli et al., Chapter 12 in this book), that will result in a physiological understanding of the processes determining yield. Indeed, it is necessary to define through a physiological understanding the important traits we need to relate to molecular markers, particularly those addressed to investigating QTL. Moreover, localization and colocalization of QTL need physiological explanation. The work of Teulat, Monneveux, et al. (1997) suggests that even if some parameters are strongly correlated, finding QTL for only one trait is not sufficient to detect all the candidate regions that might be involved in the control of the correlated traits. For barley, epinastic interactions between several QTL and between QTL and other markers were detected only in the water stress treatment, suggesting that some chromosal regions might be involved in the regulation of the expression of the traits under water stress (Teulat, Monneveux, et al., 1997).

REFERENCES

Acevedo, E. (1993). Potential of carbon isotope discrimination as a selection criterion in barley breeding. In *Stable Isotopes and Plant Carbon/Water Relations*, eds. J.R. Ehleringer, A.E. Hall, and G.D. Farquhar. New York: Academic Press, pp. 399-417.

Albrecht, S.L., C.L. Douglas Jr., E.L. Klepper, P.E. Rasmussen, R.W. Rickman, R.W. Smile, D.E. Wilkins, and D.J. Wysocki (1995). Effects of foliar methanol applications on crop yield. *Crop Science* 35: 1642-1646.

Al-Karaki, G.N. (1997). Barley response to salt stress at varied levels of phosphorus. *Journal of Plant Nutrition* 20: 1635-1643.

Araus, J.L. (1996). Integrative physiological criteria associated with yield potential. In *Increasing Yield Potential in Wheat: Breaking the Barriers*, eds. M.P. Reynolds, S. Rajaram, and A. McNab. Mexico City, Mexico, DF: CIMMYT, pp. 150-166.

Araus, J.L., T. Ali Dib, and M.M. Nachit (1998). Some insights about morphophysiological traits associated with cereal yield increases in Mediterranean environments. In *SEWANA (South Europe, West Asia and North Africa) Durum Research Network Proceedings of the SEWANA Durum Network Workshop*, March 20-23, 1995, eds. M.M. Nachit, M. Baum, E. Porceddu, P. Monneveux, and E. Picard. Aleppo, Syria: ICARDA, Aleppo, pp. 139-158.

Araus J.L., J. Bort, S. Ceccarelli, and S. Grando (1997). Relationship between leaf structure and carbon isotope discrimination in field grown barley. *Plant Physiology and Biochemistry* 35: 533-541.

Araus, J.L. and R. Buxó (1993). Changes in carbon isotope discrimination in grain cereals from the north-western Mediterranean Basin during the past seven millenia. *Australian Journal of Plant Physiology* 20: 117-128.

Araus, J.L., J. Casadesús, and J. Bort (2001). Recent tools for screening of physiological traits determining yield. In *Application of Physiology in Wheat Breeding*, eds. M. Reynolds, I. Ortiz-Monasterio, and A. McNab. Mexico City, Mexico, DF: CIMMYT, pp. 59-76.

Araus, J.L., A. Febrero, R. Buxo, M.D. Camalich, D. Martin, F. Molina, M.O. Rodriguez-Ariza, and I. Romagosa (1997). Changes in carbon isotope discrimination in grain cereals from different regions of the western Mediterranean Basin during the past seven millennia. Palaeoenvironmental evidence of a differential change in aridity during the late Holocene. *Global Change Biology* 3: 107-118.

Araus, J.L., G.A. Slafer, and I. Romagosa (1999). Durum wheat and barley yields in antiquity estimated from ^{13}C discrimination of archaeological grains: A case study from the western Mediterranean Basin. *Australian Journal of Plant Physiology* 26: 345-352.

Arnau, G. and P. Monneveux (1995). Physiology and genetics of terminal water stress tolerance in barley. *Journal of Genetics and Breeding* 49: 327-332.

Arnau, G., P. Monneveux, D. This, and L. Alegre (1997). Photosynthesis of six barley genotypes as affected by water stress. *Photosynthetica* 34: 67-76.

Ashley, D.A. and H.R. Boerma (1989). Canopy photosynthesis and its association with seed yield in advanced generations of a soybean cross. *Crop Science* 29: 1042-1045.

Austin, R.B., M.C. Cantero-Martínez, J.L. Arrue, E. Playan, and P. Cano-Marcellán (1998). Yield-rainfall relationships in cereals cropping systems in the Ebro river valley of Spain. *European Journal of Agronomy* 8: 239-248.

Austin, R.B., E. Playan, and J. Gimeno (1998). Water storage in soils during fallow: Prediction of the effects of rainfall pattern and soil conditions in the Ebro valley of Spain. *Agricultural Water Management* 36: 213-231.

Azevedo, R.A., R.M. Alas, R.J. Smith, and P.J. Lea (1998). Response of antioxidant enzymes to transfer from elevated carbon dioxide to air and ozone fumigation, in the leaves and roots of wild-type and a catalase-deficient mutant of barley. *Physiologia Plantarum* 104: 280-292.

Bacon, M.A., S. Wilkinson, and W.J. Davies (1998). PH-regulated leaf cell expansion in droughted plants is abscisic acid dependent. *Plant Physiology* 118: 1507-1515.

Bergmann, H., B. Machelett, and V. Leinhos (1994). Effect of natural amino alcohols on the yield of essential amino acids and the amino acid pattern in stressed barley. *Amino Acids Vienna* 7: 327-331.

Bertholdsson, N.O. (1998). Selection methods for malting barley consistently low in protein concentrations. *European Journal of Agronomy* 9: 213-222.

Borel, C., T. Simmonneau, D. This, and F. Tardieu (1997). Stomatal conductance and ABA concentration in the xylem sap of barley lines of contrasting genetic origins. *Australian Journal of Plant Physiology* 24: 607-615.

Bort, J., J.L. Araus, H. Hazzam, S. Grando, and S. Ceccarelli (1998). Relationships between early vigour, grain yield, leaf structure and stable isotope composition in field grown barley. *Plant Physiology and Biochemistry* 36: 889-897.

Bort, J., H.R. Brown, and J.L Araus (1995). Lack of C_4 photosynthetic metabolism in ears of C_3 cereals. *Plant Cell and Environment* 18: 897-702.

Bort, J., H.R. Brown, and J.L. Araus (1996). Refixation of respiratory CO_2 in the ears of C_3 cereals. *Journal of Experimental Botany* 47: 1567-1575.

Bort, J., A. Febrero, T. Amaro, and J.L. Araus (1994). Role of awns on ear water use efficiency and grain weight in barley. *Agronomie* 2: 133-139.

Bouzerzour, H. and M. Dekhili (1995). Heritabilities, gains from selection and genetic correlations for grain yield of barley grown in two contrasting environments. *Field Crops Research* 41: 173-178.

Briantais, J.M., J. Dacosta, Y. Goulas, J.M. Ducruet, and I. Moya (1996). Heat stress induces in leaves an increase of the minimum level of chlorophyll fluorescence, Fo: A time-resolved analysis. *Photosynthesis Research* 48: 189-196.

Bridger, G.M., H.R. Klinck, and D.L. Smith (1995). Timing and rate of ethephon application to two-row and six-row spring barley. *Agronomy Journal* 87: 1198-1206.

Bukhov, N.G., N. Boucher, and R. Carpentier (1997). The correlation between the induction kinetics of the photoacoustic signal and chlorophyll fluorescence in barley leaves is governed by changes in the redox state of the photosystem II acceptor side. A study under atmospheric and high CO_2 concentrations. *Canadian Journal of Botany* 75: 1399-1406.

Bulman, P. and D.L. Smith (1993). Yield and yield component response of spring barley to fertilizer nitrogen. *Agronomy Journal* 85: 226-231.

Campbell, S.A. and T.J. Close (1997). Dehydrins: Genes, proteins, and associations with phenotypic traits. *New Phytologist* 137: 61-67.

Cantero-Martínez, C., J.M. Villar, I. Romagosa, and E. Fereres (1995a). Growth and yield responses of two contrasting barley cultivars in a Mediterranean environment. *European Journal of Agronomy* 4: 317-326.

Cantero-Martínez, C., J.M. Villar, I. Romagosa, and E. Fereres (1995b). Nitrogen fertilization of barley under semi-arid rainfed conditions. *European Journal of Agronomy* 4: 309-316.

Casano, L.M., K.M. Martín, and B. Sabater (1994). Sensitivity of superoxide dismutase transcript levels and activities of oxidative stress is lower in mature-senescent than in young barley leaves. *Plant Physiology* 106: 1033-1039.

Cattivelli, L., G. Delogu, V. Terzi, and A.M. Stanca (1994). Progress in barley breeding. In *Genetic Improvements of Field Crops: Current Status and Development*, ed. G.A. Slafer. New York: Marcel Dekker, pp. 95-181.

Ceccarelli, S. (1994). Specific adaptation and breeding for marginal conditions. *Euphytica* 77: 205-219.

Ceccarelli, S. and S. Grando (1993). From conventional breeding to molecular biology. In *Proceedings of the International Crop Science I*, eds. D.R. Buxton, R. Shibles, R.A. Forsberg, B.L. Blad, K.H. Asay, G.M. Paulsen, and R.F. Wilson. Madison, WI: Crop Science Society of America, pp. 533-537.

Ceccarelli, S. and S. Grando (1996). Drought as a challenge for the plant breeder. *Plant Growth Regulation* 20: 149-155.

Ceccarelli, S., S. Grando, and A. Impiglia (1998). Choice of selection strategy in breeding barley for stress environments. *Euphytica* 103: 307-318.

Close, T.J. (1996). Dehydrins: Emergence of a biochemical role of a family of plant dehydration proteins. *Physiologia Plantarum* 97: 795-803.

Craufurd, P.Q., R.B. Austin, E. Acevedo, and M.A. Hall (1991). Carbon isotope discrimination and grain yield in barley. *Field Crops Research* 27: 301-313.

Daeter, W. and W. Hartung (1995). Stress-dependent redistribution of abscisic acid (ABA) in *Hordeum vulgare* L. leaves: The role of epidermal ABA metabolism, tonoplastic transport and the cuticle. *Plant Cell and Environment* 18: 1367-1376.

De Ruiter, J.M. and I.R. Brooking (1996). Effect of sowing date and nitrogen on dry matter and nitrogen partitioning in barley. *New Zealand Journal of Crop and Horticultural Science* 24: 65-76.

Desimone, M., A. Henke, and E. Wagner (1996). Oxidative stress induces partial degradation of the large subunit of ribulose-1,5-bisphosphate carboxylase/oxygenase in isolated chloroplasts of barley. *Plant Physiology* 111: 789-796.

Desimone, M., E. Wagner, and U. Johanningmeier (1998). Degradation of active oxygen modified ribulose-1,5-bisphosphate carboxylase/oxygenase by chloroplastic proteases requires ATP-hydrolysis. *Planta* 205: 459-466.

Dodd, I.C. and W.J. Davies (1996). The relationship between leaf growth and ABA accumulation in the grass leaf elongation zone. *Plant Cell and Environment* 19: 1047-1056.

Dofing, S.M. (1995). Phenological development-yield relationships in spring barley in a subarctic environment. *Canadian Journal of Plant Science* 75: 93-97.

Ellis, R.P. and B. Marshall (1998). Growth, yield and grain quality of barley (*Hordeum vulgare* L.) in response to nitrogen uptake. II. Plant development and rate of germination. *Journal of Experimental Botany* 49: 1021-1029.

Evans, L.T. (1989). Opportunities for increasing the yield potential of wheat. In *The Future Development of Maize and Wheat in the Third World*. Mexico City, Mexico DF: CIMMYT, pp. 79-93.

Fay, P., D.T. Mitchell, and B.A. Osborne (1996). Photosynthesis and nutrient-use efficiency of barley in response to low arbuscular mycorrhizal colonization and addition of phosphorus. *New Phytologist* 132: 425-433.

Febrero, A., A. Blum., I. Romagosa, and J.L. Araus (1992). Relationships between carbon isotope discrimination in field grown barley and some physiological traits of juvenile plants in growth chambers. Abstracts Supplement of the First International Crop Science Congress, Ames, IA: p. 26.

Febrero, A., T. Bort, J. Voltas, and J.L. Araus (1994). Grain yield, carbon isotope discrimination and mineral content in mature kernels of barley, under irrigated and rainfed conditions. *Agronomie* 2: 127-132.

Febrero, A., S. Fernandez, J.L. Molina-Cano, and J.L. Araus (1998). Yield, carbon isotope discrimination, canopy reflectance and epidermal conductance of barley isolines of differing glaucousness. *Journal of Experimental Botany* 49: 1575-1581.

Filella, I., T. Amaro, J.L. Araus, and J. Peñuelas (1996). Relationship between photosynthetic radiation-use efficiency of barley canopies and the photochemical reflectance index. *Physiologia Plantarum* 96: 211-216.

Forster, B.P., J.R. Russell, R.P. Ellis, L.L. Handley, D. Robinson, C.A. Hackett, R. Waugh, D.C. Gordon, R. Keith, and W. Powell (1997). Locating genotypes and genes for abiotic stress tolerance in barley: A strategy using maps, markers and the wild species. *New Phytologist* 137: 141-147.

Fricke, W. and T.J. Flowers (1998). Control of leaf cell elongation in barley: Generation rates of osmotic pressure and turgor, and growth-associated water potential gradients. *Planta* 206: 53-65.

Fricke, W., A. James, S. McDonald, and L. Mattson-Djos (1997). Why do leaves and leaf cells of N-limited barley elongate at reduced rates? *Planta* 202: 522-530.

Fukai, T.S. (1995a). Growth and yield response of barley and chickpea to water stress under three environments in southeast Queensland. I. Light interception, crop growth and grain yield. *Australian Journal of Agricultural Research* 46: 17-33.

Fukai, T.S. (1995b). Growth and yield response of barley and chickpea to water stress under three environments in southeast Queensland. III. Water use efficiency, transpiration efficiency and soil evaporation. *Australian Journal of Agricultural Research* 46: 49-60.

Gallagher, J.N., P.V. Biscoe, and R.K. Scott (1975). Barley and its environment. V. Stability of grain weight. *Journal of Applied Ecology* 12: 319-336.

García del Moral, L.F., A. Sopena, J.L. Montoya, P. Polo, J. Voltas, P. Codesal, J.M. Ramos, and J.L. Molina-Cano (1998). Image analysis and chemical composition of the barley plant as predictors of malting quality in Mediterranean environments. *Cereal Chemistry* 75: 755-761.

Gavuzzi, P., F. Rizza, M. Palumbo, R.G. Campanile, G.L. Ricciardi, and B. Borghi (1997). Evaluation of field and laboratory predictors of drought and heat tolerance in winter cereals. *Canadian Journal of Plant Science* 77: 523-531.

Gómez-Macpherson, H. and R.A. Richards (1995). Effect of sowing time on yield and agronomic characteristics of wheat in south-eastern Australia. *Australian Journal of Agricultural Research* 46: 1381-1399.

Gontia, A.S., P.K. Nigam, and S.K. Dwivedi (1994). Energy utilization of barley (*Hordeum vulgare* L.) genotypes under various planting patterns. *Bionature* 14: 63-67.

Gorny, A.G. (1996). Evaluation of the response to limited water, nitrogen and phosphorus supply in spring barley genotypes selected for vigorous seminal roots. *Journal of Applied Genetics* 37: 11-27.

Grando, S. and S. Ceccarelli (1995). Seminal root morphology and coleoptile length in wild *(Hordeum vulgare* ssp. *spontaneum)* and cultivated *(Hordeum vulgare* ssp. *vulgare)* barley. *Euphytica* 86: 73-80.

Grundy, A.C., N.D. Boatman, and W.R.J. Froud (1996). Effects of herbicide and nitrogen fertilizer application on grain yield quality of wheat and barley. *Journal of Agricultural Science* 126: 379-385.

Harinasut, P., K. Tsutsui, T. Takabe, M. Nomura, T. Takabe, and S. Kishitani (1996). Exogenous glycinebetaine accumulation and increased salt-tolerance in rice seedlings. *Bioscience, Biotechnology and Biochemistry* 60: 366-368.

Havaux, M. and F. Tardy (1999). Loss of chlorophyll with limited reduction of photosynthesis as an adaptative response of Syrian barley land races to high-light and heat stress. *Australian Journal of Plant Physiology* 26: 569-578.

Havaux, M., F. Tardy, and Y. Lemoine (1998). Photosynthetic light-harvesting function of carotenoids in higher-plant leaves exposed to high light irradiances. *Planta* 205: 242-250.

Hay, R.K.M. (1995). Harvest index: A review of its use in plant breeding and crop physiology. *Annals of Applied Biology* 126: 197-216.

Hay, R.K.M. (1999). Physiological control of growth and yield in wheat: Analysis and synthesis. In *Crop Yield, Physiology and Processes*, eds. D.L. Smith and C. Hamel. Berlin: Springer-Verlag, pp. 1-38.

Hendrix, J.E., J.C. Lindin, D.H. Smith, and C.W. Ross (1986). Influence of assimilate partitioning during inflorescence development on grain numbers in winter wheat. In *Phloem Transport*, eds. J. Cronshaw, W.J. Lucas, and R.T. Gianquinta. New York: AR Liss, pp. 557-559.

Holtorf, H., S. Reinbothe, C. Reinbothe, B. Bereza, and K. Apel (1995). *Proceedings of the National Academy of Sciences, USA* 92: 3254-3258.

Huang, J. and R.E. Redmann (1995). Responses of growth, morphology, and anatomy to salinity and calcium supply in cultivated and wild barley. *Canadian Journal of Botany* 73: 1859-1866.

Huang, J. and R.E. Redmann (1996). Carbon balance of cultivated and wild barley under salt stress and calcium deficiency. *Photosynthetica* 32: 23-35.

Hubick, K.T. and G.D. Farquhar (1989). Carbon isotope discrimination and the ratio of carbon gained to water lost in barley cultivars. *Plant Cell and Environment* 13: 795-804.

Ishitani, M., T. Nakamura, S.Y. Han, and T. Takabe (1995). Expression of the betaine aldehyde dehydrogenase gene in barley in response to osmotic stress and abscisic acid. *Plant Molecular Biology* 27: 307-315.

Isla, R., R. Aragüés, and A. Royo, (1998). Validity of various physiological traits as screening criteria for salt tolerance in barley. *Field Crops Research* 58: 97-107.

Isla, R., A. Royo, and R. Aragüés (1997). Field screening of barley cultivars to soil salinity using a sprinkler and a drip irrigation system. *Plant and Soil* 197: 105-117.

Jackson, P., M. Robertson, M. Cooper, and G. Hammer (1996). The role of physiological understanding in plant breeding; from a breeding perspective. *Field Crops Research* 49: 11-39.

Jamieson, P.D., R.J. Martin, and G.S. Francis (1995). Drought influences on grain yield of barley, wheat, and maize. *New Zealand Journal of Crop Horticultural Science* 23: 55-66.

Jamieson, P.D., R.J. Martin., G.S. Francis, and D.R. Wilson (1995). Drought effects on biomass production and radiation-use efficiency in barley. *Field Crops Research* 43: 77-86.

Kafawin, O.M. (1998). Seed size and water potential effects on germination and early growth of two barley cultivars. *Dirasat Agricultural Sciences* 25: 335-342.

Kasumov, N.A., Z.I. Abbasova, and G. Gunduz (1998). Effects of salt stress of the respiratory components of some plants. *Turkish Journal of Botany* 22: 389-396.

Kernich, G.C., G.M. Halloran, and R.G. Flood (1995). Variation in developmental patterns of wild barley (*Hordeum spontaneum* L.) and cultivated barley (*H. vulgare* L.). *Euphytica* 82: 105-115.

Khan, M.J. and E.P. Glenn (1996). Yield and evapotranspiration of two barley varieties affected by sodium chloride salinity and leaching fraction in lysimeter tanks. *Communications in Soil Science and Plant Analysis* 27: 157-177.

Kishitami, S., K. Watanabe, S. Yasuda, K. Arakawa, and T. Tabake (1994). Accumulation of glycinebetaine during cold acclimation and freezing tolerance in leaves of winter and spring barley plants. *Plant Cell Environment* 17: 89-95.

Kramell, R., R. Atzorn, G. Schneider, O. Miersch, C. Brueckner, J. Schmidt, G. Sembdner, and B. Parthier (1995). Occurrence and identification of jasmonic acid and its acid conjugates induced by osmotic stress in barley leaf tissue. *Journal of Plant Growth Regulation* 14: 29-36.

Krasnovsky, A.A. Jr. (1993). Singlet molecular oxygen and primary mechanisms of photo-oxidative damage of chloroplasts: Studies based on detection of oxygen and pigment phosphorescence. *Proceedings of the Royal Society of Edinburg, Section B, Biological Sciences* 102: 219-235.

Krol, M., M.D. Spangfort, N.P.A. Huner, G. Oquist, P. Gustafsson, and S. Jansson (1995). Chlorophyll a/b-binding proteins, pigment conversions, and early light-induced proteins in a chlorophyll b-less barley mutant. *Plant Physiology* 107: 873-883.

Lambers, H., III F.S Chapin, and T.L. Pons (1998). *Plant Physiological Ecology.* New York: Springer.

Larcher, W. (1995). *Physiological Plant Ecology: Ecophysiology and Stress Physiology of Functional Groups,* Third Edition. Berlin: Springer.

Lehmann, J., R. Atzorn, C. Brueckner, S. Reinbothe, J. Leopold, C. Wasternack, and B. Parthier (1995). Accumulation of jasmonate, abscisic acid, specific transcripts and proteins in osmotically stressed barley leaf segments. *Planta* 197: 156-162.

Leopold, J., B. Hause, J. Lehmann, A. Graner, B. Parthier, and C. Wasternack (1996). Isolation, characterization and expression of a cDNA coding for a jasmonate-inducible protein of 37 kDa in barley leaves. *Plant Cell and Environment* 19: 675-684.

Liang, Y., O. Shen, Z. Shen, and T. Ma (1996). Effects of silicon on salinity tolerance of two barley cultivars. *Journal of Plant Nutrition* 19: 173-183.

Liu, L., D.C.I. Gitzi, and J.W. McClure (1995). Effects of UV-B on flavonoids, ferulic acid, growth and photosynthesis in barley primary leaves. *Physiologia Plantarum* 93: 725-733.

López, M.V. and J.L. Arrue (1997). Growth, yield and water use efficiency of winter barley in response to conservation tillage in a semi-arid region of Spain. *Soil Tillage Research* 44: 35-54.

López-Castañeda, C. and R.A. Richards (1994). Variation in temperate cereals in rainfed environments. III. Water use and water use efficiency. *Field Crops Research* 39: 85-98.

López-Castañeda, C., R.A. Richards, and G.D. Farquhar (1995). Variation in early vigour between wheat and barley. *Crop Science* 35: 472-479.

López-Castañeda, C., R.A. Richards, G.D. Farquhar, and R.E. Williamson (1996). Seed and seedling characteristics contributing to variation in early vigor among temperate cereals. *Crop Science* 36: 1257-1266.

Loss, S.P. and K.H.M. Siddique (1994). Morphological and physiological traits associated with wheat yield increases in Mediterranean environments. *Advances in Agronomy* 52: 229-276.

Lu, Z. and P.M. Neumann (1998). Water-stressed maize, barley and rice seedlings show species diversity in mechanisms of leaf growth inhibition. *Journal of Experimental Botany* 49: 1945-1952.

Macnicol, P.K. and P. Raymon (1998). Role of phosphoenolpyruvate carboxylase in malate production by the developing barley aleurone layer. *Physiologia Plantarum* 103: 132-138.

Makela, P., J. Mantila, R. Hinkkanen, E. Pehu, and S.P. Peltoten (1996). Effect of foliar applications of glycinebetaine on stress tolerance, growth, and yield of spring cereals and summer turnip rape in Finland. *Journal of Agronomy and Crop Science* 176: 223-234.

Manderscheid, R. and H.J. Weigel (1995). Do increasing atmospheric CO_2 concentrations contribute to yield increases of German crops? *Journal of Agronomy and Crop Science* 175, 73-82.

Mano, Y., H. Nakazumi, and K. Takeda (1996). Varietal variation in and effects of some major genes on salt tolerance at the germination stage in barley. *Breeding Science* 46: 227-233.

Mayeux, H.S., H.B. Johnson, H.W. Polley, and S.R. Malone (1997). Yield of wheat across a subambient carbon dioxide gradient. *Global Change Biology* 3: 269-278.

Metodiev, M.V., T.D. Tsonev, and L.P. Popova (1996). Effect of jasmonic acid on the stomatal and nonstomatal limitation of leaf photosynthesis in barley leaves. *Journal of Plant Growth Regulation* 15: 75-80.

Miralles, D.J. and R.A. Richards (2000). Duration of the stem elongation period influences the number of fertile florets in wheat and barley. *Australian Journal of Plant Physiology* 27: 931-940.

Mitchell, J.H., S. Fukai, and M. Cooper (1996). Influence of phenology on grain yield variation among barley cultivars grown under terminal drought. *Australian Journal of Agricultural Research* 47: 757-774.

Mogensen, V.O. (1992). Effect of drought on growth rate of grains of barley. *Cereal Research Communications* 20: 225-231.

Montane, M.H., F. Tardy, K. Kloppstech, and M. Havaux (1998). Differential control of xanthophylls and light-induced stress proteins, as opposed to light-harvesting chlorophyll a/b proteins, during photosynthetic acclimation of barley leaves to light irradiance. *Plant Physiology* 118: 227-235.

Mulholland, B.J., C.R. Black, I.B. Taylor, J.A. Roberts, and J.R. Lenton (1996). Effect of soil compaction on barley (*Hordeum vulgare* L.) growth. I. Possible role for ABA as a root-sourced chemical signal. *Journal of Experimental Botany* 47: 539-549.

Mulholland, B.J., I.B. Taylor, C.R. Black, and J.A. Roberts (1996). Effect of soil compaction on barley (*Hordeum vulgare* L.) growth. II. Are increased xylem sap ABA concentrations involved in maintaining leaf expansion in compacted soils? *Journal of Experimental Botany* 47: 551-556.

Munns, R., D.P. Schachtman, and A.G. Condon (1995). The significance of a two-phase growth response to salinity in wheat and barley. *Australian Journal of Plant Physiology* 22: 561-569.

Muñoz, P., J. Voltas, J.L. Araus, E. Igartua, and I. Romagosa (1998). Changes in adaptation of barley releases over time in northeastern Spain. *Plant Breeding* 117: 531-535.
Nakamura, T., M. Ishitani, P. Harinasut, M. Nomura, T. Takabe, and T. Takabe (1996). Distribution of glycinebetaine in old and young leaf blades of salt-stressed barley plants. *Plant Cell Physiology* 37: 873-877.
Naylor, R.E.L., D.T. Stokes, and S. Matthews (1998). Biomass, shoot uniformity and yield of winter barley. *Journal of Agricultural Science* 131: 13-21.
Nogués, S., L. Alegre, J.L. Araus, L. Pérez-Aranda, and R. Lannoye (1994). Modulated chlorophyll fluorescence and photosynthetic gas exchange as rapid screening methods for drought tolerance in barley genotypes. *Photosynthetica* 30: 465-474.
Nomura, M., Y. Muramoto, S. Yasuda, T. Takabe, and S. Kishitani (1995). The accumulation of glycinebetaine during cold acclimation in early and late cultivars of barley. *Euphytica* 83: 247-250.
Passioura, J.B. (1996). Drought and drought tolerance. *Plant Growth Regulation* 20: 79-83.
Pearce, R.S., M.A. Dunn, J.E. Rixon, P. Harrison, and M.A. Hughes (1996). Expression of cold-inducible genes and frost hardiness in the crown meristem of young barley (*Hordeum vulgare* L. cv. Igri) plants grown in different environments. *Plant Cell and Environment* 19: 275-290.
Peltoten, J., A. Virtanen, and E. Haggren (1995). Using a chlorophyll meter to optimize nitrogen fertilizer application for intensively-managed small-grain cereals. *Journal of Agronomy and Crop Science* 174: 309-318.
Peñuelas, J., R. Isla, I. Filella, and J.L. Araus (1997). Visible and near-infrared reflectance assessment of salinity effects on barley. *Crop Science* 37: 198-202.
Pessarakli, M. (ed.) (1994). *Handbook of Plant and Crop Stress.* New York: Marcel Dekker.
Pessarakli, M. (ed.) (1999). *Handbook of Plant and Crop Stress,* Second Edition, Revised and Expanded. New York: Marcel Dekker.
Pessarakli, M. and H. Fardad (1995). Nitrogen (total and ^{15}N) uptake by barley and wheat under two irrigation regimes. *Journal of Plant Nutrition* 18: 2655-2667.
Poetter, E., J. Beator, and K. Kloppstech (1996). The expression of mRNAs for light-stress proteins in barley: Inverse relationship of mRNA levels of individual genes within the leaf gradiant. *Planta* 199: 314-320.
Popova, L.P. (1998). Fluridone and light-affected chloroplast ultrastructure and ABA accumulation in drought-stressed barley. *Plant Physiology and Biochemisry* 36: 313-319.
Popova, L.P., T.D. Tsonev, G.N. Lazova, and Z.G. Stoinova (1996). Drought- and ABA-induced changes in photosynthesis of barley plants. *Physiologia Plantarum* 96: 623-629.
Pospisil, P. and J.U. Naus (1998). Theoretical simulation of temperature induced increase of quantum yield of minimum chlorophyll fluorescence PHIF(o). *Journal of Theoretical Biology* 193: 125-130.
Pospisil, P., J. Skotnica, and J. Naus (1998). Low and high temperature dependence of minimum Fo and maximum Fm chlorophyll fluorescence in vivo. *Biochimica et Biophysica Acta* 1363: 95-99.

Read, J.J., R.C. Johnson, B.F. Carver, and S.A. Quarrie (1991). Carbon isotope discrimination, gas exchange and yield of spring wheat selected for ABA content. *Crop Science* 31: 139-146.

Reuber, S., J.F. Bornman, and G. Weissenboeck (1996). A flavonoid mutant of barley (*Hordeum vulgare* L.) exhibits increased sensitivity to UV-B radiation in the primary leaf. *Plant Cell and Environment* 19: 593-601.

Reynolds, M.P. and W.H. Pfeiffer (2000). Applying physiological strategies to improve yield potential. In *Durum Wheat Improvement in the Mediterranean Region: New Challenges,* eds. C. Royo, M.M. Nachit, N. Di Fonzo, and J.L. Araus. Options méditerranéennes Serie A 40: 95-103.

Richards, R.A. (1996a). Defining selection criteria to improve yield under drought. *Plant Growth Regulation* 20: 157-166.

Richards, R.A. (1996b). Increasing the yield potential of wheat: Manipulating sources and sinks. In *Increasing Yield Potential in Wheat: Breaking the Barriers,* eds. M.P. Reynolds, S. Rajaram, and A. McNab. Mexico City, Mexico, DF: CIMMYT, pp. 134-149.

Romagosa, I. and J.L. Araus (1991). Genotype-environment interaction for grain yield and ^{13}C discrimination in barley. *Barley Genetics* VI: 563-567.

Royo, A. and R. Aragüés (1995). Effect of salinity on various morpho-physiological characters and grain yield in barley. *Investigación Agraria Producción y Protección Vegetales* 10: 71-84.

Saebo, A. and L.M. Mortensen (1996). Growth, morphology and yield of wheat, barley and oats grown at elevated atmospheric CO_2 concentration in a cool, maritime climate. *Agriculture, Ecosystems and Environment* 57: 9-15.

Sanada, Y., H. Ueda, K. Kuribayashi, T. Andoh, F. Hayashi, N. Tamai, and K. Wada (1995). Novel light-dark change of proline levels in halophyte (*Mesembryanthemum crystallium* L.) and glycophyte (*Hordeum vulgare* L. and *Triticum aestivum* L.) leaves and roots under salt stress. *Plant Cell Physiology* 36: 965-970.

Savin, R. and M.E. Nicolas (1996). Effects of short periods of drought and high temperature on grain growth and starch accumulation of two malting barley cultivars. *Australian Journal of Plant Physiology* 23: 201-210.

Savin, R., P.J. Stone, and M.E. Nicolas (1996). Responses of grain growth and malting quality of barley to short periods of high temperature in field studies using portable chanbers. *Australian Journal of Plant Physiology* 47: 465-477.

Schittenhelm, S., J.A. Okeno, and W. Friedt (1996). Prospects of agronomic improvement in spring barley based on a comparison of old and new germplasm. *Journal of Agronomy and Crop Science* 176: 295-303.

Sharma, P.N. and A. Tripathi (1994). Water relations and photosynthesis in barley plants grown under conditions of deficiency and toxicity of iron. *Indian Journal of Plant Physiology* 37: 85-88.

Sharratt, B.S. (1998). Barley yield and evapotranspiration governed by tillage practices in interior Alaska. *Soil Tillage Research* 46: 225-229.

Simmonds, N.W. (1996). Yields of cereal grain and protein. *Experimental Agriculture* 32: 351-356.

Slafer, G.A. and J.L. Araus (1998). Improving wheat responses to abiotic stresses. In *Proceedings of the 9th International Wheat Genetics Symposium,* Volume 1, ed. A.E. Slinkard. Saskatoon: University of Saskatchewan, pp. 201-213.

Slafer, G.A., J.L. Araus, and R.A. Richards (1999). Promising traits for future breeding to increase wheat yield. In *Wheat: Ecology and Physiology of Yield Determination,* eds. E.H. Satorre and G.A. Slafer. Binghamton, NY: Food Products Press, pp. 379-415.

Slafer, G.A. and G.C. Kernich (1996). Have changes in yield (1900-1992) been accompanied by a decreased yield stability in Australian cereal production? *Australian Journal of Agricultural Research* 47: 323-334.

Slafer, G.A., E.H. Satorre, and F.H. Andrade (1994). Increases in grain yield in bread wheat from breeding and associated physiological changes. In *Genetic Improvements of Field Crops: Current Status and Development,* ed. G.A. Slafer. New York: Marcel Dekker, pp. 1-68.

Smith, D.L., M. Dijak, P. Bulman, B.L. Ma, and C. Hamel (1999). Barley: Physiology of yield. In *Crop Yield, Physiology and Processes,* eds. D.L. Smith and C. Hamel. Berlin: Springer-Verlag, pp. 67-107.

Stefanov, D., I. Yordanov, and T. Tsonev (1996). Effect of thermal stress combined with different irradiance on some photosynthetic characteristics of barley (*Hordeum vulgare* L.) plants. *Photosynthetica* 32: 171-180.

Takeuchi, T.S. and J.P. Thornber (1994). Heat-induced alterations in thylakoid membrane protein composition in barley. *Australian Journal of Plant Physiology* 21: 759-770.

Tambussi, E.A., J. Casadesús, and J.L. Araus (2000). Spectroradiometrical evaluation of photosynthetic efficiency in durum wheat subjected to drought. *Options Méditerranéennes.*

Tardieu, F. and T. Simmonneau (1998). Variability among species of stomatal control under fluctuating soil water status and evaporative demand: Modelling isohydric and anisohydric behaviours. *Journal of Experimental Botany* 49: 419-432.

Tardy, F., A. Créach, and M. Havaux (1998). Photosynthetic pigment concentration, organization and interconversions in a pale green Syrian land ace of barley (*Hordeum vulgare* L. Tadmor) adapted to harsh climatic conditions. *Plant Cell Environment* 21: 479-489.

Tardy, F. and M. Havaux (1997). Thylakoid membrane fluidity and thermostability during the operation of the xanthophyll cycle in higher-plant chloroplasts. *Biochimica et Biophysica Acta* 1330: 179-193.

Teulat, B., P. Monneveux, J. Werey, C. Borries, I. Souyris, A. Charrier, and D. This (1997). Relationships between relative water content and growth parameters under water stress in barley: A QTL study. *New Phytologist* 137: 99-107.

Teulat, B., D. Rekika, M.M. Nachit, and P. Monneveux (1997). Comparative osmotic adjustments in barley and tetraploid wheats. *Plant Breeding* 116: 519-523.

Tjus, S.E., B.L. Moller, and H.V. Scheller (1998). Photosystem I is an early target of photoinhibition in barley illuminated at chilling temperatures. *Plant Physiology* 116: 755-764.

Tsonev, T.D., G.N. Lazova, Z.G. Stoinova, and L.P. Popova (1998). A possible role for jasmonic acid in adaptation of barley seedlings to salinity stress. *Journal of Plant Growth Regulation* 17: 153-159.

Vallelian, B.L., E. Mosinger, J.P. Metraux, and P. Schweizer (1998). Structure, expression and localization of a germin-like protein in barley (*Hordeum vulgare* L.) that is insolubilized in stressed leaves. *Plant Molecular Biology* 37: 297-308.

Volis, S., S. Mendlinger, W.L. Olsvig, U.N. Safriel, and N. Orlovsk (1998). Phenotypic variation and stress resistance in core and peripheral populations of *Hordeum spontaneum*. *Biodiversity and Conservation* 7: 799-813.

Volkmar, K.M. and C. Chang (1995). Influence of hydrophilic gel polymers on water relations and growth and yield of barley and canola. *Canadian Journal of Plant Science* 75: 605-611.

Voltas, J., I. Romagosa, and J.L. Araus (1998). Growth and final weight of central and lateral barley grains under Mediterranean conditions as influenced by sink strength. *Crop Science* 38: 84-89.

Voltas, J., I. Romagosa, A. Lafarga, A.P. Armesto, A. Sombrero, and J.L. Araus (1999). Genotype by environment interaction for grain yield and carbon isotope discrimination of barley in Mediterranean Spain. *Australian Journal of Agricultural Research* 50: 1263-1271.

Voltas, J., I. Romagosa, P. Muñoz, and J.L. Araus (1998). Mineral accumulation, carbon isotope discrimination and indirect selection for grain yield in two-rowed barley grown under semiarid conditions. *European Journal of Agronomy* 9: 147-155.

Wallace, D.H. and W. Yan (1998). *Plant Breeding and Whole-System Crop Physiology*. Wallingford, UK: CAB International.

Watanabe, N. (1998). A method to distinguish leaf colour variation in Syrian barley. *Journal of Genetics and Breeding* 52: 289-293.

Watanabe, N., J. Naruse, R.B. Austin, and C.L. Morgan (1995). Variation in thylakoid proteins and photosynthesis in Syrian land races of barley. *Euphytica* 82: 213-220.

Wraith, J.M., J.M. Baker, and T.K. Blake (1995). Water uptake resumption following soil drought: A comparison among four barley genotypes. *Journal of Experimental Botany* 46: 873-880.

Yasseen, B.T. and B.K.S. Al Maamari (1995). Further evaluation of the resistance of black barley to water stress: Preliminary assessment for selecting drought resistant barley. *Journal of Agronomy and Crop Science* 174: 9-19.

Zelitch, I. (1982). The close relationship between net photosynthesis and crop yield. *BioScience* 32: 796-802.

Zhang, W., F. Diao, B. Yu, and Y. Liu (1998). H^+-ATPase and H^+-transport activities in tonoplast vesicles for barley roots under salt stress and influence of calcium and abscisic acid. *Journal of Plant Nutrition* 21: 447-458.

Zhong, G.Y. and J. Dvorak (1995). Evidence for common genetic mechanisms controlling the tolerance of sudden salt stress in the tribe Triticeae. *Plant Breeding* 114: 297-302.

Chapter 12

Genetic Bases of Barley Physiological Response to Stressful Conditions

Luigi Cattivelli
Paolo Baldi
Cristina Crosatti
Maria Grossi
Giampiero Valè
Antonio Michele Stanca

ADAPTATION TO STRESS ENVIRONMENTS

Genetic adaptation implies the shaping of population gene pools in response to environmental challenges due to climate and soil. Temperatures, rainfall, nutrient availability, and presence of toxic compounds have limited the diffusion of the different species. Barley is grown either in the northern countries close to the polar circle (i.e., Finland and even Iceland) or up to 4,500 m above sea level in the Himalayan mountains, as well as at the limits of the desert where the average rainfall is between 200 and 300 mm per year. Such a great diffusion, despite differences in the climatic conditions, already suggests that the barley gene pool should contain characters for wide environmental adaptability and good stress resistance.

In order to survive, plants have developed the ability to modify their metabolism, adapting themselves to the environment. Under conditions of stress, one of the strategies that plants have adopted is to slow down growth. This allows, for example, the conservation of energy for defense purposes. That plants continuously modify their metabolism to accommodate environmental changes implies the existence of sensors, which monitor changes in the environment, and signaling pathways, which transduce this information to activate cell response. Thus, barley has been able to colonize extremely hostile environments, underlying the sophisticated mechanisms it has evolved to coordinate growth requirements with mechanisms for survival. Genetic variability plays a primary role in determining positive adap-

---

This work was supported by the European Union project GEN-RES PL98-104.

tation to environmental stresses and, hence, in supporting the spread of various barley genotypes to extreme climatic conditions (Cattivelli et al., 1994). In barley, plant growth habit and heading date are the basic traits involved in plant adaptation to environments because they allow synchronizing the plant life cycle with seasonal changes. However, adaptation to environments involves also the ability of plants to cope with stress factors such as frost, drought, and low soil fertility. Compared to other cereals, barley is well adapted to poorly fertile soils or drought environments, thanks to its high nitrogen utilization efficiency Delogu et al., 1998) and water-use efficiency (WUE) (Good and Bell, 1980), although it is less frost resistant than other winter cereals.

Genetics

The ability of plants to withstand physical stresses has been generally considered as a polygenic trait, although specific loci are known to play an important role. Traditional genetics approaches were used to describe characters such as the heading date or growth habit and their involvement in plant adaptation to the environment. More recently, the application of molecular marker technology to the analysis of quantitative traits has led to the identification of a relatively small number of quantitative trait loci (QTL) having a major effect on the ability of the plants to survive under stress conditions.

Loci Controlling Heading Date

The genetics of the factors controlling heading date is complicated by the marked interaction of genotype and environment. Three main factors control heading in cereals: cold requirement (vernalization), photoperiod, and earliness. Vernalization response is strongly affected by the photoperiod, whereas photoperiod response can be masked or canceled by vernalization (Takahashi and Yasuda, 1971). Earliness per se has been extensively studied in spring genotypes. Various dominant and recessive genes play an instrumental role in heading and length of plant cycle, thereby determining varying levels of earliness: *Ea, Ea2, Ea4, Ea5, Ea6,* and *Ea7* (Hockett and Nilan, 1985). Also, a known gene factor—*ea*—controls earliness irrespective of light and temperature conditions (Gustafsson et al., 1982). QTL analysis of heading date was carried out to dissect the interaction between heading date and environment in the cross Dicktoo x Morex. Under long-day conditions, flowering time was determined by two QTL localized on chromosome 2(2H) (in the approximate region where the *Ea* locus was positioned; Hayes et al., 1993) and 7(5H), whereas under short-day conditions heading date was controlled by a different set of QTL identified on barley chromosomes 1(7H), 3(3H), 5(1H), and 7(5H). The interaction among all these loci controls the flowering time under field conditions (Pan et al., 1994). A good level of earliness can prove to be an effective breeding strategy for enhancing the yield stability of barley in drought areas. High grain yields in such regions can be achieved by using early genotypes whose heading date coincides with the end of the rainy season—the escape strategy (Gallagher et al., 1987). It is known, however, that in fertile conditions earliness is not correlated with grain yield.

Loci Controlling Growth Habit

Although influenced by temperature and photoperiod, growth habit is controlled by genetic factors. Winter habit depends on the presence of the dominant allele at locus *Sh* and of the recessive alleles at loci *sh2* and *sh3*. All the other allele combinations among these three genes are found in spring genotypes. The loci *Sh*, *Sh2*, and *Sh3* are respectively located on chromosomes 4(4H), 7(5H), and 5(1H). The homozygote genotype *shsh* is epistatic with respect to the recessive alleles *sh2* and *sh3* and evinces a facultative behavior toward the spring habit when sown in spring. The *Sh3* allele is epistatic with respect to alleles *Sh* and *sh2*. Without vernalization and in long-day conditions, all the *Sh3Sh3* cultivars are essentially spring types. *Sh2*, which is epistatic vis-à-vis with alleles *Sh* and *Sh3*, has a series of multiple alleles that induce several spring-to-winter variants (Cattivelli et al., 1994). Fowler et al. (1996) have reported that the loci controlling vernalization requirement in wheat and in rye are responsible for the duration of the expression of cold-regulated genes, demonstrating a relationship between growth habit, frost resistance, and expression of the genes involved in cold acclimation (see the following).

Loci Controlling Frost Resistance

It is known that winter cultivars are generally hardier than their spring counterparts, although winter barley is less hardy than winter wheat or rye. Genetic analyses have indeed found that a QTL for winter survival on barley chromosome 7(5H) is associated with the *Sh2* locus (Hayes et al., 1993) and with QTL for heading date and vernalization response under long-day conditions (Pan et al., 1994). These results are probably due to genetic linkage rather than to pleiotropic effects; indeed, recombinants between vernalization requirement and winter survival traits have been described (Doll et al., 1989). Restriction fragment length polymorphism (RFLP) analysis of the homeologous 5A chromosome of wheat has proved that vernalization requirement and frost resistance are controlled by two different, but tightly linked loci (*Vrn-A1* and *Fr1*, respectively) (Galiba et al., 1995). In wheat, the availability of chromosome substitution lines allowed the identification of chromosomes carrying loci with a relevant role in frost resistance. Thus, when the 5A chromosome of the frost-sensitive variety 'Chinese Spring' was replaced by the corresponding chromosome of the frost-resistant 'Cheyenne' variety, the frost tolerance of 'Chinese Spring' was greatly increased (Sutka, 1981; Veisz and Sutka, 1989). This phenomenon was also true in the opposite direction; namely, when the 5A chromosome of 'Chinese Spring' was substituted with the corresponding chromosome originated from a highly frost-sensitive *Triticum spelta* accession, the frost resistance of the recipient 'Chinese Spring' decreased (Galiba et al., 1995). Because of its large effect on frost resistance, molecular marker-assisted selection for the *Vrn-A1-Fr1* 5A chromosome interval has been proposed as a tool to improve cold hardiness of cultivars (Storlie et al., 1998). The *Vrn-A1* locus has been found to form a homeologous series with *Vrn-B1* (formerly *Vrn2*) on chromosome 5B and *Vrn-D1* (formerly *Vrn3*) on chromosome 5D (Snape et al., 1997). Comparison of a common set of RFLP markers suggested that the *Vrn-A1* locus is homeologous to *Vrn-H1* (formerly *Sh2*), located on barley chromosome arm 5HL (Laurie et al., 1995), and to *Vrn-R1* (formerly *Sp1*), located on rye chromosome arm 5RL (Plaschke et al., 1993).

Loci Controlling Drought Tolerance

Many morphological and physiological traits are found to be linked to drought resistance, including tillering, root development, plant vigor, leaf water potential, relative water content (RWC), stomata size, membrane stability, osmotic adjustment, desiccation tolerance, leaf rolling, waxiness, leaf temperature, and the accumulation of metabolites or hormones such as proline, betaine, and abscisic acid (ABA). In the C_3 plant, carbon isotope composition ($\delta^{13}C$) is an indicator of the intracellular to atmospheric partial pressure of CO_2 (p_i/p_a) and, therefore, of WUE (Farquhar and Richardson, 1984). Changes in p_i/p_a, and thus in $\delta^{13}C$, can arise from changes in the balance between leaf stomatal conductance and photosynthetic capacity. If intrinsic photosynthetic capacity of leaves is increased, WUE could be improved without limitation of yield potential (Austin et al., 1990; Romagosa and Araus, 1991). $\delta^{13}C$ was employed as a technique to compare the WUE of wheat, barley, and wheat/barley disomic chromosome addition lines. The results identified the genetic factors controlling the barley higher WUE on chromosome 4(4H) (Handley et al., 1994). The genetic bases of the relation between RWC and growth parameters under drought conditions have been dissected by Teulat et al. (1997). A major QTL controlling RWC was positioned together with a QTL for leaf number on chromosome 1(7H), explaining the negative genetic correlation found between RWC and leaf area index under water stress. Low area of mature leaves implies a lower water loss by transpiration. Great attention has also been given to the relation between osmotic adjustment and drought tolerance (Blum, 1989). Osmotic adjustment is defined as a decrease in osmotic potential within the cell due to solute accumulation during a condition of declining leaf water potential. At low soil moisture, osmotic adjustment maintains cell turgor. Several QTL regions related to variation in osmotic adjustment have been mapped on the barley genome. Notably, the osmotic adjustment QTL located on chromosome 1(7H) was found to be colinear with a QTL for the same character mapped on the rice chromosome 8, and with a locus for osmoregulation on the wheat chromosome 7A, suggesting the presence of a conserved region for osmotic adjustment in cereals (Teulat et al., 1998). Increased tissue ABA concentration is thought to play a key role in drought response. ABA mediates many components of the plant stress response, from stomata closure to gene expression. A major locus affecting drought-induced ABA production was mapped on the long arm of the 5A chromosome of wheat in the vicinity of the locus controlling frost resistance, suggesting a genetic linkage between ABA accumulation and stress tolerance (Quarrie et al., 1994).

Loci Controlling Salt Tolerance

A plant growing in soil with a high level of salt must have a transmembrane transport system that maintains a $K^+:Na^+$ ratio above the external ratio. Salt tolerance in cereals is known to be associated with the control of shoot Na^+ content; tolerant lines have more efficient systems to exclude Na^+ from their shoots. Loci involved in salt tolerance have been identified on chromosomes 4(4H) and 7(5H) of *Hordeum vulgare* and on $1H^{ch}$, $4H^{ch}$, and $5H^{ch}$ of *H. chilense* (Forster et al., 1990). The gene pool of *H. spontaneum* may also represent an interesting source of new loci for salt tolerance, and several QTL were detected on chromosomes 1(7H), 4(4H), 5(1H), and 6(6H) (Ellis et al., 1997). A single locus *(Kna1)* on chromosome 4D of bread wheat was shown to control Na^+ against K^+ discrimination, and, indeed, in a saline environment, bread wheat (genomes AABBDD) accumulates less Na^+ and more K^+ than durum wheat (genomes AABB) (Dubcovsky et al., 1996).

Genes

The ability of barley to withstand stress situations is mediated by specific sets of genes that modify the cell metabolism, making cells able to cope with adverse conditions. Fluctuations in water availability, changes of temperature, and variations in solute concentration are common in all climates. All these stresses can damage plants, causing changes in cell volume and membrane shape, disruption of water potential gradients, physical damages to the membranes, and protein degradation. Some of these damages are common to most, if not all, environmental stresses, whereas others are specific: As an example, dehydration is caused by water stress as well as by cold stress and salt stress; most stresses, including biotic stresses, cause the formation of activated oxygen species (AOS), such as superoxide anion

(O_2^-) or hydrogen peroxide (H_2O_2), and a certain rate of protein damage. Specific changes are, for example, modifications in membrane fluidity due to temperature stress.

Although it is becoming increasingly clear that plants can efficiently sense stress and mobilize the appropriate defense responses, it is not so clear how this information is obtained and transduced to the cell defense machinery. The molecular mechanisms leading to the plant response involve three steps (see Figure 12.1): (1) perception of external changes, (2) transduction of the signal to the nucleus, and (3) gene expression. Although in recent years a number of genes whose expression is induced or enhanced by environmental changes have been cloned (Bray, 1993; Waters et al., 1996; Hughes and Dunn, 1996), little is known about the regulatory mechanism

FIGURE 12.1. Schematic Representation of the Mechanisms Controlling Plant Response to Abiotic Stresses

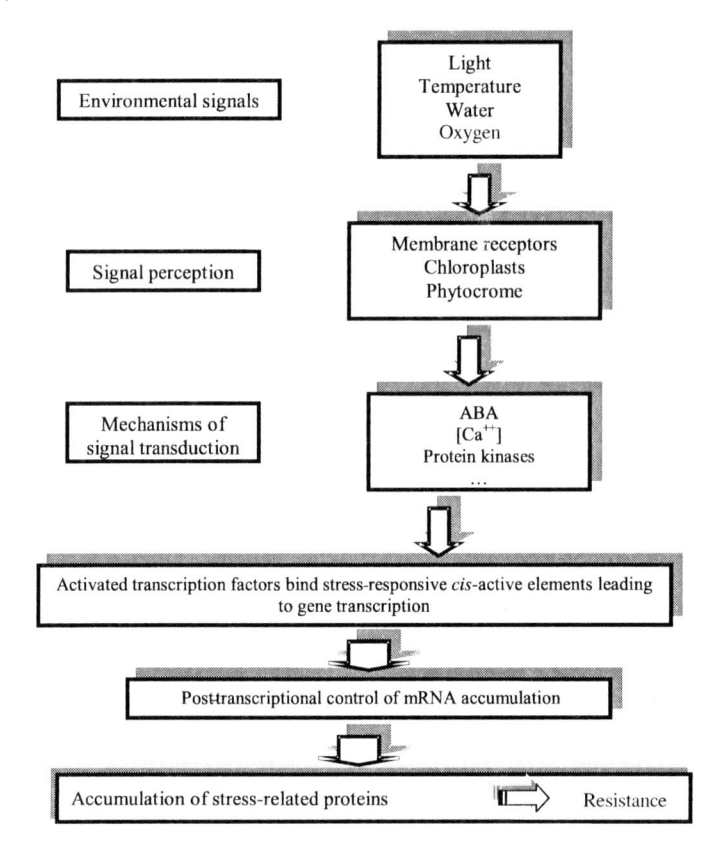

Environmental signals	Light Temperature Water Oxygen
Signal perception	Membrane receptors Chloroplasts Phytocrome
Mechanisms of signal transduction	ABA [Ca^{++}] Protein kinases ...

Activated transcription factors bind stress-responsive *cis*-active elements leading to gene transcription

Posttranscriptional control of mRNA accumulation

Accumulation of stress-related proteins ⇒ Resistance

controlling the stress responses. This is mainly because all stress responses are multigenic traits involving many genes that may have either redundant or additive effects and may interact with one another in different and complex ways. One of the primary targets to better understand plant stress responses is to clarify the functions of the stress-induced genes and how they are regulated by external changes. Most of these studies have been carried out using either barley or *Arabidopsis* as model plants, although it is generally believed that most of the molecular pathways controlling plant responses to environmental factors are well conserved in all plant species.

Signal Perception

The precise nature of the primary sensor of most environmental stimuli is still not known. Nevertheless, all receptors identified so far in plants (i.e., receptors for ethylene, red light, blue light, and calcium) are located in membranes. One of the best-characterized classes of membrane receptors is formed by protein kinases. The binding of the ligand causes a conformational change of the receptor, triggering kinase activity and the subsequent signal transduction through protein kinase cascades. These receptor-like protein kinases (RLPKs) contain characteristic domains: the extracellular leucine-rich repeat (LRR) motif, the single membrane-spanning domain, and the cytoplasmic protein kinase domain (Braun and Walker, 1996). LRR motifs are also present in many genes conferring resistance to pathogens. Recently, an *Arabidopsis* gene induced early after dehydration, low temperatures, and high salt and coding for an RLPK was isolated, suggesting a possible role of this class of genes in the perception of environmental stimuli (Hong et al., 1997).

In recent years, increased evidence has pointed to the role of the chloroplast as primary stress sensor. In fact, plants must constantly balance energy absorbed through the photosynthetic apparatus with energy utilized through metabolism. Almost all environmental stresses have the potential to upset this balance, leading to an overproduction of electrons that, in turn, can damage the reaction center of photosystem II (PSII), and in particular the D1 protein, and may lead to the perturbation and inhibition of photosynthetic electron transport. This phenomenon is known as photoinhibition of photosynthesis (Andersson et al., 1992). The mechanisms leading to an inhibition of electron transport through PSII are not known, but the multiple reduction-oxidation (redox) reactions in PSII have the potential to create AOS, leading to the degradation of the D1 protein (Russell et al., 1995). Some authors have suggested that the redox state of PSII might be the first component in a redox sensing/signaling pathway, acting synergically with other signal transduction pathways to integrate stress response. Gray et al. (1997) found that the expression of a wheat cold-induced gene *(Wcs19)* is correlated with the reduction state of PSII. The redox state of PSII has been proposed to regulate also the transcription and translation of *psbA,* a gene coding for D1. Danon and

Mayfield (1994) proposed a model in which light would activate the translation of *psbA* by changes in the redox potential generated during photosynthesis. In barley, it has been shown that chloroplast has the ability to control the transcription of several genes according to the light conditions. In darkness, an increased level of adenosine diphosphate (ADP) led to the phosphorylation of the transcription factor (called PGTF) controlling the activity of the *psbD* chloroplast gene coding for the D2 protein of the PSII reaction center. Phosphorylation of PGTF inhibits the binding of PGTF to the promoter of *psbD*, leading to reduced gene transcription (Kim et al., 1999). The experimental evidence just described provides a direct link between the quantity of light absorbed by the photosystems and the synthesis of PSII core proteins by coupling photosynthetic reducing power to gene regulation. Such mechanisms would allow the appropriate provision of gene products according to the environmental or developmental demands.

Mutations affecting chloroplast development modify the expression of cold-regulated genes. In barley, the cold-induced expression of the *cor14b* gene (a sequence coding for a chloroplast-localized protein) of the *blt14* gene family, as well as of a gene coding for a dehydrin protein *(paf93)*, was strongly reduced in plants carrying the albino mutation a_n, further supporting the role of the chloroplast in the perception of stress signals. Notably, the barley albino mutant a_n was not able to harden when exposed to low temperature, providing direct evidence of the relationship between expression of cold-regulated genes and the development of cold hardening. Failure of cold acclimation in the mutant cannot be ascribed merely to the absence of photosynthetic activity, since etiolated wild-type plants accumulated cold-regulated (COR) messenger (m) RNAs and improved frost resistance when exposed to cold (Grossi et al., 1998).

It has also been reported that light modulates the stress response; that is, the expression of the cold-regulated barley gene *cor14b* is temperature dependent, although enhanced by red or blue light, suggesting the involvement of phytochrome or blue light photoreceptors in the signal cascade events leading to gene activation (Crosatti et al., 1999).

Taken together, the data presented earlier suggest that plant stress response is the result of a complex interaction among different signaling pathways, each probably involving specific sensors and different cell compartments. All these signals must be transduced to the nucleus in order to trigger gene expression and the appropriate cell response.

Signal Transduction

Various molecules have been proposed to be involved in stress signal transduction pathways. This is probably because, for each stress, different pathways are activated at the same time in the cell. The interaction and reciprocal modulation among these signaling cascades, which are probably connected with one another as a net, determine the specific stress response of the cell.

One of the first molecules involved in stress signaling is the plant hormone ABA. It is well known that environmental stresses induce a considerable increase in endogenous ABA levels in barley (Grossi et al., 1995; Murelli et al., 1995) as well as in all plants. Exogenous application of ABA under nonstress conditions leads to a certain degree of protection against freezing stress (Heino et al., 1990). ABA, therefore, has been proposed as a necessary mediator in triggering most of the physiological and adaptive stress responses. The role of ABA in signal transduction has been analyzed mainly by using ABA-insensitive mutants. Among them, maize *vp1* and different *Arabidopsis abi* genes have been extensively characterized. The *vp1* and *abi3* genes are expressed in seeds and encode homologous transcriptional activators. The ABI1 and ABI2 gene products appear to function mainly in vegetative tissues and are implicated in signaling drought and cold stress (reviewed by Shinozaki and Yamaguchi-Shinozaki, 1996). The *abi1* gene encodes a protein related to type 2C protein serine/ threonine phosphatases with a putative Ca^{2+}-binding domain at the N-terminus (Leung et al., 1994). In barley, two ABA mutants have been described—an ABA-deficient mutant, which produces a low level of ABA in response to water stress (Walker-Simmons et al., 1989), and an ABA-insensitive mutant (Raskin and Ladyman, 1988)—although the precise role of these mutations in the response to environmental stresses is still not clear.

A number of stress-inducible genes not under the control of ABA are also known, indicating that different stress signaling pathways exist. In barley, besides a number of ABA-regulated genes (the most common are the dehydrins, Close et al., 1989), there are also drought-responsive (Grossi et al., 1995) or cold-responsive (Cattivelli and Bartels, 1990) genes whose expression is not affected by ABA. A comparative study between the drought and the cold molecular responses showed that, although ABA is the key hormone in drought response (application of exogenous ABA induces the expression of most, but not all, genes induced by dehydration), its role in cold response is limited (only a few cold-regulated genes are controlled by ABA) (Grossi et al., 1992).

In recent years, Ca^{2+} was found to be involved as a second messenger in the regulation of many plant responses to environmental stimuli (Webb et al., 1996). Cytoplasmic Ca^{2+} concentration often shows significant changes in plant cells under the influence of various stress signals, such as touch (Knight et al., 1991), wind (Knight et al., 1992), cold (Polisensky and Braam, 1996; Knight et al., 1996), oxidative stress (Price et al., 1994), anoxia (Sedbrook et al., 1996), and heat shock (Gong et al., 1998). In alfalfa as well as in *Arabidopsis,* it has been demonstrated that a Ca^{2+} influx acts also as a component of the signal transduction leading to gene activation at low temperature. Addition of calcium chelators prevents cold acclimation and the expression of cold-regulated genes, whereas addition of calcium ionophore induces the expression of genes also at 25°C (Monroy and Dhindsa, 1995; Tahtiharju et al., 1997). Calcium has been proposed to act also in relation to ABA, and some effects of ABA are mediated by an ABA-induced change in $[Ca^{2+}]_{cyt}$. Evidence suggests that ABA induces transient variations in $[Ca^{2+}]_{cyt}$, involving the activation of Ca^{2+} channels to permit Ca^{2+} influx across the plasma membrane and Ca^{2+} release from intracellular stores (Ward et al., 1995; Gilroy and Jones, 1992). In barley aleurone cells, the expression of some genes responsive to ABA (RAB) has been shown to be dependent on the presence of Ca^{2+}. At a constant ABA level with increasing extracellular Ca^{2+}, an increasing RAB mRNA expression was noticed in barley aleurone protoplasts. On the contrary, RAB gene expression was inhibited by blockers of the Ca^{2+} channels as well as by Ca^{2+} antagonists (Van der Mulen et al., 1996). This evidence suggests a role for Ca^{2+} in the ABA signal transduction pathways.

Protein phosphorylation and dephosphorylation are key steps in the signaling processes of all plants. Numerous genes encoding protein kinases and phosphatases have been cloned. Several calcium-dependent protein kinases (CDPKs) have been identified in *Arabidopsis* (Hrabak et al., 1996). These proteins have a serine/threonine protein kinase catalytic domain, an autoinhibitory domain, and a Ca^{2+}-binding domain. Thus, CDPKs in plants seem to be capable of detecting changes in the cytoplasmic concentration of free calcium and of inducing the cellular response. Sheen (1996) was able to connect specific CDPKs to stress signal transduction pathways and also to link CDPK function with gene expression. The promoter of the stress-responsive barley gene *HVA1* was fused to green fluorescent protein to monitor stress signaling in maize protoplasts. Expression of this reporter system was induced, similarly to the *HVA1* gene in barley, by ABA and stress conditions such as drought, cold, and salinity. In the absence of these stimuli, expression could be induced also by increasing the cellular Ca^{2+} level. Coexpression of the catalytic domains of two CDPKs, ATCDPK1 and ATCDPK1a, activated the *HVA1* promoter, bypassing the stress signals. Furthermore, the activation was diminished by protein phosphatase 2C, which is capable of blocking the ABA response (Sheen, 1996). Notably, mRNAs encoding these ATCDPKs were present at very low levels in control plants but were enhanced rapidly in response to stress conditions (Urao et al., 1994).

Gene Activation

Most of the described transduction pathways lead to gene activation. One of the approaches to understanding how stress-activated genes are regulated is to analyze their promoters in order to characterize the sequences (*cis*-acting elements) requested for gene activation and to identify transcription factors that bind to these sequences. One of the best-characterized *cis*-acting elements is the ABA-responsive element (ABRE), which contains the palindromic motif CACGTG with the G-box ACGT core element. The ACGT-like elements have been observed in a number of plant genes regulated by various environmental and physiological changes (Giuliano et al., 1988). In many cases, ABRE alone is not sufficient for gene activation, but an additional element (coupling element) must be present. Different types of coupling elements have been identified. An ABRE in the barley *Amy32b* α-amylase promoter has been shown to allow ABA induction only in the presence of a coupling element, known as the O2S element, that interacts with the ABRE within tight positional constraints. A second coupling element has been identified in the promoter of the barley *HVA22* gene (Shen and Ho, 1995). This element acts together with a G-box type ABRE in conferring high ABA induction. Another coupling element has been identified in a second ABA-inducible barley gene, *HVA1* (Shen et al., 1996). These two barley coupling elements are not completely interchangeable, indicating specific control of gene expression.

Other *cis*-acting elements involved in the activation of stress-induced/ABA-independent genes have also been identified. The TACCGACAT DNA element works as a drought-responsive element (DRE) in the *Arabidopsis* genes *lti78/rd29a* and *lti76/rd29b* (Yamaguchi-Shinozaki and Shinozaki, 1994). The *rd29a* DRE sequence is activated by an ABA-independent pathway working either in cold or drought stress. The promoter of *rd29a* contains both the DRE and the ABRE, suggesting that the two signal transduction pathways interact to control gene expression during drought stress. Indeed, the activation of *rd29a* is ABA independent in the first few hours after dehydration but becomes ABA dependent in the later stages of expression. In barley, the low-temperature expression of the *blt4.9* gene coding for a nonspecific lipid transfer protein was shown to be mediated by the novel *cis*-acting element CCGAAA (Dunn et al., 1998).

Posttranscriptional control of stress-responsive genes is also known. In barley, particularly, the low-temperature-responsive genes are regulated either at the level of gene activation or at the level of mRNA stability (Dunn et al., 1994), and low-temperature-dependent protein factors are involved to modulate mRNA stability (Phillips et al., 1997). The activity of the low-temperature-dependent protein factors can also be modified by the presence of green chloroplasts. Indeed, etiolated barley plants accumulate the mRNAs corresponding to the cold-regulated gene *blt14* at detectable levels already at 22°C. When the same plants are exposed to cold in the absence of light, an increased mRNA accumulation above the level present in green, cold-treated plants can be detected (Grossi et al., 1998). Several low-temperature-induced RNA-binding proteins, which might stabilize or activate mRNA, have been isolated in *Arabidopsis* and barley (Carpenter et al., 1994; Dunn et al., 1996).

Different alleles at the barley regulatory locus controlling the expression of cold-regulated genes could also be postulated, since many barley cold-responsive genes have a different threshold induction temperature, depending on whether they originate from resistant (winter) or susceptible (spring) genotypes. Crosatti et al. (1996) have shown that winter genotypes have a higher threshold induction temperature for a specific COR protein than do spring ones. Similar results were obtained for many other cold-regulated genes (Grossi et al., 1998).

Stress-Related Genes Isolated in Barley

It has been well demonstrated that modification in gene expression is a common feature in all plants subjected to environmental stresses. Some components involved in signal perception and transduction are expected to be similar among different plant species and stresses (described earlier). Notably, a wide range of species express a common set of genes and proteins in response to the same stimulus. Although the function of many of these genes is still unknown, we can assume that, on the basis of correlative evidence, these proteins play an active role in the plant response to stress. In barley, a number of stress-related genes have been cloned. According to their sequence similarity, expression behavior, and/or putative function, these genes can be divided into several groups, as listed in Table 12.1.

Drought-Responsive Genes: The Lea Genes

Many of the genes induced by dehydration, ABA treatment, salt, and osmotic conditions belong to the late embryogenesis abundant (LEA) gene family. In addition to stress conditions, LEA proteins are also accumulated in the embryo during desiccation. The function of most of them is still unknown, even if their extremely hydrophilic characteristics suggest an osmoprotective role. They have been divided into three classes based on their conserved amino acid motifs.

Genes belonging to *Lea* class 1, first isolated from cotton seeds *(Lea D19)*, encode for highly hydrophilic proteins with a very high content of glycine and charged residues. They contain one to four copies of a conserved 20 amino acid repeat, although in the barley B19-related LEA proteins the 20 amino acid sequence was found only as a monomer. In addition, they also share N-terminal (GETWPGGTGGK) and C-terminal (EGIDIDESKF) consensus amino acid motifs. Among the members of the barley embryo-specific *Lea B19* gene family, a high degree of conservation in the untranslated regions was also found between *B19.4* and *B19.3* and between *B19.1* and *Em* complementary (c) DNA. The *B19* genes have similar regulation during embryo development, although at a different level, but each member of the gene fam-

TABLE 12.1. Stress-Related Genes Isolated in Barley

Gene	Stress-Related Expression	Observation	Reference
Drought-Responsive Genes			
B19 gene family	In immature embryos after ABA, salt, and mannitol treatment	Lea group 1 (Em-like); expressed during development of embryo desiccation tolerance	Espelund et al., 1992; Hollung et al., 1994
Dhn1	Drought, ABA	Lea group 2; mapped on chr. 5H	Close et al., 1989
Dhn2	Drought, ABA	Lea group 2; mapped on chr. 5H	Close et al., 1989
Dhn3	Drought, ABA	Lea group 2; mapped on chr. 6H	Close et al., 1989
Dhn4	Drought, ABA	Lea group 2; mapped on chr. 6H	Close et al., 1989
Dhn5	Drought, ABA, cold	Lea group 2; mapped on chr. 6H	Van Zee et al., 1995
Dhn6	Drought, ABA	Lea group 2; mapped on chr. 4H	Close, 1996
Paf93	Drought, cold	Lea group 2; mapped on chr. 6H	Cattivelli and Bartels, 1990; Grossi et al., 1995
ABA2	In shoots during drought, ABA, cold	Lea group 2; mapped on chr. 5H	Gulli et al., 1995
ABA3	In shoots during drought, ABA, cold	Lea group 2; mapped on chr. 6H	Gulli et al., 1995
HVA1	In seedlings during drought, salt, ABA, and cold treatment (depending on developmental stage)	Lea group 3; expressed in the aleurone during development of embryo desiccation tolerance	Hong et al., 1988; Sutton et al., 1992; Hong et al., 1992
PG22-69	Modulated by ABA	Homologous to aldose reductase; expressed during development of embryo desiccation tolerance	Bartels et al., 1991
BADH	Induced during drought; modulated by ABA in roots	Betaine aldehyde dehydrogenase	Ishitani et al., 1995
ABA8	Induced during drought and ABA treatment	Homologous to aldose reductase	Gulli et al., 1995

TABLE 12.1 *(continued)*

Gene	Stress-Related Expression	Observation	Reference
Heat Shock Genes			
Hvhsp17	Heat stress	In transgenic tobacco, expressed in the vascular bundles of the stem and petioles after heat stress, heavy metal, and ABA treatment	Marmiroli et al., 1993; Raho et al., 1996
Hsp18	Heat stress		Marmiroli et al., 1993
Hsp26	Heat stress	Encodes for a chloroplast-localized protein	Kruse et al., 1993
Hsp70	Heat stress	Expressed also during leaf development	Kruse et al., 1993
Hsp90	Heat stress	Induced also during fungal infection	Wolthen-Larsen et al., 1993
Low-Temperature-Responsive Genes			
blt4 gene family	Expressed in leaves during cold, drought, and ABA treatment; *blt4.9* strongly expressed in epidermal cells of cold-acclimated emerging leaves	Codes for nonspecific lipid transfer protein; possibly involved in wax synthesis or secretion; mapped on chr. 2H	Hughes et al., 1992; Dunn et al., 1991; White et al., 1994
blt14 gene family	Expressed during cold acclimation		Dunn et al., 1990; Cattivelli and Bartels, 1990; Grossi et al., 1998
Blt63	Cold	Codes for elongation factor 1α	Dunn et al., 1993
blt101 gene family	Expressed during cold acclimation in the perivascular layers of the transition zone of the crown	The protein has a leader sequence for the secretory pathway	Goddard et al., 1993; Pearce et al., 1998
Blt801	Induced by cold and ABA treatment	Codes for an RNA-binding protein	Dunn et al., 1996
Cor14b	Induced by cold and enhanced by light	Expressed only in leaves; encodes for a chloroplast-localized protein; mapped on chr. 2H	Cattivelli and Bartels, 1990; Crosatti et al., 1995, 1999

Gene	Stress-Related Expression	Observation	Reference
Tmc-ap3	Constitutively expressed; enhanced by cold	Expressed only in leaves; encodes for a chloroplast-localized protein; mapped on chr. 1H	Baldi et al., 1999
Other Stress-Regulated Genes			
Ldh	Expressed during anaerobic stress	Lactate dehydrogenase	Hondred and Hanson, 1990
Adh	Expressed during anaerobic stress	Alcohol dehydrogenase	Good et al., 1988
AlaAT-2	Expressed in roots during anaerobic stress	Alanine aminotransferase	Muench and Good, 1994
FDH	Induced in roots during Fe deficiency and anaerobic stress	Formate dehydrogenase	Suzuki et al., 1998
Hb	Induced in roots during anaerobic stress	Homologous to hemoglobin; expressed in the aleurone layer	Taylor et al., 1994
Germin-like	Enhanced during salt stress		Hurkman and Tanaka, 1996
Early light induced	Enhanced during photoxidative conditions	Encodes for a chloroplast-localized protein	Montané et al., 1997
Ndh-A	Enhanced during photoxidative conditions	Chloroplast gene	Martin et al., 1996
Ndh-F	Enhanced during photoxidative conditions	Plastid genes coding for a 70 kDa subunit of the NAD(P)H dehydrogenase	Catala et al., 1997
HvHAK1	Constitutively expressed; enhanced by K+ starvation	Expressed only in roots	Santa-Marìa et al., 1997
Ltp7a2b	Enhanced in presence of cadmium or ABA	Nonspecific lipid transfer protein	Hollenbach et al., 1997

ily is differentially regulated by ABA or osmotic stress (Espelund et al., 1992).

The dehydrins represent class 2 of the *Lea* gene family. Dehydrins are the main group of proteins induced by drought, salt stress, cold acclimation, embryo development, and ABA in barley, as well as in many other species. Dehydrins have a highly conserved lysine (Lys)-rich stretch of 15 amino acids (EKKGIMDKIKEKLPG) known as the "K segment." In addition, many dehydrins contain also a stretch of serine (Ser), the "S segment," which can be phosphorylated, and a further consensus amino acid sequence (-V/T-DEYGNP), the "Y segment." There are five types of dehydrins in respect to the YSK consensus sequences. The Y_nSK_2 dehydrins have one to three Y segments, one S segment, and two K segments, and they are strongly induced by dehydration. The K_n dehydrins have only the K segment and are induced during cold acclimation. The K_nS dehydrins are induced by cold and drought and contain several K segments that begin with the consensus E(H/Q)KEG rather than EKKG and a S segment near the C-terminus. The SK_n and Y_2K_n dehydrins have been found localized in the nucleus and cytoplasm in embryonic tissues and only in the cytoplasm near the shoot and root apexes, respectively (Close, 1996).

Studies have been conducted to find the correlation between dehydrins and freezing tolerance. A linkage map based on doubled haploid genotypes was developed from the cross between the winter cultivar Dicktoo and the spring cultivar Morex and used to map both dehydrin genes and QTL for frost tolerance. Genes *dnh1* and *dnh2* were located on barley chromosome 7(5H); *dhn6* on chromosome 4(4H); and *dhn3, dhn4,* and *dhn5* on chromosome 6(6H). The loci for *dnh1* and *dhn2* are in the confidence interval of the major QTL controlling freezing tolerance, although these genes are not induced by low temperature, while the major cold-induced dehydrin, *dhn5,* maps on a different chromosome. These observations suggest that neither *dhn1* nor *dhn2* is the major determinant of the frost tolerance QTL on chromosome 7 (Van Zee et al., 1995).

An interesting observation concerning the role of the dehydrins in stress tolerance was made with the wheat gene *wcor410* encoding for an acidic dehydrin. This protein accumulates in the vicinity of the plasma membrane of the cell in the vascular transition area, and it was found associated with the development of freezing tolerance in wheat and other members of the Gramineae tribe (Danyluk et al., 1998). The authors suggest that this protein has a role in preventing the destabilization of the plasma membrane during freezing or dehydrative conditions. The three members of the *wcor410* gene family have been mapped on the long arms of the homeologous group 6 chromosomes of hexaploid wheat, while the regulation of the *Wcor* gene family appears to lie on chromosome 5A.

Class 3 members of the *Lea* genes contain sequences homologous to the cotton *D11* gene isolated from dormant seeds. All the proteins of this family contain a tandem repeat motif of 11 amino acids that may form an amphiphilic α-helix structure (Dure et al., 1989; Dure, 1993). In barley, a single gene *(HVA1)* belonging to this class has been isolated. The *HVA1* gene is expressed in aleurone layers during embryo desiccation, although its expression is rapidly induced in young seedlings by ABA, dehydration, salt, and cold (Hong et al., 1992).

Among the genes up-regulated in embryos during the development of desiccation tolerance or in vegetative tissues during the stress-induced response, several genes encoding for products with enzymatic activity have also been cloned. The aldose reductase homologous gene from barley is expressed in the embryo, developmentally regulated, and ABA modulated (Bartels et al., 1991). Aldose reductase is an enzyme of the polyol pathway involved in the accumulation of sorbitol. It has been proposed that sorbitol has a role in osmoregulation.

In response to drought or salt stress, some plants, such as barley, increase the accumulation of glycinebetaine, which is the last step of betaine synthesis. The reaction is catalyzed by betaine aldehyde dehydrogenase (BADH). The mRNA levels of *BADH* increased eight- and twofold in leaves and roots, respectively, when barley plants were subjected to high-salt and drought conditions. Treatment with ABA was less effective in the accumulation of the *BADH* mRNA than was osmotic stress, suggesting that the expression of *BADH* mRNA may require only indirect action of ABA (Ishitani et al., 1995).

Heat Shock (HS) Genes

The HS proteins (HSPs) are highly conserved among diverse organisms, from bacteria to eukaryotes to plants, and they have been traditionally divided into two classes: high molecular weight HSPs (60-110 kDa), including HSP110, HSP90, HSP70, and HSP60, and low molecular weight (LMW) HSPs (15-30 kDa). The LMW HSPs are grouped in multigene families with conserved sequences and common cellular localization: classes I and II in the cytoplasm, class III in the chloroplast, and class IV in the endomembrane system. Some HSPs are constitutively expressed or developmentally regulated, indicating a role in the normal cell life cycle besides conferring protection from heat stress. For instance, it has been found that an HSP70 cognate protein is involved in the correct packaging of the zein storage proteins (Marocco et al., 1991). The expression of HS proteins was investigated during development of barley. The 32 and 30 kDa proteins have been characterized as precursors of the plastidic proteins of 26 kDa. The mRNAs of the 26 kDa and of the 70 kDa proteins were found to increase after heat stress, although the same proteins were also present in two-day-old control plants. At later stages only HSP70 was observed. Different levels of expression were found in different segments of the leaf (Kruse et al., 1993). HSPs may be expressed also in stress conditions different from heat shock. Young spinach seedlings exposed to 5°C for two days showed increased synthesis of 79 kDa HSPs, already present at 20°C (Neven et al., 1992). This group of proteins belongs to the 70 kDa HSP family, suggesting that cold treatment could involve some HSPs for protein folding or assembly processes. Low temperature, but not heat shock or water stress, enhanced the expression of a gene coding for a 70 kDa HS cognate (HSC70) of spinach (Anderson et al., 1994).

In barley, cDNA and genomic clones coding for the 18 and 17 kDa HSPs have been identified and characterized. In ten-day-old plants, the expression of the *Hvhsp18* gene was triggered after 10 to 15 min of heat treatment at 40°C (Marmiroli et al., 1993). The analysis of the corresponding genomic clone revealed two heat shock elements (HSEs). The barley *Hvhsp17* gene promoter was used to transform tobacco plants. The chimeric pHS/GUS fusion was found to be induced by heat treatment, ABA, and some metal ions. Furthermore, the expression was strictly confined to the vascular bundles of the xylematic component in the stem and petioles (Raho et al., 1996). Another barley cDNA clone was identified as a member of the 90 kDa HSP family (HSP90). The transcript was identified during infection attempts by the powdery mildew fungus, but it was found to accumulate rapidly in leaves after heat shock treatment (Wolthen-Larsen et al., 1993).

Low-Temperature-Responsive Genes

The development of the acclimation process to low temperatures, leading to frost-tolerant tissue, requires the expression of a number of COR genes (Cattivelli and Bartels, 1989). In barley, more than 20 cDNA clones whose expression is affected by low temperatures have been isolated (Cattivelli and Bartels, 1990; Dunn et al., 1990, 1991; Goddard et al., 1993; Hughes et al., 1992, Baldi et al., 1999). The accumulation of COR mRNAs depends primarily on the low-temperature stimulus, but it can also be modulated by other factors, such as application of ABA, drought stress, or light. The analysis of the expression pattern reveals that COR mRNAs reach their steady state within two to three days after exposure to cold, whereas when the plants are moved from low temperatures to 20°C the level of COR mRNAs drops in a few hours (Cattivelli and Bartels, 1990). Their expression and the accumulation of the corresponding proteins have been correlated with frost resistance (Crosatti et al., 1996; Pearce et al., 1996).

The sequence and expression analysis of the COR genes allows the identification of several gene classes. The members of the *blt14* gene family are posttranscriptionally up-regulated (Dunn et al., 1994) in response to cold, but not to drought or ABA. Up to now, five *blt14*-related genes have been isolated; they are expressed in different plant tissues and show different threshold induction temperatures and genotype-dependent induction kinetics (Grossi et al., 1998). In situ hybridization analysis of COR gene expression showed that *blt14* genes (as well as another COR gene, *blt101*) are strongly expressed in perivascular cell layers in the vascular transition zone of cold-acclimated barley crowns, although they are also present in other organs and tissues (Pearce et al., 1998).

A putative function can be assigned to several barley COR genes on the basis of their sequence similarities. The *blt4* COR gene family encodes for nonspecific lipid transfer proteins (LTPs). Although nonspecific LTPs do not bind specifically to lipids, they are known to transfer lipids between donor and acceptor membranes in vitro. All members of the *blt4* gene family have an extracellular transport consensus signal peptide in the N-terminus, suggesting a possible involvement in wax synthesis or secretion (White et al., 1994). The mRNAs corresponding to *blt4.9* were strongly expressed in epidermal cells of barley cold-acclimated emerging leaves (Pearce et al., 1998); therefore, a possible role in reducing stress-induced water loss can be assumed. The COR gene *blt63* is a member of a multigene family encoding for the protein synthesis elongation factor 1α (Dunn et al., 1993), while the *blt801* clone encodes for an RNA-binding protein. The sequence of BLT101 shows an N-terminal amino acid stretch with a consensus RNA-binding domain and a C-terminal domain with repeated glycine residues interspersed with tyrosine and arginine (Dunn et al., 1996).

Specific COR proteins have been found to be localized in the chloroplasts. The first COR gene isolated with such characteristics was *cor14b* (formerly *pt59;* Crosatti et al., 1995). The accumulation of COR14 occurs only at low temperatures, but it is enhanced after even brief exposure of the plants to light. COR14 is stored in amounts only slightly greater in the resistant cultivars than in the susceptible ones, although the former have a higher induction temperature threshold of COR14 than the latter (Crosatti et al., 1996). This fact may represent an evolutionary advantage, enabling the resistant varieties in the field to prepare for the cold well ahead of the susceptible ones. The COR gene *tmc-ap3* encodes for a putative chloroplastic amino acid selective channel protein. The gene *tmc-ap3* is expressed at a low level under normal growing temperatures, although its expression is strongly enhanced after cold treatment. A positive correlation between the expression of *tmc-ap3* and frost tolerance was found either among barley cultivars or among cereal species. Western analysis showed that the *cor tmc-ap3* gene product is localized in the chloroplastic outer envelope membrane, supporting its putative function. These results suggest that an increased amount of a chloroplastic amino acid selective channel protein could be required for cold acclimation in cereals (Baldi et al., 1999).

Other Stress-Related Genes

In plants subjected to anaerobic stress, there is a shift in carbohydrate metabolism from the oxidative to the fermentative pathway, with the induction of the enzymes, mainly alcohol dehydrogenase (ADH), associated with the flow of carbon into glycolysis and alcoholic fermentation (Dennis et al.,

1992). In barley, three ADH loci are known: *Adh1* and *Adh2* are closely linked on chromosome 4(4H), whereas the third locus, *Adh3*, is located on chromosome 6(6H); the corresponding DNA sequences have been isolated (Good et al., 1988; Trick et al., 1988). *Adh1* is constitutively expressed in developing seeds, while all isoenzymes are strongly induced in root cells under anoxia conditions (Harberd and Edwards, 1983; Mayne and Lea, 1984). In hypoxically treated barley roots, two cDNA were isolated that are homologous to vertebrate and bacterial lactate dehydrogenase (LDH) (Hondred and Hanson, 1990). A cDNA clone coding for the alanine aminotransferase (AlaAT) was also isolated in barley roots under hypoxic conditions (Muench and Good, 1994). AlaAT is the enzyme that catalyzes the production of alanine, and alanine concentration has been shown to increase under hypoxia in all plants. A barley hemoglobin gene has been found in isolated aleurone and root tissues under anaerobic conditions. The deduced amino acid sequence shows 71 percent identity with a nonlegume hemoglobin gene (Taylor et al., 1994). There is consensus for six specific residues involved in rapid oxygen rebinding between barley and *Parasponia andersonii*. Two hypotheses have been made regarding the role of the hemoglobin gene during anoxia: it could facilitate diffusion of oxygen in root tissues or may be an oxygen sensor causing the switch from aerobic to anaerobic metabolism.

Formate dehydrogenase, an enzyme of anaerobic metabolism, has been found in iron-deficient barley roots. Sequence analysis revealed 21 amino acids at the N-terminus, similar to a transit peptide targeted to mitochondria; an NAD^+-binding site; and the formate-binding site (Suzuki et al., 1998). The expression of this gene suggests that iron deficiency may cause modifications similar to those induced by anoxia, probably through changes in heme biosynthesis.

Plants exposed to high salt concentration may accumulate germin-like proteins. Germins are proteins specifically induced during the germination of the embryo. Germin-like proteins, spherulins, were also found in *Physarum plasmodium* during starvation, osmotic stress, and changes of temperature. When barley seedlings were grown in the presence of 200 mM NaCl, an increased synthesis of two 26 kDa polypeptides (pI 6.3 and 6.5) was detected. The sequence of their N-terminals revealed homology to the germin proteins (Hurkman et al., 1991). The expression of the barley gene coding for a germin-like protein was found to be developmentally regulated and modulated by NaCl in young plants (Hurkman and Tanaka, 1996).

Plants exposed to low temperature and high light show a reduction in photosynthetic capacity, an effect known as photoinhibition. Expression of the early light-induced protein (ELIP) genes is correlated with the photoinactivation of PSII, the degradation of the D1 protein, and changes in pigment content. ELIPs are small proteins (13.5-17 kDa), localized in the stroma of chloroplast in the vicinity of the D1 protein, and they have been found to be accumulated during greening of etiolated seedlings in several species (Grimm et al., 1989). These observations suggest that they could play a role during the assembly or repair of PS II. The highest protein levels were found in plants exposed to low temperature and high light intensities (500 to 1500 $\mu mol \cdot m^{-2} \cdot s^{-1}$) in a long-term experiment, whereas there was only a transient increase of their expression under short-term stress (Montané et al., 1997). Isolated ELIPs in native form revealed that these proteins bind chlorophyll and lutein. These proteins represent unique chlorophyll-binding proteins with a transient function during light stress; it has been proposed that ELIPs protect the developing photosynthetic apparatus from overexcitation (Adamska, 1997; Krol et al., 1999).

Plastid genes can also be involved in the adaptation to photooxidative conditions. The *nhd-A* plastid gene encodes for a polypeptide localized in the tylakoid membrane when mature senescent leaves are incubated under photooxidative conditions. The authors suggest that this protein may be involved in the protection of chloroplast during photooxidative stress (Martin et al., 1996). The *ndh-F* gene encodes for the 70 kDa subunit of the plastid NAD(P)H-dehydrogenase complex. This protein was found in tylakoids of green tissues, but not in etiolated leaves. The levels of the protein decreased with aging but increased during senescence. The accumulation in young tissues is stimulated by photooxidative stress, suggesting a role in protection of electron transport. Nevertheless, the presence of this protein in nonphotosynthetic tissues led also to the hypothesis that the NAD(P)H-dehydrogenase complex may have a role in chlororespiration (Catala et al., 1997).

Specific genes are also involved in the interaction between plant roots and soil compounds. K^+ is an essential nutrient because it is involved in enzyme activity, osmoregulation, and normal growth and development. Several K^+ transporters have been cloned in different organisms: *Escherichia coli,* yeast, fungi, and plant species. In barley, the *HvHAK1* multigene family encodes for proteins homologous to *SoHAK1* and *Kup* (the K^+ transporter genes of *Schwanniomyces occidentalis* and *E. coli,* respectively). The *HvHAK1* cDNA conferred high-affinity uptake to a K^+ uptake-deficient yeast (Santa-Marìa et al., 1997). Accumulation of the *HvHAK1* mRNA has been detected only in roots, with a higher level of expression in K^+-starved plants. *HvHAK1* belongs to a gene family, and the mapping in different genomes of Triticeae species revealed hybridization signals on chromosome groups 2, 3, 4, and 6 (Santa-Marìa et al., 1997).

The barley gene *Ltp7a2b,* expressed in leaf epidermal cells of barley, encodes for a nonspecific LTP of 12.3 kDa, with a putative signal peptide characteristic for targeting to the secretorty pathway (Hollenbach et al., 1997). The level of the *Ltp7a2b* mRNA decreased during aging of leaves, although its expression was stimulated when seedlings were grown in the presence of cadmium. The expression of *Ltp7a2b* was higher in leaves containing elevated levels of ABA or with thicker waxy layers of cuticle. These observations suggest that this protein may play a role in the transfer of cutin from the site of synthesis to the cuticle (Hollenbach et al., 1997).

IMPROVING ABIOTIC STRESS TOLERANCE IN BARLEY

Recent developments in plant molecular biology have allowed the identification of many genes involved in plant adaptation to hostile environments, although many unsolved questions still remain. Genetic studies as well as molecular analysis point out the complexity of plant stress tolerance, a typical polygenic trait. Improvement of stress resistance will be a crucial factor in the near future to increase yield stability and to expand the growing area of barley as well as of other crops. This objective can be achieved through either exploiting the genetic variability naturally present in the *H. vulgare* and *H. spontaneum* gene pools or introducing new or modified genes in the barley genome via plant transformation. The first approach will be facilitated by the utilization of molecular markers to assist selection, although the feasibility of such an approach will largely depend on the number of genomic regions involved in the control of a given stress resistance. Based on actual knowledge, improvement of frost resistance could be easily performed with molecular marker selection targeting the only large QTL region detected on chromosome 7(5H) (Hayes et al., 1993; Storlie et al., 1998). On the contrary, selection for drought tolerance will be very difficult due to the great number of traits and corresponding QTL involved (Teulat et al., 1997, 1998). Exploitation of natural genetic variability, although essential for future plant breeding, may not be sufficient to improve significantly the stress tolerance of barley. Cultivated barley has a wide environmental adaptation, although other species are more frost resistant (i.e., rye) or more drought or salt tolerant (i.e., *H. spontaneum* or wild species of the *Hordeum* genus) than barley. In this scenario, plant biotechnology is expected to play an important role in the future due to its ability to transfer important genes into virtually all crops. Transformation is now also available in barley, and important traits have been already modified via genetic engineering (i.e., resistance to pathogens). Molecular improvement of stress resistance can be achieved by modifying or enhancing already existing stress response mechanisms. Small molecules, such as amino acids or soluble sugars, are normally accumulated in barley under stress (Murelli et al., 1995; Pesci, 1988). Much experimental evidence suggests a correlation between proline and drought resistance (Hanson et al., 1979) or between betaine and freezing tolerance (Allard et al., 1998), although selection for high proline or betaine accumulation under stress conditions has rarely led to improvement in stress tolerance. Molecular biology has explored the role of osmolytes by engineering the metabolic pathways controlling the synthesis of soluble sugars (i.e., plants overproducing trehalose showed an improved resistance to drought; Holmstrom et al., 1996), proline (Kishor et al., 1995), or glycinebetaine. Rice plants with the ability to synthesize glycinebetaine were established by intoducing the *codA* gene for choline oxidase. The transgenic plants showed an increased tolerance to salt and cold stress. Notably, the low glycine-betaine content makes it unlikely that the transgenic phenotype is due to increased osmotic adjustment; the most probable mechanism for the action of glycinebetaine could be the stabilization of the structure and function of the proteins (Sakamoto and Murata, 1998). These results suggest that a high- stress-induced accumulation of osmolytes is not necessarily effective to achieve stress tolerance; indeed, the best results were obtained when a moderate level of osmolytes was constitutively accumulated in the cells.

Many genes controlling simple characters are well known, while genes controlling polygenic traits, such as stress tolerance, are only rarely available. The first efforts to dissect the molecular mechanisms involved in stress resistance led to the isolation of many genes induced or up-regulated upon exposure to stress conditions, such as cold-regulated (Cattivelli and Bartels, 1990) or drought-induced genes (Close et al., 1989). Current evidence suggests that a higher expression of stress-induced genes is correlated with higher stress resistance. Under winter field conditions, frost-resistant winter barley cultivars accumulate higher amounts of cold-regulated proteins than do spring ones (Giorni et al., 1999). On the contrary, barley albino mutants defective in their chloroplast development show reduced expression of all cold-regulated genes, and they do not became hardened (Grossi et al., 1998). Despite this general evidence, stress resistance is not expected to improve significantly by transferring only one of the stress-induced genes isolated so far because each of them represents only a single component of a very complex molecular response. An exception is the barley *HVA1* gene that confers tolerance to water deficit and salinity when transferred in rice. Plants accumulating HVA1 protein at a high level show a higher growth rate under stress conditions compared with nontransformed rice, suggesting that the LEA protein coded by *HVA1* plays a major role in protecting cells from dehydration (Xu et al., 1996). Nevertheless, similar results have not been obtained when other members of the LEA gene family have been used for plant transformation, indicating that different LEA proteins may have different roles in plant stress response.

Many results are expected from recent works that aim to identify regulatory genes able to trigger the stress response pathway(s) leading to the expression of most of the genes involved in a given molecular response. A first example in this direction was the isolation of the gene coding for the heat shock transcription factor *(HSF)* responsible for the control of the transcription of all heat shock genes. A transgenic *Arabidopsis* expressing the chimeric protein between HSF and β-glucuronidase showed a constitutively active HSF leading to a constitutive synthesis of heat shock proteins and, therefore, to enhanced thermotolerance (Lee et al., 1995). Several recent results have shown that it is also possible, having the appropriate genes, to improve drought tolerance in model plants. *Arabidopsis* transformed with *CBF1* or *DREB1A,* two highly homologous genes coding for transcription factors, showed improved drought, salt, and freezing tolerance (Jaglo-Ottosen et al., 1998; Kasuga et al., 1999). Expression of *DREB1A* under the control of a stress-induced promoter gave rise to minimal effects on plant growth while providing a significant improvement in stress resistance (Kasuga et al., 1999). The next step will be the identification of the homologous regulatory gene(s) in barley in order to reproduce the resistant phenotype in this important crop. It is becoming clear that full knowledge of the regulatory system leading to stress response and adaptation will be crucial to identifying useful genes for improvement of stress resistance. This goal is expected to be achieved in the coming years, either through the identification and characterization of mutants or through the isolation of barley genes homologous to the regulatory genes already isolated in other species.

MOLECULAR RESPONSE TO PATHOGENS

Barley, as any other crop, is subjected to the attack of a wide range of fungal and viral pathogens, leading to yield reduction. Breeding for pathogen resistance has therefore been considered an important task to limit yield losses, and new resistance genes, frequently derived from wild relatives, have been introgressed into cultivated barley (Jørgensen, 1992 a,b; Moseman et al., 1990; Kintzios and Fischbeck, 1996). The availability of molecular marker-based genomic maps has led, first, to the tagging of many disease resistance genes (see Figure 12.2) and, second, to the development of molecular marker-assisted selection strategies to select resistant cultivars on the basis of their genotype rather than their phenotype (Graner, Foroughi-Wehr, and Tekauz, 1996; Graner, Bauer, et al., 1996; Graner and Tekauz, 1996).

Genes for Resistance

Resistance to Powdery Mildew

Powdery mildew, caused by the fungal pathogen *Blumeria graminis* f. sp. *hordei,* is one of the most damaging diseases of intensive barley cultivation; various barley major resistance genes, most of them behaving as race specific toward this pathogen, have been tagged by molecular markers.

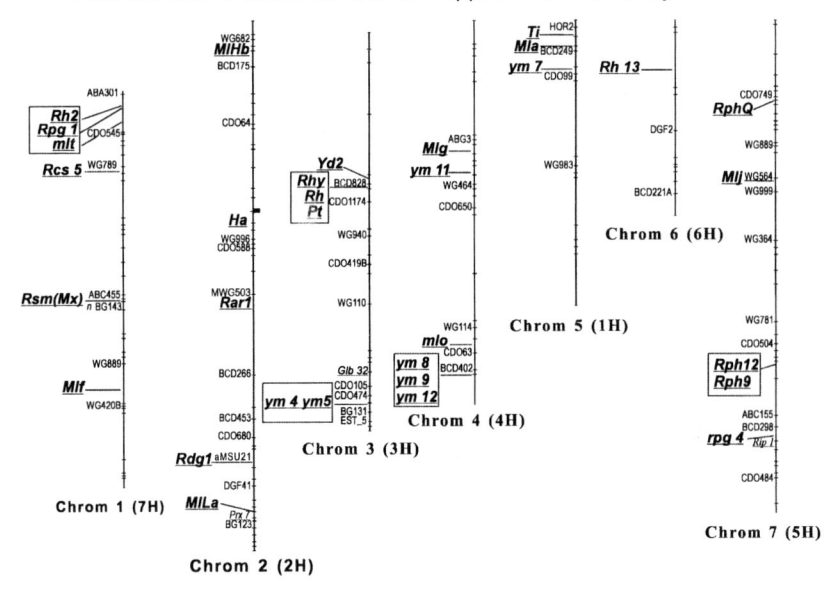

FIGURE 12.2. Resistance Genes Mapped onto the Barley Genome

Mla is a multiallelic powdery mildew resistance locus consisting of several dominant-acting alleles determining resistance specificity toward different races of *B. graminis* f. sp. *hordei;* at least 31 different alleles, with different degrees of dominance, have been identified on the basis of their race specificity (Jahoor and Fischbeck, 1987; Jørgensen, 1994; Kintzios et al., 1995; Kintzios and Fischbeck, 1996). Most of them have been incorporated into European germplasm from accessions of *H. spontaneum*. The *Mla* locus maps onto chromosome 5S(1HS) (Jahoor et al., 1993).

A multiallelic resistance locus, *mlo* is defined by a series of recessive alleles, each of them conferring resistance to all tested isolates of *B. graminis* f. sp. *hordei* (Jørgensen, 1992a), with the exception of a Japanese powdery mildew isolate (Lyngkjaer et al., 1995). Although most of the resistance alleles have been isolated after mutagenesis of susceptible *Mlo* barley cultivars, *mlo* genotypes can also occur spontaneously. The locus identified by this resistance gene maps onto chromosome 4L(4HL) (Hinze et al., 1991). The *mlo* resistance gene has been recently cloned using a positional cloning approach and the wild-type *Mlo* susceptible allele showed to encode for a membrane-spanning protein (Büschges et al., 1997). It is well known that *mlo* resistance is conferred by the absence of the MLO wild-type protein; nevertheless, the function of *Mlo* and its role in determining powdery mildew resistance are still unclear. Two models have been proposed (Büschges et al., 1997): The first hypothesis originates from the observation that plants carrying the resistance allele *mlo* spontaneously show leaf cell death phenotype and cell wall apposition even in the absence of pathogen attack (Wolter et al., 1993). *Mlo* could therefore have a negative control function in leaf cell death, suppressing the default cell suicide program in foliar tissues. This hypothesis is supported by two lines of evidences: first, the more efficient *mlo* resistance alleles are those exhibiting the more pronounced necrosis, and, second, mutants at unlinked *Ror1* and *Ror2* loci, two loci required for *mlo* resistance (Freialdenhoven et al., 1996), abolished the leaf cell death phenotype. In the second hypothesis, the MLO protein acts as a negative regulator in the plant defense response; in this case, the cell death observed in the *mlo* genotypes would represent the result of an enhanced activation of defense response arrays. This hypothesis is supported by the chronological order of the onset of defense-related events in *mlo* genotypes in the absence of the pathogen, and by the recent observations that the inhibition of phenylalanine ammonia-lyase activity in combination with complexing phosphate compounds led to the suppression of *mlo5*-mediated resistance, to the penetration of an avirulent *mlo* powdery mildew isolate, and to an enhanced penetration efficiency of a virulent *mlo* powdery mildew isolate (Lyngkjaer et al., 1997).

Resistance conferred by the barley *Mlg* resistance locus is determined by a semidominant allele that confers race-specific resistance to powdery mildew. The *Mlg* locus has been mapped to chromosome 4(4H) (Görg et al., 1993).

Three additional genes conferring race-specific resistance to powdery mildew have been derived from *H. spontaneum;* these genes were identified as *mlt, Mlf,* and *Mlj* and mapped onto chromosomes 1S(7HS), 1L(7HL), and 7L(5HL), respectively (Schönfeld et al., 1996). Other powdery mildew resistance genes have been introgressed into cultivated barley from *H. laevigatum* [*Ml(La);* chromosome 2L(2HL); Giese et al., 1993] and from *H. bulbosum* [*MlHb;* chromosome 2S(2HS); Graner, Bauer, et al., 1996]. Besides these identified major genes, additional loci conferring quantitative resistance (QTL) to powdery mildew have been identified, although the percentage of variance explained by these QTL was always under 20 percent (Heun, 1992; Saghai Maroof et al., 1994; Backes et al., 1995; Thomas et al., 1995).

The activity of some powdery mildew resistance genes is sometimes affected by the action of other unlinked genes. Two *Ror* genes were identified as required for the activities associated with the resistance specified by *mlo*. Mutations at the *Ror* genes restore the penetration of powdery mildew in the host cells (Freialdenhoven et al., 1996). Also, the resistance specified by dominant genes requires additional factors encoded by *Rar* genes *(Rar1* and *Rar2)*. Mutants at the *Rar1* locus suppress most of the resistant specificities controlled by *Mla* as well as by other loci (Jørgensen, 1996). It has also been shown that functional alleles of the *Rar* genes are required for triggering hypersensitive cell death and the accumulation of pathogenesis-related proteins that normally occurs during the powdery mildew resistance response (Freialdenhoven et al., 1994). Furthermore, interaction analyses performed with *Ror* and *Rar* genes in genetic background testing of race-specific as well as non-race-specific powdery mildew resistances demonstrated that race-specific resistance, does not depend on gene functions required for non-race-specific resistance, and vice versa (Peterhansel et al., 1997). The *Rar1* gene locus has been mapped to the long arm of barley chromosome 2(2H) (Freialdenhoven et al., 1994); the high-resolution genetic mapping of the *Rar1* locus (Lahaye, Shirasu, and Schulze-Lefert, 1998) together with the identification of YAC clones encompassing the *Rar1* gene (Lahaye, Hartmann, et al., 1998) open the way for gene isolation via positional cloning.

Resistance to Rusts

Several sources of resistance to leaf rust caused by *Puccinia hordei* were identified from *H. spontaneum* accessions (Moseman et al., 1990; Jin et al., 1995). In a comparative study of leaf rust resistance between cultivated and wild barley, it was shown that sources of resistance are rather limited in *H. vulgare* but are more common in *H. spontaneum* (Jin et al., 1995) as a consequence of the long history of coevolution between the host species and the leaf rust pathogen. Sources of partial resistance to barley leaf rust have also been identified within Ethiopian barley land races, and a barley accession of unknown pedigree from CIMMYT (International Maize and Wheat Improvement Center, Mexico) is known to possess a single dominant gene *(RphQ)* effective in conferring resistance to both leaf and stem rust *(P. graminis)* (Poulsen et al., 1995). Recent results indicate that *RphQ* is either allelic or closely linked to the leaf rust resistance locus *Rph2;* the *RphQ* locus has been mapped to the centromeric region of chromosome 7(5H) (Borovkova et al., 1997).

To date, 14 genes for resistance to *P. hordei* (*Rph* genes) have been identified in barley and its progenitor *H. spontaneum* (Jin et al., 1996); some of them have also been mapped. *Rph4* has been placed onto the chromosome 5(1H); *Rph 10* and *Rph11* (derived from *H. spontaneum*) were mapped onto chromosomes 3(3H) and 6(6H), respectively; *Rph2, Rph9,* and *Rph12* were mapped onto chromosome 7(5H) (Borovkova et al., 1997; Borovkova et al., 1998). QTL for partial resistance to leaf rust, explaining more than 50 percent of the phenotypic variance, as well as isolate-specific QTL were also mapped (Qi et al., 1998).

Genes for resistance to barley stem rust pathogen *(P. graminis* f. sp. *tritici)* have also been identified and mapped; the stem rust resistance genes *Rpg4* and *Rpg1* have been located on chromosomes 7L(5HL) and 1S(7HS), respectively (Borovkova et al., 1995; Kilian et al., 1994), and a high resolution genetic map of the *Rpg1* chromosomal region was established (Kilian et al., 1997). Loci conferring quantitative resistance to barley stripe rust pathogen *(P. striiformis* f. sp. *hordei)* have been recovered, and some of them have been introgressed into new barley genetic backgrounds by using a molecular marker-assisted selection procedure (Chen et al., 1994; Toojinda et al., 1998).

Resistance to Other Fungal Diseases

Scald or leaf blotch *(Rhynchosporium secalis)* resistance genes have been identified and located on the barley genome: *Rhy* and *Rh* have been mapped onto chromosome 3L(3HL) (Barua et al., 1993, Graner and Tekauz,1996), *Rh2* onto the short arm of chromosome 1(7HS) (Schweizer et al., 1995), and *Rh13* onto chromosome 6S(6HS) (Abbott et al., 1995). Loci controlling quantitative resistance to scald were also detected (Thomas et al., 1995; Backes et al., 1995).

A major gene conferring resistance to net blotch *(Pyrenophora teres* f. sp. *teres)* at the seedling stage *(Pt)* has been mapped onto chromosome 3L(3HL) in close proximity to the scald resistance gene *Rh* (Graner, Foroughi-Wehr, and Tekauz, et al., 1996). Two major QTL conferring resistance at the seedling stage were identified on chromosomes 4(4H) and 6L(6HL). In addition, seven QTL for net blotch resistance were identified in an adult plant; these QTL together accounted for 67.6 percent of the phenotypic variance and were mapped onto chromosomes 1S(7HS), 2S(2HS), 3S(3HS), 3L(3HL), 4(4H), 6S(6HS), and 7S(5HS) (Steffenson et al., 1996).

Resistance to spot blotch (caused by *Cochliobolus sativus*) at the seedling stage has been found to be governed by a single gene *(Rcs 5)* mapped onto the distal region of chromosome 1S(7HS). Two QTL, together explaining 70.1 percent of the phenotypic variance, confer resistance at the adult plant stage: the largest QTL effect was assigned to chromosome 5S(1HS), while the other mapped to chromosome 1S(7HS), near the seedling resistance locus (Steffenson et al., 1996).

A QTL accounting for 58.5 percent of the phenotypic variance for resistance to leaf stripe (caused by *Pyrenophora graminea*) was detected in the subcentromeric region of the long arm of chromosome 1L(7HL); furthermore, a minor QTL explaining 29.3 percent of the variance was assigned to chromosome 2S(2HS) (Pecchioni et al., 1996). A semidominant major gene conferring resistance toward *P. graminea, Rdg1,* has been located on chromosome 2L(2HL) at a distance of about 20 percent recombination from the powdery mildew resistance locus *MlLa* (Thomsen et al., 1997).

Resistance to typhula blight caused by the fungal pathogen *Typhula incarnata* is ruled by a single resistance locus *(Ti)* located in the distal portion of barley chromosome 5S(1HS) (Graner, Bauer, et al., 1996).

Resistance to Viral Diseases

The aphid-transmitted barley yellow dwarf virus (BYDV) is considered the most economically important virus of cereals. Environmental factors and genetic background interact in the determination of resistance or tolerance to the different strains of this pathogen. The *Yd2* gene has been the most effective means of providing tolerance against BYDV in cultivated barley (Delogu et al., 1995). The *Yd2* gene has been located in the centromeric region of chromosome 3L(3HL) (Collins et al., 1996). A codominant polymerase chain reaction (PCR)-based marker closely linked to this gene has also been developed and provides a useful tool for selecting segregants carrying *Yd2* in barley breeding programs (Paltridge et al., 1998). Barley yellow mosaic virus (BaYMV) and barley mild mosaic virus (BaMMV) are the two major viral pathogens of winter barley in Europe and Japan. The transmission of virus particles to the root cells of susceptible plants is mediated by the soilborne fungus *Polymyxa graminis* under winter low-temperature conditions. Genes carrying complete immunity *(ym4, ym5, ym9, ym11)* as well as those conferring partial resistance *(ym7, ym8, ym12)* have been identified (Graner, Bauer, et al., 1996). The *ym4* recessive resistance locus mediates complete immunity against BaMMV and BaYMV-type 1 but is uneffective against BaYMV-type 2. This gene has been mapped to barley chromosome 3L(3HL) (Graner and Bauer, 1993). High-resolution genetic mapping has established syntenic relationships with the homeologous rice chromosome 1 for the *ym4* chromosomal region (Bauer et al., 1996). A PCR-based codominant sequence-tagged site (STS) marker was developed to include selection for *ym4* in marker-assisted breeding programs (Tuvesson et al., 1998). The *ym5* resistance gene has been positioned in close proximity to the *ym4* (Graner, Bauer, et al., 1996), whereas other recessive resistance genes conferring resistance to BaMMV *(ym8, ym9, ym11, ym12)* are located on the long arm of chromosome 4(4HL); molecular markers closely linked to these genes are also available for marker-assisted breeding programs (Graner, Bauer, et al., 1996; Bauer et al., 1997).

A single gene conferring resistance toward the seed-transmitted barley stripe mosaic virus (BSMV), *RsmMx*, has been mapped onto the centromeric region of the short arm of chromosome 1(7HS) (Edwards and Steffenson, 1996).

Resistance to Other Pathogens

A source of partial resistance to bacterial leaf streak caused by *Xanthomonas campestris* pv. *hordei* is controlled by three QTL located on chromosomes 3(3H) and 7(5H). The resistance locus on the long arm of chromosome 3 appears to be a major gene (El Attari et al., 1998).

A number of resistance genes against cereal cyst nematode (CCN), *Heterodera avenae*, are well known, and one of them, the major dominant

Ha2 resistance gene, has been mapped onto the centromeric region of the long arm of chromosome 2(2H) (Kretschmer et al., 1997). Another major locus controlling CCN resistance, designated as *Ha4*, has been recently identified on the long arm of chromosome 7(5HL) (Barr et al., 1998).

Few data are available for genes conferring resistance to aphids: a source of resistance toward the Russian wheat aphid was identified (Nieto-Lopez and Blake, 1994). A QTL with large effect controlling resistance to cereal aphids was detected in the distal region of the short arm of chromosome 1(7H), while a second QTL was detected on chromosome 5 (Moharramipour et al., 1997).

Evolution of New Resistance Genes in Host Plants

Populations of fungal and bacterial pathogens are able to generate virulent biotypes, and if such biotypes became predominant in the field, the resistance gene would become ineffective. Thus, the ability of plant species to survive over evolutionary time depends on their ability to generate useful diversity at resistance gene loci. A growing number of natural plant resistance genes have been cloned and grouped into classes on the basis of their structural features (Backer et al., 1997; Hammond-Kosack and Jones, 1996). A common feature of most proteins characterized to date, encoded by *R* genes, is the presence of long stretches of LLR, thought to be involved in recognition of pathogen-derived signal molecules by binding and thus mediating recognition specificity (Jones and Jones, 1997). The repeated nature of the LRR region in resistance genes offers the possibility of unequal crossing-over within alleles, leading to changes in the number of repeats. Duplications and deletions of LRR-encoding sequences have been observed in several members of the tomato *Cf* gene family (conferring resistance to the fungal pathogen *Cladosporium fulvum*) (Dixon et al., 1996, 1998), and in the *Arabidopsis RPP5* gene (conferring resistance to *Peronospora parasitica*) (Parker et al., 1997). In this direction, recombination events that vary the LRR copy number could represent a mechanism for the generation of new resistance specificities able to recognize different ligands. Indeed, a recent work describing the different alleles of the *L flax* gene, conferring resistance to the rust fungus *Melamspora lini,* reported that variations in both amino acid sequence and length contribute to the generation of new resistance specificities (Ellis et al., 1999).

The contributions of classical genetic studies and of recent molecular analyses have documented many instances in which plant resistance genes are not randomly distributed in the genome but commonly occur as tightly linked clusters. Different loci in these clusters often encode resistance against different races of the same pathogen species (Crute and Pink, 1996; Bent, 1996). For both tightly linked and unlinked but related genes, recombination that generates duplication followed by divergence is the most likely source of new recognition specificities (Bent, 1996). The barley chromosomal region containing the *Mla* locus for mildew resistance can be considered a chromosomal region where a cluster of related sequences is located. To date, 31 different alleles, with different degrees of dominance, have been identified on the basis of their race specificity toward powdery mildew (Jørgensen, 1994; Kintzios et al., 1995). Evidence indicates that this region has evolved through repeated duplications, as suggested by the presence of the hordein loci *Hor1* and *Hor2*, flanking, respectively, on the downside and upside of the *Mla* locus. The hordein loci are complex loci, containing about 6 to 8 (for *Hor1*) and 13 (for *Hor2*) gene copies (Wettstein-Knowles, 1992). Furthermore, most of the markers closely linked to this locus are multicopy RFLP markers (Schüller et al., 1992; Schönfeld et al., 1996), suggesting that inter- or intrachromosomal duplication of sequences has occurred within these loci. The recovery of susceptible progeny after crosses between lines carrying different alleles supports the occurrence of interallelic recombination (Jahoor et al., 1993). Such interallelic recombination could potentially provide (although not demonstrated) the mechanism for the generation of new alleles with novel specificities (Crute and Pink, 1996).

Notably, several duplicated RFLP markers linked to the *Mla* locus on chromosome 5(1H) are also in linkage with two other powdery mildew resistance genes, *mlt* and *Mlf*, that were mapped onto chromosome 1S(7HS) and 1L(7HL), respectively, suggesting an evolutionary interrelationship among *mlt*, *Mlf*, and *Mla* loci for mildew resistance (Schönfeld et al., 1996).

It is well known that resistance genes in barley are not evenly distributed on the seven chromosomes but tend to form clusters; these have been classified as heterospecific clusters, if composed of genes with different specificities, and homospecific clusters, if all genes confer resistance against different races of the same pathogen. Besides *Mla*, homospecific gene clusters include two *ym* clusters conferring resistance to BaMMV, located on chromosomes 3L(3HL) and 4L(4HL). Some evidence also indicates that genes conferring resistance to *P. hordei* (*Rph* genes), on chromosome 7(5H), could constitute a cluster of linked genes (Jin et al., 1996; Borovkova et al., 1998).

Homologue Relationships Between Resistance Genes

Genomes of the grass species are known to share a tightly conserved gene order (synteny). Comparative genetic mapping carried out with different cereal genomes has demonstrated colinearity of molecular markers as well as of genes (cDNAs) among wheat, barley, rye, oat, maize, and rice (Devos and Gale, 1997).

Given these relationships between genomes, candidates for homeologous resistance loci were proposed. Among them, homeology was suggested for barley and wheat loci harboring the powdery mildew resistance genes *mlt* of barley and *pm5* of wheat, both located on the short arm of the homeologous chromosome group 7 and both recessively inherited. *Mlf* of barley and *Pm1*, *Pm9*, and *Pm18* of wheat are also syntenic on the long arm of homeologous chromosomes group 7. Homeology has also been suggested among *Mlj* of barley, *Pm2* of wheat, and *Pm7* of rye, all positioned on the homeologous chromosome group 5, and for *Mla* in barley and *Pm3* in wheat, positioned on the homeologous chromosome group 1S (discussed in Schönfeld et al., 1996).

Since rice has been proposed as a model plant for cereal genomes, given its small genome size and the low amount of repetitive DNA (Kurata et al., 1994), application of synteny and colinearity suggests the possible use of the rice genome as a first step toward the isolation of agronomically important genes from cereals with larger genomes (Izawa and Shimamoto, 1996; Devos and Gale, 1997). A high level of colinearity was observed between expressed sequence tag (EST) cDNA clones of rice chromosome 1 and the Triticeae homeologous group 3 (Kurata et al., 1994). This information allowed the identification of a rice chromosome 1 marker cosegregating with an RFLP marker of barley chromosome 3, tightly associated with *ym4* controlling resistance to BaMMV and to BaYMV-strain 1. This evidence represents the starting point for a map-based cloning approach for the rice *ym4* orthologous gene (Graner and Bauer, 1993; Bauer et al., 1996).

The relationships among cereal genomes are not always conserved; some examples show a break in colinearity between barley (and wheat) and rice. The *Lrk10* wheat gene encodes for a protein with three domains: an extracellular domain, a transmembrane domain, and a serine/threonine kinase domain. This gene has been mapped onto the same locus, on wheat chromosome 1AS, of the *Lr10* gene conferring resistance to the pathogen *Puccinia recondita* (Feuillet et al., 1997). Two probes were derived from *Lrk10* (a DNA fragment corresponding to the kinase domain of the protein and a second one corresponding to the putative extracellular domain) and used to map the homeologous loci in both wheat and barley genomes (Gallego et al., 1998; Pecchioni, Arru, et al., 1999). In barley, the kinase probe identified two homeologous loci on chromosomes 1HS and 3HS, while the extracellular domain probe mapped onto only the homeologous position on chromosome 1HS. In rice, the two probes identified only one locus on chromosome 1, corresponding to the tip of Triticeae group 3S chromosome, suggesting the absence of colinearity at the locus on chromosome 1S (Gallego et al., 1998). A break in synteny between barley and rice genomes was also observed in the region of barley chromosome 2 harboring the *Rar1* locus, necessary for most of the powdery mildew resistance specified at the *Mla* locus (Lahaye, Shirasu, and Schulze-Lefert, 1998; Lahaye, Hartman, et al., 1998), as well as in the region of barley chromosome 7, where the stem rust resistance gene *Rpg1* is positioned (Kilian et al., 1995).

An extensive analysis of the organization of resistance gene homologues (RGHs) in the genomes of the cereal species (rice, barley, and foxtail millet) was recently performed. RGHs were obtained from barley and rice by PCR amplification of domains (nucleotide binding site [NBS] and LRR) conserved in the major class of disease resistance genes of dicotyledonous plants (Leister et al., 1998). The results suggested that a rapid rearrangement of NBS-LRR genes has occurred in the genomes analyzed. This was supported by different experimental evidence, including the variation in copy number of some RGHs within different genotypes of barley and rice (intraspecific copy number variation possibly due to interchromosomal duplications or deletions); the observation that most of the rice and barley RGH sequences represent highly dissimilar NBS-LRR genes (leading to a substantial absence of interspecific cross-hybridization signals); and the frequent nonsyntenic map position of the loci detected with the few RGH probes showing interspecific cross-hybridization signals. Two possible mechanisms were proposed to explain the lack of synteny among NBS-LRR genes, namely, a rapid sequence divergence and species-specific ectopic recombination events of RGH loci. These mechanisms could be explained on the basis of a more rapid evolution of the monocot NBS-LRR genes in comparison to the rest of the tested monocot genomes (Leister et al., 1998).

From Resistance Genes to Resistant Phenotypes

Upon specific recognition of pathogens, plants respond by activating a battery of defense reactions. Intercellular transduction pathways involve AOS, benzoic acid, salicylic acid, jasmonic acid, protein phosphorylation/dephosphorylation, and ion influx/efflux, leading to the activation of whole-plant defense mechanisms (Hammond-Kosack and Jones, 1996). Some of these metabolic changes as well as the activation of defense genes have been described also in the section on barley response to abiotic stresses.

Activated Oxygen Species (AOS) and Salicylic Acid (SA)

In different plant-pathogen interactions, the generation of AOS has been observed, and this represents probably a key step in the plant defense response. Different roles have been proposed for AOS in the defense response, such as a direct antimicrobial effect, a reinforcement of the cell walls mediated by oxidative cross-linking of cell wall proteins, a triggering of localized hypersensitive response (HR), an increased activity of enzymes required for SA biosynthesis, and an effect in the activation of plant defense genes mediated by the activation of transcription factors (reviewed in Hammond-Kosack and Jones, 1996; Mehdy et al., 1996). A direct role of H_2O_2 in conferring disease resistance has been demonstrated by engineering plants with the H_2O_2-generating glucose oxidase of *Aspergillus niger*. Transgenic potato showed resistance to the bacterial pathogen *Erwinia carotovora* and an enhanced resistance to *Phytophthora infestans* due to the H_2O_2-mediated activation of the array of defense genes (Wu et al., 1997).

A typical resistance response of barley to powdery mildew infection is the formation of cell wall apposition, termed papillae, directly beneath the site of attempted fungal penetration; these papillae are thought to represent a physical reinforcement of the host cell wall. In barley cells carrying the resistance gene *Mla3* and following interaction with powdery mildew, a localized accumulation of H_2O_2 and cross-linking of proteins (process supported by the presence of H_2O_2) in epidermal cells undergoing HR and around papillae have been demonstrated (Thordal-Christensen et al., 1997). Powdery mildew-related accumulation of H_2O_2 was detected in barley genotypes carring the resistance genes *Mlg*, *Mla12*, and *mlo5*, either at the level of papillae or in the cytosol vesicles near the papillae. It has been suggested that the function of these vesicles is to target H_2O_2 and other defense-related compounds to the papillae. This behavior was not observed in cells that were successfully penetrated by the pathogen, confirming a substantial role of H_2O_2 in the barley response to powdery mildew (Huckelhoven et al., 1999). Generation of H_2O_2 may occur through activity of the oxalate oxidase, an H_2O_2-generating enzyme induced during barley-powdery mildew interaction. A gene coding for an epidermis/papilla-specific oxalate oxidase-like protein involved in H_2O_2 generation has been recently identified (Wei et al., 1998), and two oxalate oxidase barley genes, induced during powdery mildew infection, have recently been cloned, leading to a model hypothesis in which oxalate oxidase plays a central role in the signal transduction pathway regulating HR (Zhou et al., 1998).

In many plant-pathogen systems, it has been demonstrated that SA, derived from the phenylpropanoid pathway, plays a key role as a signaling factor in the induction of plant disease resistance (Ryals et al., 1996, Camp et al., 1998). Many functions in plant defense to pathogens have been assigned to SA: enhancer of HR cell death, signal in the process of systemic acquired resistance (SAR), and signal in the induction of arrays of defense genes. A lower concentration of SA is normally required for SAR in comparison to that required for SA-mediated HR. In the latter action, SA is an inhibitor of catalase and ascorbate peroxidase leading to the accumulation of H_2O_2. In barley, SA treatment does not induce the accumulation of PR proteins, and the level of SA does not increase in tissues challenged with *Etysiphe graminis* f. sp. *hordei* and *E. graminis* f. sp. *tritici*. An increased level of SA was, nevertheless, observed after inoculation with *Pseudomonas syringae* pv. *syringae,* demonstrating that SA accumulation in barley is pathogen specific (Vallelian-Binschedler et al., 1998).

PR and Other Defense-Related Proteins

During pathogen attack, many defense-related genes are coordinately induced. There are genes coding for enzymes of biosynthetic pathways, such as phenylalanine ammonia-lyase (the key enzyme of the phenylpropanoid pathway); for proteins involved in modifications of the cell wall (callose synthase, hydroxyproline-rich glycoproteins, thionins, lipid transfer proteins, and peroxidase); and for other proteins with antimicrobial activity, generally classified as PR proteins (Bryngelsson and Collinge, 1992).

The role of several of these proteins isolated from barley has been elucidated, either by testing their capacity to inhibit pathogens in vitro or by the overexpression of the corresponding genes in transgenic plants. Genes coding for enzymes with antifungal properties, β-1,3-glucanases and chitinases, and for a ribosome-inactivating protein are known to be accumulated in barley during pathogen attacks (Freialdenhoven et al., 1994; Boyd et al., 1995; Valè et al., 1998). The genomic position of most of these pathogen-responsive genes do not comap with known resistance genes or QTL, confirming that they do not respresent *R* genes (Pecchioni, Valè, et al., 1999). Nevertheless, when these genes were cooverexpressed in transgenic tobacco plants, a synergistic protective effect against the fungal pathogen *Rhizoctonia solani* was detected (Jach et al., 1995).

Thionins (Bohlmann et al., 1988) and LTPs (García-Olmedo et al., 1995) represent a group of small, basic, cysteine-rich antimicrobial compounds. The induction of leaf thionin genes has been detected in barley inoculated with powdery mildew (Bohlmann et al., 1988; Boyd et al., 1995), *Septoria nodorum* (Titarenko et al., 1993), and barley leaf stripe (Valè et al., 1994). The toxicity of thionins could be derived from their action in the permeabilization of cell membranes, followed by inhibition of macromolecular synthesis, or from their interference with thioredoxin-mediated redox reactions (García-Olmedo et al., 1992). Expression of the barley thionin gene in transgenic tobacco led to enhanced protection against the bacterial pathogens *P. syringae* pv. *syringae* and *P. syringae* pv. *tabaci* (Carmona et al., 1993).

LTPs are associated with the cell wall and possess in vitro-demonstrated antifungal and antibacterial activity. The genes encoding LTPs are induced in some plant-pathogen interactions, whereas in others the expression of these genes is repressed (García-Olmedo et al., 1995). The mechanism responsible for the toxicity of LTPs is not well understood, but it is thought to be different from that of thionins, as suggested by the differences observed in activity spectra (Molina and García-Olmedo, 1997). Transgenic *Arabidopsis* plants expressing the barley *LTP2* gene showed an enhanced level of resistance following inoculation with the bacterial pathogen *P. syringae* pv. *tomato*. Furthermore, exogenous application of LPT2 protein on tobacco leaves inoculated with *P. syringae* pv. *tabaci* eliminates the symptoms of disease, while transgenic tobacco plants expressing LPT2 reduce the growth of the same pathogen (Molina and García-Olmedo, 1997).

Barley plants challenged with powdery mildew show an accumulation of peroxidase transcripts, which could be involved in the polymerization of lignin, suberin, and hydroxyproline-rich glycoproteins during the formation of the structural barriers represented by the papillae (Thordal-Christensen et al., 1992). One of these pathogen-induced peroxidase genes *(pBH6-301)* has been recently used for the transformation of tobacco plants; although the encoded protein was correctly processed, as demonstrated from the secretion of the protein in the extracellular space, an infection assay with the tobacco powdery mildew fungus *Erysiphe cichoracearum* failed to indicate that the transgenic plants had achieved an enhanced level of resistance (Kristensen et al., 1997).

Phytoalexins are important defense compounds, and several plants synthesize the stilbene-type phytoalexin resveratrol when attacked by pathogens (Schröder et al., 1988). Stilbene biosynthesis requires only the presence of stilbene synthase being the precursor molecules present in plants (Hain et al., 1990). The biosynthesis of resveratrol was engineered in tobacco plants by the constitutive expression of the terminal biosynthetic enzyme stilbene synthase, and the transgenic plants exhibited enhanced resistance to necrotrophic fungus *Botrytis cinerea* (Hain et al., 1993). The expression of resveratrol has recently been engineered also in barley and wheat by the expression of the stilbene synthase gene (Leckband and Lorz, 1998). Transgenic barley plants tested for resistance against the fungal pathogen *B. cinerea* revealed an increased level of resistance.

Mechanisms to generate transgenic plants resistant to viral diseases are based on several different approaches (reviewed in Baulcombe, 1996); the most known involve the utilization of the viral coat protein (CP), whose presence in the host cells would inhibit the virion disassembly, leading to an increased resistance against the virus (Baulcombe, 1996). Transgenic plants expressing the CP genes of three isolates of BYDV and one isolate of the same virus have been recently obtained with oat and barley plants, respectively (McGrath et al., 1997). When tested with the respective viral isolates, the transgenic oat and barley plants showed high levels of resistance, and, at least in barley, the level of resistance achieved correlated with the expression level of the CP gene.

New Breeding Strategies for Disease Resistance

The study of the mechanisms underlying plant resistance to pathogens has led to the tagging of a great number of chromosomal loci harboring resistance genes and to the cloning of resistance and defense genes. The chromosomal localization of resistance genes as well as the identification of closely linked molecular markers are leading to the application of marker-assisted selection procedures during the development of breeding programs. This will open the way to the pyramiding in a single genotype of multiple resistance genes with specificity for a given pathogen; this approach will exploit the minimal likelihood that a single pathogen would simultaneously lose the corresponding avirulence genes, increasing in this way the durability of plant resistance genes.

The availability of isolated genes offers the possibility to use genetic transformation as a tool for resistance breeding. As suggested by Lamb et al. (1992), strategies to engineer plants against pathogens could be based on (1) cloned resistance genes or chimeric resistance genes; (2) genes involved in regulatory mechanisms, i.e., genes leading to the accumulation of compounds involved in signal transduction; (3) genes involved in biosynthetic pathways, i.e., those of phytoalexin biosynthesis; and (4) defense mechanisms based on a single gene. The only barley resistance gene cloned so far is the powdery mildew resistance gene *mlo* (Büschges et al., 1997); however, many other barley resistance genes have been tagged, and high-resolution mapping projects are underway, creating the basis for the isolation of the genes by using a positional cloning approach. Such work will make available other barley resistance genes during the coming years. New sources of resistance could be produced through the creation of new resistance sequences by combining domains of different cloned resistance genes in order to change or to amplify the range of pathogen recognition capacity. Defense mechanisms based on a single gene are well known, and many genes coding for a single protein, either from plant or from nonplant sources, have been shown to significantly enhance tolerance to pathogens (Broekaert et al., 1997; García-Olmedo et al., 1996). Moreover, the antimicrobial activities of several compounds frequently show synergistic effects, supporting the advantage of using combinatorial expression systems in transgenic plants.

CONCLUSIONS

Recent advances in the molecular understanding of resistance toward pathogens or physical stresses have produced a number of molecular tools (either markers or genes) that will contribute to the improvement of barley adaptation to stressful conditions. Barley breeding in the near future will integrate traditional methods with molecular breeding tools. DNA-based strategies will improve the efficiency of selection for monogenic as well as for polygenic traits. The use of molecular markers will move the focus of selection from phenotype to genotype, therefore reducing the negative effects of the environment and increasing the selection response. Recent achievements in barley transformation will allow further improvement of elite barley genotypes with traits that cannot be modified through traditional cross- and selection-based methods.

REFERENCES

Abbott, D.C., E.S. Lagudah, and A.H.D. Brown (1995). Identification of RFLPs flanking a scald resistance gene in barley. *Journal of Heredity* 86: 152-154.

Adamska, I. (1997). ELIPs—Light-induced stress proteins. *Physiologia Plantarum* 100: 794-805.

Allard, F., M. Houde, M. Krol, A. Ivanov, N.P.A. Huner, and F. Sarhan (1998). Betaine improves freezing tolerance in wheat. *Plant and Cell Physiology* 11: 1194-1202.

Anderson, J.V., L.G. Neven, Q. B. Li, D.W. Haskel, and C.L. Guy (1994). A cDNA encoding the endoplalmic reticulum-luminal heat shock protein from spinach (*Spinacia oleracea* L.). *Plant Physiology* 104: 303-304.

Andersson, B., A.H. Salter, I. Virgin, I. Vass, and S. Styring (1992). Photodamage to photosystem II-Primary and secondary events. *Journal of Photochemical and Photobiology, Series B, Biology* 15: 15-31.

Austin, R.B., P.Q. Craufurd, M.A. Hall, E. Acevedo, B. da Silveira Pinheiro, and E.C.K. Ngugi (1990). Carbon isotope discrimination as a means of evaluating drought resistance in barley, rice and cowpea. *Bulletin de la Societe Botanique de France* 137: 21-30.

Backer, B., P. Zambryski, B. Staskawicz, and S.P. Dinesh-Kumar (1997). Signaling in plant-microbe interactions. *Science* 276: 726-733.

Backes, G., A. Graner, B. Foroughi-Wher, G. Fishbeck, G. Wenzel, and A. Jahoor (1995). Localization of quantitative trait loci (QTL) for agronomic important characters by the use of a RFLP map in barley (*Hordeum vulgare* L.). *Theoretical and Applied Genetics* 90: 294-302.

Baldi, P., M. Grossi, N. Pecchioni, G. Valè, and L. Cattivelli (1999). High expression level of a gene coding for a chloroplastic amino acid selective channel protein is correlated to cold acclimation in cereals. *Plant Molecular Biology* 41: 233-243.

Barr, A.R., K.J. Chalmers, A. Karakousis, J.M. Kretschmer, S. Manning, R.C.M. Lance, J. Lewis, S.P. Jeffries, and P. Langridge (1998). RFLP mapping of a new cereal cyst nematode resistance locus in barley. *Plant Breeding* 117: 185-187.

Bartels, D., K. Engelhardt, R. Roncarati, K. Schneider, A.M. Rotter, and F. Salamini (1991). An ABA and GA modulated gene expressed in the barley embryo encodes an aldose reductase related protein. *EMBO Journal* 10: 1037-1043.

Barua, U.M., K.J. Chalmers, C.A. Hackett, W.T.B. Thomas, W. Powell, and R. Waugh (1993). Identification of RAPD markers linked to a *Rynchosporium secalis* resistance locus in barley using near-isogenic lines and bulked segregant analysis. *Heredity* 71: 177-184.

Bauer, E., T. Lahaye, P. Schulze-Lefert, T. Sasaki, and A. Graner (1996). High resolution mapping and rice synteny around the *ym4* virus resistance locus on chromosome 3L. In *Proceedings of the Fifth International Oat Conference and Seventh International Barley Genetics Symposium,* eds. G. Scoles and B. Rossnagel. Saskatoon: University of Saskatchewan Extension Press, pp. 317-319.

Bauer, E., J. Weyen, A. Schiemann, A. Graner, and F. Ordon (1997). Molecular mapping of novel resistance genes against barley mild mosaic virus (BaMMV). *Theoretical and Applied Genetics* 95: 1263-1269.

Baulcombe, D.C. (1996). Mechanisms of pathogen-derived resistance to viruses in transgenic plants. *Plant Cell* 8: 1833-1844.

Bent, A.F. (1996). Plant disease resistance genes: Function meets structure. *Plant Cell* 8: 1757-1771.

Blum, A. (1989). Osmotic adjustment and growth in barley genotypes under drought stress. *Crop Science* 29: 230-233.

Bohlmann, H., S. Clausen, S. Behnke, H. Giese, C. Hiller, U. Reimann-Philipp, G. Schrader, V. Barkholt, and K. Apel (1988). Leaf-specific thionins of barley—A novel class of cell wall proteins toxic to plant-pathogenic fungi and possibly involved in the defense mechanism of plants. *EMBO Journal* 7: 1559-1565.

Borovkova, I.G., Y. Jin, and B.J. Steffenson (1998). Chromosomal location and genetic relationship of leaf rust resistance genes *Rph9* and *Rph12* in barley. *Phytopathology* 88: 76-80.

Borovkova, I.G., Y. Jin, B.J. Steffenson, A. Kilian, T.K. Blake, and A. Kleinhofs (1997). Identification and mapping of a leaf rust resistance gene in barley line Q21861. *Genome* 40: 236-241.

Borovkova, I.G., B.J. Steffenson, Y. Jin, J.B. Rasmussen, A. Kilian, A. Kleinhofs, B.G. Rossnagel, and K.N. Kao (1995). Identification of molecular markers linked to the stem rust resistance gene *rpg4* in barley. *Phytopathology* 85: 181-185.

Boyd, L.A., P.H. Smith, E.M. Foster, and J.K.M. Brown (1995). The effects of allelic variation at the *Mla* resistance locus in barley on the early development of *Erysiphe graminis* f. sp. *hordei* and host response. *Plant Journal* 7: 959-968.

Braun, D.M. and J.C. Walker (1996). Plant transmembrane receptors: New pieces in the signaling puzzle. *Trends in Biotechnological Science* 21: 70-73.

Bray, E.A. (1993). Molecular responses to water deficit. *Plant Physiology* 103: 1035-1040.

Broekaert, W.F., B.P.A. Cammue, M.F.C. DeBolle, K. Thevissen, G.W. DeSamblanx, and R.W. Osborn (1997). Antimicrobila, peptides from plants. *Critical Reviews in Plant Science* 16: 297-323.

Bryngelsson, T. and D.B. Collinge (1992). Biochemical and molecular analyses of the response of barley to infection by powdery mildew. In *Barley, Genetics, Biochemistry, Molecular Biology and Biotechnology*, ed. P.R. Shewry. Wallingford, UK: C.A.B. International, pp. 459-480.

Büschges, R., K. Hollricher, R. Panstruga, G. Simons, M. Wolter, A. Frijters, R. van Daelen, T. van der Lee, P. Diergaarde, J. Groenendijk, et al. (1997). The barley *Mlo* gene: A novel control element of plant pathogen resistance. *Cell* 88: 695-705.

Campvan, W., M. van Montagu, and D. Inzè (1998). H2O2 and NO: Redox signals in disease resistance. *Trends in Plant Science* 3: 330-334.

Carmona, M.J., A. Molina, J.A. Fernandez, J.J. Lopez-Fando, and F. García-Olmedo (1993). Expression of the thionin gene from barley in tobacco confers enhanced resistance to bacterial pathogens. *Plant Journal* 3: 457-462.

Carpenter, C.D., J.A. Kreps, and A.E. Simon (1994). Genes encoding glycine-rich *Arabidopsis thaliana* proteins with RNA-binding motifs are influenced by cold treatment and an endogenous circadian rhythm. *Plant Physiology* 104: 1015-1025.

Catala, R., B. Sabater, and A. Guera (1997). Expression of the plastid *ndh-F* gene product in photosynthetic and non-photosynthetic tissues of developing barley seedlings. *Plant and Cell Physiology* 38: 1382-1388.

Cattivelli, L. and D. Bartels (1989). Cold-induced mRNAs accumulate with different kinetics in barley coleoptiles. *Planta* 178: 184-188.

Cattivelli, L. and D. Bartels (1990). Molecular cloning and characterization of cold-regulated genes in barley. *Plant Physiology* 93: 1504-1510.

Cattivelli, L., G. Delogu, V. Terzi, and A.M. Stanca (1994). Progress in barley breeding. In *Genetic Improvement of Field Crops,* ed. G.A. Slafer. New York: Marcel Dekker, pp. 95-181.

Chen, F.Q., D. Prehn, D. Mulrooney, A. Corey, and H. Vivar (1994). Mapping genes for resistance to barley stripe rust *(P. striiformis* f. sp. *hordei). Theoretical and Applied Genetics* 88: 215-219.

Close, T.J. (1996). Dehydrins: Emerge of a biochemical role of a family of a plant dehydration proteins. *Physiologia Plantarum* 97: 795-803.

Close, T.J., A.A. Kortt, and P.M. Chandler (1989). A cDNA-based comparison of dehydration-induced proteins (dehydrins) in barley and corn. *Plant Molecular Biology* 13: 95-108.

Collins, N.C., N.G. Paltridge, C.M. Ford, and R.H. Symons (1996). The *Yd2* gene for barley yellow dwarf virus resistance maps close to the centromere on the long arm of barley chromosome 3. *Theoretical and Applied Genetics* 92: 858-864.

Crosatti, C., E. Nevo, A.M. Stanca, and L. Cattivelli (1996). Genetic analysis of the accumulation of COR14 proteins in wild *(Hordeum spontaneum)* and cultivated *(Hordeum vulgare)* barley. *Theoretical and Applied Genetics* 93: 975-981.

Crosatti, C., P. Polverino de Laureto, R. Bassi, and L. Cattivelli (1999). The interaction between cold and light controls the expression of the cold-regulated barley gene *cor14b* and the accumulation of the corresponding protein. *Plant Physiology* 119: 671-680.

Crosatti, C., C. Soncini, A.M. Stanca, and L. Cattivelli (1995). The accumulation of a cold-regulated chloroplastic protein is light-dependent. *Planta* 196: 458-463.

Crute, I.R. and D.A.C. Pink (1996). Genetics and utilization of pathogen resistance in plants. Plant Cell 8: 1747-1755.

Danon, A. and S.P. Mayfield (1994). Light regulated translation of chloroplast messenger RNAs through redox potential. *Science* 266: 1717-1719.

Danyluk, J., A. Perron, M. Houde, A. Limin, B. Fowler, N. Benhamou, and F. Sarhan (1998). Accumulation of an acidic dehydrin in the vicinity of the plasma membrane during cold acclimation of wheat. *Plant Cell* 10: 623-638.

Delogu, G., L. Cattivelli, N. Pecchioni, D. De Falcis, T. Maggiore, and A.M. Stanca (1998). Nitrogen uptake and use efficiency during growth and development of winter barley and winter wheat. *European Journal of Agronomy* 9: 11-20.

Delogu, G., L. Cattivelli, M. Snidaro, and A.M. Stanca (1995). The *Yd2* gene and enhanced resistance to barley yellow dwarf virus (BYDV) in winter barley. *Plant Breeding* 114: 417-420.

Dennis, E.S., M. Olive, R. Dolferus, A. Millar, W.J. Peacock, and T.L. Setter (1992). Biochemistry and molecular biology of the anaerobic response. In *Inducible Plant Proteins*, ed. J.L. Wray. Cambridge: Cambridge University Press, pp. 231-245.

Devos, K.M. and M.D. Gale (1997). Comparative genetics in the grasses. *Plant Molecular Biology* 35: 3-15.

Dixon, M.S., K. Hatzixanthis, D.A. Jones, K. Harrison, and J.D.G. Jones (1998). The tomato *Cf-5* disease resistance gene and six homologs show pronounced allelic variation in leucine-rich repeat copy number. *Plant Cell* 10: 1915-1925.

Dixon, M.S., D.A. Jones, J.S. Keddie, C.M. Thomas, K. Harrison, and J.D.G. Jones (1996). The tomato *Cf-2* disease resistance locus comprises two functional genes encoding leucine-rich repeat proteins. *Cell* 84: 451-459.

Doll, H., V. Hahr, and B. Sogaard (1989). Relationship between vernalization requirement and winter hardiness in double haploid of barley. *Euphytica* 42: 209-213.

Dubcovsky, J., G.S. Maria, E. Epstein, M.C. Luo, and J. Dvorak (1996). Mapping of the K^+/Na^+ discrimination locus in wheat. *Theoretical and Applied Genetics* 92: 448-454.

Dunn, M.A., K. Brown, R.L. Lightowlers, and M.A. Hughes (1996). A low-temperature-responsive gene from barley encodes a protein with single stranded nucleic acid binding activity which is phosphorylated in vitro. *Plant Molecular Biology* 30: 947-959.

Dunn, M.A., N.J. Goddard, L. Zhang, R.S. Pearce, and M.A. Hughes (1994). Low-temperature-responsive barley genes have different control mechanisms. *Plant Molecular Biology* 24: 879-888.

Dunn, M.A., M.A. Hughes, R.S. Pearce, and P.L. Jack (1990). Molecular characterization of a barley gene induced by cold treatment. *Journal of Experimental Botany* 41: 1405-1413.

Dunn, M.A., M.A. Hughes, L. Zhang, R.S. Pearce, A.S. Quigley, and P.L. Jack (1991). Nucleotide sequence and molecular analysis of the low-temperature-induced cereal gene, *blt4*. *Molecular and General Genetics* 229: 389-394.

Dunn, M.A., A. Morris, P.L. Jack, and M.A. Hughes (1993). A low-temperature-responsive translation elongation factor 1α from barley (*Hordeum vulgare* L). *Plant Molecular Biology* 23: 221-225.

Dunn, M.A., A.J. White, S. Vural, and M.A. Hughes (1998). Identification of promoter elements in a low-temperature-responsive gene (*blt4.9*) of barley (*Hordeum vulgare* L). *Plant Molecular Biology* 38: 551-564.

Dure, L., III (1993). A repeating 11-mer amino acid motif and plant desiccation. *The Plant Journal* 3: 363-369.

Dure, L., III, M. Crouch, J. Harada, T.-H.D. Ho, J. Mundy, R.S. Quatrano, T. Thomas, and Z.R. Sung (1989). Common amino acid sequence domains among the LEA proteins of higher plants. *Plant Molecular Biology* 12: 475-486.

Edwards, M.C. and B.J. Steffenson (1996). Genetics and mapping of barley stripe mosaic virus resistance in barley. *Phytopathology* 86: 184-187.

El Attari, H., A. Rebai, P.M. Hayes, G. Barrault, G. Dechamp-Guillaume, and A. Sarrafi (1998). Potential of doubled-haploid lines and localization of quantitative trait loci (QTL) for partial resistance to bacterial leaf streak *(Xanthomonas campestris* pv. *hordei)* in barley. *Theoretical and Applied Genetics.* 96: 95-100.

Ellis, J.G., G.J. Lawrence, J.E. Luck, and P.N. Dodds (1999). Identification of regions in alleles of the flax rust resistantce gene *L* that determine differences in gene-for-gene specificity. *Plant Cell* 11: 495-506.

Ellis, R., B.P. Forster, R. Waugh, N. Bonar, L.L. Handley, D. Robinson, D.C. Gordon, and W. Powell (1997). Mapping physiological traits in barley. *New Phytology* 137: 149-157.

Espelund, M., S. Saeboe-Larssen, D.W. Hughes, G.A. Galau, F. Larsen, and K. Jakobsen (1992). Late embryogenesis-abundant genes encoding proteins with different numbers of hydrophilic repeats are regulated differentially by abscisic acid and osmotic stress. *The Plant Journal* 2: 241-252.

Farquhar, G.D. and R.A. Richardson (1984). Isotopic composition of plants correlates with water use efficiency of wheat genotypes. *Australian Journal of Plant Physiology* 11: 539-552.

Feuillet, C., G. Schachermayr, and B. Keller (1997). Molecular cloning of a new receptor-like kinase gene encoded at the *Lr10* disease resistance locus of wheat. *The Plant Journal* 11: 45-52.

Forster, B.P., M.S. Phillips, T.E. Miller, E. Baird, and W. Powell (1990). Chromosome location of genes controlling tolerance to salt (NaCl) and vigour in *Hordeum vulgare* and *H. chilense. Heredity* 65: 99-107.

Fowler, D.B., L.P. Chauvin, A.E. Limin, and F. Sarhan (1996). The regulatory role of vernalization in the expression of low-temperature-induced genes in wheat and rye. *Theoretical and Applied Genetics* 93: 554-559.

Freialdenhoven, A., C. Peterhänsel, J. Kurt, F. Kreuzaler, and P. Schulze-Lefert (1996). Identification of genes required for the function of non-race-specific *mlo* resistance to powdery mildew in barley. *Plant Cell* 8: 5-14.

Freialdenhoven, A., B. Scherag, K. Hollricher, D.B. Collinge, H. Thordal-Christensen, and P. Schulze-Lefert (1994). *Nar1* and *Nar2,* two loci required for *Mla$_{12}$*-specified race-specific resistance to powdery mildew in barley. *Plant Cell* 6: 983-994.

Galiba, G., S.A. Quarrie, J. Sutka, A. Morgounov, and J.W. Snape (1995). RFLP mapping of the vernalization *(Vrn1)* and frost resistance *(Fr1)* genes on chromosome 5A of wheat. *Theoretical and Applied Genetics* 90: 1174-1179.

Gallagher, L.W., M. Belhadri, and A. Zahour (1987). Interrelationships among three major loci controlling heading date of spring barley when grown under short day lengths. *Crop Science* 27: 155-160.

Gallego, F., C. Feuillet, M. Messmer, A. Penger, A. Graner, M. Yano, T. Sasaki, and B. Keller (1998). Comparative mapping of the two wheat leaf rust resistance loci *Lr1* and *Lr10* in rice and barley. *Genome* 41: 328-336.

García-Olmedo, F., A. Molina, A. Segura, and M. Moreno (1995). The defensive role of nonspecific lipid-transfer proteins in plants. *Trends in Microbiology* 3: 72-74.

García-Olmedo, F., A. Molina, A. Segura, M. Moreno, A. Castagnaro, E. Titarenko, P. Rodriguez-Palenzuela, M. Pineiro, and I. Diaz (1996). Engineering plants against pathogens: A general strategy. *Field and Crop Research* 45: 79-84.

García-Olmedo, F., G. Salcedo, R. Sanchez-Monge, C. Hernandez-Lucas, M.J. Carmona, J.J. Lopez-Fando, J.A. Fernandez, L. Gomez, J. Royo, F. García-Maroto, et al. (1992). Trypsin-amylase inhbitors and thionins: Possible defence proteins from barley. In *Barley, Genetics, Biochemistry, Molecular Biology and Biotechnology,* ed. P.R. Shewry. Wallingford, UK: C.A.B. International, pp. 335-350.

Giese, H., A.G. Holm-Jensen, H.P. Jensen, and J. Jensen (1993). Localization of the *Laevigatum* powdery mildew resistance gene to barley chromosome 2 by the use of RFLP markers. *Theoretical and Applied Genetics* 85: 897-900.

Gilroy, S. and R.L. Jones (1992). Gibberellic acid and abscisic acid coordinately regulate cytoplasmic calcium and secretory activity in barley aleurone protoplasts. *Proceedings of the National Academy of Sciences, USA* 89: 3591-3595.

Giorni, E., C. Crosatti, P. Baldi, M. Grossi, C. Marè, A.M. Stanca, and L. Cattivelli (1999). Cold-regulated gene expression during winter in frost tolerant and frost susceptible barley cultivars grown under field conditions. *Euphytica 106: 149-157.*

Giuliano, G., E. Pichersky, V.S. Malik, M.P. Timko, P.A. Scolnick, and A.R. Cashmore (1988). An evolutionary conserved protein binding sequence upstream of a plant light-regulated gene. *Proceedings of the National Academy of Sciences, USA* 85: 7089-7093.

Goddard, N.J., M.A. Dunn, L. Zhang, A.J. White, P.L. Jack, and M.A. Hughes (1993). Molecular analysis and spatial expression pattern of a low-temperature-specific barley gene, *blt101*. *Plant Molecular Biology* 23: 871-879.

Gong, M., A.H. van der Luit, M.R. Knight, and A.J. Trewavas (1998). Heat-shock changes in intracellular Ca^{2+} level in tobacco seedlings in relation to thermotolerance. *Plant Physiology* 116: 429-437.

Good, A.G., L.E. Pelcher, and W.L. Crosby (1988). Nucleotide sequence of a complete barley alcohol dehydrogenase 1 cDNA. *Nucleic Acids Research* 16: 7182.

Good, N.E. and D.H. Bell (1980). Photosynthesis, plant productivity, and crop yield. In *The Biology of Crop Productivity,* ed. P.S. Carlson. New York: Academic Press, pp. 3-51.

Görg, R., K. Hollricher, and P. Schulze-Lefert (1993). Functional analysis and RFLP-mediated mapping of the *Mlg* resistance locus in barley. *The Plant Journal* 3: 857-866.

Graner, A. and E. Bauer (1993). RFLP mapping of the *ym4* virus resistance gene in barley. *Theoretical and Applied Genetics* 86: 689-693.

Graner, A., E. Bauer, J. Chojecki, A. Tekauz, A. Kellerman, G. Proeseler, M. Michel, V. Valkov, G. Wenzel, and F. Ordon (1996). Molecular mapping of genes for disease resistance in barley. In *Proceedings of the Fifth International Oat Conference and Seventh International Barley Genetics Symposium,* eds. G. Scoles and B. Rossnagel. Saskatoon: University of Saskatchewan Extension Press, pp. 253-255.

Graner, A., B. Foroughi-Wehr, and A. Tekauz (1996). RFLP mapping of a gene in barley conferring resistance to net blotch *(Pyrenophora teres)*. *Euphytica* 91: 229-234.

Graner, A. and A. Tekauz (1996). RFLP mapping in barley of a dominant gene conferring resistance to scald *(Rhynchosporium secalis)*. *Theoretical and Applied Genetics* 93: 421-425.

Gray, G.R., L.P. Chauvin, F. Sarhan, and N.P.A. Huner (1997). Cold acclimation and freezing tolerance: A complex interaction of light and temperature. *Plant Physiology* 114: 467-474.

Grimm, B., E. Kruse, and K. Kloppstech (1989). Transiently expressed early light-inducible thylakoid proteins share transmembrane domains with light-harvesting chlorophyll binding proteins. *Plant Molecular Biology* 13: 583-593.

Grossi, M., L. Cattivelli, V. Terzi, and A.M. Stanca (1992). Modification of gene expression induced by ABA, in relation to drought and cold stress in barley shoots. *Plant Physiology and Biochemistry* 30: 97-103.

Grossi, M., E. Giorni, F. Rizza, A.M. Stanca, and L. Cattivelli (1998). Wild and cultivated barleys show differences in the expression pattern of a cold-regulated gene family under different light and temperature conditions. *Plant Molecular Biology* 38: 1061-1069.

Grossi, M., M. Gulli, A.M. Stanca, and L. Cattivelli (1995). Characterization of two barley genes that respond rapidly to dehydration stress. *Plant Science* 105: 71-80.

Gulli, M., E. Maestri, H. Hartings, G. Raho, C. Perrotta, K.M. Devos, and N. Marmiroli (1995). Isolation and characterization of abscisic acid inducible genes in barley seedlings and their responsiveness to environmental stress. *Plant Physiology (Life Science Advance)* 14: 89-96.

Gustafsson, Å., I. Dormling, and U. Lundqvist (1982). Gene, genotype and barley climatology. *Biological Zentralbatten* 101: 763-782.

Hain, R., B. Bieseler, H. Kindl, G. Schröder, and R.H. Stöcker (1990). Expression of a stilbene synthase gene in *Nicotiana tabacum* results in synthesis of the phytoalexin resveratrol. *Plant Molecular Biology* 15: 325-335.

Hain, R., H.-J. Reif, E. Krause, R. Langebartels, H. Kindl, B. Vornam, W. Wiese, E. Schmelzer, P.H. Schreier, R.H. Stöcker, and K. Stenzel (1993). Disease resistance results from foreign phytoalexin expression in a novel plant. *Nature* 361: 153-156.

Hammond-Kosack, K.E. and J.D.G. Jones (1996). Resistance gene-dependent plant defense responses. *Plant Cell* 8: 1773-1791.

Handley, L.L., E. Nevo, J.A. Raven, R. Martines-Carrasco, C.M. Scrimgeour, H. Pakniyat, and B.P. Forster (1994). Chromosome 4 controls potential water use efficiency (δ^{13}) in barley. *Journal of Experimental Botany* 45: 1661-1663.

Hanson, A.D., C.E. Nelsen, A.R. Pedersen, and E.H. Everson (1979). Capacity for proline accumulation during water stress in barley and its implications for breeding for drought resistance. *Crop Science* 19: 489-493.

Harberd, N.P. and K.J.R. Edwards (1983). Further studies on the alcohol dehydrogenases in barley: Evidence for a third alcohol dehydrogenase locus and data on the effect of an alcohol dehydrogenase-1 null mutation in homozygous and in heterozygous conditions. *Genetical Research* 41: 109-116.

Hayes, P.M., T. Blake, T.H.H. Chen, S. Tragoonrung, F. Chen, A. Pan, and B. Liu (1993). Quantitative trait loci on barley (*Hordeum vulgare* L.) chromosome-7 associated with components of winterhardiness. *Genome* 36: 66-71.

Heino, P., G. Sandman, V. Lang, K. Nordin, and E.T. Palva (1990). Abscisic acid deficiency prevents developing of freezing tolerance in *Arabidopsis thaliana* (L.) Hheynh. *Theoretical and Applied Genetics* 79: 801-806.

Heun, M. (1992). Mapping quantitative powdery mildew resistance of barley using a restriction fragment length polymorphism map. *Genome* 35: 1019-1025.

Hinze, K., R.D. Thompson, E. Ritter, F. Salamini, and P. Schulze-Lefert (1991). Restriction fragment length polymorphism-mediated targeting of the *ml-o* resistance locus in barley *(Hordeum vulgare)*. *Proceedings of the National Academy of Sciences, USA* 88: 3691-3695.

Hockett, E.A., and R.A. Nilan (1985). Genetics. In *Barley,* ed. D.C. Rasmusson. Madison, WI: ASA, CSSA, SSSA, pp. 187-230.

Hollenbach, B., L. Schreiber, W. Hartung, and K.J. Dietz (1997). Cadmium leads to stimulated expression of the lipid transfer protein genes of barley: Implication for the involvement of lipid transfer proteins in wax assembly. *Planta* 203: 9-19.

Hollung, K., M. Espelund, and K.S. Jakobsen (1994). Another Lea B19 gene (Group 1 Lea) from barley containing a single 20 amino acid hydrophilic motif. *Plant Molecular Biology* 25: 559-564.

Holmstrom, K.-O., E. Mantyla, B. Welin, A. Mandal, and E.T. Palva (1996). Drought tolerance in tobacco. *Nature* 379: 683-684.

Hondred, D. and A.D. Hanson (1990). Hypoxically inducible barley lactate dehydrogenase: cDNA cloning and molecular analysis. *Proceedings of the National Academy of Sciences, USA* 87: 7300-7304.

Hong, B., R. Barg, and T.-O. Ho (1992). Developmental and organ-specific expression of an ABA- and stress-induced protein in barley. *Plant Molecular Biology* 18: 663-674.

Hong, B., S. Uknes, and T.-H.D. Ho (1988). Cloning and characterization of a cDNA encoding a mRNA rapidly induced by ABA in barley aleurone layers. *Plant Molecular Biology* 11: 495-506.

Hong, S.W., J.H. Jon, J.M. Kwak, and H.G. Nam (1997). Identification of a receptor-like protein kinase gene rapidly induced by abscisic acid, dehydration, high salt and cold treatments in *Arabidopsis thaliana*. *Plant Physiology* 113: 1203-1212.

Hrabak, E.M., L.J. Dickmann, J.S. Satterlee, and M.R. Sussman (1996). Characterization of eight new members of the calmodulin-like domain protein kinase gene family from *Arabidopsis thaliana*. *Plant Molecular Biology* 31: 405-412.

Huckelhoven, R., J. Fodor, C. Preis, and K.-H. Kogel (1999). Hypersensitive cell death and papillae formation in barley attacked by powdery mildew fungus are associated with hydrogen peroxide but not with salicylic acid accumulation. *Plant Physiology* 119: 1251-1260.

Hughes, M.A. and M.A. Dunn (1996). The molecular biology of plant acclimation to low temperature. *Journal of Experimental Botany* 47: 291-305.

Hughes, M.A., M.A. Dunn, R.S. Pearce, A.J. White, and L. Zhang (1992). An abscisic acid responsive low temperature barley gene has homology with a maize phospholipid transfer protein. *Plant Cell and Environment* 15: 861-866.

Hulbert, S.H. (1997). Structure and evolution of the *Rp1 complex* conferring rust resistance in maize. *Annual Review of Phytopathology* 35: 293-310.

Hurkman, W. J. and C.T. Tanaka (1996). Effect of salt stress on germin gene expression in barley roots. *Plant Physiology* 110: 971-977.

Hurkman, W. J., H.P. Tao, and C.K. Tanaka (1991). Germin-like polypeptides increase in barley roots during salt stress. *Plant Physiology* 97: 366-374.

Ishitani, M., T. Nakamura, S.Y. Han, and T. Takabe (1995). Expression of the betaine aldehyde dehydrogenase gene in barley in response to osmotic stress and abscisic acid. *Plant Molecular Biology* 27: 307-315.

Izawa, T. and K. Shimamoto (1996). Becoming a model plant: The importance of rice to plant science. *Trends* in *Plant Science* 1: 95-99.

Jach, G., B. Görnhardt, J. Mundy, J. Logemann, E. Pinsdorf, R. Leah, J. Schell, and C. Maas (1995). Enhanced quantitative resistance against fungal disease by combinatorial expression of different barley antifungal proteins in transgenic tobacco. *The Plant Journal* 8: 97-109.

Jaglo-Ottosen, K.R., S.J. Gilmour, D.G. Zarka, O. Schabenberger, and M.F. Thomashow (1998). *Arabidopsis CBF1* overexpression induces COR genes and enhances freezing tolerance. *Science* 280: 104-106.

Jahoor, A. and G. Fischbeck (1987). Genetical studies of resistance to powdery mildew in barley lines derived from *Hordeum spontaneum* collected in Israel. *Plant Breeding* 99: 265-273.

Jahoor, A., A. Jacobi, M.E. Schuller, and G. Fischbeck (1993). Genetical and RFLP studies at the *Mla* locus conferring powdery mildew resistance in barley. *Theoretical and Applied Genetics* 85: 713-718.

Jin, Y., G.H. Cui, B.J. Steffenson, and J.D. Franckowiak (1996). New leaf rust resistance genes in barley and their allelic and linkage relationships with other *Rph* genes. *Phytopathology* 86: 887-890.

Jin, Y., B.J. Steffenson, and H.E. Bockelman (1995). Evaluation of cultivated and wild barley for resistance to pathotypes of *Puccinia hordei* with wide virulence. *Genetic Resources and Crop Evolution* 42: 1-16.

Jones, D.A. and J.D.G. Jones (1997). The role of leucine-rich repeat proteins in plant defenses. *Advances in Botanical Research* 24: 89-167.

Jørgensen, J.H. (1992a). Discovery, characterization and exploitation of *Mlo* powdery mildew resistance in barley. *Euphytica* 63: 141-152.

Jørgensen, J.H. (1992b). Sources and genetics of resistance to fungal pathogens. In *Barley, Genetics, Biochemistry, Molecular Biology and Biotechnology,* ed. P.R. Shewry. Wallingford, UK: C.A.B. International, pp. 441-457.

Jørgensen, J.H. (1994). Genetics of powdery mildew resistance in barley. *Critical Reviews in Plant Science* 13: 97-119.

Jørgensen, J.H. (1996). Effect of three suppressors on the expression of powdery mildew resistance in barley. *Genome* 39: 492-498.

Kasuga, M., S. Miura, K. Yamaguchi-Shinozaki, and K. Shinozaki (1999). Improving plant drought, salt and freezing tolerance by gene transfer of a single stress-inducible transcription factor. *Nature Biotechnology* 17: 287-291.

Kilian, A., J. Chen, F. Han, B.J. Steffenson, and A. Kleinhofs (1997). Towards map-based cloning of the barley stem rust resistance genes *Rpg1* and *rpg4* using rice as an intergenomic cloning vehicle. *Plant Molecular Biology* 35: 187-195.

Kilian, A., D.A. Kudrna, A. Kleinhofs, M. Yano, N. Kurata, B. Steffenson, and T. Sasaki (1995). Rice-barley synteny and its application to saturation mapping of the barley Rpg1 region. *Nucleic Acid Research* 23: 2729-2733.

Kilian, A., B.J. Steffenson, M.A. Saghai Maroof, and A. Kleinhofs (1994). RFLP markers linked to the durable stem rust resistance gene *Rpg1* in barley. *Molecular Plant-Microbe Interaction* 7: 298-301.

Kim, M., D.A. Christopher, and J.E. Mullet (1999). ADP-dependent phosphorylation regulates association of a DNA-binding complex with the chloroplast psbD blue-light-responsive promoter. *Plant Physiology* 119: 663-670.

Kintzios, S. and G. Fischbeck (1996). Identification of new sources for resistance to powdery mildew in *H. spontaneum*-derived winter barley lines. *Genetic Resources and Crop Evolution* 43: 25-31.

Kintzios, S., A. Jahoor, and G. Fischbeck (1995). Powdery-mildew-resistance genes *Mla29* and *Mla32* in *H. spontaneum*-derived winter barley lines. *Plant Breeding* 114: 265-266.

Kishor, P.B.K., Z. Hong, G.-H. Miao, C.-A.A. Hu, and D.P.S. Verma (1995). Overexpressing of delta^1-pyrroline-5-carboxylate synthetase increases proline production and confers osmotolerance in transgenic plants. *Plant Physiology* 108: 1387-1394.

Knight, H., A.J. Trewavas, and M.R. Knight (1996). Cold calcium signaling in *Arabidopsis* involves two cellular pools and a change in calcium signature after acclimation. *Plant Cell* 8: 489-503.

Knight, M.R., A.K. Campbell, S.M. Smith, and A.J. Trewavas (1991). Transgenic plant aequorin reports the effect of touch and cold-shock and elicitors on cytoplasmic calcium. *Nature* 352: 524-526.

Knight, M.R., S.M. Smith, and A.J. Trewavas (1992). Wind-induced plant motion immediately increases cytosolic calcium. *Proceedings of the National Academy of Sciences, USA* 89: 4967-4971.

Kretschmer, J.M., K.J. Chalmers, S. Manning, A. Karakousis, A.R. Barr, A.K.M.R. Islam, S.J. Logue, Y.W. Choe, S.J. Barker, R.C.M. Lance, and P. Langridge (1997). RFLP mapping of the *Ha2* cereal cyst nematode resistance gene in barley. *Theoretical and Applied Genetics* 94: 1060-1064.

Kristensen, B.K., J. Brandt, K. Bojsen, H. Thordal-Christensen, K.B. Kerby, D.B. Collinge, J.D. Mikkelsen, and S.K. Rasmussen (1997). Expression of a defence-related intercellular barley peroxidase in transgenic tobacco. *Plant Science* 122: 173-182.

Krol, M., A.G. Ivanov, S. Joansson, K. Kloppstech, and N.P.A. Huner (1999). Greening under high light or cold temperature affects the level of xantophyll-cycle pigments, early light inducible proteins and light-harvesting polypeptides in wild type barley and in the chlorina f2 mutant. *Plant Physiology* 120: 193-203.

Kruse, E., Z. Liu, and K. Kloppstech (1993). Expression of heat shock proteins during development of barley. *Plant Molecular Biology* 23: 111-122.

Kurata, N., G. Moore, Y. Nagamura, T. Foote, M. Yano, Y. Minobe, and M.D. Gale (1994). Conservation of genome structure between rice and wheat. *Bio/Technology* 12: 276-278.

Lahaye, T., S. Hartmann, S. Topsch, A. Freialdenhoven, M. Yano, and P. Schulze-Lefert (1998). High resolution genetic and physical mapping of the *Rar1* locus in barley. *Theoretical and Applied Genetics* 97: 526-534.

Lahaye, T., K. Shirasu, and P. Schulze-Lefert (1998). Chromosome landing at the barley *Rar1* locus. *Molecular and General Genetics* 260: 92-101.

Lamb, C.J., J. Ryals, E.R. Ward, and R.A. Dixon (1992). Emerging strategies for enhancing crop resistance to microbial pathogens. *Bio/Technology* 10: 1436-1445.

Laurie, D.A., N. Pratchett, J.H. Bezant, and J.W. Snape (1995). RFLP mapping of five major genes and eight quantitative trait loci controlling flowering time in a winter x spring barley (*Hordeum vulgare* L.) cross. *Genome* 38: 575-585.

Leckband, G. and H. Lorz (1998). Transformation and expression of a stilbene synthase gene of *Vitis vinifera* L. in barley and wheat for increased fungal resistance. *Theoretical and Applied Genetics* 96: 1004-1012.

Lee, J.H., A. Hubler, and F. Schoffl (1995). Derepression of the activity of genetically engineered heat shock factor causes constitutive synthesis of heat shock proteins and increased thermotolerance in transgenic *Arabidopsis*. *Plant Journal* 8: 603-612.

Leister, D., J. Kurth, D.A. Laurie, M. Yano, T. Sasaki, K. Devos, A. Graner, and P. Schulze-Lefert (1998). Rapid reorganization of resistance gene homologues in cereal genomes. *Proceedings of National Academy of Sciences, USA* 95: 370-375.

Leung, J., M. Bouvier-Durand, P.C. Moris, D. Guerrier, F. Chefdor, and J. Giraudat (1994). *Arabidopsis* ABA-responsive gene *abi1*: Features of a calcium-modulated protein phosphatase. *Science* 264: 1448-1452.

Lyngkjaer, M.F., T.L.W. Carver, and R.J. Zeyen (1997). Suppression of resistance to *Erysiphe graminis* f. sp. *hordei* conferred by the *mlo5* barley powdery mildew resistance gene. *Physiological and Molecular Plant Pathology* 50: 17-34.

Lyngkjaer, M.F., H.P. Jensen, and H. Ostergard (1995). A Japanese powdery mildew isolate with exceptionally large infection efficiency on *mlo*-resistant plants. *Plant Pathology* 44: 786-790.

Marmiroli, N., A. Pavesi, G. Di Cola, H. Hartings, G. Raho, M.R. Conte, and C. Perrotta (1993). Identification, characterization, and analysis of cDNA and genomic sequences encoding two different small heat shock proteins in *Hordeum vulgare*. *Genome* 36: 1111-1118.

Marocco, A., A.L. Santucci, S. Cerioli, M. Motto, N. DiFonzo, R. Thompson, and F. Salamini (1991). Three high-lysine mutations control the level of ATP binding HSP 70-like proteins in the maize endosperm. *Plant Cell* 3: 507-515.

Martin, M., L.M. Casano, and B. Sabater (1996). Identification of the product of ndhA gene as a tylakoid protein synthesized in response to photooxidative treatment. *Plant and Cell Physiology* 37: 293-298.

Mayne, R.G. and P.J. Lea (1984). Alcohol dehydrogenase in *Hordeum vulgare*: Changes in isoenzyme level under hypoxia. *Plant Science Letter* 37: 73-78.

McGrath, P.F., J.R. Vincent, C.H. Lei, W.P. Pawlowski, K.A. Torbert, W. Gu, H.F. Kaeppler, Y. Wan, P.G. Lemaux, H.R. Rines, et al. (1997). Coat-protein-mediated resistance to isolates of barley yellow dwarf in oats and barley. *European Journal of Plant Pathology* 103: 695-710.

Mehdy, M.C.,Y.K. Sharma, K. Sathasivan, and N.W. Bays (1996). The role of activated oxygen species in plant disease resistance. *Physiologia Plantarum* 98: 365-374.

Moharramipour, S., H. Tsukumi, K. Sato, and H. Yoshida (1997). Mapping resistance to cereal aphids in barley. *Theoretical and Applied Genetics* 94: 592-596.

Molina, A. and F. García-Olmedo (1997). Enhanced tolerance to bacterial pathogens caused by the transgenic expression of barley lipid transfer protein LPT2. *The Plant Journal* 12: 669-675.

Monroy, A.F. and R.S. Dhindsa (1995). Low-temperature signal transduction: Induction of cold acclimation-specific genes of alfalfa by calcium at 25°C. *Plant Cell* 7: 321-331.

Montané, M.H., S. Dreyer, C. Triantaphyllides, and K. Kloppstech (1997). Early light-inducible proteins during long-term acclimation of barley to photooxidative stress caused by light and cold: High level of accumulation by posttranscriptional regulation. *Planta* 202: 293-302.

Moseman, J.G., E. Nevo, and M.A. El-Morshidy (1990). Reactions of *Hordeum spontaneum* to infection with two cultures of *Puccinia hordei* from Israel and United States. *Euphytica* 49: 169-175.

Muench, D.G. and A.G. Good (1994). Hypoxically inducible barley alanine aminotransferase: cDNa cloning and expression analysis. *Plant Molecular Biology* 24: 417-427.

Murelli, C., F. Rizza, F. Marinone Albini, A. Dulio, V. Terzi, and L. Cattivelli (1995). Metabolic changes associated with cold-acclimation in contrasting genotypes of barley. *Physiologia Plantarum* 94: 87-93.

Neven, L.G., D.W. Haskell, C.L. Guy, N. Denslow, P.A. Klein, L.G. Green, and A. Silverman (1992). Association of 70 kilodalton heat shock cognate proteins with acclimation to cold. *Plant Physiology* 99: 1362-1369.

Nieto-Lopez, R.M. and T.K. Blake (1994). Russian wheat aphid resistance in barley: Inheritance and linked molecular markers. *Crop Science* 34: 655-659.

Paltridge, N.G., N.C. Collins, A. Bendahmane, and R.H. Symons (1998). Development of YLM, a codominant PCR marker closely linked to the *Yd2* gene for resistance to barley yellow dwarf disease. *Theoretical and Applied Genetics* 96: 1170-1177.

Pan, A., P.M. Hayes, F. Chen, T.H.H. Chen, T. Blake, S. Wright, I. Karsai, and Z. Bedö (1994). Genetic analysis of the components of winterhardiness in barley (*Hordeum vulgare* L.). *Theoretical and Applied Genetics* 89: 900-910.

Parker, J.E., M.J. Coleman, V. Szabò, L.N. Frost, R. Schmidt, E.A. van der Biezen, T. Moores, C. Dean, M.J. Daniels, and J.D.G. Jones (1997). The *Arabidopsis* downy mildew resistance gene *RPP5* shares similarity to the Toll and interleukin-1 receptors with *N* and *L6*. *Plant Cell* 9: 879-894.

Pearce, R.S., M.A. Dunn, J. Rixon, P. Harrison, and M.A. Hughes (1996). Expression of cold-inducible genes and frost hardiness in the crown meristem of young barley (*Hordeum vulgare* L. cv. Igri) plants grown in different environments. *Plant and Cell Environment* 19: 275-290.

Pearce, R. S., C.E. Houlston, K.M. Atherton, J.E. Rixon, P. Harrison, M.A. Hughes, and M.A. Dunn (1998). Localization of expression of three cold-induced genes, *blt101*, *blt4.9* and *blt14*, in different tissues of the crown and developing leaves of cold-acclimated cultivated barley. *Plant Physiology* 117: 787-795.

Pecchioni, N., L. Arru, L. Bellini, A.M. Stanca, and G. Valè (1999). The wheat leaf rust resistance gene *Lrk10* maps to barley chromosome 1HS. *Journal of Genetics and Breeding*, 53:113-119.

Pecchioni, N., P. Faccioli, H. Toubia-Rahme, G. Valè, and V. Terzi (1996). Quantitative resistance to barley leaf stripe *(Pyrenophora graminea)* is dominated by one major locus. *Theoretical and Applied Genetics* 93: 97-101.

Pecchioni, N., G. Valè, H. Toubia-Rahme, P. Faccioli, V. Terzi, and G. Delogu (1999). Barley-*Pyrenophora graminea* interaction: QTL analysis and gene mapping. *Plant Breeding* 118: 29-35.

Pesci, P. (1988). Ion fluxes and abscisic acid-induced proline accumulation in barley leaf segments. *Plant Physiology* 86: 927-930.

Peterhansel, C., A. Freialdenhoven, J. Kurth, R. Kolsch, and P. Schulze-Lefert (1997). Interaction analyses of genes required for resistance responses to powdery mildew in barley reveal distinct pathways leading to leaf cell death. *Plant Cell* 9: 1397-1409.

Phillips, J.R., M.A. Dunn, and M.A. Hughes (1997). MRNA stability and localisation of the low temperature-responsive gene family *blt14*. *Plant Molecular Biology* 33: 1013-1023.

Plaschke, J., A. Borner, D.X. Xie, R.D.M. Koebner, R. Schlegel, and M.D. Gale (1993). RFLP mapping of genes affecting plant height and growth habit in rye. *Theoretical and Applied Genetics* 85: 1049-1054.

Polisensky, D.H. and J. Braam (1996). Cold-shock regulation of the *Arabidopsis* TCH genes and the effects of modulating intracellular calcium levels. *Plant Physiology* 111: 1271-1279.

Poulsen, D.M.E., R.J. Henry, R.P. Johnston, J.A.G. Irwin, and R.G. Rees (1995). The use of bulk segregant analysis to identify a RAPD marker linked to leaf rust resistance in barley. *Theoretical and Applied Genetics* 91: 270-273.

Price, A.H., A. Taylor, S.J. Ripley, A. Griffiths, A.J. Trewavas, and M.R. Knight (1994). Oxidative signals in tobacco increase cytosolic calcium. *Plant Cell* 6: 1301-1310.

Qi, X., R.E. Niks, P. Stam, and P. Lindhout (1998). Identification of QTLs for partial resistance to leaf rust *(Puccinia hordei)* in barley. *Theoretical and Applied Genetics* 96: 1205-1215.

Quarrie, S.A., M. Gulli, C. Calestani, A. Steed, and N. Marmiroli (1994). Location of a gene regulation drought-induced abscisic acid production on the long arm of the chromosome 5A of wheat. *Theoretical and Applied Genetics* 89: 794-800.

Raho, G., E. Lupotto, H. Hartings, A.P. Della Torre, C. Perrotta, and N. Marmiroli (1996). Tissue-specific expression and environmental regulation of the barley *Hvhsp17* gene promoter in transgenic tobacco plants. *Journal of Experimental Botany* 47: 1587-1594.

Raskin, I. and J.A.R. Ladyman (1988). Isolation and characterization of a barley mutant with abscisic-acid-insensitive stomata. *Planta* 173: 73-78.

Romagosa, I. and J.L. Araus (1991). Genotype-environment interaction for grain yield and $\delta^{13}C$ discrimination in barley. In *Proceedings of the 6th International Barley Genetics Symposium 1991*, Volume I, Helsingborg, Sweden, pp. 563-567.

Russell, A.W., C. Critchley, S.A. Robinson, L.A. Franklin, G.G.R. Seaton, W.-S. Chow, J. Anderson, and C.B. Osmond (1995). Photosystem II regulation and dynamics of the chloroplast D1 protein in *Arabidopsis* leaves during photosynthesis and photoinhibition. *Plant Physiology* 107: 943-952.

Ryals, J.A., U.H. Neuenschwander, M.G. Willits, A. Molina, H.-Y. Steiner, and M. Hunt (1996). Systemic acquired resistance. *Plant Cell* 8: 1809-1819.

Saghai, Maroof, M.A., Q. Zhang, and R.M. Biyashev (1994). Molecular marker analysis of powdery mildew resistance in barley. *Theoretical and Applied Genetics* 88: 733-740.

Sakamoto, A. and N. Murata (1998). Metabolic engineering of rice leading to biosynthesis of glycinebetaine and tolerance to salt and cold. *Plant Molecular Biology* 38: 1011-1019.

Santa-María, G.E., F. Rubio, J. Dubcovsky, and A. Rodriguez-Navarro (1997). The HAK1 gene of barley is a member of a large gene family and encodes a high-affinity potassium transporter. *Plant Cell* 9: 2281-2289.

Schönfeld, M., A. Ragni, G. Fischbeck, and A. Jahoor (1996). RFLP mapping of three new loci for resistance genes to powdery mildew *(Erysiphe graminis* f. sp. *hordei)* in barley. *Theoretical and Applied Genetics* 93: 48-56.

Schröder, G., J.W.S. Brown, and J. Schröder (1988). Molecular analysis of resveratrol synthase: cDNA, genomic clones and relationship with chalconsynthase. *European Journal of Biochemistry* 172: 161-169.

Schüller, C., G. Backes, G. Fischbeck, and A. Jahoor (1992). RFLP markers to identify the alleles on the *Mla* locus conferring powdery mildew resistance in barley. *Theoretical and Applied Genetics* 84: 330-338.

Schweizer, G.F., M. Baumer, G. Daniel, H. Rugel, and M.S. Röder (1995). RFLP markers linked to scald *(Rynchosporium secalis)* resistance gene *Rh2* in barley. *Theoretical and Applied Genetics* 90: 920-924.

Sedbrook, J.C., P.J. Kronebusch, G.G. Borisy, A.J. Trewavas, and P.H. Masson (1996). Transgenic AEQUORIN reveals organ-specific cytosolic calcium resposes to anoxia in *Arabidopsis thaliana* seedlings. *Plant Physiology* 111: 243-257.

Sheen, J. (1996). Ca^{2+}-dependent protein kinases and stress signal transduction in plants. *Science* 274: 1900-1902.

Shen, Q. and T.-H.D. Ho (1995). Functional dissection of an abscisic acid (ABA)-inducible gene reveals two independent ABA-responsive complexes each containing a G-box and a novel *cis*-acting element. *Plant Cell* 7: 295-307.

Shen, Q., P.N. Zhang, and T.-H.D. Ho (1996). Modular nature of abscisic acid (ABA) response complexes: Composite promoter units that are necessary and sufficient for ABA induction of gene expression in barley. *Plant Cell* 8: 1107-1119.

Shinozaki, K. and K. Yamaguchi-Shinozaki (1996). Molecular responses to drought and cold stress. *Current Opinion in Biotechnology* 7: 161-167.

Snape, J.W., A. Semikhodskii, L. Fish, R.N. Sarma, S.A. Quarrie, G. Galiba, and J. Sutka (1997). Mapping frost tolerance loci in wheat and comparative mapping with other cereals. *Acta Agronomica Hungarica* 45: 265-270.

Steffenson, B.J., P.M. Hayes, and A. Kleinhofs (1996). Genetics of seedling and adult plant resistance to net blotch *(Pyrenophora teres* f. sp. *teres)* and spot blotch *(Cochliobolus sativus)* in barley. *Theoretical and Applied Genetics* 92: 552-558.

Storlie, E.W., R.E. Allan, and M.K. Walker Simmons (1998). Effect of the Vrn1-Fr1 interval on cold hardiness levels in near-isogenic wheat lines. *Crop Science* 38: 483-488.

Sutka, J. (1981). Genetic studies of frost resistance in wheat. *Theoretical and Applied Genetics* 59: 145-152.

Sutton, F., X. Ding, and D.G. Kenefrik (1992). Group 3 Lea genes *HVA1* regulation by cold acclimation and deacclimation in two barley cultivars with varying freeze resistance. *Plant Physiology* 99: 338-340.

Suzuki, K., R. Itai, K. Suzuki, H. Nakanishi, N.K. Nishizawa, E. Yoshimura, and S. Mori (1998). Formate dehydrogenase, an enzyme of anaerobic metabolism, is induced by iron deficiency in barley roots. *Plant Physiology* 116: 725-732.

Tahtiharju, S., V. Sangwan, A.F. Monroy, R.S. Dhindsa, and M. Borg (1997). The induction of kin genes in cold-acclimating *Arabidopsis thaliana: Evidence of a role for calcium. Planta* 203: 442-447.

Takahashi, R. and S. Yasuda (1971). Genetics of earliness and growth habit in barley. In *Barley Genetics II. Proceedings of the Second International Barley Genetics Symposium,* ed. R.A. Nilan. Pullman: Washington State University Press, pp. 388-408.

Taylor, E.R., X.Z. Nie, A.W. MacGregor, and R.D. Hill (1994). A cereal haemoglobin gene is expressed in seed and root tissues under anaerobic conditions. *Plant Molecular Biology* 24: 853-862.

Teulat, B., P. Monneveux, J. Wery, C. Borries, I. Souyris, A. Charrier, and D. This (1997). Relationship between relative water content and growth parameters under water stress in barley: A QTL study. *New Phytology* 137: 99-107.

Teulat, B., D. This, M. Khairallah, C. Borries, C. Ragot, P. Sourdille, P. Leroy, P. Monneveux, and A. Charrier (1998). Several QTLs involved in osmotic-adjustment trait variation in barley. *Theoretical and Applied Genetics* 96: 688-698.

Thomas, W.T.B., W. Powell, R. Waugh, K.J. Chalmers, U.M. Barua, P. Jack., V. Lea, B.P. Forster, J.S. Swantson, R.P. Ellis, et al. (1995). Detection of quantitative trait loci for agronomic, yield, grain and disease characters in spring barley *(Hordeum vulgare* L.). *Theoretical and Applied Genetics* 91: 1037-1047.

Thomsen, S.B., H.P. Jensen, J. Jensen, J.P. Skou, and J.H. Jørgensen (1997). Localization of a resistance gene and identification of sources of resistance to barley leaf stripe. *Plant Breeding* 116: 455-459.

Thordal-Christensen, H., J. Brandt, B. Ho Cho, S.K. Rasmussen, P.L. Gregersen, V. Smedegaard-Petersen, and D.B. Collinge (1992). cDNA cloning and characterization of two barley peroxidase transcripts induced differentially by the powdery mildew fungus *Erysiphe graminis*. *Physiological and Molecular Plant Pathology* 40: 395-409.

Thordal-Christensen, H., Z. Zhang, Y. Wei, and D.B. Collinge (1997). Subcellular localization of H_2O_2 in plants. H_2O_2 accumulation in papillae and hypersensitive response during barley-powdery mildew interaction. *The Plant Journal* 11: 1187-1194.

Titarenko, E., J. Hargreaves, J. Keon, and S.J. Gurr (1993). Defence-related gene expression in barley coleoptile cells following infection by *Septoria nodorum*. In *Mechanisms of Plant Defense Responses*, eds. B. Fritig and M. Legrand. Dordrecht, the Netherlands: Kluwer Academic Publishers, pp. 308-311.

Toojinda, T., E. Baird, A. Booth, L. Broers, P.M. Hayes, W. Powell, W. Thomas, H. Vivar, and G. Young (1998). Introgression of quantitative trait loci (QTLs) determining stripe rust resistance in barley: An example of marker-assisted line development. *Theoretical and Applied Genetics* 96: 123-131.

Trick, M., E.S. Dennis, K.J.R. Edwards, and N.J. Peacock (1988). Molecular analysis of the alcohol dehydrogenase gene family of barley. *Plant Molecular Biology* 11: 147-160.

Tuvesson, S., L. Von Post, R. Ohlund, P. Hagberg, A. Graner, S. Svitashev, M. Schehr, and R. Elovsson (1998). Molecular breeding for the BaMMV/BaYMV resistance gene *ym4* in winter barley. *Plant Breeding* 117: 19-22.

Urao, T., T. Katagiri, T. Mizoguchi, K. Yamaguchi-Shinozaki, N. Hayashida, and K. Shinozaki (1994). Two genes that encode Ca^{2+}-dependent protein kinases are induced by drought and high salt stress in *Arabidopsis thaliana*. *Molecular and General Genetics* 224: 331-340.

Valè, G., M. Aragona, E. Torrigiani, L. Cattivelli, M. Montigiani, A.M. Stanca, and A. Porta-Puglia (1998). Characterization of a hypovirulent insertional mutant of *Pyrenophora graminea* and analysis of the barley defence response after inoculation. *Plant Pathology* 47: 657-664.

Valè, G., E. Torrigiani, A. Gatti, G. Delogu, A. Porta-Puglia, G. Vannacci, and L. Cattivelli (1994). Activation of genes in barley roots in response to infection by two *Drechslera graminea* isolates. *Physiological and Molecular Plant Pathology* 44: 207-215.

Vallelian-Bindschedler, L., J.P. Metraux, and P. Schweizer (1998). Salicylic acid accumulation in barley is pathogen specific but not required for defense-gene activation. *Molecular Plant-Microbe Interaction* 11: 702-705.

Van der Mulen, R.M., K. Visser, and M. Wang (1996). Effects of modulation of calcium levels and calcium fluxes on ABA-induced gene expression in barley aleurone. *Plant Science* 117: 75-78.

Van Zee, K., F.Q. Chen, P.M. Hayes, T.J. Close, and T.H.H. Chen (1995). Cold-specific induction of a dehydrin gene family member in barley. *Plant Physiology* 108: 1233-1239.

Veisz, O. and J. Sutka (1989). The relationships of hardening period and the expression of frost resistance in chromosome substitution lines of wheat. *Euphytica* 43: 41-45.

Walker-Simmons, M., D.A. Kudrna, and R.L. Warner (1989). Reduced accumulation of ABA during water stress in a molybdenum cofactor mutant of barley. *Plant Physiology* 90: 728-733.

Ward, J.M., Z.M. Pei, and J.I. Schroeder (1995). Roles of ion channels in initiation of signal transduction in higher plants. *Plant Cell* 7: 833-844.

Waters, E.R., G.J. Lee, and E. Vierling (1996). Evolution, structure and function of the small heat-shock proteins in plants. *Journal of Experimental Botany* 47: 325-338.

Webb, A.A.R., M.R. McAinsh, J.E. Taylor, and A.M. Hetherington (1996). Calcium ions as intracellular second messengers in higher plants. *Advanced Botanical Research* 22: 45-96.

Wei, Y.D., Z.G. Zhang, C.H. Andersen, E. Schmelzer, P.L. Gregersen, D.B. Collinge, V. Smedegaard-Petersen, and H. Thordal-Christensen (1998). An epidermis/papilla-specific oxalate oxidase-like protein in the defence response of barley attacked by the powdery mildew fungus. *Plant Molecular Biology* 36: 101-112.

Wettstein-Knowles von, P. (1992). Cloned and mapped genes: Current status. In *Barley, Genetics, Biochemistry, Molecular Biology and Biotechnology,* ed. P.R. Shewry. Wallingford, UK: C.A.B. International, pp. 73-98.

White, A.J., M.A. Dunn, K. Brown, and M.A. Hughes (1994). Comparative analysis of genomic sequence and expression of a lipid transfer protein gene family in winter barley. *Journal of Experimental Botany* 45: 1885-1892.

Wolter, M., K. Hollricher, F. Salamini, and P. Schulze-Lefert (1993). The *mlo* resistance alleles to powdery mildew infection in barley trigger a developmentally controlled defence mimic phenotype. *Molecular and General Genetics* 239: 122-128.

Wolthen-Larsen, H., J. Brandt, D.B. Collinge, and H. Thordal-Christensen (1993). A pathogen-induced gene of barley encodes a HSP90 homologue showing striking similarity to vertebrate forms resident in the endoplasmic reticulum. *Plant Molecular Biology* 21: 1097-1108.

Wu, Y., J. Kuzuma, E. Maréchal, R. Graeff, H.C. Lee, R. Foster, and N.-H. Chua (1997). Abscisic acid signaling through cyclic ADP-ribose in plants. *Science* 278: 2126-2130.

Xu, D., X. Duan, B. Wang, B. Hong, T.-H.D. Ho, and R. Wu (1996). Expression of late embryogenesis abundant protein gene, *HVA1,* from barley confers tolerance to water deficit and salt stress in transgenic rice. *Plant Physiology* 110: 249-257.

Yamaguchi-Shinozaki, K. and K. Shinozaki (1994). A novel *cis*-acting element in an *Arabidopsis* gene is involved in responsiveness to drought, low temperature or high-salt stress. *Plant Cell* 6: 251-264.

Zhou, F.S., Z.G. Zhang, P.L. Gregersen, J.D. Mikkelsen, E. deNeergaard, D.B. Collinge, and H. Thordal-Christensen (1998). Molecular characterization of the oxalate oxidase involved in the response of barley to the powdery mildew fungus. *Plant Physiology* 117: 33-41.

Chapter 13

Physiological Changes Associated with Genetic Improvement of Grain Yield in Barley

L. Gabriela Abeledo
Daniel F. Calderini
Gustavo A. Slafer

INTRODUCTION

Genetic improvement of grain yield potential has been mainly the result of an empirical selection approach of trial and error, with yield per se being the dominant trait targeted for selection (Loss and Siddique, 1994). This methodology, which was notoriously successful in most crops during most of the second half of the twentieth century (see Slafer, Satorre, and Andrade, 1994), seems to be hardly efficient enough to face the challenge of increasing crop production to satisfy the expected requirements in the near future (see Byrnes and Bumb, 1998). This is particularly true at least for crops that in the past supported a strong selection pressure for this trait. As a consequence, the use of physiological traits as selection critera could be useful for future genetic improvement in major cereals (Shorter, Lawn, and Hammer, 1991; Reynolds, Rajaram, and Sayre, 1999; Slafer, Araus, and Richards, 1999). Genetic gains in grain yield are the result of breeding effects on grain yield potential, stress tolerance, and disease resistance (see Sayre et al., 1998). This chapter focuses on the effect of plant breeding progress on barley yield potential in an attempt to identify traits that are associated with yield gains. Understanding the causes that permitted yield potential increases in the past may help identify successful traits to be used in current and future breeding programs (Calderini, Reynolds, and Slafer, 1999).

Different approaches have been used for identifying the contributions of genetic improvement to grain yield. Those most extensively discussed in the literature are (1) the analysis of data corresponding to experimental networks carried out over long periods of time, and (2) the comparison of cultivars released at different times but grown under the same experimental conditions (Slafer, Satorre, and Andrade, 1994).

The first approach involves the comparison of cultivars grown in different years and sites, with yield expressed in relative values with respect to a check cultivar. This approach is most common in the literature because this kind of information is frequently available in experimental stations throughout the world. However, these data do not allow analyzing traits related to grain yield improvement because yield is frequently the only measured trait. In addition, biotic stresses (e.g., pests, diseases) are infrequently prevented or controlled. According to Slafer and Andrade (1991), this creates two important biases in the assessments: (1) it impedes quantifying genetic gains in grain yield potential independently of gains in diseases resistance or tolerance, and (2) as key cultivars are usually kept as checks for several years, they expectedly become more susceptible to diseases than newer cultivars, causing overestimation of the genetic gains in potential yield.

The other approach compares old and new cultivars under the same growing conditions, mostly in the absence of stresses; thus, the biases mentioned previously could be prevented. Although this approach appears to be better than the use of network data with check cultivars (Slafer, Satorre, and Andrade, 1994), its major constraint is the limitations imposed by the growing environment, i.e., crop management and year-to-year variability. The inconvenience that genotype × environment interactions are frequently significant for assessing yield gains (e.g., Cox et al., 1988; Perry and D'Antuono, 1989) is reduced as the number of environments in which the studies are conducted increases.

This chapter outlines the genetic gains in barley yield, highlighting the traits that were responsible for them. For this purpose, we mainly consider experiments conducted under field conditions, comparing simultaneously cultivars released at different times. Then, the relationship between genetic improvement and grain yield is analyzed, integrating the associated changes in crop development, plant height, dry-matter production and partitioning, numerical yield components (grain number, grain weight), as well as nitrogen partitioning and grain quality. The last section of this chapter briefly compares breeding effects on barley yields with those on other small-grain cereals.

GENETIC GAINS IN GRAIN YIELD

Several authors have reported the influence of genetic improvement on the total gain in barley grain yield. Silvey (1986) found that about 33 percent of the increase in barley yield in the United Kingdom from the 1940s through the 1980s could be attributed to genetic improvement. Similar conclusions were reached by Strand (1994), who found that cultivar improvement contributed 40 percent to total yield increases in Norway. These genetic contributions to farm yield gains were slightly lower than those assessed for wheat, about 50 percent, during the twentieth century (see references in Slafer, Satorre, and Andrade, 1994).

The magnitude of genetic gains varied for different countries. Italy showed the highest rate of more than 7 $g \cdot m^{-2} \cdot y^{-1}$ (see Table 13.1), but this extremely high rate could be due to the very short and recent period considered by Martiniello et al. (1987): the last quarter of the twentieth century has been recognized as the most successful breeding period for grain yield of small-grain cereals (Slafer, Satorre, and Andrade, 1994). Therefore, considering data from experiments that analyzed genetic gains during most of the century, Spain had the highest genetic gain (about 4 $g \cdot m^{-2} \cdot y^{-1}$), and the lowest values were reported for North America (see Table 13.1).

The magnitude of genetic gains estimated in absolute terms not only reflects the efficiency of the breeding processes but also the effect of the environmental conditions under which these gains were achieved (i.e., the better the environment, the higher the gains tend to be for the same set of cultivars; Perry and D'Antuono, 1989). Thus, to assess genetic gains in relative terms prevents much of this effect (Slafer and Andrade, 1991) and reveals more genuine differences among countries. Most of the countries showed relative genetic gains of about 0.40 and 0.45 percent per year (see Table 13.1). Spain still exhibited one of the most successful breeding programs for grain yield improvement over a long period of time, whereas for the United States results depended on the study considered (see Table 13.1).

TABLE 13.1. Absolute and Relative Genetic Gains Quantified As the Slope of the Linear Regression Between Absolute or Relative Grain Yield and the Year of Release of the Cultivars in Each Experiment for Different Countries

Country	Period	GENETIC GAIN		Source
		Absolute ($g \cdot m^{-2} \cdot y^{-1}$)	Relative (%/y)	
Canada	1910-1988	1.8	0.43	Bulman, Mather, and Smith (1993)
	1910-1987	2.0	0.48	Jedel and Helm (1994a)
Italy	1960-1980s	7.4	1.10	Martiniello et al. (1987)
Norway	1960-1992	2.8	(#)	Strand (1994)
Spain	1937-1993	4.1	0.76	Muñoz et al. (1998)
United Kingdom	1880-1980	1.9	0.39	Riggs et al. (1981)
United States	1920-1978	2.7	0.87	Wych and Rasmusson (1983)
	1920-1984	1.6	0.40	Boukerrou and Rasmusson (1990)

(#) Data insufficient to assess the relative genetic gain

Differences between countries based on markedly different lengths of the analyzed periods must be considered cautiously, due to trends in yield were not linear over time. Riggs et al. (1981) showed that the rate of yield improvement during the second half of this century was approximately three times greater than the corresponding rate for the previous analyzed period in their study (1880-1953). Trends in grain yield in other studies including cultivars released over long periods agree with those reported by Riggs et al. (1981) for England (see Figure 13.1). Therefore, studies including cultivars released after the 1950s would expectedly have higher gains.

Relative yields (i.e., the yield of each cultivar as a percentage of the average yield of the experiment; Calderini and Slafer, 1999) could be useful for analyzing together the data corresponding to different countries. This pro-

FIGURE 13.1. Relationship Between Grain Yield and the Year of Release of Cultivars in Canada, Spain, the United Kingdom, and the United States

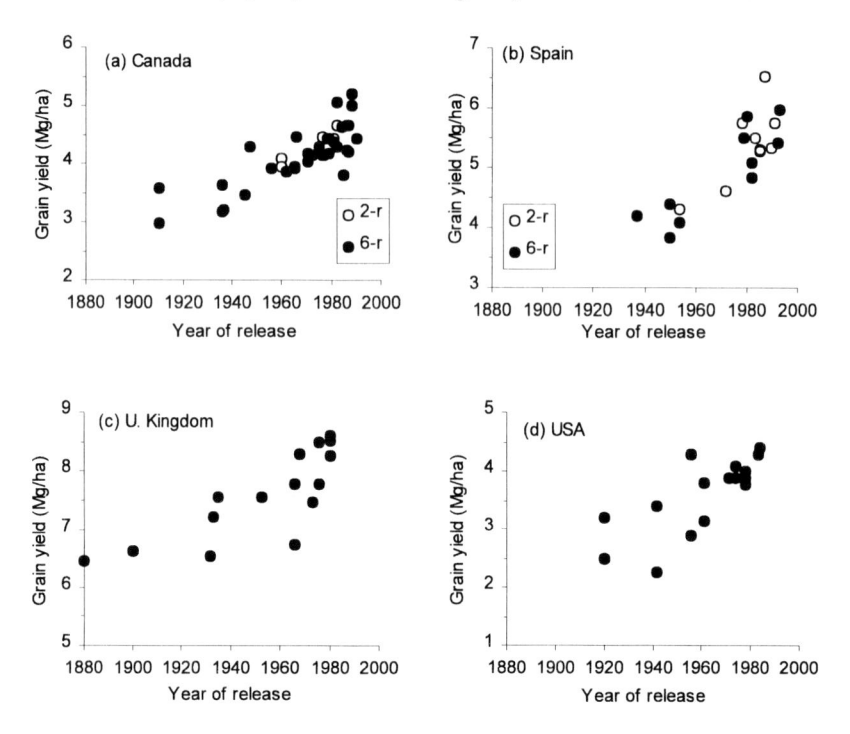

Source: (a) Bulman, Mather, and Smith, 1993; Jedel and Helm, 1994a; (b) Muñoz et al., 1998; (c) Riggs et al., 1981; (d) Wych and Rasmusson, 1983; Boukerrou and Rasmusson, 1990.

cedure showed that breeding progress in grain yield was only marginally successful until 1941 ($p < 0.05$), and then the relative genetic gain was approximately 0.6 percent per year, on average, for the evaluated countries (see Figure 13.2).

Another source of variation that should be considered in analyzing barley breeding progress is spike morphology (two- and six-rowed barleys), because estimates of genetic gains seemed to be different. For example, the genetic gain in two-rowed barleys was higher than that in six-rowed cultivars (7.4 and 5.2 $g \cdot m^{-2} \cdot y^{-1}$, respectively) in Italy (Martiniello et al., 1987). The corresponding figures were 4.1 and 3.3 $g \cdot m^{-2} \cdot y^{-1}$ (in two- and six-rowed cultivars, respectively) in Spain (Muñoz et al., 1998). However, this trend was not confirmed in Canada (Jedel and Helm, 1994a), where six-rowed cultivars showed genetic gain (2.2 $g \cdot m^{-2} \cdot y^{-1}$) similar to that of two-rowed barleys (2.0 $g \cdot m^{-2} \cdot y^{-1}$). In Figure 13.3, these genetic gains are shown as relative to the mean yield of each group of barley in the corresponding experiment. These differences among countries might be associated with the varying importance of each group of genotypes in European and North American countries, which could modify the selection pressure on two- and six-rowed barleys.

FIGURE 13.2. Relationship Between Relative Grain Yield (Cultivar Yield As a Percentage of the Average Yield of the Experiment) and Year of Release of Cultivars in Different Countries

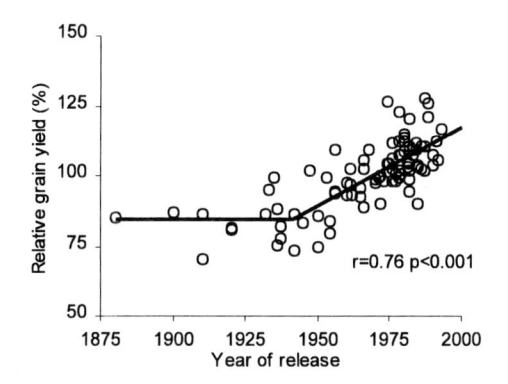

Source: Data correspond to experiments carried out in Canada (Bulman, Mather, and Smith, 1993; Jedel and Helm, 1994a), Spain (Muñoz et al., 1998), the United Kingdom (Riggs et al., 1981), and the United States (Wych and Rasmusson, 1983; Boukerrou and Rasmusson, 1990).

Note: The correlation coefficient corresponds to the linear regression of the bilineal model.

FIGURE 13.3. Relative Genetic Gains of Two- and Six-Rowed Barleys in Canada, Italy, and Spain

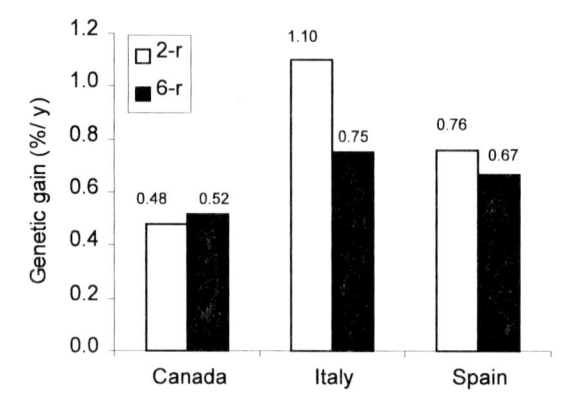

[Bar chart showing genetic gain (%/y) for Canada, Italy, and Spain with 2-r and 6-r bars: Canada 2-r = 0.48, 6-r = 0.52; Italy 2-r = 1.10, 6-r = 0.75; Spain 2-r = 0.76, 6-r = 0.67]

Source: (Canada) Jedel and Helm, 1994a; (Italy) Martiniello et al., 1987; (Spain) Muñoz et al., 1998.

PHYSIOLOGICAL CHANGES ASSOCIATED WITH GENETIC GAINS IN GRAIN YIELD

Crop Development

Even though the duration of the crop cycle affects grain yield, there were not clear effects of barley breeding on the length of the period between sowing (or emergence) and anthesis or maturity, with a few exceptions (see Table 13.2). For example, modern Italian cultivars matured earlier than the older cultivars (Martiniello et al., 1987). This agrees with the trend shown for the United Kingdom, where modern barleys also matured earlier than their predecessors (Riggs et al., 1981). However, even though these trends were significant, their slopes were small: one hundred years of barley breeding reduced time to anthesis by just seven days in the United Kingdom, in association with a decrease in the final number of leaves on the main stem, from 9.1 to 8.3, and a reduction in the rate of leaf emergence of about 0.018 leaves per day. On the other hand, breeding programs in the United States resulted in a trend (though only significant at $p < 0.25$) of slightly faster development, advancing anthesis by one to two days during the period 1920 to 1978, but without any effect on time to maturity (Wych and Rasmusson, 1983). These slight trends for the United States were not supported in a more recently published study with newer cultivars, as they tended to have slightly later maturity dates (Boukerrou and Rasmusson, 1990). Similarly, barley cultivars selected to represent different periods of genetic improve-

TABLE 13.2. Coefficient of Regression (Slope) for the Relationship Between Days to Anthesis or Maturity and the Year of Release of the Cultivars, and the Corresponding Coefficients of Correlation (r) for Cultivars Released in Different Eras in Canada, the United Kingdom, and the United States

Country	Period	Days to Anthesis		Days to Maturity		Source
		Slope (d·y^{-1})	r	Slope (d·y^{-1})	r	
Canada	1910-1988	+0.01	+0.07 NS	−0.01	−0.13 NS	Bulman, Mather, and Smith (1993)
	1910-1982	+0.07	+0.55**	+0.09	+0.59**	Jedel and Helm (1994b)
United Kingdom	1880-1980	−0.07	−0.64**	(#)	(#)	Riggs et al. (1981)
United States	1920-1978	−0.03	−0.57 NS	(#)	(#)	Wych and Rasmusson (1983)
	1920-1984	−0.01	−0.16 NS	+0.04	+0.62*	Boukerrou and Rasmusson (1990)

*$p < 0.1$, **$p < 0.01$
NS = not significant
(#) Data insufficient to assess coefficient of regression to days to maturity

ment in Spain did not show a consistent association between year of release of the cultivars and days to anthesis (Muñoz et al., 1998).

For Canadian barleys, whereas Bulman, Mather, and Smith (1993) reported no consistent historical trend for date of anthesis and date of maturity, Jedel and Helm (1994b) found that days to anthesis and maturity lengthened by 0.075 d·y^{-1} and by 0.088 d·y^{-1}, and days to stem elongation were also extended, a trait that was positively correlated with grain yield (Jedel and Helm, 1994b).

It is clear, therefore, that no consistent modifications in phenology took place over time or across countries, and wherever the trend was examined, the magnitude of the change was small or negligible. We could conclude, then, that genetic improvement of barley yields was not associated with specific changes in developmental pattern.

Plant Height

Lodging is one of the most important problems of barley under high-yielding growing conditions. Losses due to lodging may often add up to 30 to 40 percent of the grain yields (Pinthus, 1973; Stapper and Fischer, 1990), so improvement in lodging resistance has been an important criterion in plant breeding programs. The simplest morphological trait most frequently selected for reducing lodging in cereals has been plant height. Most of the studies considered in the present chapter have evaluated the effect of breeding on plant height (Riggs et al., 1981; Wych and Rasmusson, 1983; Bulman, Mather, and Smith, 1993; Jedel and Helm, 1994a).

Results from all countries showed that breeding programs reduced barley height consistently, as most modern cultivars were always significantly shorter than older cultivars. However, the dynamics of stem length reduction varied among countries. In England, a continuous reduction in height was found during the last century (see Figure 13.4a), for which Riggs et al. (1981) showed a significant trend toward height reduction of about 0.6 cm·y^{-1} ($r = 0.84$; $p < 0.001$). On the other hand, this relationship did not show a linear trend in the United States (see Figure 13.4b), exhibiting a break point around 1960. Therefore, plant height reduction between 1920 and 1960 was greater than during the following 30 years. Furthermore, Jedel and Helm (1994b) found that plant height decreased between 0.15 and 0.30 cm·y^{-1} in Canada. In Italy, plant height also decreased, 23 and 15 percent in two- and six-rowed genotypes, respectively (Martiniello et al., 1987). As a consequence of all these reductions, the average height of modern barley cultivars is between 75 and 95 cm in all countries analyzed.

This implies that, in the future, it is likely that breeders would not further reduce plant height, as the stature of modern barleys is within the range that would optimize yield (Richards, 1992). In fact, the lack of further shorten-

FIGURE 13.4. Relationship Between Plant Height of Cultivars Released at Different Times and the Year of Release of the Cultivars in the United Kingdom and the United States

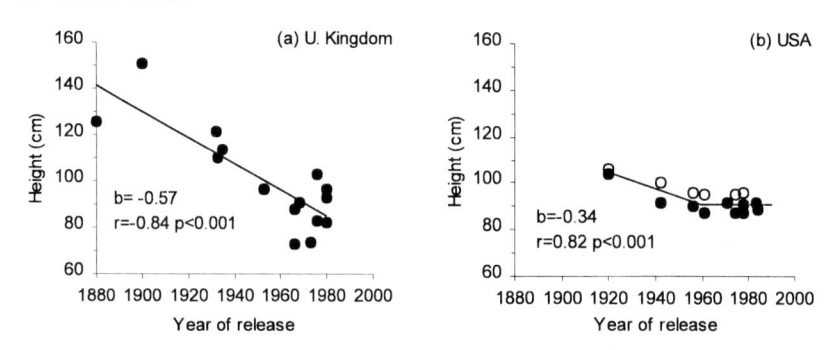

Source: (a) Riggs et al., 1981; (b) closed circles, Wych and Rasmusson, 1983; open circles, Boukerrou and Rasmusson, 1990.

ing of stem length in the United States after the 1960s (see Figure 13.4b) seems to support this suggestion: barleys released by midcentury there were already within the range of plant heights that would optimize yields.

As the quantification of damage caused by lodging is made using different and subjective scales, studies in which the effect of genetic improvement on lodging was analyzed used different semiquantitative scales. This enables us to compare quantitatively values reported for different countries, but a strong trend to increase lodging resistance was found in all cases.

Dry-Matter Production and Partitioning

High-yielding varieties should possess higher biomass production and/or higher partitioning toward the harvested organs (i.e., higher harvest index) than their low-yielding predecessors. Breeding efforts in the United States achieved an increase in barley yield partially through increasing the vegetative biomass (see Table 13.3), but the increase in grain yield was larger than that in vegetative biomass, at rates of 15.7 ($p < 0.01$) and 6.8 kg·ha^{-1}·y^{-1} ($p < 0.05$), respectively. This is in agreement with the conclusions reported for the United Kingdom, where a weak association between total aboveground biomass and grain yield was observed by Riggs et al. (1981), though they found a slight increase in biomass in the most modern, highest-yielding cultivars.

On the other hand, vegetative biomass has not been modified over the years in Canada (Bulman, Mather, and Smith, 1993; Jedel and Helm, 1994a) or in Italy. Therefore, aboveground biomass was generally not associated with the year of release of the cultivars (see Figure 13.5), and, consequently, most gains in grain yield were independent of biomass gains (see Table 13.3).

Grain yield was strongly correlated, on the other hand, with harvest index (see Figure 13.6), which confirms that the genetic increase in grain yield was mostly due to higher partitioning of biomass to reproductive organs

TABLE 13.3. Correlation Coefficient (r) Between Grain Yield and Vegetative Biomass for Different Countries

Country	Period	r	Source
Canada	1910-1987	0.01 NS	Bulman, Mather, and Smith (1993)
	1910-1987	0.27 NS	Jedel and Helm (1994a)
Italy	1900-1980s	0.32 NS	Martiniello et al. (1987)
United Kingdom	1880-1980	0.35 NS	Riggs et al. (1981)
United States	1920-1978	0.97**	Wych and Rasmusson (1983)
	1920-1984	0.56*	Boukerrou and Rasmusson (1990)

* $p < 0.1$, ** $p < 0.001$
NS = not significant

FIGURE 13.5. Relationship Between Aboveground Biomass and Year of Release of the Cultivars in Various Countries

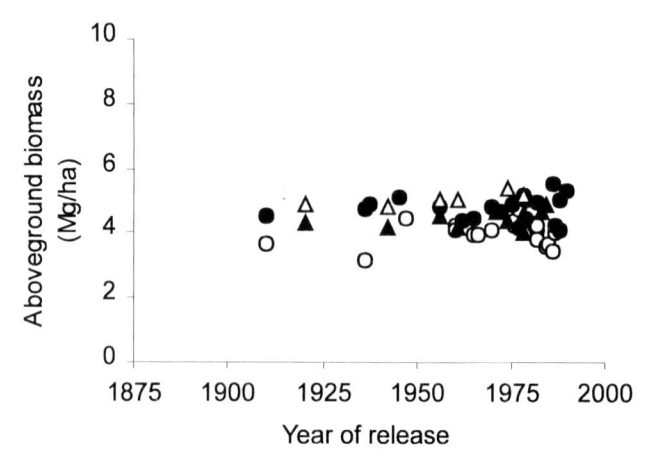

Source: Closed circles, Bulman, Mather, and Smith, 1993, and open circles, Jedel and Helm, 1994a (Canada); open squares, Riggs et al., 1981 (United Kingdom); open triangles, Wych and Rasmusson, 1983, and closed triangles, Boukerrou and Rasmusson, 1990 (United States).

(Hay, 1995). Most of the modern cultivars reached harvest index values of 50 percent, which suggests that future improvement of grain yield may not rely on further increasing harvest index (see Austin et al., 1980).

Numerical Yield Components

Grain yield has traditionally been divided into two main components: grains/m^2 and individual grain weight. Numbers of spikes/m^2 and number of grains per spike are the primary factors considered in analyzing the former main component.

Most studies reported a positive correlation between grain yield and number of grains/m^2 (see Table 13.4), and in the case of six-rowed barleys in Italy, the improvement in number of grains/m^2 was strong enough to maintain a positive rate of yield improvement against a significantly negative trend in individual grain weight (see Table 13.4). However, for two-rowed cultivars released in different eras in Italy and in the United States, a positive and significant association between grain yield and grain weight was found. In addition, in these countries, grain yield was more strongly associated with grain weight than with grain number (see Table 13.4).

The increase in grain number through genetic improvement was more closely associated with the number of spikes/m^2 than with the number of

FIGURE 13.6. Relationship Between Grain Yield and Harvest Index in Cultivars Released in Different Eras in Canada, Italy, the United Kingdom, and the United States

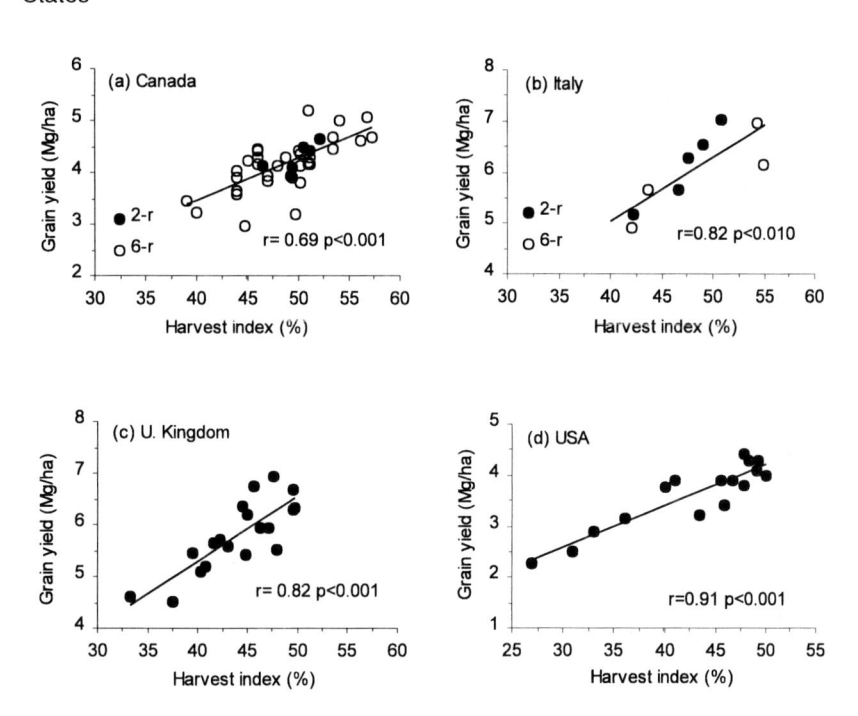

Source: (a) Bulman, Mather, and Smith, 1993; Jedel and Helm, 1994a; (b) Martiniello et al., 1987; (c) Riggs et al., 1981; (d) Wych and Rasmusson, 1983; Boukerrou and Rasmusson, 1990.

grains per spike, except in six-rowed barleys (see Table 13.4). In the United Kingdom, old and modern barleys differed in their number of tillers per plant. The increase in the number of spikes/m^2 was at a rate of 2.3 spikes/m^2 per year between 1880 and 1980 (Riggs et al., 1981). Whereas in North America, barley cultivars that represented the period from 1920 to 1978 (Wych and Rasmusson, 1983) differed significantly in spikes/m^2 (with a rate of 0.9 spikes/m^2 per year), although Boukerrou and Rasmusson (1990) reported that no trend associated with time for this component. Estimation of breeding progress achieved in Italian winter barley showed that modern two- and six-rowed cultivars had more tillers/m^2 than the old cultivars (Martiniello et al., 1987). In six-rowed cultivars, the number of grains per spike increased ten units, from the end of the nineteenth century to the 1980s, and this was,

TABLE 13.4. Correlation Coefficient (r) Between Grain Yield and Its Numerical Components for Cultivars Released at Different Eras in Different Countries

Country		Period	Between Grain Yield and		Between number of grains/ m^2 and		Source
			Grains/m^2	Grain Weight	Spikes/m^2	Grains/Spike[1]	
Canada	2-rowed	1960-1982	0.48*	−0.68 NS	0.96****	0.18 NS	Jedel and Helm (1994b)
	6-rowed	1910-1987	0.75*	0.43 NS	−0.78****	0.24 NS	
Italy	2-rowed	1900-1980s	0.42 NS	0.90**	0.88***	0.69 NS	Martiniello et al. (1987)
	6-rowed	1900-1980s	0.98***	−0.87**	0.94**	0.99****	
United Kingdom		1880-1980	0.79****	0.22 NS	0.89****	0.02 NS	Riggs et al. (1981)
United States		1920-1978	0.84**	0.91***	0.86**	0.59 NS	Wych and Rasmusson (1983)
		1920-1984	0.20 NS	0.84***	0.59*	0.18 NS	Boukerrou and Rasmusson (1990)

*$p < 0.1$, **$p < 0.05$, ***$p < 0.01$, ****$p < 0.001$
NS = not significant

apparently, the only case where number of grains/m^2 was correlated with number of grains per spike (Martiniello et al., 1987; $y = 696x + 8547$, $r = 0.99$, $p < 0.001$).

Improvement of grain yield associated with grain weight showed a rate of 0.06 mg/grain per year in the United States (Wych and Rasmusson, 1983, $r = 0.96$, $p < 0.001$; Boukerrou and Rasmusson, 1990, $r = 0.82$, $p < 0.05$). Similarly, in Italy, two-rowed cultivars had significant increases in grain weight, but the six-rowed genotypes decreased their grain weight by about 25 percent (see Table 13.5).

Nitrogen Partitioning and Grain Quality

Nitrogen is one of the major environmental factors affecting biomass production and final grain yield and quality. However, very few studies have evaluated the effect of barley breeding on nitrogen economy. As far as we are aware, just three studies corresponding to two countries have analyzed the influence of genetic improvement on nitrogen uptake and partitioning in barley under field conditions, i.e., two conducted in Canada (Bulman, Mather, and Smith, 1993; Jedel and Helm, 1994a) and one conducted in the United States (Wych and Rasmusson, 1983).

Grain quality is not just restricted to nitrogen content of grains. The evaluation of malting quality includes various parameters (Molina-Cano, 1989). From an economical point of view, the most important is malt extract (Molina-Cano, 1987). This character is generally closely and positively associated with grain weight, but frequently negatively related to protein content of grains (Hsi and Lambert, 1954, and see Savin and Molina-Cano, Chapter 19 in this book).

Grain protein concentration was not consistently affected by genetic improvement during the twentieth century in the United States (Wych and Rasmusson, 1983). In Canada, however, this trait was higher in older than in modern cultivars (Bulman, Mather, and Smith, 1993; Jedel and Helm, 1994a). It is interesting to note that grain nitrogen concentration followed a quadratic pattern, with the concentration being more affected during the past years (see Figure 13.7a). The initial value of grain protein concentration remained constant or even slightly increased until approximately 1940, after which it sharply decreased. This could likely be related to the shape of the relationship between grain yield and the year of release of cultivars shown earlier (see Figure 13.1). In addition, genetic improvement tended to increase diastatic power and alpha-amylase content as well as the percentage of plump grains and malt extract (see Figure 13.7b; Wych and Rasmusson, 1983).

Unfortunately, total nitrogen absorbed by the crop was scarcely studied. Studies evaluating this trait were carried out in Canada and the United States, and the authors found that genetic improvement increased total nitrogen absorbed by the crop (Wych and Rasmusson, 1983; Bulman, Mather,

TABLE 13.5. Grain Yield, Number of Grains/m^2, and Grain Weight, and Their Percentage of Change in the Oldest and Newest Cultivars Released in Different Eras in Different Countries

Country	Period	Grain Yield (Mg/ha)			Number of Grains/m^2			Grain Weight (mg)			Source
		Oldest Cultivars	Newest Cultivars	Change (%)	Oldest Cultivars	Newest Cultivars	Change (%)	Oldest Cultivars	Newest Cultivars	Change (%)	
Canada 2-4 rowed	1960-1982	4.0	4.4	9	12,167	15,908	24	32.5	33.9	4	Jedel and Helm (1994b)
6-rowed	1910-1987	3.1	4.4	30	18,770	21,226	12	43.8	40.7	−8	
Italy 2-rowed	1900-1980s	5.7	6.6	14	12,385	13,555	9	45.7	48.9	7	Martiniello et al. (1987)
6-rowed	1900-1980s	4.9	6.6	25	9,685	15,860	39	50.8	41.3	−23	
United Kingdom	1880-1980	3.9	5.2	25	9,885	13,729	28	39.3	38.0	−3	Riggs et al. (1981)
United States	1920-1978	2.5	3.6	31	15,008	16,500	9	27.1	29.9	9	Wych and Rasmusson (1983)
	1920-1984	3.3	4.1	20	15,933	16,695	5	35.3	37.8	7	Boukerrou and Rasmusson (1990)

Note: Genotypes were classified as oldest and newest ones, i.e., those released in the first and last thirds, respectively, of the breeding period considered in each study.

FIGURE 13.7. Malting Quality Traits

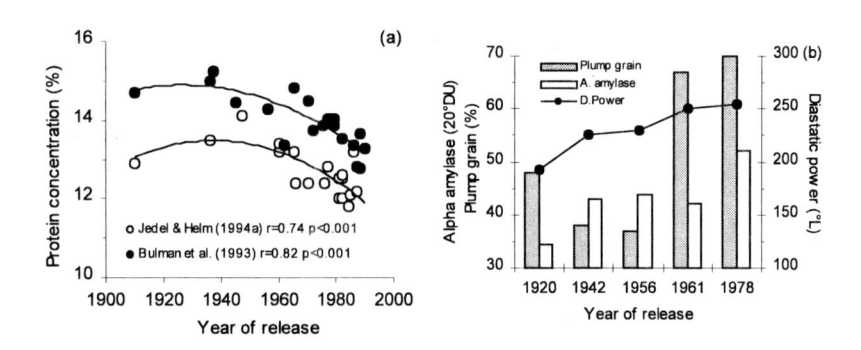

Source: (a) open circles, Bulman, Mather, and Smith, 1993; closed circles, Jedel and Helm, 1994a; (b) Wych and Rasmusson, 1983.

Note: (a) Relationship between percentage of total grain protein concentration and the year of release of cultivars in Canada and the United States. (b) Comparison of alpha-amylase, plump grain, and diastatic power in cultivars released in different eras in the United States.

and Smith, 1993). Nitrogen partitioning to the grain was not altered with time, despite the fact that newer cultivars accumulated more grain nitrogen (Bulman, Mather, and Smith, 1993; Wych and Rasmusson, 1983). Therefore, dry-matter partitioning shows a far greater increase at maturity than nitrogen partitioning. This difference is clearly shown by the ratio between nitrogen harvest index and dry-matter harvest index throughout the breeding process (see Figure 13.8).

Only one study measured the postheading nitrogen uptake (considered as the difference between total plant nitrogen at maturity and total plant nitrogen at heading), but this trait was not related to the year of release of the cultivars (Bulman, Mather, and Smith, 1993). Although quality of malting barley could be as important as yield, there is little knowledge about how it changed with breeding. For example, only one of the studies included in this chapter analyzed malt extract in cultivars released in different eras (Wych and Rasmusson, 1983), and the authors found clear trends throughout breeding. Apart from this, there is no doubt that the consequences of changes in nitrogen economy with genetic improvement of yield need to be investigated more intensively, with the aim of identifying opportunities and difficulties for future breeding programs (Cregan and Van Berkum, 1984).

FIGURE 13.8. Changes in Nitrogen Harvest Index (NHI) to Harvest Index (HI) Ratio with the Year of Release of the Cultivars

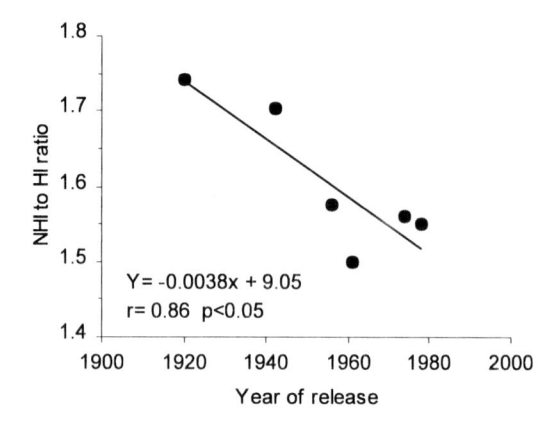

[Scatter plot: NHI to HI ratio vs Year of release; $Y = -0.0038x + 9.05$; $r = 0.86$, $p < 0.05$]

Source: Ratio data from Wych and Rasmusson, 1983.

COMPARISON BETWEEN BREEDING PROGRESS REACHED IN BARLEY AND OTHER SMALL-GRAIN CEREALS

The analysis of plant breeding impact on grain yield and related traits, as discussed earlier, has received different efforts depending on the crop. For example, many studies have been carried out to determine plant breeding effects on wheat (see Feil, 1992; Loss and Siddique, 1994; Slafer, Satorre, and Andrade, 1994; Calderini, Reynolds, and Slafer, 1999). On the other hand, breeding effects on barley physiology have received much less attention than those on wheat, and even less research has been conducted on oat (see Peltonen-Saino, 1994), triticale, and rye. In addition, to the best of our knowledge, no studies have reported on the effects of plant breeding on these crops in a single experiment. This lack of information and lack of balance between research on different cereals should be kept in mind throughout this section because a partial evaluation of plant breeding programs could strongly influence the conclusions. However, to analyze comparatively genetic gain effects on different cereals would provide an interesting opportunity to check general assumptions about plant breeding effects and, more important, to set the bases for a future wider analysis to compare the effects of plant breeding on crop physiology. Due to the lack of studies evaluating the effects of genetic gains on different small-grain cereals in the same experiment, a useful approach to compare them could be the analysis of data from countries where plant breeding has been evaluated in barley and, at least, one more small-grain cereal crop. A literature survey showed that this information is available for Canada, Finland, Italy, the United Kingdom, and the United States. In addition, to evaluate the effect of plant breeding, complementary information could be obtained from independent experiments involving cultivars barlley, wheat, and oat released in different eras.

Genetic gains showed differences among countries (as discussed earlier) and crops (see Figure 13.9). Among countries, Italy showed the greatest increase, with both barley and wheat reaching about fourfold the values found in other countries, and data from Finland's cultivars were among the lowest values (see Figure 13.9). The highest gains found in Italy could be explained by the shorter and later period analyzed by Martinello et al. (1987) for barley and complemented by the same period analysis reported for wheat by Canevara et al. (1994), i.e., from 1960 to the 1980s. Among crops, plant breeding showed a greater increase in barley than in wheat in three out of the five analyzed countries (see Figure 13.9). In addition, genetic gains in oat

FIGURE 13.9. Absolute and Relative Genetic Gains Reported for Barley, Wheat, and Oat in Canada, Finland, Italy, the United Kingdom, and the United States

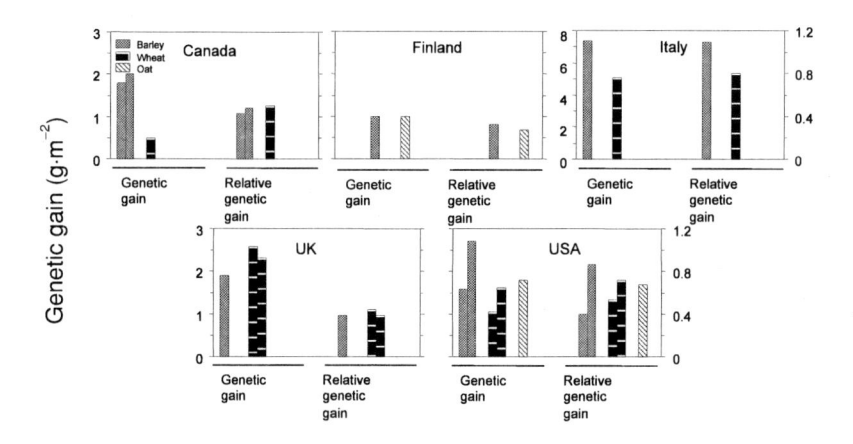

Source: (Canada) Bulman, Mather, and Smith, 1993; Jedel and Helm, 1994a; Hucl and Baker, 1987; (Finland) Peltonen-Sainio, 1990; Peltonen-Sainio and Peltonen, 1994; (Italy) Martinello et al., 1987; Canevara et al., 1994; (United Kingdom) Austin et al., 1980; Austin, Ford, and Morgan, 1989; Riggs et al., 1981; (United States) Wych and Rasmusson, 1983; Boukerrou and Rasmusson, 1990; Deckerd, Busch, and Kofoid, 1985; Cox et al., 1988; Wych and Stuthman, 1983.

Note: The genetic gain scale (g·m^{-2}) for Italy is different.

seemed to be similar or slightly higher than in wheat, depending on the study (see Figure 13.9d).

Most differences among cereals were associated with the environmental conditions of the study in which genetic gains were assessed and, consequently, were removed using relative genetic gains (see Figure 13.9), although some differences remained.

The reduction of plant height was a clear effect of breeding programs on both barley and wheat (e.g., Riggs et al., 1981; Austin et al., 1980; Slafer and Andrade, 1989). However, this effect was less clear in oat, as contrasting results have been reported. Thus, whereas Peltonen-Sainio (1994) and Lewes (1977) found a trend of plant height reduction in Finland and the United Kingdom, respectively, Wych and Stuthman (1983) did not find any trend in oats released in different eras in the United States. This lack of plant height reduction could be attributed to the fact that the oat straw is also harvested (see Wych and Stuthman, 1983).

In wheat, plant breeding reduced plant height as a continuous trend in most of the countries throughout the twentieth century, with the unique exception of India, where there was no plant height reduction (nor grain yield increase) before the green revolution (see Sinha et al., 1981; Kulshrestha and Jain, 1982; Calderini, Reynolds, and Slafer, 1999). This was not the case in barley, as shown earlier (see Figure 13.4). These differences between plant breeding programs could be due to the fact that plant height in wheat was 130 cm or higher at the beginning of the century, whereas this trait was more variable in barley (about 140 and 100 cm in cultivars from the United Kingdom and the United States, respectively). In oat, plant height of about 140 cm was reported in the study carried out by Mac Key (1988); however, the Finnish study showed that old cultivars (released in 1921) were only 90 cm tall (Peltonen-Sainio, 1994). In the United States, this trait was about 110 cm in the oldest cultivar (released in 1923) considered by the study carried out by Wych and Stuthman (1983). Modern cultivars show plant height values between 75 and 85 cm in barley, wheat, and oat, with the exception of oat in the United States, at about 110 cm (see Figure 13.4; Calderini, Reynolds, and Slafer, 1999; Peltonen-Sainio, 1994; Wych and Stuthman, 1983).

It is generally agreed that harvest index has been the main trait related to grain yield increase in small-grain cereals (see Figure 13.10). This was shown for barley (see earlier) and has been reported for wheat (Loss and Siddique, 1994; Slafer, Satorre, and Andrade, 1994; Calderini, Reynolds, and Slafer, 1999) and oat (Wych and Stuthman, 1983; Peltonen-Sainio, 1994). Therefore, and taking into account the effect of plant breeding on plant height, the increase in grain yield might have been due to the decrease in stem dry matter releasing carbohydrates to be allocated into reproductive organs. Similar values in stem weight decrease and grain yield increase were found in wheat (Calderini, Dreccer, and Slafer, 1995). In agreement,

FIGURE 13.10. Relationship Between Grain Yield and Harvest Index of Barley, Wheat, and Oat in Different Countries

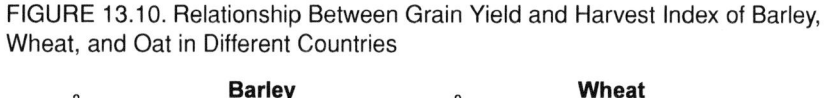

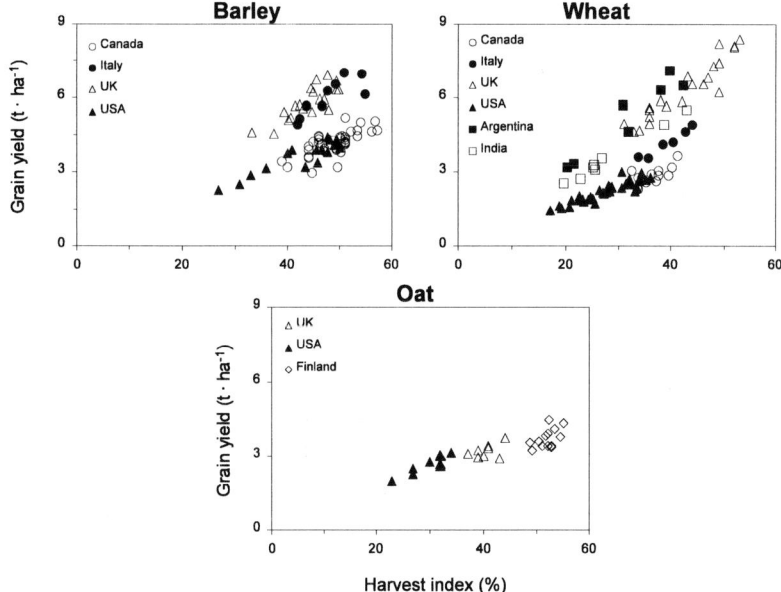

Source: (barley) Bulman, Mather, and Smith, 1993; Jedel and Helm, 1994a; Martinello et al., 1987; Riggs et al., 1981; Wych and Rasmusson, 1983; Boukerrou and Rasmusson, 1990; (wheat) Hucl and Baker, 1987; Canevara et al., 1994; Austin, Ford, and Morgan, 1989; Cox et al., 1988; Calderini, Dreccer, and Slafer, 1995; Kulshrestha and Jain, 1982; (oat) Wych and Stuthman, 1983; Peltonen-Sainio and Karjalainen, 1991; Mac Key, 1988.

negative associations between harvest index and vegetative biomass were reported for the United Kingdom and Australia (Austin, Ford, and Morgan, 1989; Siddique, Kirby, and Perry, 1989).

As shown earlier, most of the plant height values of barley, wheat, and oat are within the range of plant heights (between 70 and 100 cm) at which grain yield is maximized in wheat (Richards, 1992; Miralles and Slafer, 1995). If this is also true for other small-grain cereals, the conscious or unconscious strategy of increasing harvest index by reducing plant height, which was a remarkably successful strategy in the past, seems unlikely to keep yielding the same success (see Slafer, Araus, and Richards, 1999; Reynolds, Rajaram, and Sayre, 1999). In addition, harvest index values of modern cultivars (see Figure 13.10) are near the theoretical limit of biomass partitioning (about 60 percent) calculated by Austin et al. (1980). Therefore, increasing crop biomass should be the aim of future plant breeding programs.

In the past, biomass was almost not affected by plant breeding of small-grain cereals, with a few exceptions. In barley, data shown in Figure 13.5 confirm this, and the positive association between grain yield and biomass found by Wych and Rasmusson (1983) seems to be more an exception than a rule. In wheat, biomass was not constantly modified by breeding, as shown by several reviews (see Feil, 1992; Loss and Siddique, 1994; Slafer, Satorre, and Andrade, 1994; Calderini, Reynolds, and Slafer, 1999). Very few exceptions have been found in which the increase of grain yield was associated with biomass, and most of them showed that the increase in biomass was responsible for less than 20 percent of the genetic gain in grain yield (see Calderini, Reynolds, and Slafer, 1999). In oat, although a positive association ($r = 0.77$; $p < 0.05$) between aboveground biomass and the year of release was found (Wych and Stuthman, 1983), this increase (12 percent) only partially explained yield gains (40 percent), which were, in fact, closely associated with harvest index gains (see Figure 13.10). On the other hand, grain yield of modern cultivars released in Finland was more associated with biomass increase (50 percent) than with harvest index gain (20 percent). Probably this was due to the higher harvest index of old cultivars (about 49 percent), and consequently the more narrow range in which to increase it (modern cultivars have harvest index values of about 60 percent), found in Finland (Peltonen-Sainio, 1990) as compared to oats from the United Kingdom and the United States (see Figure 13.10).

Numerical components of grain yield showed less agreement among crops. Undoubtedly, grain number/m^2 was the component that better explained yield increases in wheat (Loss and Siddique, 1994; Slafer, Satorre, and Andrade, 1994; Calderini, Reynolds, and Slafer, 1999), whereas grain weight has not been markedly modified by wheat breeding (Austin et al., 1980; Waddington et al., 1986). In addition, some results even showed that individual grain weight was reduced by genetic improvements (Slafer and Andrade, 1989; Loss et al., 1989). Only a few cases have shown increases in this trait during the twentieth century (Cox et al., 1988) or the last part of it (from 1987) (Calderini, Dreccer, and Slafer, 1995). Therefore, not only was grain weight much less affected than grain number by wheat breeding, but also changes in grain weight were not consistent among studies. The general behavior in barley (see Table 13.4) is similar to that described for wheat but less conclusive (due to the lower number of cases and less clear relationships in the cases reported). Probably this difference with respect to wheat could be due to the importance of grain weight and size for quality in malting barleys (see Savin and Molina-Cano, Chapter 19 in this book). In oat, results reported contradictory effects of plant breeding. For example, grain number was consistently increased in Finland, while no association was found between grain yield and grain weight (Peltonen-Sainio and Peltonen, 1994). On the other hand, in the United States, a clear association between grain yield and grain weight ($r = 0.86$; $p < 0.01$) was reported by Wych and Stuthman (1983), with no trends in grain number throughout the analyzed period. This could be due to the fact that plant height was not modified in these oat crops (Wych and Stuthman, 1983), a process that has been found to increase grains/m^2 (see earlier).

In the few studies that analyzed nitrogen uptake in barley and oat, a positive association between this trait and the year of release of cultivars was found ($r = 0.92$; $p < 0.05$ and $r = 0.93$; $p < 0.001$ for barley and oat, respectively; Wych and Rasmusson, 1983; Wych and Stuthman, 1983). On the contrary, most of the studies conducted in wheat concluded that breeding did not consistently modify the amount of nitrogen absorbed by the crop (Fischer and Wall, 1976; Paccaud, Fossati, and Cao, 1985; Feil and Geisler, 1988; Slafer, Andrade, and Feingold, 1990; Peltonen-Sainio and Peltonen, 1994; Calderini, Torres León, and Slafer, 1995), and just few studies found increases in nitrogen uptake in wheat (e.g., Ortiz-Monasterio et al., 1997). Despite these differences among crops, modern cultivars showed a higher nitrogen-use efficiency (i.e., the yield produced per unit of nitrogen absorbed) than their older counterparts. Among crops, a weak association between aboveground nitrogen uptake and the year of release of cultivars was found in oat ($r = 0.63$; $p < 0.10$; Wych and Stuthman, 1983). In addition, most of these studies concluded that nitrogen harvest index was increased by plant breeding. As genetic variation in total nitrogen uptake in these crops has been recognized by different authors, future breeding programs could exploit this variability to increase the biomass production in new cultivars.

CONCLUSIONS

This chapter described physiological changes associated with breeding progress in barley during this century. The main conclusion is that plant breeding has increased grain yield potential at a rate higher than 0.5 percent per year since the 1940s. This was primarily because genotypes increased their dry-matter partitioning to grains (i.e., harvest index). Modern cultivars also exhibited shorter stems, likely due to breeders selecting for lodging resistance and grain yield. The optimum height for yield in wheat is between 70 and 100 cm (Richards, 1992; Miralles and Slafer, 1995). If this were also true for barley (and oat), modern cultivars would have already a stature close to the optimum, and further reductions would not maintain increases in potential yield (Slafer, Araus, and Richards, 1999).

Improvements in yield have mainly involved an increase in the number of grains per unit area, which is associated with an increase in the number of spikes per unit area. Grain weight has been especially important in the studies from the United States; this trait requires more attention because grain size is central to quality.

REFERENCES

Austin R. B., Bingham J., Blackwell R. D., Evans L. T., Ford M. A., Morgan C. L., and Taylor M. (1980). Genetic improvement in winter wheat yields since 1900 and associated physiological changes. *Journal of Agricultural Science*, Cambridge 94: 675-689.

Austin R. B., Ford M. A., and Morgan C. L. (1989). Genetic improvement in the yield of winter wheat: A further evaluation. *Journal of Agricultural Science*, Cambridge, 112: 295-301.

Boukerrou L. and Rasmusson D. D. (1990). Breeding for high biomass yield in spring barley. *Crop Science* 30(1): 31-35.

Bulman P., Mather D. E., and Smith D. L. (1993). Genetic improvement of spring barley cultivars grown in eastern Canada from 1910 to 1988. *Euphytica* 71(1): 35-48.

Byrnes B. H. and Bumb B. L. (1998). Population growth, food production and nutrient requirements. *Journal of Crop Production* 2: 1-27.

Calderini D. F., Dreccer M. F., and Slafer G. A. (1995). Genetic improvement in wheat yield and associated traits: A re-examination of previous results and the latest trends. *Plant Breeding* 114: 108-112.

Calderini D. F., Reynolds M. P., and Slafer G. A. (1999). Genetic gains in wheat yield and main physiological changes associated with them during the 20th century. In *Wheat: Ecology and Physiology of Yield Determination*, eds. E. H. Satorre and G. A. Slafer. Binghamton, NY: Food Products Press, pp. 351-377.

Calderini D. F. and Slafer G. A. (1999). Has yield stability changed with genetic improvement of wheat yield? *Euphytica* 107: 51-59.

Calderini D. F., Torres León S., and Slafer G. A. (1995). Consequences of wheat breeding on nitrogen and phosphorus yield, grain nitrogen and phosphorus concentration and associated traits. *Annals of Botany* 76: 315-322.

Canevara M. G., Romani M., Corbellini M., Perenzin M., and Borghi B. (1994). Evolutionary trends in morphological, physiological, agronomical and qualitative traits of *Triticum aestivum* L. cultivars bred in Italy since 1900. *European Journal of Agronomy* 3: 175-185.

Cox T. S., Shroyer J. P., Ben-Hui L., Sears R. G., and Martin T. J. (1988). Genetic improvement in agronomic traits of hard red winter wheat cultivars from 1919 to 1987. *Crop Science* 28: 756-760.

Cregan P. B. and Van Berkum P. (1984). Genetics of nitrogen metabolism and physiological/biochemical selection for increased grain crop productivity. *Theoretical and Applied Genetics* 67: 97-111.

Deckerd E. L., Busch R. H., and Kofoid K. D. (1985). Physiological aspects of spring wheat improvement. In *Exploitation of Physiological and Genetic Variability to Enhance Crop Productivity,* eds. J. E. Harper, L. E. Schrader, and R. W. Howell. Rockland, MD: American Society of Plant Physiology, pp. 45-54.

Feil B. (1992). Breeding progress in small grain cereals—A comparison of old and modern cultivars. *Plant Breeding* 108: 1-11.

Feil B. and Geisler G. (1988). Untersuchungen zur Bildung und Verteilung der Biomasse bei alten und neuen deutschen Sommerweizensorten. *Journal of Agronomy and Crop Science* 161: 148-156.

Fischer R. A. and Wall P. C. (1976). Wheat breeding in Mexico and yield increases. *Journal of the Australian Institute of Agricultural Science* 42: 139-148.

Hay R. K. M. (1995). Harvest index: A review of its use in plant breeding and crop physiology. *Annals of Applied Biology* 126: 197-216.

Hsi C. H. and Lambert J. W. (1954). Inter- and intra-annual relationships of some agronomic and malting quality characters of barley. *Agronomy Journal* 46: 470-474.

Hucl P. and Baker R. J. (1987). A study of ancestral and modern Canadian spring wheats. *Canadian Journal of Plant Science* 67: 87-97.

Jedel P. and Helm J. H. (1994a). Assesment of western Canadian barleys of historical interest. I. Yield and agronomic traits. *Crop Science* 34(4): 922-927.

Jedel P. and Helm J. H. (1994b). Assesment of western Canadian barleys of historical interest. II. Morphology and phenology. *Crop Science* 34(4): 927-932.

Kulshrestha V. P. and Jain H. K. (1982). Eighty years of wheat breeding in India: Past selection pressures and future prospects. *Zeitschrift fur Pflanzenzüchtung.* 89: 19-30.

Lewes D. A. (1977). Yield improvement in spring oats. *Journal of Agricultural Science, Cambridge* 89: 751-757.

Loss S. P., Kirby E. J. M., Siddique K. H. M., and Perry M. W. (1989). Grain growth and development of old and modern Australian wheats. *Field Crops Research,* 21: 131-146.

Loss S. P. and Siddique K. H. M. (1994). Morphological and physiological traits associated with wheat yield increases in Mediterranean enviroments. *Advances in Agronomy* 52: 229-276.

Mac Key J. (1988). Shoot:root interactions in oats. In *Proceedings of the 3rd International Oat Conference,* eds. B. Mattsson and R. Lyhagen. Svalöf AB, Sweden: Lund, pp. 340-344.

Martiniello P., Delogu G., Oboardi M., Boggini G., and Stanca A. M. (1987). Breeding progress in grain yield and selected agronomic characters of winter barley (*Hordeum vulgare* L.) over the last quarter of a century. *Plant Breeding* 99: 289-294.

Miralles D. J. and Slafer G. A. (1995). Yield, biomass and yield components in dwarf, semi-dwarf and tall isogenic lines of spring wheat under recommended and late sowing dates. *Plant Breeding* 114: 392-396.

Molina-Cano J. L. (1987). The EBC barley and malt committee index for the evaluation of malting quality in barley and its use in breeding. *Plant Breeding* 98: 249-256.

Molina-Cano J. L. (1989). Fabricación de malta y cerveza. Calidad cervecera de la cebada. In *La cebada. Morfología, fisiología, genética, agronomía y usos industriales,* ed. J. L. Molina-Cano. Madrid: Ediciones Mundi-Frensz, pp. 213-216.

Muñoz P., Voltas J., Araus J. L., Igartua E., and Romagosa I. (1998). Changes over time in the adaptation of barley releases in north-eastern Spain. *Plant Breeding* 117: 531-535.

Ortiz-Monasterio J. I., Sayre K. D., Rajaram S., and McMahon M. (1997). Genetic progress in wheat yield and nitrogen use efficiency under four nitrogen rates. *Crop Science* 37: 398-904.

Paccaud F. X., Fossati A., and Cao H. S. (1985). Breeding for yield and quality in winter wheat: Consequences for nitrogen uptake and partitioning efficiency. *Zeitschrift fur Pflanzenzuchtung* 94: 89-100.

Peltonen-Sainio P. (1990). Genetic improvement in the structure of oat stands in northern growing conditions during this century. *Plant Breeding* 104: 340-345.

Peltonen-Sainio P. (1994). Productivity of oats: Genetic gains and associated physiological changes. In *Genetic Improvement of Field Crops,* ed. G.A. Slafer. New York: Marcel Dekker, pp. 69-94.

Peltonen-Sainio P. and Karjalainen R. (1991). Genetic yield improvement of cereal varieties in northern agriculture since 1920. *Acta of Agricultural Scandinavia* 41: 267-273.

Peltonen-Sainio P. and Peltonen J. (1994). Progress since the 1930s in breeding for yield, its components, and quality traits of spring wheat in Finland. *Plant Breeding* 113: 177-186.

Perry M. W. and D'Antuono M. F. (1989). Yield improvement and associated characteristics of some Australian spring wheat cultivars introduced between 1860 and 1982. *Australian Journal of Agricultural Research* 40: 457-472.

Pinthus M. J. (1973). Lodging in wheat, barley and oats: The phenomenon, its causes, and preventive measures. *Advances in Agronomy* 25: 209-263.

Reynolds M. P., Rajaram S., and Sayre K. D. (1999). Physiological and genetic changes of irrigated wheat in the post-green revolution period and approaches for meeting projected global demand. *Crop Science* 39: 1611-1621.

Richards R. A. (1992). The effect of dwarfing genes in spring wheat in dry environments. I. Agronomic characteristics. *Australian Journal of Agricultural Research* 43: 517-523.

Riggs T. J., Hanson P. R., Start N. D., Miles D. M., Morgan C. L., and Ford M. A. (1981). Comparison of spring barley varieties grown in England and Wales between 1880 and 1980. *Journal of Agricultural Science, Cambridge* 97: 599-610.

Sayre K. D., Singh R. P., Huerta-Espino J., and Rajaram S. (1998). Genetic progress in reducing losses to leaf rust in CIMMYT-derived Mexican spring wheat cultivars. *Crop Science* 38: 654-659.

Shorter R., Lawn R. J., and Hammer G. L. (1991). Improving genotypic adaptation in crops—A role for breeders, physiologists and modellers. *Experimental Agriculture* 27: 155-175.

Siddique K. H. M., Kirby E. J. M., and Perry M. W. (1989). Ear to stem ratio in old and modern wheats: Relationship with improvement in number of grains per ear and yield. *Field Crops Research* 21: 59-78.

Silvey V. (1986). The contribution of new varieties to cereal yields in England and Wales between 1947 and 1983. *Journal of the National Institute of Agricultural Botany* 17: 155-168.

Sinha S. K., Aggarwal P. K., Chaturvedi G. S., Koundal K. P., and Khanna-Chopra, R. (1981). A comparison of physiological and yield characters in old and new wheat varieties. *Journal of Agricultural Science* 97: 233-236.

Slafer G. A. and Andrade F. H. (1989). Genetic improvement in bread wheat (*Triticum aestivum* L.) yield in Argentina. *Field Crops Research* 21: 289-296.

Slafer G. A. and Andrade F. H. (1991). Changes in physiological attributes of the dry matter economy of bread wheat *(Triticum aestivum)* through genetic improvement of grain yield potential at different regions of the world: A review. *Euphytica* 58: 37-49.

Slafer G. A., Andrade F. H., and Feingold F. E. (1990). Genetic improvement of bread wheat (*Triticum aestivum* L.) in Argentina: Relationships between nitrogen and dry matter. *Euphytica* 50: 63-71.

Slafer G. A., Araus J. L., and Richards R. A. (1999). Physiological traits that increase the yield potential of wheat. In *Wheat: Ecology and Physiology of Yield Determination,* eds., E. H. Satorre and G. A. Slafer. Binghamton, NY: Food Products Press, pp. 379-415.

Slafer G. A., Satorre E. H., and Andrade F. H. (1994). Increases in grain yield in bread wheat from breeding and associated physiological changes. In *Genetic Improvement of Field Crops,* ed. G. A. Slafer. New York: Marcel Dekker, pp.1-68.

Stapper M. and Fischer R. A. (1990). Genotype, sowing date and planting spacing influence on high-yielding irrigated wheat in southern New South Wales. I. Phasic development, canopy growth and spike production. *Australian Journal of Agricultural Research* 41: 997-1019.

Strand E. (1994). Yield progress and the sources of yield progress in Norwegian small grain production 1960-92. *Norsk-Landbruksforsking* 8(2): 111-126.

Waddington S. R., Ransom J. K., Osmanzai M., and Saunders D. A. (1986). Improvement in the yield potential of bread wheat adapted to northwest Mexico. *Crop Science* 26: 698-703.

Wych R. D. and Rasmusson D. C. (1983). Genetic improvement in malting barley cultivars since 1920. *Crop Science* 23: 1037-1040.

Wych R. D. and Stuthman D. D. (1983). Genetic improvement in Minnesota-adapted oat cultivars released since 1923. *Crop Science* 23: 879-881.

Chapter 14

Spectral Vegetation Indices As Nondestructive Indicators of Barley Yield in Mediterranean Rain-Fed Conditions

Jordi Bort
Jaume Casadesús
José Luis Araus
Stefania Grando
Salvatore Ceccarelli

INTRODUCTION

For many decades, the classical empirical approach, in which grain yield per se is used as the main selection criterion, has been extensively followed by cereal breeders (Loss and Siddique, 1994). However, the rates of breeding progress varied according to the environment. Whereas the genetic gain in high-yielding environments has been particularly successful, it has been much slower in other areas, such as the Mediterranean, where barley is one of the main cereal crops, especially in the most drought-prone areas (Slafer et al., 1994). The empirical breeding approach has provided only modest yield increases in this region, where drought during the last part of the crop cycle is a major factor limiting cereal yield (Ceccarelli and Grando, 1996).

The use of morphological and physiological traits as indirect selection criteria for grain yield is an alternative breeding approach. However, the limited success (to date) of this approach may be due to a lack of understanding of the physiological factors most directly involved in determining yield, in addition to the absence of proper methods for evaluating them in a rapid, routine manner (Blum, 1988; Loss and Siddique, 1994; Richards, 1996). Among the most promising screening methods, some allow a quick screening of the physiological traits that are able to integrate the performance of the crop either over time (i.e., during the plant cycle) or at the highest levels of organization (i.e., whole plant, canopy) (Araus, 1996;

This study was supported by the CICYT research projects PB97-0865 and AGF 99-0611-C03-03 (Spain).

Richards, 1996). The yield of a crop during a given period of time and under particular growing conditions is determined by three major processes or integrative traits: the interception of incident solar irradiance by the canopy, the conversion of the intercepted radiant energy to potential chemical energy, and the harvest index (Hay and Walker, 1989). The first depends on the photosynthetic area of the canopy, while the second relies on the overall photosynthetic efficiency of the crop. These two processes, which are responsible for the overall crop biomass, are more affected than the harvest index by drought and other related stresses (Sinclair, 1988; Slafer et al., 1994; Cattivelli et al., 1994) typical of Mediterranean growing conditions.

In principle, the spectra reflected by crop canopies at different wavelengths through the photosynthetically active radiation (PAR) and near infrared radiation (NIR) regions of the electromagnetic spectrum provide rapid, nondestructive, simultaneous estimations of the two processes determining crop biomass at the canopy level (Field et al., 1994; Araus, 1996; Peñuelas and Filella, 1998). Such estimations require spectral reflectance indices, which are formulations based on simple operations between reflectances at given wavelengths (such as, for example, ratios and differences), on the derivative catching of sudden changes in reflectance across wavelengths (for example, around 700 nm, i.e., the red edge position [REP]), or on the (absolute) reflectance values at specific wavelengths.

To date, spectroradiometrical indices have been used mainly for the assessment of characteristics associated with the development of the (total) photosynthetic area of the canopy. The most widespread spectral vegetation index is the normalized difference vegetation index (NDVI) followed by the simple ratio (SR) (Wiegand and Richardson, 1990a,b; Baret and Guyot, 1991; Price and Bausch, 1995). These indices have been correlated positively with photosynthetic area and biomass in small-grain cereals such as barley (Elliott and Regan, 1993; Filella et al., 1996; Masoni et al., 1996; Peñuelas, Isla, et al., 1997; Bertholdsson, 1999), bread wheat (Fernández et al., 1994; Field et al., 1994; Bellairs et al., 1996), and durum wheat (Aparicio et al., 2000). On the other hand, these indices can saturate as the total chlorophyll content on a soil area basis increases, and they can be affected by external factors such as plant architecture, incident radiation intensity, or soil background (Filella and Peñuelas, 1995). To solve this, several indices—REP, dRE (maximum amplitude of the first derivative of the reflectance spectra in the red edge), sdRE (sum of amplitudes of the first of the reflectance spectra in the same range), SqdRE (indicator of the magnitude of the discontinuity of the reflectance spectra around 700 nm)—have been established, based on the evaluation of the red edge (RE), which is the range of wavelengths (between 680-780 nm) with the highest change in reflectance when the wavelength is increased from red to near infrared (Filella and Peñuelas, 1995).

Leaf area duration also affects the total photosynthetic area and thus the radiation accumulated by the canopy, especially during the grain filling of barley and other cereals grown in Mediterranean conditions, since terminal stresses may reduce leaf area duration and no new photosynthetic area is formed, or because early maturing genotypes can escape terminal stresses. Early senescence of photosynthetic tissues is characterized by chlorophyll degradation with increases in the ratios of total carotenoids to chlorophyll a and chlorophyll b to a. In this context, spectroradiometrical indices able to provide information on the relative changes in these pigments may be appropriate indicators of the duration of photosynthetic organs. Thus, the normalized total pigment to chlorophyll a ratio index (NPCI) has been defined as a reflectance index to evaluate the ratio of total carotenoids to chlorophylla (Peñuelas, Gamon, et al., 1993). The structural independent pigment index (SIPI) is calculated using the wavelengths that showed the best semi-empirical estimation of the carotenoids:chlorophyll a ratio (Peñuelas, Filella, and Baret, 1995). The ratio of chlorophyll a to b (Chla/b) can also be assessed spectroradiometrically (Blackburn, 1998), and the normalized phaeophytinization index (NPQI) can be used as an indicator of chlorophyll degradation (Peñuelas, Filella, Lloret, et al., 1995).

The photosynthetic radiation-use efficiency (the second component in the identity determining yield) of the PAR absorbed by the canopy has been correlated with the photochemical reflectance index (PRI) (Gamon et al., 1992, 1997; Peñuelas, Filella, and Gamon, 1995; Filella et al., 1996). Thus, the changes in photosynthetic radiation-use efficiency induced by factors such as nutritional state and midday reduction in different species and functional types may be followed using the PRI (Filella et al., 1996, Gamon et al., 1997). The PRI is a valuable index, since it tracks changes in the xantophyll pigments (mainly zeaxanthin) associated with the de-epoxidation state of the xanthophyll cycle (Filella et al., 1996). Owing to the importance of drought stress in Mediterranean conditions, the water index (WI) was also evaluated, since it has been reported to be a reliable indicator of water status (Peñuelas, Filella, et al., 1993, Peñuelas and Filella, 1998), especially in barley (Peñuelas, Isla, et al., 1997).

The aforementioned evidence highlights the advantages of spectroradiometrical techniques for agronomic and plant breeding purposes. However, most of these studies have been carried out on a reduced number of genotypes, mostly combining plants grown in different environments, which does not provide a clear idea as to whether such techniques are useful for breeding purposes. In addition, the lack of field portable spectroradiometers has restricted the application of this technique to breeding programs, in which the routine assessment of large amounts of germplasm is necessary. In this context, we aimed to study the performance of the aforementioned set of indices, obtained during a single observation date, by assessing genotype differences in yield and related parameters. Whereas drought stress is the main constraint to barley production in Mediterranean conditions, cultivation in rain-fed conditions (i.e., without supplementary irrigation) is the usual managing practice for this crop. Therefore, the evaluation of these spectral reflectance indices was carried out in two Mediterranean rain-fed environments with contrasting water availability. As Wiegand et al. (1991) pointed out, if most uncertainty in yield prediction by spectral reflectance indices is site dependent, then it is necessary to relate yield versus these indices across environments differing in water availability within the production area.

MATERIAL AND METHODS

Plant Material and Growing Conditions

Five groups of lines, each consisting of 116 to 132 entries of two- and six-row barley (*Hordeum vulgare* L.), were cultivated in 1999 in rain-fed conditions in two experimental stations of ICARDA (International Center for Agricultural Research in the Dry Areas): Tel Hadya, Syria (headquarters of ICARDA), and Breda, the Netherlands. In addition, each group of lines was also rain-fed cultivated in the field of a collaborating farmer. Thus, group 1 was sown at farmer location 1, group 2 at location 2, and so on. All the trials were located in northwest Syria. Soil characteristics of the two experimental stations and the farmer fields have been reported elsewhere (Araus et al., 1997, and Ceccarelli and Grando, 1999, respectively). The two experimental stations showed consistent differences in rainfall, ranging from moderately to very dry conditions: accumulated rainfall from September 1998 to June 1999 was 309 mm in Tel Hadya and 197 mm in Breda. Differences in evapotranspiration between sites were minor. Rainfall in the five farmer trials was in the same range. For locations 1 (Tel Tafer) and 2 (Ebla), the accumulated rainfall during the same period was 311 mm and 355 mm, respectively, whereas for locations 3, 4, and 5 (Tel Brak, Jurn El-Aswad, and Baylonan), it ranged between 190 mm (site 4) and 163 mm (site 5).

Each group of lines was assembled based on the selection conducted in 1998 by ICARDA breeders and collaborating farmers (Ceccarelli et al., 2000). Entries numbered 120, 132, 116, 116, and 128 for groups 1 to 5, respectively. They included a set of nearly 50 lines common to the five groups, whereas the remaining lines were selected to perform better either in moderately dry (groups 1 and 2) or in very dry environments (groups 3, 4, and 5). Lines were grown in an unreplicated row-column design, with 30 columns and four rows, which included a systematic check, different for each group, every ten entries. Trials in the two experimental stations and in the farmer fields were sown in late and mid-November 1998, respectively, in 12 m^2

plots (eight rows 20 cm apart). Plants emerged about two weeks later. Sowing density was around 300 viable seeds/m^2.

The Tel Hadya trial was fertilized with N at 60 kg·ha^{-1} and with P_2O_5 at 40 kg·ha^{-1} before planting, whereas the Breda trial was not fertilized. Farmer trial 1 received N at 35 kg·ha^{-1} before planting and an additional 35 kg·ha^{-1} as a top dressing, farmer trial 2 was top dressed with N at 50 kg·ha^{-1}, and farmer trial 3 received N at 35 kg·ha^{-1} before planting, whereas farmer trials 4 and 5 were not fertilized. No treatments were done to control pests, diseases, or weeds, which were not a problem. Plots from all the trials were mechanically harvested, and grain yield was determined from the central six rows of each plot. Thousand kernel weight was then measured. In addition, harvest index and grain yield were evaluated in the farmer trials. For the same set of genotypes cultivated the year before (1998) at Tel Hadya and Breda (Ceccarelli and Grando, 1999), grain yield and harvest index were measured at both sites, and days from planting to heading only at Tel Hadya.

Spectral Reflectance Measurements

Radiometric measurements were carried out in the Tel Hadya and Breda trials during the last week of April 1999. Plants were around middle grain filling, which corresponds to the stages 75 (medium milk-grain stage) to 85 (soft dough-grain stage) of Zadoks' decimal code (Zadoks et al., 1974). Only the plots without lodging were measured. Canopy reflectance was detected with a narrow-bandwidth, visible/near infrared, portable field spectroradiometer fitted with an 18° field of view optics (model FieldSpec UV/VNIR, Analytical Spectral Devices, Inc., Boulder, Colorado). This instrument detects 512 continuous bands (with a sampling interval of 1.4 nm) from 350 to 1,050 nm, which covers the visible and near infrared portion of the spectrum. Individual scans were remotely triggered from a portable computer and saved to disk for subsequent analysis. The measurements were performed at midday, in cloudless conditions. All spectral measurements were taken with the sensor at a zenith angle of 90°, with the field of view optic mounted on a tripod at 1.5 m above the soil, and with the radiometer in a nadir orientation. Three spectral reflectance measurements (of 1-2 s each) were taken at each plot, which was the average of five scans. The reflectance spectrum was calculated in real time as the ratio of the reflected versus the incident spectra on the canopy, where the incident spectrum was periodically obtained from the light reflected by a white reference panel, commercially available as Spectralon (Labsphere, NorthSutton, New Hampshire), with a lambertian-like surface. Reflectance of the reference was periodically measured on each of the five plots. Radiometric indices were calculated from spectral reflectances measured, i.e., reflectance by the canopy and reflectance by the soil background. The latter was homogeneous within each site and similar in the two sites.

Spectroradiometrical Indices

Photosynthetic Area Indices

The SR and NDVI were defined from canopy reflectances after Hall et al. (1990) as SR = R770/R680 and NDVI = (R770 − R680)/(R770 + R680), since the small difference between the two wavelengths minimizes the effect of the soil background. The REP was defined as the wavelength in the range 680 to 780 nm with the maximum change in reflectance. dRE was the maximum amplitude in the first derivative of the reflectance spectra in this range, whereas sdRE was the sum of those amplitudes (Filella and Peñuelas, 1995). We also introduced SqdRE as an indicator of the magnitude of the discontinuity of the spectra around 700 nm. This index magnifies the suddenness of the change in reflectance and is calculated as $\Sigma[si *ABS(si)]$), where si is the slope of the spectrum at each wavelength between 680 and 780 nm. R680 refers to the absorbance pick of the chlorophyll. At this wavelength, the reflectance is inversely proportional to the amount of green area and very sensitive to changes in chlorophyll content but saturated at very low chlorophyll concentrations (10 $\mu g \cdot cm^{-2}$). R550 is less sensitive to changes in chlorophyll content, but it is not saturated at such lower concentrations. R800 and R900 are used as references; at these wavelengths, neither carotenoids nor chlorophylls absorb, and these reflectances are affected by only the structure of the canopy. At 900 nm, the absorption by water is null.

Senescence Indices

The NPCI was defined after Peñuelas, Gamon, et al. (1993), with a change in the sign: NPCI = (R680 − R430)/(R680 + R430). Therefore, NPCI will be positively correlated with the ratio of total carotenoids to chlorophyll a. The SIPI was calculated following Peñuelas, Filella, and Baret (1995) as SIPI = (R800 − R445)/(R800 − R680), which minimizes the disrupting effects of leaf surface and mesophyll structure on the ratio of carotenoids to chlorophyll a. The NPQI was NPQI = (R415 − R435)/(R415 + R435) (Peñuelas, Filella, Lloret, et al., 1995). NPQI shows more negative values as more degradation of chlorophylls. Based on Blackburn (1998), we defined the index Chla/b = R650/R685.

Radiation-Use Efficiency Indices

The PRI was defined following Peñuelas, Filella, and Gamon (1995) as PRI = (R531 − R570)/(R531 + R570), which has been reported to be positively correlated with photosynthetic radiation-use efficiency (PRUE), and negatively with xantophyll levels. Thus, PRI is expected to be lower in stressed conditions.

Water Status Indices

The water index (WI) was defined as WI = R970/R900 (Peñuelas et al., 1996), where high WI indicates more water stress (Peñuelas et al., 1996). However, the inverse formulation (R900/R970) has also been used in other references (Peñuelas, Piñol, et al., 1997; Peñuelas and Filella, 1998), where low WI indicates more water stress.

Statistical Analysis

Spectroradiometrical indices, grain yield, and thousand grain weight data were spatially analyzed following the method of residual maximum likelihood (REML), using the ASREML package. Several models were fitted to the data. We chose the model that provided the best linear unbiased estimates for entries, and thus the highest F value together with the smallest error residue components. Pearson correlation coefficients were used to study the relationship between radiometric indices and agronomic variables. Environment and genotype effects on the traits studied were assessed through ANOVA analysis. The percentage of grain yield variation, explained by the progressive addition of spectral reflectance indices, was determined by the coefficient of determination of a stepwise fitting. A principal components analysis was conducted in order to summarize the possibly redundant information contained in the 16 spectroradiometrical indices calculated from each reflectance spectrum. The principal components were calculated using the whole set of data considering the two sites. The analysis of variance, the principal components analysis, and the stepwise analysis were carried out using the SPSS 9.01 software (SPSS, Inc.).

RESULTS

The growing conditions were markedly different between the two experimental stations. Mean grain yield across genotypes for the five groups of lines was around 1,000 kg·ha^{-1} at Breda and about 3.5 times higher at Tel Hadya in rain-fed conditions (see Table 14.1). Kernel weight was almost 30 percent higher in Tel Hadya than in Breda. Mean NDVI, SR, and sdRE of Breda doubled those of Tel Hadya, and PRI was one-third higher, whereas SIPI and R680 decreased about one- and two-thirds, respectively. dRE increased about 60 percent, and SqdRE was almost seven times lower, whereas the mean values for the other indices were similar in the two environments. The environment (site) had no significant effects on either WI or NPQI. Genotype effect was significant for grain yield, kernel parameters, and all the spectroradiometrical indices, except WI, NPQI, and REP (see Table 14.1). Genotype by environment (i.e., site) interaction was significant for all traits except WI, REP, and dRE. For grain yield, the percentage of the

TABLE 14.1. Agronomic and Spectroradiometrical Parameters Measured on a Collection of Barley Cultivars Grown in Syria in Two Rain-Fed Environments (Breda and Tel Hadya) During 1999

	BREDA		TEL HADYA		PERCENTAGE OF THE SUM OF SQUARES		
	Mean	SD	Mean	SD	Genotype	Site	Genotype x Site
SR	1.770	0.228	3.387	1.034	13.8**	35.7**	9.8**
NDVI	0.276	0.056	0.524	0.096	9.2*	47.3**	4.9**
NPCI	0.599	0.019	0.476	0.063	11.6**	40.3**	9.5**
SIPI	1.915	0.242	1.289	0.132	8.6**	47.0**	4.8*
NPQI	−0.109	0.025	−0.105	0.046	17.7 ns	0.1 ns	28.0**
PRI	−0.148	0.007	−0.103	0.016	8.1**	50.5**	4.3**
R550	0.093	0.006	0.099	0.025	33.8**	2.0*	31.7**
R680	0.146	0.013	0.093	0.024	12.0**	41.1**	7.2**
Chla/b	0.974	0.014	1.092	0.058	7.8**	45.5**	8.8**
WI	0.999	0.064	1.033	0.126	10.2 ns	0.2 ns	18.5 ns
REP	734.021	24.900	714.386	20.297	17.8 ns	15.6**	9.8 ns
dRE¶	4.634	1.909	7.376	2.100	17.6**	23.2**	13.0 ns
SdRE	0.113	0.024	0.210	0.057	12.7**	39.4**	10.6**
SqdRE¶	7.441	0.004	0.970	0.542	17.0**	32.4**	15.5**
R800	0.265	0.020	0.312	0.060	23.3**	18.5**	23.3**
R900	0.283	0.022	0.341	0.063	22.7**	22.1**	18.7**
GY99	1041.10	222.41	3573.24	522.20	2.5**	59.0**	1.9**
KW99	34.03	5.01	41.83	4.13	28.6**	30.6**	6.7**
DHTH98			121.1	5.8			
HI98	0.386	0.078	0.413	0.080	30.9**	0.3**	18.7**
GY98	941.120	183.00	3239.87	630.52	3.2**	54.4**	3.4**
HI99FF	0.378	0.160					
GY99FF	1689.15	1150.80					

Note: In addition, agronomic data from a farmer field trial as well as from a previous year (1998) trial at both sites are also included. Spectroradiometrical parameters were measured around mid-grain filling. Data in Breda from groups of lines 1, 2, 4, and 5 (see Material and Methods). Data in Tel Hadya from groups of lines 1, 2, 3, 4, and 5. GY99 and KW99: grain yield and kernel weight during 1999. DHTH98: days to heading at Tel Hadya during 1998. HI98 and GY98: harvest index and grain yield during 1998. HI99FF and GY99FF: harvest index and grain yield at the farmer fields during 1999. Left: mean and standard deviation. Right: percentage of the sum of squares of the analyses of variance explained by the main factors in the analyses: genotype, site, and the interaction between them.

¶ Mean and standard deviation (SD) must be divided by 1,000 to obtain actual values.
**p < 0.001, *p < 0.01
ns = not significant

overall sum of squares of the ANOVA explained by the genotype effect was minor (see Table 14.1). In contrast, the effect of genotype was similar to that of site for kernel weight and for some of the spectroradiometrical indices.

The analysis of variance was also performed in order to determine which indices showed significant differences among groups of lines selected for specific adaptation to contrasting growing conditions, as well as for the interactions between the lines and the two environments (see Table 14.2). The groups were clearly affected by the growing environment. At Tel Hadya, many indices showed smaller differences across the groups compared to Breda. When the groups grow under moderate stress, they can hardly be identified by their spectral sign, although the differences in grain yield during 1999 were significant. On the other hand, under severe stress, many spectroradiometrical indices showed significant differences among all groups of lines, as well as in grain yield during 1999. Grain yield during 1998 only showed differences between the two sources of groups, selected for different growing conditions. Thus, groups 1 and 2, selected for performance in moderately dry environments, were not different from each other, but they were significantly different from groups 4 and 5, which were selected for performance under severe stress. This may be because 1998 was much drier than 1999.

When the groups of lines are considered separately, the correlation coefficients between spectroradiometrical indices and grain yield were generally higher in Tel Hadya than in Breda (see absolute values of correlation coefficients in Figure 14.1). Thus, for Breda, NPCI was the index best correlated with yield, both within each group or considering all groups (see Table 14.3). SqdRE was the second index best correlated (negatively) with yield within each group (see Figure 14.1), but the correlation was lower (and positive) when all the lines of the four groups were considered (see Table 14.3). These indices at Breda (NPCI and SqdRE) were also among those most frequently chosen in the stepwise linear regression approaches within groups of lines (data not shown), addressed to explain genotype differences in yield. When all groups were considered, NPCI and SqdRE were, respectively, positively and negatively correlated with yield (see Table 14.4). Vegetation indices such as SR and NDVI were also correlated (negatively) with yield within groups. The correlation between SqdRE, SR, and NDVI and yield can be explained by the much stronger (negative) correlation between these indices and kernel weight (see Table 14.3), the yield component last defined during grain filling. However, the correlation between kernel weight and NPCI was still significant, but much lower, within each group of lines and absent, at both sites, when all groups were considered (see Table 14.3). Kernel weight was positively correlated with grain yield in all the groups of genotypes, but more weakly for the set of groups combined. On the other hand, many of the spectroradiometrical indices measured at Breda also

TABLE 14.2. Statistical Separation of Means by Duncan's Comparison Test Within the Factor Group of Lines for Every Agronomic and Spectroradiometrical Index Measured

	Two sites	Breda	Tel Hadya	Group of lines		Two sites	Breda	Tel Hadya
GY99	bc	d	b	1	R550	a	a	a
	c	c	c	2		ab	c	a
	c	b	d	4		b	b	b
	a	a	a	5		b	a	a
KW99	b	a	b	1	R680	a	a	a
	a	a	a	2		bc	c	a
	c	c	a	4		b	b	a
	b	b	a	5		c	d	a
SR	a	d	a	1	Chla/b	b	d	a
	ab	c	a	2		a	c	a
	b	b	b	4		c	b	c
	a	a	a	5		b	a	b
NDVI	b	d	a	1	WI	a	a	c
	b	c	ab	2		c	b	d
	b	b	c	4		b	c	a
	a	a	b	5		a	d	b
NPCI	c	d	b	1	REP	c	b	b
	a	c	a	2		d	c	c
	b	b	b	4		b	a	a
	b	a	b	5		a	b	a
SIPI	a	a	c	1	dRE	c	b	a
	b	b	bc	2		d	b	b
	d	c	a	4		b	a	b
	c	d	b	5		a	a	a
NPQI	a	a	b	1	sdRE		c	a
	c	b	c	2			b	a
	a	c	a	4		b	a	b
	b	d	a	5		a	a	a
PRI	a	a	a	1	SqdRE	b	c	a
	b	a	b	2		b	b	a
	ab	b	a	4		b	a	b
	a	b	a	5		a	a	a
				1	R800	a	c	a
				2		b	d	a
				4		b	a	b
				5		a	b	a
DHTH98			ab	1	R900	ab	c	a
			b	2		c	d	a
			a	4		bc	a	b
			a	5		a	b	a
HI98	d	ab	c	1	HI99FF	c		
	c	c	b	2		d		
	b	bc	a	4		b		
	a	a	a	5		a		
GY98	b	b	b	1	GY99FF	d		
	b	b	b	2		c		
	a	a	a	4		b		
	ab	a	a	5		a		

Note: GY99 and KW99: grain yield and kernel weight during 1999. DHTH98: days to heading at Tel Hadya during 1998. HI98 and GY98: harvest index and grain yield during 1998. HI99FF and GY99FF: harvest index and grain yield at the farmer fields during 1999. Data were analyzed combining both sites (Breda and Tel Hadya) and for each site separately. Different letters indicate statistically significant differences ($p < 0.05$) between means of groups for each parameter.

FIGURE 14.1. Variation of the Correlation Coefficient Between Yield and Spectroradiometrical Indices Across Groups of Barley Genotypes from Different Origins, Grown in Syria in Two Environments

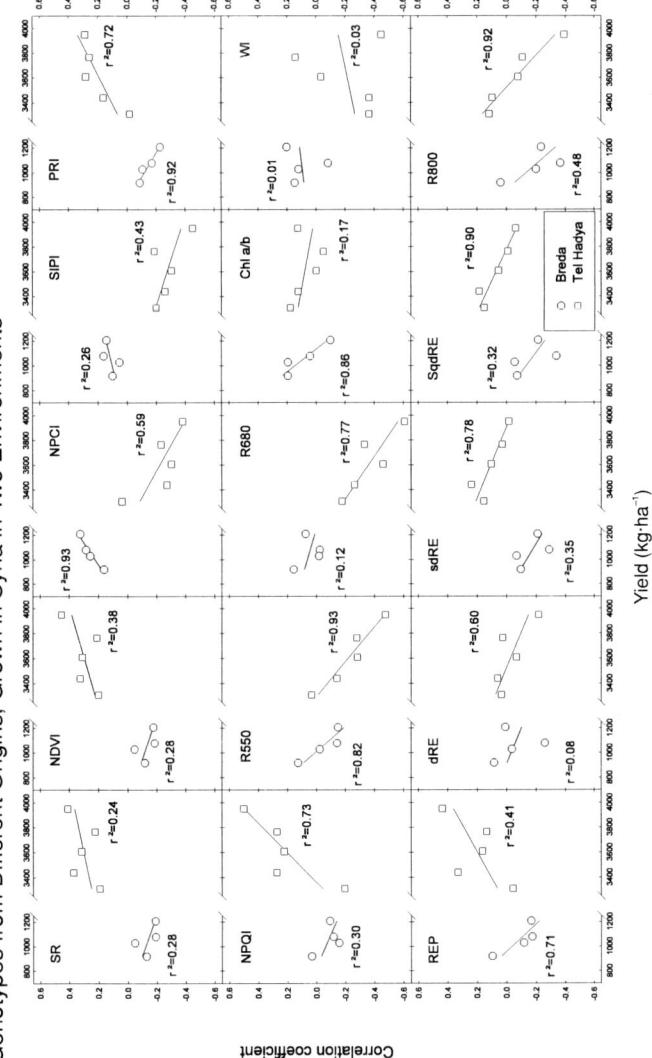

Note: Breda (circles) and Tel Hadya (squares). Spectroradiometrical indexes were measured around mid-grain filling. Data in Breda from groups of lines 1, 2, 4, and 5 (see Material and Methods). Data in Tel Hadya from groups of lines 1, 2, 3, 4, and 5. The trend of variation of the correlation coefficient of each index is shown by a regression line with its r^2 value for each environment.

TABLE 14.3. Correlation Coefficients Between the Different Spectroradiometrical Indices Measured Around Mid-Grain Filling During the 1999 Season at Two Sites (Breda and Tel Hadya) and Agronomic Parameters Measured During the 1998 and 1999 Seasons

	DHTH98		HI98		GY98		KW99		GY99	
	Breda	Tel Hadya	Breda	Tel Hadya	Breda	Tel Hadya	Breda	Tel Hadya	Breda	Tel Hadya
R550	−0.127	−0.276	0.023	−0.343	0.055	−0.366	0.053	−0.102	−0.133	−0.234
R680	−0.243	−0.401	0.049	−0.192	0.041	−0.245	0.424	−0.072	−0.235	−0.391
WI	−0.194	0.053	−0.142	0.088	−0.105	0.171	0.245	−0.025	−0.391	−0.086
NDVI	0.273	0.309	−0.001	−0.095	0.057	−0.014	−0.556	−0.065	0.230	0.356
SR	0.267	0.299	−0.010	−0.083	0.045	−0.004	−0.565	−0.075	0.217	0.370
PRI	0.129	0.353	−0.187	−0.054	−0.153	0.093	−0.323	−0.115	−0.231	0.242
SIPI	−0.273	−0.307	−0.006	0.087	−0.066	0.011	0.521	0.048	−0.223	−0.316
NPCI	−0.122	−0.309	0.132	0.055	0.252	−0.046	0.069	0.040	0.459	−0.255
NPQI	−0.051	0.273	−0.029	0.229	−0.174	0.285	0.396	0.063	−0.440	0.112
Chla/b	0.144	−0.025	0.031	−0.272	0.147	−0.299	−0.413	−0.028	0.397	0.076
REP	0.067	0.246	0.026	0.254	0.200	0.260	−0.177	−0.019	0.042	0.166
dRE	0.187	0.060	0.080	−0.227	0.267	−0.146	−0.333	−0.169	0.257	0.009
sdRE	0.258	0.087	0.011	−0.285	0.087	−0.215	−0.555	−0.138	0.185	0.166
SqdRE	0.258	0.060	0.006	−0.287	0.112	−0.221	−0.553	−0.145	0.184	0.109
R800	0.164	−0.082	0.056	−0.355	0.156	−0.307	−0.424	−0.166	0.059	0.012
R900	0.139	−0.045	−0.007	−0.346	0.182	−0.250	−0.408	−0.202	0.042	−0.017
DHTH98	1		0.166	0.269	0.009	0.305	−0.190	−0.058	0.073	0.315
HI98			1		0.308	0.262	0.165	0.207	0.245	0.242
GY98					1		0.124	0.090	0.342	0.290
KW99							1		0.109	0.156
GY99									1	

Note: DHTH98: days to heading at Tel Hadya during 1998. HI98 and GY98: harvest index and grain yield during 1998. GY99 and KW99: grain yield and kernel weight during 1999. All groups of lines (see Material and Methods) were combined for each site. All shaded coefficients were significant at 0.01.

TABLE 14.4. Stepwise Linear Regressions Between Barley Yield During 1999 and Spectroradiometrical Indices Measured Around Mid-Grain Filling the Same Year

BREDA ALL GROUPS OF LINES					
Indices	r^2	Sign in the equation	Indices	r^2	Sign in the equation
DHTH98	0.005	+	NPCI	0.213	+
NPCI	0.231	+	Chla/b	0.303	+
Chla/b	0.309	–	SqdRE	0.369	–
SqdRE	0.388	–	SIPI	0.377	–
WI	0.395	+	WI	0.383	+
NDVI	0.405	+			
dRE	0.412	+			

TEL HADYA ALL GROUPS OF LINES					
Indices	r^2	Sign in the equation	Indices	r^2	Sign in the equation
DHTH98	0.139	+	NPQI	0.132	+
NPQI	0.215	+	R680	0.174	–
R680	0.228	–			

Note: For the two sites (Breda and Tel Hadya), all groups of lines were combined. Stepwise regressions were conducted introducing first days to heading at Tel Hadya during 1998 (DHTH98) and then the spectroradiometrical indices (left half of the table) or allowing only the spectroradiometrical indices (right half of the table) to be chosen by the stepwise approach. For each site, indices are introduced from top to bottom in the order they were chosen by the model. Increment of r^2 is shown as each index is introduced in the model. The sign of the coefficient of each index in the equation line is also shown.

correlated with the number of days from emergence to heading measured the year before at Tel Hadya. The correlation between days to heading (DH) and grain yield was only significant at Tel Hadya (see Table 14.3), although DH was not chosen by the stepwise analysis at Breda or at Tel Hadya (see Table 14.4).

For Tel Hadya, most indices, except those associated with the red edge (REP, dRE, sdRE, and SqdRE), as well as Chla/b were significantly correlated with yield within each group (see Figure 14.1). The index that showed the best correlation with yield was R680 (negatively related), followed by NDVI and SR (both positively related). However, R680 and NPQI were the most frequently chosen in the stepwise linear regressions of spectradiometrical indices with yield within groups and when the five groups of lines were combined (see Table 14.4). The correlations between NPCI, NDVI, and SR and kernel weight and between kernel weight and grain yield were in most cases (groups of lines) not significant. Thus, kernel weight does not account for the significant correlations between these spectroradiometrical indices and yield. Days to heading (measured at the same site but the year before) correlated better than yield with the spectroradiometrical indices. In addition, days to heading also correlated (positively) with grain yield for all groups at Tel Hadya.

Correlation coefficients within each group between yield and the spectroradiometrical indices assayed at Breda were frequently different and even had a different sign from those at Tel Hadya. They were negative at Breda and positive at Tel Hadya for SR, NDVI, NPQI, PRI, REP, sdRE, and SqdRE and positive at Breda and negative at Tel Hadya for NPCI, SIPI, WI, and R680 (see Figure 14.1).

On the other hand, for each combination of spectroradiometrical index and site, the correlation between yield and the coefficients of correlation between yield and index across the 4 (Breda) and 5 (Tel Hadya) groups of lines assayed was studied (coefficients of correlation in the vertical axis and the mean yield in the horizontal axis). Different patterns, shown by the regression lines drawn in Figure 14.1, were observed within each site. Thus, correlation coefficients between yield and the spectroradiometrical indices SR, NDVI, NPQI, PRI, and REP became more negative in Breda and more positive in Tel Hadya as the mean grain yield across the groups of lines increased. In contrast, correlation coefficients were more positive in Breda and more negative in Tel Hadya for NPCI and SIPI as the yield increased. For other indices, such as R550, R680, Chla/b, dRE, and sdRE, coefficients of correlation became more negative as the mean yield increased at both sites. Therefore, the correlation increased as the mean yield increased within each site, regardless of the sign of the correlation between yield and these spectroradiometrical indices.

The combination of spectroradiometrical indices significantly contributing to explain differences in grain yield was assessed through an stepwise linear regression (see Table 14.4). For Breda, the combination of two to four indices explained about 25 percent of the genotype variability in yield within groups of lines (data not shown), whereas five indices accounted for up to 38 percent when the four groups were combined. The contribution of DH was minor, as observed when DH was included in the multilinear approach. For Tel Hadya, the results were similar, although the variation in performance among groups was higher. When all the groups were put together, NPQI and R680 accounted for only up to 17 percent, while the addition of DH increased this percentage to 23 percent. However, when this stepwise analysis is performed with the individual indices, the redundant effect of some indices cannot be avoided. A principal components analysis established groups of indices associated by their similar or redundant effect. The four first principal components explained up to 95 percent of the variation (see Table 14.5).

DISCUSSION

Photosynthetic area indices such as SR, NDVI, and sdRE were higher at the wetter site, reflecting the differences in green biomass (Hall et al., 1990; Peñuelas, Isla, et al., 1997; Araus et al., 2001; Aparicio et al., 2000). In addition, PRI, associated with the radiation-use efficiency, was also higher at the wetter site, which again suggests that Tel Hadya was less stressed (Filella et al., 1996; Peñuelas, Isla, et al., 1997). Nevertheless, PRI may also be positively affected by differences in biomass (Aparicio et al., 2000), which also agrees with the higher PRI found at Tel Hadya. In contrast, senescence indices such as NPCI and SIPI decreased, NPQI increased, and the index Chla/b also tended to increase slightly at the wetter site. This clearly shows that plants were more senescent (i.e., closer to the end of their cycle) at the more stressed site (Peñuelas, Gamon, et al., 1993; Peñuelas, Filella, and Baret, 1995; Peñuelas, Filella, Lloret, et al., 1995).

Correlations Between Individual Indices and Grain Yield

Some of the spectroradiometrical indices evaluated in this work have been positively correlated (either on a linear or logarithmic basis) with total green biomass, leaf area index, green area index, and the fraction of PAR absorbed by the canopy in small-grain cereals such as bread wheat and barley (Baret and Guyot, 1991; Field et al., 1994; Price and Bausch, 1995; Bellairs et al., 1996; Fernández et al., 1994; Peñuelas, Isla, et al., 1997). However, these studies have been carried out in a reduced number of genotypes, which does not provide a clear insight on the use of such techniques for breeding purposes. Here, for a very large collection of genotypes, significant correlations between some of the spectroradiometrical indices and grain yield were

TABLE 14.5. Correlation Coefficients Between the Four First Principal Components (PC1, PC2, PC3, and PC4) Obtained by a Principal Components Analysis Combining All Spectroradiometrical Data from Breda and Tel Hadya, and the Collection of Spectroradiometrical Indices from Which the Components Were Obtained

Index	PC1	PC2	PC3	PC4
R550	0.380	0.782	0.426	−0.155
R680	−0.757	0.501	0.398	−0.039
WI	0.005	−0.575	0.750	−0.201
NDVI	0.947	−0.211	−0.224	0.051
SR	0.908	−0.226	−0.196	0.050
PRI	0.899	−0.346	−0.036	−0.115
SIPI	−0.895	0.184	0.232	−0.088
NPCI	−0.887	0.382	−0.141	0.112
NPQI	−0.023	−0.758	0.592	−0.132
Chla/b	0.950	0.116	−0.097	−0.073
REP	−0.355	−0.216	0.299	0.837
dRE	0.784	0.111	0.214	0.443
sdRE	0.986	0.123	0.009	0.033
SqdRE	0.948	0.216	0.118	0.027
R800	0.807	0.489	0.286	0.021
R900	0.830	0.350	0.402	0.004
Accumulated variance	60.6	77.2	88.5	94.9

Note: Significant ($p < 0.05$) coefficients are included within shaded cells. The accumulated percentage of variation explained by the successive addition of principal components is shown in the last row.

found. However, the sign of the coefficients frequently changed from one site to another, owing to the contrasting growing conditions of these sites. Whereas the yield at Breda corresponded to a very poor environment in the limit of the agroecological cultivation area of barley (Slafer et al., 1994; Ceccarelli and Grando, 1999), that of Tel Hadya was more than three times higher. The contrasting levels of stress during growth led to different plant responses. Thus, the negative correlation of the vegetation indices (SR and NDVI) with yield within groups of lines at the poorer site (see Figure 14.1) suggests that these lines which mature earlier (or have less transpiring area) are the most productive. Vegetation indices indicate the amount of green area, and measured at grain filling, these genotypes show less green area. In contrast, in the better environment, these genotypes with more green area are the most productive. The negative correlation at Tel Hadya between yield and both R680 and R550 also supports this conclusion. The amount of green biomass at anthesis has been positively correlated with yield in nonextreme growing conditions (Slafer et al., 1994; Smith et al., 1999), since total kernel number per unit of ground area depends on the amount of leaf area. For the other indices, the opposite results in the two sites can be explained on the same basis: higher yield of the early maturing lines at the very poor site and of the lines with higher green biomass at the less stressed. Thus, at Breda, those genotypes which showed higher NPCI and SIPI and lower NPQI, and therefore the most senescent (Peñuelas, Gamon, et al., 1993; Peñuelas, Filella, and Baret, 1995; Peñuelas, Filella, and Gamon, 1995), were the most productive. At Tel Hadya, the less senescent lines were the most productive. A later senescence is associated with the presence of (a usually larger) green area for longer. On the other hand, the more senescent lines show lower photosynthetic efficiency and thus lower PRI, and vice versa (Peñuelas, Filella, and Gamon, 1995). Finally, the less senescent lines usually have higher water content and, thus, lower WI (Peñuelas, Filella, et al., 1993; Peñuelas and Filella, 1998).

Categories of Spectral Indices

To classify the set of spectroradiometrical indices studied according to their common physiological features, principal components analysis was performed. Based on the high positive correlation with the indices associated with green biomass (NDVI, SR, sdRE, SqdRE), the first principal component (PC1) could be considered as an indicator of standing green biomass. The second component (PC2) is negatively correlated with PRI and NPQI and positively correlated with reflectances at several wavelengths (R550, R680, R800, and R900). This high reflectance at all wavelengths associated with low PRUE and a high rate of chlorophyll degradation may be regarded as a function of the extent of highly reflecting dry biomass (straw, dry awns, etc.) associated with earlier maturity. In contrast, the correlation of the third component (PC3) with NPQI switches to positive, while the correlation with the simple reflectances remains positive. These high reflectances, associated with a low rate of chlorophyll degradation, might be interpreted as a function of certain developmental or structural characteristics of the canopies, such as green biomass that is suffering some level of stress associated with long-duration genotypes. Several factors can be involved, such as the extent of late (i.e., secondary) tillering, the presence of planophile leaves, or some characteristics of the spikes (number, size, color, angle). WI shows significant correlations with PC2 and PC3 (see Table 14.5), but the physiological basis is not clear. Neither the environment nor the genotype factors had significant effect on WI (see Table 14.1), probably because of the strong correlation between greenness and moisture content (Peñuelas, Isla, et al., 1997). WI has been found to be highly correlated with plant water content in several species of trees, shrubs, crops, and grasses (Peñuelas, Isla, et al., 1997), although in some instances it may be a better indicator of plant water status related to cell wall elasticity (Peñuelas et al., 1996). During grain filling, upper leaves of the canopy become senescent, not only because less water is available to the crop but because translocation proceeds toward the growing grains. Thus, changes in reflectance might be due to changes in canopy structure or geometry associated with moderate loss of water and changes in turgor in leaves and stems. The fourth component (PC4) is mainly represented by REP and dRE, which may be associated with leaf chlorophyll and nitrogen (Filella and Peñuelas, 1995).

Relationships Between Categories of Indices and Yield

The effect of each principal component on grain yield and other agronomic variables was studied using the correlation coefficients between principal components and agronomic variables (see Table 14.6) and the coefficient of each component in the stepwise regression line for each agronomic variable (see Tables 14.7 and 14.8). For the two sites, PC1 was highly and positively correlated with yield, while for the other three components the correlation was lower and negative. However, these relationships may result from the strong environmental interaction between the two sites (see Table 14.1).

However, when the correlation was performed separately for each site, the effect of some components is clearly different (opposite in some cases) depending on the site. For example, the correlation between PC2 and yield (1998 and 1999 seasons) or HI (1998 and 1999 at farmer fields) is positive in Breda but negative in Tel Hadya (see Table 14.6). Something similar happens with the coefficients for the stepwise regression for grain yield, where the coefficient of PC1 is negative in Breda and positive in Tel Hadya (see Tables 14.7 and 14.8).

In Breda, PC1 is positively correlated with DHTH (see Table 14.6), which can imply that the green biomass measured spectroradiometrically was not just indicative of the overall biomass produced by each genotype but biased by a phenological component, i.e., the persistence of green biomass when the measurements were performed (mid-grain filling). Therefore, the biomass may vary according to the phenology of each genotype. The stepwise regression for GY99 in Breda suggests that high yields are associated with low PC1 and PC3 and with high PC2 and PC4 (see Tables 14.7 and 14.8). This may be interpreted as higher yield in genotypes with lower green biomass (PC1) and earlier maturity (PC2), where lower green biomass could also be associated with earlier phenology or with higher kernel weight. Early maturation has been proposed as one of the most suitable strategies to improve yield in very stressed Mediterranean environments Loss and Sidiqque, 1994). Alternatively, genotypes with lower transpiring biomass can avoid terminal drought by controlling consumption of the available water.

TABLE 14.6. Correlation Coefficients Between the Principal Components and Some Agronomic Variables

	BREDA + TEL HADYA				BREDA				TEL HADYA			
	PC1	PC2	PC3	PC4	PC1	PC2	PC3	PC4	PC1	PC2	PC3	PC4
GY99	0.81	−0.27	−0.15	−0.08	0.17	0.34	−0.38	0.21	0.19	−0.19	−0.13	0.23
KW99	0.47	−0.24		−0.26	−0.57	−0.19	0.24	−0.32	−0.13			
DHTH98	0.08	−0.22		0.18	0.27		−0.13	0.15	0.13	−0.32		0.25
HI98		−0.24	−0.10			0.13				−0.26	−0.34	0.13
GY98	0.72	−0.35	−0.11	−0.08		0.21		0.25	−0.18	−0.39		0.14
HI99FF			0.16	0.27	0.41	0.50	−0.26	0.41		−0.23	0.42	
GY99FF			0.09	0.38	0.49	0.60	−0.46	0.58		−0.23	0.42	

Note: GY99 and KW99: grain yield and kernel weight during 1999. DHTH98: days to heading at Tel Hadya during 1998. HI98 and GY98: harvest index and grain yield during 1998. HI99FF and GY99FF: harvest index and grain yield at the farmer fields during 1999. The principal components (PC1, PC2, PC3, and PC4) were calculated combining all spectroradiometrical data from Breda and Tel Hadya. Coefficients were calculated combining data from Breda and Tel Hadya, and separately for each site. In order to contrast the sign of the coefficients, shaded cells indicate negative coefficients. Empty cells correspond to variables showing nonsignificant correlations.

In Tel Hadya, the stepwise regression for GY99 suggests that higher yields are associated with lower PC2 and PC3 and with higher PC1 and PC4 (see Tables 14.7 and 14.8). This may be interpreted as higher yield in genotypes with higher biomass (PC1) and longer grain filling (negative PC2). Despite the positive effect of PC1 on grain yield in 1999, the coefficient inthe stepwise regression for HI99FF and GY99 at the farmer fields was negative, since drought is more severe in some farmer fields than in Tel Hadya. In contrast, PC1 was not chosen in the stepwise regression for GY98 (see Tables 14.7 and 14.8), although they were negatively correlated (see Table 14.6), probably because 1998 was drier than 1999 (see mean yields in Table 14.1). Therefore, the physiological traits represented by PC1 increase yield in moderately dry conditions but decrease it in more severely dry conditions. Even in Tel Hadya in 1999, the most favorable environment, the effect of PC1 on kernel weight was negative.

PC3 has a negative effect on GY at both sites (see Table 14.6). However, the higher PC3 measured in Tel Hadya is associated with higher GY and HI in the farmer fields, where it is the first principal component to be chosen

TABLE 14.7. Coefficients of Each Principal Component in Stepwise Regressions Against Some Agronomic Variables

	BREDA + TEL HADYA					BREDA					TEL HADYA				
	r^2	PC1	PC2	PC3	PC4	r^2	PC1	PC2	PC3	PC4	r^2	PC1	PC2	PC3	PC4
GY99	0.76	1089	−358	−206	−110	0.24	−180	140	−146	36	0.16	215	−100	−84	91
KW99	0.35	2.87	−1.44	−0.34	−1.6	0.33	−9.9	1.4			0.02	−0.68			
DHTH98	0.09	0.47	−1.24	−0.37	1.02	0.09	6.4	−2.5			0.19	2.21	−1.72	−0.54	0.91
HI98	0.07		−0.02	−0.01	0.005	0.014		0.38			0.14	−0.02	−0.02		
GY98	0.66	896	−433	−141	−101	0.075	−66	76		35	0.15		−188		
HI99FF	0.10			0.025	0.042	0.31		0.143	−0.04	0.032	0.26	−0.04	−0.02	0.058	
GY99FF	0.16	77		104	427	0.589		874	410	−664	0.25	−275	−154	417	

Note: GY99 and KW99: grain yield and kernel weight during 1999. DHTH98: days to heading at Tel Hadya during 1998. HI98 and GY98: harvest index and grain yield during 1998. HI99FF and GY99FF: harvest index and grain yield at the farmer fields during 1999. The principal components (PC1, PC2, PC3, and PC4) were calculated combining all spectroradiometrical data from Breda and Tel Hadya. Coefficients were calculated combining data from Breda and Tel Hadya, and separately for each site. In addition, the overall determination coefficient (r^2) of the regression line is included for each analysis. In order to contrast the sign of the coefficients, shaded cells indicate negative coefficients. Empty cells correspond to variables not included in the model because they were not chosen by the stepwise approach.

TABLE 14.8. Stepwise Regressions, Using the First Four Principal Components (PC1, PC2, PC3, and PC4) Calculated from the Set of Spectroradiometrical Indexes Measured, Against Some Agronomic Variables

	BREDA				TEL HADYA			
	First	Second	Third	Fourth	First	Second	Third	Fourth
GY99	PC3 −0.148	PC2 +0.205	PC1 −0.220	PC4 +0.246	PC4 +0.054	PC1 +0.083	PC2 −0.127	PC3 −0.168
KW99	PC1 −0.321	PC2 +0.331			PC1 −0.017			
DHTH98	PC1 0.071	PC2 −0.094			PC2 −0.105	PC1 −0.172	PC3 −0.186	PC4 0.198
HI98	PC2 +0.016				PC2 −0.114	PC1 −0.139		
GY98	PC4 +0.060	PC2 +0.072	PC1 −0.080		PC2 −0.150			
HI99FF	PC2 +0.250	PC4 +0.292	PC3 −0.310		PC3 +0.177	PC1 −0.233	PC2 −0.260	
GY99FF	PC2 +0.357	PC4 +0.477	PC3 −0.591		PC3 +0.175	PC2 −0.226	PC1 −0.252	

Note: GY99 and KW99: grain yield and kernel weight during 1999. DHTH98: days to heading at Tel Hadya during 1998. HI98 and GY98: harvest index and grain yield during 1998. HI99FF and GY99FF: harvest index and grain yield at the farmer fields during 1999. The principal components were calculated combining all spectroradiometrical data from Breda and Tel Hadya. Then, a separate stepwise analysis was run for each site combining all groups of lines, allowing only principal components to be chosen into the model. The figures shown are r^2 values of the model as each component enters in first, second, third, or fourth position, with the sign of the principal component coefficient in the regression line.

in the stepwise regressions (see Tables 14.7 and 14.8). The physiological traits associated with this component are not clear. High reflectance in most wavelengths, together with low degradation of chlorophylls, could result from a canopy structure with higher secondary tillering, which affects yield negatively (Slafer et al., 1994; Loss and Siddique, 1994; Smith et al., 1999). Leaf shape and angle may also be associated with the pattern of spectral variation described by PC3. The third principal component would thus be higher in genotypes with large horizontal leaves, which may be associated with lower water-use efficiency and yield (Araus et al., 1993). Finally, the fourth component has a positive correlation with yield at both sites and might be associated with some nutritional trait. However, its effect is small.

CONCLUSIONS

For most indices, the sign of the correlation coefficient with yield was opposite at each site. Nevertheless, and irrespective of the site, the strength of the correlation between each index and yield within each group of lines increased as the mean yield of the group increased. In general, individual indices correlated better at the drier site. The spectroradiometrical features, as well as the physiological traits represented by them, which can be associated with genotypes with higher yields, vary according to the level of stress. The more severe the drought, the more efficient the genotypes with lower biomass and earlier maturity, whereas these traits are inefficient or negative in moderate drought. A spectroradiometrical feature (included in the third principal component) that is correlated with lower yields in both environments might be associated with canopy architecture but needs to be further investigated.

REFERENCES

Aparicio, N., V. Villegas, J. Casadesús, J.L. Araus, and C. Royo (2000). Spectral reflectance indices for assessing durum wheat biomass, green area, and yield under Mediterranean conditions. *Agronomy Journal* 92: 83-91.

Araus, J.L. (1996). Integrative physiological criteria associated with yield potential. In *Increasing yield potential in wheat: Breaking the barriers,* eds. M.P. Reynolds, S. Rajaram, and A. McNab. Mexico City, Mexico: CIMMYT, pp. 150-166.

Araus, J.L., T. Amaro, Y. Zuhair, and M.M. Nachit (1997). Effect of leaf structure and water status on carbon isotope discrimination in field-grown durum wheat. *Plant Cell and Environment* 20: 1484-1494.

Araus, J.L., J. Casadesús, and J. Bort (2001). Recent tools for screening of physiological traits determining yield. In *Applications of Physiology in Wheat Breeding,* eds. M. Reynolds, I. Ortiz-Monasterio, and A. McNab. Mexico City, Mexico, DF: CIMMYT, pp. 59-76.

Araus, J.L., M.P. Reynolds, and E. Acevedo (1993). Leaf posture, grain yield, growth, leaf structure and carbon isotope discrimination in wheat. *Crop Science* 33: 1273-1279.

Baret, F. and G. Guyot (1991). Potentials and limits of vegetation indices for LAI and APAR estimation. *Remote Sensing of Environment* 35: 161-173.

Bellairs, M., N.C. Turner, P.T. Hick, and R.C.G. Smith (1996). Plant and soil influences on estimating biomass of wheat in plant breeding plots using spectral radiometers. *Australian Journal of Agricultural Reserach* 47: 1017-1034.

Bertholdsson, N.O. (1999). Characterisation of malting barley cultivars with more or less stable grain protein content under varying environmental conditions. *European Journal of Agronomy* 10: 1-8.

Blackburn, G.A. (1998). Spectral indexes for estimating photosynthetic pigment concentrations—A test using senescent tree leaves. *International Journal of Remote Sensing* 19: 657-675.

Blum, A. (1988). *Breeding for stress environments.* Boca Raton, FL: CRC Press, p. 223.

Cattivelli, L., D. Giovanni, V. Terzi, and A.M. Stanca (1994). Progress in barley breeding. In *Genetic improvement of field crops,* ed. G.A. Slafer. New York: Marcel Dekker, pp. 95-182.

Ceccarelli, S. and S. Grando (1996). Drought as a challenge for the plant breeder. *Plant Growth Regulation* 20: 149-155.

Ceccarelli, S. and S. Grando (1999). Barley improvement. Germplasm Program Cereals, annual report for 1998. Aleppo, Syria: ICARDA.

Ceccarelli, S., S. Grando, R. Tutwiler, J. Baha, A.M. Martini, H. Salahieh, A. Goodchild, and M. Michael (2000). A methodological study on participatory barley breeding. I. Selection phase. *Euphytica* 111: 91-104.

Elliott, G.A. and K. L. Regan (1993). Use of reflectance measurements to estimate early cereal biomass production on sandplain soils. *Australian Journal of Experimental Agriculture* 33: 179-83.

Fernández, S., D. Vidal, E. Simón, and L. Solé-Sugranes (1994). Radiometric characteristics of *Triticum aestivum* cv. Astral under water and nitrogen stress. *International Journal of Remote Sensing* 15: 1867-1884.

Field, C.B., J.A. Gamon, and J. Peñuelas (1994). Remote sensing of terrestrial photosynthesis. In *Ecophysiology of photosynthesis,* eds. E.D. Schulze and M.M. Caldwell. Berlin: Springer-Verlag, pp. 511-528.

Filella, I., T. Amaro, J.L. Araus, and J. Peñuelas (1996). Relationship between photosynthetic radiation-use efficiency of barley canopies and the photochemical reflectance index (PRI). *Physiologia Plantarum* 96: 211-216.

Filella, I. and J. Peñuelas (1995). The red edge position and shape as indicators of plant chlorophyll content, biomass and hydric status. *International Journal of Remote Sensing* 15: 1459-1470.

Gamon, J.A., J. Peñuelas, and C.B. Field (1992). A narrow-waveband spectral index that tracks diurnal changes in photosynthetic efficiency. *Remote Sensing of Environment* 41: 35-44.

Gamon, J.A., L. Serrano, and J. Surfus (1997). The photochemical reflectance index: An optical indicator of photosynthetic radiation-use efficiency across species, functional types and nutrient levels. *Oecologia* 112: 492-501.

Hall, F.G., K.F. Huemmrich, and S.N. Goward (1990). Use of narrow-band spectra to estimate the fraction of absorbed photosynthetically active radiation. *Remote Sensing of Environment* 34: 273-288.

Hay, R.K.M. and A.J. Walker (1989). *An introduction to the physiology of crop yield.* Harlow, England: Longman Scientific and Technical, p. 250.

Loss, S.P. and K.H.M. Siddique (1994). Morphological and physiological traits associated with wheat yield increases in Mediterranean environments. *Advances in Agronomy* 52: 229-276.

Masoni, A., L. Ercoli, and M. Mariotti (1996). Spectral properties of leaves deficient in iron, sulphur, magnesium and manganese. *Agronomy Journal* 88: 937-943.

Peñuelas, J. and I. Filella (1998). Visible and near-infrared reflectance techniques for diagnosing plant physiological status. *Trends in Plant Science* 3: 151-156.

Peñuelas, J., I. Filella, and F. Baret (1995). Semiempirical indices to assess carotenoids/chlorophyll a ratio from leaf spectral reflectance. *Photosynthetica* 31: 221-230.

Peñuelas, J., I. Filella, C. Biel, L. Serrano, and R. Savé (1993). The reflectance at the 950-970 nm region as an indicator of plant water status. *International Journal of Remote Sensing* 14: 1887-1905.

Peñuelas, J., I. Filella, and J.A. Gamon (1995). Assessment of photosynthetic radiation-use efficiency with spectral reflectance. *New Phytologist* 131: 291-296.

Peñuelas, J., I. Filella, P. Lloret, F. Muñoz, and M. Vilajeliu (1995). Reflectance assessment of plant mite attack on apple trees. *International Journal of Remote Sensing* 16: 2727-2733.

Peñuelas, J., I. Filella, L. Serrano, and R. Savé (1996). Cell wall elasticity and water index (R970 nm/R900 nm) in wheat under different nitrogen availabilities. *International Journal of Remote Sensing* 17: 373-382.

Peñuelas, J., J.A. Gamon, K.L. Griffinand, and C.B. Field (1993). Assessing type, biomass, pigment composition and photosynthetic efficiency of aquatic vegetation from spectral reflectance. *Remote Sensing of Environment* 46: 110-118.

Peñuelas, J., R. Isla, I. Filella, and J.L. Araus (1997). Visible and near-infrared reflectance assessment of salinity effects on barley. *Crop Science* 37: 198-202.

Peñuelas, J., J. Piñol, R. Ogaya, and I. Filella (1997). Estimation of plant water concentration by the reflectance water index WI (R900/R970). *International Journal of Remote Sensing* 18: 2869-2875.

Price, J.C. and W.C. Bausch (1995). Leaf-area index estimation from visible and near-infrared reflectance data. *Remote Sensing of Environment* 52: 55-65.

Richards, R.A. (1996). Defining selection criteria to improve yield under drought. *Plant Growth Regulation* 20: 157-166.

Sinclair, T.R. (1988). Selecting crops and cropping systems for water-limited environments. In *Drought research priorities for the dryland tropics,* eds. F.D. Bidinger and C. Johansen. Patancheru, India: ICRISAT, pp. 87-94.

Slafer, G., E. Satorre, and F.H. Andrade (1994). Increases in grain yield in bread wheat from breeding and associated physiological changes. In *Genetic improvements of field crops,* ed. G.A. Slafer. New York: Marcel Dekker, pp. 1-68.

Smith, D.L., M. Dijak, P. Bulman, B.L. Ma, and C. Hamel (1999). Barley: Physiology of yield. In *Crop Yield, Physiology and Processes,* eds. D.L. Smith and C. Hamel. Berlin and Heidelberg: Springer-Verlag, pp. 67-107.

Wiegand, C.L. and A.J. Richardson (1990a). Use of spectral vegetation indices to infer leaf area, evapotranspiration and yield. I. Rationale. *Agronomy Journal* 82: 623-629.

Wiegand, C.L. and A.J. Richardson (1990b.) Use of spectral vegetation indices to infer leaf area, evapotranspiration and yield. II. Results. *Agronomy Journal* 82: 630-636.

Wiegand, C.L., A.J. Richardson, D.E. Escobar, and A.H. Gerbermann (1991). Vegetation indices in crop assessments. *Remote Sensing of Environment* 35: 105-119.

Zadoks, J.C., T.T. Chang, and C.F. Konzak (1974). A decimal code for the growth stage of cereals. *Euphytica Bulletin* Longman Scientific and Technical. 7: 42-52.

Chapter 15

Choosing Genotype, Sowing Date, and Plant Density for Malting Barley

Daniel F. Calderini
M. Fernanda Dreccer

INTRODUCTION

The main objectives of cropping systems for malting barley are both grain yield and grain quality. Although both traits are determined during the crop cycle, important decisions that will strongly affect them should be made before sowing. Among others, the choice of genotype, sowing date, and plant density are central for successfully combining the genotype potential for yield and quality with the environmental availability of resources. As was shown earlier (see García del Moral, Miralles, and Slafer, Chapter 10 this book), the barley growth cycle can be divided into phenological stages with different impacts on grain yield and quality. The length of these stages and the rates of growth and nitrogen uptake by the crop within each stage are dependent on the complex interaction among genotype, sowing date and plant density, other management practices, and climate and soil conditions. Undoubtedly, in extensive production systems, it is not possible (and perhaps not even necessary) to provide each stage of the crop cycle with the optimal environmental factors. Therefore, a compromising solution is generally reached that consists of making presowing decisions so that critical crop stages for the definition of yield and quality are given a preferential environment. In this context, the present chapter discusses different strategies for making some of the most important decisions that farmers face at presowing.

CHOOSING GENOTYPE

The choice of genotype can be based on different criteria. Among the traits that are often taken into account when choosing a cultivar are grain yield, grain quality (of its different aspects), disease and pest resistance, competitive ability against weeds, drought resistance, lodging susceptibility, and lack of preharvest sprouting. At the same time, at the level of the cropping system, the length of the growing season and the practice of double-cropping per season are important considerations for evaluating the best genotype to be used by farmers.

Genotypes are generally grouped in three categories according to their time of sowing: winter, mediterranean, and spring cultivars. The use of a particular group will be associated with climatic considerations, mainly the length of the growing season and freezing temperatures during winter. Another criterion for grouping is to distinguish between two- and six-rowed cultivars, depending on the number of fertile florets per spikelet. In this case, the choice of either two- or six-rowed barley is strongly connected with malt and brewing industry requirements, and the farmer's choice is generally based on market aspects, e.g., prices and costs, feasibility of selling, and industry contracts. In addition, two-rowed malting barley has been traditionally used by farmers in Europe and other regions of the world (e.g., Australia and South America), whereas six-rowed barley has been more common in North America. The first step toward choosing the best genotype from the broad range of possibilities defined by the combination of these two groupings (winter and spring barley or two-rowed and six-rowed barley) is to define clearly and precisely the agroclimatic characteristics of the crop production system and its socioeconomical constraints. Though socioeconomical constraints are clearly involved in crop decisions, this chapter will focus on only the ecological and physiological aspects. On this basis, the concept of "ideotype" (see Donald, 1968) could be useful for defining a genotype, considering its interaction with the environment (see Rasmusson, 1991). For example, for a long cropping season with abundant rain and wind during grain filling, we would require a high-tillering, long-season cultivar, with high grain yield potential and medium grain protein concentration (between 11 and 12 percent), medium length of grain-filling period, and high resistance to powdery mildew and lodging. Probably it is not easy to find information to fulfill the required "simple ideotype," but we consider clarifying the most important traits for each barley production system to be helpful in the process of choosing a genotype. In the sections that follow, we will discuss how to choose a genotype according to the most relevant traits.

Genotypic Effects on Grain Yield

Grain yield is dependent on the genotype and the environment in which this genotype is growing. For this reason, to choose a genotype considering grain yield as the central attribute will necessarily involve the consideration of the genotype by environment (GE) interaction. The study of this interaction can be approached in different ways, from simple evaluations that show the grain yield response of different genotypes for several environments (Finlay and Wilkinson, 1963), to the more sophisticated multiplicative models (e.g., additive main effects and multiplicative interaction model

[AMMI]; see Voltas et al., Chapter 9 in this book) and crop simulation models (see Ritchie and Alagarswamy, Chapter 16 in this book). Although a more comprehensive understanding of GE interactions is possible with the use of some analytical approaches, such as the AMMI analysis, these statistical techniques are mostly useful for breeders and researchers. For most of the realistic agronomic situations, information is generally restricted to methods describing postdictively genotypic performance across environments. Thus, aiming to describe a way to deal with GE interactions when, in practice, a cultivar has to be chosen for a particular condition, we propose the use of the Finlay and Wilkinson (1963) approach and its subsequent modifications (Eberhart and Russell, 1966). To use this index, it is important to keep in mind (1) that the environmental restrictions should be similar between the target region and the environment where the index was obtained, and (2) that the predictability of the index decreases with the number of factors involved in the yield restriction.

Finlay and Wilkinson (1963) basically proposed studying the GE interaction using the regression coefficient of the relationship between the yield of each genotype and the average yield of all the genotypes evaluated in the same environment under a wide range of conditions, usually sites and years, i.e., the environmental index (see Figure 15.1). Thus, they assume a direct measure of genotype responsiveness for better and worse growing conditions. However, as these authors found different patterns of GE interaction, they suggested that the better variety would have a high mean yield and a regression coefficient of 1. To choose a genotype, the responsiveness of cultivars to different environmental indices could be crucial once the environment in which the crop will be located has been defined. Hypothetical patterns of responsiveness of three different cultivars are shown in Figure 15.1, where the environment modifies the ranking of the genotypes. Similar

FIGURE 15.1. Relationship Between Grain Yield and Environmental Index for Three Hypothetical Barley Cultivars A, B, and C

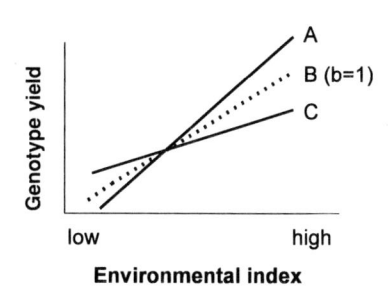

Note: The slope (b) of cultivar B is 1.

patterns can be found in the literature for barley and other cereals (see Baker, 1988; Acevedo, 1991; Ceccarelli, 1991). For example, if the probable environment for the next growing season could be identified as high yielding, cultivar A should be the higher yielding for this environment. On the other hand, cultivar C might be the best yielding in low-yielding environments. In addition, according to the response of these three genotypes, a range of environments exists where none of them shows a higher yield, and the choice of the genotype could be focused on the analysis of other traits (as discussed in the following).

To consider additional parameters, instead of "responsiveness" alone, could improve the evaluation. For instance, Eberhart and Russell (1966) proposed including deviation from the regression as a measurement of confidence of the fitness of the linear regression used by Finlay and Wilkinson (1963). For this purpose, the mean square of the regression should be included in the analysis. As mentioned earlier, the use of the GE interaction (see Figure 15.1) is a general approach for choosing genotype, but the environmental index integrates many different climatic, soil, and biotic conditions. When it is possible to find data of grain yield in the same area for different years, the wide concept of environment could be divided into (1) predictable environmental variation and (2) unpredictable environmental variation, according to Lin, Binns, and Lefkovitch (1986). This analysis assumes that year variation is higher (and less predictable) than regional variation. Therefore, to estimate the response of different cultivars to predictable and unpredictable variation, the use of mean squares of the year:location (y:l) ratio was proposed (see Lin, Binns, and Lefkovitch, 1986).

In Argentina, the approach used by Finlay and Wilkinson (1963) and complemented by both mean square of the regression (see Table 15.1) and the mean square of the y:l ratio (Lúquez, 1990) has been successfully applied for evaluating the response of different wheat cultivars and choosing a genotype. Thus, cultivars with high grain yield average, high responsiveness (coefficient b), and low mean square of the regression (Eberhart and Russell, 1966) or mean square of the y:l ratio (Lin, Binns, and Lefkovitch, 1986) will be chosen for high-yielding conditions, while those with high-grain yield average, low responsiveness *(b),* and low mean square of the regression or mean square of the y:l ratio will be chosen for low-yielding conditions.

In addition, the usefulness of a relative approach for the study of GE interaction has been discussed by Francis and Kannenberg (1978) and by Austin and Arnold (1989). Francis and Kannenberg (1978) evaluated genotype stability using the coefficient of variation instead of responsiveness (b). In their analysis, the authors divided the response of genotypes as (1) high yielding and small variation, (2) high yielding and large variation, (3) low yielding and small variation, and (4) low yielding and large variation. At the

TABLE 15.1. Analysis of the Relationship Between Grain Yield and Environmental Index for Four Wheat Cultivars

Cultivar	Grain yield (kg·ha^{-1})	b	MS
Trigomax 200	4,334 a	1.24	109,350
Las Rosas	3,912 ab	1.00	59,077
Buck Pucará	3,556 bcd	0.78	161,726
Klein Criollo	3,452 cd	0.82	76,571

Source: Matinuzzi, Machado, and Paulucci (1990).
Note: Parameters of the regression analysis as grain yield responsiveness (b) and mean squares of the regression analysis (MS) are included. Different letters show statistical differences ($p < 0.05$) between grain yields of cultivars.

same time, more recently developed multiplicative models, e.g., the AMMI model (see Voltas et al., Chapter 9 in this book), increase the number of methodologies that analyze the GE interaction, and they could provide interesting approaches for understanding the complexity of this interaction. However, the approach proposed earlier (see Figure 15.1 and Tabel 15.1) is still a simple and powerful way to choose genotype, and this methodology could be readily available for most farmers and crop advisors.

Genotypic Effects on Grain Quality

Grain quality in malting barley has many different parameters to be considered, such as grain size, extract yield, protein concentration in grains, amylase activity, dormancy, and various others (see Molina-Cano, 1989; MacGregor, 1992; Savin and Molina-Cano, Chapter 19 in this book). The value and importance of these traits are closely related to the objective of the crop production system, i.e., to achieve a level of quality that satisfies both the malting and brewing industries. From an economical viewpoint, the most important criterion is extract yield (i.e., the total extractable material that is likely to be obtained from a given malt; see Bamforth and Barclay, 1993). However, protein content of grains is also an important criterion, though proteins have ambiguous implications because they are required for biochemical processes and, at the same time, they have a negative relationship with extract yield (see MacLeod, 1979). This forces farmers to produce barley grains within a quite narrow range of protein content. Clear differences among genotypes have been found in grain protein concentration and other quality traits (see Moll, 1979; Bulman, Mather, and Smith, 1993).

Nevertheless, these traits are also variable among years and locations (Schildbach and Burbidge, 1992). In other words, as with yield, the quality of barley grains for malting is strongly modified by the environment. For example, Schildbach and Burbidge (1992) found that percentage of grains larger than 2.8 mm and β-amylase activity showed a greater degree of heritability (22 and 28 percent, respectively) than did crude protein (4.6 percent), extract (9.7 percent), and Kolbach index (6.4 percent). These results highlight the difficulties in choosing genotype according to quality parameters.

A possible tool for choosing a genotype based on its quality is to analyze the response of quality traits in different environments. An interesting approach consists of evaluating quality traits in response to those environmental parameters which have the greatest effect upon them. For example, it is commonly accepted that grain weight (and grain size) is associated with average temperature during the grain-filling period (see Chowdhury and Wardlaw, 1978), and final grain weight could respond differently to average grain-filling temperature according to genotype (Calderini et al., 1999b). Therefore, a genotype having this trait could be chosen by analyzing the relationship between grain weight and the average temperature during the grain-filling period. The interaction between other traits, such as extract yield, and environmental factors is actually poorly understood. Thus, the rational analysis of genotype response as proposed for grain weight might not be possible until greater knowledge of these factors is available. However, to systematize information about, e.g., extract yield, the average variability in different environments could be a useful empirical criterion for choosing genotypes based on this trait.

Most barley cultivars have low dormancy values to allow immediate malting, thus reducing costs and deterioration resulting from long storage until dormancy is terminated. However, this low dormancy might lead to preharvest sprouting, which is an important cause of grain quality damage. These cultivars might be troublesome when crops face rainy conditions during grain filling. Cultivar differences in dormancy level have been reported in barley (Benech-Arnold et al., 1999) and could be useful for choosing genotypes, taking into account the more probable weather conditions during grain filling.

Choosing Genotypes to Cope with Abiotic and Biotic Stresses

Abiotic stresses (e.g., drought and heat stress) and biotic stresses (e.g., diseases, pests, and weeds) have a major impact on barley grain yield and quality. In many cases, these stresses are associated with growing areas or production systems and it is possible to calculate the probability of occurrence (see Wardlaw and Wrigley, 1994). General evaluations for choosing genotypes to cope with one of these particular stresses, as commented earlier, could be complemented by other, more specific considerations.

Regarding abiotic constraints, drought stress is among the most important, taking into account the large proportion of subhumid or semiarid regions where barley is produced (see Poehlman, 1985). The negative effect of drought stress on grain yield is related to the timing of this stress and its intensity (see Savin and Molina-Cano, Chapter 19 in this book). A common assumption is that barley is better adapted to water stress than is wheat (Acevedo, 1991; López Castañeda, and Richards, 1994). Several authors (e.g., Anderson and Reinbergs, 1985; Poehlman, 1985; Acevedo, 1991; López Castañeda and Richards, 1994) have shown that this characteristic is mainly based on an avoidance strategy due to earlier grain maturity. However, other causes could also be mentioned (see López Castañeda and Richards, 1994; González, Martín, and Ayerbe, 1999). These suggest that genotypes selected under drought (or at least under some degree of water shortage) would be the best choice for subhumid/semiarid regions or areas where water stress conditions could occur during the growing season. However, Cooper et al. (1997), working with wheat, showed that lines selected for yield in water-stressed nurseries do not perform better than other lines selected in nonstressed nurseries. In addition, Slafer and Araus (1998) concluded that, except under very low grain yield environments, high-yielding cultivars selected in high-yielding environments would also show the best yield response in medium- to low-yielding environments. For this reason, use of the Finlay and Wilkinson (1963) analysis could still be useful in choosing genotypes to be placed in environments with probable water stress. Nevertheless, for very low-yielding environments (lower than 1.5 t·ha^{-1}), choosing cultivars selected under water stress conditions could possibly be a useful strategy. Local cultivars or land races have been found to be the highest-yielding ones under drought stress in rain-fed Mediterranean areas of Syria and Turkey (Ceccarelli, 1984; Tahir et al., 1994).

One more issue that has occupied breeders and researchers focused on low rainfall areas was to identify groups of genotypes with advantages for reaching higher yields in these environments. Despite the common assumption that long crop cycle cultivars use much water for vegetative growth that would be better used for grain filling, different evaluations of wheat carried out by crop simulation models showed that both long and short crop cycle cultivars reached similar yields at current sowing dates in low rainfall areas or seasons (Stapper and Harris, 1989; Connor and Loomis, 1991; Savin et al., 1995). In addition, long crop cycle cultivars showed yield advantages over short crop cycle cultivars at early sowing dates (Connor and Loomis, 1991; Savin et al., 1995) or under better water availability (Stapper and Harris, 1989; Savin et al., 1995). On the contrary, short crop cycle cultivars performed better than long crop cycle ones when the crops were sown late

(Connor and Loomis, 1991). López Castañeda and Richards (1994) found a negative relationship between grain yield and time to anthesis in barley and other small-grain cereals when these crops faced terminal drought. Because of the characteristic variability of low rainfall environments, yield stability could be an important criterion for choosing genotype. Acevedo (1991) showed that six-rowed barley cultivars have lower stability, calculated as proposed by Eberhart and Russell (1966), than two-rowed barleys. In addition, two-rowed land races showed higher stability than two-rowed improved barleys (Acevedo, 1991). Other studies carried out on wheat also showed that plant breeding decreased yield stability (Calderini and Slafer, 1999).

Heat is another important abiotic stress, making heat tolerance an objective for plant breeding (see Reynolds, Balota, et al., 1994). It has been found that traits such as membrane thermostability (measured in flag leaf) and canopy temperature depression show good association with grain yield under hot conditions (Reynolds, Balota, et al., 1994). Therefore, these traits could be useful for choosing genotypes to be sown in hot environments. In addition to the effect of heat stress on grain yield, high temperatures also modify grain quality by reducing grain weight or modifying grain protein content (see Randall and Moss, 1990; Savin and Molina-Cano, Chapter 19 in this book). Special attention has been paid lately to the effect of short periods (three to five days) of heat stress during grain filling on grain quality of wheat and barley (see Blumenthal et al., 1994; Stone and Nicolas, 1994; Savin and Nicolas, 1999). Important variability of grain yield and quality under heat stress conditions has been found in wheat (Stone and Nicolas, 1995). Although the extent of the cultivar survey made in wheat (see Stone and Nicolas, 1994) has not been paralleled in barley, cultivar variability has been found in response to heat shock stress in barley (Savin and Nicolas, 1996). For this reason, cultivar evaluations in response to heat conditions are important for choosing genotypes to be sown under probable heat stress conditions.

In winter barley, considerable damage can be caused by extremely low temperatures at different stages of development. Variation in the tolerance to cold exists and is related to different morphological, cytological, and metabolic characteristics, e.g., the ability to form a deep crown (Tahir et al., 1994), cell size (Limin and Fowler, 1999), and differences in fatty acid metabolism among genotypes (Murelli et al., 1995), respectively. These traits could be important in the near future for breeding programs aimed to obtain cold tolerance and to choose genotypes. For other abiotic stresses, such as high soil contents of aluminum, boron, or saline, genetic variability in tolerance has been found (e.g., Schaller et al., 1981; Nable, 1988; Jana, 1991; Minella and Sorrells, 1992; Yau, Hamblin, and Ryan, 1994). In addition, tolerance to these stresses has actually been enhanced by the introgression of genetic variability from promising wild relatives in barley and wheat (Jana, 1991; Mujeeb-Kazi, 1995).

Biotic stresses can be grouped into the categories of diseases, pests, and weeds. Considering the potential effect of diseases and pests (e.g., barley yellow dwarf virus, powdery mildew, net blotch, and Russian aphid), choosing a genotype should be done according to plant tolerance/resistance. In line with this idea, Little and Doodson (1972) proposed a rating scale for mildew, and Tekauz (1985) for net blotch. Environmental conditions clearly affect the development of diseases and pests and their impact on the crop. Therefore, cultural practices such as sowing date, fertilization, irrigation, and plant density should be considered the complement for the control of these biotic stresses. Evaluations of cultivar tolerance/resistance (from susceptible to highly resistant) are commonly available in different countries (e.g., Tekauz, 1990; Ordon and Friedt, 1994; Barreto et al., 1998). Moreover, public agricultural research agencies provide yearly rankings of cultivar resistance. However, not only the resistance evaluation of cultivars should be considered. In addition, the possibility of disease or plague escape by choosing the barley cycle length (short-, medium-, or long-season cultivar) can be helpful, as well as information about probable weather conditions (see Carmona, Moschini, and Conti, 1997).

Weeds are a particular case of biotic stress because they compete for soil resources that are also taken up by the plant, and their relative position in the canopy undermines light capture by the crop. Thus, processes involved in competition are different from those involved in the plant-disease or plant-plague interaction. From the point of view of the cultivar choice, the competitive ability of the crop to cope with weeds could be important in evaluating different cultivars. It has been shown that cultivars of barley, wheat, and oat have different competitive abilities (Satorre and Snaydon, 1991), and plant breeders are taking this into account in selection procedures (e.g., Valentine, 1982; Powell et al., 1985). At the same time, some agronomists advocate the use of competitive cultivars as a component of weed control systems (e.g., Hakansson, 1983; Challaiah et al., 1986; Wicks et al., 1986). Satorre (1988) showed that the highest-yielding cultivars of barley, wheat, and oat could be at the same time the most competitive, probably reflecting their greater ability to capture scarce resources. In addition, Reynolds, Acevedo, et al. (1994), found that the higher the grain yield in pure stands, the lower the increase in grain yield when competition was reduced in wheat. Therefore, as high-yielding cultivars can maintain their ability to capture resources under adverse conditions (see Slafer and Araus, 1998), it is possible to choose high-yielding and highly-competitive cultivars. Still, when choosing cultivars based on their competitive ability, it is important to account for environmental effects on crop and weed interaction (see Radosevich, Holt, and Ghersa, 1997). For example, in studies in which crop-weed competition was evaluated under low and high rates of nitrogen fertilization, it was shown that mustard improved its competitive ability against barley (Liebman and Robichaux, 1990) and wheat (Guglielmini and Satorre, 2000). Therefore, competitive ability should be considered in relation to the level of inputs of the crop production system (see Legere and Samson, 1999). Finally, some traits have been found associated with a higher competitive ability. In barley, a positive association between competitive ability and seed size was shown (Satorre, 1988).

Lodging is an important cause of yield losses and has long been a problem in cereal cultivation (Pinthus, 1973). It has been estimated that lodging can reduce yields of barley and wheat up to 40 percent, depending on its severity and on the time of its occurrence (Sisler and Olson, 1951; Day and Dickson, 1958, Stapper and Fischer, 1990). At the same time, negative consequences on grain quality have been reported, and sprouting has also been found to occur more frequently in lodged crops (Pinthus, 1973; Briggs, 1990). Lodging severity is dependent on (1) the size of the forces to which the stem is subjected, (2) the bending strength of the stem and its resistance to buckling, and (3) the anchorage strength of the root system (Crook and Ennos, 1994). In addition, timing of lodging affects its severity, as lodging at the grain milk stage was found to produce higher yield losses than lodging during other moments just after anthesis (Jedel and Helm, 1991). As lodging susceptibility has a clear relationship with plant height (Pinthus, 1973; Crook and Ennos, 1994), and this trait is easily evaluated, plant breeding has focused on increasing plant resistance to lodging by shortening plant height. Thus, plant breeding programs decreased plant height in most winter cereals, such as barley (Riggs et al., 1981; Abeledo, Calderini, and Slafer, Chapter 13 in this book), oat (Peltonen-Sainio, 1990), and wheat (Calderini, Reynolds, and Slafer, 1999). As cultivar differences in lodging resistance were also found to be significantly associated with internode diameter (see Pinthus, 1973), it is possible that the higher lodging resistance shown by modern six-rowed cultivars is the consequence of both shorter culm length and higher culm diameter (see Jedel and Helm, 1994). In addition, six-rowed cultivars had significantly greater culm diameter than two-rowed (more susceptible to lodging) barley in this study (Jedel and Helm, 1994). This characteristic could be helpful in choosing cultivars to cope with environments with high lodging probability.

CHOOSING SOWING DATE

It is generally accepted that within the crop growing season the delay of sowing date has a negative effect on grain yield. This penalty for yield has been observed widely in such winter cereals as oat (Peltonen-Sainio, 1996) and wheat (see Ortiz Monasterio, Dhillon, and Fischer, 1994; Dennett,

1999), as well as in other crops, e.g., maize, soybean, and sunflower (Andrade, 1995; Otegui et al., 1995). In barley, different studies have shown the importance of early sowing dates for maximizing grain yield in both spring (e.g., Kirby, 1969; Zubriski, Vasey, and Norum, 1970; McFadden, 1970; Ciha, 1983) and winter barleys (e.g., Green, Furmston, and Ivins, 1985). In addition, the importance of early sowing date has been shown in tropical climates also (Okafor, 1988). Experiments carried out in Argentina found a decrease in grain yield of about 25 kg·ha^{-1} per day when sowing date was delayed between mid-June and mid-October, and this decrease was similar in both two- and six-rowed Mediterranean barleys (see Figure 15.2). Green, Furmston, and Ivins (1985) found a decrease in grain yield of 0.43 percent for every day sowing was delayed after mid-September for two-rowed winter barley sown in England. In the experiments carried out in Argentina and England, the number of grains/m^2 was the yield component that better explained final grain yield ($r = 0.98, p < 0.001; r = 0.97, p < 0.001$, for the studies carried out in Argentina and England, respectively). Decreases in grain yield by delaying sowing date could be associated with a shorter crop cycle. Thus, the longer a crop is able to grow, the greater its biomass (see Evans, 1993). Dofing (1997) found a close association ($r = 0.83, p < 0.001$) between grain yield and thermal time units accumulated between sowing and heading in 24 barley genotypes. In addition, the author did not find a significant association between the grain-filling period and grain yield.

Although early sowing date is a useful criterion for choosing sowing date, the positive relationship between earliness and grain yield could be modified in places (or years) where abiotic (e.g., water shortage early in the

FIGURE 15.2. Grain Yield of Two-Rowed (B1215) and Six-Rowed (B1614) Cultivars Sown in Buenos Aires (Argentina) on Four Sowing Dates

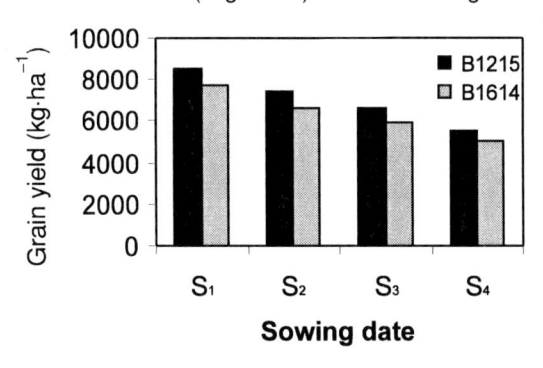

Source: Calderini et al. (1998) pp. 2-5).
Note: Sowing dates were June 20 (S_1), July 21 (S_2), August 28 (S_3), and October 11 (S_4) during the 1995-1996 growing season.

growing season or freezing temperatures at heading) or biotic (e.g., disease and pest damage) constraints have a high probability of occurring. At the same time, another restriction for this general principle is to know how much the sowing date could be advanced to increase grain yield potential considering cultivar diversity (long-, medium-, and short-season cultivars) and also the effect of sowing date on grain quality. In addition, although a longer whole crop cycle (or sowing to anthesis period) has been recognized as a high-grain-yielding condition (Dofing, 1997), it is accepted that the crop cycle phenophases do not have the same impact on grain yield and quality (see Fischer, 1985; García del Moral, Miralles, and Slafer, Chapter 10 in this book; Savin and Molina-Cano, Chapter 19 in this book). For these reasons, to choose the optimal sowing date for the crop growing season, the following schematic analysis could be useful: (1) identify the developmental stages during which grain yield and quality are determined, (2) match the key stages of development for barley grain yield and quality with the best potential environment, and (3) identify environmental restrictions (e.g., low soil temperature at sowing, frost damage at heading) to reaching the previous potential environment for the key developmental stages. Finally, once the key stages are located in the growing season, the sowing date can be inferred according to the timing of these stages.

Identifying the Developmental Stages During Which Grain Yield and Quality Are Determined

These aspects of the barley crop are widely discussed in other chapters of the present book; thus, to cope with the objective of choosing sowing date only some highlighted ideas will be mentioned here. Most of the studies trying to explain yield differences in barley and other cereals have shown that grain number, instead of grain weight, is the component better associated with yield (see Evans, 1993, and references cited therein). For this reason, except in particular cases such as late-season drought, the period during which grain number is determined should be considered central for choosing sowing date to maximize grain yield. On the other hand, grain weight determination could be more directly associated with grain quality. From this point of view, although grain number is the result of structures (e.g., plants per square meter, spikes per plant, spikelets per spike, etc.) that develop and grow during different phenophases (see García del Moral, Miralles, and Slafer, Chapter 10 in this book), and considering that a longer period between emergence and heading has been shown as a condition for higher grain number (Kitchen and Rasmusson, 1983) and yield (Dofing, 1997), the key stage for grain number determination should be more precisely defined for a successful planting date choice. Kitchen and Rasmusson (1983) found a high correlation between the length of both the spike initiation and spike growth periods and grain number ($r = 0.87$, $p < 0.01$;

$r = 0.90, p < 0.01$, respectively). In addition, in experiments carried out under controlled conditions, Miralles, Richards, and Slafer (2000) have shown that fertile florets per spike and grains per square meter were increased linearly when the period between triple mound and heading was extended from 35 to 55 days. More precisely, preliminary results suggest that the period surrounding booting might be responsible for differences in final grain number in barley (Miralles and Ariznabarreta, unpublished data). Therefore, this period should be matched with the best environment for maximizing grain number.

Grain quality seems to be more clearly determined within a phenophase (i.e., between anthesis and physiological maturity; see García del Moral, Miralles, and Slafer, Chapter 10 in this book) than is grain number. Therefore, grain quality determination is mostly restricted to the postanthesis period and the environmental conditions during that time. Although considerations for choosing sowing date focus on the anthesis to physiological maturity period, it is necessary to keep in mind that the most important proportion of nitrogen (about 60-70 percent) is taken up by the crop prior to anthesis (Bulman and Smith, 1994; McTaggart and Smith, 1995), and, thus, what has occurred previous to heading partially determines grain protein concentration. In addition, recent studies revealed that the assumption that grain weight and quality are restricted to the grain-filling period could be simplistic because the period during which carpels of florets are growing is also important for defining grain weight potential (Scott et al., 1983; Calderini et al., 1999a). Finally, the period during which grains are merely losing water content (i.e., after physiological maturity) could also modify grain quality due to preharvest sprouting.

Matching the Key Stages of Development for Barley Grain Yield and Quality with the Best Potential Environment

Many factors determine the development and growth of a genotype (see Slafer and Rawson, 1994; García del Moral, Miralles, and Slafer, Chapter 10 in this book). However, a practical first step for assessing the likely environment in which the key phenological stages of a barley genotype will be inserted could be to consider those factors which determine grain yield potential and, at the same time, are difficult or impossible to modify in a farmer system, i.e., solar radiation and temperature. These factors, integrated as the photothermal quotient (see Fischer, 1985), have been used successfully for predicting the number of grains per square meter around anthesis in wheat (Fischer, 1985; Slafer and Savin, 1991; Magrin et al., 1993; Ortiz Monasterio, Dhillon, and Fischer, 1994), maize (Andrade, Otegui, and Vega, 2000), and sunflower (Cantagallo, Chimenti, and Hall, 1997). The cause of this association is that crops, and especially spikes, are source limited during this period (see Fischer, 1984, 1985). Therefore, environmental conditions that permit a greater radiation interception by the crop during this critical period would set a higher grain number. Although it is important to recognize that the critical experiment for evaluating the usefulness of photothermal quotient has still not been carried out in barley, recent experiments (Miralles, Richards, and Slafer, 2000) conducted under controlled conditions showed that the higher the spike weight at heading (at least between 0.1 and 0.3 g per spike), the greater the number of fertile florets per spike. Therefore, to place the period during which the spikes are growing fast (around booting initiation) at a higher photothermal quotient could be a valuable strategy for maximizing grain yield.

On the other hand, to match the grain-filling period with the best environmental conditions for grain quality by choosing sowing date is at least theoretically possible. However, to combine the best environmental conditions for both grain yield and grain quality as independent traits seems to be difficult because they are not completely independent, since some environmental requirements overlap. For this reason, a compromise must be made.

During the grain-filling period, the most important factor that modifies grain quality and might not be modified in most field conditions is temperature. Therefore, it could be convenient to analyze how temperature should be considered from the viewpoint of sowing date determination. A negative relationship between final grain weight and the average temperature for the grain-filling period has been found in barley (MacLeod and Duffus, 1988). Therefore, high average temperature during grain filling could decrease barley quality by diminishing grain size and weight. At the same time, it is possible that grain protein concentration could be increased, taking into account differences between dry matter and nitrogen dynamics in the grains (see Jenner, Ugalde, and Aspinal, 1991; Savin and Nicolas, 1996; Stone and Savin, 1999). In addition, to the effect of temperatures on grain nitrogen concentration, other grain quality traits could also be affected by high temperatures, as those higher than 30°C modify the balance between different proteins (see Wardlaw and Wrigley, 1994). Thus, maximum temperatures higher than 30°C during grain filling, just for a few days (commonly named heat shock effect), also affect grain quality (Blumenthal et al., 1994; Wardlaw and Wrigley, 1994; Stone and Nicolas, 1994) by modifying both final grain size and weight (Stone and Nicolas, 1995; Savin, Stone, and Nicolas, 1996; Calderini et al., 1999b) and starch and protein composition (Jenner, 1994; Wardlaw and Wrigley, 1994; Stone and Nicolas, 1995; Savin, Stone, and Nicolas, 1996). Therefore, to choose sowing date considering its effect on grain quality for malting, mild temperatures during the grain-filling period and maximum temperatures below 30°C would be optimum. Such conditions imply high grain size and weight (contributing at the same time to higher grain yield) and a good balance between proteins. This "optimum temperature condition" for grain quality could be changed if the objective is just high grain protein concentration. Correll et al. (1994) modeled, by multiple regression procedures, the barley grain protein concentration in South Australia considering climatic data (i.e., total rainfall from July to September and consecutive days in November with temperatures >35°C). This procedure could be useful for estimating the probability of grain protein concentration based on historical climatic records and its interaction with sowing date. In addition, the risk of negative effects of heat shock temperatures on other quality traits (see Savin and Molina-Cano, Chapter 19 in this book) could also be evaluated.

Identifying Environmental Restrictions to Reaching the Potential Environment for the Key Developmental Stages

Once the potential environment for the key developmental stages has been identified, and the key stages have been located within the growing season, the following step is to analyze whether other environmental restrictions for maximizing grain yield and quality exist. Important climatic constraints for grain yield and quality are frost damage between heading and physiological maturity and low water availability or low temperature at sowing. Thus, environmental restrictions for barley should be considered according to the region in which the crop will be grown. One of the ways to solve the puzzle of matching key stages for crop grain yield and quality with the potential environment and its restrictions could be to employ a probability-based approach. An example of this approach for choosing sowing date is described here for an important barley-growing area of Argentina (Tres Arroyos, Buenos Aires Province; 38°20'S, 60°15'W; elevation 115 m above sea level). For this area, the probability of the occurrence of some important climatic factors was evaluated weekly for choosing sowing date (see Figure 15.3):

- Frozen risk (mean temperature < 0°C): to protect the plant sensitivity between heading and physiological maturity
- Photothermal quotient equal to or higher than 1 (MJ·m^{-2} · °Cday^{-1}, base temperature = 0°C): to place the period between booting initiation and seven days after heading under the best conditions for grain number determination
- Rainfall equal to or higher than 10 mm in less than a week: to assess sufficient water availability for germination and emergence of the crop
- 100°Cday accumulated in two weeks: to evaluate the delay of the period between sowing and emergence
- Temperatures higher than 30°C: to avoid heat shock temperatures during grain quality determination, i.e., between anthesis and physiological maturity

FIGURE 15.3. Probability of the Occurrence of Various Events in Each Week After the Mean Earliest Frozen Date of the Year

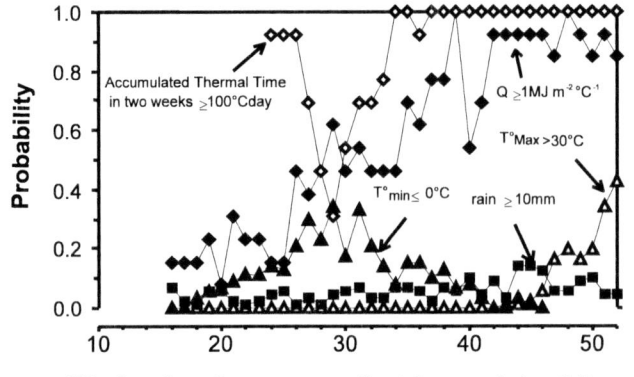

Note: The events are photothermal quotient > 1 (closed diamonds), maximum temperature > 30°C (open triangles), rainfall >10 mm (closed squares), minimum temperatures < 0°C (closed triangles), and accumulated thermal time units (Tb = 0°) ≥100°C during two weeks (open diamonds). Data were calculated weekly from climatic records of 25 years corresponding to an important barley-growing area of Argentina (Tres Arroyos, Buenos Aires Province; 38°20'1S, 60°15'1W; elevation 115 m above sea level).

Based on this analysis, which could be complemented with mean climatic values of the historical data set (e.g., mean photothermal quotient), optimal time of heading could be determined, and, then, according to thermal time units between emergence and heading for the chosen genotype (taking into account the interaction between vernalization, photoperiod, and thermal time; see García del Moral et al., Chapter 10 in this book), the time of sowing is calculated (see Figure 15.3). For example, heading date is planned to occur at 42 weeks after the earliest frozen date in Figure 15.3, and sowing date will be accordingly calculated depending on the period between sowing and heading for the chosen genotype. It is important to highlight here that time of heading could be modified depending on the farmer's option of frozen risk. In addition, other environmental constraints such as water stress should also be included in this analysis by evaluating rainfall probability or, more precisely, potential and actual evapotranspiration. On the other hand, biotic constraints such as disease could be analyzed by models that consider humidity and temperature for calculating disease risk (see Carmona, Moschini, and Conti, 1997). Weeds could also affect sowing date choice. For example, in Argentina, 1.5 million hectare have been reported to be infested with wild oat (Cattena et al., 1998). This weed has an important negative effect on barley yield and, as a contaminant, on grain quality. Therefore, yield and quality of barley will depend on competition between the crop and weed. The effect of different sowing dates that favor the crop's competitive ability, taking into account the wild oat population dynamic (see Scursoni, Benech-Arnold, and Hirchoren, 1999), i.e., the timing and number of plants of each cohort, could be evaluated using a similar approach to that shown in Figure 15.3.

Another interesting approach for choosing sowing date could be the use of models that simulate the development and growth of the crop. These models have been used for evaluating management practices in maize (Otegui, Ruiz, and Petruzzi, 1996), sorghum (Muchow, Hammer, and Carberry, 1991), and wheat (Stapper and Harris, 1989; Savin et al., 1995). For example, in barley, the QBAR model was used for evaluating different sowing dates in northeastern Australia (Goyne et al., 1996). Although all these examples clearly show the usefulness of crop simulation models for farming use, these models have still not incorporated the effect of some abiotic (e.g., heat shock effect) or biotic (e.g., diseases, weeds) constraints to evaluate properly the effect of sowing date on grain yield and quality. For this reason, even when choosing sowing date by crop simulation models, the use of probabilistic approaches, as described earlier, could be an important complement.

CHOOSING PLANT DENSITY

Grain yield can be described by the following relationship:

$$GY = B \cdot HI$$

where grain yield per square meter (GY) is the product of biomass per square meter (B) and harvest index (HI).

It could be accepted that harvest index is a somewhat fixed trait for a given genotype except under certain conditions, e.g., terminal drought stress (López Castañeda and Richards, 1994) and extremely high plant density (Donald and Hamblin, 1976). Therefore, the more common way to reach a higher yield is through greater biomass production, and to do this, the crop must capture more resources. Plant density is a management practice that strongly modifies the ability of the crop to capture resources (mainly solar radiation, water, and nutrients). For this reason, plant density should be chosen according to the resources available to the crop; thus, the crop response to plant density could be analyzed as a competition process. Under potential conditions (without water and nutrient limitation and at optimum sowing date) within a barley-growing area, the amount of solar radiation intercepted by the crop is the main determinant of biomass production and, consequently, yield. The manipulation of plant density directly affects the evolution of the ground cover at the beginning of the growing season and, therefore, the ability of the crop to intercept solar radiation (see Figure 15.4).

In the experiment shown in Figure 15.4, the crop had lower ability to intercept solar radiation at a rate of 140 plants/m^2 than at 300 plants/m^2. This was the cause of differences in aboveground biomass found at heading (862 versus 1,161 g/m^2 in the low- and high-density treatments, respectively), as well as the lower yield in the crop sown at a rate of 140 plants/m^2 (4,927 versus 6,179 g/m^2 in the low- and high-density treatments, respectively). These results agree with previous reports that showed changes in plant density modified intercepted radiation and, consequently, aboveground biomass at ear emergence (McDonald, 1990) and aboveground biomass and yield at harvest (Kirby, 1969; Martin and Field, 1987).

The Relationship Between Grain Yield and Plant Density

Several authors have used mathematical models to explain the effect of plant density on yield (e.g., Holliday, 1960; Willey and Heath, 1969; Ratkowsky, 1983). Although parabolic models have been proposed (Hol-

FIGURE 15.4. Relationship Between Solar Radiation Intercepted by the Crop and Days After Sowing at 140 (Open Triangles) and 300 (Closed Circles) Plants Per Square Meter in a Two-Rowed Barley Cultivar (B1215)

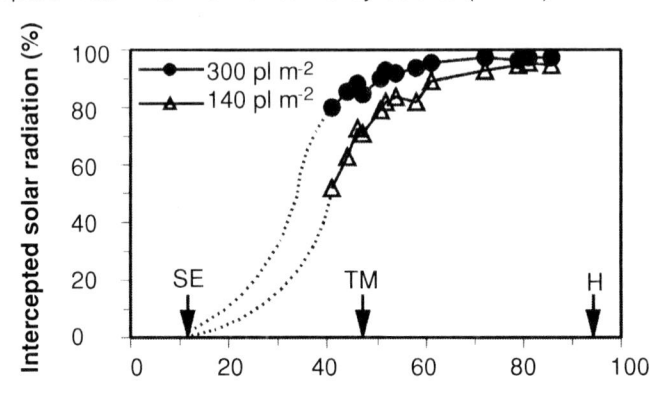

Source: Dreccer, Miralles, and Calderini (unpublished data).
Note: This experiment was carried out in Buenos Aires during the 1996-1997 growing season. Arrows indicate the dates of seedling emergence (SE), triple mound (TM), and heading (H). Dashed lines show calculated fitness for the period between seedling emergence and 40 days after sowing.

liday, 1960), the response of grain yield to plant density could be better described by asymptotic relationships (see Stippers, 1983; Firbank and Watkinson, 1990; Satorre, 1999) (see Figure 15.5). A more difficult task has been to quantify this asymptotic relationship. For example, the threshold value of plant density at which yield starts to decrease in barley (see Figure 15.5) is controversial in the literature. Kirby (1969) showed increases in yield up to 200 plants/m^2 depending on the sowing date. On the other hand, Dofing (1996) showed responses in yield up to 140 plants/m^2. In addition, sowing rates of 500 to 700 plants/m^2 are commonly used at high latitudes (Peltonen and Peltonen-Sainio, 1997), whereas sowing rates between 200 and 300 plants/m^2 are usually used at medium latitudes, as in Argentina (Dreccer, Miralles, and Calderini, 1997; Satorre and Slafer, 1999). The difficulty in explaining the relationship between yield and plant density is that this relationship is the result of the interaction between the availability of resources to produce biomass, and consequently grain yield, and the genotype. For this reason, the shape of the relationship that describes the response of yield to plant density showed in Figure 15.5 could be modified (e.g., maximum grain yield, minimum threshold of plant density at which yield starts to decrease) even without water and nutrient limitation. Climatic factors such as temperature and solar radiation deeply influence the response showed in Figure 15.5. For example, better conditions for crop growth (high irradiance and fresh temperatures) will move the asymptotic relationship (see Figure 15.5) up and to the right.

FIGURE 15.5. Relationship Between Yield and Plant Density, Where "a" Is the Maximum Yield and "c" the Plant Density Level at Which Maximum Yield Is Reached

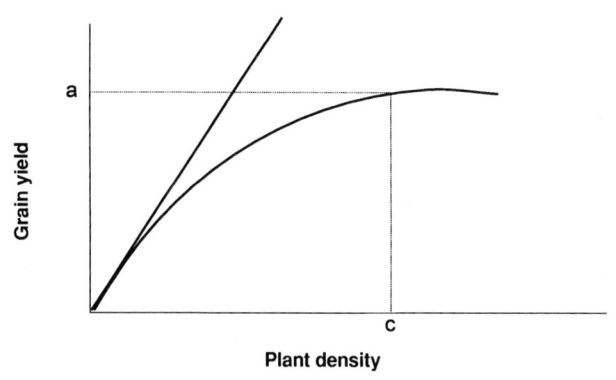

Note: The linear association shows the hypothetical relationship when there is no competition between plants.

Generally, the supply of limiting resources positively modifies the relationship between yield and plant density (see Satorre, 1999). On the contrary, the reduction of nitrogen or water availability produces a lower maximum yield, a lower threshold of plant density for reaching this maximum yield, and a lower rate at which yield responds to increases in plant density. Therefore, when the crop faces low nitrogen or water availability, plant density should be reduced with respect to potential conditions. This strategy will reduce the competition level to which plants are exposed. For this reason, when water shortage increases during the growing season, as commonly occurs in Mediterranean climates, crops are sown at low densities to provide as much water stored in the soil as possible per plant. The disadvantage of this strategy is that water is lost through direct evaporation from the soil in the beginning of the season because leaf growth to cover the soil occurs slowly (see French and Schultz, 1984).

Sowing date might affect barley yield-density relationships (see Kirby, 1969) at sowing rates that may or may not produce competition between plants (Harper, 1977). As noted earlier, the delay of sowing date has a negative effect on grain number and grain yield because the crop is exposed to longer days and higher temperatures that increase the developmental rate, thus reducing the crop cycle (see García del Moral, Miralles, and Slafer, Chapter 10 in this book). This general effect of sowing date modifies most of the grain yield components (e.g., spikes per plant, grains per spike, and individual grain weight). One component, spikes per plant, is reduced due to a shorter tillering period. The direct consequence of this is a lower number of tillers per plant, which affects both solar radiation interception and the number of fertile tillers per plant. These components could be at least partially compensated for by increasing plant density at sowing. In this case, the interaction between sowing date and plant density modifies both the availability of resources, by reducing the period during which the crop can take up resources, and the ability of the crop to capture them, by increasing the demand for resources by a higher number of plants per square meter. An example of this is found at high latitudes, where, due to the short period for tillering, high seeding rates are commonly used (e.g., 500-700 seeds/m^2; Peltonen and Peltonen-Sainio, 1997).

The Effect of Plant Tillering and Planting Arrangement on the Grain Yield-Plant Density Relationship

In small-grain cereals as barley, the ability of the crop to capture resources could be modified by its capability for tillering. At a low sowing rate, the number of tillers per plant is maximized, but this number is strongly reduced when the sowing rate is increased. Therefore, in crops sown at low density, tillers are essential to capture solar radiation that affects grain yield (as shown earlier). In addition, the threshold value of plant density at which maximum grain yield is reached could be modified by the tillering capability of the genotype. For example, Dofing (1996), working with two pairs of near-isogenic barley lines differing in tiller type, showed that uniculm lines reached a lower maximum yield than conventional tillering lines, up to 320 plants/m^2, and showed a higher threshold value for plant density. Other authors found a similar response to seeding rate between two- versus six-rowed cultivars (Jedel and Helm, 1995) or uniculm versus tillered cultivars (McDonald, 1990) in environments with restricted water availability. These results show the difficult task of calculating general threshold values of plant density to obtain maximum yields. However, based on the asymptotic relationship between grain yield and plant density, the availability of resources, and genotype characteristics (e.g., tillering capability), it is possible to envisage the level of plant density for coping with the expected environment where the barley crop will be grown.

Planting arrangement of crops at a given plant density is another management practice to be considered in analyzing the relationship between yield and plant density. However, it is generally accepted that planting arrangement has a lower impact on the relationship between grain yield and plant density than other management practices, such as sowing date or nitrogen availability, and planting arrangement is often not significant if plant densities are above the threshold value to maximize grain yield (Auld, Kemp, and Medd, 1983). Planting arrangements can be described by their rectangularity (distance between rows and distance between plants in a row). Although rectangularity of 1:1 is considered the best arrangement for reaching the highest yield (Holliday, 1963; Auld, Kemp, and Medd, 1983), barley is generally sown in rows with clumped arrangement of plants. Several authors have reported that row spacing between 10 and 15 cm produces higher grain yields than wider row spacing (20-25 cm) in winter cereals (e.g., Holliday, 1963; Panda et al., 1996). Satorre (1999) has suggested that the extent to which rectangularity affects yield depends on the plasticity of individual plants and the environmental conditions. Connor and Loomis (1991) proposed that uniform spacing allows roots to reach the perimeter of interplant space in the shortest time; in contrast, crowded plants in widely spaced rows suffer early competition, restricting water use and early growth. Therefore, under low rainfall conditions, water is distributed over a longer period with an arrangement of higher rectangularity. In addition, when the crop faces competition from weeds, increasing plant density and reducing rectangularity can increase competitive ability of the crop against the weeds (Flemming, Young, and Ogg, 1988; Satorre and Arias, 1990).

Other Important Effects of Plant Density

Although it is not expected that grain yield could be modified above the threshold value of plant density by a direct effect of sowing rate on yield, grain yield in Figure 15.5 should be considered as "potential harvestable yield," since high plant densities could decrease yield by indirect effects. For example, under high plant densities, competition between crop plants is increased. Therefore, greater crop density tends both to reduce biomass accumulated in stems and to increase plant height (Pinthus, 1973), which increases lodging risk. On the other hand, higher plant density could produce a wet environment in the canopy, thus favoring diseases.

Compared with the effect on grain yield, little attention has been paid to the effect of plant density on grain quality for malting. The clearest consequence of increasing plant density is the negative effect on individual grain weight (Kirby, 1969; Lauer, 1991; Dofing, 1996). At the same time, increasing plant density seems to decrease grain plumpness (Lauer, 1991; Hanson and Lukach, 1992) and the difference in grain plumpness between different categories of tillers (Lauer, 1991), resulting in a more heterogeneous seed lot. This response may be genotype dependent, as it has been observed that plant density affects differently the way in which yield is partitioned between yield components in two- versus six-rowed barley cultivars (Jedel and Helm, 1995) or uniculm versus tillered cultivars (McDonald, 1990). Last, but not least, it is important to note that low plant densities can also affect the crop yield indirectly by creating more opportunities for weed growth (Doll, 1997).

CONCLUDING REMARKS

Farmers and crop advisors must make decisions regarding processes in most agricultural production systems. In malting barley, grain yield and quality strongly depend on the success of properly choosing genotype, sowing date, and plant density. This involves combining the genotype potential for yield and quality with the environmental availability of resources. For choosing genotype, a simple and effective methodology based on the response of cultivars to different environmental indices, as proposed by Finlay and Wilkinson (1963) and enhanced by Eberhart and Russell (1966), seems to be useful. This analysis provides a quantitative approach to the response of cultivars to a given genotype-season combination. This could also be complemented by other important considerations, including more specific effects of biotic and abiotic factors on grain yield and quality. Sowing date is another important presowing decision. As a general rule, anticipating sowing date has a clearly positive effect on grain yield and grain quality as well. The extent to which it is possible to follow this general rule and how to consider particular stresses (e.g., frozen risk) that might affect grain yield and quality can be approached probabilistically. When choosing sowing date, the critical periods for determining grain yield and quality should be matched with the best environmental conditions. For this objective, the probability of the occurrence of favorable climatic conditions (e.g., high photothermal quotient) and climatic constraints (e.g., frozen risk, high maximum temperatures) can be evaluated based on historical climatic records. Finally, the choice of plant density is based on the asymptotic relationship between grain yield and plant density, which accounts for the availability of resources to the crop. Therefore, the grain yield-plant density relationship is viewed as a competition process in which a higher availability of resources would increase the threshold value of plant density at which maximum yield is reached. On the other hand, tillering capability of the cultivar and sowing date clearly modify this threshold value. In addition, plant arrangement could also affect (though with a lower impact) the choice of plant density.

REFERENCES

Acevedo, E. (1991). Morphophysiological traits of adaptation of cereals to Mediterranean environments. In *Improvement and Management of Winter Cereals Under Temperature, Drought and Salinity Stresses,* eds. E. Acevedo, E. Fereres, C. Gimenez, and J.P. Srivastava. Madrid: Instituto Nacional de Investigaciones Agrarias, pp. 85-96.

Anderson, M.K. and E. Reinbergs (1985). Barley breeding. In *Barley,* ed. D.C. Rasmusson. Madison, WI: ASA, pp. 231-268.

Andrade, F.H. (1995). Analysis of growth and yield of maize, sunflower and soybean grown at Balcarce, Argentina. *Field Crops Research* 41: 1-12.

Andrade, F.H., M.E. Otegui, and C. Vega (2000). Intercepted radiation at flowering and kernel number in maize. *Agronomy Journal* 92: 92-97.

Auld, B.A., D.R. Kemp, and R.W. Medd (1983). The influence of spatial arrangement on grain yield of wheat. *Australian Journal of Agricultural Research* 34: 99-108.

Austin, R. B. and M.H. Arnold (1989). Variability in wheat yields in England: Analysis and future prospects. In *Variability in Grain Yields,* eds. J.R. Anderson and P.B.R. Hazell. Baltimore, MD: Johns Hopkins University Press, pp. 100-106.

Baker, R.J. (1988). Tests for crossover genotype-environmental interactions. *Canadian Journal of Plant Science* 68: 405-410.

Bamforth, C.W. and A.H.P. Barclay (1993). Malting technology and the uses of malt. In *Barley: Chemistry and Technology,* eds. A.W. MacGregor and R.S. Bhatty. St. Paul, MN: American Association of Cereal Chemists, pp. 297-354.

Barreto, D., M. Carmona, M. Ferrazini, and B. Pérez (1998). Reaction of *Hordeum distichum* to *Pyrenophora teres*. *Annual Wheat Newsletter* 44: 7-9.

Benech-Arnold R.L., M.C. Giallorenzi, J. Frank, and V. Rodriguez (1999). Termination of hull-imposed dormancy in barley is correlated with changes in embryonic ABA content and sensitivity. *Seed Science Research* 9: 39-47.

Blumenthal, C., C.W. Wrigley, I.L. Batey, and E.W.R. Barlow (1994). The heat-shock response relevant to molecular and structural changes in wheat yield and quality. *Australian Journal of Plant Physiology* 21: 901-909.

Briggs, K.G. (1990). Studies of recovery from artificially induced lodging in several six-row barley cultivars. *Canadian Journal of Plant Science* 70: 173-181.

Bulman, P., D.E. Mather, and D.L. Smith (1993). Genetic improvement of spring barley cultivars grown in eastern Canada from 1910 to 1988. *Euphytica* 71: 35-48.

Bulman, P. and D.L. Smith (1994). Post-heading nitrogen uptake, retranslocation, and partitioning in spring barley. *Crop Science* 34: 977-984.

Calderini, D.F., L.G. Abeledo, R. Savin, and G.A. Slafer (1999a). Carpel size and temperature in pre-anthesis modify potential grain weight in wheat. *Journal of Agricultural Science (Cambridge)* 132: 453-460.

Calderini, D.F., L.G. Abeledo, R. Savin, and G.A. Slafer (1999b). Final grain weight in wheat as affected by short periods of high temperature during pre- and post-anthesis under field conditions. *Australian Journal of Plant Physiology* 26: 453-458.

Calderini, D.F., D.J. Miralles, M.F. Dreccer, L.G. Abeledo, and A. Lorenzo (1998). Rendimiento y componentes del rendimiento en genotipos de cebadas de 2 y 6 hileras. In *Proceedings IV Congreso Nacional de Trigo, Mar del Plata, Argentina*, pp. 2-5.

Calderini, D.F., M.P. Reynolds, and G.A. Slafer (1999). Genetic gains in wheat yield and associated physiological changes during the twentieth century. In *Wheat: Ecology and Physiology of Yield Determination*, eds. E.H. Satorre and G.A. Slafer. Binghamton, NY: Food Products Press, pp. 351-377.

Calderini, D.F. and G.A. Slafer (1999). Has yield stability changed with genetic improvement of wheat yield? *Euphytica* 107: 51-59.

Cantagallo, J.E., C.A. Chimenti, and A.J. Hall (1997). Number of seeds per unit area in sunflower correlates well with a photothermal quotient. *Crop Science* 37: 1780-1786.

Carmona, M.A., R.C. Moschini, and H.A. Conti (1997). Meteorological factors influencing the incidence of barley scald and its spatial distribution over the argentine pampas region. *Journal of Plant Pathology* 79: 203-209.

Cattena, H., P. Avalle, F. Strimmer, and M. Durante (1988). Control químico de *Avena fatua* en trigo *(Triticum aestivum)*. In *Proceedings XI Reunión Argentina sobre la maleza y su control, Córdoba, Argentina*, pp. 63-78.

Ceccarelli, S. (1984). Utilization of landraces and *Hordeum spontaneum* in barley breeding for dry areas at ICARDA. *Rachis* 3: 8-11.

Ceccarelli, S. (1991). Selection for specific environments or wide adaptability? In *Improvement and Management of Winter Cereals Under Temperature, Drought and Salinity Stresses*, eds. E. Acevedo, E. Fereres, C. Gimenez, and J.P. Srivastava. Madrid: Instituto Nacional de Investigaciones Agrarias, pp. 227-237.

Challaiah, O.C., Burnside, G.A. Wicks, and V.A. Johnson (1986). Competition between winter wheat *(Triticum aestivum)* cultivars and downy brome *(Bromus tectorum)*. *Weed Science* 34: 689-693.

Chowdhury, S.I. and I.F. Wardlaw (1978). The effect of temperature on kernel development in cereals. *Australian Journal of Agricultural Research* 29: 205-223.

Ciha, A.J. (1983). Seeding rate and seeding date effects on spring seeded small grain cultivars. *Agronomy Journal* 75: 795-799.

Connor, D.J. and Loomis, R.S. (1991). Strategies and tactics for water-limited agriculture in low rainfall Mediterranean climates. In *Improvement and Management of Winter Cereals Under Temperature, Drought and Salinity Stresses,* eds. E. Acevedo, E. Fereres, C. Gimenez, and J.P. Srivastava. Madrid: Instituto Nacional de Investigaciones Agrarias, pp. 441-465.

Cooper, M., R.E. Stucker, I.H. DeLacy, and B.D. Harch (1997). Wheat breeding nurseries, target environments, and indirect selection for grain yield. *Crop Science* 37: 1168-1176.

Correll, R., J. Butler, L. Spouncer, and C. Wrigley (1994). The relationship between grain-protein content of wheat and barley and temperatures during grain filling. *Australian Journal of Plant Physiology* 21: 869-873.

Crook, M.J. and Ennos, A.R. (1994). Stem and root characteristics associated with lodging resistance in four winter wheat cultivars. *Journal of Agricultural Science (Cambridge)* 123: 167-174.

Day, A.D. and A.D. Dickson (1958). Effect of artificial lodging on grain and malt quality of fall-sown irrigated barley. *Agronomy Journal* 50: 338-340.

Dennett, M.D. (1999). Effects of sowing date and the determination of optimum sowing date. In *Wheat: Ecology and Physiology of Yield Determination,* eds. E.H. Satorre and G.A. Slafer. Binghamton, NY: Food Products Press, pp. 123-140.

Dofing, S.M. (1996). Near-isogenic analysis of uniculm and conventional-tillering barley lines. In *Proceedings of the Fifth International Oat Conference and the Seventh International Barley Genetics Symposium,* eds. A. Slinkard, G. Scoles, and B. Rossnagel. Saskatoon: University of Saskatchewan Extension Press, pp. 617-619.

Dofing, S.M. (1997). Ontogenic evaluation of grain yield and time to maturity in barley. *Agronomy Journal* 89: 685-690.

Doll, H. (1997). The ability of barley to compete with weeds. *Biological Agriculture and Horticulture* 14: 43-51.

Donald, C.M. (1968). The breeding of crop ideotypes. *Euphytica* 17: 385-403.

Donald, C.M. and J. Hamblin (1976). The biological yield and harvest index of cereals as agronomic and plant breeding criteria. *Advances in Agronomy* 28: 361-405.

Dreccer, M.F., D.J. Miralles, and D.F. Calderini (1997). Ventajas de las cebadas de seis hileras. *Super Campo* 33: 96-97.

Eberhart, S.A. and W.A. Russell (1966). Stability parameters for comparing varieties. *Crop Science* 6: 36-40.

Evans, L.T. (1993). Increases in yield: Trends and limits. In *Crop Evolution, Adaptation and Yield,* ed. L.T. Evans. Cambridge: Cambridge University Press, pp. 269-316.

Finlay, K.W. and G.N. Wilkinson (1963). The analysis of adaptation in a plant-breeding program. *Australian Journal of Agricultural Research* 14: 342-354.

Firbank, L.G. and A.R. Watkinson (1990). On the effects of competition: From monocultures to mixtures. In *Perspectives on Plant Competition,* eds. J.B. Grace and D. Tilman. San Diego: Academic Press, pp. 165-192.

Fischer, R.A. (1984). Wheat. In *Potential Productivity of Field Crops under Different Environments,* eds. W.H. Smith and S.J. Banta. Los Baños, CA: IRRI, pp. 129-153.

Fischer, R. A. (1985). Number of kernels in wheat crops and the influence of solar radiation and temperature. *Journal of Agricultural Science (Cambridge)* 105: 447-461.

Flemming, G.F., F.L. Young, and A.C. Ogg (1988). Competitive relationships among winter wheat *(Triticum aestivum)*, jointed goatgrass *(Aegilops scylindrica)*, and downy brome *(Bromus tectorum)*. *Weed Science* 36: 479-486.

Francis, T.R. and L.W. Kannenberg (1978). Yield stability in short-season maize. I. A descriptive method for grouping genotypes. *Canadian Journal of Plant Science* 58: 1029-1034.

French, R.J. and J.E. Schultz (1984). Water use efficiency of wheat in a Mediterranean-type environment. 1. Relation between yield, water use and climate. *Australian Journal of Agricultural Research* 35: 743-764.

González, A., I. Martín, and L. Ayerbe (1999). Barley yield in water-stress conditions. The influence of precocity, osmotic adjustment and stomatal conductance. *Field Crops Research* 62: 23-34.

Goyne, P.J., H. Meinke, S.P. Milroy, G.L. Hammer, and J.M. Hare (1996). Development and use of a barley crop simulation model to evaluate production management strategies in north-eastern Australia. *Australian Journal of Agricultural Research* 47: 997-1015.

Green, C.F., D.T. Furmston, and J.D. Ivins (1985). Time of sowing and the yield of winter barley. *Journal of Agricultural Science (Cambridge)* 104: 405-411.

Guglielmini, A.C. and E.H. Satorre (2000). Nitrogen fertilizer application and competitive balance between spring wheat (*Triticum aestivum* L.) and volunteer oil seed rape (*Brassica napus* L.). *Ecología Australia* 10: 133-142.

Hakansson, S. (1983). Competition and production in short-lived crop-weed stands: Density effects. Report 127. Alnarp, Sweden: The Swedish University of Agricultural Sciences, Department of Plant Husbandry.

Hanson, B.K. and J.R. Lukach (1992). Barley response to planting rate in North Dakota. *North Dakota Farm Research* 49: 14-19.

Harper, J.L. (1977). *Population Biology of Plants*. London: Academic Press.

Holliday, R. (1960). Plant population and crop yield. *Field Crop Abstracts* 13: 159-167.

Holliday, R. (1963). The effect of row width on the yield of cereals. *Field Crop Abstracts* 16: 71-81.

Jana, S. (1991). Strategies for breeding salt-tolerant wheat and barley. In *Improvement and Management of Winter Cereals Under Temperature, Drought and Salinity Stresses*, eds. E. Acevedo, E. Fereres, C. Gimenez, and J.P. Srivastava. Madrid: Instituto Nacional de Investigaciones Agrarias, pp. 351-371.

Jedel, P.E. and J.H. Helm (1991). Lodging effects on a semidwarf and two standard barley cultivars. *Agronomy Journal* 83: 158-161.

Jedel, P.E. and J.H. Helm (1994). Assessment of western Canadian barleys of historical interest. II. Morphology and phenology. *Crop Science* 34: 927-932.

Jedel, P.E. and J.H. Helm (1995). Agronomic response to seeding rate of two- and six-rowed barley cultivars in central Alberta. *Canadian Journal of Plant Science* 75: 315-320.

Jenner, C.F. (1994). Effects of exposure of wheat ears to high temperature on dry matter accumulation and carbohydrate metabolism in the grain of two cultivars. II. Carry-over effects. *Australian Journal of Plant Physiology* 18: 179-190.

Jenner, C.F., D.T. Ugalde, and D. Aspinal (1991). The physiology of starch and protein deposition in the endosperm of wheat. *Australian Journal of Plant Physiology* 18: 211-226.

Kirby, E.J.M. (1969). The effect of sowing date and plant density on barley. *Annals of Applied Biology* 63: 513-521.

Kitchen, B.M. and D.C. Rasmusson (1983). Duration and inheritance of leaf initiation, spike initiation, and spike growth in barley. *Crop Science* 23: 939-943.

Lauer, J.G. (1991). Barley tiller response to plant density and ethephon. *Agronomy Journal* 83: 968-973.

Legere, A. and N. Samson (1999). Relative influence of crop rotation, tillage, and weed management on weed associations in spring barley cropping systems. *Weed Science* 47: 112-122.

Liebman, M. and R. Robichaux (1990). Competition by barley and pea against mustard: Effects on resource acquisition, photosynthesis and yield. *Agricultural Ecosystems Environment* 31: 155-172.

Limin, A.E. and D.B. Fowler (1999). Inheritance of cell size and low-temperature tolerance in wheat (*Triticum aestivum* L.). In *Proceedings of the ASA, CSSA and SSSA Annual Meeting*, eds. H.H. Cheng and V.B. Cardwell. Salt Lake City, UT: p. 71.

Lin, C.S., M.R. Binns, and L.P. Lefkovitch (1986). Stability analysis: Where do we stand? *Crop Science* 26: 894-900.

Little, R. and J.K. Doodson (1972). The reaction of spring barley cultivars to mildew, their disease resistance rating and an interim report on their yield response to mildew control. *Journal of the National Institute of Agricultural Botany* 12: 447-455.

López Castañeda, C. and R.A. Richards (1994). Variation in temperate cereals in rainfed environments. I. Grain yield, biomass and agronomic characteristics. *Field Crops Research* 37: 51-62.

Lúquez, J. (1990). Aplicaciones del método de estabilidad de tipo IV al rendimiento de trigo pan (*Triticum aestivum* L.). *Proceedings II Congreso Nacional de Trigo, Pergamino, Argentina* II: 114-120.

MacGregor, A.W. (1992). Evaluation of barley malting quality. In *Proceedings of the Sixth International Barley Genetics Symposium*, Volume II, eds. L. Munck, K. Kirkegaard, and B. Jensen. Helsingborg, Sweden: The Organizing Committee of the Nordic Countries, pp. 969-978.

MacLeod, A.M. (1979). The physiology of malting. In *Brewing Science* (Volume 1), ed. J.R.A. Pollock. Reading, England: Academic Press, pp. 1-143.

MacLeod, L.C. and C.M. Duffus (1988). Temperature effects on starch granules in developing barley grains. *Journal of Cereal Science* 8: 19-37.

Magrin, G.O., A.J. Hall, C. Baldy, and M.O. Grondona (1993). Spatial and interannual variations in the photothermal quotient: Implications for the potential kernel number of wheat crops in Argentina. *Agricultural and Forest Meteorology* 67: 29-41.

Martin, M.P.L.D. and R.J. Field (1987). Competition between vegetative plants of wild oat *(A. fatua)* and wheat *(Triticum aestivum* L.). *Weed Research* 27: 119-124.

Matinuzzi, H.J., N. Machado, and P. Paulucci (1990). Respuesta al ambiente en cultivares de trigo pan *(Triticum aestivum* L.). *Proceedings II Congreso Nacional de Trigo, Pergamino, Argentina* II: 134-142.

McDonald, G.K. (1990). The growth and yield of uniculm and tillered barley over a range of sowing rates. *Australian Journal of Agricultutral Research* 41: 449-461.

McFadden, A.D. (1970). Influence of seeding dates, seeding rates and fertilizers on two cultivars of barley. *Canadian Journal of Plant Science* 50: 693-699.

McTaggart, I.P. and K.A. Smith (1995). The effect of rate, form and timing of fertilizer N on nitrogen uptake and grain N content in spring malting barley. *Journal of Agricultural Science* 125: 341-353.

Minella, E. and M.E. Sorrells (1992). Aluminum tolerance in barley: Genetic relationships among genotypes of diverse origin. *Crop Science* 32: 593-598.

Miralles, D.J., R.A. Richards, and G.A. Slafer (2000). Duration of the stem elongation period influences the number of fertile florets in wheat and barley. *Australian Journal of Plant Physiology* 27: 931-940.

Molina-Cano, J.L. (1989). *La cebada. Morfología, fisiología, genética, agronomía y usos industriales.* Madrid: Mundi-Prensa.

Moll, M. (1979). Analysis and composition of barley malt. In *Brewing Science* (Volume 1), ed. J.R.A. Pollock. Reading, England: Academic Press, pp. 1-143.

Muchow, R.C., G.L. Hammer, and P.S. Carberry (1991). Optimizing crop and cultivar selection in response to climatic risk. In *Climatic Risk in Crop Production: Models and Management for the Semiarid Tropics and Subtropics,* eds. R.C. Muchow and J.A. Bellamy. London: CAB International, pp. 235-262.

Mujeeb-Kazi, A. (1995). Interespecific crosses: Hybrid production and utilization. In *Utilizing Wild Grass Biodiversity in Wheat Improvement: 15 Years of Wild Cross Research at CIMMYT,* eds. A. Mujeeb-Kazi and G.P. Hettel. Mexico City, Mexico, DF: CIMMYT Research Report No. 2, pp. 14-21.

Murelli, C., F. Rizza, F. Marinone Albinini, A. Dulio, V. Terzi, and L. Cattivelli (1995). Metabolic changes associated with cold-acclimation in contrasting barley genotypes. *Physiologia Plantarum* 94: 87-93.

Nable, R.O. (1988). Resistance to boron toxicity amongst barley and wheat cultivars: A preliminary examination of the resistance mechanism. *Plant and Soil* 112: 45-52.

Okafor, Ll. (1988). Barley yield and responses to planting date, seed rate and N fertilizers in the Chad basin of Nigeria. *Tropical Agriculture* 3: 249-253.

Ordon, F. and W. Friedt (1994). Agronomic traits of exotic barley germplasms resistant to soil-borne mosaic-inducing viruses. *Genetic Resources and Crop Evolution* 41: 43-46.

Ortiz Monasterio, J.I., S.S. Dhillon, and R.A. Fischer (1994). Date of sowing effects on grain yield and yield components of irrigated spring wheat cultivars and relationships with radiation and temperature in Ludhiana, India. *Field Crops Research* 37: 169-184.

Otegui, M.E., M.G. Nicolini, R.A. Ruiz, and P. Dodds (1995). Sowing date effects on grain yield components for different maize genotypes. *Agronomy Journal* 87: 29-33.

Otegui, M.E., R.A. Ruiz, and D. Petruzzi (1996). Modeling hybrid and sowing date effects on potential grain yield of maize in a humid temperate region. *Field Crops Research* 47: 167-174.

Panda, S.C., A. Pattanaik, B.S. Rath, R.K. Tripathy, and B. Behera (1996). Response of wheat *(Triticum aestivum)* to crop geometry and weed management. *Indian Journal of Agronomy* 41: 553-557.

Peltonen, J. and P. Peltonen-Sainio (1997). Breaking uniculm growth habit of spring cereals at high latitudes by crop management. II. Tillering, grain yield and yield components. *Journal of Agronomy and Crop Science* 178: 87-95.

Peltonen-Sainio, P. (1990). Genetic improvements in the structure of oat stands in northern growing conditions during this century. *Plant Breeding* 104: 340-345.

Peltonen-Sainio, P. (1996). Sowing time effects on growth duration and formation and realization of yield potential of oat in northern growing conditions. *Cereal Research Communications* 24: 223-229.

Pinthus, M.J. (1973). Lodging in wheat, barley, and oats: The phenomenon, its causes, and preventive measures. *Advances in Agronomy* 25: 210-263.

Poehlman, J.M. (1985). Adaptation and distribution. In *Barley,* ed. D.C. Rasmusson. Madison, WI: ASA, pp. 1-17.

Powell, W., P.D.S. Caligari, P.H. Goudappel, and W.T.B. Thomas (1985). Competitive effects in monocultures and mixtures of spring barley *(Hordeum vulgare)*. *Theoretical and Applied Genetics* 71: 443-450.

Radosevich, S., J. Holt, and C. Ghersa (1997). *Weed ecology. Implications for management.* New York: John Wiley and Sons.

Randall, P.J. and H.J. Moss (1990). Some effects of temperature regime during grain filling on wheat quality. *Australian Journal of Agricultural Research* 41: 603-617.

Rasmusson, D.C. (1991). A plant breeder's experience with ideotype breeding. *Field Crops Research* 26: 191-200.

Ratkowsky, D.A. (1983). *Non Linear Regression Modelling. A Unified Practical Approach.* New York: Marcel Dekker.

Reynolds, M.P., E. Acevedo, K.D. Sayre, and R.A. Fischer (1994). Yield potential in modern wheat varieties: Its association with a less competitive ideotype. *Field Crops Research* 37: 149-160.

Reynolds, M.P., M. Balota, M.I.B. Delgado, I. Amani, and R.A. Fischer (1994). Physiological and morphological traits associated with spring wheat yields under hot, irrigated conditions. *Australian Journal of Plant Physiology* 21: 717-730.

Riggs, T.J., P.R. Hanson, N.D. Start, D.M. Miles, C.L. Morgan, and M.A. Ford (1981). Comparison of spring barley varieties grown in England and Wales between 1880 and 1980. *Journal of Agricultural Science (Cambridge)* 97: 599-610.

Satorre, E.H. (1988). The competitive ability of spring cereals. PhD thesis. Reading, England: University of Reading, 262 pp.

Satorre, E.H. (1999). Plant density and distribution as modifiers of growth and yield. In *Wheat: Ecology and Physiology of Yield Determination,* eds. E.H. Satorre and G.A. Slafer. Binghamton, NY: Food Products Press, pp. 141-159.

Satorre, E.H. and S.P. Arias (1990). Competencia entre trigo *(Triticum aestivum)* y malezas. III. El efecto de la densidad del cultivo y la maleza. *Proceedings II Congreso Nacional de Trigo, Pergamino, Argentina* IV: 1-10.

Satorre, E.H. and G.A. Slafer (1999). Wheat production systems of the Pampas. In *Wheat: Ecology and Physiology of Yield Determination,* eds. E.H. Satorre and G.A. Slafer. Binghamton, NY: Food Products Press, pp. 141-159.

Satorre, E.H. and R.W. Snaydon (1991). A comparison of root and shoot competition between spring cereals and *Avena fatua. Weed Research* 32: 45-55.

Savin, R. and M.E. Nicolas (1996). Effects of short periods of drought and high temperature on grain of two malting barley cultivars. *Australian Journal of Plant Physiology* 23: 201-210.

Savin, R. and M.E. Nicolas (1999). Effects of timing of heat stress and drought on growth and quality of barley grains. *Australian Journal of Agricultural Research* 50: 357-364.

Savin, R., E.H. Satorre, A.J. Hall, and G.A. Slafer (1995). Assessing strategies for wheat cropping in the monsoonal climate of the Pampas using the CERES-wheat simulation model. *Field Crops Research* 42: 81-91.

Savin, R., P.J. Stone, and M.E. Nicolas (1996). Responses of grain growth and malting quality of barley to short periods of high temperature in field studies using portable chambers. *Australian Journal of Agricultural Research* 47: 465-477.

Schaller, C.W., J.A. Berdeque, C.W. Dennet, R.A. Richards, and M.D. Winslow (1981). Screening the world barley collection for salt tolerance. In *Proceedings of the Fourth International Barley Genetics, Symposium* Volume II, eds. L. Munck, K. Kirkegaard, and B. Jensen. Edinburg, Scotland: pp. 22-29.

Schildbach, R. and M. Burbidge (1992). Barley varieties and their malting and brewing qualities. In *Proceedings of the Sixth International Barley Genetics Symposium.* Helsingborg, Sweden: The Organizing Committee of the Nordic Countries, pp. 953-967.

Scott, R.W., M. Appleyard, G. Fellowers, and E.J.M. Kirby (1983). Effect of genotype and position in the ear on carpel and grain growth and mature grain weight of spring barley. *Journal of Agricultural Science (Cambridge)* 100: 383-391.

Scursoni, J., R.L. Benech-Arnold, and H. Hirchoren (1999). Demography of wild oat in barley crops: Effect of crop, sowing rate, and herbicide treatment. *Agronomy Journal* 91: 478-485.

Sisler, W.W. and P.J. Olson (1951). A study of methods of influencing lodging in barley and the effect of lodging upon yield and certain quality characteristics. *Scientific Agriculture* 31: 117-186.

Slafer, G.A. and J.L. Araus (1998). Improving wheat responses to abiotic stresses. *Proceedings of the Ninth International Wheat Genetics Symposium,* ed. A.E. Slinkard. Saskatoon: University of Saskatchewan Extension Press, pp. 201-213.

Slafer, G.A. and H.M. Rawson (1994). Sensitivity of wheat phasic development to major environmental factors: A re-examination of some assumptions made by

physiologists and modellers. *Australian Journal of Plant Physiology* 21: 393-426.
Slafer, G.A. and R. Savin (1991). Developmental base temperature in different phenological phases of wheat (*Triticum aestivum* L.). *Journal of Experimental Botany* 42: 1077-1082.
Stapper, M. and R.A. Fischer (1990). Genotype, sowing date and planting spacing influence on high-yielding irrigated wheat in southern New South Wales. I. Phasic development, canopy growth and spike production. *Australian Journal of Agricultural Research* 41: 997-1019.
Stapper, M. and H.C. Harris (1989). Assessing the productivity of wheat genotypes in a Mediterranean climate, using a crop simulation model. *Field Crops Research* 20: 129-152.
Stippers, C.J.T. (1983). An alternative approach to the analysis of mixed cropping experiments. I. Estimation of competition effects. *Netherland Journal of Agricultural Science* 31: 1-11.
Stone, P.J. and M.E. Nicolas (1994). Wheat cultivars vary widely in their responses of grain yield and quality to short periods of post-anthesis heat stress. *Australian Journal of Plant Physiology* 21: 887-900.
Stone, P.J. and M.E. Nicolas (1995). A survey of the effects of high temperature during grain filling on yield and quality of 75 wheat cultivars. *Australian Journal of Agricultural Research* 46: 475-492.
Stone, P.J. and R. Savin (1999). Grain quality and its physiological determinants. In *Wheat: Ecology and Physiology of Yield Determination,* eds. E.H. Satorre and G.A. Slafer. Binghamton, NY: Food Products Press, pp. 85-120.
Tahir, M., V. Shevtsov, H. Pashayani, A. Ottekin, H. Tosun, and T. Akar (1994). Stress tolerance in winter and facultative barley. *Rachis* 13: 5-11.
Tekauz, A. (1985). A numerical scale to classify reactions of barley to *Pyrenophora teres. Canadian Journal of Plant Pathology* 7: 181-183.
Tekauz, A. (1990). Determination of barley cultivar reaction to *Pyrenophora graminea* using disease nurseries. *Canadian Journal of Plant Pathology* 12(1): 57-62.
Valentine, J. (1982). Variation in monoculture and in mixture for grain yield and other characters in spring barley. *Annals of Applied Biology* 101: 127-141.
Wardlaw, I.F. and C.W. Wrigley (1994). Heat tolerance in temperate cereals: An overview. *Australian Journal of Plant Physiology* 21: 695-703.
Wicks, G.A., R.E. Ramsel, P.T. Nordquist, J.W. Schmidt, and O.C. Challaiah (1986). Impact of wheat cultivars on establishment and suppression of summer annual weeds. *Agronomy Journal* 78: 59-62.
Willey, R.W. and S.B. Heath (1969). The quantitative relationship between plant population and crop yield. *Advances in Agronomy* 21: 281-321.
Yau, S.K., J. Hamblin, and J. Ryan (1994). Phenotypic variation in boron toxicity tolerance in barley, durum and bread wheat. *Rachis* 13: 20-25.
Zubriski, T.C., E.H. Vasey, and E.B. Norum (1970). Influence of nitrogen and potassium fertilizers and dates of seeding on yield and quality of malting barley. *Agronomy Journal* 62: 216-219.

Chapter 16

Computer Simulation of Barley Growth, Development, and Yield

Joe T. Ritchie
Gopalsamy Alagarswamy

INTRODUCTION

Crop simulation models and systems analysis are becoming integral parts of decision-making processes in agriculture. This is especially true in rain-fed agriculture, which often requires tactical and conditional decisions during the cropping season, such as the midseason crop management correction. Reducing plant population in response to some weather-related catastrophes, such as severe drought, is one such example. Any attempt to analyze such conditional decisions through agricultural experimentation in a variable climate requires conducting experiments for several years to characterize adequately the variability. In practice, this approach is time-consuming, costly, and difficult to implement. Alternatively, dynamic crop simulation models based on plant processes and mechanisms can be used to analyze conditional "what if" questions in variable climatic conditions, both locally (site specific) and globally.

The barley crop simulation model (Otter-Nacke et al., 1991), similar to the wheat (Ritchie and Otter-Nacke, 1985; Otter-Nacke, Godwin, and Ritchie, 1987; Ritchie, 1991c) and maize (Jones and Kiniry, 1986) models, is a dynamic and plant process-oriented crop simulation model. It is designed to simulate the effects of barley cultivars, planting density, weather, soil, water, and nitrogen on growth, development, and yield. It was developed as a joint effort by the International Center for Agricultural Research in the Dry Areas (ICARDA), the International Fertilizer Development Cen-

---

We gratefully acknowledge S. Otter-Nacke and D. C. Godwin for their efforts in the model development and testing. We thank S. Daroub, Michigan State University, for critical reading of the manuscript. The model development work was supported by a grant from the International Center for Agricultural Research in the Dry Areas (ICARDA) and the International Benchmark Sites Network for Agrotechnology Transfer (IBSNAT).

ter (IFDC), and Michigan State University. This collaborative effort evolved through mutual interests in utilizing the modeling approach to improve barley management practices and facilitate technology transfer in the 150 to 300 mm rainfall areas of the West Asia and North Africa (WANA) region. In the WANA region, barley is the most important feed and forage crop supporting sheep production, an important source of dietary protein and a capital base for resource-poor farmers. The barley model is not an explanatory model for improving the understanding of how plants work. It is intended to be a functional model useful as a within-year crop decision-making tool and as a multiyear risk assessment analysis tool to evaluate alternate crop management strategies. On a regional basis, the model is intended to be useful for yield forecasting and analysis of various policy decisions regarding crop production and natural resource conservation.

To accurately simulate barley growth, development, and yield, the model accounts for the following three processes: (1) phasic development describing the duration of growth stages as influenced by genotypes and weather; (2) biomass accumulation and partitioning to different plant parts; and (3) grain yield and yield components. To accomplish these primary purposes, it was necessary to exclude several potentially important yield-limiting factors, such as phosphorus, pests and diseases, soil salinity, poor soil aeration, and catastrophic weather events. Excluding these factors in the present model does not minimize their importance, nor does it imply that they are too complex to model. A particular limiting factor could be added to the general model framework as a subroutine, when it is sufficiently developed.

Crop simulation models use diverse simulation procedures with different levels of details to predict crop growth, grain yield, and yield components. These procedures are categorized into three major groups: (1) assimilate production or source dependence (Williams, Jones, and Dyke, 1984); (2) potential pool size for assimilate or sink dependence (Mass and Arkin, 1980); and (3) a combination of source and sink dependence. Ritchie (1991b) discussed the advantages and limitations of using these approaches in various crop simulation models to simulate grain yield. The CERES family of models uses the combination of sink and source evaluation along with potentially mobile assimilates (reserves) when the current source cannot match the sink demand. In some instances, when the sink demand exceeds the source, stored reserves can meet the sink demand. The major advantage of using the source-sink and reserve approach in simulation models is that the yield predictions are not dependent on a predetermined harvest index. In developing the CERES-based family of crop models, the following general principles were considered.

The potential yield of a crop is generally determined by the crop duration when water and nutrient supplies are nonlimiting and temperatures are optimum during the growing season. When the duration is longer, the productivity is higher in several crops (Tanaka, 1983; Akita, 1989). Selecting for shorter-duration crops may impose a yield penalty. This is especially true in temperate crops. The increase in barley yields in England between 1977 and 1983 was due to the shift from spring to autumn sowing, leading to a longer growing season (Silvey, 1986). Yield is greatly influenced by the length of period from when the stem and inflorescence start to grow until the end of grain filling. Hence, control of crop duration is important to achieve an optimum yield.

The plant life cycle is divided into several growth stages. The phasic development deals with timing and duration of different growth stages as affected by genetics of the plants and the environment in which they are grown. Due to genetic diversity in photoperiod sensitivity, vernalization, and cold tolerance, barley cultivars are grown in a wide range of environments both north and south of the equator. Unlike high-value commercial crops, barley is generally grown as a rain-fed crop where the water supply during the season is highly variable. Under such growing conditions, it is essential to sow short-duration barley cultivars to reduce crop failure caused by terminal drought. A simulation model should include the quantitative aspects of phasic development because of genetic diversity and diversity of environments in which barley is grown. Hence, in simulating barley yield, the model focuses on three important areas: growth duration, growth rate, and the extent to which these two processes are influenced by soil water and nitrogen status.

The model calculates phasic development to drive the model through time. It calculates the number and appearance of leaves on the main stem, number of tillers, and number of grains on a plant. The morphological development is simulated separately, but it is closely coupled with phasic development and growth. Expansion growth of leaves and stems is simulated separately from mass growth because the expansion growth is considered "sink" driven, primarily by the temperature of expanding tissue. Mass growth is considered to be the "source" necessary to fill the expanding tissue and also provides assimilates to the root system for expansion and maintenance. Mass growth is driven by radiation interception by the leaves. The various aspects of development and growth are evaluated separately to accommodate the logic of partitioning assimilates into different plant parts according to well-established principles. Some principles used in the model for assimilate partitioning during different phases of development are as follows:

1. During vegetative growth, shoots have a higher priority than roots for assimilates as long as the supply of water and nutrients in the soil is

adequate. When water or nutrient supply is limited, then roots have a higher priority than shoots.
2. During grain development, grains are the dominant sink for assimilates. Assimilates for filling the grains are derived from current photosynthesis and stored assimilates. Deficiencies of water and nutrients have little effect on the ability of materials to be transported to the grain.

Many of the equations used to calculate growth, development, and yield in CERES barley are presented in the following two sections, as they appear in the FORTRAN source codes.

PHASIC DEVELOPMENT

Plant development is controlled by the thermal environment to which plants are exposed. The rate of development is directly proportional to the temperature between the base temperature (T_b) and a maximum temperature of 26°C. This linear relation between development and temperature allows the use of a thermal time concept in describing plant development. The concept of thermal time use in phasic development and sources of possible errors in calculating thermal time are described by Ritchie and NeSmith (1991). The phyllochron concept is used to describe plant development and is expressed as the thermal time (°Cd) between successive leaf tip appearance. The phyllochron is a measure of the rate of development of plant leaves that allows systematic tracking of plant development. Thermal time and phyllochron are vital in simulating phenological events in the plant's life cycle.

The growth stages in the model are numbered 1 to 9 (see Table 16.1). The active plant growing stages are stages 1 to 5, and the remaining stages describe events involved in crop management. The developmental stages are logically divided by the partitioning of biomass to specific plant organs that grow actively during each of the stages.

Fallow or Presowing: Stage 7

This stage is used for simulating the soil water or nutrient balance during the fallow periods before sowing. Simulating soil water balance during this period is useful when the initial soil water at the time of sowing is unknown. Long-term simulations or crop sequence simulations require simulation of soil water balance during this stage. If the soil water balance is running during this stage, the soil is assumed to be uniformly dry when the previous crop is harvested or uniformly wet when there is sufficient rainfall during the fallow period.

TABLE 16.1. Growth Stages of Barley As Used in Model

Stage	Description	Growing Plant Parts
7	Fallow or presowing	
8	Sowing to germination	
9	Germination to emergence	Roots and coleoptile
1	Emergence to terminal spikelet initiation	Roots and leaves
2	Terminal spikelet initiation to end of leaf growth and beginning of ear growth	Roots, leaves, and stems
3	End of leaf growth and beginning of ear growth to end of preanthesis ear growth	Roots, leaves, and ear
4	End of preanthesis ear growth to beginning of grain filling	Roots, stems, and grains
5	Grain filling	Grains and roots
6	End of grain filling to harvest	

Sowing to Germination: Stage 8

Seeds usually imbibe water rapidly. Simulating germination, or actually water imbibition, takes one day unless the soil water content is at the lower limits of plant-extractable soil water or soil temperature is below 0°C. It is unlikely that seeds will be sown in cold soil. However, it is possible that seeds may be sown in dry soil and will not germinate until rain or irrigation wets the soil.

Germination to Emergence: Stage 9

Both soil temperature and depth of sowing of seeds control timing of seedling emergence. The soil water content will not influence emergence, as it is assumed that if soil water is sufficient to promote germination of seeds, then it is also sufficient to promote seedling emergence. Temperature influences time of emergence. The depth of sowing influences emergence through its effect on the time necessary for the expansion of the coleoptile of the emerging seedling to reach the soil surface. The thermal time for emergence after water imbibition is expressed as follows (all equation variables are defined in the appendix):

$$P9 = 80.0 + 10.2 * SDEPTH$$

Emergence to Terminal Spikelet Initiation: Growth Stage 1

The thermal time for this growth stage can be highly variable depending on genotype and environment. Temperature, vernalization, and photoperiod influence the duration of this stage. Genetic variations of cultivars for vernalization and photoperiod cause variation in total thermal time required to complete this stage. Vernalization is assumed to occur at temperatures from 0 to 18°C (Trione and Metzger, 1970). The optimum temperature is assumed to be in the range of 0 to 7°C, while temperatures from 7 to 18°C have a decreasing influence on vernalization. Daily minimum and maximum temperatures are used to calculate the daily vernalization effectiveness factor, with a value ranging from 0 to 1 (see Figure 16.1). Summing the vernalization effectiveness factor (CUMVD) each day gives the total vernalization days.

Fifty vernalization days are assumed to be sufficient for complete vernalization of all the cultivars (see Figure 16.2). The genetic variability for vernalization is considered in the model by the genetic coefficient P1V to calculate its influence on the growth of plants in growth stage 1. The P1V (K value in Figure 16.2) is used to express the vernalization requirement for both spring and winter barley. The spring barley genotypes have a low sensitivity to vernalization.

Young seedlings are devernalized when exposed to high temperature. The relative development rates for three contrasting genotypes are shown in Figure 16.2, where Maris Mink represents a spring genotype and Igri a winter one. Data for developing these relationships are usually obtained from controlled environment chambers in which seedlings are exposed to effective vernalizing temperatures for varying durations and then transferred to more normal growing temperatures. If CUMVD is less than ten days and daily maximum temperature exceeds 30°C, then the number of vernalization days is decreased by 0.5 days/degree of temperature above 30°C. If CUMVD is greater than ten days, no devernalization is assumed to occur.

FIGURE 16.1. Relationship Between Plant Crown Temperature and the Relative Vernalization That Barley Plants Receive Each Day

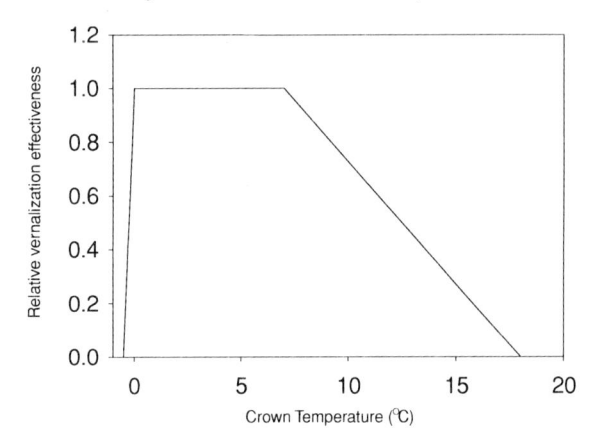

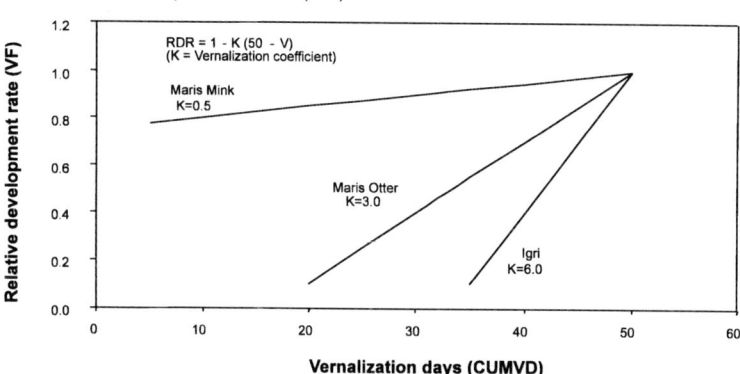

FIGURE 16.2. The Relationship to Predict the Influence of Vernalization Days on the Relative Development Rate (VF)

[Graph showing RDR = 1 - K (50 - V) (K = Vernalization coefficient); Maris Mink K=0.5; Maris Otter K=3.0; Igri K=6.0; x-axis: Vernalization days (CUMVD) 0-60; y-axis: Relative development rate (VF) 0.0-1.2]

Short photoperiod can also delay plant development because barley is a quantitative long-day plant. Day lengths shorter than 20 h can delay development, and this delay is dependent on photoperiod sensitivity of the genotype used. The genetic variation for photoperiod sensitivity is expressed as a genetic coefficient, P1D (C in Figure 16.3). In the model, photoperiod length includes civil twilight besides the normal day length and is calculated from latitude and time of year. The relative development rate concept is used and its value ranges from 0 to 1. The relative sensitivity of different barley genotypes to photoperiod as determined from controlled climate experiments is presented in Figure 16.3.

Both vernalization days and photoperiod are used to modify the accumulation of thermal time in growth stage 1. The relative development rates for vernalization (VF) and for photoperiod (DF) are calculated using the genetic coefficients P1V and P1D, respectively. The minimum value of VF and DF is then multiplied by the thermal time to reduce the rate of thermal time accumulation. The equation for calculating the duration follows:

$$TDU = \Sigma \ DTT \overset{300}{\underset{0}{*}} [AMIN1(VF,DF)]$$

When the thermal time accumulation reaches 300°Cd, the duration of growth stage 1 is completed.

Terminal Spikelet Initiation to End of Leaf Growth and Beginning of Ear Growth: Growth Stage 2

This stage starts at terminal spikelet initiation and is considered to be strictly under temperature control, requiring about three phyllochrons from

FIGURE 16.3. Relationship of Photoperiod (P) and Relative Development Rate (DF) for Various Barley Genotypes

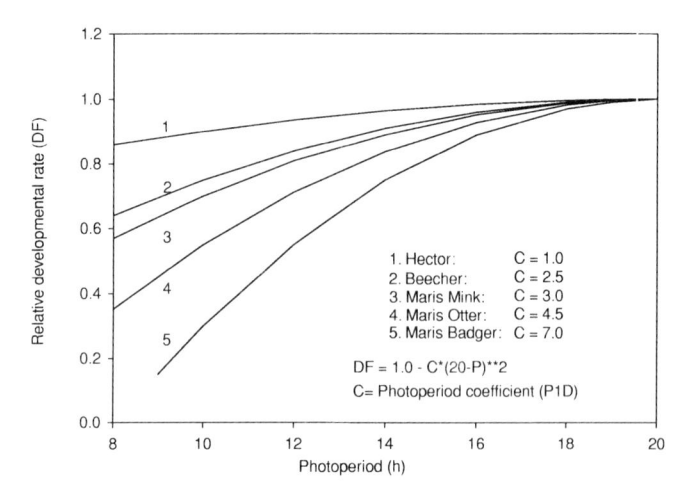

1. Hector: $C = 1.0$
2. Beecher: $C = 2.5$
3. Maris Mink: $C = 3.0$
4. Maris Otter: $C = 4.5$
5. Maris Badger: $C = 7.0$

$$DF = 1.0 - C*(20-P)**2$$
C = Photoperiod coefficient (P1D)

terminal spikelet initiation to flag leaf appearance. The phyllochron value used in the model is 75°Cd. A recent field study by Sharratt (1999) has confirmed this phyllochron value for barley.

End of Leaf Growth and Beginning of Ear Growth to End of Preanthesis Ear Growth: Growth Stage 3

During this stage, the ear develops rapidly and is a major sink for assimilates produced by the plant. The duration of this stage is equivalent to two phyllochrons, even though no new leaves are formed during this stage.

End of Preanthesis Ear Growth to Beginning of Grain Filling: Growth Stage 4

Flowering of the barley plant occurs during this growth stage. It is assumed to take 200°Cd for plants to go from maximum ear size and volume to the time when linear grain mass accumulation begins. No other aboveground alternate sinks develop during this stage under normal growing conditions. The plants accumulate assimilates as stored carbohydrates for translocation to the developing kernels during the kernel-filling stage.

Grain Filling: Growth Stage 5

The size of the kernel is determined during this stage. The thermal time for this stage varies slightly among genotypes and is input as a genetic coefficient, P5. Kernel filling stage usually begins two to ten days after flowering and is associated with a rapid and linear increase in the kernel weight.

End of Grain Filling to Harvest: Growth Stage 6

This stage is reserved for calculating time from physiological maturity to harvest. This stage provides an opportunity for specific model users who need to consider possible yield reduction associated with delay in harvesting the crop after it reaches maturity.

GROWTH AND ORGAN DEVELOPMENT

The objective of the growth routine is threefold: (1) to calculate the leaf area of plants as the site of biomass production through the conversion of carbon dioxide and light energy to biomass and as a water evaporating surface, (2) to partition the biomass produced between different plant parts, and (3) to calculate grain yield from the number of grains filled and the average grain weight. The development of the growth subroutine was a major challenge because partitioning of assimilates is a dynamic process controlled by phenological development, requiring several feedback mechanisms, especially those related to root-shoot partitioning.

Dry-Matter Production

Monteith (1977) demonstrated that cumulative seasonal light interception by crops of barley, potatoes, sugar beets, and apples grown with adequate soil water supply in a temperate climate was linearly and quantitatively related to biomass production. The slope of the overall relation for these well-fertilized crops without water stress was known as radiation-use efficiency (RUE). Although the calculated relationship for the different crops had different intercepts, considerable similarity existed. Evans (1993) developed a similar relationship between dry mass produced at harvest and radiation intercepted throughout the growing season for several tropical crops. Radiation values used by Monteith (1977) were for wavelengths in the photosynthetically active radiation (PAR) range and assumed to be 50 percent of the solar radiation.

In the model, an RUE value of 2.96 $g \cdot MJ^{-1}$ of intercepted PAR by plants is used to estimate the daily assimilate production. Total solar radiation is a daily input variable in the weather file and is converted into PAR by multiplying the input values by 0.5.

A major difficulty associated with verifying an intercepted radiation-biomass production relationship has been the lack of knowledge concerning the fraction of assimilates partitioned to the roots. Although some investigations have established an RUE relationship, including root weight to aboveground biomass, considerable uncertainty still remains in calculating the fraction of assimilates partitioned to the roots because of root death, root respiration, and root exudates. Several studies (Barber and Martin, 1976; Martin, 1977; Sauerbeck and Johnen, 1977; Shadan, 1980) have reported that rather large losses of assimilates partitioned to roots through root exudations, sloughing, and other mechanisms. If these processes do indeed involve significant amounts of assimilate loss, then most reported measurements of root weight are low estimates of the amount of assimilates transported to the roots. In attempting to develop a reasonably representative root system and the required amount of assimilates transported to the roots, a somewhat higher RUE value than usually reported is used in the model. The daily potential biomass produced by plants is simulated by the following equation:

$$Y1 = 1.5 - 0.768 \cdot (ROWSPC \cdot 0.01)**2 \cdot PLTPOP]**0.1$$
$$PCARB = 0.5 \cdot RUE \cdot SRAD/PLTPOP \cdot [1.0 - EXP(-Y1 \cdot LAI)]$$

Two factors can reduce potential biomass production: nonoptimal temperatures and water deficit. A weighted daytime temperature is calculated from the input minimum (TEMPMN) and maximum (TEMPMX) daytime temperatures for use in the photosynthesis temperature reduction factor (PRFT), where the optimum daytime temperature is considered to be 18°C. The daytime temperature (T) is approximated by this relationship:

$$T = 0.25 \cdot TEMPMN + 0.75 \cdot TEMPMX$$

The reduction in photosynthesis due to temperature is then:

$$PRFT = 1 - 0.0025 \cdot (T - 18)**2$$

If PRFT values are less than 0, then it is set equal to 0 (see Figure 16.4).

Water deficit reduces dry-matter production rates below their potential whenever extraction of soil water falls below the potential transpiration rate calculated for the crop in the water balance subroutine (Ritchie, 1998). Similarly, nitrogen deficit reduces dry-matter production rates. The minimum values of the two reduction factors, SWFAC (calculated in the water balance subroutine; details in Ritchie, 1998) and NSTRESS (calculated in the nitrogen subroutine; details in Godwin and Singh, 1998), are then used with PRFT to calculate the actual biomass (CARBO) production:

$$CARBO = PCARB \cdot [AMIN1 (SWFAC, NFAC) \cdot PRFT]$$

In the model, the respiration rates are assumed to be proportional to the carbon fixed by photosynthesis and, hence, are not calculated independently but rather are incorporated into the calculation of PCARB.

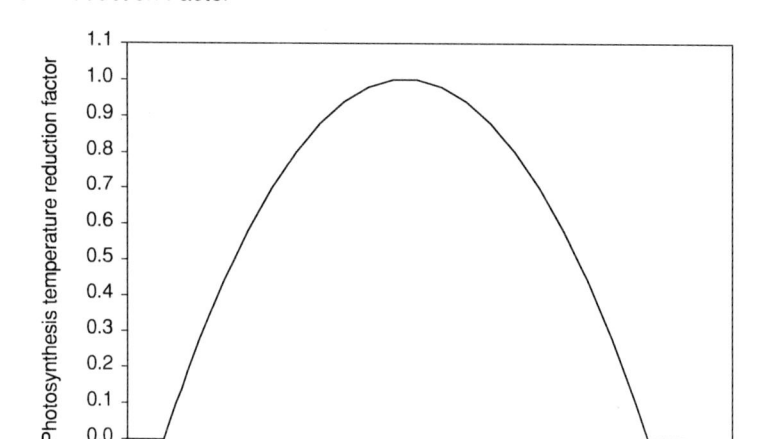

FIGURE 16.4. The Relationship Between Daytime Temperature and Photosynthesis Reduction Factor

[Graph: Photosynthesis temperature reduction factor (y-axis, 0.0 to 1.1) versus Daytime temperature (°C) (x-axis, -5 to 45). Curve peaks at ~1.0 near 17-18°C, reaches zero near -5°C and 40°C.]

The remainder of the growth components involve the partitioning of assimilates to individual plant organs during the five phasic developmental stages when the plants are growing. The various plant parts that are actively growing in each growth stage are given in Table 16.1. The CARBO is converted to individual plant growth by dividing its value by the input plant population (PLTPOP). Modeling on a single-plant basis assumes that plants are reasonably homogeneous and allows more flexibility in using the model with a wide range of plant populations.

Growth Stage 1

Development of main stem leaf area. Plant leaf area acts as both an assimilation and a transpiring surface. Modeling leaf area is one of the important aspects of a model. It influences light interception, plant transpiration, and dry-matter production. Leaf area expansion is the component of plant growth most sensitive to deficits of water and nutrients. Leaf expansion growth is assumed to be more sensitive to plant water deficits than to the photosynthesis process. Thus, drought reduces leaf expansion growth more than photosynthesis. This reduction in expansion growth without a concomitant decrease in photosynthesis can increase the specific leaf weight and the proportion of assimilates partitioned to the roots. The model accounts for these dynamic and differential plant responses by using separate water deficit functions for reducing leaf expansion growth and rate of photosynthesis. These soil water deficit functions are calculated from the estimated soil-root system limited uptake rate and the potential transpiration rate (Ritchie, 1998).

The daily increase in plant leaf area results from the growth of leaves on both the main stem and the tillers. The area of the leaves on the main stem is calculated as a priority over the tiller leaf area. The leaf area of a plant is the product of the rates of leaf appearance and expansion of the growing leaves. One leaf blade at a time is assumed to be expanding on a stem during vegetative growth. Kirby, Appleyard, and Fellowes (1982) demonstrated that while two leaves are in an active expansion stage, only one of these leaves is contributing to new visible blade growth. The growth of the other leaf occurs within the sheath of the uppermost visible leaf and, hence, does not contribute to the leaf area available to intercept solar radiation.

The leaf appearance rate is a linear function of temperature between the range of about 0° and 26°C. Thus, leaves appear at predictable intervals using temperature data. The leaf appearance interval, the phyllochron (PHINT), is also used as a basis for determining leaf area growth. Both the rates of leaf appearance and expansion are controlled by temperature in a reasonably similar fashion. Thus, leaf area growth rate can be calculated as a function of temperature without separately considering the individual leaf appearance and expansion rates.

The expansion rate of the first leaves on the plant is less than that of leaves that appear later due to the ability of the plant to acquire assimilates to support larger leaves after more green leaf area is produced. The first few leaves of the plant, therefore, are smaller. The plant leaf area growth rate (PLAGMS) for main stem leaves is determined from the following equation:

$$PLAGMS = 7.5 * CUMPH ** 0.5 * AMIN1 (SWFAC, EGFT, NFAC) * TI$$

From PLAGMS, total plant leaf area growth rate (PLAG) is determined:

$$PLAG = PLAGMS * (0.3 + 0.7 *TILN)$$

SWFAC and NFAC are the two major constraints that were used to influence the potential leaf area growth rate. SWFAC is calculated in the soil water balance routine while NFAC is calculated from the N subroutine. The leaf area to weight ratio (AWR) is first calculated to determine the mass of leaves that can grow in a given day. The AWR includes all of the aboveground biomass of the plant, which consists of only leaves, since the stem has a negligible contribution to plant biomass in growth stage 1. The AWR for early vegetative growth is described by this equation:

$$AWR = (150 - 0.075 *TDU) * 1.1$$

AWR can change more than the previous equation implies, but we found it difficult to allow AWR to vary with the environmental conditions along with other dynamic factors affecting partitioning. If AWR is allowed to vary with environmental conditions, then it is difficult to control tiller development in the model. In growth stage 1, the model allows AWR to vary with the aging of the plant. AWR is not used in later stages of plant growth.

The potential leaf growth (GROLF, grams per plant) is the mass of assimilate required to support daily expansion growth in the plant and is expressed by this equation:

$$GRORT = PLAG/AWR$$

The root growth is then assumed to consist of the remainder of the daily assimilate supply:

$$GRORT = CARBO - GROLF$$

If GRORT is more than 35 percent of CARBO in stage 1 growth, the plant is assumed to have an adequate assimilate supply for leaf growth to proceed at its potential rate (Gregory et al., 1978). If GRORT is less than 35 percent of CARBO, then a feedback mechanism in the model reduces GROLF to 65 percent of CARBO and constrains GRORT to 35 percent of CARBO. When this happens, a new value for PLAG is determined by this equation:

$$PLAG = GROLF * AWR$$

The total cumulative plant leaf area (PLA) is then updated by adding the daily leaf growth:

$$PLA = PLA + PLAG$$

The previous considerations are used to account for the competition for assimilates between the main stem and the tillers. Early in growth stage 1, seedlings have a reservoir of carbohydrates available for leaf and root growth in the seed endosperm. The model supplements the supply of assimilates with the carbohydrates from seed storage until the seed endosperm storage capacity is depleted.

Leaf senescence. Leaf senescence is coupled with plant leaf development. Barley plants have four green leaves on each stem. Assuming that no other stresses cause early leaf senescence, senescence begins in the oldest leaves after four leaves have already been formed. The plant maintains the four newest leaves as green leaves while the other leaves senesce. This type of senescence can be explained by both the shading of the older, lower leaves and the upper, newer leaves and by physical damage caused by the older leaves as a result of stem expansion and leaf sheath splitting. The plant leaf area loss rate (PLALR) is determined by this equation:

$$PLALR = [PLSC\ (LN - 4) - PLSC\ (LN - 5)] * TI$$

In the presence of soil water and nitrogen deficits, the plant leaf area senescence is increased by SWFAC and NFAC through their influence on PLALR. The senescence leaf area (SENLA) is determined from the following equation:

$$SENLA = SENLA + PLALR$$

From PLA and SENLA, the green leaf area index (LAI) is determined using the following equation (where the constant 10^{-4} converts leaf area units from cm^2 to m^2):

$$LAI = (PLA - SENLA) * PLTPOP * 10^{-4}$$

These equations are used to calculate senescence throughout the life cycle of the crop. Thus, when PLA reaches its maximum value at the end of leaf growth, senescence causes a gradual decline in LAI.

Tillering. The potential rate of tiller formation depends on the thermal time after seedling emergence, as described in the concepts presented earlier for leaf development. After three phyllochrons (CUMPH), potential tiller production is assumed to be proportional to the leaf number:

$$TC1 = -2.5 + CUMPH, \text{ when } CUMPH > 3$$

The actual number of tillers formed is assumed to be controlled by the available assimilates. The competition for assimilates between tillers is evaluated using the number of tillers per square meter (TPSM). The tiller production is based on competition limitations and is calculated by the following empirical equations:

$$TPSM = PLTPOP * TILN$$
$$TC2 = 2.56E{-}7 * (3000 - TPSM)^{**}3$$

TILN) is updated by the rate of daily tiller formation and is calculated using the least of SWFAC, NFAC, and the factor accounting for solar radiation < 10 MJ·m^{-2} per day(LIF1) and tiller production ratios (TC1 and TC2) using the following equation:

$$TILN = TILN + TI * AMIN1\ (SWFAC, NFAC, LIF1) *$$
$$AMIN1\ (TC1, TC2)$$

For early tiller growth, TC1 is usually the rate-limiting factor. As more and more tillers are formed, TC2 becomes the rate-limiting factor due to increased competition among the tillers. SWFAC and NFAC can also reduce tillering. If water deficit or assimilate competition causes TILN to be less than one, it is assumed that the plant population does not decrease. If TPSM is greater than 1,000 at the end of stage 1 growth, TILN is reduced to $1,000/m^2$. This is needed for later calculations of TILN and is based on the assumption that no more than 1,000 tillers/m^2 would be able to produce ears at the maturity stage. Such reduction in TILN is needed during growth stage 2, where a reduction in tiller number due to assimilate competition is considered.

Tiller death. Tiller death results from an insufficient supply of assimilates to support existing tiller growth and maintenance. Assimilate demand by a single tiller is calculated with the assumption that the demand is proportional to the rate at which the stem elongation can occur. This single-stem elongation rate can vary considerably among plant genotypes. Dwarf or semidwarf genotypes show less demand per stem than genotypes with large stems. The rate of potential biomass gain for a single tiller is determined from thermal time and the weight of a single tiller (TILSW), updated by the following equation:

$$TILSW = TILSW + G3 * 1.18 * 10E-5 * SUMDTT ** 2$$

Once the single-tiller demand for assimilates is determined, the model provides a time-delayed balance with the assimilate supply to determine the number of tillers that can elongate under the existing assimilate constraints. This is accomplished by deriving a daily ratio (RTSW) between total plant stem weight (STMWT) and potential plant stem weight (TILSW · TILN).

$$RTSW = STMWT/(TILSW * TILN)$$

The time-delayed reduced tiller number is updated with this equation:

$$TILN = TILN - [TILN * DTT * 0.005 * (1 - RTSW)]$$

Growth Stage 2: Stem Growth

The number of tillers developed in growth stage 1 usually exceeds the number of tillers that can eventually survive to produce ears. As the potential sinks exceed the assimilate supply, it is essential to properly partition available assimilates among the stems, leaves, and roots. The partitioning of assimilates to the aboveground parts of the plant is expressed by the following relationship:

$$PTF = 0.7 + 0.1 * AMIN1 (SWFAC, NFAC, LIF1)$$

In the absence of water deficit, PTF equals 0.8. When the most severe water deficit exists, PTF equals 0.7. After assimilates are partitioned to the aboveground portion of the plant, the remainder is allocated to the roots. Once PTF is determined, that fraction must be partitioned between the stems and the leaves. After terminal spikelet formation at the end of growth stage 1, the fraction of assimilates partitioned to the stem gradually increases from near zero to about 0.5 by the end of this stage. The rate of whole-plant stem growth (GROSTM) is expressed by the following equation:

$$GROSTM = (0.15 + 0.12 * DTT/PHINT) * CARBO * PTF$$

GROLF is then determined:

$$GROLF = CARBO * PTF - GROSTM$$

During this stage, the new leaf area is considered to be proportional to the new leaf weight. This proportionality constant, 127 $cm^{-2} * g^{-1}$, assumes only leaf blade area, while the leaf weight includes both leaf blades and sheaths. The total leaf area (PLA) is then updated:

$$PLA = PLA + GROLF * 127.0$$

Growth Stage 3: Preanthesis Ear Growth

During this stage of development, plant growth continues to the end of preanthesis ear growth and has a duration equivalent of two phyllochrons. The major growing parts during this period are the stem, ear, and roots. No distinction is made between the weight of the ear and that of the stem due to the difficulty of knowing exactly when the major portion of ear growth is initiated and the relatively short expansion period of ear growth. Ear and stem weights are combined into the single variable STMWT.

The PTF is assumed to be constant unless there is a soil water deficit. The partitioning of assimilates to the aboveground parts of the plant is similar to that used for stem growth in growth stage 2:

$$PTF = 0.75 + 0.1 * AMIN1 (SWFAC, NFAC, LIF1)$$

This equation reflects a 5 percent increase in the portion of assimilates partitioned to the top parts of the plant, as compared to growth stage 2. The range of PTF is from 0.85 under conditions of no water deficit to 0.75 under conditions of the most severe water deficit. As in growth stage 2, the remaining assimilate portion is allocated to the roots after partitioning assimilates to the top portions of the plant.

The rate of potential biomass gain for a single tiller during growth stage 3 is a linear function of thermal time multiplied by genotype coefficient G3. TILSW is updated by the following equation:

$$\text{TILSW} = \text{TILSW} + \text{G3} * (0.6 + 0.00266 * \text{SUMDTT})$$

Similar to growth stage 2, the single-tiller assimilate demand is balanced against the assimilate supply and a reduction in tiller number is possible:

$$\text{TILN} = \text{TILN} - [\text{TILN} * \text{DTT} * 0.005 * (1 - \text{RTSW})]$$

The minimum stem weight (SWMIN) is calculated at the end of growth stage 3. It represents the minimum stem weight when the stem reserve assimilates are relocated during grain filling. When the assimilate supply from photosynthesis is not sufficient to meet the demand of potential grain growth rate, the stem carbohydrate reserves are used to meet this demand. The stem weight may then decrease, but only to the estimated SWMIN.

PLALR due to senescence in growth stage 3 is as follows:

$$\text{PLALR} = 0.0003 * \text{DTT} * \text{GPLA}$$

The senescence leaf area is then updated:

$$\text{SENLA} = \text{SENLA} - \text{PLALR}$$

Growth Stage 4: End of Ear Expansion to the Beginning of Grain Filling

The duration of growth stage 4 is 200°Cd. During this stage, the plant has no aboveground organs that are actively expanding. Carbohydrates are temporarily stored in stems during this stage, resulting in increased stem weight. The reserve carbohydrates can be later translocated to the grain. This stage is important for establishing grain yield because stem plus ear weight is assumed to be proportional to the number of grains to be filled. Plants reach anthesis during this stage. Of the daily biomass produced, 90 percent is partitioned to the plant top unless there is water deficit. Under conditions of water deficit, PTF can be reduced to 80 percent.

For simplicity, all of the weight increase in the plant top is assumed to occur in the stem and ear during this stage. The PLALR and the SENLA values are calculated similar to those for growth stage 3, except that the rate is faster as a result of leaf aging:

$$\text{PLALR} = 0.0006 \, \text{DTT} * \text{GPLA}$$

Soil water deficit can increase the senescence rate. The number of kernels per plant (GPP) is assumed to be proportional to the stem plus ear weight at the end of growth stage 4. A genotype coefficient, G1, is used to account for genetic differences among barley genotypes for the number of kernels produced per unit weight of stem plus ear at anthesis:

$$GPP = STMWT * G1$$

The value obtained for GPP from this calculation is not altered during the remainder of the plant's life cycle unless severe water deficits cause the abortion of some of the grains during the grain-filling stage. The adjustment of GPP for abortion will be discussed later.

Growth Stage 5: Grain Filling

At the beginning of this stage, the base temperature for thermal time determination is changed from 0 to 1°C for the subsequent thermal time determination. Several difficulties arise in establishing the proper rate of growth for plants during the grain-filling stage when the leaves are aging. Root growth is thought to be small during this stage. Some studies (Bidinger, Musgrave, and Fischer, 1977; Austin et al., 1980) indicated that part of the assimilate supply used for grain filling comes from carbohydrates stored in other plant parts, such as stems, when there is severe water deficit. Total aboveground biomass production rates decline during the grain-filling stage compared to growth rates in earlier stages. Furthermore, maintenance respiration increases due to larger plant size. The assumption throughout all previous growth stages that respiration is proportional to gross photosynthesis as altered by the temperature reduction factor is less accurate for the grain-filling stage.

Plant growth rate may be controlled by assimilate demand during grain filling (Evans, Wardlaw, and Fischer, 1975). Another possibility is that more assimilates are partitioned to roots when the demand for grain filling is low. With these dynamic feedback mechanisms and the uncertainty about the amount of translocation to the grain, the scaling of daily photosynthesis is difficult during grain filling. To estimate the amount of assimilates derived from photosynthesis and the amount from stored assimilates, the following relationship is used (during this stage, the fraction of assimilates partitioned to the aboveground plant parts is assumed to be somewhat dependent on the stored assimilate supply, as indicated by the stem weight):

$$PTF = SWMIN/STMWT * 0.35 + 0.65$$

This relationship allows more mass to be partitioned to roots when there is ample stored assimilates in the plant top. When the stored assimilates are exhausted for grain filling, all of the assimilates from current photosynthesis are used for grain filling, i.e., when SWMIN = STMWT. The PTF value ranges from 0.7 at the beginning of growth stage 4 to 1.0 at the end of growth stage 5. As in other stages of development, the fraction of assimilates not used for aboveground growth is partitioned to the roots.

Aging of the leaves influences assimilate production and sink demand and is approximated using an equation that reduces the original calculated value of daily biomass growth (CARBO):

$$CARBO = CARBO * [1 - (1.2 - 0.8 * SWMIN/STMWT) * (SUMDTT + 100)/(P5 + 100)]$$

The leaf aging effect is determined by the (SUMDTT + 100)/(P5 + 100) ratio in the equation, and the sink demand is inferred indirectly through the SWMIN/STMWT ratio. Green leaf area reduction (PLALR) in growth stage 5 is approximated by the following relationship as the plant approaches maturity:

$$PLALR = GPLA * 2 * SUMDTT/P5**2$$

The kernel growth rate is calculated on a single-kernel basis. It is assumed that all kernels grow at the same rate. A relative kernel-filling rate is first calculated as a function of temperature. The grain-filling rate is assumed to be independent of temperature for air temperatures 17°C or higher. The relative rate of grain filling (RGFILL) between 0 and 17 is approximated by this equation:

$$RGFILL = 0.065 * TEMPM$$

This relative grain-filling rate has values between 0 and 1. The daily potential grain growth (GROGRN) is calculated from RGFILL, GPP, and a genotype coefficient, G2, using the following equation (where 0.001 is a factor to convert mass units from mg to g):

$$GROGRN = RGFILL * GPP * G2 * 0.001$$

Plant water deficits influence grain filling through a reduction in assimilate supply. While calculating the actual rate of daily grain growth (mg/kernel), if the daily aboveground growth rate (CARBO) plus stored assimilate supply is adequate to support growth of all the kernels, GROGRN is not reduced. Since the stem is assumed to be the storage organ for assimilates, the daily stem growth is the difference between supply and demand. GROSTM is calculated by the following equation:

$$GROSTM = CARBO * PTF - GROGRN$$

Depending on the state of reserves in the stem and the daily assimilate supply, the value of GROSTM may be either positive or negative. The updated stem weight is then calculated:

$$STMWT = STMWT + GROSTM$$

In the model, STMWT is controlled between a maximum value and a minimum value (SWMIN). If STMWT is greater than SWMIN, the rate of kernel growth is equal to the calculated potential rate. If STMWT is depleted to SWMIN, the grain-filling rate is equal to the daily biomass partitioned to the plant top (CARBO * PTF). Reduction in grain growth usually occurs near the end of the grain-filling stage because reserves are depleted and the daily assimilation rate is low. This occurs primarily because of reduced green leaf area due to leaf senescence. Grain weight (GRNWT) is updated daily using GROGRN.

$$GRNWT = GRNWT + GROGRN$$

The grain yield per unit area at the end of growth stage 5 is calculated by multiplying GRNWT by PLTPOP. If water deficit prevented individual kernel weights from attaining a minimum weight of 20 mg, individual kernel weight is assumed to be 20 mg and the original kernel number (GPP) is reduced to provide the calculated whole-plant grain weight. The reduction in the kernel number is intended to account for kernel abortion due to severe water deficits during most of the grain-filling period.

Development of the Root System

Root length density is needed to calculate root water absorption and to evaluate the soil water deficit factors that decrease various plant processes. Throughout each stage of plant development, some assimilates are partitioned to the roots (GRORT). To simulate a root system it is necessary to convert root mass (GRORT) to root length. To account for biomass losses due to root exudations and sloughing, we assume that 60 percent of the biomass partitioned to the roots is actually involved in structural root weight gains. Furthermore, an additional 0.5 percent of the total root mass is assumed to be lost daily through respiration and death. The root weight is not, however, directly used in the model and is included with the model output for those interested in comparing the model result with field measurements. The conversion of GRORT to root length is done using the factor of 1.05×10^{-4} cm root length per gram of root (Gregory et al., 1978). Based on this approximation, the daily new root length (RLNEW) is calculated using the following equation (where 1.05 is a constant that converts the daily biomass partitioned to the roots of an individual plant to cm of root length/cm^2 of soil):

$$RLNEW = GRORT * PLANTS * 1.05$$

Because the model is not three-dimensional, the root length calculations assume that the roots are equally distributed throughout the soil area. This assumption is not valid for young plants whose roots are concentrated near the seed. In some instances, this assumption can cause erroneous predictions.

The downward movement of roots is assumed to be proportional to the daily thermal time in the same way as leaf development is affected, except when the soil is dry at the depth where roots are growing downward or when the plant itself is experiencing a water deficit. When there is no water deficit, the downward rate of root growth is assumed to be 0.22 cm/°Cd. The downward rate of root growth is used to update root depth (RTDEP):

$$RTDEP = RTDEP + DTT \cdot 0.22 * AMIN1 (SWDF1 * 2.0, SWDF)$$

The factor SWDF has a value of 1 when there is at least 25 percent of the total extractable soil water (ESW) at the depth where the roots are growing. If the ESW is less than 25 percent, then:

$$SWDF = 4 * [SW(L) - LL(L)]/ESW(L)$$

The distribution of new root growth within the soil layers is evaluated by calculating a relative root length density factor (RLDF) at each soil depth increment [DLAYR(L)] where the roots are growing:

$$RLDF(L) = SWDF * WR(L) * DLAYR(L)$$

The WR(L) is soil input for each depth increment that depends on soil properties and ranges in value from 0 to 1. It represents the relative preference for root growth at different depths if the soil water content and nutrient level are not below threshold values. For most soils, values for WR(L) are usually near 1 for the top 30 cm and then decline to near zero at 2 m.

The distribution of roots at various soil depths is affected by several management factors. Abundant root formation occurs at the crown of the plant, usually causing a higher density of roots near the soil surface. The upper soil layers generally contain higher amounts of plant nutrients and organic matter and promote more root development. Roots usually penetrate the upper soil layers more easily because these layers have been disturbed by tillage operations. A root length density factor [RLDF(L)] is calculated for each soil depth increment down to the lower boundary of the root zone (RTDEP) to obtain the distribution of new root growth with depth:

$$RLDF(L) = WR(L) * DLAYR(L) * AMIN (SWDF, RNFAC)$$

The value of SWDF is a limitation on root growth due to water shortage in a layer and has been defined for root depth routines, and RNFAC is a limitation on root growth due to nitrogen shortage. As with SWDF, RNFAC has a value ranging from 1 for no limitation to near 0 for a strong limitation on root growth.

$$RNFAC = 1.0 - [1.17 * EXP(-0.15 * TOTN)]$$

The sum of the values of RLDF(L) for each layer with roots is calculated as a total root length density factor (TRLDF). The fraction of new root growth at each soil depth then becomes a normalized factor RLDF(L)/TRLDF. The root length density RLV(L) for each depth increment is then updated (where the last term of the equation calculates a 1 percent reduction in RLV(L) to account for root loss by death, sloughing, and other factors):

RLV(L) = RLV(L) + [RLDF(L)/TRLDF] * [RLNEW/DLAYR(L)]
− 0.01 * RLV(L).

POTENTIAL MODEL IMPROVEMENT

Since the development of the model, some new approaches to phenology and cold tolerance have been suggested for possible model performance.

Phenology

The phenology subroutine uses a linear and optimum response function to describe leaf tip appearance as a function of temperature. The slope of the linear response up to the optimum temperature is the rate of leaf tip appearance, and the inverse of the slope is the phyllochron. Considerable evidence suggests that both barley and wheat leaf appearance have a nonlinear response to temperature. Other evidence supports that photoperiod modifies the temperature response of leaf development. The rate of change in daylength at seedling emergence is reported to influence the phyllochron in wheat and barley (Baker, Gallagher, and Monteith, 1980; Kirby and Perry, 1987). Masle et al. (1989) reported that photothermal time correlated better with the leaf appearance rate than with thermal time alone. However, such relationships could not be reproduced in growth chamber studies (Cao and Moss, 1989) or in other field studies (Baker et al., 1986, Hotsonyame and Hunt, 1997). Some studies also indicated combined and interactive effects of temperature and photoperiod. A constant phyllochron approach is used in the model to predict leaf appearance. However, Cao and Moss (1989) reported that leaf appearance rate increased with daylength increases at a given temperature, indicating that the phyllochron decreased as daylength increased for any temperature. Sharratt (1999) has shown that the phyllochron values were 80°Cd for an early sown crop and 60°Cd for a late-sown crop in Alaska, where daylengths exceed 18 h. Slafer and Rawson (1997) suggested that photoperiod may have little or no effect on phyllochron beyond an optimum daylength of 18 h for wheat. Based on this, Sharratt (1999) indicated that shifts in air temperature, and possibly not the daylength, have caused changes in phyllochron. Jamie, Cutforth, and Ritchie (1998) proposed a nonlinear beta function to describe the combined and interactive effects of temperature and photoperiod on leaf tip appearance in winter wheat and barley. This nonlinear model has four genetic coefficients for each cultivar to predict the combined effects of temperature and daylength on leaf appearance. These types of nonlinear functions and variable rather than constant phyllochron values may have to be utilized to improve phenology modeling in order to account for the combined effects of temperature and photoperiod on leaf appearance rate. There seems to be consistent evidence that wheat and barley leaf appearance rates have an approximate optimum temperature above which further increases in temperature do not result in more rapid leaf appearance. In barley and wheat, this threshold temperature appears to be between 13 and 16°C (Ritchie and NeSmith, 1991; Jamie, Cutforth, and Ritchie, 1998). Variations in reported phyllochron values probably contain errors if not corrected for this threshold temperature, unless air temperatures during the study were below the threshold value.

Low-Temperature Tolerance

Although low-temperature tolerance was not discussed in the earlier description of the model, a procedure is used to kill all or part of the plant population due to low temperatures. The procedure for barley is the same as described for wheat in Ritchie (1991c). The procedure took no account of genetic variation in cold tolerance. Exposure to low temperature produces changes in morphological, biochemical, and physiological characters in winter cereals. These changes are highly correlated with the cold tolerance of plants. In order to investigate the production risks of winter cereals exposed to low temperature, Fowler, Limin, and Ritchie (1999) proposed quantitative procedures to describe the acclimation, dehardening, and damage due to cold temperature in winter wheat. This low-temperature model has been field validated for winter cereals grown in Canada. Furthermore, the potential model improvement includes a genetic coefficient to distinguish among genotypes varying in tolerance to low temperature.

EVALUATION OF THE MODEL

Evaluation by Model Developers

Before using the model as a decision-making tool, it is important to evaluate adequately the performance of the model for its ability to simulate field crop conditions. To evaluate the model under different growing conditions, a database was assembled from published experiments (Rothamsted, United Kingdom) as well as from unpublished experiments. The database represented a wide range of growing conditions, including short-season spring crops, environments with limited water availability, and regions with temperature extremes during the growing season (Ritchie, 1991a). It included a total of 115 experiment × treatment combinations and covered a time span of 19 years (1967-1986). A range of experimental locations situated in latitudes ranging from 35°N (Syria) to 56°N (Scotland) was included (see Table 16.2). The genetic coefficients for the varieties used are given in Table 16.3. The evaluation of the model requires comparing simulated values with the observed values from independent experiments. The root mean square error (RMSE), which measures the variation between simulated and observed values, was used to evaluate the model performance.

Evaluation of Phasic Development

Accurate simulation of phasic development is a primary requirement for any simulation model because the allocation of assimilates to various plant parts depends on the phenology of the plant. The predicted and observed ear emergence for various experiments is shown in Figure 16.5. The data for spring and fall season barley are reported separately. Fall-sown crops show a substantial agreement between predicted and observed ear emergence dates (RMSE of 9.5 days, coefficient of variation [CV] of 7 percent). The data fell into two groups representing two distinct environments: data showing emergence prior to day 130 were from low-latitude experimental sites in Syria, whereas the data showing emergence after day 150 were from experiments at high-latitude sites situated in Germany and Scotland. For the spring-sown crops, the model predicted ear emergence slightly earlier for

TABLE 16.2. The List of Locations and Barley Varieties Used in Experiments in the Model Evaluation

Location/Country	Latitude	Longitude	Time	Variety Used
Jindres, Syria	36.43°	36.73°	1981, 1983	A. Abiad, Beecher
Breda, Syria	35.93°	37.17°	1983	A. Abiad, Beecher
Tel Hadya, Syria	36.02°	36.93°	1980-1981	Beecher
Khanasser, Syria	35.08°	37.50°	1980, 1982	Beecher
Stuttgart, Germany	48.77°	9.18°	1983	Optima, IGRI
Rothamsted, UK	51.82°	0.36°	1967, 1972, 1976, 1979	Maris Badger, Julia
Cambridge, UK	52.18°	0.13°	1977	Maris Mink
The Murrays, UK	55.88°	2.83°	1979-1981	IGRI, Video, Golden Promise, Maris Otter, Maris Mink, Georgie
Mandan, ND, USA	46.82°	100.88°	1983-1986	Azure, Bedford, Bumper, Hector, Larker, Robust, Summit
Bozeman, MO, USA	45.78°	108.50°	1977	Maris Badger

TABLE 16.3. Relative Genetic Coefficient Values for the Barley Varieties Used in the Evaluation Experiments

No.	Variety Name	PIV	PID	P5	G1	G2	G3
1	A. Abiad	0.5	3.0	4.0	3.5	4.0	2.0
2	Beecher	0.5	2.5	4.0	2.2	3.5	4.0
3	Optima	6.0	3.0	4.0	3.0	2.5	3.0
4	IGRI	6.0	3.0	2.0	2.0	4.0	3.0
5	Maris Badger	0.5	7.0	5.0	4.0	5.0	3.0
6	Julia	0.5	7.0	5.0	3.5	3.0	2.0
7	Maris Mink	0.5	3.0	2.0	5.0	3.0	2.0
8	Video	6.0	3.0	2.0	4.0	5.0	2.5
9	Golden Promise	0.5	4.0	2.0	5.0	3.0	2.0
10	Maris Otter	3.0	4.5	2.0	5.0	3.0	2.0
11	Georgie	0.5	3.0	2.0	3.0	4.0	2.0
12	Azure	0.5	1.0	3.5	4.0	3.0	4.0
13	Bedford	0.5	1.0	3.5	5.0	3.0	2.0
14	Bumper	0.5	1.0	3.5	4.0	3.0	4.0
15	Hector	0.5	1.0	3.5	4.0	3.0	2.0
16	Larker	0.5	1.0	3.5	4.0	3.0	4.0
17	Robust	0.5	1.0	3.5	4.0	3.0	4.0
18	Summit	0.5	1.0	3.5	4.0	3.0	2.0

Note: Relative coefficient values are converted to absolute values based on the following equations:

P1V = Vernalization coefficient. Relative amount of development that is slowed for each day of unfulfilled vernalization, assuming that 50 days of vernalization is sufficient for all barley cultivars. $P1V = P1V \cdot 0.0054545 + 0.0003$.

P1D = Photoperiod coefficient. Relative amount of development that is slowed when plants are grown in a photoperiod 1 h shorter than optimum (which is considered to be 20 h). $P1D = P1D \cdot 0.002$.

P5 = Grain-filling duration coefficient. Degree-days above a base of 1°C after anthesis to maturity. $P5 = 300 + P5 \cdot 40.0$.

G1 = Kernel number coefficient (no./g). Kernel number per unit weight of stem (less leaf blades and sheaths) plus spike at anthesis. $G1 = 5.0 + G1 \cdot 5.0$.

G2 = Kernel weight coefficient. Kernel-filling rate under optimal conditions (mg · day^{-1}). $G2 = 0.65 + G2 \cdot 0.35$.

G3 = Spike number coefficient. Nonstressed dry weight of a single stem (excluding leaf blades and sheaths) and spike when elongation ceases (g). $G3 = G3 \cdot 0.35 - 0.005$.

most of the ranges shown (RMSE of 8 days, CV of 5 percent). Similar trends in maturity date predictions for fall- and spring-sown crops were also observed (data not provided). Taking into account the diversity of environments, genotypes, and crop management practices, it may be concluded that the model simulates phasic development reasonably well.

Evaluation of Biomass Accumulation

The observed and simulated biomass for both fall- and spring-sown crops are given in Figure 16.6. The scatter of points for the spring-sown crop (RMSE of 1542 kg·ha^{-1}, CV of 21 percent) is somewhat greater compared to that for the fall-sown crop (RMSE of 1670 kg·ha^{-1}, CV of 18 percent). The model overpredicts biomass values in low-yielding environments, but the predictions were reasonable in high-yielding environments. There could be two possible reasons for this. The model may not accurately account for the severe environmental stresses that the crops experience in low-yielding environments, for example, severe water deficit conditions. Another possible reason could be that some factors not accounted for in the model, such as pest damages and disease, are limiting yield. At high-yielding environments show reasonable agreement between simulated and observed data. This indicates that the model simulates the leaf area index, solar radiation interception, and subsequent conversion to biomass reasonably well when crop growing condi-

FIGURE 16.5. Predicted versus Observed Ear Emergence Day for (A) Fall-Sown and (B) Spring-Sown Barley

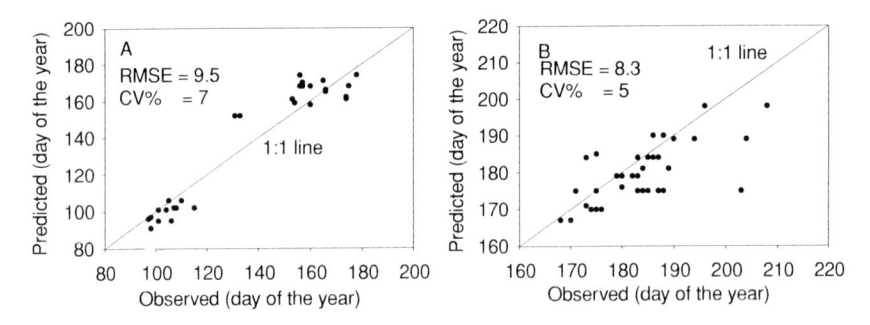

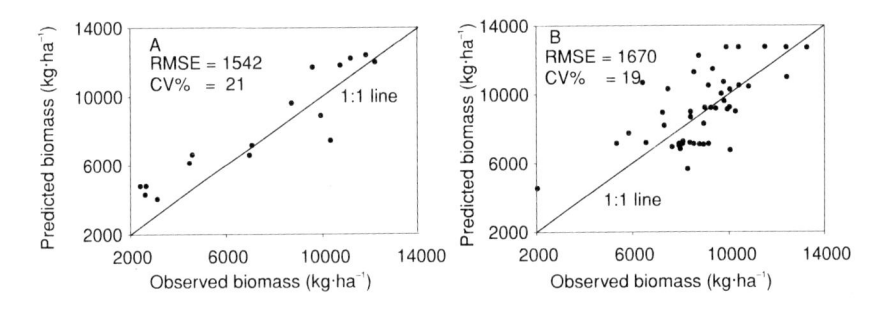

FIGURE 16.6. Predicted versus Observed Aboveground Biomass for (A) Fall-Sown and (B) Spring-Sown Barley

tions are favorable for expression of genetic potential of the geno-type grown in that environment.

Evaluation of Yield and Yield Components

The model was evaluated in a range of grain yield environments varying from 2,000 to 12,000 kg·ha^{-1}. The simulated and observed yields for fall- and spring-sown crops are given in Figure 16.7. A more systematic departure of points from the 1:1 line can be observed, unlike the random distribution of points observed for the biomass curve. For the fall-sown crop, the model tends to underpredict the grain yield in high-yielding environments (RMSE of 1780 kg·ha^{-1} and a high CV of 32 percent). The model predictions for the spring-sown crop were better compared to those for the fall-sown crop (RMSE of 1125 kg·ha^{-1}, CV of 25 percent). The relation between simulated and observed grain number for both fall- and spring-sown crops is presented in Figure 16.8. The model underestimates grain number for the fall-sown

FIGURE 16.7. Predicted versus Observed Grain Yield for (A) Fall-Sown and (B) Spring-Sown Barley

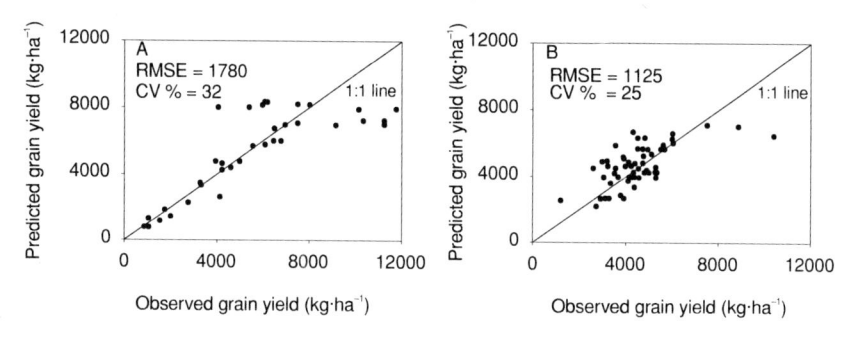

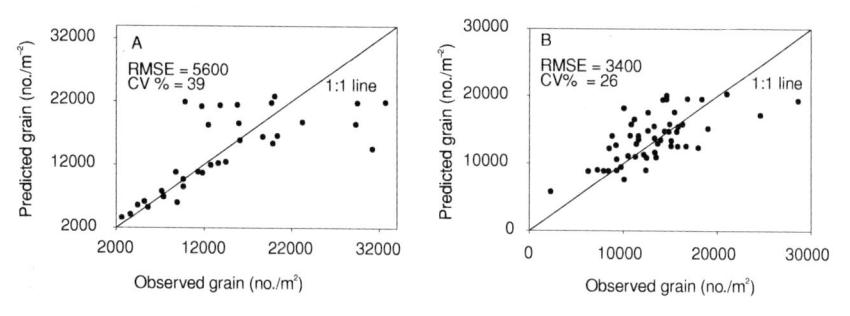

FIGURE 16.8. Predicted versus Observed Grain (no/m^2) for (A) Fall-Sown and (B) Spring-Sown Barley

crop (RMSE of 5600 and a high CV of 39 percent) in high-yielding environments (see Figure 16.8A). Since grain numbers are highly correlated with grain yield, it is necessary to predict accurately grain number. The grain yield simulation in the model is closely linked to tiller number and stem weight at the time of anthesis. Since the model predictions for biomass were reasonable, it is necessary to evaluate the model further with additional data sets for the relationships used to partition biomass into stems and ears, especially during ear development. This will perhaps help to improve the predictions of grain number, leading to better prediction of grain yield.

Evaluation of Modeling Responses to Crop Management

Crop simulation models are increasingly used as decision-making tools in agriculture. Hence, it is necessary to evaluate the ability of the models to simulate plant responses to management decisions involving, for example, irrigation and nitrogen fertilizer input. Data from two irrigation experiments conducted in Rothamsted were used to evaluate the model's capability to simulate crop responses to irrigation. The simulated and observed responses to various irrigation treatments are shown in Figure 16.9. The model is able to simulate reasonably the response of plants to irrigation.

Two experiments in the evaluation database involved the use of different nitrogen (N) rates as a management variable. These experiments used three or more N treatments and the same irrigation management for each treatment. Figure 16.10 shows the simulated and observed response curves. The model overpredicted yield for low-N treatments in both experiments. The initial soil N content in the different soil profile layers, which are model-input conditions, was not measured, and the estimated values were likely too high. It may also be possible that the effects of soil nitrogen deficit on growth are not fully accounted for in the model or that the rate of mineral-

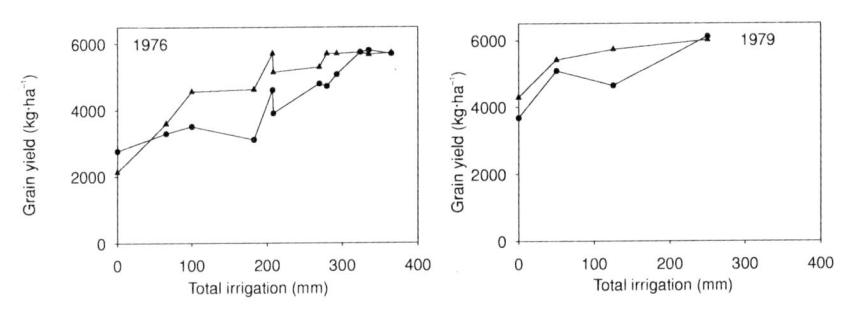

FIGURE 16.9. Predicted (▲) and Observed (•) Responses to Irrigation Application for Experiments Conducted in Rothamsted, England, with Barley Cultivar Julia

FIGURE 16.10. Predicted (▲) and Observed (●) Responses to Nitrogen Application Rates in (A) Cambridge, England, and (B) Bozeman, Montana

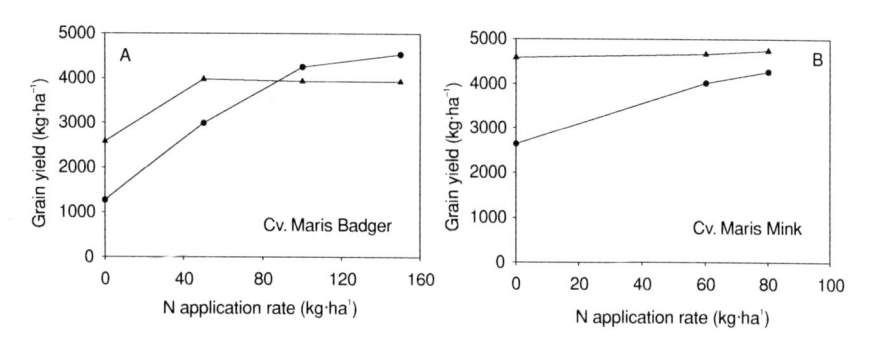

ization of organic N may be too high. There is a further need to evaluate model performance under low-N conditions.

The forgoing evaluation sections indicate that the model is capable of reasonably simulating barley phasic development and biomass production. The model underestimates grain number and, consequently, grain yield in high-yielding environments for fall-sown crops. The evaluation exercise also indicated that the model does not accurately simulate the effects of N stress in low-N situations, which is the general crop growing condition for barley. These problem areas need to be confirmed and the model needs to be improved with additional validation.

Evaluation by Independent Model Users

Evaluation in Argentina

Travasso and Magrin (1998) in Argentina independently validated the model. Data sets ($n = 83$) from experiments in a wide range of environmental conditions covering different sowing dates, years (1988 to 1993), cultivars (five), fertility, soils, and climates were used to evaluate the model. The genetic coefficients for the five cultivars were obtained from dates of planting (June to September) for experiments conducted in Buenos Aires. The differences in planting dates contributed to differences in crop season length (85 to 164 days), yield (8,000 to 16,000 kg·ha^{-1}), grain number (8,000 to 16,000 no./m^2), and mean grain weight (0.033 to 0.06 g). The validation data sets came from experiments conducted in the main barley growing region of Argentina (southern and western portions of Buenos Aires). The calibration of genetic coefficients among genotypes aided in differentiating the genotypes by their dependence on growth conditions in regard to yield either during early or late growth phases. This result indicated the sensitivities of genotypes to the types of water deficit that can occur either early or late in the growing season.

Results from Travasso and Magrin (1998), using all of the 83 data sets, indicated a mean absolute error of 288 kg·ha^{-1} (8 percent of mean observed yield) and an RMSE of 379 kg·ha^{-1} (11.7 percent of mean observed yield). Besides simulating grain yield correctly, the model is also able to simulate the grain number, with an RMSE of 1,075 grains/m^2 (16 percent of mean value), and grain size, with an RMSE of 0.0058 g (13 percent of mean value). This model validation work has proved that the model is able to predict yield estimates correctly and hence can be used as a research tool to assess the management practices required to obtain best yields for barley production in Argentina.

Evaluation in Morocco

Hanchane, El Mourid, and Karrou (1994) evaluated the model in the semiarid conditions of Morocco for phenology and yield under both irrigated and nonirrigated regimes. The model predicted the days to flowering and maturity with reasonable accuracy in both drier and irrigated conditions for cultivars Arig 8 and ACSAD 60. The model simulated yield and yield components reasonably well under irrigated conditions. However, the model overestimated total biomass, grain yield, and its components in the nonirrigated regime. The authors concluded that though the model is promising for use as a research tool under irrigated conditions, it needs to be further evaluated in diverse and drier crop growing conditions.

APPENDIX: GLOSSARY OF TERMS

AMIN1: FORTRAN function to select the minimum of the stress factors.

AWR: Leaf area to weight ratio.

CARBO: Daily biomass production (g/plant).

CUMPH: Cumulative phyllochron intervals.

CUMVD: Cumulative number of vernalization days.

DF: Day length factor.

DLAYR(L): Thickness of soil profile layer (cm).

DTT: Daily thermal time (°Cd).

DUL(L): Drained upper limit.

EGFT: Temperature factor accounting for extremes of mean temperature below 17°C on leaf expansion growth.

ESW: Total extractable soil water.

G1: Genotype coefficient for kernel number per unit weight of stem plus ear at anthesis.

G2: Genotype coefficient for kernel filling rate under optimum conditions (mg/seed).

G3: Genotype coefficient for potential dry weight of a single stem plus ear when elongation ceases under optimum conditions.

GPLA: Plant green leaf area (cm^2/day).

GPP: Grains/kernels per plant.

GRNWT: Grain weight (g/plant).

GROGRN: Potential grain growth rate (g/plant per day).

GROLF: Potential leaf growth (g/plant per day).

GRORT: Potential root growth (g/plant per day).

GROSTM: Potential stem growth (g/plant per day).

L: Depth at which roots are growing.

LAI: Leaf area index (m^2 leaf/m^2 land area).

LIF1: Factor accounting for effects of solar radiation < 10 $MJ \cdot m^{-2}$/day.

LL(L): Lower limit.

LN: Leaf number on the main stem accumulated since emergence.

NFAC: Nitrogen factor.

NSTRESS: Nitrogen stress factor.

PCARB: Potential biomass production ($g \cdot m^{-2}$).

P5: Duration of grain filling (°Cd).

PHINT: Phyllochron leaf appearance interval.

PLA: Cumulative leaf area (cm^2/plant).

PLAG: Total plant leaf area growth rate (cm^2/day).

PLAGMS: Leaf area growth rate for main stem (cm^2/day).

PLALR: Plant leaf area loss rate (cm^2/day).

PLSC: Cumulative leaf area when each main stem leaf reaches full size (cm^2/plant).

PLTPOP: Plant population (no./m^2).

P1V: Vernalization coefficient.

P1D: Photoperiod coefficient.

PRFT: Temperature reduction factor for photosynthesis.

PTF: Top fraction of the plant.

RGFILL: Relative rate of grain filling (0-1).

RLDF(L): Root length density factor at depth increment L (0-1).

RLNEW: Daily new root length (cm root/cm^2 land).

RLV(L): Root length density for soil layer L (cm root/cm^3 soil).

RMSE: Root mean square error.

RNFAC: Limitation on root growth due to nitrogen shortage.

ROWSPC: Row spacing (cm).

RTDEP: Root depth (cm).

RTSW: Daily ratio between total plant stem weight and potential plant stem weight (g/plant).

RUE: Radiation use efficiency (g·m^{-2}·MJ^{-1}).

SDEPTH: Depth of seeding in soil (cm).

SENLA: Area of leaf that senesces in a day (cm^2/plant).

SRAD: Solar radiation (MJ·m^{-2}·d^{-1}).

STMWT: Total stem weight (g/plant).

SUMDTT: The sum of growing degree-days for phenological stage (°Cd).

SWDF: Water deficit factor for the deepest soil layer where roots are growing.

SWDF1: Plant deficit factor determined in the soil water balance routine.

SWFAC: Soil water deficit factor.

SW(L): Soil water content at depth increment L.

SWMIN: Minimum stem weight (g/plant).

T: Daytime temperatures.

TC1: Tiller coefficient 1.

TC2: Tiller coefficient 2.

TDU: Thermal development unit.

TEMPM: Mean temperature (°C).

TEMPMX: Maximum temperature (°C).

TEMPMN: Minimum temperature (°C).

TI: Daily phyllochron fraction (DTT/PHINT).

TILN: Tiller number (no./plant).

TILSW: Weight of a single tiller (g).

TOTN: Total nitrogen in a particular soil layer.

TPSM: Number of tillers per square meter.

TRLDF: Total root length density factor (sum of RLDF for all layers).

VF: Vernalization factor.

WR(L): Weighting factor for soil depth (L) to determine new root growth distribution (unitless).

Y1: Light extinction effect varying slightly with spacing between rows and plant population (unitless).

REFERENCES

Akita, S. (1989). Improving yield potential in tropical rice. In *Progress in irrigated rice research*. Los Banos, the Philippines: IRRI, pp. 41-73.

Austin, R.B., C.L. Morgan, M.A. Ford, and R.D. Blackwell (1980). Contributions to grain yield from pre-anthesis assimilation in tall and dwarf barley phenotypes in two contrasting seasons. *Annals of Botany* 45: 309-319.

Baker, C.K., J.N. Gallagher, and J.L. Monteith (1980). Day length change and leaf appearance in winter wheat. *Plant Cell and Environment* 3: 285-287.

Baker, J.T., P.J. Pinter Jr., R.J. Reginato, and E.T. Kanemasu (1986). Effects of temperature on leaf appearance in winter wheat cultivars. *Agronomy Journal* 78: 605-613.

Barber, D.A. and J.K. Martin (1976). The release of organic substances by cereal roots into soil. *New Phytologist* 76: 69-80.

Bidinger, F.R., R.B. Musgrave, and R.A. Fischer (1977). Contribution of stored pre-anthesis assimilate to grain yield in wheat and barley. *Nature* 270: 431-433.

Cao, W. and D.N. Moss (1989). Temperature and day length interaction on phyllochron in wheat and barley. *Crop Science* 29: 1021-1025.

Evans, L.T. (1993). *Crop evolution, adaptation and yield*. Cambridge: Cambridge University Press.

Evans, L.T., I.F. Wardlaw, and R.A. Fischer (1975). Wheat. In *Crop physiology: Some case histories,* ed. L.T. Evans. Cambridge: Cambridge University Press, pp. 101-149.

Fowler, D.B., A.E. Limin, and J.T. Ritchie (1999). Low-temperature tolerance in cereals: Model and genetic interpretation. *Crop Science* 39: 626-633.

Godwin, D.C. and U. Singh (1998). Nitrogen balance and crop response to nitrogen in upland and low land cropping systems. In *Understanding options for agricultural production*, eds. G.Y. Tsuji, G. Hoogenboom, and P.K. Thornton. Dordrecht: Kluwer Academic Publishers, pp. 55-77.

Gregory, P.J., M. McGowan, P.V. Biscoe, and B. Hunter (1978). Water relations of winter wheat. 1. Growth of root system. *Journal of Agricultural Science (Cambridge)* 91: 91-102.

Hanchane, M., M. El Mourid, and M. Karrou (1994). Evaluation du Modele CERES-Orge pour les zones semi-arides du Maroc. *Proceedings of the international workshop on agro-ecological characterization, ICARDA, Aleppo, Syria,* ed. S. Verma, April 19-22, 1994.

Hotsonyame, G.K. and L.A. Hunt (1997). Sowing date and photoperiod effects on leaf appearance rate in field-grown wheat. *Canadian Journal of Plant Science* 77: 23-31.

Jamie, Y.W., H.W. Cutforth, and J.T. Ritchie (1998). Interaction of temperature and day length on leaf appearance rate in wheat and barley. *Agricultural and Forest Meteorology* 92: 241-249.

Jones, C.A. and J.R. Kiniry (1986). *CERES-Maize: A simulation model of maize growth and development*. College Station, Texas: Texas A&M University Press.

Kirby, E.J.M., M. Appleyard, and G. Fellowes (1982). Effect of sowing date on the temperature response of leaf emergence and leaf size in barley. *Plant Cell and Environment* 5: 477-484.

Kirby, E.J.M. and M.W. Perry (1987). Leaf emergence rates of wheat in a Mediterranean environment. *Australian Journal of Agricultural Research* 38: 455-464.

Martin, J.K. (1977). Factors influencing the loss of carbon from the wheat roots. *Soil Biology and Biochemistry* 9: 1-7.

Masle, J., G. Doussinault, G.D. Farquhar, and B. Sun (1989). Foliar stage in wheat correlates better to photothermal time than to thermal time. *Plant Cell and Environment* 12: 235-247.

Mass, S.J. and J.F. Arkin (1980). TAMW: A wheat growth and development simulation model. *Research center program and model documentation no. 80-3.* Temple, TX: Blackland Center, Texas Agricultural Experiment Station.

Monteith, J.L. (1977). Climate and the efficiency of crop production in Britain. *Philosophical Transactions of the Royal Society of London* B281: 277-294.

Otter-Nacke, S., D.C. Godwin, and J.T. Ritchie (1987). *Testing and validating the CERES-Wheat model in diverse environments*. AgRISTRAS Publ. No. Y.M-15-00407. Springfield, VA: NITS.

Otter-Nacke, S., J.T. Ritchie, D.C. Godwin, and U. Singh (1991). *CERES Barley V. 2.10: User's guide for barley growth simulation model*. Aleppo, Syria: ICARDA; East Lansing, MI: Michigan State University; Muscle Shoals, AL: IFIC; Honolulu: University of Hawaii.

Ritchie, J.T. (1991a). *Development of a barley yield simulation model*. Final progress report prepared for agreement number 86-CRSR-2-2867 with ICARDA. Aleppo, Syria: ICARDA.

Ritchie, J.T. (1991b). Specifications of the ideal model for predicting crop yields. In *Climatic risk in crop production: Models and management for the semi-arid tropics and subtropics*, eds. R.C. Muchow and J.A. Bellamy. Proceedings of international symposium, St. Lucia, Brisbane, Queensland, Australia, July 2-6, 1990. Wallingford, UK: C.A.B. International, pp. 97-122.

Ritchie, J.T. (1991c). Wheat phasic development. In *Modelling plant and soil systems*, eds. J. Hanks and J.T. Ritchie. Madison, WI: ASA, pp. 31-54.

Ritchie, J.T. (1998). Soil water balance and plant water stress. In *Understanding options for agricultural production*, eds. G.Y. Tsuji, G. Hoogenboom, and P.K. Thornton. Dordrecht: Kluwer Academic Publishers, pp. 41-54.

Ritchie, J.T. and D.S. NeSmith (1991). Temperature and crop development. In *Modelling plant and soil systems*, eds. J. Hanks and J.T. Ritchie. Madison, WI: ASA, pp. 5-29.

Ritchie, J.T. and S. Otter-Nacke (1985). Description and performance of CERES-Wheat: A user-oriented wheat yield model. In *ARS wheat yield project*. ARS-38. Springfield, VA: NTIS, pp. 159-175.

Sauerbeck, D.R. and B.G. Johnen (1977). Root formation and decomposition during plant growth. In *Soil organic matter studies*, Volume 1. Vienna: IAEA, pp. 141-148.

Shadan, M.M. (1980). Fixation, translocation and root exudation of ^{14}C-labeled assimilate by two genotypes of *Phaseolus vulgaris* L. subjected to root anaerobiosis. MS thesis. East Lansing: Michican State University.

Sharratt, B. (1999). Thermal requirement for barley maturation and leaf development in interior Alaska. *Field Crops Research* 63: 179-184.

Silvey, V. (1986). The contribution of new varieties to cereal yields in England and Wales between 1947 and 1983. *Journal of National Institute of Agricultural Botany* 17: 155-168.

Slafer, G.A. and H.M. Rawson (1997). Phyllochron in wheat affected by photoperiod under two temperature regimes. *Australian Journal of Plant Physiology* 24: 151-158.

Tanaka, A. (1983). Physiological aspects of productivity in field crops. In *Potential productivity in field crops under different environments*, Los Banos, the Phillipines: IRRI, pp. 61-80.

Travasso, M.T. and G.O. Magrin (1998). Utility of CERES-Barley under Argentine conditions. *Field Crops Research* 57: 329-333.

Trione, E.J. and R.J. Metzger (1970). Wheat and barley vernalization in a precise temperature gradient. *Crop Science* 10: 390-392.

Williams, J.R., C.A. Jones, and P.T. Dyke (1984). A modeling approach to determining the relationship between erosion and soil productivity. *Transactions of the American Society of Agricultural Engineers* 27: 129-144.

Chapter 17

Bases of Preharvest Sprouting Resistance in Barley: Physiology, Molecular Biology, and Environmental Control of Dormancy in the Barley Grain

Roberto L. Benech-Arnold

INTRODUCTION

Dormancy is the failure to germinate because of some internal block that prevents the completion of the germination processes (Black, Butler, and Hughes, 1987). For completeness it should be added that dormant seeds cannot germinate in the same conditions (e.g., water, air, temperature) under which nondormant seeds do so. Although the adaptive significance of dormancy is quite evident for plants living in the "wild," dormancy has been seen always as a complication in seeds from plants that are grown as crops. Indeed, nobody would tolerate uncertainty about the emergence of the crop because of the possible existence of dormancy in the lot of seeds to be used for sowing. For that reason, crops that originally must have had dormancy have undergone heavy selection against dormancy throughout their domestication process. In some cases, this selection against dormancy has gone too far and the seeds are fully germinable even prior to crop harvest. This situation, combined with rainy or damp conditions prevailing during the last stages of maturation, leads to germination in the mother plant, a phenomenon that is better known as preharvest sprouting. This highly undesirable phenomenon has a wide range of consequences, all of them adverse. In terms of seed viability, these consequences range from the immediate loss of seed viability upon subsequent desiccation when the embryo has grown beyond the point at which it loses its desiccation tolerance, to a dramatic reduction in seed longevity when embryo growth has not progressed that far (Del Fueyo, Marcaida, and Benech-Arnold, 1999). In addition, the initiation

---

I want to thank Maria Cristina Giallorenzi and Verónica Rodriguez for their help with the figures.

of the germination process triggers the synthesis of enzymes that promote reserve hydrolysis; this results not only in a reduction of grain yield due to carbohydrate respiration but also in the generation of a favorable environment for saprophytic attack (Castor and Frederiksen, 1977).

Although preharvest sprouting could be prevented by using genotypes that lose their dormancy a few weeks after crop harvest, the problem still persists mainly because, due to the paucity of our knowledge on the genetic, physiological, and environmental control of dormancy, it is very difficult to adjust the timing of dormancy loss to a precise and narrow "time window" (i.e., neither as early as to expose the crop to the risk of preharvest sprouting, nor as late as to have a dormant seed lot at the time of the next sowing). In malting barley, the task is even harder and breeders are compelled to work at the "edge of a knife." Indeed, the malting process itself requires grain germination; hence, a low dormancy level at harvest is a desirable characteristic because the grain can be malted immediately after crop harvest, thus avoiding costs and deterioration resulting from grain storage until dormancy is terminated. The possibility of solving the compromise between obtaining cultivars with low dormancy at harvest, but not with such an anticipated termination of dormancy that leads to sprouting risks, requires a thorough knowledge of the mechanisms determining dormancy release in the maturing grain. Moreover, it is essential to understand how those mechanisms are genetically and environmentally controlled.

This chapter is devoted to discussing the existing information for barley and other cereals on the physiology and genetics of dormancy inception, maintenance, and loss, with the final aim of understanding the nature of preharvest sprouting susceptibility in barley. Also, the chapter is intended to analyze the perspectives for obtaining cultivars that, without having a long-lasting dormancy, could present resistance to preharvest sprouting.

PHYSIOLOGY OF DORMANCY IN BARLEY

Where Is Dormancy Located in Cereal Grains?

Dormancy inception occurs very early in cereals. Embryos are usually fully germinable from the early stages of development (i.e., 15-20 DAP [days after pollination]) if isolated from the entire grain and incubated in water (Walker-Simmons, 1987; Benech-Arnold, Fenner, and Edwards, 1991; Benech-Arnold, Giallorenzi, et al., 1999); the entire grain, however, reaches full capacity of germination well after it has been acquired by the embryo. This coat (endosperm plus pericarp)-imposed dormancy is the barrier preventing untimely germination, and its duration depends on the genotype and on the environment experienced during maturation and beyond. In summary, though cases of embryo dormancy have been reported for barley and other cereal crops (Norstog and Klein, 1972; Black, Butler, and Hughes, 1987), sprouting-susceptible cultivars are those whose coat-imposed dormancy is terminated well before harvest maturity. In barley, the glumellae (the hull) adhering to the caryopsis represents a further constraint for embryo germination in addition to that already imposed by endosperm plus pericarp (Corbineau and Come, 1980). In a recent study, Benech-Arnold, Giallorenzi, et al. (1999) followed the dynamics of the release from dormancy imposed by the different structures surrounding the embryo, in grains from a sprouting-susceptible cultivar (B 1215) and a sprouting-resistant one (Quilmes Palomar). As expected, embryos from both varieties germinated precociously from early stages of development if excised from the entire grain (see Figure 17.1a). In both cultivars, dormancy imposed by endosperm plus pericarp was steadily overcome at a similar rate throughout development (see Figure 17.1b). However, although caryopses presented low dormancy from well before physiological maturity (PM, defined as the moment when the grain has attained maximum dry weight), the presence of the hull prevented grain germination prior to that stage. Hull-imposed dormancy started to be removed from PM onward, with a rate that was different depending on the cultivar: in B 1215 grains, this restriction was removed abruptly, whereas in Q. Palomar ones, the removal occurred at a lower rate (see Figure 17.1c). This difference was instrumental for the determination of a "time window" going from PM to harvest maturity (HM, defined as the moment when the grains have reached 12 percent of humidity on a fresh basis), within which one cultivar (B 1215) behaves as susceptible to sprouting and the other one (Q. Palomar) as resistant. In other words, the different behavior associated with sprouting presented by these two cultivars was related to the extent to which the presence of the hull prevents germination, and not to different dormancy levels of the caryopses. This is substantially different from what has been reported for wheat and sorghum, for which differences in dormancy level leading to contrasting sprouting behavior are inherent to the rate at which dormancy imposed by endosperm plus pericarp is lost (Walker-Simmons, 1987; Steinbach et al., 1995).

Dormancy imposed by the hull is a common feature in grasses (Simpson, 1990; Bewley and Black, 1994). It has been suggested that high polyphenoloxidase activity existing in the barley hull results in oxygen deprivation for the embryo (Lenoir, Corbineau, and Come, 1986). The way in which oxygen influences germination of dormant seeds is largely unknown, but it has been hypothesized that oxygen concentration might determine the rate at which germination inhibitors are catabolized (Neil and Horgan, 1987). This proposition is strongly supported by recent results, presented by Wang et al. (1998), showing that the dormancy breaking effect of a strong oxidant, such as hydrogen peroxide, is through a reduction in the endogenous level of the germination inhibitor abscisic acid (ABA). The question arising is, How can this mechanism operate differentially throughout development and among genotypes presenting contrasting sprouting behavior?

FIGURE 17.1. Germination Indices (GI) of (a) Embryos, (b) Dehulled Caryopses, and (c) Grains from a Sprouting-Susceptible (B/1215) and a Sprouting-Resistant (Q. Palomar) Cultivar, Harvested at Different Days After Pollination (DAP) and Incubated at 20º C for 12 Days

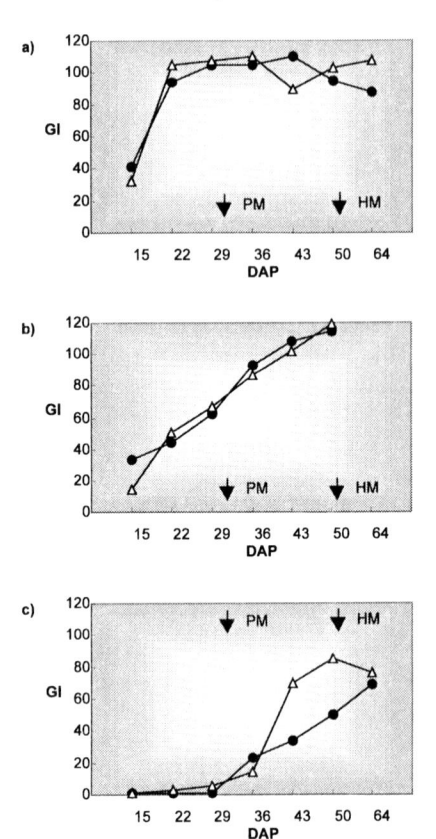

Source: From Benech-Arnold, Giallorenzi, et al., 1999. Reproduced with permission of CABI Publishing. This figure was originally published in Benech-Arnold R.L., Giallorenzi M.C., Frank J., and Rodriguez V. (1999). Termination of hull-imposed dormancy in barley is correlated with changes in embryonic ABA content and sensitivity. *Seed Science Research* 9, 39-47.

Note: The arrows indicate the moments when physiological maturity (PM) and harvest maturity (HM) took place. The GI considers both the final germination percentage and the velocity of germination; a maximum value of 120 was attained when all the embryos, dehulled caryopses, or grains had germinated within the first day after incubation.

Hormonal Regulation of Dormancy in the Barley Grain

Research on the mechanisms of dormancy in the developing seeds of many species suggests a strong involvement of the phytohormone ABA (King, 1982; Fong, Smith, and Koehler, 1983; Karssen et al., 1983; Walker-Simmons, 1987; Black, 1991; Benech-Arnold, Fenner, and Edwards, 1991, 1995; Benech-Arnold et al., 1995; Steinbach et al., 1995; Steinbach, Benech-Arnold, and Sánchez, 1997). Indeed, ABA-deficient or -insensitive mutants of *Arabidopsis* and maize precociously germinate (Robichaud, Wong, and Sussex, 1980; Karssen et al., 1983), and application of the ABA-synthesis inhibitor fluridone has been shown to reduce dormancy in developing seeds of some species (Fong, Smith, and Koehler, 1983; Xu, Coulter, and Bewley, 1991; Steinbach, Benech-Arnold, and Sánchez, 1997). In cereals, the imposition of dormancy on the embryo by the structures that surround it has been suggested to be mediated by the high levels of endogenous ABA existing in the embryo during grain development (Walker-Simmons, 1987; Steinbach et al., 1995). ABA content in the embryo is usually low until 15 DAP (Walker-Simmons, 1987; Steinbach et al., 1995; Benech-Arnold, Giallorenzi, et al., 1999). From that moment onward, ABA content increases, coinciding with the acquisition by the embryo of the capacity to germinate if isolated from the rest of the grain; hence, one possibility is that precocious germination would be prevented by the surrounding structures by impeding the leaching of ABA outside the embryo (Bewley and Black, 1994). ABA content has been reported to peak at around PM and to decline afterward (Goldbach and Michael, 1976; Walker-Simmons, 1987; Quarrie, Tuberosa, and Lister, 1988; Morris, Jewer, and Bowles, 1991; Steinbach et al., 1995; Benech-Arnold, Giallorenzi, et al., 1999). In sprouting-susceptible and sprouting-resistant barley cultivars, ABA embryonic content is usually similar until PM, and maximum ABA content occurs prior to PM (see Figure 17.2). However, immediately after PM, a dramatic reduction in embryonic ABA content can be observed in embryos from the sprouting-susceptible B 1215, coinciding with the aforementioned abrupt termination of hull-imposed dormancy that takes place in these grains after PM (see Figure 17.1c); in Q. Palomar embryos, in contrast, ABA content is kept at high levels for longer (i.e., until 43 DAP).

Wang, Heimovaara-Dijkstra, and Van Duijn (1995) also found differences in ABA content between dormant and nondormant barley genotypes, but as a result of differential de novo synthesis after some hours of incubation of the mature grains.

The role of changes in embryo responsiveness to ABA has been suggested as a key one for controlling dormancy in cereals and other species (Robichaud, Wong, and Sussex, 1980; Walker-Simmons, 1987; Corbineau, Poljakoff-Mayber, and Come, 1991; Steinbach et al., 1995; Benech-Arnold et al., 2000), and differences in sensitivity to ABA of mature embryos has

FIGURE 17.2. ABA Content in Embryos from a Sprouting-Susceptible (B 1215) and Sprouting-Resistant (Q. Palomar) Cultivar, Harvested at Different Days After Pollination (DAP)

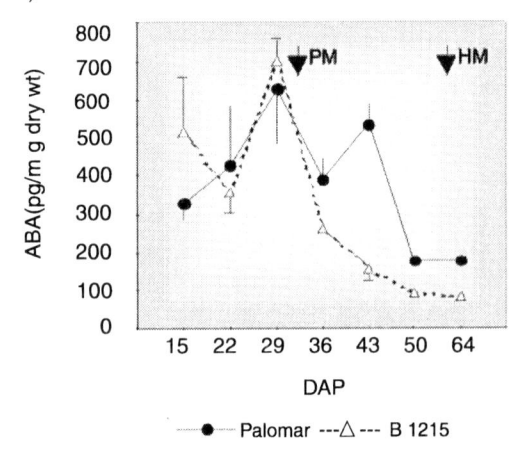

Source: From Benech-Arnold, Giallorenzi, et al., 1999. Reproduced with permission of CABI Publishing. This figure was originally published in Benech-Arnold R.L., Giallorenzi M.C., Frank J., and Rodriguez V. (1999). Termination of hull-imposed dormancy in barley is correlated with changes in embryonic ABA content and sensitivity. *Seed Science Research* 9, 39-47.

been reported for barley genotypes presenting contrasting dormancy (Van Beckum, Libbenga, and Wang, 1993; Wang et al., 1994; Wang, Heimovaara-Dijkstra, and Van Duijn, 1995; Visser et al., 1996). In the B 1215–Q. Palomar system, termination of dormancy is correlated also with changes in embryo sensitivity to ABA (see Figure 17.3). In an interesting paper, Visser et al. (1996) showed that the low embryo sensitivity to ABA exhibited by a barley cultivar with no dormancy was related, not to alterations in the ABA transduction pathway, but to a high rate of degradation of the hormone in the outside walls of the embryo. As a general conclusion based on the existing information, Wang, Heimovaara-Dijkstra, and Van Duijn (1995) proposed that release from dormancy of barley grains should be through (1) a reduction in the ABA sensitivity of the embryo, (2) a reduction in ABA embryonic content, (3) a higher ability of ABA to diffuse out of the embryo, and (4) reduced de novo ABA synthesis during incubation.

Resuming the idea that the hull impedes embryo germination because it interferes with ABA oxidation (or metabolization) through oxygen deprivation, it could be argued now that release from hull-imposed dormancy occurs because oxygen in high concentrations is not necessary when the ger-

FIGURE 17.3. Germination Indices (GI) of Embryos from a Sprouting-Susceptible (B 1215) and a Sprouting-Resistant (Q. Palomar) Cultivar, Harvested at Different Days After Pollination (DAP), After 12 Days of Incubation at 20°C in the Presence of 50 µM ABA.

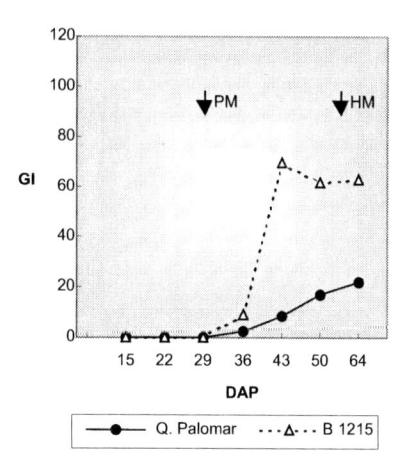

Source: From Benech-Arnold, Giallorenzi, et al., 1999. Reproduced with permission of CABI Publishing. This figure was originally published in Benech-Arnold R.L., Giallorenzi M.C., Frank J. and Rodriguez V. (1999). Termination of hull-imposed dormancy in barley is correlated with changes in embryonic ABA content and sensitivity. *Seed Science Research* 9, 39-47.

Note: The arrows indicate the moments when physiological maturity (PM) and harvest maturity (HM) took place. The GI considers both the final germination percentage and the velocity of germination; a maximum value of 120 was attained when all the embryos had germinated within the first day after incubation.

mination inhibitor (i.e., ABA) is no longer present or when sensitivity to the inhibitor is very low. It has been shown also that gibberellic acid (GA) strongly improves germination in hypoxia of oat seeds presenting hull-imposed dormancy (Lecat, Corbineau, and Come, 1992). This is broadly consistent with results, presented by Benech-Arnold, Giallorenzi, et al. (1999), showing that inhibition of GA synthesis with paclobutrazol lowers the germination capacity of the grains when they are surrounded by the glumellae, without altering the germination capacity of the naked caryopses. These results further support the proposition that hull-imposed dormancy is hormonally regulated. The germination-promoting effect of gibberellins (GAs) is well documented for mature seeds of a number of species (Lona, 1956; Karssen et al., 1989; Hilhorst, 1995; Karssen, 1995), but it has been recently demonstrated that the degree of dormancy in developing seeds also depends on the extent to which ABA action is counterbalanced by the effect of GAs (Steinbach, Benech-Arnold, and Sánchez, 1997; Benech-Arnold et al., 2000). It might be, then, that the low dormancy presented by grains from sprouting-susceptible cultivars arises also from a high GA content or sensitivity that counteracts more effectively the inhibitory effect imposed by ABA. Nevertheless, the way in which GAs determines dormancy level in developing seeds is by far less understood than in the case of ABA. Hence, much work is required if a deep knowledge of the hormonal nature of dormancy in the barley grain is intended.

Dormancy and the Production of α-Amylase

During germination, the growing embryo synthesizes and secretes GAs into the starchy endosperm. Gibberellins diffuse into the aleurone to trigger the synthesis of hydrolytic enzymes such as α-amylase. This enzyme starts the degradation of the starchy endosperm, thus aggravating the damage already imposed by the initiation of embryo growth when germination has been triggered prior to crop harvest. This role of GAs as promoter of the postgerminative event "production of α-amylase" should not be confounded with the pregerminative role of GAs referred to in the preceding section. Also, the involvement of ABA in inhibiting the expression of α-amylase genes has been shown for barley and other cereals (Mundy, 1984; Jacobsen and Beach, 1985; Robertson et al., 1989; Van Beckum, Libbenga, and Wang, 1993; Jacobsen, Gubler, and Chandler, 1995; Pagano et al., 1997), thus recreating the aforementioned antagonism between ABA and GA for deciding the grain dormancy level. Moreover, the acquisition of the capacity to produce α-amylase activity was reported to be strongly related to changes in embryonic sensitivity to ABA (Benech-Arnold, Giallorenzi, et al., 1999). Dormancy of the barley grain has been correlated with GA responsiveness of the isolated aleurone layer (Schuurink, Sedee, and Wang, 1992); however, it is worth emphasizing that changes in both ABA and GA responsiveness for the production of α-amylase are more likely consequences of changes in dormancy rather than the cause of those changes. Nevertheless, the acquisition of GA responsiveness of the aleurone may be delayed with respect to the process of dormancy relief. For example, Benech-Arnold, Giallorenzi, et al. (1999) reported that by the time grains from a couple of cultivars presenting contrasting rates of exit from dormancy had reached a similar germination capacity, the ability to produce α-amylase was still low in the genotype that had been released later from dormancy. Interestingly, embryo sensitivity to ABA in that cultivar was not as low yet as that observed for the cultivar that had been released earlier from dormancy. Hence, a pair of cultivars with equally low dormancy at harvest would be subjected to a different level of damage due to preharvest sprouting, so long as one of them has not acquired yet the capacity of producing enough α-amylase (i.e., because of a low GA responsiveness of the aleurone) to generate an important degradation of the endosperm.

Understanding the physiological and genetic control of α-amylase production is of paramount importance in malting barley, since the malting process itself depends on the production of this enzyme. In fact, much research has been carried out in relation to this topic over the years. However, since a deep analysis of the control mechanisms for the production of this enzyme is beyond the scope of this chapter, the reader is directed to some other comprehensive reviews (Briggs, 1992; Jacobsen, Gubler, and Chandler, 1995).

The Expression of Dormancy in the Barley Grain

Except for the case of seeds that present full dormancy and consequently do not germinate at any temperature, dormancy is commonly expressed at certain temperatures. Vegis (1964) introduced the concept of degrees of relative dormancy from the observation that, as dormancy is released, the temperature range permissive for germination widens, until germination is maximal under a wide thermal range. This seems to be the case for dormant cereal grains: whereas in summer cereals such as sorghum, dormancy is not expressed at high temperatures (i.e., 30°C) (Benech-Arnold et al., 1995; Benech-Arnold, Enciso, and Sánchez, 1999), in winter cereals such as wheat and barley, it is not expressed at low temperatures (i.e., 10°C or lower) (Bewley and Black, 1994; Gosling et al., 1981; Mares, 1984; Black, Butler, and Hughes, 1987; Walker-Simmons, 1988). It should be emphasized that the depressed germination, which in winter cereals occurs when temperatures exceed certain values is truly an expression of dormancy, and not an inevitable effect of temperature on germination, for it does not take place in isolated embryos or in grains that have after-ripened (Mares, 1984). Moreover, it has been shown in wheat that isolated embryos incubated at high temperatures (i.e., 25-30°C) are more effectively inhibited by ABA than embryos incubated at lower temperatures (i.e., 10-15°C) (Walker-Simmons, 1988). Similarly, it was observed for barley grains that, so long as they are released from dormancy throughout development and maturation, they are able to germinate at higher temperatures (Benech-Arnold, Giallorenzi, et al., 1999). This lack of expression of dormancy at low temperatures in barley grains implies that in years when damp conditions are combined with low air temperatures around harvest time, both resistant and susceptible cultivars might be expected to sprout.

GENETICS AND MOLECULAR BIOLOGY OF DORMANCY IN BARLEY

Early genetic investigations (Buraas and Skinnes, 1984) revealed that seed dormancy in Scandinavian barleys was governed by several recessive, nucleoplasmic loci with high heritability. Genetic control of barley seed dormancy has also been studied by means of quantitative trait loci (QTL) mapping (Ullrich et al., 1993; Takeda, 1996). A saturated molecular marker linkage map based on the six-row Steptoe/Morex (S/M) mapping population has been developed (Kleinhofs et al., 1993) and extensively used for QTL analysis by the North American Barley Genome Mapping Project (Hayes et al., 1993; Han et al., 1996; Romagosa et al., 1996). Steptoe is a six-row feed barley with high levels of dormancy (Muir and Nilan, 1973). Morex is a six-row malting type that does not express dormancy (Rasmusson and Wilcoxson, 1979). Four regions of the barley genome on chromosomes 1(7H), 4(4H), and 7(5H) were associated with most of the differential genotypic expression for dormancy in the S/M cross (Ullrich et al., 1996; Oberthur et al., 1995; Han et al., 1996; Larson et al., 1996). They were designated SD1 to SD4 by Han et al. (1996) and accounted for approximately 50, 15, 5, and 5 percent of the phenotypic differences, respectively, in germination following several postharvest periods. In an early study, Livers (1957, cited by Romagosa et al., 1999) found some evidence that one or more postharvest dormancy *(phd)* genes may be located on chromosome 7, which is where two of the S/M QTL are located. Takeda (1996), using QTL analysis with the Harrington/TR306 (H/T) population, identified one region each on chromosomes 5(1H) and 7(5H) that controlled dormancy. The chromosome 7(5H) H/T QTL coincides with the S/M QTL SD2 on the end of the long arm and was suggested to be allelic. In a recent study, Romagosa et al. (1999) investigated the individual effects of the S/M SD QTL on dormancy during seed development and after-ripening. With this aim, three pairs each of doubled haploid lines (DHLs), derived from Steptoe/Morex F_1s with the MM, SS, SS MM, and SS SS genotypes at the SD1 and SD2 QTL and fixed M genotypes (MM MM) at the SD3 and SD4 QTL, were identified by restriction fragment length polymorphism (RFLP) analysis. Morex and genotype MM SS MM MM were the first to start losing dormancy throughout development; the other genotypes remained dormant until the end of seed development (see Figure 17.4a). Similarly, Morex and genotypes MM SS MM MM and MM SS SS SS had completely lost dormancy after 30 days of after-ripening, whereas other genotypes presented a pattern of exit from dormancy that progressively resembled that observed for the highly dormant Steptoe (see Figure 17.4b). Since the presence of the Steptoe allele at SD1 on chromosome 7(5H) delayed exit from dormancy, the authors concluded that SD1 is the most important QTL in determining the time of dormancy release.

The gene *Vp1* encodes a transcription factor whose involvement in the control of embryo sensitivity to ABA has been evidenced since the isolation of maize *vp1* mutants that are insensitive to ABA and present viviparity (McCarty et al., 1991). Preharvest sprouting in cereals is very similar pheno-

FIGURE 17.4. Germination Percentage of Various Barley Genotypes During (a) Seed Development and (b) After-Ripening After Crop Harvest

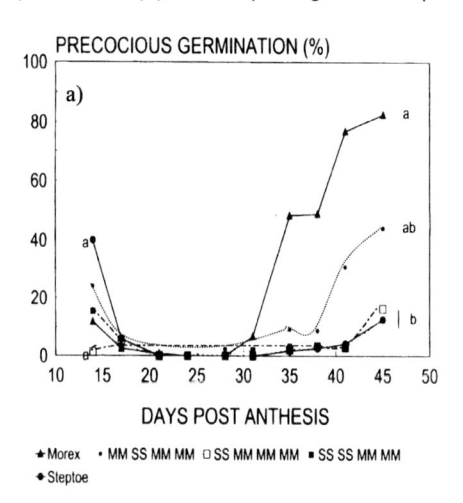

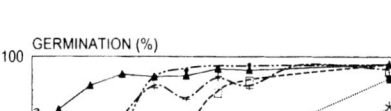

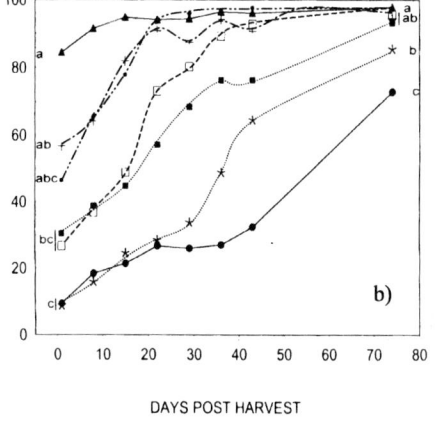

Source: From Romagosa et al., 1999.

Note: Genotypic means followed by the same letter are not statistically significant according to Duncan test (a) ($\alpha < 0.05$) or LSD test (b) ($\alpha < 0.05$). MM (Morex) and SS (Steptoe) designations refer to the genotypes of the pair of flanking markers for the four seed dormancy (SD) QTL in order: SD1, SD2, SD3, SD4, e.g., MM SS MM MM.

typically to the *vp1* mutation in maize, raising the interesting possibility that preharvest sprouting in barley and other cereals is caused, in part, by the physiological disruption of the *Vp1* function. Genes homologous to *vp1* from barley (Hollung et al., 1997) and other Gramineae, such as rice (Hattori, Terada, and Hamasuna, 1994), sorghum (Carrari et al., 2000), and wild oat (Jones, Peters, and Holdsworth, 1997), have been cloned and sequenced and, in some cases, close correlations between *Vp1* expression and dormancy were found (Jones, Peters, and Holdsworth, 1997; Carrari et al., 2000). In a study carried out with sorghum, Lijavetzky et al. (2000) found a QTL that explained more than 50 percent of the phenotypic variation in a segregating F_2 generated by crossing two varieties with contrasting dormancy; this QTL was linked to the RFLP marker UMC3 that, in maize, is linked to the gene *Vp1*. Interestingly, *Vp1* expression during imbibition of immature grains correlated closely with their dormancy level (Carrari et al., 2001). If in this case *Vp1* turns out to map within the QTL pointed out by Lijavetzky et al. (2000), then this would be the first time for a cereal that a genetic analysis revealed the participation of a gene controlling a physiological mechanism (i.e., embryo sensitivity to ABA) that is known to be in part responsible for the different dormancy level of sorghum cultivars with contrasting behavior in regard to preharvest sprouting. Unfortunately, none of the genetic studies carried out so far for barley has been able to identify the involvement of genes that could conceivably control any of the physiological mechanisms described in previous sections.

Protein kinases often act in the transduction of external signals and could have a role in the effects of environmental conditions on expression of dormancy. For that reason, a protein kinase mRNA (PKABA1), which accumulates in mature wheat seed embryos and which is responsive to applied ABA was cloned, and its expression was analyzed during imbibition of dormant and nondormant wheat seeds. When dormant seeds are imbibed, embryonic PKABA1 mRNA levels remain high for as long as the seeds are dormant, whereas they decline and disappear in embryos of germinating seeds (Anderberg and Walker-Simmons, 1991; Holappa and Walker-Simmons, 1995). The role of this kinase in dormant seeds is currently under characterization, but a potential role of phosphorylation-dependent responses in maintenance of seed dormancy is also supported by recent characterization of the *abi1* mutant of *Arabidopsis* (ABA insensitive with no dormancy) (Meyer, Leube, and Grill, 1994). The participation of this protein kinase in maintaining dormancy of the barley grain remains to be investigated. Differences in gene expression in imbibed dormant and nondormant caryopses of *Avena fatua* (wild oats) have been determined through the technique of differential display (Li and Foley, 1994, 1995; Johnson et al., 1995). Monitoring gene expression in dormant and nondormant caryopses of barley through differential display could eventually evince yet unknown physiological and biochemical mechanisms controlling dormancy, provided the function of genes that are differentially expressed is finally elucidated.

In summary, both genetics and molecular biology studies could aid in the search for barley cultivars that, without having a long-lasting dormancy, could present resistance to preharvest sprouting. However, complementarity with physiological studies is essential if such a goal is intended. The most profitable genetic investigations would be those which, for example, through QTL analysis, point out the participation of genes with known physiological function. If in the end the phenotype happens to correlate well with some characteristic of that gene (i.e., differences between phenotypes in terms of gene expression timing, sequence, regulation, etc.), then the possibilities for manipulating the system are high. Work with recently isolated barley dormancy mutants promises a good amount of revelations (Molina-Cano et al., 1999).

ENVIRONMENTAL CONTROL OF DORMANCY IN BARLEY

The effects of parental environment on dormancy level of seeds have been reported for a wide range of species (for reviews, see Fenner, 1991; Wulff, 1995). Some well-defined patterns emerge, however, with certain environmental factors tending to have similar effects in different species. Low dormancy is generally associated with high temperatures, short days, red light, drought, and nutrient availability during seed development (Walker-Simmons and Sesing, 1990; Fenner, 1991; Benech-Arnold, Fenner, and Edwards, 1991, 1995; Gate, 1995). Sprouting susceptibility is determined mainly by the genotype; but, as in other species, it is known that dormancy of the barley grain can be influenced also by the environment experienced by the mother plant (Khan and Laude, 1969; Nicholls, 1982; Reiner and Loch, 1976; Schuurink, Van Beckum, and Heidekamp, 1992; Cochrane, 1993; see also Auranen, 1995, for references). In cultivars with very fast dormancy release after physiological maturity, or in those with long-lasting dormancy, this environmental influence might not affect their sprouting behavior. In other words, the former will behave always as sprouting susceptible while the latter will be always sprouting resistant. However, in cultivars with intermediate behavior, changes in the speed of exit from dormancy after physiological maturity, as affected by the environment, may determine that, in some years, those cultivars behave as sprouting resistant and in others as sprouting susceptible.

Among the different factors acting on the mother plant, temperature appears to be the main one responsible for year-to-year variation in grain dormancy of a genotype. Therefore, knowledge on how temperature modulates dormancy release may help to predict sprouting susceptibility and, together with meteorological information, give an estimate of sprouting risk in cultivars with intermediate sprouting behavior. Evidence suggests that temperature might be effective only within a sensitivity period during grain filling (Kivi, 1966; Reiner and Loch, 1976; Buraas and Skinnes, 1985). For example, Reiner and Loch (1976) determined that low temperatures during the first half of grain filling combined with high temperatures during the second half results in a low dormancy level of the grain and, presumably, in preharvest sprouting susceptibility. Conversely, high temperatures during the first half combined with low temperatures during the second one produced the highest dormancy levels. The authors established a linear relationship between the ratio of the temperatures prevailing in both halves of the filling period and the dormancy level of the grains three weeks after harvest. This model has since been used by the German malting industry to predict dormancy level in the malting barley harvest lots.

In a recent work, Rodriguez et al. (1999) identified a time window within the grain-filling period of cultivar Q. Palomar, with sensitivity to temperature for the determination of dormancy. This time window was found to go from 300°C days to 350°C days after heading (accumulated over a base temperature of 5.5°C). A positive linear relationship was established between the average temperature perceived by the crop during this time window and the germination index of the grains 12 days after PM (see Figure 17.5).

FIGURE 17.5. The Relationship Between Temperature Experienced by the Crop in the Sensitivity Window Going from 300 to 350°C day After Heading ($Tm_{300-350}$) and the Germination Index of Grains Harvested 12 Days After Physiological Maturity (GI [12 DAPM]) and Incubated at 20°C

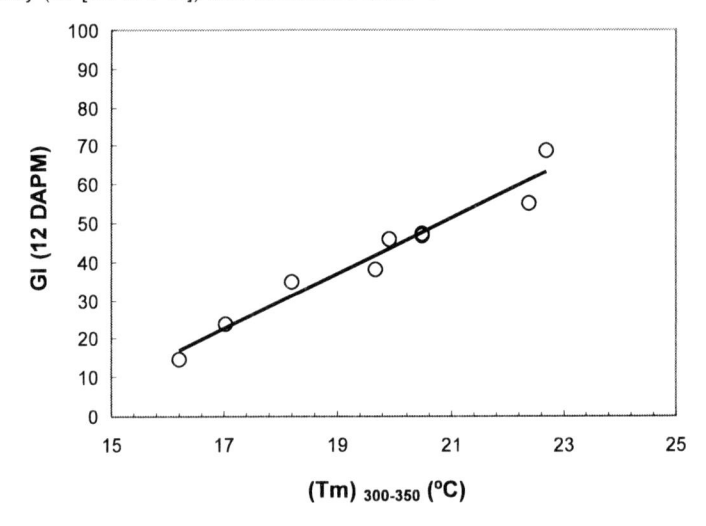

Source: From Rodriguez et al., 2001.

Twelve days after PM is approximately halfway between physiological and harvest maturity; grain germination index measured at this stage is a good estimate of the rate at which the grains are being released from dormancy after PM. According to this model, the higher the temperature experienced during the sensitivity time window, the faster the rate at which grains will be released from dormancy after PM and, consequently, the lower the dormancy level prior to crop harvest. Such a situation, combined with a forecast of heavy rains for the forthcoming days, implies a risk for the crop, and the farmer could decide to anticipate the harvest. Conversely, low temperatures experienced by the crop during the sensitivity window would result in a high dormancy level prior to harvest, making the crop resistant to sprouting. This model was successfully validated against data collected from commercial plots 700 km away from the site where the model was produced (Rodriguez et al., 2001). However, it was also noted from validation that temperature explains only one dimension of the observed variability in dormancy. Indeed, some other unknown factor(s) was (were) responsible for displacing up or down the relationship between temperature and dormancy (Rodriguez et al., 2001). Current efforts are directed toward identifying this (these) factor(s) and to quantify its (their) effects. Also, it is intended to expand the use of this model to other commercial cultivars.

Comprehensive knowledge of how the environment controls dormancy of the barley grain will not result in the development of sprouting-resistant cultivars. However, it could aid in developing efficient agronomic practices for reducing the incidence of this adversity.

CONCLUDING REMARKS

The task of adjusting the timing of exit from dormancy of commercial cultivars to the necessities of both the farmers and the malting industry does not seem to be an easy one. However, an adequate knowledge of the physiology and the genetics of dormancy in the barley grain should help to solve the compromise between obtaining cultivars with low dormancy at harvest, but not with such an anticipated termination of dormancy that leads to sprouting. Although much progress has been made in recent years, we are as yet far from having detailed knowledge of the physiology and the genetics of dormancy. It is worth emphasizing that studies linking genetics (and molecular biology) with physiology appear to be the most promising ones. For example, if genes controlling sensitivity to hormones (either ABA or GA) are finally identified and their participation in the control of dormancy is eventually evidenced, then efforts should be directed toward understanding the regulation of those genes. It would not be surprising to find out that the action of genes controlling, for example, sensitivity to ABA, is cancelled after the grain has undergone desiccation (Kermode, 1995). If the transduction pathway is finally understood, then it should not be very difficult to manipulate the timing of such a cancellation. This is just speculation to illustrate how molecular studies oriented by physiological studies could yield tools for the production of cultivars with a precise timing of dormancy release. The recent isolation of barley dormancy mutants (Molina-Cano et al., 1999) is an important step toward that aim.

The development of agronomic practices derived from our knowledge of dormancy could help to reduce the incidence of sprouting until new cultivars with controlled exit from dormancy are developed. This chapter exemplified how a comprehensive assessment of the effects of temperature during seed development on dormancy can be used for deciding management practices. It is quite evident, however, that other factors in addition to temperature modulate the resistance to sprouting of barley crops. So long as these factors are identified and their effects quantified, decisions on management practices will be made on even more solid bases.

REFERENCES

Anderberg, R.J. and M.K. Walker-Simmons (1991). Isolation of a wheat cDNA clone for an abscisic acid-inducible transcript with homology to protein kinases. *Proceedings of the National Academy of Sciences, USA* 89: 10183-10187.

Auranen, M. (1995). Preharvest sprouting and dormancy in malting barley in northern climatic conditions. *Acta Agriculturae Scandinavica* 45: 89-95.

Benech-Arnold, R.L., S. Enciso, and R.A. Sánchez (1999). Fluridone stimulus of dormant sorghum seeds germination at low temperatures is not accompanied by changes in ABA content. In preharvest *Sprouting in Cereals 1998* (Part II), ed. D. Weipert. Detmold, Germany: Association of Cereal Research, pp. 76-80.

Benech-Arnold, R.L., S. Enciso, R.A. Sánchez, F. Carrari, L. Perez-Flores, N. Iusem, H.S. Steinbach, D. Lijavetzky, and R. Bottini (2000). Involvement of ABA and GAs in the regulation of dormancy in developing sorghum seeds. In *Seed Biology: Advances and Applications,* eds. M. Black, K.J. Bradford, and J. Vázquez Ramos. Oxon, UK: C.A.B. International, pp. 101-111.

Benech-Arnold, R.L., M. Fenner, and P.J. Edwards (1991). Changes in germinability, ABA levels and ABA embryonic sensitivity in developing seeds of *Sorghum bicolor* induced by water stress during grain filling. *New Phytologist* 118: 339-347.

Benech-Arnold, R.L., M. Fenner, and P.J. Edwards (1995). Influence of potassium nutrition on germinability, ABA content and embryonic sensitivity to ABA of developing seeds of *Sorghum bicolor* (L.) Moench. *New Phytologist* 130: 207-216.

Benech-Arnold, R.L., M.C. Giallorenzi, J. Frank, and V. Rodriguez (1999). Termination of hull-imposed dormancy in barley is correlated with changes in embryonic ABA content and sensitivity. *Seed Science Research* 9: 39-47.

Benech-Arnold, R.L., G. Kristof, H.S. Steinbach, and R.A. Sánchez (1995). Fluctuating temperatures have different effects on embryonic sensitivity to ABA in *Sorghum* varieties with contrasting preharvest sprouting susceptibility. *Journal of Experimental Botany* 46: 711-717.

Bewley, J.D. and M. Black (1994). *Seeds: Physiology of Development and Germination* (Second Edition). New York: Plenum Press.

Black, M. (1991). Involvement of ABA in the physiology of developing and mature seeds. In *Abscisic Acid Physiology and Biochemistry*, ed. W.J. Davies. Oxford. Bios Scientific Publishers Limited, pp. 99-124.

Black, M., J. Butler, and M. Hughes (1987). Control and development of dormancy in cereals. In *Proceedings of the IV Symposiun on Pre-Harvest Sprouting in Cereals*, ed. D. Mares. Boulder, CO: Westview Press, pp. 379-392.

Briggs, D.E. (1992). Barley germination: Biochemical changes and hormonal control. In *Barley: Genetics, Biochemistry, Molecular Biology and Biotechnology*, ed. P.R. Shewry. Oxford, UK: C.A.B. International, Alden Press, pp. 369-401.

Buraas, T. and H. Skinnes (1984). Genetic investigations on seed dormancy in barley. *Hereditas* 101: 235-244.

Buraas, T. and H. Skinnes (1985). Development of seed dormancy in barley, wheat and triticale under controlled conditions. *Acta Agriculturae Scandinavica* 35: 233-244.

Carrari, L., J. Perez-Flores, D. Lijavetzky, S. Enciso, R.A. Sanchez, R.L. Benech-Arnold, and N. Iusem (2001). Cloning and expression of a sorghum gene with homology to maize *vp1*: Its potential involvement in preharvest sprouting resistance. *Plant Molecular Biology* 45: 631-640.

Castor, L.L. and R.A. Frederiksen (1977). Seed moulding of grain sorghums caused by *Fusarium* and *Curvularia* (abstract). *Proceedings of the American Phytopathological Society* 4: 151.

Cochrane, M.P. (1993). Effects of temperature during grain development on the germinability of barley grains. *Aspects of Applied Biology* 36: 103-113.

Corbineau, F. and D. Come (1980). Quelques caractéristiques de la dormance du caryopse d'Orge (*Hordeum vulgare* variété Sonja). *Comptes Rendues de l'Academie des Sciences de Paris, Série D.* 280: 547-550.

Corbineau, F., A. Poljakoff-Mayber, and D. Come (1991). Responsiveness to abscisic acid of embryos of dormant oat *(Avena sativa)* seeds: Involvement of ABA-inducible proteins. *Physiologia Plantarum* 83: 1-6.

Del Fueyo, P.A., V. Marcaida, and R.L. Benech-Arnold (1999). El pregerminado en granos de cebada afecta su longevidad potencial. Una evaluación cuantitativa. In *Abstracts from the III Congreso Latinoamericano de Cebada,* Montevideo, Uruguay: INIA, p. 52.

Fenner, M. (1991). The effects of the parent environment on seed germinability. *Seed Science Research* 1: 75-84.

Fong, F., J.D. Smith, and D.E. Koehler (1983). Early events in maize seed development. *Plant Physiology* 73: 899-901.

Gate, P. (1995). Ecophysiologie de la germination sur pied. *Perspectives Agricoles* 204: 22-29.

Goldbach, H. and G. Michael (1976). Abscisic acid content of barley grains during ripening as affected by temperature and variety. *Crop Science* 16: 797-799.

Gosling, P.G., R.A. Butler, M. Black, and J.M. Chapman (1981). The onset of germination ability in developing wheat. *Journal of Experimental Botany* 32: 621-627.

Han, F., S.E. Ullrich, J.A. Claney, V. Jitkov, A. Kilian, and I. Romagosa (1996). Verification of barley seed dormancy loci via linked molecular markers. *Theoretical and Applied Genetics* 92: 87-91.

Hattori, T., T. Terada, S.T. Hamasuna (1994). Sequence and functional analyses of the rice gene homologous to the maize *Vp1*. *Plant Molecular Biology* 24: 805-810.

Hayes, P.M., B.H. Liu, S.J. Knapp, F. Chen, B. Jones, T. Blake, J. Franckowiak, D. Rasmusson, M. Sorrells, S.E. Ullrich, et al. (1993). Quantitative trait locus effects and environmental interaction in a sample of North American barley germplasm. *Theoretical and Applied Genetics* 87: 392-401.

Hilhorst, H.W.M. (1995) A critical update on seed dormancy. I. Primary dormancy. *Seed Science Research* 5: 61-73.

Holappa, L.D. and M.K. Walker-Simmons (1995). The wheat abscisic acid-responsive protein kinase mRNA, PKABA1, is up-regulated by dehydration, cold temperature and osmotic stress. *Plant Physiology* 108: 1203-1210.

Hollung, K., M. Espelund, K. Schou, and K.S. Jakobsen (1997). Developmental, stress and ABA modulation of mRNA levels for bZip transcription factors and *Vp1* in barley embryos and embryo-derived suspension cultures. *Plant Molecular Biology* 35: 561-571.

Jacobsen, J.V. and L.R. Beach (1985). Control of transcription of alpha-amylase and r-RNA genes in barley aleurone protoplasts by gibberellin and abscisic acid. *Nature* 316: 275-277.

Jacobsen, J.V., F. Gubler, and P.M. Chandler (1995). Gibberellin action in germinated cereal grains. In *Plant Hormones. Physiology, Biochemistry and Molecular Biology*, ed. P. Davies. Dordrecht, the Netherlands: Kluwer Academic Publishers, pp. 246-271.

Johnson, R.R., H.J. Cranston, M.E. Chaverra, and W.E. Dyer (1995). Characterization of cDNA clones for differently expressed genes in embryos of dormant and non-dormant *Avena fatua* L. caryopses. *Plant Molecular Biology* 28: 113-122.

Jones, H.D., N.C.B. Peters, and M.J. Holdsworth (1997). Genotype and environment interact to control dormancy and differential expression of the *VIVIPAROUS-1* homologue in embryos of *Avena fatua*. *Plant Journal* 12: 911-920.

Karssen, C.M. (1995). Hormonal regulation of seed development, dormancy, and germination studied by genetic control. In *Seed Development and Germination*, eds. J. Kigel and G. Galili. New York: Marcel Dekker, Inc., pp. 333-350.

Karssen, C.M., D.L.C. Brinkhorst-Van der Swan, A.E. Breekland, and M. Koorneef (1983). Induction of dormancy during seed development by endogenous abscisic acid: Studies on abscisic acid deficient genotypes of *Arabidopsis thaliana* (L.). *Planta* 157: 158-165.

Karssen, C.M., S. Zagorski, J. Kepczynski, and S P.C. Groot (1989). Key role for endogenous gibberellins in the control of seed germination. *Annals of Botany* 63: 71-80.

Kermode, A.R. (1995). Regulatory mechanisms in the transition from seed development to germination: Interactions between the embryo and the seed environment. In *Seed Development and Germination,* eds. J. Kigel and G. Galili. New York: Marcel Dekker, Inc., pp. 273-332.

Khan, R.A. and H.M. Laude (1969). Influence of heat stress during seed maturation on germinability of barley seed at harvest. *Crop Science* 9: 55-58.

King, R.W. (1982). Abscisic acid in seed development. In *The Physiology and Biochemistry of Seed Development, Dormancy and Germination,* ed. A.A. Kahn, Amsterdam: Elsevier Biomedical Press, pp. 157-181.

Kivi, E. (1966). The response of certain preharvest climatic factors on sensitivity to sprouting in the ear of two-row barley. *Acta Agriculturae Fennica* 107: 228-246.

Kleinhofs, A., A. Kilian, M.A. Saghai Maroof, R.M. Biyashev, P.M. Hayes, F.Q. Chen, N. Lapitan, A. Fenwich, T.K. Blake, V. Kanazin, et al. (1993). A molecular, isozyme and morphological map of the barley *(Hordeum vulgare)* genome. *Theoretical and Applied Genetics* 86: 705-712.

Larson S., G. Bryan, W. Dyer, and T. Blake (1996). Evaluating gene effects of a major barley seed dormancy QTL in reciprocal back-cross populations. *Journal of Quantitative Trait Loci* (online) 2: <http://probe.nalusda.gov:8000/otherdocs/jqtl/jqtl1996-04/larson15a.htm>.

Lecat, S., F. Corbineau, and D. Come (1992). Effects of gibberellic acid on the germination of dormant oat *(Avena sativa* L.) seeds as related to temperature, oxygen and energy metabolism. *Seed Science and Technology* 20: 421-433.

Lenoir, C., F. Corbineau, and D. Come (1986). Barley *(Hordeum vulgare)* seed dormancy as related to glumella characteristics. *Physiologia Plantarum* 68: 301-307.

Li, B. and M.E. Foley (1994). Differential polypeptide patterns in imbibed dormant and after-ripened *Avena fatua* embryos. *Journal of Experimental Botany* 45: 275-279.

Li, B. and M.E. Foley (1995). Cloning and characterization of differentially expressed genes in imbibed dormant and after-ripened *Avena fatua* embryos. *Plant Molecular Biology* 29: 823-831.

Lijavetzki, D., M.C. Martinez, F. Carrari, H.E. Hopp (2000). QTL analysis and mapping of preharvest sprouting resistance in sorghum. *Euphytica* 112: 125-135.

Livers, R. W. (1957). Linkage studies with chromosomal translocation stocks in barley. PhD thesis (Dissertation Abstract AAT 5801125). St. Paul: University of Minnesota.

Lona, F. (1956). L'acido gibberéllico determina la germinazione del semi di Lactuca scariola in fase di scotoinhibizione. *Ateneo Pamense* 27: 641-644.

Mares, D. (1984). Temperature dependence of germinability of wheat *(Triticum aestivum)* grain in relation to preharvest sprouting. *Australian Journal of Agricultural Research* 35: 115-128.

McCarty, D.R., T. Hattori, C.B. Carson, V. Vasil, M. Lazar, and I.K. Vasil (1991). The *viviparous-1* developmental gene of maize encodes a novel transcriptional activator. *Cell* 66: 895-905.

Meyer, K., M.P. Leube, and E. Grill (1994). A protein phosphatase 2C involved in ABA signal transduction in *Arabidopsis thaliana*. *Science* 264: 1452-1455.

Molina-Cano, J.L., A. Sopena, J.S. Swanston, A.M. Casas, M.A. Moralejo, A. Ubieto, I. Lara, A.M. Perez-Vendrell, and I. Romagosa (1999). A mutant induced in the malting barley cv. Triumph with reduced dormancy and ABA response. *Theoretical and Applied Genetics* 98: 347-355.

Morris, P.C., P.C. Jewer, and D.J. Bowles (1991). Changes in water relations and endogenous abscisic acid content of wheat and barley grains and embryos during development. *Plant, Cell and Environment* 14: 443-446.

Muir, C.E. and R.A. Nilan (1973). Registration of Steptoe barley. *Crop Science* 13: 770.

Mundy, J. (1984). Hormonal regulation of alpha-amylase inhibitor synthesis in germinating barley. *Carlsberg Research Communications* 48: 81-90.

Neil, S.J. and R. Horgan (1987). Abscisic acid and related compounds. In *The Principles and Practice of Plant Hormone Analysis*, eds. L. Rivier and A. Crozier. London: Academic Press, pp. 111-167.

Nicholls, P.B. (1982). Influence of temperature during grain growth and ripening of barley on the subsequent response to exogenous gibberellic acid. *Australian Journal of Plant Physiology* 9: 373-383.

Norstog, K. and R.M. Klein (1972). Development of cultured barley embryos. II. Precocious germination and dormancy. *Canadian Journal of Botany* 50: 1887-1894.

Oberthur, L., W. Dyer, S.E. Ullrich, and T.K. Blake (1995). Genetic analysis of seed dormancy in barley (*Hordeum vulgare* L.). *Journal of Quantitative Trait Loci* (online) 1: <http://probe.nalusda.gov:8000/otherdocs/jqtl/jqtl1995-05/dormancy.html>.

Pagano, E.A., R.L. Benech-Arnold, M. Wawrzkiewics, and H.S. Steinbach (1997). Alpha-amylase activity in developing sorghum caryopses from sprouting resistant and susceptible varieties: The role of ABA and GAs on its regulation. *Annals of Botany* 79: 13-17.

Quarrie, S.A., R. Tuberosa, and P.G. Lister (1988). Abscisic acid in developing grains of wheat and barley genotypes differing in grain weight. *Plant Growth Regulation* 7: 3-17.

Rasmusson, D.C. and R.D. Wilcoxson (1979). Registration of Morex barley. *Crop Science* 19: 293.

Reiner, L. and V. Loch (1976). Forecasting dormancy in barley—Ten years experience. *Cereal Research Communication* 4: 107-110.

Robertson, M., M.K. Walker-Simmons, D. Munro, and R.D. Hill (1989). Induction of alpha-amylase inhibitor synthesis in barley embryos and young seedlings by abscisic acid and dehydration stress. *Plant Physiology* 91: 415-420.

Robichaud, C.S., J. Wong, and I.M. Sussex (1980). Control of in vitro growth of viviparous embryo mutants of maize by abscisic acid. *Developmental Genetics* 1: 325-330.

Rodriguez, V., J. González Martín, P. Insausti, and R.L. Benech-Arnold (1999). Un método para predecir la susceptibilidad al brotado en el cultivo de cebada, utilizando la temperatura durante el llenado de los granos. In *Abstracts from the III Congreso Latinoamericano de Cebada*. Montevideo, Uruguay: INIA, p. 59.

Rodriguez, V., J. González Martín, P. Insausti, J.M. Margineda, and R.L. Benech-Arnold (2001). preharvest sprouting susceptibility in barley correlates positively with mean temperature perceived during grain filling. *Agronomy Journal* 93 (in press).

Romagosa, I., F. Han, J.A. Clancy, and S.E. Ullrich (1999). Individual locus effects on dormancy during seed development and after ripening in barley. *Crop Science* 39: 74-79.

Romagosa, I., S.E. Ullrich, F. Han, and P.M. Hayes (1996). Use of additive main effects and multiplicative interaction model in QTL mapping for adaptation in barley. *Theoretical and Applied Genetics* 93: 30-37.

Schuurink, R.C., N.J.A. Sedee, and M. Wang (1992). Dormancy of the barley grain is correlated with gibberellic acid responsiveness of the isolated aleurone layer. *Plant Physiology* 100: 1834-1839.

Schuurink, R.C., J.M.M. Van Beckum, and F. Heidekamp (1992). Modulation of grain dormancy in barley by variation of plant growth conditions. *Hereditas* 117: 137-143.

Simpson, G.M. (1990). *Seed Dormancy in Grasses*. Cambridge: Cambridge University Press.

Steinbach, H.S., R.L. Benech-Arnold, G. Kristof, R.A. Sánchez, and S. Marcucci Poltri (1995). Physiological basis of preharvest sprouting resistance in *Sorghum bicolor* (L.) Moench: ABA levels and sensitivity in developing embryos of sprouting resistant and susceptible varieties. *Journal of Experimental Botany* 45: 701-709.

Steinbach, H.S., R.L. Benech-Arnold, and R.A. Sánchez (1997). Hormonal regulation of dormancy in developing sorghum seeds. *Plant Physiology* 113: 149-154.

Takeda, K. (1996). Varietal variation and inheritance of seed dormancy in barley. In *Pre-Harvest Sprouting in Cereals 1995*, eds. K. Noda and D. Mares. Osaka, Japan: Center for Academic Societies, pp. 205-212.

Ullrich, S.E., F. Han, T.K. Blake, L.E. Oberthur, W.E. Dyer, and J.A. Clancy (1996). Seed dormancy in barley: Genetic resolution and relationship to other traits. In *Pre-Harvest Sprouting in Cereals 1995*, eds. K. Noda and D. Mares. Osaka, Japan: Center for Academic Societies, pp. 157-163.

Ullrich, S.E., P.M. Hayes, W.E. Dyer, T.K. Blake, and J.A. Clancy (1993). Quantitative trait locus analysis of seed dormancy in 'Steptoe' barley. In *Pre-Harvest Sprouting in Cereals 1992*, eds. M.K. Walker-Simmons and J.L. Ried. St. Paul, MN: American Society of Cereal Chemistry, pp. 136-145.

Van Beckum, J.M.M., K.R. Libbenga, and M. Wang (1993). Abscisic acid and gibberellic acid-regulated responses of embryos and aleurone layers isolated from dormant and non-dormant barley grains. *Physiologia Plantarum* 89: 483-489.

Vegis, A. (1964). Dormancy in higher plants. *Annual Review of Plant Physiology* 15: 185-224.

Visser, K., A.P.A. Visser, M.A. Cagirgan, J.W. Kijne, and M. Wang (1996). Rapid germination of a barley mutant is correlated with a rapid turnover of abscisic acid outside the embryo. *Plant Physiology* 111: 1127-1133.

Walker-Simmons, M.K. (1987). ABA levels and sensitivity in developing wheat embryos of sprouting resistant and susceptible cultivars. *Plant Physiology* 84: 61-66.

Walker-Simmons, M.K. (1988). Enhancement of ABA responsiveness in wheat embryos by high temperature. *Plant, Cell and Environment* 11: 769-775.

Walker-Simmons, M.K. and J. Sesing (1990). Temperature effects on embryonic abscisic acid levels during development of wheat grain dormancy. *Journal of Plant Growth Regulation* 9: 51-56.

Wang, M., R. Bakhuizen, S. Heimovaara-Dijkstra, M.J. Zeijl, M.A. De Vries, J.M. Van Beckum, and K.M.C. Sinjorgo (1994). The role of ABA and GA in barley grain dormancy: A comparative study between embryo dormancy and aleurone dormancy. *Russian Journal of Plant Physiology* 41: 577-584.

Wang, M., S. Heimovaara-Dijkstra, and B. Van Duijn (1995). Modulation of germination of embryos isolated from dormant and non-dormant grains by manipulation of endogenous abscisic acid. *Planta* 195: 586-592.

Wang, M., R.M. van der Meulen, K. Visser, H.-P. Van Schaik, B. Van Duijn, and A.H. de Boer (1998). Effects of dormancy-breaking chemicals on ABA levels in barley grain embryos. *Seed Science Research* 8: 129-137.

Wulff, R.D. (1995). Environmental maternal effects on seed quality and germination. In *Seed Development and Germination,* eds. J. Kigel and G. Galili. New York: Marcel Dekker, Inc., pp. 491-505.

Xu, N., K.M Coulter, and J.D. Bewley (1991). Abscisic acid and osmoticum prevent germination of developing alfalfa embryos, but only osmoticum maintains the synthesis of developmental proteins. *Planta* 182: 382-390.

Chapter 18

The Proteins of the Mature Barley Grain and Their Role in Determining Malting Performance

Peter R. Shewry
Helen Darlington

INTRODUCTION

The mature grain of barley grown under "normal" agricultural conditions in the United Kingdom contains about 7 to 12 percent total protein (calculated as total organic nitrogen [N] × 5.7), the precise proportion varying with N supply (Kirkman, Shewry, and Miflin, 1982). Protein is therefore a minor component when compared with starch, which accounts for about 70 to 80 percent of the mature grain mass. Nevertheless, it is a major determinant of the quality of the grain for the two major end uses, as feed for livestock and as raw material for malting, brewing, and distilling.

A high proportion of the barley grown in Western Europe is used as feed for pigs and poultry, which are monogastric animals lacking the rumen bacteria capable of interconverting amino acids, as in ruminants such as cattle. It is therefore important to provide a balance of nutritionally essential amino acids in the diet. The well-documented deficiency of barley grain in lysine and, to a lesser extent, in threonine (see Munck, 1972; Bright and Shewry, 1983) led in the 1960s and 1970s to the search for high-lysine phenotypes present in germplasm collections or induced by mutation breeding. Although this work has not led to the production of commercially grown high-lysine barley cultivars, it has nevertheless provided a considerable stimulus to the analysis of barley grain protein (see Shewry, Williamson, and Kreis, 1987; Munck, 1992).

Whereas the role of grain protein in determining nutritional quality is well-defined, its role in malting quality is more obscure. As long ago as 1930, Bishop showed that high levels of grain protein were disadvanta-

---

IACR receives grant-aided support from the Biotechnology and Biological Sciences Research Council of the United Kingdom.

geous due to a negative correlation with hot water extract (Bishop, 1930a). He also showed that this effect varied with the cultivar, and that increases in respiration and rootlet growth contributed to malting losses. Subsequent studies have suggested that the negative effect of total protein on malting may relate to the increased proportion of hordeins that occurs under conditions of high N availability (Bishop, 1930b; Kirkman, Shewry, and Miflin, 1982) but have failed to provide a convincing explanation for this, as discussed in the section Barley Grain Proteins and Malting Quality.

Malting quality may also be affected by other proteins synthesized during grain development. The most notable of these is β-amylase, which is synthesized only in the developing grain (see β-amylase), and α-glucosidase, which is present in the quiescent grain and synthesized de novo on germination (see Briggs, 1992). This contrasts with most other enzymes involved in cell and reserve digestion that are synthesized de novo only during germination (notably α-amylases and β-glucanases).

The mature barley grain contains many proteins with different structures and properties. It is clearly impossible to provide a complete description within the limits of this chapter, so we will focus on proteins present in the mature grain in significant quantities that play a role in determining the end-use properties of the grain. These can broadly be classified into three groups: enzymes and enzyme inhibitors, storage proteins, and protective proteins. However, these proteins are not evenly distributed within the grain, and it is therefore necessary initially to outline the structure of the mature barley grain.

BARLEY GRAIN STRUCTURE IN RELATION TO PROTEIN DISTRIBUTION

The mature barley grain consists of two distinct parts, the embryo and endosperm, surrounded by the fused testa (seed coat) and pericarp (a maternal tissue) and the lemma and palea, which form the husk (see Figure 18.1). The embryo is diploid and formed from fusion of the egg nucleus with one pollen nucleus. It comprises the embryonic axis (plumule and radicle surrounded by the coleoptile and coleorhiza, respectively), which develops during germination to give the seedling, and a single cotyledon, called the scutellum. The latter is a source of hydrolytic enzymes during germination but also contains storage compounds: oil droplets and protein bodies containing 7S globulins (Burgess and Shewry, 1986). A second fertilization event also occurs: the second pollen nucleus joins with two polar nuclei in the egg cell to form the triploid endosperm. The endosperm subsequently differentiates to form two tissues in the mature grain, the aleurone and the starchy endosperm. The aleurone in barley comprises two or three layers of

FIGURE 18.1. Idealized Section of a Mature Barley Grain

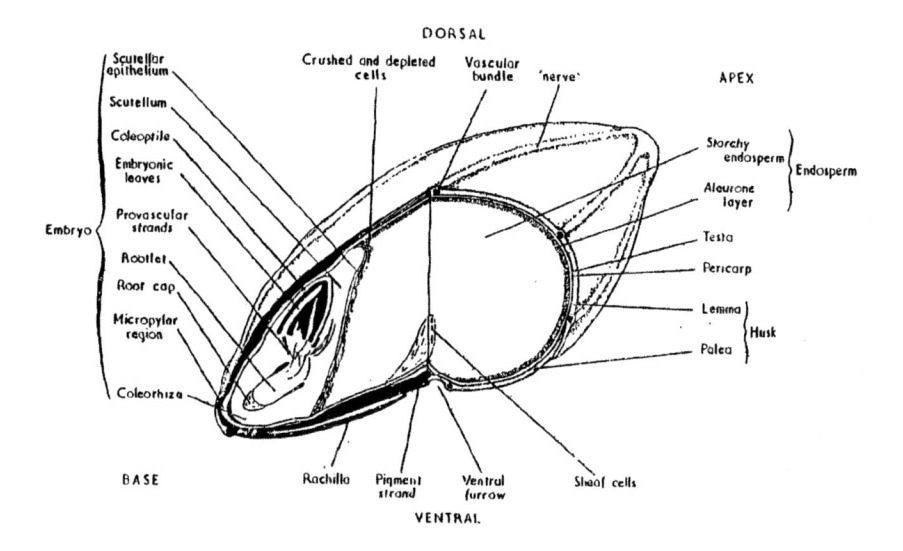

Source: From Briggs, 1973, p. 2. Reprinted with permission of Academic Press, London.

thick-walled cells that surround the starchy endosperm. They remain alive in the mature grain and, similiar to the scutellum, contribute to the synthesis and secretion of hydrolytic enzymes during germination. They also resemble the scutellum in storing oil and 7S globulin storage proteins (Yupsanis et al., 1990). The starchy endosperm comprises the bulk of the mature grain and consists of large cells packed with deposits of two types of storage compound: starch and hordein storage proteins. These cells die during the later stages of grain maturation but contain stored enzymes (β-amylase, α-glucosidase) that contribute to starch digestion during germination.

Some differences also exist in the compositions of cells in different parts of the starchy endosperm, with the outer subaleurone cells being smaller and with higher contents of storage protein. Hordeins are not present in the embryo and aleurone, while 7S globulins are not present in the starchy endosperm. However, the distributions of other proteins vary, with the chymotrypsin inhibitor CI-2 being present in the aleurone and starchy endosperm, and β-amylase being present in both these tissues and in the embryo (see Shewry and Kreis, 1992).

BARLEY GRAIN PROTEINS

Hordein Storage Proteins

The hordeins show a high level of polymorphism, with multiple components being present within a single genotype and a high level of variation in the numbers and properties (i.e., M_r, charge, pI) of components among the different genotypes. This is illustrated by Figure 18.2, which shows SDS-PAGE (second-division segregation–polyacrylamide gel electrophoresis) patterns of hordeins from a range of cultivars. This polymorphism has allowed the use of hordeins as markers for determining the distinctness of varieties and the identity and purity of seed samples (see Shewry et al., 1979).

Determination of the amino acid sequences (deduced mainly from complementary [c] DNA and genomic DNA) and the genetics of the individual hordeins has allowed them to be classified into four groups, two major

FIGURE 18.2. SDS-PAGE of Total Hordein Fractions from Single Seeds of Six Cultivars of Barley to Illustrate Polymorphism

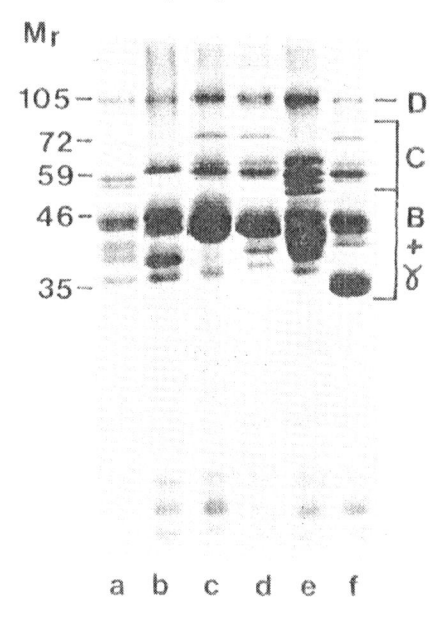

Source: Taken from Bunce et al., 1986, p. 422. Reprinted with permission.
Note: The different cultivars are (a) Athos, (b) Keg, (c) Jupiter, (d) Hoppel, (e) Igri, and (f) Sundance.

(B and C) and two minor (D and γ) (see Table 18.1 and Figure 18.3) (Shewry, 1993). Each group appears to be encoded by a single gene locus on chromosome 1H(5), although minor components may be encoded by remote genes on the same chromosome. The demonstration that all hordeins have a common evolutionary origin (Kreis et al., 1985; Shewry, 1993) indicates that various *Hor* loci were derived from a single locus and then separated by translocation to other locations on the chromosome.

Comparison of the sequences of the hordein proteins (see Figure 18.3) shows that all contain extensive regions of repeated sequences that have similar proline- and glutamine-rich motifs in B, C, and γ-hordeins. In contrast, three unrelated motifs are present in D hordein. Whereas C hordein consists essentially of repetitive sequences with only short nonrepetitive domains at the *N*- and *C*-termini (of 12 and 6 residues, respectively), the B, γ-, and D hordeins all have extensive nonrepetitive domains. Formation of interchain disulfide bonds allows B, D, and some γ-hordeins to form high M_r polymers, whereas the C hordeins contain no cysteine residues and are present only as monomers.

Extensive structural studies have shown that the repetitive sequences in C hordeins form an unusual spiral structure based on a mixture of β-reverse turns and poly-L-proline II structure, and the repetitive sequences in the other types of hordein proteins may form related structures (see Shewry, Miles, and Tatham, 1994). In contrast, the nonrepetitive sequences in B, γ-, and D hordeins probably adopt globular conformations that are rich in α-helix.

TABLE 18.1. Characterization of the Major Groups of Hordein Proteins

Group	Amount (% total)	Partial amino acid composition* (mol %)	Polymer formation?	Loci
B	70-80	35 Q, 30 P, 2.5 C, 0.6 M, 0.5 K	Yes	Hor2
C	10-20	40 Q, 30 P, 9 F, 0 C, 0.2 K, 0.2 M	No	Hor1
D	2-4	30 Q, 12 P, 16 G, 1.5 C, 1 K, 8 T	Yes	Hor3
γ	less than 5**	30 P, 20 Q, 3 C, 1 M, 1.5 K	Yes/No	Hor5 (Hrd F)

*Partial amino acid compositions based on data by Shewry (1993) with means of $γ_3$ and $γ_1 + γ_2$ for γ-hordeins. Standard single-letter abbreviations for amino acids are used: C, cysteine; F, phenylalanine; G, glycine; K, lysine; M, methionine; P, proline; Q, glutamine; T, threonine.

**Amounts of γ-hordeins not determined precisely.

FIGURE 18.3. Schematic Depiction of the Amino Acid Sequences of B, C, D, and γ-Hordeins

S-RICH

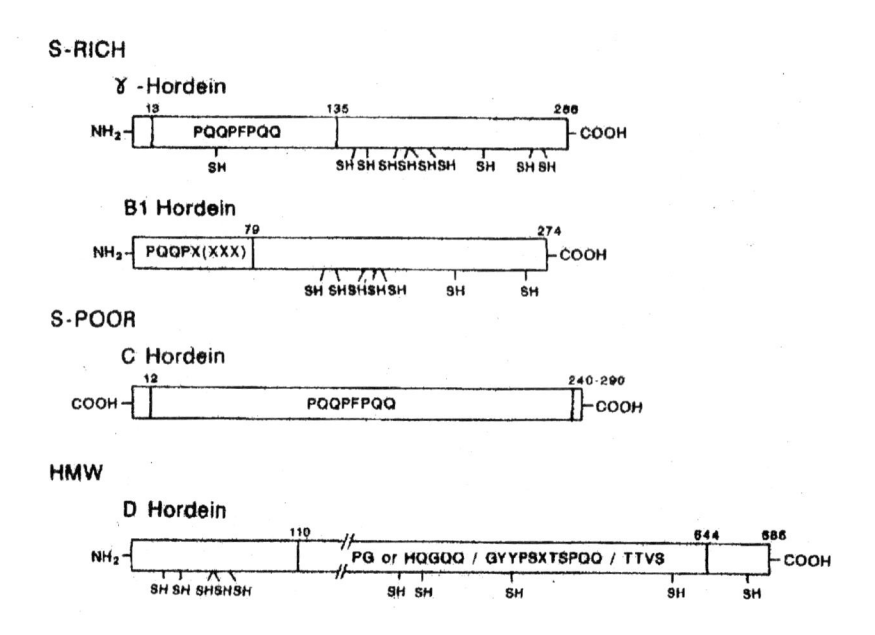

γ-Hordein

```
           13                   135                          286
NH₂ —[ ]— PQQPFPQQ |—————————————————|— COOH
              SH       SH SH SHSHSHSH   SH    SH SH
```

B1 Hordein

```
            79                                           274
NH₂ —[ ]— PQQPX(XXX) |——————————————————————————|— COOH
                       SH SHSHSHSH       SH           SH
```

S-POOR

C Hordein

```
        12                                          240-290
COOH —[ ]—————————— PQQPFPQQ ———————————————[ ]— COOH
```

HMW

D Hordein

```
                    110                                       644      686
NH₂ —[ ]——————|—// PG or HQGQQ / GYYPSXTSPQQ / TTVS //|————[ ]— COOH
     SH SH SHSHSH          SH SH          SH           SH      SH
```

Note: SH indicates positions of cysteine residues.

Protective Proteins

In common with other seeds, the mature barley grain contains a range of proteins that may contribute to a broad but nonspecific resistance to pathogenic microorganisms and/or invertebrate pests. Inhibitors of α-amylase and proteinases were initially identified in barley in the late 1960s and early 1970s, notably by Mikola and co-workers in Helsinki (see, for example, Kirsi and Mikola, 1971; Mikola and Kirsi, 1972). However, more recent work, facilitated by molecular cloning, has shown that these belong to at least four separate families of inhibitors, defined on the basis of their amino acid sequences and the mechanistic classes of their target enzymes (see Shewry, 1999). Despite considerable attempts to identify endogenous target enzymes, only one has so far been identified, discussed in the following.

The barley α-amylase 2 isoenzyme is inhibited by the bifunctional barley α-amylase/subtilisin inhibitor (usually called BASI). BASI is an M_r 19,600 protein that interacts with α-amylase 2 and subtilisin at nonoverlapping sites on opposite sides of the molecule (Vallée et al., 1998), α-Amylase 2 is one of the two major α-amylase isoforms synthesized during malting and germination, and the high affinity of BASI for this enzyme (Ki [inhibitor constant] 2.2×10^{-10} M at pH 8 and 37°C [Abe, Sidenius, and Svensson, 1993]) suggests that the inhibitor acts as an endogenous regulator, perhaps to prevent starch degradation during premature germination. In addition, its activity against bacterial subtilisin indicates a second role in conferring resistance.

The chymotrypsin inhibitors CI-1 and CI-2 were initially identified in the high-lysine line Hiproly, where increases in their amounts contribute to the high-lysine phenotype (Hejgaard and Boisen, 1980). The authors reported the presence of about 9.5 and 11.5 g lysine per 100 g protein in CI-1 and CI-2, respectively, and these high contents have been borne out by subsequent studies. Molecular cloning demonstrated that both CI-1 and CI-2 consisted of at least two isoforms, with the mature proteins comprising 83 or 84 residues with M_r of about 9,000 (Williamson et al., 1987; Williamson, Forde, and Kreis, 1988). CI-1 and CI-2 differ from almost all other inhibitors that have so far been characterized in lacking intrachain disulfide bonds, which has made them an attractive subject for protein engineering and folding studies (see, for example, Campbell, 1992). CI-1 and CI-2 do not appear to have any antinutritional impact and may again contribute to wide spectrum resistance. In the long term, their main impact on grain utilization will probably be as sources of high lysine for improved animal feed, either as wild-type proteins or after manipulation to insert additional residues of lysine and other nutritionally limiting amino acids.

The cereal inhibitor family is widely distributed in cereals and includes inhibitors of trypsin, exogenous α-amylase, bifunctional inhibitors, and forms that lack biological activity. The subunit M_r varies from about 12,000 to 16,000, with monomeric, dimeric, and tetrameric forms, with some forms also being glycosylated (Carbonero and García-Olmedo, 1999). Barley contains monomeric inhibitors of trypsin (BTI-CMe) and α-amylase (BMAI-1), the latter being glycosylated and a major allergen in baker's asthma. It also contains homodimeric (BDAI-1) and tetrameric (BTAI) α-amylase inhibitors, the latter comprising two copies of the CMd subunit and one copy each of the CMa and CMb subunits.

Protein Z is a major grain albumin estimated to be present at levels of 1.5 to 2.5 mg/g in commercial cultivars. However, it is increased about twofold in Hiproly and, together with CI-1, CI-2, and β-amylase, contributes to the high-lysine phenotype (Hejgaard and Boisen, 1980). It is highly resistant to denaturation and proteolysis and was the first well-defined protein of barley origin found to survive the malting and brewing processes to be present in beer, where it is called antigen-1 and may contribute to foam stability and/or haze formation (Hejgaard, 1977; Hejgaard and Kaersgaard, 1983; Sorensen et al., 1993). When purified from barley, protein Z consists of four antigenically related forms with M_r of about 40,000 and pI of 5.56 to 5.80 (Hejgaard, 1982). Protein Z belongs to the serpin family of serine proteinase inhibitors, which are widely distributed in animal tissues. Although early attempts to demonstrate inhibitory activity failed due to technical problems, recent work has shown that isoforms of protein Z purified from barley or expressed in *Escherichia coli* are able to inhibit serine proteinases, notably α-chymotrypsin (Dahl, Rasmussen, and Hejgaard, 1996; Dahl et al., 1996).

In addition to these families of inhibitors, barley grain also contains a range of other proteins that may contribute to resistance. These include enzymes such as endochitinase and β-glucanase, type 1 ribosome inactivating protein, and low M_r cysteine-rich proteins such as thionins and nonspecific lipid transfer protein (see Shewry, 1993; Shewry and Lucas, 1997). Most of these proteins do not have any known impact on the quality of the grain for processing or feed, the exception being lipid transfer protein (LTP).

LTP was initially identified as the product of a cDNA cloned from aleurone and was thought to have inhibitory activity (Mundy and Rogers, 1986). Hence, it was called "probable amylase/protease inhibitor," or PAPI. It was subsequently shown to have sequence homology with a well-characterized family of nonspecific phospholipid transfer proteins from other species (Bernhardt and Somerville, 1989) and to be capable of transferring phosphatidyl choline from liposomes to mitochondria in vitro (Breu et al., 1989). Nevertheless, it is now generally considered to have a protective role, possibly by binding to and destabilizing membranes of pathogens. Its role in malting and brewing is discussed under Surface-Active Proteins and Foam Stablization.

β-*Amylase*

The vast majority of enzymes present in the developing barley grain play no role in reserve mobilization during germination or malting. An exception is β-amylase [(1-4)-α-D-glucan maltohydrolase], which is a major component of the diastatic activity of malted grain. β-Amylase also acts as a storage protein, in that the amount increases with N availability (Giese and Hejgaard, 1984), and it also contributes to the high-lysine phenotype of Hiproly (discussed earlier).

β-Amylase is present in the mature grain as a mixture of free and latent (bound) forms, the latter being extracted in the presence of a reducing agent or after proteolytic digestion with papain. Both the free and bound forms appear to consist of the same protein, and both accumulate during development, particularly in the aleurone layer and starchy endosperm. β-Amylase is synthesized as a cytosolic protein, and some evidence suggests that conversion of the free form to the bound form results from association with the periphery of starch granules during grain desiccation (Hara-Nishimura,

Nishimura, and Daussant, 1986). The mature protein consists of 535 amino acids with M_r of about 60,000 (Kreis et al., 1987), and release of the bound form during germination results from partial proteolysis at the C-terminus, generating additional enzymically active isoforms with M_r of about 58,000; 56,000; and 54,000.

Starch Granule Proteins

Grain texture (i.e., hardness or softness) of wheat and barley is determined by the adhesion of "matrix" proteins to the surface of the starch granule. In wheat, this property appears to be determined by the puroindolines, a group of low M_r ($\cong$ 15,000) proteins that bind to the surface of the starch granules of soft but not hard wheat grains, providing a "nonstick" surface. Two types of puroindoline (pin a and b) are present in bread wheat and have similar structures, including tryptophan-rich motifs that could be involved in starch binding. Hardness appears to result from mutations that affect either the amount of pin a or the amino acid sequence, and presumably also the binding properties, of pin b (Giroux and Morris, 1997, 1998). However, the precise mechanism of this is not known, and the pins also exhibit lipid-binding properties that could contribute to the phenomenon (Dubreil, Compoint and Marion, 1997).

Barley also shows genetically determined differences in texture, which may be determined by a genetic locus on chromosome 5H that is homologous with the hardness *(Ha)* locus on chromosome 5D of bread wheat (Chalmers et al., 1993). Puroindoline homologues are also present in barley, and we have recently purified a pin-b homologue and used PCR (polymerase chain reaction) amplification to isolate corresponding cDNAs (Shewry and Darlington, unpublished results). However, the role of barley puroindolines in determining grain texture is still uncertain.

BARLEY GRAIN PROTEINS AND MALTING QUALITY

Hordein, Gel Protein, and Hot Water Extract

Advances in our ability to separate and characterize hordeins led in the late 1970s and 1980s to attempts to relate differences in their composition, properties, and distribution to malting performance. In 1979, Baxter and Wainwright (1979a) separated hordein fractions from 16 varieties by acid gel electrophoresis and suggested that better quality varieties had lower proportions of fast-moving B hordein components. However, this correlation was not confirmed by Shewry et al. (1980) or Riggs et al. (1983), who analyzed more extensive collections of cultivars by SDS-PAGE and acid gel electrophoresis, respectively. Baxter and Wainwright (1979b) also drew attention to a characteristic of hordeins that could also affect malting performance: the ability of the B, D, and some γ-hordeins to form high M_r polymers stabilized by interchain disulfide bonds. Baxter and Wainwright (1979b) extracted hordeins in two sequential fractions, in 70 percent (v/v) ethanol (monomeric components) and 70 percent (v/v) ethanol + 2-mercaptoethanol (subunits of polymeric components). They showed that the proportions of these fractions varied with genotype and environment, and that the polymeric fraction was more resistant to proteolysis during malting. They also showed that the spent grains contained a higher proportion of polymeric hordeins than malts, suggesting that some polymerization occurs by disulfide exchange during mashing. Based on these observations, they proposed that the polymeric hordeins could entrap starch granules, resulting in poor digestion. These results were supported by Shewry et al. (1981), who showed small differences between the proportions of polymeric hordeins in good- and poor-quality malting cultivars.

When barley flour is stirred with 1.5 percent SDS, the unextractable polymeric hordeins can be recovered as a "gel protein" fraction following centrifugation. SDS-PAGE of this fraction after reduction of disulfide bonds showed that it comprised B and D hordeins, being particularly enriched in the latter (Smith and Lister, 1983). Furthermore, the amount of N present in the gel protein fraction showed an inverse correlation with malting quality as defined by hot water extract (HWE) of malt ($r = -0.757$, $p = <0.001$, $n = 84$) (Smith and Lister, 1983). These results supported the earlier work of Baxter and Wainwright (1979b) and have since been confirmed and extended by Skerritt and Janes (1992).

The high proportion of D hordein in gel protein indicates that this component may play a specific role in the formation of the high M_r gel protein polymers, analagous to the role played by the related high molecular weight (HMW) subunits in forming high M_r wheat gluten polymers (see Shewry et al., 1995). The formation of highly cross-linked polymers would also be facilitated by the distribution of cysteine residues throughout the D hordein protein (see the previous discussion under Hordein Storage Proteins). It therefore came as no surprise when Howard et al. (1996) reported that the amount (mg/g flour) of D hordein present in three varieties of barley grown under five nutrient regimes was negatively correlated with malt extract, the effect being independent of cultivar. This clearly established a role of D hordein in determining malting quality and provided a rationale for the long-established effect of grain N on quality. However, more recent work has cast doubt on this role.

In order to determine the relationship between D hordein, gel protein, and malting performance, Brennan et al. (1998) compared the composition and performance of a series of near-isogenic pairs of lines that differed in the absence (−) or presence (+) of D hordein. These were produced by crossing a naturally occurring D hordein null line identified by screening a barley germplasm collection with European lines, followed by eight backcrosses with the European parent. Comparisons of six pairs of lines showed a statistically significant effect on gel protein (means of 0.2903 and 0.1288 g/g in D hordein + and − lines, respectively, $p = 0.003$) but no effect on HWE or other malting parameters. Similarly, large-scale malting and brewing of selected lines also failed to show any differences (Smith and co-workers, unpublished data).

A further factor that may influence the relationship between hordein and malting is the distribution of hordeins, and in particular of the polymeric hordeins, within the grain. Millet, Montembault, and Autran (1991) compared the distribution of hordeins in pearling fractions produced by sequential abrasion of one cultivar each of good and poor malting quality. They showed that a higher proportion of hordein was present in the outer region of the poor-quality cultivar, and that more of this was present as polymers. We have since carried out similar studies of the distribution of total N, hordein, gel protein, and hordein groups in pearling fractions from two good-quality (Britannia, Chariot) and two poor-quality (Derkado, Hart) cultivars (Shewry et al., 1996). The distributions were broadly similar, with gel protein being concentrated in the outer part of the starchy endosperm (fractions 2-4) compared with the aleurone layer (fraction 1) or central starchy endosperm (fraction 5).

In conclusion, several lines of evidence indicate that hordein has an effect on malting performance, which may be related to the total amount of hordein and to differences in its composition, properties, and distribution within the grain. However, these effects are complex and may be affected by the environment (notably nutrition) and genetic background (i.e., other malting characteristics of the line), making it difficult to draw firm conclusions as to the mechanisms.

β-*Amylase and Diastatic Power*

Diastatic power represents the combined activity of four enzymes, α-amylase, β-amylase, α-glucosidase, and limit dextrinase, with β-amylase being the second most active (after α-amylase). Diastatic power is particularly important when brewing with high levels of starch adjunct, and some evidence suggests that β-amylase activity may be limiting under such conditions (see Evans et al., 1997). Furthermore, it is not sufficient to select for lines with increased total levels of β-amylase activity, as part of the problem resides in the heat lability of the enzyme. Whereas α-amylase is heat stable, β-amylase activity is reduced during kilning and the bulk of the remaining activity is lost during a typical mash at 65°C (Palmer, 1989; Bamforth and Quain, 1989). Kihara, Kaneko, and Ito (1998) have demonstrated that cultivars of barley vary in the extent to which their β-amylase enzymes are heat stable, with modern Japanese varieties showing a higher level of stability than European, North American, or Australian varieties. It is also possible to increase the thermostability of barley β-amylase by mutagenesis, with seven single mutations giving increases in T_{50} (the temperature at which 50 percent of the initial activity was lost during a 30-minute period), ranging from 0.8 to 3.2°C, and an increase of 11.6°C when the seven mutations were combined (Yoshigi et al., 1995). Alternatively, heat-stable forms of β-amylase may be obtained from thermophilic microorganisms such as *Clostridium thermosulfurogenes* (Kitamoto et al., 1988). Kihara et al. (1997) have recently reported the production of transgenic barley expressing heat-stable β-amylase under the control of a barley β-amylase promoter, but the origin and characteristics of the enzyme were not stated.

Surface-Active Proteins and Foam Stabilization

Foam stability is an important character that may be enhanced by hydrophobic peptides but adversely affected by lipids. Sorensen et al. (1993) divided a beer foam fraction into HMW and low molecular weight (LMW) components by gel filtration on Sephadex G75. The HMW fraction comprised mainly carbohydrate but did contain about 10 percent protein, the major component of which was protein Z. In contrast, the LMW fraction comprised mainly protein, over half of which was LTP-1. In addition, hordein and glutelin protein fragments were also present. Foam assays showed that the LMW fraction was mainly responsible for foam potential, and the HMW fraction for foam half-life.

Barley LTP-1 binds acyl groups in a hydrophobic cavity (Heinemann et al., 1996). However, the form of LTP-1 present in beer appears to have been subtly modified during the malting and brewing process from that present in barley (Sorensen et al., 1993; Bech et al., 1995). Thus, although the two forms have identical amino acid sequences, they differ in pI, M_r, and immunoreactivity. The modified protein is also more efficient at stabilizing foam. The presence of LTP-1 in beer may result from its high level of resistance to proteolysis. Jones and Marinac (1995) showed that LTP-1 was able to inhibit the major cysteine and serine endoproteinases in malt and suggested that it could affect proteolysis during malting. However, Davy et al. (1999) recently showed that native LTP-1 does not inhibit the major malt cysteine endoproteinase EP-B, but that the denatured protein behaves as a competitive inhibitor. Denaturation of the protein during wort boiling could therefore lead to inhibition of the malt proteinases.

Wheat puroindolines also bind polar lipids to stabilize foams (Dubreil, Compoint, and Marion, 1997). Consequently, the barley homologues may also be expected to contribute to foam stabilization in beer, providing they survive degradation during malting and brewing.

Puroindolines and Grain Texture

Two distinct phenomena can result in textural differences in barley grain. Grain hardness or softness is a genetically determined phenomenon that is probably based on differences in the degree of adhesion of "matrix proteins" to the surface of the starch granule, as discussed earlier. However, the barley grains of all varieties also exhibit textural differences that are primarily, if not wholly, environmentally determined. Thus, individual grains may be vitreous or floury (also called steely or mealy) or exhibit regions of both textural types. Steeliness appears to be favored under conditions of high temperature and N availability and is associated with poor malting quality (Palmer and Harvey, 1977; Chandra, Proudlove, and Baxter, 1999). However, the molecular basis for this phenomenon is not known, and there is no obvious way to control the extent of steeliness apart from controlling the growth conditions, which is clearly not feasible in the field.

Although hardness (i.e., genetically determined differences in texture) has been neglected in barley compared to wheat, recent evidence suggests that it may affect malting performance. Because hardness relates to starch/protein adhesion, it is possible to discriminate between hard and soft kernels by scanning electron microscopy of endosperm cells after freeze fracture. Brennan et al. (1996) carried out comparative analyses of two cultivars of good malting quality and two of poor quality. In the good-quality cultivars, a clean separation between the granules and matrix proteins was observed with little or no damage to the granule integrity. In contrast, fracture often occurred across the granules of poor-quality cultivars, and more proteinaceous material appeared to be associated with the starch granule surface. These differences were associated with variations in milling energy (a measure of hardness or vitreousness; see Swanston and Ellis, Chapter 5 in this book), with the good-quality cultivars having lower values than the poor-quality cultivars (599-714 compared with 765-845 joules/mg).

The role of barley puroindolines in determining grain texture is still unclear (Darlington et al., 2000). However, it may be possible to develop soft-textured barleys with good malting performance by expressing genes for wheat puroindolines a and/or b in the developing grain.

CONCLUSIONS

The proteins deposited in the developing barley grain may affect the malting quality in various ways. Hordeins, in the form of gel protein, and starch granule proteins may affect the physical structure, preventing or facilitating the access of enzymes to the starch granules during malting and brewing, while β-amylase contributes directly to diastatic activity, and surface-active proteins to foam stabilization of the final product. Although these effects may still be poorly understood in detail, the availability of transformation systems provides an opportunity to explore their mechanisms and to make direct improvements.

REFERENCES

Abe, J., U. Sidenius, and B. Svensson (1993). Arginine is essential for the α-amylase inhibitory activity of the α-amylase/subtilisin inhibitor (BASI) from barley seeds. *Biochemical Journal* 293: 151-155.

Bamforth, C.W. and D.E. Quain (1989). Enzymes in brewing and distilling. In *Cereal Science and Technology*, ed. G.H. Palmer. Aberdeen, Scotland: Aberdeen University Press, pp. 326-366.

Baxter, E.D. and T. Wainwright (1979a). Hordein and malting quality. *Journal of American Society of Brewing Chemists* 37: 8-12.

Baxter, E.D. and T. Wainwright (1979b). The importance in malting and mashing of hordein proteins with a relatively high sulphur content. In *Proceedings of the 17th European Brewery Convention Congress, Berlin.* Dordrecht, The Netherlands: European Brewing Convention, DSW, pp. 131-143.

Bech, L.M., P. Vaag, B. Heinemann, and K. Breddam (1995). Throughout the brewing process barley lipid transfer protein 1 (LTP1) is transformed into a more foam-promoting form. In *Proceedings of the 25th European Brewery Convention, Brussels.* Oxford, UK: IRL Press, Oxford University Press, pp. 561-568.

Bernhardt, W.R. and C.R. Somerville (1989). Coidentity of putative amylase inhibitors from barley and finger millet with phospholipid transfer proteins inferred from amino acid sequence homology. *Archives of Biochemistry and Biophysics* 269: 695-697.

Bishop, L.R. (1930a). The Institute of Brewing Research scheme. 1. The prediction of extract. *Journal of the Institute of Brewing* 36: 421-434.

Bishop, L.R. (1930b). The nitrogen content and quality of barley. *Journal of the Institute of Brewing* 36: 352-369.

Brennan, C.S., N. Harris, D. Smith, and P.R. Shewry (1996). Structural differences in the mature endosperms of good and poor malting barley cultivars. *Journal of Cereal Science* 24: 171-177.

Brennan, C.S., D.B. Smith, N. Harris, and P.R. Shewry (1998). The production and characterisation of *Hor 3* null lines of barley provides new information on the relationship of D hordein to malting performance. *Journal of Cereal Science* 28: 291-299.

Breu, V., F. Guerbette, J.C. Kader, C.G. Kannangara, B. Svensson, and P. Von Wettstein-Knowles (1989). A 10 kD barley basic protein transfers phosphatidylcholine from liposomes to mitochondria. *Carlsberg Research Communications* 54: 81-84.

Briggs, D.E. (1973). Hormones and carbohydrate metobolism in germinating cereal grains. In *Biosynthesis and Its Control in Plants,* ed. B.V. Milborrow. London: Academic Press, pp. 219-277.

Briggs, D.E. (1992). Barley germination: Biochemical changes and hormonal control. In *Barley: Genetics, Biochemistry, Molecular Biology and Biotechnology,* ed. P.R. Shewry. Wallingford, UK: CAB International, pp. 369-401.

Bright, W.J. and P.R. Shewry (1983). Improvement of protein quality in cereals. In *CRC Critical Reviews in Plant Sciences.* Boca Raton, FL: CRC Press, pp. 49-93.

Bunce, N.A.C., B.G. Forde, M. Kreis, and P.R. Shewry (1986). DNA restriction fragment length polymorphism at hordein loci: Application to identifying barley cultivars. *Seed Science and Technology* 14: 419-429.

Burgess, S.R. and P.R. Shewry (1986). Identification of homologous globulines from embryos of wheat, barley, rye and oats. *Journal of Experimental Botany* 37: 1863-1871.

Campbell, A.F. (1992). Protein engineering of the barley chymotrypsin inhibitor 2. In *Plant Protein Engineering,* eds. P.R. Shewry and S. Gutteridge. Sevenoaks, UK: Edward Arnold, pp. 257-268.

Carbonero, P. and F. García-Olmedo (1999). A multigene family of trypsin/α-amylase inhibitors from cereals. In *Seed Proteins,* eds. P.R. Shewry and R. Casey. Dordrecht: Kluwer Academic Publishers, pp. 617-634.

Chalmers, K.J., U.M. Barua, C.A. Hackett, W.T.B. Thomas, R. Waugh, and W. Powell (1993). Identification of RAPD markers linked to genetic factors controlling the milling energy of barley. *Theoretical and Applied Genetics* 87: 314-320.

Chandra, G.S., M. Proudlove, and E.D. Baxter (1999). The structure of barley endosperm—An important determinant of malt modification. *Journal of the Science of Food and Agriculture* 79: 37-46.

Dahl, S.W., S.K. Rasmussen, and J. Hejgaard (1996). Heterologous expression of three plant serpins with distinct inhibitory specificities. *Journal of Biological Chemistry* 271: 25083-25088.

Dahl, S.W., S.K. Rasmussen, L.C. Petersen, and J. Hejgaard (1996). Inhibition of coagulation factors by recombinant barley serpin BSZx. *FEBS Letters* 394: 165-168.

Darlington, H.F., L. Tesci, N. Harris, D. Griggs, I. Cantrell, and P.R. Shewry (2000). Starch granule associated proteins in barley and wheat. *Journal of Cereal Science* 32: 21-29.

Davy, A., Ib. Svendsen, L. Bech, D. Simpson, and V. Cameron-Mills (1999). LTP does not represent a cysteine endoprotease inhibitor in barley grains. *Journal of Cereal Science* 30: 237-244.

Dubreil, L., J.-P. Compoint, and D. Marion (1997). Interaction of puroindolines with wheat flour polar lipids determines their foaming properties. *Journal of Agricultural and Food Chemistry* 45: 108-116.

Evans, D.E., L.C. MacLeod, J.K. Eglinton, C.E. Gibson, X. Zhang, W. Wallace, J.H. Skerritt, and R.C.M. Lance (1997). Measurement of beta-amylase in malting barley (*Hordeum vulgare* L.) I. Development of a quantitative ELISA for β-amylase. *Journal of Cereal Science* 26: 229-239.

Giese, H. and J. Hejgaard (1984). Synthesis of salt-soluble proteins in barley. Pulse-labeling study of grain filling in liquid-cultured detached spikes. *Planta* 161: 172-177.

Giroux, M.J. and C.J. Morris (1997). A glycine to serine change in puroindoline-b is associated with wheat grain hardness and low levels of starch-surface friabilin. *Theoretical and Applied Genetics* 95: 857-864.

Giroux, M.J. and C.J. Morris (1998). Wheat grain hardness results from highly conserved mutations in the friabilin components of puroindoline-a and b. *Proceedings of the National Academy of Sciences, USA* 95: 6262-6266.

Hara-Nishimura, I., M. Nishimura, and J. Daussant (1986). Conversion of free β-amylase to bound β-amylase on starch granules in the barley endosperm during desiccation phase of seed development. *Protoplasma* 134: 149-153.

Heinemann, B., K.V. Andersen, P.R. Nielsen, L.M. Bech, and F.M. Poulsen (1996). Structure in solution of a four-helix lipid binding protein. *Protein Science* 5: 13-23.

Hejgaard, J. (1977). Origin of a dominant beer protein: Immunochemical identity with a β-amylase-associated protein from barley. *Journal of the Institute of Brewing* 83: 94-96.

Hejgaard, J. (1982). Purification and properties of protein Z—A major albumin of barley endosperm. *Physiologia Plantarum* 54: 174-182.

Hejgaard, J. and S. Boisen (1980). High lysine proteins in Hiproly barley breeding: Identification, nutritional significance and new screening methods. *Hereditas* 93: 311-320.

Hejgaard, J. and P. Kaersgaard (1983). Purification and properties of the major antigenic beer protein of barley origin. *Journal of the Institute of Brewing* 89: 402-410.

Howard, K.A., K.R. Gayler, H. Eagles, and G.M. Halloran (1996). The relationship between D hordein and malting quality in barley. *Journal of Cereal Science* 24: 47-56.

Jones, B.L. and L.A. Marinac (1995). Barley LTP1 (PAPI) and LTP2: Inhibitors of green malt cysteine endoproteases. *Journal of the American Society of Brewing Chemists* 53: 194-195.

Kihara, M., T. Kaneko, and K. Ito (1998). Genetic variation of β-amylase thermostability among varieties of barley, *Hordeum vulgare* L., and relation to malting quality. *Plant Breeding* 117: 425-428.

Kihara, M., Y. Okada, H. Kuroda, K. Sacki, and K. Ito (1997). Generation of fertile transgenic barley synthesizing thermostable β-amylase. *Journal of the Institute of Brewing* 103: 153.

Kirkman, M.A., P.R. Shewry, and B.J. Miflin (1982). The effect of nitrogen nutrition on the lysine content and protein composition of barley seeds. *Journal of the Science of Food and Agriculture* 33: 115-127.

Kirsi, M. and J. Mikola (1971). Occurrence of proteolytic inhibitors in various tissues of barley. *Planta* 96: 281-291.

Kitamoto, N., H. Yamagata, T. Kato, N. Tsukagoshi, and S. Udaka (1988). Cloning and sequencing of the gene encoding thermophilic β-amylase of *Clostridium thermosulfurogenes*. *Journal of Bacteriology* 170: 5848-5854.

Kreis, M., B.G. Forde, S. Rahman, B.J. Miflin, and P.R. Shewry (1985). Molecular evolution of the seed storage proteins of barley, rye and wheat. *Journal of Molecular Biology* 183: 499-502.

Kreis, M., M. Williamson, B. Buxton, J. Pywell, J. Hejgaard, and I. Svendsen (1987). Primary structure and differential expression of β-amylase in normal and mutant barleys. *European Journal of Biochemistry* 169: 517-525.

Mikola, J. and M. Kirsi (1972). Differences between endospermal and embryonal trypsin inhibitors in barley, wheat and rye. *Acta Chemica Scandinavica* 26: 787-795.

Millet, M.O., A. Montembault, and J.-C. Autran (1991). Hordein compositional differences in various anatomical regions of the kernel between two different barley types. *Sciences des Alimentations* 11: 155-161.

Munck, L. (1972). Improvement of nutritional value in cereals. *Hereditas* 72: 1-128.

Munck, L. (1992). The case of high-lysine barley breeding. In *Barley: Genetics, Biochemistry, Molecular Biology and Biotechnology*, ed. P.R. Shewry. Wallingford, UK: CAB International, pp. 573-601.

Mundy, J. and J.C. Rogers (1986). Selective expression of a probable amylase-protease inhibitor in barley aleurone cells: Comparison to the barley amylase-subtilisin inhibitor. *Planta* 169: 51-63.

Palmer, G.H. (1989). Cereals in malting and brewing. In *Cereal Science and Technology*, ed. G.H. Palmer. Aberdeen, Scotland: Aberdeen University Press, pp. 61-242.

Palmer, G.H. and A.E. Harvey (1977). The influence of endosperm structure on the behaviour of barleys in the sedimentation test. *Journal of the Institute of Brewing* 83: 295-299.

Riggs, T.J., M. Sanada, A.F. Morgan, and D.F. Smith (1983). Use of acid gel electrophoresis in the characterisation of "B" hordein protein in relation to malting quality and mildew resistance in barley. *Journal of the Science of Food and Agriculture* 34: 576-586.

Shewry, P.R. (1993). Barley seed proteins. In *Barley: Chemistry and Technology*, eds. J. MacGregor and R. Bhatty. St. Paul, MN: American Association of Cereal Chemists, pp. 131-197.

Shewry, P.R. (1999). Enzyme inhibitors of seeds: Types and properties. In *Seed Proteins*, eds. P.R. Shewry and R. Casey. Dordrecht: Kluwer Academic Publishers, pp. 587-615.

Shewry, P.R., C. Brennan, A.S. Tatham, T. Warburton, R. Fido, D. Smith, D. Griggs, I. Cantrell, and N. Harris (1996). The development, structure and composition of the barley grain in relation to its end use properties. In *Cereals 96— Proceedings of the 46th Australian Cereal Chemistry Conference*, September 1996, ed. C. Wrigley. Sydney: Cereal Chemistry Division, pp. 158-162.

Shewry, P.R., A.J. Faulks, S. Parmar, and B.J. Miflin (1980). Hordein polypeptide pattern in relation to malting quality and the varietal identification of malted grain. *Journal of the Institute of Brewing* 86: 138-141.

Shewry, P.R. and M. Kreis (1992). Tissue-specific gene expression in the developing barley seed. In *Barley Genetics VI. Proceedings of the Sixth International*

Barley Genetics Symposium, Volume 2, ed. L. Munck. Copenhagen: Munksgaard International Publishers, pp. 725-735.

Shewry, P.R. and J.A. Lucas (1997). Plant proteins that confer resistance to pests and pathogens. *Advances in Botanical Research* 26: 135-192.

Shewry, P.R., M.J. Miles, and A.S. Tatham (1994). The prolamin storage proteins of wheat and related cereals. *Progress in Biophysics and Molecular Biology* 16: 37-59.

Shewry, P.R., H.M. Pratt, A.J. Faulks, S. Parmar, and B.J. Miflin (1979). The storage protein (hordein) polypeptide pattern of barley (*Hordeum vulgare* L.) in relation to varietal identification and disease resistance. *Journal of the National Institute of Agricultural Botany* 15: 34-50.

Shewry, P.R., A.S. Tatham, F. Barro, P. Barcelo, and P. Lazzeri (1995). Biotechnology of breadmaking: Unraveling and manipulating the multi-protein gluten complex. *Bio/Technology* 13: 1185-1190.

Shewry, P.R., M.S. Williamson, and M. Kreis (1987). Effects of mutant genes on the synthesis of storage components in developing barley endosperms. In *Developmental Mutants in Higher Plants,* eds. H. Thomas and D. Grierson. Cambridge: Cambridge University Press, pp. 95-118.

Shewry, P.R., M.S. Wolfe, S.E. Slater, S. Parmar, A.J. Faulks, and B.J. Miflin (1981). Barley storage proteins in relation to varietal identification, malting quality and mildew resistance. In *Proceedings of the 4th International Barley Genetics Symposium,* ed. R.N.H. Whitehouse. Edinburgh: Edinburgh University Press, pp. 596-603.

Skerritt, J.H. and P.W. Janes (1992). Disulphide-bonded "gel protein" aggregates in barley: Quality-related differences in composition and reductive dissociation. *Journal of Cereal Science* 16: 219-235.

Smith, D.B. and P.R. Lister (1983). Gel forming proteins in barley grain and their relationship with malting quality. *Journal of Cereal Science* 1: 229-239.

Sorensen, S.B., L.M. Bech, M. Muldbjerg, T. Beenfeldt, and K. Breddam (1993). Barley lipid transfer protein 1 is involved in beer foam formation. *MBAA Technical Quarterly* 30: 136-145.

Vallée, F., A. Kadziola, Y. Bourne, M. Juy, K.W. Rodenburg, B. Svensson, and R. Haser (1998). Barley α-amylase bound to its endogenous protein inhibitor BASI: Crystal structure of the complex at 1.9 Å resolution. *Structure* 6: 649-659.

Williamson, M.S., J. Forde, B. Buxton, and M. Kreis (1987). Nucleotide sequence of barley chymotrypsin inhibitor-2 (CI-2) and its expression in normal and high-lysine barley. *European Journal of Biochemistry* 165: 99-106.

Williamson, M.S., J. Forde, and M. Kreis (1988). Molecular cloning of two isoinhibitor forms of chymotrypsin inhibitor-1 (CI-1) from barley endosperm and their expression in normal and mutant barleys. *Plant Molecular Biology* 10: 521-535.

Yoshigi, N., Y. Okada, H. Maeba, H. Sahara, and T. Tamaki (1995). Construction of a plasmid used for the expression of a sevenfold-mutant barley β-amylase with increased thermostability in *Escherichia coli* and properties of the seven-fold-mutant β-amylase. *Journal of Biochemistry* 118: 562-567.

Yupsanis, T., S.R. Burgess, P.J. Jackson, and P.R. Shewry (1990). Characterization of the major protein component from aleurone cells of barley (*Hordeum vulgare* L.). *Journal of Experimental Botany* 41: 385-392.

Chapter 19

Changes in Malting Quality and Its Determinants in Response to Abiotic Stresses

Roxana Savin
José Luis Molina-Cano

INTRODUCTION

Barley is the fourth most important cereal in terms of world production, with a total annual production output ranging between 2 and 2.5 million Mg since 1990 (Food and Agriculture Organization [FAO] statistics, 2000). Its major use is for animal feed, followed by malt for brewing and direct human consumption. The use of barley for making fermented beverages is indicated in the earliest agricultural records (Dickson, 1979). Barley is the primary cereal used in the production of malt. Historically, this may have resulted, in part, from its availability compared to that of other cereals, but there are other important reasons for its use. Although wheat and rye also produce α- and β-amylase, which are more efficient together than separately in hydrolyzing starch, only barley has a tightly adhering lemma and palea. These structures protect the embryo during grain handling and the coleoptile during malting, resulting in more uniform germination. In addition, steeped barley grain has a firmer consistency than wheat or rye and can be handled at high moisture levels with less risk of damage (Dickson, 1979; Rasmusson, 1985). The lemma and palea, or hulls, also serve as an aid for filtering the brewing mash. The production of malt, and subsequently beer, from barley grain is a biotechnological process that is markedly dependent on the initial raw material, i.e., the barley grain used to produce malt (Cattivelli et al., 1994).

Variation in malting quality from year to year has been identified as a major problem in the brewing industry. Although variable environmental conditions during grain filling are commonly accepted to be the cause of that

---

We wish to thank Stuart Swanston and Gustavo Slafer for critical reading of this chapter and for their suggestions.

variation, there appears to be little information on the relationships between grain weight, quality, and the climatic conditions during barley grain growth. The reports in the literature on how weather conditions affect grain composition are usually unclear. Most of the research had sowing dates, locations, or years as treatments; the weather conditions during grain-filling were not properly defined; and the final results were usually a measure of interactions among several (including some undetermined) factors.

Among abiotic stresses, high temperature and drought are two of the major environmental factors affecting grain yield and possibly quality (Conroy et al., 1994; Paulsen, 1994; Reynolds et al., 1994). Periods of high temperature and drought are quite common during the grain-filling period in many cereal-growing regions of the world (Stone and Nicolas, 1994; Savin, 1996). Maximum day temperatures frequently rise above 30°C and can reach 40°C (Wardlaw and Wrigley, 1994). Moreover, climatic changes such as general warming and greater incidences of periodic high temperatures and drought are also likely to influence growth and development, reducing grain yield and possibly quality (Conroy et al., 1994).

Most of the experiments concerned with the effects of high temperatures and drought during grain filling on grain growth and quality of temperate cereals have been carried out on wheat. Little is known about the effects of high temperature and drought on grain growth, grain composition, and malting quality in barley (Jenner, 1994). In addition, because the commercial requirements of grain quality are different for wheat (i.e., bread/noodle quality, with grains milled to produce flour) and barley (i.e., malting quality, for which grains must germinate to produce malt), it is not possible to draw many conclusions concerning the effects of high temperature on malting quality from wheat studies.

The work presented in this chapter aims to synthesize the knowledge of the effects of these stresses by examining the responses of grain components and grain quality to short periods of high temperature and drought during the grain-filling period.

PHASES OF GRAIN GROWTH

Prior to fertilization, the embryo sac of most angiosperms contains four different types of cells, i.e., the egg cell, the two synergids, the antipodals, and the central cell (Bosnes, Weideman, and Olsen, 1992). The early events taking place in the barley embryo sac after fertilization have been described in detail by several workers (Pope, 1937; Cass and Jensen, 1970; Mogensen, 1982; Engell, 1989).

The fertilization of the barley ovule gives rise to two tissues, the diploid embryo and the triploid endosperm. The embryo differentiates further into the embryonic axis (which gives rise to the seedling) and a single cotyledon, the scutellum, which contains some storage reserves but also functions as a source of enzymes for digestion of the endosperm reserves during germination (Kreis and Shewry, 1992). The endosperm differentiates further into the starchy endosperm and the aleurone layer. The former is the major storage tissue and consists at maturity of dead cells distended by starch granules in a protein matrix. The aleurone consists of up to four layers of small cells surrounding the starchy endosperm that remain alive during maturation and produce hydrolytic enzymes during germination (Kreis and Shewry, 1992).

The growth of the caryopsis after anthesis can be divided into two main stages, cell division and grain filling (Jenner, Ugalde, and Aspinall, 1991). Cell division commences at fertilization and is complete within approximately 20 to 25 days, depending on the temperature. Grain filling commences five to ten days after anthesis and occupies the rest of the period until the grain ripens. Grain filling is the deposition of polymeric products in the cells and organelles formed during the cell division phase.

The duration of grain growth depends strongly on temperature (Jenner, 1994) and environmental stresses such as drought (Nicolas, Gleadow, and Dalling, 1984). The duration of 50 days (see Figure 19.1) would correspond to a day/night temperature regime of 18/13°C, which is close to the optimum for wheat and presumably also for barley (Chowdhury and Wardlaw, 1978). Expressing the timing of events on a relative scale (see Figure 19.1) allows easier comparison between experiments differing, for example, in temperature regimes. The lower part of Figure 19.1 shows the events that take place during the cell division and grain-filling phases in the endosperm of the barley grain.

Cell Division

The first division of the primary endosperm nucleus occurs about 14 h after fertilization (Engell, 1989). Four days later, after several cycles of nuclear division, the multinucleate syncytium is formed, i.e., a multinucleate mass of protoplasm without cell walls. At approximately five days after anthesis, the syncytium becomes vacuolar, and soon after that, the cellularization stage is initiated by the appearance of the cell walls separating the nuclei in the endosperm syncytium (Bosnes, Weideman, and Olsen, 1992; see Figure 19.1). This process first appears in the syncytium adjacent to the ventral crease, then spreads laterally along the endosperm wings to the dorsal side of the grain (Bosnes, Weideman, and Olsen, 1992).

By the time the cellularization stage is complete, the differentiation of aleurone cells is already apparent in the ventral crease area (Bosnes, Weideman, and Olsen, 1992). In the starchy endosperm, cell division ceases between 14 and 20 days after anthesis in wheat (Evers, 1970; Wardlaw, 1970)

FIGURE 19.1. Dry-Matter Accumulation for Different Parts of a Barley Grain from Anthesis to Maturity

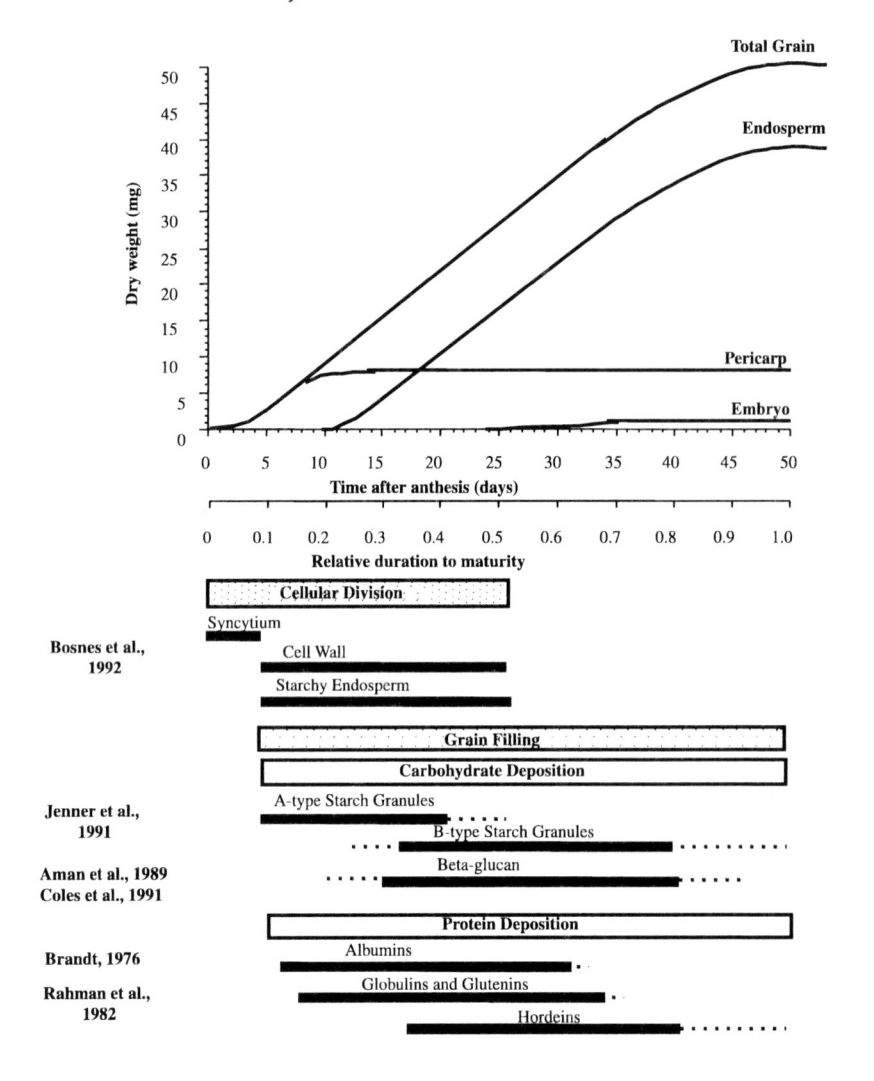

Source: Adapted from Savin (1996, p. 7).

Note: The lower part of the figure shows the timing of different events that occur in the endosperm. Stippled bars indicate subphases of the grain growth period. Solid bars represent the timing of different events. Dotted lines indicate unclear start or finish times of particular events.

and between 14 and 30 days after anthesis in barley (Cochrane and Duffus, 1981; Bosnes, Weideman, and Olsen, 1992). Expressing these results on a relative scale indicates that cell division in the starchy endosperm ceases between 0.35 and 0.5 (see Figure 19.1), depending on the cultivar (Gleadow, Dalling, and Halloran, 1982; Cochrane and Duffus, 1983), environmental conditions (Wardlaw, 1970; Brocklehurst, Moss, and Williams, 1978), and differences inherent in cell-counting methods (Duffus and Cochrane, 1992). The maximum number of endosperm cells varies, depending on cultivar (Brocklehurst, 1977; Gleadow, Dalling, and Halloran, 1982; Cochrane and Duffus, 1983), environmental conditions (Wardlaw, 1970; Brocklehurst, Moss, and Williams, 1978), and grain position on the ear (Singh and Jenner, 1982).

The maximum number of endosperm cells appears to be closely related to final grain weight in wheat (Brocklehurst, 1977) and in barley (Cochrane and Duffus, 1983). It is therefore of interest to know what factors control the rate and duration of endosperm cell division and what determines whether a cell differentiates into a starchy endosperm cell or into an aleurone cell (Duffus and Cochrane, 1992).

A rapid increase in water content of the grain occurs during the phase of endosperm cell division (Sofield, Wardlaw, et al., 1977), and the endosperm volume also increases rapidly (Renwick and Duffus, 1987). However, the increase in volume is due more to an increase in the number of endosperm cells than to an increase in the volume of individual cells during the period of cell division (Briarty, Hughes, and Evers, 1979). By the end of the cell division phase, the grain has achieved its maximum length and further increase in size is by cell enlargement.

Grain Filling

More than 80 percent of the increase in individual cell size occurs after the end of cell division (Hughes, 1976, cited by Nicolas, Gleadow, and Dalling, 1984). Cell volume continues to increase until around 35 days after anthesis without any change in grain water content (Sofield, Evans, et al., 1977). The volume and dry weight of the grain reach a maximum at about the same time, and this is followed by a rapid decrease in water content and grain volume that corresponds to grain maturation (Jennings and Morton, 1963; Sofield, Wardlaw, et al., 1977). The buildup of lipids, which is observed toward maturity in the chalazal zone of the grain, may be partly responsible for restricting the supply of water to the endosperm (Sofield, Wardlaw, et al., 1977b) but is unlikely to be the only cause of grain maturation (Nicolas, Gleadow, and Dalling, 1984).

Carbohydrate Accumulation

Starch accounts for most (75 to 85 percent) of grain dry weight at maturity, and, consequently, yield largely reflects starch accumulation in the endosperm (Jenner, Ugalde, and Aspinall, 1991). Starch is deposited in amyloplasts as granules with an ordered crystalline structure. Two distinct populations of starch granules are present in the mature endosperm of wheat and barley, i.e., large, lenticular-shaped granules (A-type) and small, spherical granules (B-type). The A-type granules appear a few days after anthesis, and their formation stops before the end of cell division (see Figure 19.1). Consequently, the endosperm cells formed last contain fewer A-type granules than do older cells (Jenner, Ugalde, and Aspinall, 1991). B-type starch granules first appear at about 15 days after anthesis. Although early work suggested that B-type granules bud off from the larger A-type granules (Buttrose, 1960), more recently, Williams and Duffus (1977) have found that the small amyloplasts do not arise from the large amyloplasts but constitute a separate population. At maturity, B-type granules greatly outnumber the A-type granules, typically by a ratio of 10:1 (Jenner, Ugalde, and Aspinall, 1991). The B-type granules are much smaller (<10 µm) than the A-type granules (20-45 µm), and they generally account for less than 30 percent of the total weight of accumulated starch (Jenner, Ugalde, and Aspinall, 1991).

The starch of mature barley grains contains about 25 percent amylose and 75 percent amylopectin (Morrison, Milligan, and Azudin, 1984), although waxy varieties contain between 2 and 10 percent amylose (Tester et al., 1991) and high-amylose types can contain more than 38 percent amylose (Morrison, Milligan, and Azudin, 1984; Swanston and Ellis, Chapter 5, and Ullrich, Chapter 6, this book). Banks, Greenwood, and Muir (1973) showed that amylose content increases steadily during grain growth. Thus, it seems likely that amylopectin synthesis predominates at the early stages of grain growth. The increase in amylose content might be caused by a change in the relative composition of all granules, or it may be due to changes in relative numbers of the A- and B-type starch granules (Duffus and Cochrane, 1992).

The nonstarch polysaccharides found in mature barley grains include fructans, cellulose, (1→3, 1→4)-β-D-glucan (subsequently referred to as β-glucan), arabinoxylan, and glucomannan (Duffus and Cochrane, 1992; Swanston and Ellis, Chapter 5, and Ullrich, Chapter 6, this book). Over 96 percent of total grain cellulose is present in the husk, with very little cellulose in the endosperm cell walls. The cell walls of the starchy endosperm contain approximately 70 percent β-glucan and 20 percent arabinoxylan, whereas the aleurone cell walls contain about 26 percent β-glucan and 67 percent arabinoxylan (Duffus and Cochrane, 1992). Some evidence suggests that β-glucan deposition occurs relatively late in grain growth, since dry-matter accumulation is almost complete before most of the β-glucan is synthesized (Aman, Graham, and Tilly, 1989; Coles, Jamieson, and Haslemore, 1991).

Protein Accumulation

Proteins account for about 10 percent of the dry weight of mature barley grain and are usually classified into several solubility groups (Osborne, 1924), i.e., albumins (water soluble), globulins (soluble in dilute saline), prolamins (soluble in alcohol-water mixtures), and glutelins (soluble only in dilute acid or alkali). The prolamins, called hordeins, are the major storage proteins in the barley grain. The albumins, globulins, and glutelins consist predominantly of structural and metabolic proteins, although they also include some storage proteins (Kreis and Shewry, 1992; Shewry and Darlington, Chapter 18 in this book).

Most of the albumins are accumulated during early stages of development, while globulins, glutelins, and particularly hordeins accumulate later in grain development (Brandt, 1976). Globulins and glutelins have been detected early in grain filling (Rahman, Shewry, and Miflin, 1982), but hordeins can be detected by SDS-PAGE (second-division segregation–polyacrylamide gel electrophoresis) only at about 14 days after anthesis, when the endosperm has reached about 10 percent of its final weight (Rahman, Shewry, and Miflin, 1982) (see Figure 19.1). Since hordeins exert an important influence on the technological characteristics of the mature barley grain, there has been much recent interest in the mechanisms controlling their synthesis and deposition (Rahman, Shewry, and Miflin, 1982; Kreis and Shewry, 1992; Shewry, 1993; Swanston et al., 1997). Hordeins account for 30 to 50 percent of total protein at maturity. Hordeins comprise two main subfractions, i.e., B hordein (sulfur rich, average molecular weight [MW] 40,000) and C hordein (sulfur poor, average MW 70,000) and two minor subfractions, i.e., γ hordein (sulfur rich, average MW 15,000) and D hordein (MW 100,000) (Rahman, Shewry, and Miflin, 1982; Howard et al., 1996). The B and C hordeins account for 75 to 90 and 10 to 20 percent, respectively, while γ and D hordeins form less than 5 percent of the total hordein fraction (Kreis and Shewry, 1992; Howard et al., 1996; Shewry and Darlington, Chapter 18 in this book).

The developing barley grain also contains a number of enzymes usually associated with starch degradation (Swanston and Ellis, Chapter 5 in this book). These include α- and β-amylases, α-glucosidase, and limit dextrinase (Duffus and Cochrane, 1993). MacGregor and Dushnicky (1989) have shown that two groups of α-amylase are present in different parts of the caryopsis. The group designated α-amylase 1 is found in the pericarp early in grain development, and its activity falls rapidly thereafter. α-Amylase 1 is also synthesized in the germinating embryo. The other group, α-amylase 2, is produced by the aleurone during seedling growth. Distinct isoforms can also be detected within each α-amylase group (Fincher and Stone, 1993), and up to 12 α-amylase isoforms, in total, have been resolved by isoelectric focusing (Simon and Jones, 1988).

In ungerminated barley grain, β-amylases may account for 1 to 2 percent of total protein in the starchy endosperm (Fincher and Stone, 1993; Swanston and Ellis, Chapter 5 in this book; and Shewry and Darlington, Chapter 18 in this book). In contrast to the majority of hydrolytic enzymes responsible for endosperm mobilization in barley, the β-amylases are synthesized exclusively during grain matuation in the starchy endosperm, rather than after the initiation of germination in the aleurone or scutellum (Fincher and Stone, 1993).

Limit dextrinase and α-glucosidase are present in small amounts in ungerminated grains (Fincher and Stone, 1993). During germination, rapid synthesis of these enzymes is observed, and both the aleurone layer and the embryo appear to be involved.

BARLEY GRAIN COMPOSITION

The major components of a barley grain are starch (50 to 70 percent of dry weight), protein (9 to 12 percent of dry weight), and β-glucans (3 to 5 percent of dry weight) (MacGregor and Fincher, 1993; Swanston and Ellis, Chapter 5 in this book). Some variations in these proportions can be found in the literature, due to differences in genotype (Kenn, Dagg, and Stuart, 1993; Allan et al., 1995), environmental conditions (Morgan and Riggs, 1981; Coles, Jamieson, and Haslemore, 1991), and measuring techniques, particularly for starch (McDonald and Stark, 1988; MacGregor and Fincher, 1993). The emphasis in the present chapter is to determine whether, in addition to their effect on yield, periods of high temperature and drought during grain filling can alter the composition of the barley grain, which could, in turn, alter malting quality.

The different components of a barley grain are synthesized in sequence, and sometimes synthetic pathways overlap (see Figure 19.1). Consequently, if a short period of high temperature or drought occurs at a certain point in the grain-filling period, it may affect one or more components that are being synthesized concurrently and result in a different composition of the mature grain.

Effect of Brief Periods of High Temperature

Table 19.1 summarizes the experiments from the literature that have been carried out with short periods of high temperature during grain filling in barley. This type of stress reduces grain weight by 5 to 30 percent, depending on the cultivar, time of exposure, and duration of the stress.

Pooling the data from some of the experiments presented in Table 19.1 showed a general positive relationship between grain weight and starch con-

TABLE 19.1. Reductions in Individual Grain Weight (IGW) Caused by Short Periods of High Temperature During Grain Filling in Barley

IGW (%)	Factors investigated	Temperature day/night (°C)	Duration days	Duration h·d⁻¹	Type of experiment	Reference
25-28*	Genotype	30/30	7	24	Chamber	[1]
5-9	Timing of exposure	35/25	5	5	Chamber	[2]
4-23	Duration of exposure	40/15	5-10	6	Glasshouse	[3]
13-24	Genotype	40/20	5	6	Field	[4]
34		40/16	5	6	Chamber	[5]
4-13	Temperature regime	40/16	5	6	Chamber	[6]
30*		35/25	3	12	Chamber	[7]
13-24	Genotype	35/25	3	12	Chamber	[8]
5-24	Timing of exposure	40/15	5	6	Chamber	[9]

*Starch dry weight per endosperm

[1] MacLeod and Duffus (1988)
[2] MacNicol et al. (1993)
[3] Savin and Nicolas (1996)
[4] Savin, Stone, and Nicolas (1996)
[5] Savin et al. (1997a)
[6] Savin et al. (1997b)
[7] Wallwork et al. (1998a)
[8] Wallwork et al. (1998b)
[9] Savin and Nicolas (1999)

tent per grain ($R^2 = 0.74$, $P < 0.001$). This is consistent with the main effect of heat stress operating through the metabolism of starch in the grains (see Figure 19.2a). Inspecting the data more closely, it was clear that two sets of data are differentiated (triangles and circles versus squares) in the content of starch per grain, but within each group the association between starch content and grain weight was maintained (see Figure 19.2a). However, when starch was considered as a percentage of the total dry-matter of the grain, the relative change in the two parameters, in response to the treatments, was different. In both data sets, starch was slightly but consistently more sensitive than total dry-matter. Consequently, the percentage of starch tended to decrease as grain dry-matter was reduced by heat stress (see Figure 19.2b).

Starch is the major source of fermentable sugars during malting, and its properties are critical to achieving optimum malt extract values (Allan et al., 1995). Starch in the endosperm is arranged in granules that follow a bimodal distribution with two clear peaks, one for large, A-type granules and the other for small, B-type granules, as shown in Figure 19.1. In some of the experiments reported in Table 19.1, there were reductions due to heat stress in

FIGURE 19.2. Relationship Between Starch in (a) Absolute and (b) Relative Content and Grain Weight for Heat Stress Experiments

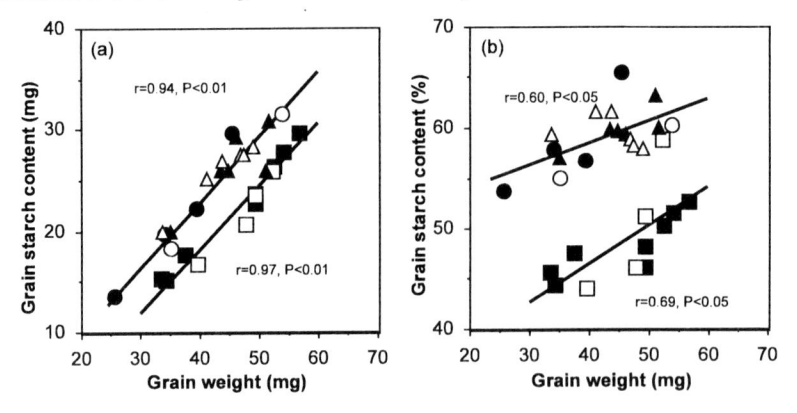

Source: The data are from some of the experiments reported in Table 19.1: ● Savin, Stone, and Nicolas (1996); ○ Savin et al. (1997a); ■ Savin et al. (1997b); □ Savin and Nicolas (1999); ▲ Wallwork et al. (1998b); △ Savin and Nicolas (1996).

the number of both A- and B-type starch granules. The reductions in grain weight were, however, more related to the reduction in the number of B-type ($R^2 = 0.70, P < 0.001$) than A-type starch granules ($R^2 = 0.19$, not significant [ns]). This does not mean that the B-type granules are intrinsically more sensitive than the A-type but is a reflection of the timing of the stress (i.e., after 15 days after anthesis), which coincided with the initiation of B-type starch granules (see Figure 19.1). For example, when high temperatures were applied early enough, i.e., ten days after anthesis (Savin and Nicolas, 1999), the number of A-type starch granules was more affected by heat stress than that of B-type granules.

Some genotypic differences were found in the diameter and volume of A-type granules among malting barley cultivars (Allan et al., 1995; Savin, 1996) and may be related to levels of malt extract. It is suggested that barley samples with a greater proportion of larger diameter starch granules could potentially produce more fermentable sugars, as they would have a greater percentage of starch by volume (Allan et al., 1995).

It is commonly accepted that accumulation of starch is more sensitive to high temperature than the accumulation of nitrogen (Bhullar and Jenner, 1985; Jenner, Ugalde, and Aspinall, 1991), which frequently determines increases in grain nitrogen percentage. Most of the experiments described in Table 19.1 showed decreases in grain nitrogen proportion when grain weight was reduced as a consequence of heat stress (see Figure 19.3a). However,

FIGURE 19.3. Relationship Between Grain Nitrogen in (a) Absolute and (b) Relative Content and Grain Weight for Heat Stress Experiments

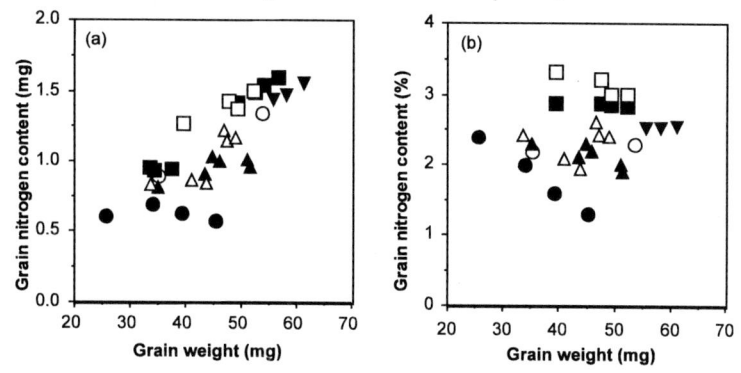

Source: The data are from some of the experiments reported in Table 19.1: ● Savin, Stone, and Nicolas (1996); ○ Savin et al. (1997a); ■ Savin et al. (1997b); □ Savin and Nicolas (1999); ▲ Wallwork et al. (1998b); △ Savin and Nicolas (1996); ▼ MacNicol et al. (1993).

the relative amount of nitrogen did not show consistent differences between treatments and across studies (see Figure 19.3b). In only one experiment (Savin, Stone, and Nicolas, 1996) was a significant increase in nitrogen percentage found when heat stress was applied (see Figure 19.3b, closed circles), as nitrogen accumulation appeared not to have been affected in this particular case (see Figure 19.3a).

It has been reported that the concentration of β-glucans in the grain is dependent on both genotypic factors and environmental conditions (Stuart, Loi, and Fincher, 1988; Coles, Jamieson, and Haslemore, 1991; Kenn, Dagg, and Stuart, 1993; Swanston et al., 1997). The absolute content of β-glucan in the grains was reduced in response to short periods of very high maximum temperature. Thus, a positive relationship between β-glucan and final grain weight was observed as a consequence of heat stress (see Figure 19.4a). As the relative effect of these treatments on both β-glucan and grain weight was similar, the percentage of β-glucan did not consistently change when grain weight was reduced by heat stress (see Figure 19.4b). However, in some particular studies (Savin et al., 1997a, b; Savin and Nicolas, 1999), a trend to reduce grain β-glucan percentage with heat stress occurred together with a dramatic reduction in the duration of grain-filling. It is possible that the reduction in β-glucan content was related to the reduced duration of the grain-filling period, since β-glucan deposition in the grain occurs relatively late during grain filling (Swanston et al., 1997).

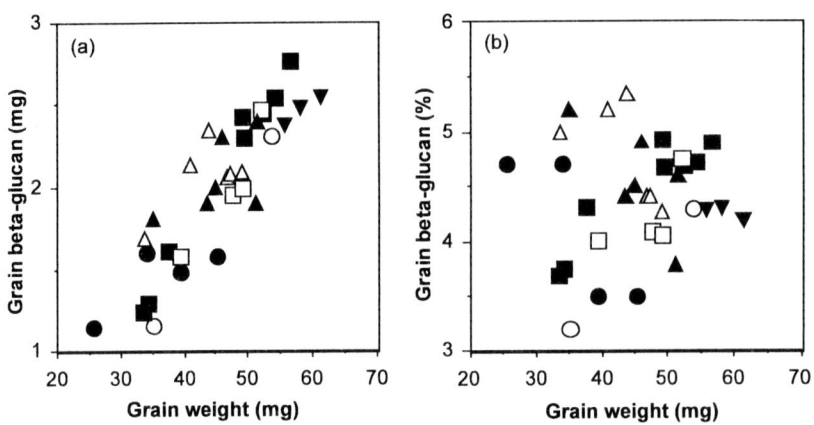

FIGURE 19.4. Relationship Between Grain β-Glucan in (a) Absolute and (b) Relative Content and Grain Weight for Heat Stress Experiments

Source: The data are from some experiments reported in Table 19.1; symbols as in Figure 19.3.

Effect of Brief Periods of Drought

Little is known about the effects of postanthesis drought on grain quality in cereals. One problem with many drought experiments is that the intensity and timing of the stress, relative to grain growth, are often poorly defined, so that results are difficult to interpret or compare. Only five reported studies appear to have examined the effects of drought on malting quality (see Table 19.2).

Once again, the effect observed on starch content, measured in absolute values, was offset by changes in total dry matter, so the percentage of starch showed no consistent relationship with grain weight when affected by drought (see Figure 19.5). This conclusion should, however, be viewed with caution, as only two experiments analyzed starch content.

When severe (Savin and Nicolas, 1996) or mild (Savin and Nicolas, 1999) droughts were applied, there was a tendency for β-glucan content in the grains to be reduced. This is in agreement with results reported under field (Stuart, Loi, and Fincher, 1988; Coles, Jamieson, and Haslemore, 1991) and growth cabinet conditions (MacNicol et al., 1993). However, Morgan and Riggs (1981) found that barley extract viscosity (an indicator of β-glucans) increased with drought applied from heading onward. The amount of grain β-glucan was strongly correlated with grain weight under drought (see Figure 19.6a; $R^2 = 0.87$, $P < 0.001$), but grain β-glucan percentage did not show a clear response under drought (see Figure 19.6b).

TABLE 19.2. Reductions in Individual Grain Weight (IGW) Caused by Short Periods of Drought During Grain Filling in Barley

IGW (%)	Factors investigated	Temperature day/night (°C)	Intensity days	Intensity RWC (%)	Type of experiment	Reference
3-6	Timing of exposure	24*	3	nr	Chamber	[1]
12	Timing of exposure	15*	nr	nr	Field	[2]
7-29	Timing of exposure	21/16	4	22-51	Chamber	[3]
20	Genotype	20/15	10	50	Glasshouse	[4]
2-17	Timing of exposure	20/15	10	70-80	Glasshouse	[5]

*Mean temperature of the whole period
RWC = relative water content
nr = not reported
[1] Morgan and Riggs (1981)
[2] Coles, Jamieson, and Haslemore, (1991)
[3] MacNicol et al. (1999)
[4] Savin and Nicolas (1996)
[5] Savin and Nicolas (1999)

FIGURE 19.5. Relationship Between Starch in (a) Absolute and (b) Relative Content and Grain Weight Under Drought

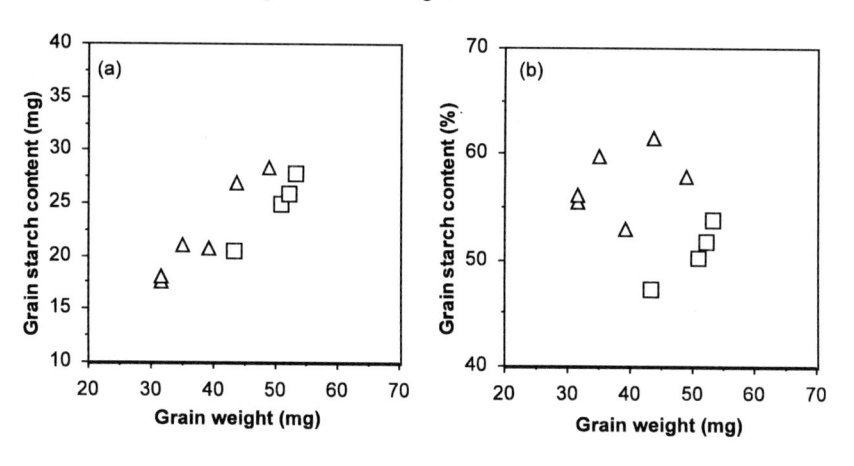

Source: Data are from the experiments reported in Table 19.2: △ Savin and Nicolas (1996b); □ Savin and Nicolas (1999).

FIGURE 19.6. Relationship Between Grain ß-Glucan in (a) Absolute and (b) Relative Content and Grain Weight Under Drought

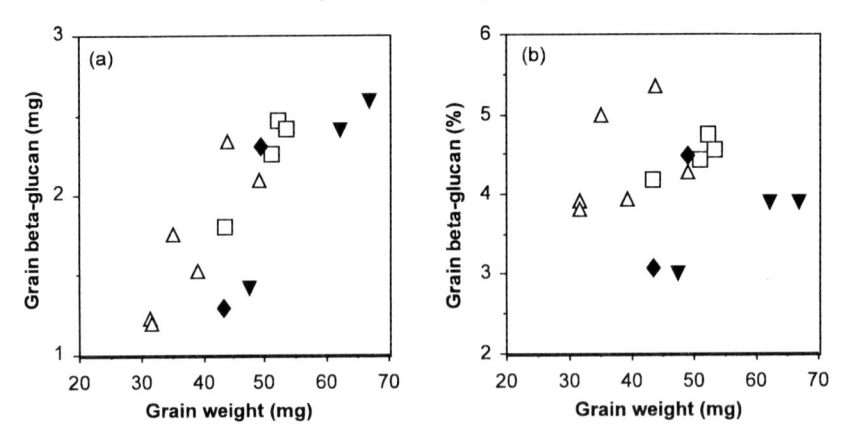

Source: Data are from the experiments reported in Table 19.2: □ Savin and Nicolas (1999); △ Savin and Nicolas (1996); ▼ MacNicol et al. (1993); ◆ Coles, Jamieson, and Haslemore (1991).

Most of the experiments reported in Table 19.2 found grain nitrogen content, in absolute terms, to be similar in the controls and the drought treatments. However, across all the experiments there was a linear relationship between grain nitrogen content (in mg) and grain weight (see Figure 19.7a; $R^2 = 0.45, P < 0.1$), but in most cases, grain nitrogen percentage was similar under drought conditions (see Figure 19.7b).

To summarize briefly, short periods of high temperature and drought not only reduced the final grain weight but also altered the main components of the barley grain (see Figures 19.2-19.7). It is important, however, to determine if, and to what extent, these changes relate to changes in malting quality.

QUALITY SPECIFICATIONS

There is no simple, clear group of variables that is unanimously regarded as defining the quality of the grain or malt (Molina-Cano, 1987). Rather, quality requirements in malting barley represent a consensus of the specifications required by commercial brewers to produce their products in an efficient manner consistent with desired product properties or traditional methodologies (Burger and LaBerge, 1985). Moreover, breweries have different specifications, and there are certain ranges for each quality factor within which each brewer can operate by adjusting conditions or by tolerating

FIGURE 19.7. Relationship Between Grain Nitrogen in (a) Absolute and (b) Relative Content and Grain Weight Under Drought

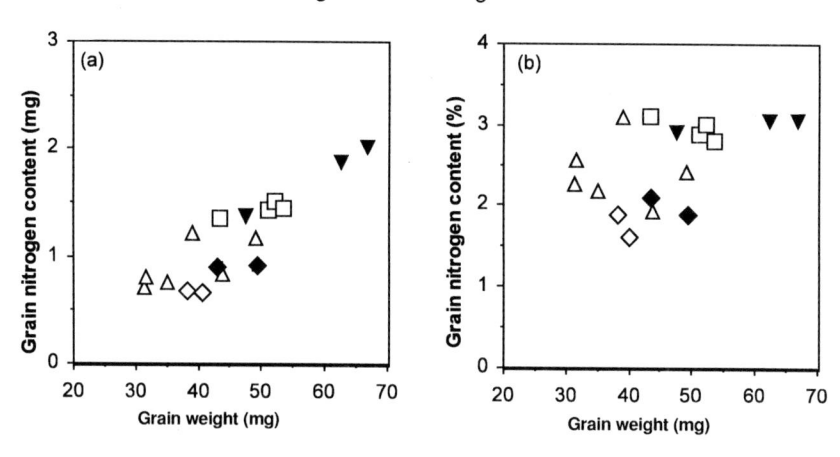

Source: Data are from the experiments reported in Table 19.2: □ Savin and Nicolas (1999); △ Savin and Nicolas (1996); ▼ MacNicol et al. (1993); ◆ Coles, Jamieson, and Haslemore (1991); ◇ Morgan and Riggs (1981).

some variation in the product (Burger and LaBerge, 1985). The number of actual specifications can vary from 8 to 35 (Cole, 1993). In this chapter, only a few of the most important parameters of malting quality are taken into account. These include malt extract, diastatic power and β-glucan content in the malt and starch, as well as nitrogen and β-glucan content in the grain. The effects of the treatments on these components of the grain were discussed in the previous section and are considered here only as they relate to malt extract. From the characteristic of the malt, malt extract is undoubtedly accepted as the principal indicator of malting quality (Molina-Cano, 1987; Coles, Jamieson, and Haslemore, 1991).

From the experiments reported in Tables 19.1 and 19.2, malt extract varied from around 80 to 55 percent, most of the data being between 65 and 68 percent. These values are below what would be desirable for the industry (Molina-Cano, 1989; Henry, 1990). However, it is not uncommon to have values within this range, even from field-grown barley (e.g., Stuart, Loi, and Fincher, 1988; Eagles et al., 1995). In addition, it is appropriate to note that the methodology to determine malt extract used in some of these experiments (Institute of Brewing [IOB], 1982) utilizes a unithermal extraction and may give lower values than commercial malting (Henry and McLean, 1984; Skerritt and Janes, 1992) or other methods (European Brewing Convention [EBC], 1975).

RESPONSES OF SOME ASPECTS OF GRAIN QUALITY TO HIGH TEMPERATURE AND DROUGHT

Relationship Between Malt Extract and Grain Starch Concentration

There was a clear trend between malt extract and starch content under high temperatures ($R^2 = 0.64$, $P < 0.001$; see Figure 19.8a) and drought ($R^2 = 0.71$, $P < 0.05$; see Figure 19.8b) over a wide range of starch contents. However, no relationship was found between these two variables within some of the individual experiments (e.g., see open and closed squares in Figure 19.8).

Relationship Between Malt Extract and β-Glucan Degradation

As discussed earlier, the starchy endosperm cell walls represent a potential barrier, limiting access of hydrolytic enzymes to their substrates within the cells. For this reason, cell wall degradation is considered an important early event in germination or in the malting process (Stuart, Loi, and Fincher, 1988).

FIGURE 19.8. Relationship Between Malt Extract and Starch Content for the (a) Heat Stress and (b) Drought Experiments

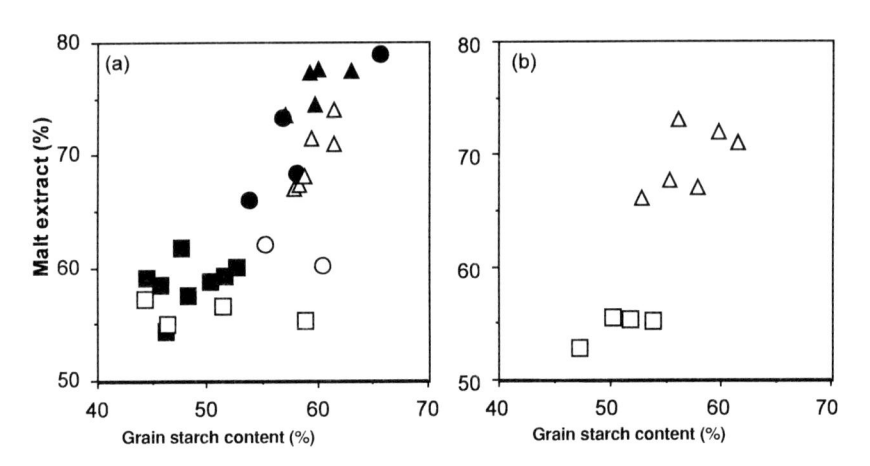

Source: The data are from some of the experiments reported in Tables 19.1 and 19.2: ● Savin, Stone, and Nicolas (1996); ○ Savin et al. (1997a); ■ Savin et al. (1997b); □ Savin and Nicolas (1999); ▲ Wallwork et al. (1998b); △ Savin and Nicolas (1996).

Stuart, Loi, and Fincher (1988) proposed that the difference between the percentage of β-glucans in grain and in malt, when the latter is given as a percentage of the former, is an estimate of the β-glucan degradation during malting. Although there was, generally, a positive relationship between malt extract and β-glucan degradation, this relationship was not statistically significant ($R^2 = 0.14$, ns; see Figure 19.9a), due to the data from Wallwork et al. (1998b), which presented high values of malt extract but lower values of β-glucan degradation (see Figure 19.9a, closed triangles). In the case of the drought experiments, there was a linear relationship between malt extract and β-glucan degradation ($R^2 = 0.87$, $P < 0.05$; see Figure 19.9b), but this was based on only two sets of data.

Relationship Between Malt Extract and Grain Nitrogen Concentration

The grain nitrogen content (percent of dry matter) in the control plants of the experiments reported in Tables 19.1 and 19.2 ranged from 1.3 (Savin, Stone and Nicolas, 1996) to 3 (MacNicol et al., 1993, Saven et al., 1997b) percent. Grain nitrogen content shouldbe around 1.6 to 1.8 percent to be widely accepted by the malting industry. It is generally reported that there is a strong negative relationship (Bishop's equation) between malt extract and nitrogen (or protein) content (Smith, 1990; Eagles et al., 1995; Howard

FIGURE 19.9. Relationship Between Malt Extract and β-Glucan Degradation for the (a) Heat Stress and (b) Drought Experiments

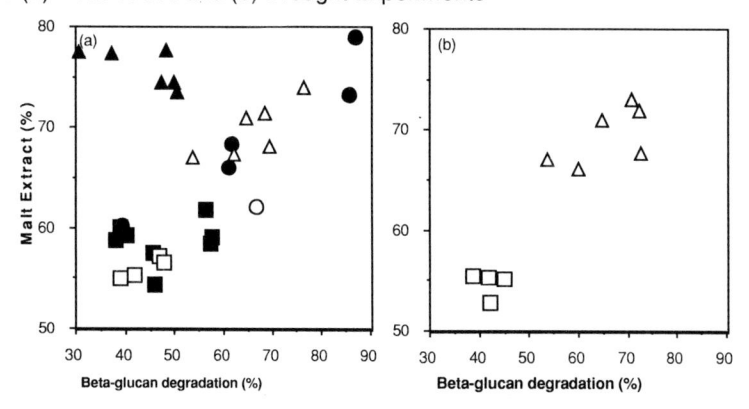

Source: The data are from some of the experiments reported in Tables 19.1 and 19.2: ● Savin, Stone, and Nicolas (1996); ○ Savin et al. (1997a); ■ Savin et al. (1997b); □ Savin and Nicolas (1999); ▲ Wallwork et al. (1998b); △ Savin and Nicolas (1996).

et al., 1996), although this negative relationship is cultivar dependent (Howard et al., 1996) and is not always significant. For example, seasonal differences in protein concentration in Mediterranean climates of Australia (Eagles et al., 1995). Similar findings about the relative influence of protein on malt extract in Spain, in contrast to Scotland, were demonstrated (Molina-Cano et al., 1995). Differences in total protein content between the two environments did not adequately account for extract yield differences. Spanish samples with 12 percent total protein content yielded more extract than their Scottish counterparts, with 10 percent. It was therefore suggested that other grain characters, including hordein and β-glucan composition, and differences in germination behavior due to the very contrasting environments could be responsible for an apparent deviation from Bishop's equation.

There was a negative linear relationship between malt extract and nitrogen content for the heat stress experiments (see Figure 19.10a; $R^2 = 0.59$; $P < 0.001$) and drought experiments (see Figure 19.10b; $R^2 = 0.30, P < 0.1$).

Across all the experiments reported in Tables 19.1 and 19.2, malt extract was consistently reduced with increases in nitrogen percentage, even in a range of relatively high grain nitrogen concentrations. However, no relationship between these parameters was found when some of the experiments were considered separately (see symbols within Figure 19.10). This

FIGURE 19.10. Relationship Between Malt Extract and Grain Nitrogen Concentration for the (a) Heat Stress and (b) Drought Experiments

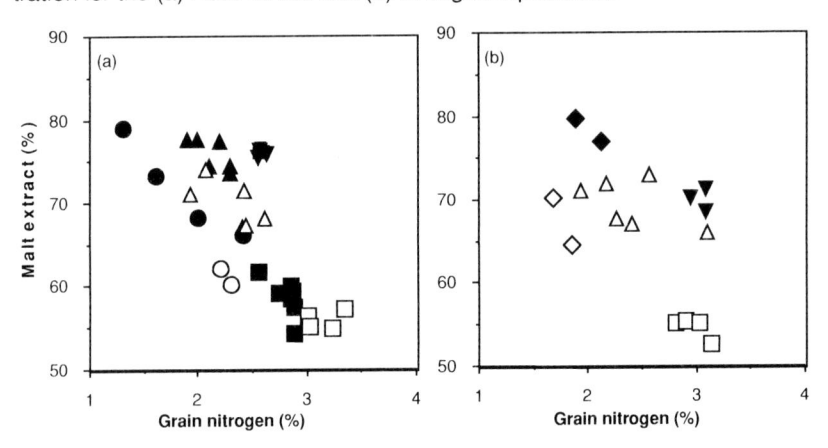

Source: The data are from some of the experiments reported in Tables 19.1 and 19.2: ● Savin, Stone, and Nicolas (1996); ○ Savin et al. (1997a); ■ Savin et al. (1997b); □ Savin and Nicolas, (1999); ▲ Wallwork et al. (1998b); △ Savin and Nicolas (1996); ▼ MacNicol et al. (1993); ♦ Coles, Jamieson, and Haslemore (1991); ◇ Morgan and Riggs (1981).

suggests that short periods of heat stress and drought do not consistently change malt extract, even though grain nitrogen content was increased in several of the experiments reported earlier.

The conclusion from the previous analysis is that grain nitrogen concentration is an important factor affecting malting quality, and that when the environmental factors influence the availability of nitrogen (as in Howard et al., 1996, and Eagles et al., 1995), malt extract is negatively correlated with nitrogen percentage. It was also found that pooling data from all the experiments reported in this chapter gave a negative relationship between malt extract (ME) and grain nitrogen concentration (N) under stress, the relationship being very similar to those published in the literature (ME = 94.85 − 11.24N, $R^2 = 0.38$, $P < 0.05$). Despite this, considering the experiments separately, it did not appear that the level of nitrogen concentration was a relevant characteristic for predicting the value of malt extract (see Figure 19.10). It is therefore proposed that the lack of consistent effects of heat stress and drought on malt extract were due to these stresses simultaneously affecting various grain components/properties that counterbalanced one another (see the following subsection).

For certain purposes, a good-quality malt will require a high capacity for the synthesis of starch-degrading enzymes during malting (Duffus and Cochrane, 1992). Diastatic power may be defined as the collective activity of all the amylolytic enzymes involved in the hydrolysis of starch to fermentable sugars during mashing (Evans and MacLeod, 1993). In fact, the term *diastatic power* encompasses the action of at least four different enzymes, i.e., the α-amylases, β-amylases, α-glucosidase, and limit dextrinase. These enzymes all have different activities, substrate specificities, and stabilities (Evans and MacLeod, 1993). Many studies have found a positive correlation between nitrogen content and diastatic power (e.g., Smith, 1990), suggesting that the production of grain high in both extract and diastatic power could be difficult (Eagles et al., 1995). With the data obtained from experiments reported in Tables 19.1 and 19.2, no significant relationship could be shown between nitrogen content and diastatic power (data not shown). In addition, there was no relationship between the reduction of malt extract and that of diastatic power with heat stress (see Figure 19.11a) and drought stress (see Figure 19.11b).

As with the relationship between malt extract and nitrogen percentage within each experiment (see Figure 19.10), there was no relationship between malt extract and diastatic power, even though a large variation in diastatic power was found within some of the experiments (see references cited in Tables 19.1 and 19.2). This could be an indication that individual components, such as starch or nitrogen content, diastatic power, and β-glucan content, cannot explain malt extract values when barley plants are exposed to short periods of very high temperature or drought. It is likely that

FIGURE 19.11. Relationship Between Malt Extract and Diastatic Power for the (a) Heat Stress and (b) Drought Experiments

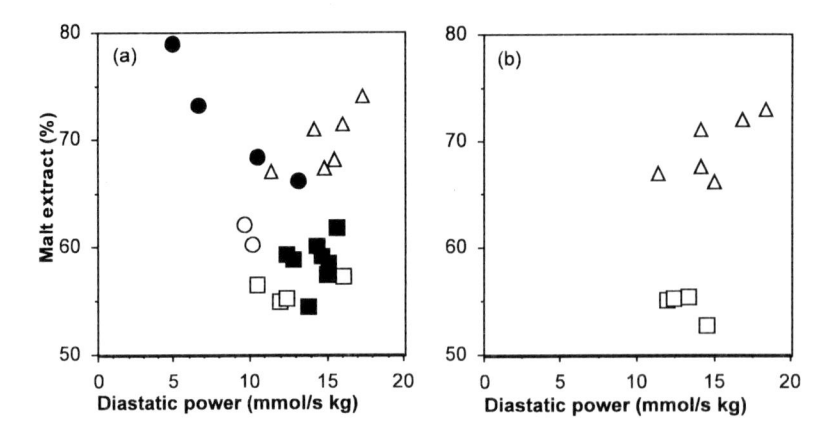

Source: The data are from some of the experiments reported in Tables 19.1 and 19.2: ● Savin, Stone, and Nicolas (1996); ○ Savin et al. (1997a); ■ Savin et al. (1997b); □ Savin and Nicolas, (1999); △ Savin and Nicolas (1996).

interactions among these quality parameters may be responsible for the limited effect of those stresses on malt extract in most experiments.

If this is true, simple regression analyses would not be the correct tool to uncover the reasons for effects (or lack of them) of heat stress and drought (or any other factor) on malt extract. Until more research can be conducted and our understanding of partial relationships substantially improved, a mathematically less precise but mechanistically more acceptable model has been developed, considering, initially, the simultaneous effects of the major factors controlling malt extract.

Malt Extract As an Expression of Multiple Interactions

An alternative approach to understanding the effect of short periods of very high temperature and drought could consider a net balance between the different parameters that play an important role in malting quality. A tentative simplified model is presented diagrammatically in Figure 19.12. With this model in mind, one can understand how the negative and positive effects of particular environmental factors on different parameters that play an important role in the determination of malt extract can shift the net balance. This can result in changes of a different magnitude (and direction) to those expected from observing individual parameters in isolation.

FIGURE 19.12. Theoretical Model Illustrating a Possible Balanced Effect of the Amount of Starch (St), β-Glucan (β-G), β-Glucanases (β-Gls), Nitrogen (N), and Diastatic Power (DP) on Malt Extract (ME)

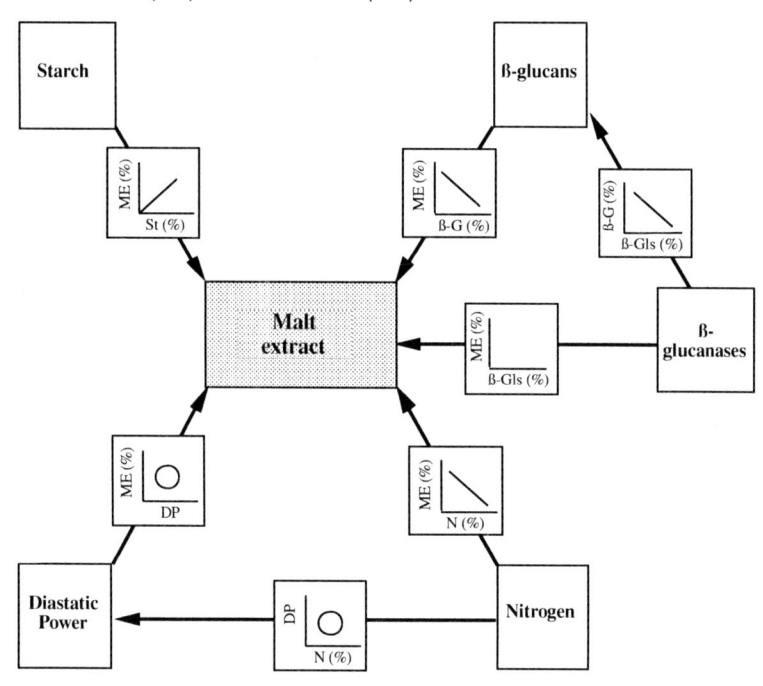

Source: Adapted from Savin (1996).

Note: All the relationships are drawn to indicate the general effect of one factor on another.

For example, in one of the experiments reported in Table 19.1 (Savin and Nicolas, 1996; Savin, 1996), there was a large effect of the heat stress treatment on grain nitrogen concentration (ranging from 2.2 to 3.1 percent), and a change of around 10 percent would therefore be expected in malt extract (following a slope of –10.5 percent ME/percent N; Smith, 1990). However, no significant changes in malt extract were observed. In the same experiment, the treatments affected the diastatic power, possibly as an indirect consequence of the increased nitrogen concentration, and that might counterbalance, at least partly, the negative effect on extract expected from the increased protein concentration. Malt β-glucan concentration was also reduced, reflecting increased β-glucan degradation during malting. All these effects combined could reasonably explain the lack of severe effects on malt extract. A similar exercise can be done with other experiments reported in Tables 19.1 and 19.2.

The major drawback of this model is that the knowledge of the nature of these relationships is still in its infancy, and clearly more research must be conducted before robust relationships can be developed and the model can be substantially improved.

FUTURE RESEARCH

This chapter has identified several issues that deserve further research:

1. A possibility to overcome drought- and heat-associated variations in grain weight and grain quality would be to modify barley genotypes genetically so as to render them tolerant to heat stress. A survey of barley genotypes under short periods of very high temperature would be a first step toward identifying existing sources of such thermotolerance in barley.
2. There is need to study further the effect of short periods of high temperature and drought during the grain-filling period under field conditions and with nitrogen levels within those required by the malting industry.
3. It would be important to investigate further starch composition (i.e., amylose and amylopectin content) and protein composition (i.e., hordeins) when short periods of high temperature and drought were imposed during grain filling. These issues require further investigation because starch and protein quality have important implications during mashing. More detailed research examining the deposition of amylose and amylopectin as well as the different hordein fractions during grain filling under short periods of high temperature and drought would be essential.

IMPLICATIONS

Short periods of high temperatures (> 35°C) and drought can have a significant effect on grain weight and then grain yield, and the amount of "maltable" grain (grains > 2.5 mm) can be considerably reduced (Savin, Stone, and Nicolas, 1996). This has practical implications for both farmers and maltsters. Moreover, these stresses significantly modified grain composition, as shown in the third section. This also has practical implications, since the stressed grains may require different malting schedules or need to be directed toward different end uses.

The question whether short periods of very high temperature and drought affect malt extract (as a measure of malting quality) does not have a simple answer. These environmental constraints can change grain composition to a certain extent, but these changes can have opposing effects on malt extract. The problem proved to be too complex to be analyzed simply by linear regression, which appears to be sufficient for nitrogen fertilization studies. This complexity could be disappointing to those who wish to formulate simple generalizations of environmental effects on malting quality. On the other hand, it could be viewed as a challenge to pursue more work to improve our knowledge in this field, with the potential to improve malting quality through the breeding of cultivars tolerant to particular environmental constraints. This knowledge will also help to develop simulation models that could potentially be used to predict the responses of grain yield and grain quality to a particular stress, thereby improving the ability of "administrators" to assess the market for grains before they are harvested.

REFERENCES

Allan, G.R., A. Chrevatidis, F. Sherkat, and I.M. Stuart (1995). The relationship between barley starch and malt extract for Australian barley varieties. In *Proceedings of the 5th Scientific and Technical Convention of the Institute of Brewing*, Victoria Falls, Zimbabwe, pp. 70-79.

Aman, P., H. Graham, and A.C. Tilly (1989). Content and solubility of mixed linked (1-3, 1-4)-β-D-glucan in barley and oats during kernel development and storage. *Journal of Cereal Science* 10:45-50.

Banks, W., C.T. Greenwood, and D.D. Muir (1973). Studies on the biosynthesis of starch granules. Part 5. Properties of the starch components of normal barley and barley with starch of high amylose content during growth. *Starch* 25:153-157.

Bhullar, S.S. and C.F. Jenner (1985). Differential responses to high temperatures of starch and nitrogen accumulation in the grain of four cultivars of wheat. *Australian Journal of Plant Physiology* 12:363-375.

Bosnes, M., F. Weideman, and O.A. Olsen (1992). Endosperm differentiation in barley wild-type and sex mutants. *The Plant Journal* 2:661-674.

Brandt, A. (1976). Endosperm protein formation during kernel development of wild type and a high-lysine barley mutant. *Cereal Chemistry* 53:890-901.

Briarty, L.G., C.E. Hughes, and A.D. Evers (1979). The developing endosperm of wheat: A stereological analysis. *Annals of Botany* 44:641-658.

Brocklehurst, P.A. (1977). Factors controlling grain weight in wheat. *Nature* 266:348-349.

Brocklehurst, P.A., J.P. Moss, and W. Williams (1978). Effects of irradiance and water supply on grain development in wheat. *Annals of Applied Biology* 90: 265-276.

Burger, W.C. and D.E. LaBerge (1985). Malting and brewing quality. In *Barley*, ed. D.C. Rasmusson. Madison, WI: American Society of Agronomy, pp. 367-401.

Buttrose, M.S. (1960). Submicroscopic development and structure of starch granules in cereal endosperms. *Journal of Ultrastructure Research* 4:231-257.
Cass, D.D. and W.A. Jensen (1970). Fertilization in barley. *American Journal of Botany* 57:62-70.
Cattivelli, L., G. Delogu, V. Terzi, and A.M. Stanca (1994). Progress in barley breeding. In *Genetic Improvement of Field Crops*, ed. G.A. Slafer. New York: Marcel Dekker, pp. 95-181.
Chowdhury, S.I. and I.F. Wardlaw (1978). The effect of temperature on kernel development in cereals. *Australian Journal of Agricultural Research* 29:205-223.
Cochrane, M.P. and C.M. Duffus (1981). Endosperm cell number in barley. *Nature* 289:399-401.
Cochrane, M.P. and C.M. Duffus (1983). Endosperm cell number in cultivars of barley differing in grain weight. *Annals of Applied Biology* 102:177-181.
Cole, N.W. (1993). Diastase-significance and control. In *Proceedings of the 6th Australian Barley Technical Symposium,* Tasmania, Australia, pp.152-154.
Coles, G.D., P.D. Jamieson, and R.M. Haslemore (1991). Effect of moisture stress on malting quality in Triumph barley. *Journal of Cereal Science* 14: 161-177.
Conroy, J.P., S. Sneweera, A.S. Basra, G. Rogers, and B. Nissen-Wooller (1994). Influence of rising atmospheric CO_2 concentrations and temperature on growth, yield and grain quality of cereal crops. *Australian Journal of Plant Physiology* 21:741-758.
Dickson, A.D. (1979). Barley for malting and food. In *Barley.* Agriculture Handbook 338. Washington, DC: U.S. Department of Agriculture, pp. 136-146.
Duffus, C.M. and M.P. Cochrane (1992). Grain structure and composition. In *Barley: Genetics, Biochemistry, Molecular Biology and Biotechnology,* ed. P.R. Shewry. Wallingford, UK: C.A.B. International, pp. 291-317.
Duffus, C.M. and M.P. Cochrane (1993). Formation of the barley grain-morphology, physiology, and biochemistry. In *Barley: Chemistry and Technology,* eds. A.W. MacGregor and R.S. Bhatty. St. Paul, MN: American Association of Cereal Chemists, pp. 31-72.
Eagles, H.A., A.G. Bedggod, J.F. Panozzo, and P.J. Martin (1995). Cultivar and environmental effects on malting quality in barley. *Australian Journal of Agricultural Research* 46:831-844.
Engell, K. (1989). Embryology of barley: Time course and analysis of controlled fertilization and early embryo formation based on serial sections. *Nordic Journal of Botany* 9:265-280.
European Brewery Convention (EBC) (1975). *Analytica,* Third edition. Dordrecht, the Netherlands: EBC.
Evans, D.E. and L.C. MacLeod (1993): Diastatic power: Definitions, description and its implications for barley breeding in Australia. In *Proceedings of the 6th Australian Barley Technical Symposium,* Tasmania, pp. 129-133.
Evers, A.D. (1970). Development of the endosperm of wheat. *Annals of Botany* 34:547-555.
Fincher, G.B. and B.A. Stone (1993). Physiology and biochemistry of germination in barley. In *Barley: Chemistry and Technology,* eds. A.W. MacGregor and R.S. Bhatty. St. Paul, MN: American Association of Cereal Chemists, pp. 247-295.

Food and Agriculture Organization (FAO) (2000). FAO Statistics. Web site: <www.fao.org>.

Gleadow, R.M., M.J. Dalling, and G.M. Halloran (1982). Variation in endosperm characteristics and nitrogen content in six wheat lines. *Australian Journal of Plant Physiology* 9:539-551.

Henry, R.J. (1990). Barley quality: An Australian perspective. *Aspects of Applied Biology* 25:5-14.

Henry, R.J. and MacLean, B.T. (1984). Rapid small-scale determination of malt extract in barley breeding. *Journal of the Institute of Brewing* 90:371-374.

Howard, K.A., K.R. Gayler, H.A. Eagles, and G.M. Halloran (1996). The relationship between D hordein and malting quality in barley. *Journal of Cereal Science* 24:47-53.

Institute of Brewing (IOB) (1982). *Institute of Brewing Recommended Methods of Analysis*. London: The Institute of Brewing.

Jenner, C.F. (1994). Starch synthesis in the kernel of wheat under high temperature conditions. *Australian Journal of Plant Physiology* 21:791-806.

Jenner, C.F., D.T. Ugalde, and D. Aspinall (1991). The physiology of starch and protein deposition in the endosperm of wheat. *Australian Journal of Plant Physiology* 18:211-226.

Jennings, A.C. and R.K. Morton (1963). Changes in carbohydrate, protein and non-protein nitrogenous compounds of developing wheat grain. *Australian Journal of Biological Science* 16:318-331.

Kenn, D.A., A.H.S Dagg, and I.M. Stuart (1993). Effect of environment and genotype on the fermentability of malt produced from four Australian barley varieties. *American Society of Brewing Chemists Journal* 51:119-122.

Kreis, M. and P.R. Shewry (1992). The control of protein synthesis in developing barley seeds. In *Barley: Genetics, Biochemistry, Molecular Biology and Biotechnology*, ed. P.R. Shewry. Wallingford, UK: C.A.B. International, pp. 319-333.

MacGregor, A.W. and L. Dushnicky (1989). α-Amylase in developing barley kernels—A reappraisal. *Journal of the Institute of Brewing* 95:29-33.

MacGregor, A.W. and G.B. Fincher (1993). Carbohydrates of the barley grain. In *Barley: Chemistry and Technology*, eds. A.W. MacGregor and R.S. Bhatty. St. Paul, MN: American Association of Cereal Chemists, pp. 73-130.

MacLeod, L.C. and C.M.. Duffus (1988). Reduced starch content and sucrose synthase activity in developing endosperm of barley plants grown at elevated temperatures. *Australian Journal of Plant Physiology* 15:367-375.

MacNicol, P.K., J.V. Jacobsen, M.M. Keys, and I.M. Stuart (1993). Effects of heat and water stress on malt quality and grain parameters of Schooner barley grown in cabinets. *Journal of Cereal Science* 18:61-68.

McDonald, A.M.L. and J.R. Stark (1988). A critical examination of procedures for the isolation of barley starch. *Journal of the Institute of Brewing* 94:125-132.

Mogensen, V.O. (1982). Effect of drought on growth rate of grains of barley. *Cereal Research Communications* 20:225-231.

Molina-Cano, J.L. (1987). The EBC barley and malt committee index for the evaluation of malting quality in barley and its use in breeding. *Plant Breeding* 98:249-256.
Molina-Cano, J.L. (1989). La cebada. *Morfología, Fisiología, Genética, Agronomía y Usos Industriales,* ed. J.L. Molina-Cano. Madrid: Ediciones Mundi-Prensa, 252 p.
Molina-Cano, J.L., T. Ramo, R.P. Ellis, J.S. Swanston, H. Bain, T. Uribe-Echeverría, and A.M. Pérez-Vendrell (1995). Effect of grain composition on water uptake by malting barley: A genetic and environmental study. *Journal of the Institute of Brewing* 101: 79-83.
Morgan, A.G. and T.J. Riggs (1981). Effects of drought on yield and on grain and malt characters in spring barley. *Journal of Science Food Agriculture* 32:339-346.
Morrison, W.R., T.P. Milligan, and M.N. Azudin (1984). A relationship between the amylose and lipid contents of starches from diploid cereals. *Journal of Cereal Science* 2:257-271.
Nicolas, M.E., R.M. Gleadow, and M.J. Dalling (1984). Effects of drought and high temperature on grain growth in wheat. *Australian Journal of Plant Physiology* 11:553-566.
Osborne, T.B. (1924). *The Vegetable Proteins.* London: Longmans, Green and Company.
Paulsen, G.M. (1994). High temperature responses of crop plants. In *Physiology and Determination of Crop Yield,* eds. K.J. Boote, J.M. Bennett, T.R. Sinclair, and G.M. Paulsen. Madison, WI: American Society of Agronomy, pp. 365-389.
Pope, M.N. (1937). The time factor in pollen tube growth and fertilization in barley. *Journal of Agricultural Research* 54:525-529.
Rahman, S., P.R. Shewry, and B.J. Miflin (1982). Differential protein accumulation during barley grain development. *Journal of Experimental Botany* 33:717-728.
Rasmusson, D.C. (1985). *Barley.* Madison, WI: American Society of Agronomy.
Renwick, F. and C.M. Duffus (1987). Factors affecting dry weight accumulation in developing barley endosperm. *Physiologia Plantarum* 69:141-146.
Reynolds, M.P., M. Balota, M.I.B. Delgado, I. Amani, and R.A. Fischer (1994). Physiological and morphological traits associated with spring wheat yield under hot, irrigated conditions. *Australian Journal of Plant Physiology* 21:717-730.
Savin, R. (1996). Effects of post-anthesis stress on grain filling and malting quality of barley. PhD Thesis. Melbourne, Australia: The University of Melbourne, 160 pp.
Savin, R. and M.E. Nicolas (1996). Effects of short periods of drought and high temperature on grain growth and starch accumulation of two malting barley cultivars. *Australian Journal of Plant Physiology* 23:201-210.
Savin, R. and M.E. Nicolas (1999). Effects of timing of heat stress and drought on grain growth and malting quality of barley. *Australian Journal of Agricultural Research* 50:357-364.
Savin, R., P.J. Stone, and M.E. Nicolas (1996). Responses of grain growth and malting quality of barley to short periods of high temperature in field studies using portable chambers. *Australian Journal of Agricultural Research* 47:465-477.

Savin, R., P.J. Stone, M.E. Nicolas, and I.F. Wardlaw (1997a). Effects of heat stress and moderately high temperature on grain growth and malting quality of barley. *Australian Journal of Agricultural Research* 48:615-624.

Savin, R., P.J. Stone, M.E. Nicolas, and I.F. Wardlaw (1997b). Effects of the temperature regime before heat stress on grain growth and malting quality of barley. *Australian Journal of Agricultural Research* 48:625-634.

Shewry, P.R. (1993). Opportunities to improve barley by genetic engineering. In *Proceedings of the 6th Australian Barley Technical Symposium*, Tasmania, pp. 1-14.

Simon, P. and R.L. Jones (1988). Synthesis and secretion of catalytically active barley α-amylase isoforms by *Xenopus oocytes* injected with barley mRNAs. *European Journal of Cell Biology* 47:213-221.

Singh, B.K. and C.F. Jenner (1982). Association between concentrations of organic nutrients in the grain, endosperm cell number and grain dry weight within the ear of wheat. *Australian Journal of Plant Physiology* 9:83-95.

Skerritt, J.H. and P.W. Janes (1992). Disulphide-bonded "gel protein" aggregates in barley: Quality-related differences in composition and reductive dissociation. *Journal of Cereal Science* 16:219-235.

Smith, D.B. (1990). Barley seed protein and its effects on malting and brewing quality. *Plant Varieties and Seeds* 3:63-80.

Sofield, I., L.T. Evans, M.G. Cook, and I.F. Wardlaw (1977). Factors influencing the rate and duration of grain filling in wheat. *Australian Journal of Plant Physiology* 4:785-797.

Sofield, I., I.F. Wardlaw, L.T. Evans, and S.Y. Zee (1977). Nitrogen, phosphorus and water contents during grain development and maturation in wheat. *Australian Journal of Plant Physiology* 4:799-810.

Stone, P.J. and M.E. Nicolas (1994). The effects of short periods of high temperature during grain filling on grain yield and quality vary widely between wheat cultivars. *Australian Journal of Plant Physiology* 21:887-900.

Stuart, I.M., L. Loi, and G.B. Fincher (1988). Varietal and environmental variations in (1-3, 1-4)-β-glucan levels and (1-3, 1-4)-β-glucanase potential in barley: Relationships to malting quality. *Journal of Cereal Science* 7:61-71.

Swanston, J.S., R.P. Ellis, A. Pérez-Vendrell, J. Voltas, and J.L. Molina-Cano (1997). Patterns of barley grain development in Spain and Scotland and their implication for malting quality. *Cereal Chemistry* 74:456-461.

Tester, R.F., J.B. South, W.R. Morrison, and R.P. Ellis (1991). The effects of ambient temperature during the grain filling period on the composition and properties of starch from four barley genotypes. *Journal of Cereal Science* 13:113-127.

Wallwork, M.A.B., S.J. Logue, L.C. MacLeod, and C.F. Jenner (1998a). Effect of high temperature during grain filling on starch synthesis in developing barley grain. *Australian Journal of Plant Physiology* 25:173-181.

Wallwork, M.A.B., S.J. Logue, L.C. MacLeod, and C.F. Jenner (1998b). Effects of a period of high temperature during grain filling on the grain growth characteristics and malting quality of three Australian malting barleys. *Australian Journal of Agricultural Research* 49:1287-1296.

Wardlaw, I.F. (1970). The early stages of grain development in wheat: Response to light and temperature in a single variety. *Australian Journal of Biological Science* 23:765-774.

Wardlaw, I.F. and C.W. Wrigley (1994). Heat tolerance in temperate cereals: An overview. *Australian Journal of Plant Physiology* 21:695-703.

Williams, J.M. and C.M. Duffus (1977). The development of endosperm amyloplasts during grain maturation in barley. *Journal of the Institute of Brewing* 84:47-50.

Index

α-amylase, 488-489, 504, 508-510
α-glucosidase enzyme, 529-530
ABA (abscisic acid)
    component of malting quality, 88
    dormancy regulation, 485-488
    germination inhibition, 89
    and growth regulators, 284-286, 290
    responsive element (ABRE), 317
    stress signaling, 315
Abeledo, L. Gabriela, 361
Abiotic stress and changes in malting quality, 523-524
    carbohydrate accumulation, 528-529
    cell division, 525-527
    effect of drought, 534-536, 537, 538-544
    effect of heat stress, 530-534, 538-544
    future research, 544
    grain composition, 530-536
    grain filling, 525, 527-530. *See also* Grain, development stage
    implications, 544-545
    multiple interactions, 542-544
    protein accumulation, 529-530
    quality specifications, 536-537
Abiotic stress tolerance, 71-75, 312, 328-330, 418-422. *See also* Stress adaptation
Abscisic acid (ABA). *See* ABA
Acid detergent fiber (ADF), 39
Acreage and production, 7
Activated oxygen species (AOS), 339-340
Adaptation, 65. *See also* Physiological mechanisms to cope with stress
Agrobacterium tumefaciens transformation, 147-150
Agronomic trait, QTL mapping, 36-37

Alagarswamy, Gopalsamy, 445
Alcohol dehydrogenase (ADH), 325-326
AMIN1, 451, 454, 456, 458, 459, 460, 465, 474
AMMI model, 214-215, 234-235
Amplified fragment length polymorphism (AFLP), 33, 129
Analysis of variance (ANOVA) of multienvironment trials, 213-214
Ancestors of cultivated barley, 15-17
Animal feed. *See* Feed for animals
Animal performance parameters to test feed barley, 120-121, 128
Aphid resistance, 336
Araus, José Luis, 269, 387
Assays. *See* Identification assays
Awns, 254, 281
AWR, 456-457, 474

β-amylase, 504, 510, 513-514
β-glucan, 97-100, 533-536, 538-539
β-glucanase, 504. *See also* Heat-stable (1,3-1,4)-β-glucanase
β-glucuronidase (GUS), 156-157
Baldi, Paolo, 307
Barley stem rust, 333-334
Barley stripe rust, 132
Barley yellow dwarf virus (BYDV), 335
Barley yellow mosaic virus (BaYMV), 42-43, 71
BASI, 508-509
Benech-Arnold, Roberto L., 481
Beta-Amylase Thermostability, 92

Betaine, 286
Betaine dehydrogenase (BADH), 286-287
Bin map, 33
Biolistic method of embryo transformation, 144-147
Biomass
   CARBO, 454, 455, 457, 460, 463, 464, 474
   dry-matter production, 453-464
   evaluation of accumulation, 470-471
   partitioning, 369-370
   PCARB, 454, 475
Biotic stress, 418, 421-422. *See also* Disease resistance; Diseases
Bort, Jordi, 387
Breeding
   disease resistance, 342-343
   doubled haploids, 180-181
   feed barley, 128-132
   GE. *See* Genotype by environment interaction
   grain yield improvements, 376-381
   major genes, 183-185
   malt barley, 101-102
   marker-assisted selection, 181-192
   methods, 179-181
   pedigree inbreeding, 179-180
   problems and success, 177-179
   shuttle, 181
   stress adaptation and potential yield, 272
   stress physiology, 287-288
Brewing. *See* Malting barley
Brown rust, 86

Calderini, Daniel F., 361, 413
Canopy and growing environment, 270-271, 388-389
CARBO, 454, 455, 457, 460, 463, 464, 474
Carbohydrates. *See also* Starch
   accumulation in grain filling, 528-529
   in feed barley, 118, 124-125

Casadesús, Jaume, 387
Casas, Ana-María, 15
Cassette vectors, 150-151
Cattivelli, Luigi, 307
Ceccarelli, Salvatore, 387
Cell division, 525-527
Cell walls, 93-100
Centers of origin, 17-20, 24-25
Centromeres, mapping of, 33-34
Cereal
   acreage, 7
   production, 6-10
Cereal cyst nematode (CCN), 335-336
Characteristics
   agronomic characteristics in genetic transformation, 169-170
   control by genes, 177
   feed genetic characteristics, 121-128
   feed physical characteristics, 117, 122-124
   feed quality, 116-121
   root characters and stress tolerance, 71-74
Chemical composition for feed barley, 118-120, 124-128
Chicken feed, 165
Chlorophyll synthesis, 273, 288-289. *See also* Photosynthesis
Chloroplasts, 313-314
Chromosomes, 31-32. *See also* Molecular mapping of genome; *specific genetic subjects*
   feed barley, 122-123
Climatic variables affecting yield, 207-208, 225-231
Coarse grains, 10-11
Cocultivation of embryos, 147-150
CO_2 concentrations and yield, 271, 279, 284
Complexly inherited traits, 131-132
Composition of grain, 530
   in abiotic stress, 530-536, 537
   chemical composition for feed barley, 118-120, 124-128
Computer simulation. *See* Crop simulation models

Crop management
  evaluation of computer model, 472-473
  physiological mechanisms to cope with stress, 292-293
Crop simulation models, 445-448
  dry-matter production, 453-464. *See also* Phasic development
  evaluation of model, 467-474
  model improvement, 466-467
  phasic development. *See* Phasic development
  principles, 447-448
  root development, 464-466
Crosatti, Cristina, 307
Cultivars
  Crest, 131
  effects of breeding, 101-102
  Golden Promise, 87-88, 143, 145, 168, 169-170
  Lewis/Baronesse, 131
  Orca, 132
  Proctor, 102
  Triumph, 89, 102
  Valier, 131
Cultivation
  for feed, 5
  history of, 1-4
  origin, 15-25
CUMPH, 456, 458, 474
CUMVD, 450, 474

Darlington, Helen, 503
Date choices for sowing, 422-424
  determination of grain yield and quality, 424-425, 471-472
  matching grain development and environment, 425-427
  probability-based approach, 427-429
Daylength sensitivity, 258
Deassimilatory processes, 275
Dehydrins, 286, 321-322
Density of plants and grain yield, 429-430
  defined, 429
  and grain yield, 430-432, 433-434

Density of plants and grain yield *(continued)*
  planting arrangement, 432-433
  tillering, 432-433
Development, 243
  cell division, 525-527
  defined, 243
  diagram of morphological changes, 247
  dry-matter production, 369-370, 453-464, 526
  environmental factors, 254-259
  genetic control of flowering time, 259-260
  grain-filling phase, 244-245, 246, 253-254, 462-464, 527-530
  leaf and spikelet initiation, 246-250, 255-256
  leaf emergence, 250-252, 255-256
  phases, 244, 245
  photoperiod responses, 258-259
  plant. *See* Phasic development
  reproductive phase, 244-246
  stages of, 243-244
  temperature responses, 255-256, 288, 317-318
  tillering, 252-253, 432-433, 458-459
  triple mound stage, 248
  vegetative phase, 244-245, 248
  vernalization responses, 256-258
DF, 451, 474
Diastatic power, 513-514
Disease resistance
  aphids, 336
  barley stem rust, 333-334
  barley stripe rust, 132
  barley yellow dwarf virus (BYDV), 335
  breeding strategies, 342-343
  brown rust, 86
  cereal cyst nematode (CCN), 335-336
  consideration in planting selection, 421
  leaf blotch, 71, 334
  leaf rust, 68-69, 71, 185, 333
  leaf streak, 335

Disease resistance *(continued)*
  leaf stripe, 334
  mildew, 70, 71, 86, 185-186, 330-333, 340
  molecular response, 330
  net blotch, 69-70, 71, 334
  new resistance genes, 336-337
  Rhynchosporium, 71, 334
  scald blotch, 334
  spot blotch, 69, 334
  typhula blight, 334
  wild barley, 66, 68-71
Diseases
  barley yellow mosaic virus (BaYMV), 42-43, 71
  effect on yield, 208
  mapping of genes, 40-43
  powdery mildew, 42
  stem rust, 41-42, 333-334
DLAYR(L), 465, 466, 474
Dormancy, 481-482, 495-496
  defined, 481
  environmental control, 493-495
  expression of, 489
  genetics and molecular biology, 489-493
  hormonal regulation, 485-488
  imposed by the hull, 483, 486-487
  location in cereal grains, 482-484
  low, 418
  physiology, 482-489
  postharvest, 89
  preharvest sprouting, 481-482, 490, 492
  production of α-amylase, 488-489
  quantitative trait lock (QTL), 89, 489-493
  time window, 482, 483, 494-495
Double cassette vectors, 150-151
Doubled haploids, 180-181
Drecher, M. Fernanda, 413
Drought stress
  and malting quality, 534-536, 537, 538-544
  and yield, 208, 290-292

Drought tolerance, 65-66
  choosing grain for planting, 419-420
  genes (*Lea* genes), 318, 321-323
  loci control, 72-73, 310-311
  physiological basis, 289, 290-292
Dry-matter production, 369-370, 453-464, 526
DTT, 451, 459, 460, 461, 465, 474
DUL(L), 474

Ear types
  six-row, 3-4, 15-16, 253, 414
  two-row, 3-4, 15-16, 253, 414
EGFT, 456, 475
Ellis, Roger P., 65, 85
Embryo transformation
  biolistic method, 144-147
  cocultivation, 147-150
Endosperm, 93-100
Environmental factors, 207-209. *See also* Stress adaptation
  climatic variables, 207-208, 225-231
  CO_2 concentrations, 271, 279, 284
  disease. *See* Disease resistance; Diseases
  dormancy, 493-495
  drought. *See* Drought stress; Drought tolerance
  effect on plant development, 254-259
  in factorial regression models, 222-231
  and grain development, 425-427
  and leaf appearance, 251, 255-256
  photosynthesis, 270-271, 273-274
  physiological. *See* Physiology of yields
  story of isogenic lines in Spain, 222, 224-231
  water stress, 271, 273, 274
Enzymes
  α-amylase, 488-489, 504, 508-510
  α-glucosidase, 529-530
  β-amylase, 504, 510, 513-514
  during grain filling, 529-530

Enzymes *(continued)*
  production, 89-93
  recombinant, 162
ESW, 465, 475
Example data set. *See* Yield of isogenic lines in Spain
Expressed sequence tags (ESTs), 129
Extract yield, 417-418

Factorial regression model in yield study, 219-231, 235-236
Fallow stage, 448-449
Feed for animals, 115-116
  animal performance parameters, 120-121, 128
  breeding, 128-132
  carbohydrates, 118, 124-125
  chemical composition, 118-120, 124-128
  crop utilization, 10-11, 115
  cultivation, 5
  feed quality characteristics, 116-121
  future projections, 132-133
  genetic characteristics, 121-128
  physical characteristics, 117, 122-124
  protein content, 119, 125-126, 503
  straw, 12
  versus malting barley, 115-116
Fertile Crescent, 17, 24-25, 67
Fertilization, 504-505, 524-525
Fertilizer. *See also* Nitrogen
  and grain quality, 277
  and yield, 271
Fine-structure mapping of QTL, 39-40
Fischbeck, Gerhard, 1
Flavonoid compounds in feed barley, 119-120
Flowering, 65, 259-260
Foam stabilization, 510, 514
Food consumption, 5, 12
Formate dehydrogenase, 326
Future of barley
  crop improvement, 77
  feed grain, 132-133

Future of barley *(continued)*
  malt quality barley, 102-104
  research, 544

G1, 461-462, 475
G2, 463, 475
G3, 459, 475
García del Moral, Luis F., 205, 243
Genes
  in barley breeding, 183-185
  homologue relationships between resistance genes, 337-339
  Lea genes, 318, 321-323
  new resistance genes, 336-337
  plastid, 327
  regulating flowering time, 259-260
  wild barley. *See* Wild barley for crop improvement
Genetic adaptation to stress, 307-308, 311-313, 343
  abiotic stress, 71-75, 312, 328-330, 418-422
  anaerobic stress, 325
  breeding strategies, 342-343
  defined, 307
  disease. *See* Disease resistance
  drought tolerance, 72-73, 310-311, 318, 321-323, 419-420
  frost resistance, 309-310, 322, 324-325, 420
  gene activation, 316-318
  genes, 319-321
  growth habit, 309
  heading date, 308-309
  heat shock (HS), 323-324, 420
  homologue relationships between resistance genes, 337-339
  low-temperature-responsive genes, 324-325, 467
  phenotypes, 339-342
  salt tolerance, 287, 293-294, 311, 323
  schematic of plant response to stress, 312
  signal perception, 313-314
  signal transduction, 314-316
  stress-related genes, 318-328

Genetic diversity for crop
    improvement, 76-77
Genetic mapping. *See* Molecular
    mapping of genome
Genetic transformation, 143-144
    agronomic characteristics, 169-170
    biolistic method, 144-147
    cocultivation, 147-150
    field testing, 166-167
    homeotic mutants, 160-162
    identification assays, 155-160
    marker-free transgenic plants,
        150-151
    micromalting, 167-169
    quality testing, 151-155
    recombinant proteins, 162-164
    tailoring heat-tolerant $(1,3-1,4)$-β-
        glucanase, 164-165
Genetics and breeding of malt quality
        attributes
    effects of breeding, 101-102
    future directions, 102-104
    history of, 85-87
    mutations, 127
    quality. *See* Malting quality
Genetics of feed barley, 121-128. *See
        also* Feed for animals
Genotype
    adaptation to stresses, 418-422
    category of sowing time, 414
    choosing for planting, 413-414
    desirable traits, 413-414
    drought-resistant, 291
    effect on grain quality, 417-418
    effect on grain yield, 414-417
    grain quality, 417-418
    grain yield, 414-417
    responsiveness, 415-416
Genotype by environment interaction
        (GE), 205-206
    barley, 206-209
    breeding implications, 231-234
    causes for yield variances, 208-209
    climatic variables, 207-208, 225-231
    crossover, 205
    disease, 208

Genotype by environment interaction
        (continued)
    drought, 208
    effect on grain yield, 414-417
    example of lines in Spain. *See* Yield
        of isogenic lines in Spain
    qualitative, 205
    SAS programs, 234-237
    soil, 208
Genotype by location (GL), 205, 233
Genotype by location by year (GLY),
        206, 233
Genotype by year (GY), 205-206, 233
Germination. *See also* Dormancy
    effect of gibberellic acid (GA),
        487-488
    indices, 484
    percentages of various genotypes,
        491
    preharvest sprouting, 481-482, 490,
        492
    transport of soluble material,
        100-101
Germins, 326
Gibberellins
    malting additive, 88, 90
    to promote germination, 487, 488
Glaucousness, 281
Glossary of terms, 474-477
GPLA, 461, 463, 475
GPP, 461, 462, 463, 475
Grain
    development stage, 244-245, 246,
        253-254, 462-464, 527-530
    number versus mass, 276-277
    quality, 75-76, 277, 417-418
    size variation, 75
    structure, 504-505
    texture, 511, 515
Grain yield
    density of plants. *See* Density of
        plants and grain yield
    determination of, 388
    effects on, 207-209, 244-246
    genetic improvement. *See* Grain
        yield, genetic improvement
        and physiological changes

Grain yield *(continued)*
  genotypic effects on, 414-417
  Mediterranean experiment. *See*
    Spectral vegetation and barley
    yield in Mediterranean rain-
    fed conditions
  and plant density, 430-432
Grain yield, genetic improvement and
    physiological changes,
    361-362, 381-382
  breeding, 376-381
  crop development, 366-367
  dry-matter production, 369-370
  genetic gains, 377-378
  nitrogen partitioning and grain
    quality, 373, 375-376
  numerical components, 370-373,
    374, 380-381
  partitioning of biomass, 369-370
  plant height, 367-369, 378-379
  worldwide gains, 362-366
Grando, Stefania, 387
GRNWT, 464, 475
GROGRN, 463, 475
GROLF, 457, 460, 475
GRORT, 457, 464, 475
Grossi, Maria, 307
GROSTM, 460, 463, 475
Growth. *See also* Development
  defined, 243
  habit controlled by loci, 309
  regulators, 284-286, 290

H. spontaneum, 15-17, 19, 20-21, 23-24.
    See also Western Mediterranean
    H. spontaneum
H. vulgare, 21-24
Han, Feng, 31
Harvest index (HI), 275
Harvesting, whole-plant, 12
Heading date controlled by loci, 308-309
Heat shock (HS) genes, 323-324

Heat stress
  adaptation, 164-165
  effect on malting quality, 530-534,
    538-544
  HS genes, 323-324
  tolerance, 420
Heat-stable (1,3-1,4)-β-glucanase
  background, 157
  in chicken feed, 165
  expression of, 166-167
  product analysis, 158
  production of recombinant proteins,
    162-163
  tailoring of, 164-165
  under control of D-hordein
    promoter, 161
Heterospecific clusters, 337
High-lysine feed, 130
History
  ancestors of cultivated barley, 15-17
  centers of origin, 17-20
  cereal production, 6-10
  cultivation, 1-4, 15-25
  genetics and breeding of malt
    quality attributes, 85-87
  utilization, 5, 10-12
Homologue relationships between
    resistance genes, 337-339
Hordeins
  in breeding malt quality barley,
    95-97, 511-513
  during grain filling, 529
  in genetic transformations, 150, 151,
    154, 162-163
  storage proteins, 506-508
Hordeum agriocrithon, 16
Hormones
  control system, 87-89
  regulation of dormancy, 485-488
Horvath, Henriette, 143
Hot water extract, 86-88, 102, 103, 512
Huang, Jintai, 143
Hull, effect on dormancy, 483, 486-487
Hull-less grain, 129-130

Identification assays
   b-glucuronidase (GUS), 156-157
   heat-stable (1,3-1,4)-β-glucanase.
      See Heat-stable (1,3-1,4)-β-glucanase
   phosphinotricin acetyl transferase (PAT) activity, 155-156
   polymerase chain reaction (PCR), 155, 156
   southern blot analysis, 157-160
   western blot analysis, 157, 166
   zymogram, 157
Ideotype, 414
Igartua, Ernesto, 15, 205
Igri x Franka maps, 32
IML (Interactive Matrix Language), 234-237
Improvement programs, 65-66. *See also* Wild barley for crop improvement
Inherited traits
   complexly, 131-132
   simply, 129-131
International Center for Agricultural Research in the Dry Areas (ICARDA), 231-232, 288, 390
International Maize and Wheat Improvement Center (CIMMYT), 232
Isoenzymes, 90, 91

Kleinhofs, Andris, 31
Kneifel, 102

L, 475
LAI, 454, 458, 475
Leaf
   area duration, 389
   blotch, 71, 334
   development, 246-252, 255-256
   expansion, 273
   rust, 68-69, 71, 185, 333
   streak disease resistance, 335
   stripe disease resistance, 334

LIF1, 458, 459, 460, 475
Light stress, 288-289
Linkage maps, 32-33, 46-52, 260
Lipid levels in feed barley, 119
LL(L), 465, 475
LN, 458, 475
Lodging resistance, 367-369, 422
Low-phytic-acid barley, 130
LTPs, 341, 510
Lysine content of feed barley, 127, 130-131

Malting, defined, 86
Malting barley
   genetics and breeding. *See* Genetics and breeding of malt quality attributes
   grain quality, 75-76
   mapping QTL, 38
   production, 9
   protein. *See* Protein
   sowing. *See* Sowing of malting barley
   utilization, 11
   versus feed barley, 115-116
Malting quality. *See also* Protein
   hormone control system, 87-89
   nitrogen partitioning, 373, 375-376
   production of enzymes, 89-93
   response to abiotic stresses. *See* Abiotic stress and changes in malting quality
   specifications, 536-537
   starch, protein, and cell walls, 93-100, 511
   traits, 375
   transport of soluble material, 100-101
   variations, 523-524
Mapping. *See* Molecular mapping of genome
Marker-assisted selection
   future opportunities, 192
   major genes, 181-186
   problems, 189-92
   for QTL deployment, 189-192
   for QTL detection, 186-189

Mass growth, 447
Mediterranean conditions. *See* Spectral vegetation and barley yield in Mediterranean rain-fed conditions; Western Mediterranean *H. spontaneum*
Metabolism modification, 307
Micromalting, 167-169
Microsatellites, 129
Mildew, 42, 70, 71, 86, 185-186, 330-333, 340
Milling energy, 97-99, 102, 103
Minerals in feed barley, 120, 127-128
Miralles, Daniel J., 243
Molecular mapping of genome, 31-32, 45
   centromeres and telomeres, 33-34
   chromosomes, 31-32
   comparative mapping among grass species, 43-44
   comparative mapping of QTL, 44-45
   comparative maps within barley, 43-45
   disease resistance genes, 40-43
   dormancy, 489-493
   linkage maps, 32-33, 46-52, 260
   physical, 34-35
   quantitative trait loci (QTL), 35-40
   synteny, 43-45
Molecular marker-assisted selection (MMAS), 129, 132
Molecular markers in selection for major genes, 181-186
Molina-Cano, José Luis, 15, 205, 523
Moralejo, Marian, 15
Morex
   malting type, 490
   Steptoe/Morex maps, 32, 33, 94, 490
Moroccan *H. spontaneum,* 19-21, 23, 24
Morphological adaptations, 281-282, 310, 365
Mutations
   ABA-insensitive, 315
   chloroplast development, 314
   in genotype by environment (GE), 209
   high-lysine, 127

Mutations *(continued)*
   homeotic, 160-162
   Hooded, 160
   malting barley, 127
   pant, 120
   waxy, 93-94

Naked barley
   history of, 2
   production, 9
Near infrared reflectance (NIR), 129
Near infrared transmission (NIT), 129
Net blotch, 69-70, 71, 334
NFAC, 454, 456, 458, 459, 460, 475
Nitrogen
   effect of heat stress and drought, 539-542
   fertilizer, 271, 277, 292
   NFAC, 454, 456, 458, 459, 460, 475
   NSTRESS, 454, 475
   partitioning and grain quality, 373, 375-376
Nonstarch polysaccharides, 528-529
North American Barley Genome Mapping Project (NABGMP), 33, 68
NSTRESS, 454, 475

Origin of cultivated barley, 15-25
Osmoprotectants, 286, 290
Osmoregulators, 286-287, 290
Osmotic adjustment, 310-311
Overview of agricultural contribution, 1-12

P1D, 451, 476
P1V, 450, 451, 476
P5, 463, 475
Pant mutations, 120
PAPI (probable amylase/protease inhibitor), 510

Partition of genotypic variation, 220-221
Partitioning of biomass, 369-370
Partitioning of nitrogen, 373, 375-376
Pathogens
    aphids, 336
    cereal cyst nematode (CCN), 335-336
Pax Romana, 17
PCARB, 454, 475
Pedigree inbreeding, 179-180
Peroxidase, 341-342
Phasic development, 447, 448
    emergence to terminal spikelet initiation (stage 1), 449-451, 455-459
    end of grain filling to harvest (stage 6), 449, 453
    end of leaf growth and start of ear growth to end of preanthesis ear growth (stage 3), 449, 452, 460-461
    end of preanthesis ear growth to start of grain filling (stage 4), 449, 452, 461-462
    evaluation of model, 468-469, 470
    fallow or presowing (stage 7), 448-449
    germination to emergence (stage 9), 449
    grain filling (stage 5), 449, 452-453, 462-464
    sowing to germination (stage 8), 449
    terminal spikelet initiation to end of leaf growth and start of ear growth (stage 2), 449, 451-452, 459-460
Phenology, 277-278, 466-467
Phenotypes, resistant, 339-342
PHINT, 456, 460, 475
Phosphinotricin acetyl transferase (PAT) activity assay, 155-156
Photoassimilation, 274
Photoinhibition of photosynthesis, 313, 326-327
Photooxidative damage, 282-284
Photoperiod responses, 258-259
    P1D, 451, 476

Photosynthesis
    area indices, 392
    photoinhibition, 313, 326-327
    PRFT, 454, 476
    and yield, 270-271, 273-274, 285
Phyllochron, 250-252, 448
    CUMPH, 456, 458, 474
    PHINT, 456, 460, 475
    TI, 477
Physical characteristics of feed barley, 117, 122-124
Physical mapping. *See* Molecular mapping of genome
Physiological mechanisms to cope with stress, 282
    ABA and growth regulators, 88, 89, 284-286, 290
    breeding implications, 287-288
    crop management, 292-293
    drought stress, 290-292, 310-311. *See also* Drought stress; Drought tolerance
    excess radiation, 282-284
    light stress, 288-289
    osmoprotectants, 286, 290
    osmoregulators, 286-287, 290
    photooxidative damage, 282-284
    salinity, 287, 293-294
    stress proteins, 286
Physiology of yields, 269-270, 294-295
    early vigor and fast development, 280-281
    factors determining yield, 270-271, 272
    grain number versus grain mass, 276-277
    grain quality, 75-76, 277, 373, 375-376
    harvest index (HI), 275
    morphological and physiological adaptations, 281-282
    phenology manipulation, 277-278
    photosynthesis, 270-271, 273-274, 285
    potential yield versus stress adaptation, 272
    present trends, 278

Physiology of yields *(continued)*
 radiation-use efficiency, 274-275, 389
 retrospective studies, 278
 sink limitation, 275-276
 stress-coping mechanisms. *See* Physiological mechanisms to cope with stress
 water-use efficiency (WUE), 278-280, 282
 yield potential, defined, 269
Phytoalexins, 342
PLA, 457, 458, 460, 475
PLAG, 456, 457, 475
PLAGMS, 456, 475
PLALR, 458, 461, 463, 475
Plant density and grain yield, 430-432
Plant height changes to increase yield, 367-369, 378-379
Planting arrangement effect on grain yield and density, 433. *See also* Sowing of malting barley
Plastid genes, 327
PLSC, 458, 475
PLTPOP, 454, 455, 458, 464, 476
Polymerase chain reaction (PCR), 155, 156
Polyphenolic compounds in feed barley, 119-120
Polysaccharide, 97, 528-529
Potential yield. *See also* Physiology of yields
 defined, 269
 versus stress adaptation, 272
Powdery mildew, 42, 70, 71, 86, 185-186, 330-333, 340
Preharvest sprouting, 481-482, 490, 492. *See also* Dormancy
PRFT, 454, 476
Probability-based approach of matching grain yield and quality with environment, 427-429
Protein, 503-504, 515-516
 accumulation in grain filling, 529-530
 β-amylase, 504, 510, 513-514
 coat (CP), 342

Protein *(continued)*
 dephosphorylation, 316
 diastatic power, 513-514
 in feed barley, 119, 125-126, 503
 foam stabilization, 510, 514
 gel, 512-513
 in grain, 373
 grain structure and protein distribution, 504-505
 hordein storage proteins, 506-508
 hordeins and malting quality, 511-513
 kinases, 313, 316, 492
 and malting performance, 503-504, 511-513
 phosphorylation, 316
 PR, 340-341
 protective, 508-510
 puroindolines, 515
 recombinant, 162-164
 starch, and cell walls, 93-100
 starch granule, 511
 surface-active, 510, 514
 Z, 509-510
Protoplasts as tool for gene studies, 152, 153
PTF, 459, 460, 461, 462, 463, 476
Pulsed field gel electrophoresis (PFGE), 34
Puroindolines, 515

Quality traits
 nitrogen partitioning and, 373, 375-376
 QTL mapping, 37-39
 specifications for malting, 536-537
Quantitative trait loci (QTL). *See also* Genetic adaptation to stress
 analysis, 122, 129
 chromosome 2(2H), 94-95
 disease resistance gene mapping, 40-43
 dormancy, 89, 489-493
 drought tolerance, 72-73, 310-311

Quantitative trait loci *(continued)*
  fine-structure mapping, 39-40
  frost resistance, 309-310, 322, 324-325
  growth habit control, 309
  heading date control, 308-309
  mapping, 35-36
  mapping for agronomic traits, 36-37
  mapping for quality traits, 37-39
  marker-assisted selection for deployment, 189-192
  marker-assisted selection for detection, 186-189
  salt tolerance control, 311

RAD and yield, 270-271
Radiation, 270-271, 282-284
Radiation use efficiency (RUE)
  defined, 476
  dry-matter production, 453, 454
  indices, 392
  and yield, 274-275, 389
Random amplified polymorphic DNA (RAPD), 33, 129
Recombinant proteins, 162-164
Restriction fragment length polymorphism (RFLP), 20-21, 22, 32
RGFILL, 463, 476
Rhynchosporium, 71, 334
Rice as model for cereal genomes, 338
Ripening of the grain, 254
Ritchie, Joe T., 445
RLDF(L), 465, 466, 476
RLNEW, 464, 466, 476
RLV(L), 466, 476
RMSE, 468-472, 474, 476
RNFAC, 465, 476
Romagosa, Ignacio, 205
Root characters and stress tolerance, 71-74
Root development model, 464-466
ROWSPC, 454, 476
RTDEP, 465, 476

RTSW, 459, 461, 476
RUE. *See* Radiation use efficiency

Salicylic acid (SA), 340
Salt stress, 287, 293-294
Salt tolerance, 287, 293-294, 311, 323
SAS programs, 234
  analysis of variance using GLM, 235
  estimates of genotypic sensitivities, 236-237
  factorial regression using GLM, 235-236
  IML (Interactive Matrix Language) for AMMI model, 234-235
Savin, Roxana, 523
Scald blotch, 334
Scottish Crop Research Institute, 74
Screening and selection of genetic traits, 129
SDEPTH, 449, 476
SDS-PAGE (second-division segregation-polyacrylamide gel electrophoresis), 506, 512
Senescence, 392, 457-458
SENLA, 458, 461, 476
Sequence tagged sites (STSs), 129
Shewry, Peter R., 503
Shuttle breeding, 181
Signal perception, 313-314
Signal transduction, 314-316
Simple sequence repeats (SSRs), 129
Simply inherited traits, 129-131
Simulation models. *See* Crop simulation models
Sink
  demand, 446
  limitation, 275-276
  strength, 274
Slafer, Gustavo A., 243, 361
Software SAS programs, 234-237
Soil
  abiotic stresses, 420-421
  evaporation, 278-279
  factors and yield, 208

Soluble material transport, 100-101
Southern blot analysis, 157-160
Sowing of malting barley, 413, 434-435
   adaptation to stresses, 418-422
   category of sowing time, 414
   date choices, 422-424
   defined, 429
   desirable traits, 413-414
   determination of grain yield and quality, 424-425
   genotype, choosing of, 413-414
   grain quality, 417-418
   grain yield, 414-417, 430-432, 433-434
   plant density, 429-430
   matching grain development and environment, 425-427
   planting arrangement, 432-433
   probability-based approach, 427-429
   tillering, 432-433
Spain study. *See* Yield of isogenic lines in Spain
Spectral vegetation and barley yield in Mediterranean rain-fed conditions, 387-390, 408
   analysis of results, 401-408
   growing conditions, 391
   photosynthetic area, 392
   plant material, 390-391
   radiation-use efficiency indices, 392
   results, 393-401, 402
   senescence, 392
   spectral reflectance measurements, 391
   spectroradiometrical indices, 392-393, 403-404
   statistical analysis, 393
   water status, 393
Spikelets, 244, 246, 248-250
Spot blotch, 69, 334
Spratt Archer grain, 75
Sprouting. *See* Dormancy; Germination
SRAD, 454, 476
Stanca, Antonio Michele, 307
Starch. *See also* Carbohydrates
   effect of drought, 538

Starch *(continued)*
   effect of heat stress, 530-534, 538
   grain content, 528
   and grain texture, 511
   granule proteins, 511
   nonstarch polysaccharides, 528-529
   protein, and cell walls, 93-100
Stem length shortened for increased yield, 368-369, 378-379
Stem rust, 41-42, 333-334
Steptoe/Morex maps, 32, 33, 94, 490
STMWT, 459, 460, 462, 463, 476
Straw for feeding ruminants, 12
stress
   abiotic stress tolerance, 71-75, 312, 328-330, 418-422
   anaerobic stress tolerance, 325
   genetic. *See* Genetic adaptation to stress
   heat, 164-165, 323-324, 420, 530-534, 538-544. *See also* Heat-stable $(1,3$-$1,4)$-β-glucanase
   malting quality. *See* Abiotic stress and changes in malting quality
   versus potential yield, 272
   root characters and, 71-74
   salt, 287, 293-294, 311, 323
   water stress, 271, 273, 274. *See also* Drought stress
Stress adaptation. *See* Genetic adaptation to stress; Physiological mechanisms to cope with stress
Stress proteins, 286
SUMDTT, 459, 461, 463, 476
Swanston, J. Stuart, 85
SWDF, 465, 476
SWDF1, 465, 476
SWFAC, 454, 456, 458, 459, 460, 476
SW(L), 465, 476
SWMIN, 461, 462, 463, 476
Synteny, 43-45

T, 454, 476
Taxonomy of wild barley, 67-68

TC1, 458, 459, 476
TC2, 458, 459, 476
TDU, 451, 456, 476
Telomeres, mapping of, 33-34
Temperature. *See also* Environmental factors
   development and temperature responses, 255-256, 288, 317-318
   during grain-filling period, 426
   effect on dormancy, 493-495
   low-temperature-responsive genes, 324-325, 467
TEMPM, 463, 477
TEMPMN, 454, 477
TEMPMX, 454, 477
Texture of grain
   and puroindolines, 515
   and starch, 511
Thermal time, 448, 474. *See also* Phasic development
Thionins, 341
Thomas, William T. B., 177
TI, 456, 458, 477
Tiller death, 459
Tillering, 252-253, 432-433, 458-459
TILN, 456, 458, 459, 461, 477
TILSW, 459, 461, 477
Time window for dormancy, 482, 483, 494-495
Timentin, 147
TOTN, 465, 477
TPSM, 458, 459, 477
Transcription factor, 153
Transformation. *See* Genetic transformation
Transgenic plants
   agrobacterium-mediated transformation, 147-150
   agronomic characteristics, 169-170
   biolistic method of transformation, 144-147
   field testing, 166-167
   homeotic genes, 160-162
   identification assays, 155-160
   marker-free, 150-151

Transgenic plants *(continued)*
   micromalting, 167-169
   origination, 144
   quality testing, 151-155
   recombinant proteins, 162-164
   regenerated, 144
   tailoring heat-tolerant (1,3-1,4)-β-glucanase, 164-165
Transpiration efficiency, 279
Transport of soluble material, 100-101
Triple mound stage, 248
TRLDF, 466, 477
Typhula blight resistance, 334

Ullrich, Steven E., 115
Utilization
   crop, 10-12
   history of, 5

Valè, Giampiero, 307
van Eeuwijk, Fred, 205
Vernalization
   CUMVD, 450, 474
   in growth stage, 449-451
   heading date, 308-309
   P1V, 450, 451, 476
   and plant development, 256-258
   VF, 451, 477
VF, 451, 477
Vigor in early stages, 280-281
Voltas, Jordi, 205
von Wettstein, Diter, 143

Water status indices, 393
Water stress and yield, 271, 273, 274. *See also* Drought stress
Water-use efficiency (WUE), 278-280, 282
Wax bloom, 281
Waxy mutations, 93-94

Web sites
   disease resistance, 71
   feed grain, 123, 124, 126
   genetic control of flowering time, 259
   genome mapping, 33, 35, 68
Weeds as biotic stress, 421-422, 428-429
Weight of grain reduced by drought, 534-536
Western blot analysis, 157, 166
Western Mediterranean
   center of origin, 24-25
   H. vulgare, 21-24
   H. spontaneum, 23-24
Wheat, history of, 3
Wild barley for crop improvement, 65-66, 77
   abiotic stress tolerance, 71-75
   adaptation, 65
   disease resistance, 68-71
   drought tolerance, 65-66
   flowering, 65
   genetic diversity, 76-77
   grain quality, 75-76
   origin, 67-68
   taxonomy, 67-68
Wong, Oi T., 143
WR(L), 465, 477

Xanthophyll cycle, 284, 289, 389

Y1, 454, 477
Yield. *See also* Grain yield; Physiology of yields
   evaluation in computer simulation, 471-472
   extract, 417-418
   grain. *See* Grain yield
   potential. *See* Potential yield
Yield of isogenic lines in Spain
   AMMI model, 214-215
   analysis of variance (ANOVA), 213-214
   assessment of adaptation, 217-219
   background, 209-213
   biplot representation, 215-217, 218
   factorial regression model, 219-231
   incorporation of environmental information, 222, 224-226
   sensitivities to environmental changes, 226-231

Zymogram, 157

Order a copy of this book with this form or online at:
http://www.haworthpressinc.com/store/product.asp?sku=4472

BARLEY SCIENCE
Recent Advances from Molecular Biology to Agronomy of Yield and Quality

_____ in hardbound at $129.95 (ISBN: 1-56022-909-8)
_____ in softbound at $69.95 (ISBN: 1-56022-910-1)

COST OF BOOKS_____

OUTSIDE USA/CANADA/
MEXICO: ADD 20%____

POSTAGE & HANDLING_____
(US: $4.00 for first book & $1.50
for each additional book)
Outside US: $5.00 for first book
& $2.00 for each additional book)

SUBTOTAL_____

in Canada: add 7% GST____

STATE TAX____
(NY, OH & MIN residents, please
add appropriate local sales tax)

FINAL TOTAL____
(If paying in Canadian funds,
convert using the current
exchange rate, UNESCO
coupons welcome.)

BILL ME LATER: ($5 service charge will be added)
(Bill-me option is good on US/Canada/Mexico orders only;
not good to jobbers, wholesalers, or subscription agencies.)

Check here if billing address is different from
shipping address and attach purchase order and
billing address information.

Signature_____

PAYMENT ENCLOSED: $_____

PLEASE CHARGE TO MY CREDIT CARD.

Visa   MasterCard   AmEx   Discover
        Diner's Club   Eurocard   JCB

Account # _____

Exp. Date _____

Signature _____

Prices in US dollars and subject to change without notice.

NAME_____
INSTITUTION_____
ADDRESS_____
CITY_____
STATE/ZIP_____
COUNTRY_____ COUNTY (NY residents only)_____
TEL_____ FAX_____
E-MAIL_____

May we use your e-mail address for confirmations and other types of information?   Yes   No
We appreciate receiving your e-mail address and fax number. Haworth would like to e-mail or fax special discount offers to you, as a preferred customer. **We will never share, rent, or exchange your e-mail address or fax number.** We regard such actions as an invasion of your privacy.

Order From Your Local Bookstore or Directly From
The Haworth Press, Inc.
10 Alice Street, Binghamton, New York 13904-1580 • USA
TELEPHONE: 1-800-HAWORTH (1-800-429-6784) / Outside US/Canada: (607) 722-5857
FAX: 1-800-895-0582 / Outside US/Canada: (607) 722-6362
E-mail: getinfo@haworthpressinc.com
PLEASE PHOTOCOPY THIS FORM FOR YOUR PERSONAL USE.
www.HaworthPress.com

FORTHCOMING AND NEW BOOKS FROM HAWORTH CROP SCIENCES

Take 20% Off Each Book! SPECIAL OFFER

BACTERIAL DISEASE RESISTANCE IN PLANTS
Molecular Biology and Biotechnological Applications
P. Vidhyasekaran, PhD, FNA
Over 550 Pages!
Presents reviews of current research but goes on to suggest future research strategies to exploit the studies in interventions with biotechnological, commercial, and field applications. Generously illustrated with figures and tables that make the data more quickly understandable.
$110.95 hard. ISBN: 1-56022-924-1.
$59.95 soft. ISBN: 1-56022-925-X.
Available Summer 2002. Approx. 582 pp. with Index.

TILLAGE FOR SUSTAINABLE CROPPING
P. R. Gajri, PhD, V. K. Arora, PhD, and S. S. Prihar, PhD
Over 200 Pages!
Provides a rational framework for tillage systems that takes into consideration soil and climatic characteristics and the availability of other edaphic inputs. This well-referenced volume also examines soil sustainability in terms of pollution, greenhouse gases, water contamination, growing production demands soil degradation, and looks at the way crops respond to tillage techniques in terms of weed growth, root growth, crop yields, and more.
$89.95 hard. ISBN: 1-56022-902-0.
$39.95 soft. ISBN: 1-56022-903-9.
Available Summer 2002. Approx. 246 pp. with Index.

CROP IMPROVEMENT
Challenges in the Twenty-First Century
Edited by Manjit S. Kang, PhD
Covering a range of essential food and fiber crops, this book addresses physiological and biochemical responses of plants to drought and heat stress, genotype-by-environment interactions, and use of best linear unbiased prediction. It also emphasizes the need to integrate molecular genetic techniques with traditional plant breeding methods to develop hardier, more productive crops.
$124.95 hard. ISBN: 1-56022-904-7.
$59.95 soft. ISBN: 1-56022-905-5.
Over 350 Pages!
2001. Available now. 390 pp. with Index.

FACULTY: ORDER YOUR NO-RISK EXAM COPY TODAY!
Send us your examination copy order on your stationery; indicate course title, enrollment, and course start date. We will ship and bill on a 60-day examination basis, and cancel your invoice if you decide to adopt! We will always bill at the lowest available price, such as our special "5+ text price." Please remember to order softcover where available. (We cannot provide examination copies of books not published by The Haworth Press, Inc., or its imprints.) (Outside US/Canada, a proforma invoice will be sent upon receipt of your request and must be paid in advance of shipping. A full refund will be issued with proof of adoption.)

Food Products Press®
An Imprint of The Haworth Press, Inc.
10 Alice Street
Binghamton, New York 13904-1580 USA

BARLEY SCIENCE
Recent Advances from Molecular Biology to Agronomy of Yield and Quality
Edited by Gustavo Slafer, PhD, Jose Luis Molina-Cano, PhD, Roxana Savin, PhD, Jose Luis Araus, PhD, and Ignacio Romagosa, PhD
Over 550 Pages!
This comprehensive book covers every aspect of barley from molecular biology to agronomy of yield and quality for malting, food, and animal feed. In addition, you will find a new technique for estimating yield differences among genotypes—the long-sought bridge between physiology and breeding.
$129.95 hard. ISBN: 1-56022-909-8.
$69.95 soft. ISBN: 1-56022-910-1.
2001. Available now. 590 pp. with Index.

IN VITRO PLANT BREEDING
Acram Taji, PhD, Prakash P. Kumar, PhD, and Prakash Lakshmanan, PhD
This comprehensive book presents the basic concepts and applied techniques of plant cell and tissue culture. This helpful book is written in clear language, illustrated with examples, schematic descriptions, and tables to make the concepts clear.
$49.95 hard. ISBN: 1-56022-907-1.
$29.95 soft. ISBN: 1-56022-908-X.
2001. Available now. 168 pp. with Index.

PLANT VIRUSES AS MOLECULAR PATHOGENS
Edited by Jawaid A. Khan, PhD, and Jeanne Dijkstra, PhD
Over 450 Pages!
Here you'll find new information about virus taxonomy, the molecular basis of virus transmission and movement, replication and gene expression of RNA/DNA viruses, and novel aspects of plant virus detection technologies and their control strategies. This is the only book to compile all of this information in a single volume!
$129.95 hard. ISBN: 1-56022-894-6.
$59.95 soft. ISBN: 1-56022-895-4.
2001. Available now. 530 pp. with Index.

PLANT GROWTH REGULATORS IN AGRICULTURE AND HORTICULTURE
Their Role and Commercial Uses
Edited by Amarjit S. Basra, PhD
Over 225 Pages!
This thorough book will give crop scientists and growers the latest information on several growth regulators, or phytohormones, that are used to improve the growth and yield of plants. Examining new approaches, such as molecular genetics, to investigate hormone root physiology, this guide will help you improve the yields of important crops and increase the efficiency of specific plants.
$94.95 hard. ISBN: 1-56022-891-1.
$39.95 soft. ISBN: 1-56022-896-2. 2000. 264 pp. with Index.

PHYSIOLOGICAL BASES FOR MAIZE IMPROVEMENT

Edited by Maria Elena Otegui, DrSci, and Gustavo A. Slafer, PhD

NEW!

Offers a thorough and concise guide to recent literature and developments about increasing the crop efficiency of corn. International experts in the field discuss and analyze how to effectively improve crop breeding and produce better and larger yields of corn through such processes as molecular genetics.
$69.95 hard. ISBN: 1-56022-889-X.
$39.95 soft. ISBN: 1-56022-911-X. 2000. 218 pp. with Index.

CROP RESPONSES AND ADAPTATIONS TO TEMPERATURE STRESS

Over 200 Pages! **NEW!**

New Insights and Approaches
Edited by Amarjit S. Basra, PhD

Examines research on such vital crops as corn, sorghum, soybeans, and rice, and contains methods and suggestions for growing crops in varying temperatures and areas of the world, not just in their natural climate.
$94.95 hard. ISBN: 1-56022-890-3.
$49.95 soft. ISBN: 1-56022-906-3. 2000. 302 pp. with Index.

Take 20% Off Each Book! SPECIAL OFFER

INTENSIVE CROPPING

Efficient Use of Water, Nutrients, and Tillage
S. S. Prihar, PhD, P. R. Gajri, PhD,
D. K. Benbi, PhD, and V. K. Arora, PhD

NEW! **Over 225 Pages!**

This book explains the need for intensive cropping and goes in-depth on the technologies and practices necessary for management of water, nutrients, and energy.
$79.95 hard. ISBN: 1-56022-881-4. 1999.
$49.95 soft. ISBN: 1-56022-899-7. 2000.
264 pp. with Index. **NOW PUBLISHED IN PAPERBACK!**

COTTON FIBERS

Developmental Biology, Quality Improvement, and Textile Processing
Edited by Amarjit S. Basra, PhD

NEW! **Over 350 Pages!**

"UNIQUE.... useful as a means of learning about cotton growth and development, and the exciting opportunities provided by genetic engineering."
—*Textile Research Journal*
$129.95 hard. ISBN: 1-56022-867-9. 1999.
$49.95 soft. ISBN: 1-56022-898-9. 2000.
388 pp. with Index. **NOW PUBLISHED IN PAPERBACK!**

CALL OUR TOLL-FREE NUMBER: 1-800-429-6784
US & Canada only / 8am–5pm ET; Monday–Friday
Outside US/Canada: + 607-722-5857
FAX YOUR ORDER TO US: 1-800-895-0582
Outside US/Canada: + 607-771-0012
E-MAIL YOUR ORDER TO US:
orders@haworthpressinc.com
VISIT OUR WEB SITE AT:
http://www.HaworthPress.com

Order Today and Save!

TITLE	ISBN	REGULAR PRICE	20%–OFF PRICE

- Discount available only in US, Canada, and Mexico and not available in conjunction with any other offer.
- Individual orders outside US, Canada, and Mexico must be prepaid by check, credit card, or money order.
- In Canada: Add 7% for GST after postage & handling. Residents of Newfoundland, Nova Scotia, New Brunswick, and Labrador, add an additional 8% for province tax.
- MN, NY, and OH residents: Add appropriate local sales tax.

Please complete information below or tape your business card in this area.

NAME _____

ADDRESS _____

_____

CITY _____

STATE _____ ZIP _____

COUNTRY _____

COUNTY (NY residents only) _____

TEL _____ FAX _____

E-MAIL _____
May we use your e-mail address for confirmations and other types of information?
( ) Yes ( ) No. We appreciate receiving your e-mail address and fax number. Haworth would like to e-mail or fax special discount offers to you, as a preferred customer. We will never share, rent, or exchange your e-mail address or fax number. We regard such actions as an invasion of your privacy.

POSTAGE AND HANDLING: If your book total is:	Add
up to $29.95	$5.00
$30.00 – $49.99	$6.00
$50.00 – $69.99	$7.00
$70.00 – $89.99	$8.00
$90.00 – $109.99	$9.00
$110.00 – $129.99	$10.00
$130.00 – $149.99	$11.00
$150.00 and up	$12.00

❑ **BILL ME LATER** ($5 service charge will be added).
(Bill-me option is not available on orders outside US/Canada/Mexico. Service charge is waived for booksellers/wholesalers/jobbers.)

Signature _____

❑ PAYMENT ENCLOSED _____
(Payment must be in US or Canadian dollars by check or money order drawn on a US or Canadian bank.)

❑ PLEASE CHARGE TO MY CREDIT CARD:
❑ AmEx ❑ Diners Club ❑ Discover ❑ Eurocard ❑ JCB ❑ Master Card ❑ Visa

Account # _____ Exp Date _____

Signature _____

May we open a confidential credit card account for you for possible future purchases? ( ) Yes ( ) No

- US orders will be shipped via UPS; Outside US orders will be shipped via Book Printed Matter. For shipments via other delivery services, contact Haworth for details. Based on US dollars. Booksellers: Call for freight charges. • If paying in Canadian funds, please use the current exchange rate to convert total to Canadian dollars. • Payment in UNESCO coupons welcome. • Please allow 3-4 weeks for delivery after publication. • Prices and discounts subject to change without notice. • Discount not applicable on books priced under $15.00.

The Haworth Press, Inc.
10 Alice Street, Binghamton, New York 13904-1580 USA